Selected Titles in This Series

45 **Nicholas M. Katz and Peter Sarnak,** Random matrices, Frobenius eigenvalues, and monodromy, 1999

44 **Max-Albert Knus, Alexander Merkurjev, and Markus Rost,** The book of involutions, 1998

43 **Luis A. Caffarelli and Xavier Cabré,** Fully nonlinear elliptic equations, 1995

42 **Victor Guillemin and Shlomo Sternberg,** Variations on a theme by Kepler, 1990

41 **Alfred Tarski and Steven Givant,** A formalization of set theory without variables, 1987

40 **R. H. Bing,** The geometric topology of 3-manifolds, 1983

39 **N. Jacobson,** Structure and representations of Jordan algebras, 1968

38 **O. Ore,** Theory of graphs, 1962

37 **N. Jacobson,** Structure of rings, 1956

36 **W. H. Gottschalk and G. A. Hedlund,** Topological dynamics, 1955

35 **A. C. Schaeffer and D. C. Spencer,** Coefficient regions for Schlicht functions, 1950

34 **J. L. Walsh,** The location of critical points of analytic and harmonic functions, 1950

33 **J. F. Ritt,** Differential algebra, 1950

32 **R. L. Wilder,** Topology of manifolds, 1949

31 **E. Hille and R. S. Phillips,** Functional analysis and semigroups, 1957

30 **T. Radó,** Length and area, 1948

29 **A. Weil,** Foundations of algebraic geometry, 1946

28 **G. T. Whyburn,** Analytic topology, 1942

27 **S. Lefschetz,** Algebraic topology, 1942

26 **N. Levinson,** Gap and density theorems, 1940

25 **Garrett Birkhoff,** Lattice theory, 1940

24 **A. A. Albert,** Structure of algebras, 1939

23 **G. Szegö,** Orthogonal polynomials, 1939

22 **C. N. Moore,** Summable series and convergence factors, 1938

21 **J. M. Thomas,** Differential systems, 1937

20 **J. L. Walsh,** Interpolation and approximation by rational functions in the complex domain, 1935

19 **R. E. A. C. Paley and N. Wiener,** Fourier transforms in the complex domain, 1934

18 **M. Morse,** The calculus of variations in the large, 1934

17 **J. M. Wedderburn,** Lectures on matrices, 1934

16 **G. A. Bliss,** Algebraic functions, 1933

15 **M. H. Stone,** Linear transformations in Hilbert space and their applications to analysis, 1932

14 **J. F. Ritt,** Differential equations from the algebraic standpoint, 1932

13 **R. L. Moore,** Foundations of point set theory, 1932

12 **S. Lefschetz,** Topology, 1930

11 **D. Jackson,** The theory of approximation, 1930

10 **A. B. Coble,** Algebraic geometry and theta functions, 1929

9 **G. D. Birkhoff,** Dynamical systems, 1927

8 **L. P. Eisenhart,** Non-Riemannian geometry, 1927

7 **E. T. Bell,** Algebraic arithmetic, 1927

6 **G. C. Evans,** The logarithmic potential, discontinuous Dirichlet and Neumann problems, 1927

(See the AMS catalog for earlier titles)

Random Matrices, Frobenius Eigenvalues, and Monodromy

American Mathematical Society

Colloquium Publications
Volume 45

Random Matrices, Frobenius Eigenvalues, and Monodromy

Nicholas M. Katz
Peter Sarnak

American Mathematical Society
Providence, Rhode Island

1991 *Mathematics Subject Classification.* Primary 11G25, 14G10, 60Fxx, 14D05; Secondary 11M06, 82Bxx, 11Y35.

ABSTRACT. The main topic of this book is the deep relation between the spacings between zeros of zeta and L-functions and spacings between eigenvalues of random elements of large compact classical groups. This relation, the Montgomery-Odlyzko law, is shown to hold for wide classes of zeta and L-functions over finite fields.

The book draws on, and gives accessible accounts of, many disparate areas of mathematics, from algebraic geometry, moduli spaces, monodromy, equidistribution, and the Weil Conjectures, to probability theory on the compact classical groups in the limit as their dimension goes to infinity and related techniques from orthogonal polynomials and Fredholm determinants. It will be useful and interesting to researchers and graduate students working in any of these areas.

Library of Congress Cataloging-in-Publication Data

Katz, Nicholas M., 1943–

Random matrices, Frobenius eigenvalues, and monodromy / Nicholas M. Katz, Peter Sarnak.

p. cm. — (Colloquium publications / American Mathematical Society, ISSN 0065-9258 ; v. 45)

Includes bibliographical references.

ISBN 0-8218-1017-0 (hardcover : alk. paper)

1. Functions, Zeta. 2. L-functions. 3. Random matrices. 4. Limit theorems (Probability theory) 5. Monodromy groups. I. Sarnak, Peter. II. Title. III. Series: Colloquium publications (American Mathematical Society) ; v. 45.

QA351.K36 1998

515′.56—dc21 98-20459

CIP

∞ The paper used in this book is acid-free and falls within the guidelines
established to ensure permanence and durability.
Visit the AMS home page at URL: http://www.ams.org/

10 9 8 7 6 5 4 3 2 1 04 03 02 01 00 99

Contents

Introduction 1

Chapter 1. Statements of the Main Results 17
- 1.0. Measures attached to spacings of eigenvalues 17
- 1.1. Expected values of spacing measures 23
- 1.2. Existence, universality and discrepancy theorems for limits of expected values of spacing measures: the three main theorems 24
- 1.3. Interlude: A functorial property of Haar measure on compact groups 25
- 1.4. Application: Slight economies in proving Theorems 1.2.3 and 1.2.6 25
- 1.5. Application: An extension of Theorem 1.2.6 26
- 1.6. Corollaries of Theorem 1.5.3 28
- 1.7. Another generalization of Theorem 1.2.6 30
- 1.8. Appendix: Continuity properties of "the i'th eigenvalue" as a function on $U(N)$ 32

Chapter 2. Reformulation of the Main Results 35
- 2.0. "Naive" versions of the spacing measures 35
- 2.1. Existence, universality and discrepancy theorems for limits of expected values of naive spacing measures: the main theorems bis 37
- 2.2. Deduction of Theorems 1.2.1, 1.2.3 and 1.2.6 from their bis versions 38
- 2.3. The combinatorics of spacings of finitely many points on a line: first discussion 42
- 2.4. The combinatorics of spacings of finitely many points on a line: second discussion 45
- 2.5. The combinatorics of spacings of finitely many points on a line: third discussion: variations on Sep(a) and Clump(a) 49
- 2.6. The combinatorics of spacings of finitely many points of a line: fourth discussion: another variation on Clump(a) 54
- 2.7. Relation to naive spacing measures on $G(N)$: Int, Cor and TCor 54
- 2.8. Expected value measures via INT and COR and TCOR 57
- 2.9. The axiomatics of proving Theorem 2.1.3 58
- 2.10. Large N COR limits and formulas for limit measures 63
- 2.11. Appendix: Direct image properties of the spacing measures 65

Chapter 3. Reduction Steps in Proving the Main Theorems 73
- 3.0. The axiomatics of proving Theorems 2.1.3 and 2.1.5 73
- 3.1. A mild generalization of Theorem 2.1.5: the φ-version 74
- 3.2. M-grid discrepancy, L cutoff and dependence on the choice of coordinates 77
- 3.3. A weak form of Theorem 3.1.6 89

3.4. Conclusion of the axiomatic proof of Theorem 3.1.6 90
3.5. Making explicit the constants 98

Chapter 4. Test Functions 101
4.0. The classes $\mathcal{T}(n)$ and $\mathcal{T}_0(n)$ of test functions 101
4.1. The random variable $Z[n, F, G(N)]$ on $G(N)$ attached to a function F in $\mathcal{T}(n)$ 103
4.2. Estimates for the expectation $E(Z[n, F, G(N)])$ and variance $\mathrm{Var}(Z[n, F, G(N)])$ of $Z[n, F, G(N)]$ on $G(N)$ 104

Chapter 5. Haar Measure 107
5.0. The Weyl integration formula for the various $G(N)$ 107
5.1. The $K_N(x, y)$ version of the Weyl integration formula 109
5.2. The $L_N(x, y)$ rewriting of the Weyl integration formula 116
5.3. Estimates for $L_N(x, y)$ 117
5.4. The $L_N(x, y)$ determinants in terms of the sine ratios $S_N(x)$ 118
5.5. Case by case summary of explicit Weyl measure formulas via S_N 120
5.6. Unified summary of explicit Weyl measure formulas via S_N 121
5.7. Formulas for the expectation $E(Z[n, F, G(N)])$ 122
5.8. Upper bound for $E(Z[n, F, G(N)])$ 123
5.9. Interlude: The $\sin(\pi x)/\pi x$ kernel and its approximations 124
5.10. Large N limit of $E(Z[n, F, G(N)])$ via the $\sin(\pi x)/\pi x$ kernel 127
5.11. Upper bound for the variance 133

Chapter 6. Tail Estimates 141
6.0. Review: Operators of finite rank and their (reversed) characteristic polynomials 141
6.1. Integral operators of finite rank: a basic compatibility between spectral and Fredholm determinants 141
6.2. An integration formula 143
6.3. Integrals of determinants over $G(N)$ as Fredholm determinants 145
6.4. A new special case: $O_-(2N + 1)$ 151
6.5. Interlude: A determinant-trace inequality 154
6.6. First application of the determinant-trace inequality 156
6.7. Application: Estimates for the numbers $\mathrm{eigen}(n, s, G(N))$ 159
6.8. Some curious identities among various $\mathrm{eigen}(n, s, G(N))$ 162
6.9. Normalized "n'th eigenvalue" measures attached to $G(N)$ 163
6.10. Interlude: Sharper upper bounds for $\mathrm{eigen}(0, s, SO(2N))$, for $\mathrm{eigen}(0, s, O_-(2N + 1))$, and for $\mathrm{eigen}(0, s, U(N))$ 166
6.11. A more symmetric construction of the "n'th eigenvalue" measures $\nu(n, U(N))$ 169
6.12. Relation between the "n'th eigenvalue" measures $\nu(n, U(N))$ and the expected value spacing measures $\mu(U(N)$, sep. $k)$ on a fixed $U(N)$ 170
6.13. Tail estimate for $\mu(U(N)$, sep. $0)$ and $\mu(\mathrm{univ}$, sep. $0)$ 174
6.14. Multi-eigenvalue location measures, static spacing measures and expected values of several variable spacing measures on $U(N)$ 175
6.15. A failure of symmetry 183
6.16. Offset spacing measures and their relation to multi-eigenvalue location measures on $U(N)$ 185

6.17. Interlude: "Tails" of measures on $\mathbb{R}^r$ 189
6.18. Tails of offset spacing measures and tails of multi-eigenvalue location measures on $U(N)$ 192
6.19. Moments of offset spacing measures and of multi-eigenvalue location measures on $U(N)$ 194
6.20. Multi-eigenvalue location measures for the other $G(N)$ 195

Chapter 7. Large N Limits and Fredholm Determinants 197
7.0. Generating series for the limit measures μ(univ, sep.'s a) in several variables: absolute continuity of these measures 197
7.1. Interlude: Proof of Theorem 1.7.6 205
7.2. Generating series in the case $r = 1$: relation to a Fredholm determinant 208
7.3. The Fredholm determinants $E(T, s)$ and $E_{\pm}(T, s)$ 211
7.4. Interpretation of $E(T, s)$ and $E_{\pm}(T, s)$ as large N scaling limits of $E(N, T, s)$ and $E_{\pm}(N, T, s)$ 212
7.5. Large N limits of the measures $\nu(n, G(N))$: the measures $\nu(n)$ and $\nu(\pm, n)$ 215
7.6. Relations among the measures μ_n and the measures $\nu(n)$ 225
7.7. Recapitulation, and concordance with the formulas in [**Mehta**] 228
7.8. Supplement: Fredholm determinants and spectral determinants, with applications to $E(T, s)$ and $E_{\pm}(T, s)$ 229
7.9. Interlude: Generalities on Fredholm determinants and spectral determinants 232
7.10. Application to $E(T, s)$ and $E_{\pm}(T, s)$ 235
7.11. Appendix: Large N limits of multi-eigenvalue location measures and of static and offset spacing measures on $U(N)$ 235

Chapter 8. Several Variables 245
8.0. Fredholm determinants in several variables and their measure-theoretic meaning (cf. [**T-W**]) 245
8.1. Measure-theoretic application to the $G(N)$ 248
8.2. Several variable Fredholm determinants for the $\sin(\pi x)/\pi x$ kernel and its $\pm$ variants 249
8.3. Large N scaling limits 251
8.4. Large N limits of multi-eigenvalue location measures attached to $G(N)$ 257
8.5. Relation of the limit measure Off μ(univ, offsets c) with the limit measures $\nu(c)$ 263

Chapter 9. Equidistribution 267
9.0. Preliminaries 267
9.1. Interlude: zeta functions in families: how lisse pure $\mathcal{F}$'s arise in nature 270
9.2. A version of Deligne's equidistribution theorem 275
9.3. A uniform version of Theorem 9.2.6 279
9.4. Interlude: Pathologies around (9.3.7.1) 280
9.5. Interpretation of (9.3.7.2) 283
9.6. Return to a uniform version of Theorem 9.2.6 283
9.7. Another version of Deligne's equidistribution theorem 287

Chapter 10. Monodromy of Families of Curves 293
10.0. Explicit families of curves with big G_{geom} 293
10.1. Examples in odd characteristic 293
10.2. Examples in characteristic two 301
10.3. Other examples in odd characteristic 302
10.4. Effective constants in our examples 303
10.5. Universal families of curves of genus $g \geq 2$ 304
10.6. The moduli space $\mathcal{M}_{g,3K}$ for $g \geq 2$ 307
10.7. Naive and intrinsic measures on $USp(2g)^{\#}$ attached to universal families of curves 315
10.8. Measures on $USp(2g)^{\#}$ attached to universal families of hyperelliptic curves 320

Chapter 11. Monodromy of Some Other Families 323
11.0. Universal families of principally polarized abelian varieties 323
11.1. Other "rational over the base field" ways of rigidifying curves and abelian varieties 324
11.2. Automorphisms of polarized abelian varieties 327
11.3. Naive and intrinsic measures on $USp(2g)^{\#}$ attached to universal families of principally polarized abelian varieties 328
11.4. Monodromy of universal families of hypersurfaces 331
11.5. Projective automorphisms of hypersurfaces 335
11.6. First proof of 11.5.2 335
11.7. Second proof of 11.5.2 337
11.8. A properness result 342
11.9. Naive and intrinsic measures on $USp(\text{prim}(n,d))^{\#}$ (if n is odd) or on $O(\text{prim}(n,d))^{\#}$ (if n is even) attached to universal families of smooth hypersurfaces of degree d in $\mathbb{P}^{n+1}$ 346
11.10. Monodromy of families of Kloosterman sums 347

Chapter 12. GUE Discrepancies in Various Families 351
12.0. A basic consequence of equidistribution: axiomatics 351
12.1. Application to GUE discrepancies 352
12.2. GUE discrepancies in universal families of curves 353
12.3. GUE discrepancies in universal families of abelian varieties 355
12.4. GUE discrepancies in universal families of hypersurfaces 356
12.5. GUE discrepancies in families of Kloosterman sums 358

Chapter 13. Distribution of Low-lying Frobenius Eigenvalues in Various Families 361
13.0. An elementary consequence of equidistribution 361
13.1. Review of the measures $\nu(c, G(N))$ 363
13.2. Equidistribution of low-lying eigenvalues in families of curves according to the measure $\nu(c, USp(2g))$ 364
13.3. Equidistribution of low-lying eigenvalues in families of abelian varieties according to the measure $\nu(c, USp(2g))$ 365
13.4. Equidistribution of low-lying eigenvalues in families of odd-dimensional hypersurfaces according to the measure $\nu(c, USp(\text{prim}(n,d)))$ 366

13.5. Equidistribution of low-lying eigenvalues of Kloosterman sums in evenly many variables according to the measure $\nu(c, USp(2n))$ 367
13.6. Equidistribution of low-lying eigenvalues of characteristic two Kloosterman sums in oddly many variables according to the measure $\nu(c, SO(2n+1))$ 367
13.7. Equidistribution of low-lying eigenvalues in families of even-dimensional hypersurfaces according to the measures $\nu(c, SO(\text{prim}(n,d)))$ and $\nu(c, O_-(\text{prim}(n,d)))$ 368
13.8. Passage to the large N limit 369

Appendix: Densities 373
AD.0. Overview 373
AD.1. Basic definitions: $W_n(f, A, G(N))$ and $W_n(f, G(N))$ 373
AD.2. Large N limits: the easy case 374
AD.3. Relations between eigenvalue location measures and densities: generalities 378
AD.4. Second construction of the large N limits of the eigenvalue location measures $\nu(c, G(N))$ for $G(N)$ one of $U(N)$, $SO(2N+1), USp(2N), SO(2N), O_-(2N+2), O_-(2N+1)$ 381
AD.5. Large N limits for the groups $U_k(N)$: Widom's result 385
AD.6. Interlude: The quantities $V_r(\varphi, U_k(N))$ and $V_r(\varphi, U(N))$ 386
AD.7. Interlude: Integration formulas on $U(N)$ and on $U_k(N)$ 390
AD.8. Return to the proof of Widom's theorem 392
AD.9. End of the proof of Theorem AD.5.2 399
AD.10. Large N limits of the eigenvalue location measures on the $U_k(N)$ 401
AD.11. Computation of the measures $\nu(c)$ via low-lying eigenvalues of Kloosterman sums in oddly many variables in odd characteristic 403
AD.12. A variant of the one-level scaling density 405

Appendix: Graphs 411
AG.0. How the graphs were drawn, and what they show 411
Figure 1 413
Figure 2 414
Figure 3 415
Figure 4 416

References 417

Introduction

In a remarkable numerical experiment, Odlyzko [**Odl**] found that the distribution of the (suitably normalized) spacings between successive zeroes of the Riemann zeta function is (empirically) the same as the so-called GUE measure, a certain probability measure on $\mathbb{R}$ arising in random matrix theory. His experiment was inspired by work of Montgomery [**Mon**], who determined the pair correlation distribution between zeroes (in a restricted range), and who noted the compatibility of what he found with the GUE prediction. Recent results of Rudnick and Sarnak [**Ru-Sar**] are also compatible with the belief that the distribution of the spacings between zeroes, not only of the Riemann zeta function, but also of quite general automorphic L-functions over $\mathbb{Q}$, are all given by the GUE measure, or, as we shall say, all satisfy the Montgomery-Odlyzko Law. Unfortunately, proving this seems well beyond range of existing techniques, and we have no results to offer in this direction.

However, it is a long established principle that problems which seem inaccessible in the number field case often have finite field analogues which are accessible. In this book we establish the Montgomery-Odlyzko Law for wide classes of zeta and L-functions over finite fields.

To fix ideas, let us consider a special case, which none the less contains all the essential phenomena, the case of curves over finite fields. Thus we consider a finite field $\mathbb{F}_q$, and a proper, smooth, geometrically connected curve $C/\mathbb{F}_q$ of genus g. [For example, if we take a homogeneous form $F(X, Y, Z)$ over $\mathbb{F}_q$ of degree d in three variables such that F and its first partial derivatives have no common zeroes in $\overline{\mathbb{F}}_q$, then the projective plane curve of equation $F = 0$ in $\mathbb{P}^2$ is such a curve, of genus $g = (d-1)(d-1)/2$.] The zeta function of $C/\mathbb{F}_q$, denoted $Z(C/\mathbb{F}_q, T)$, was first introduced by Artin [**Artin**] in his thesis. It is the basic diophantine invariant of $C/\mathbb{F}_q$, constructed out of the numbers $N_n := \mathrm{Card}(C(\mathbb{F}_{q^n}))$ of points on C with coordinates in the unique field extension $\mathbb{F}_{q^n}$ of $\mathbb{F}_q$ of each degree $n \geq 1$. T As a formal series over $\mathbb{Q}$ in one variable T, $Z(C/\mathbb{F}_q, T)$ is defined as the generating series

$$Z(C/\mathbb{F}_q, T) := \exp\left(\sum_{n\geq 1} N_n T^n/n\right).$$

One knows that in fact $Z(C/\mathbb{F}_q, T)$ is a rational function of T, of the form

$$P(T)/(1-T)(1-qT),$$

where $P(T)$ is a polynomial of degree $2g$ with $\mathbb{Z}$-coefficients. By the Riemann Hypothesis for curves over finite fields [**Weil-CA**], one knows that the reciprocal zeroes of $P(T)$ all have complex absolute value Sqrt(q), i.e., we have

$$P(T) = \prod_{j=1}^{2g}(1 - \alpha_j T), \quad \text{with } |\alpha_j|_{\mathbb{C}} = \mathrm{Sqrt}(q) \text{ for all } j.$$

We write

$$\alpha_j = \mathrm{Sqrt}(q)e^{i\varphi_j}, \qquad 0 \le \varphi_j < 2\pi.$$

Renumbering, we may assume that

$$0 \le \varphi_1 \le \varphi_2 \le \cdots \le \varphi_{2g} < 2\pi.$$

The normalized spacings between the (reciprocal) zeroes of the zeta function of $C/\mathbb{F}_q$ are the following $2g$ real numbers. The first $2g - 1$ are

$$(g/\pi)(\varphi_2 - \varphi_1), (g/\pi)(\varphi_3 - \varphi_2), \ldots, (g/\pi)(\varphi_{2g} - \varphi_{2g-1}),$$

and the last is the "wraparound" spacing

$$(g/\pi)(\varphi_1 + 2\pi - \varphi_{2g}).$$

The spacing measure $\mu = \mu(C/\mathbb{F}_q)$ attached to $C/\mathbb{F}_q$ is the probability measure on $\mathbb{R}$, supported in $\mathbb{R}_{\ge 0}$, which gives mass $1/2g$ to each of the $2g$ normalized spacings.

Before going on, we must first say what is the GUE measure on $\mathbb{R}$, cf. 1.0–2. For this, we first pick an integer $N \ge 1$, and consider the unitary group $U(N)$ of size N. Given an element A in $U(N)$, its N eigenvalues lie on the unit circle, and we form the N normalized (to have mean 1) spacings between pairs of adjacent eigenvalues, and out of these N spacings we form the probability measure on $\mathbb{R}$ which gives mass $1/N$ to each of the N normalized spacings. This measure we call $\mu(A, U(N))$, the spacing measure attached to an element A in $U(N)$. We view $A \mapsto \mu(A, U(N))$ as a measure-valued function on $U(N)$. One can make sense of the integral of this function over $U(N)$ against the total mass one Haar measure dA on $U(N)$: the result makes sense as a probability measure on $\mathbb{R}$, denoted

$$\mu(U(N)) := \int_{U(N)} \mu(A, U(N))\, dA.$$

One then shows that as N grows, the measures $\mu(U(N))$ on $\mathbb{R}$ have a limit which is again a probability measure on $\mathbb{R}$, which we denote $\mu(\text{univ})$, and call the GUE measure.[1] One shows that its cumulative distribution function

$$\mathrm{CDF}_{\mu(\text{univ})}(x) := \int_{[-\infty, x]} d\mu(\text{univ})$$

is continuous on $\mathbb{R}$. [In fact, this measure has a density, which vanishes outside $\mathbb{R}_{>0}$, and is real analytic on $\mathbb{R}_{\ge 0}$, cf. Appendix: Graphs for a picture.]

For the application to curves that we have in mind, we need to know that we can obtain the GUE measure not just from the series of unitary groups $U(N)$, but also from any of the series of compact classical groups. Indeed, suppose we are given any compact subgroup K of a given unitary group $U(N)$. We can, for each element A in K, form the spacing measure attached to A thought of as an element of $U(N)$. To remind ourselves that we do this only for elements of K, we denote this measure $\mu(A, K)$. Then we view $A \mapsto \mu(A, K)$ as a measure-valued function on K, and we integrate this function against the total mass one Haar measure dA

[1] In the physics literature, this measure often carries Wigner's name

on the compact group K. The result, denoted $\mu(K) := \int_K \mu(A, K)\,dA$, is itself a probability measure on $\mathbb{R}$.

We can perform this construction with K any of the compact classical groups, $U(N)$ or $SU(N)$ or $USp(2N)$ or $SO(2N+1)$ or $SO(2N)$ or $O(2N+1)$ or $O(2N)$ in their standard representations. We show that for $G(N)$ running over any of these series of compact classical groups, the sequence of probability measures $\mu(G(N))$ on $\mathbb{R}$ converges, as N grows, to the **same** measure $\mu(\text{univ})$, the GUE measure, that we obtained as the large N limit of the $\mu(U(N))$ measures. [The case which will be relevant to curves over finite fields will turn out to be the compact symplectic groups $USp(2N)$.]

Now let us return to a curve $C/\mathbb{F}_q$ over a finite field, of some genus g. Since the spacing measure $\mu(C/\mathbb{F}_q)$ gives each of $2g$ points mass $1/2g$, its CDF is a step function, with $2g$ jumps. So it cannot possibly be the case that $\mu(C/\mathbb{F}_q)$ is equal to the GUE measure, whose CDF is continuous. Moreover, as we shall see later in this Introduction, over any finite field there are sequences of curves of increasing genus whose spacing measures are arbitrarily close to the delta measure δ_0 supported at the origin. So it is simply **not true** that the spacing measures of **all** curves of large genus are close to the GUE measure. What we show is that "most" curves of large genus over a large finite field have their spacing measure quite close to the GUE measure, or in other words that "most" curves of sufficiently large genus over a sufficiently large finite field satisfy the Montgomery-Odlyzko Law to an arbitrary degree of precision.

To make this more precise, we need a numerical measure of how close two probability measures on $\mathbb{R}$, say μ and ν, are. For this, we use the Kolmogoroff-Smirnov discrepancy, defined as the greatest vertical distance between the graphs of their CDF's:

$$\operatorname{discrep}(\mu, \nu) := \operatorname*{Sup}_{s \text{ in } \mathbb{R}} |\operatorname{CDF}_\mu(s) - \operatorname{CDF}_\nu(s)|.$$

Notice that $\operatorname{discrep}(\mu, \nu)$ is a number which always lies in the closed interval $[0, 1]$, just because CDF's of probability measures take values in $[0, 1]$.

Now let us denote by $\mathcal{M}_g(\mathbb{F}_q)$ the set, known to be finite, consisting of all $\mathbb{F}_q$-isomorphism classes of genus g curves over $\mathbb{F}_q$. Our essential result about the spacing measures $\mu(C/\mathbb{F}_q)$ attached to curves over finite fields, and their relation to the GUE measure $\mu(\text{univ})$, is this:

Theorem (cf. 12.2.3). *We have the double limit formula*

$$\lim_{g\to\infty} \lim_{q\to\infty} (1/|\mathcal{M}_g(\mathbb{F}_q)|) \sum_{C \text{ in } \mathcal{M}_g(\mathbb{F}_q)} \operatorname{discrep}(\mu(\mathit{univ}), \mu(C/\mathbb{F}_q)) = 0.$$

More precisely, for any real $\varepsilon > 0$, there exists an integer $N(\varepsilon)$ such that for any genus $g > N(\varepsilon)$, we have the inequality

$$\lim_{q\to\infty} (1/|\mathcal{M}_g(\mathbb{F}_q)|) \sum_{C \text{ in } \mathcal{M}_g(\mathbb{F}_q)} \operatorname{discrep}(\mu(\mathit{univ}), \mu(C/\mathbb{F}_q)) \le g^{\varepsilon - 1/6}.$$

To see what this says concretely, pick a small $\varepsilon > 0$, and fix a genus $g > N(\varepsilon)$. Then for q sufficiently large, say $q > M(\varepsilon, g)$, we will have

$$(**) \qquad (1/|\mathcal{M}_g(\mathbb{F}_q)|) \sum_{C \text{ in } \mathcal{M}_g(\mathbb{F}_q)} \operatorname{discrep}(\mu(\text{univ}), \mu(C/\mathbb{F}_q)) \le 2g^{\varepsilon - 1/6}.$$

To see what this last inequality means about "most" curves, pick any two positive real numbers α and β with $\alpha + \beta = 1/6 - \varepsilon$. Denote by

$$\mathcal{M}_g(\mathbb{F}_q)(\text{discrep} > g^{-\alpha}) \subset \mathcal{M}_g(\mathbb{F}_q)$$

the set of those C in $\mathcal{M}_g(\mathbb{F}_q)$ for which

$$\text{discrep}(\mu(\text{univ}), \mu(C/\mathbb{F}_q)) > g^{-\alpha}.$$

Then we easily infer from (**) above that

$$|\mathcal{M}_g(\mathbb{F}_q)(\text{discrep} > g^{-\alpha})|/|\mathcal{M}_g(\mathbb{F}_q)| \leq 2g^{-\beta},$$

i.e., the fraction of curves in $\mathcal{M}_g(\mathbb{F}_q)$ whose discrepancy exceeds $g^{-\alpha}$ is at most $2g^{-\beta}$, provided that $g > N(\varepsilon)$ and provided that $q > M(\varepsilon, g)$. In other words, if g and then q are sufficiently large, then the probability is at least $1 - 2g^{-\beta}$ that a randomly chosen curve in $\mathcal{M}_g(\mathbb{F}_q)$ has discrepancy $\leq g^{-\alpha}$. This is the sense in which most curves of sufficiently large genus over a sufficiently large finite field have a spacing measure which is arbitrarily close to the GUE measure.

To explain how one proves such results, we must now return to a discussion of the GUE measure $\mu(\text{univ})$ and its genesis from compact classical groups $G(N)$. Suppose we take a particular $G(N)$, and an element A in $G(N)$. How close is the spacing measure $\mu(A, G(N))$ to the GUE measure? The answer is that "most" elements A of a large $G(N)$ have their spacing measures quite close to the GUE measure, as the following "law of large numbers" shows.

Theorem (cf. 1.2.6). *In any of the series of compact classical groups $G(N) = U(N)$ or $SU(N)$ or $USp(2N)$ or $SO(2N+1)$ or $SO(2N)$ or $O(2N+1)$ or $O(2N)$, we have*

$$\lim_{N\to\infty} \int_{G(N)} \text{discrep}(\mu(A, G(N)), \mu(\mathit{univ}))\, dA = 0.$$

More precisely, given $\varepsilon > 0$, there exists an integer $N(\varepsilon)$ such that for any $N > N(\varepsilon)$, we have

$$\int_{G(N)} \text{discrep}(\mu(A, G(N)), \mu(\mathit{univ}))\, dA \leq N^{\varepsilon - 1/6}.$$

We also remark that the integrand above,

$$A \mapsto \text{discrep}(\mu(A, G(N)), \mu(\text{univ})),$$

is a continuous (cf. 1.0.12) central function on $G(N)$. This remark will allow us below to apply Deligne's equidistribution theorem (cf. 9.2.6, 9.6.10, 9.7.10) in a completely straightforward way.

The connection between Theorems 12.2.3 and 1.2.6 comes about through monodromy, and Deligne's equidistribution theorem (9.6.10). Recall that the zeta function of a genus g curve $C/\mathbb{F}_q$ is of the form $P(T)/(1-T)(1-qT)$, for P a polynomial of degree $2g$ with the property that the auxiliary polynomial $P(T/\operatorname{Sqrt}(q))$ has all its roots on the unit circle. However, the polynomial $P(T/\operatorname{Sqrt}(q))$ has a bit more structure; namely, its $2g$ roots on the unit circle can be partitioned into g pairs of inverses $(\xi, 1/\xi)$ on the unit circle. One interpretation of this fact is that there exists a conjugacy class $\vartheta(C/\mathbb{F}_q)$ in the compact group $USp(2g)$ such that

$$P(T/\operatorname{Sqrt}(q)) = \det(1 - T\vartheta(C/\mathbb{F}_q)).$$

Because conjugacy classes in $USp(2g)$ are uniquely determined by their characteristic polynomials, there is a unique such conjugacy class $\vartheta(C/\mathbb{F}_q)$ in $USp(2g)$, which we call the unitarized Frobenius conjugacy class attached to $C/\mathbb{F}_q$.

Now fix an integer $N \geq 1$ and a genus $g \geq 1$. Consider a proper smooth family $\pi : \mathcal{C} \to S$ of genus g curves, parameterized by a scheme S which, for simplicity, we assume to be smooth and surjective over $\mathrm{Spec}(\mathbb{Z}[1/N])$ with geometrically connected fibres. We further assume that for every prime number p which does not divide N, the geometric monodromy group of the family of curves

$$\pi \otimes \mathbb{F}_p : \mathcal{C} \otimes \mathbb{F}_p \to S \otimes \mathbb{F}_p$$

in characteristic p is the full symplectic group $Sp(2g)$. Once we have made this assumption about the monodromy of the family, Deligne's equidistribution theorem (cf. 9.6.10) says the following. For each finite field $\mathbb{F}_q$ of characteristic not dividing N, and each point s in the finite set $S(\mathbb{F}_q)$ of $\mathbb{F}_q$-valued points of S, look at the curve $\mathcal{C}_s/\mathbb{F}_q$ named by the point s, and look at its unitarized Frobenius conjugacy classes $\vartheta(\mathcal{C}_s/\mathbb{F}_q)$ in $USp(2g)$. Then these unitarized Frobenius conjugacy classes are equidistributed in the space $USp(2g)^{\#}$ of conjugacy classes in $USp(2g)$ for the probability measure μ_{Haar} on $USp(2g)^{\#}$ which is the direct image from $USp(2g)$ of its normalized Haar measure, in the following sense: for any $\mathbb{C}$-valued continuous central function f on $USp(2g)$, we have the limit formula

$$\lim_{q\to\infty, q \text{ prime to } N} (1/|S(\mathbb{F}_q)|) \sum_{s \text{ in } S(\mathbb{F}_q)} f(\vartheta(\mathcal{C}_s/\mathbb{F}_q))$$
$$= \int_{USp(2g)^{\#}} f \, d\mu_{\mathrm{Haar}}.$$

In order to apply this to study the discrepancy for the curves over finite fields $\mathcal{C}_s/\mathbb{F}_q$ which occur in our family, we have only to apply Deligne's equidistribution theorem above to the continuous central function f on $USp(2g)$ given by

$$A \mapsto \mathrm{discrep}(\mu(A, USp(2g)), \mu(\mathrm{univ})).$$

We know from Theorem 1.2.6 quoted above that, given $\varepsilon > 0$, there is an $N(\varepsilon)$ such that for $g \geq N(\varepsilon)$, we have, for this f, the estimate

$$\int_{USp(2g)^{\#}} f \, d\mu_{\mathrm{Haar}} \leq g^{\varepsilon - 1/6}.$$

So if our family $\mathcal{C}/S$ has $g > N(\varepsilon)$, and we use Deligne's equidistribution theorem to calculate this integral, we find the estimate

$$\lim_{q\to\infty, \text{prime to } N} (1/|S(\mathbb{F}_q)|) \sum_{s \text{ in } S(\mathbb{F}_q)} \mathrm{discrep}(\mu(\mathrm{univ}), \mu(\mathcal{C}_s/\mathbb{F}_q)) \leq g^{\varepsilon - 1/6}.$$

In particular, if q is prime to N and sufficiently large, we will have

$$(**) \qquad (1/|S(\mathbb{F}_q)|) \sum_{s \text{ in } S(\mathbb{F}_q)} \mathrm{discrep}(\mu(\mathrm{univ}), \mu(\mathcal{C}_s/\mathbb{F}_q)) \leq 2g^{\varepsilon - 1/6}.$$

To see what this means about the discrepancy of "most" curves in the family $\mathcal{C}/S$, pick a pair of positive real numbers α, β with $\alpha + \beta = 1/6 - \varepsilon$. Denote by

$$S(\mathbb{F}_q)(\mathrm{discrep} > g^{-\alpha}) \subset S(\mathbb{F}_q)$$

the set of those s in $S(\mathbb{F}_q)$ for which

$$\mathrm{discrep}(\mu(\mathrm{univ}), \mu(\mathcal{C}_s/\mathbb{F}_q)) > g^{-\alpha}.$$

Then we easily infer from $(**)$ above that

$$|S(\mathbb{F}_q)(\text{discrep} > g^{-\alpha})|/|S(\mathbb{F}_q)| \leq 2g^{-\beta},$$

provided that $g \geq N(\varepsilon)$ and that q is prime to N and sufficiently large (how large depends on the particular family $\mathcal{C}/S$).

To obtain Theorem 12.2.3 stated above about $\mathcal{M}_g$, we need only replace $S(\mathbb{F}_q)$ in the above formulas by $\mathcal{M}_g(\mathbb{F}_q)$. There are some technical difficulties to be overcome in justifying this formal replacement; cf. 10.6 and 10.7 for an exhaustive discussion of these difficulties and their resolution.

Once we know that "most" curves over finite fields have their spacing measure close to the GUE measure, it is natural to ask if, given a finite field $\mathbb{F}_q$, we can exhibit a single **explicit** sequence of curves $\{C_g/\mathbb{F}_q\}_g$ with C_g of genus g, whose spacing measures $\mu(C_g/\mathbb{F}_q)$ approach the GUE measure $\mu(\text{univ})$ as $g \to \infty$, in the sense that $\lim_{g\to\infty} \text{discrep}(\mu(C_g/\mathbb{F}_q), \mu(\text{univ})) = 0$. We do not know how to do this at present.

To the extent that we can write down families (of varieties, of exponential sums, of ...) over finite fields whose geometric monodromy groups are large classical groups, we will get results similar to those for curves for the behavior of the discrepancy in these families as well. We work this out explicitly for universal families of abelian varieties (where the group is again Sp), of smooth hypersurfaces in projective space (where the group is either Sp or O), and for multi-variable Kloosterman sums (where the group is either Sp, SL or $SO(odd)$). Again in these more general cases we do not know how to write down explicit sequences of objects of the type considered whose spacing measures approach the GUE measure.

In the case of the Kloosterman sums $\text{Kl}_n(\psi, a \text{ in } \mathbb{F}_q^\times)$ there is a plausible candidate for such a sequence.

Conjecture. *Fix a finite field $\mathbb{F}_q$, fix any choice of a in $\mathbb{F}_q^\times$ and fix any choice of the nontrivial additive character ψ of $\mathbb{F}_q$. Then the spacing measure*

$$\mu(\text{Kl}_n(\psi, a \textit{ in } \mathbb{F}_q^\times))$$

attached to $\text{Kl}_n(\psi, a \text{ in } \mathbb{F}_q^\times)$, or more precisely to its L-function, tends to the GUE measure as $n \to \infty$ in the sense that

$$\lim_{n\to\infty} \text{discrep}(\mu(\text{Kl}_n(\psi, a \textit{ in } \mathbb{F}_q^\times)), \mu(\textit{univ})) = 0.$$

Suppose now that we fix an integer $N \geq 1$, and a large integer g. Suppose that we are given a curve $C/\mathbb{Z}[1/N]$ which is proper and smooth with geometrically connected fibres of genus g. For any prime p not dividing N, the reduction mod p of our curve $C/\mathbb{Z}[1/N]$ is a curve $C \otimes \mathbb{F}_p/\mathbb{F}_p$ of genus g, which has a spacing measure $\mu(C \otimes \mathbb{F}_p/\mathbb{F}_p)$. When is it reasonable to expect that for most primes p which are prime to N, the spacing measure $\mu(C \otimes \mathbb{F}_p/\mathbb{F}_p)$ is close to the GUE measure $\mu(\text{univ})$? When should we expect some other behaviour?

Given $C/\mathbb{Z}[1/N]$ as above, for every prime p not dividing N, we obtain a unitarized Frobenius conjugacy class $\vartheta(C \otimes \mathbb{F}_p/\mathbb{F}_p)$ in $USp(2g)$. When is it reasonable to expect that these classes $\vartheta(C \otimes \mathbb{F}_p/\mathbb{F}_p)$ are equidistributed in $USp(2g)^\#$, in the sense that for any $\mathbb{C}$-valued continuous central function f on $USp(2g)$ we have

$$\int_{USp(2g)} f(A)\, dA = \lim_{X\to\infty} (1/\pi(X)) \sum_{p \leq X \text{ prime to } N} f(\vartheta(C \otimes \mathbb{F}_p/\mathbb{F}_p))?$$

The **Generalized Sato-Tate Conjecture** is that this equidistribution (of the classes $\vartheta(C \otimes \mathbb{F}_p/\mathbb{F}_p)$ in $USp(2g)^{\#}$) holds whenever $C/\mathbb{Z}[1/N]$ has big arithmetic monodromy, in the sense that for every prime l, the action of $\mathrm{Gal}(\overline{\mathbb{Q}}/\mathbb{Q})$ on $H^1(C \otimes \overline{\mathbb{Q}}, \mathbb{Q}_l)$ has image which is open in the group $GSp(2g, \mathbb{Q}_l)$ of symplectic similitudes. As an immediate application of Theorem 12.2.3, we find:

Theorem. *In the notations of Theorem 1.2.6, let $\varepsilon > 0$ and $g > N(\varepsilon)$. Suppose $C/\mathbb{Z}[1/N]$ is a curve of genus g as above, which has big arithmetic monodromy. Suppose the generalized Sato-Tate conjecture holds. Then we have the inequality*

$$\lim_{X\to\infty} (1/\pi(X)) \sum_{p \le X \text{ prime to } N} \mathrm{discrep}(\mu(\vartheta(C \otimes \mathbb{F}_p/\mathbb{F}_p)), \mu(univ)) \le g^{\varepsilon - 1/6}.$$

Corollary. *Suppose for each integer g_i in an infinite subset Γ of $\mathbb{Z}_{\ge 1}$, we are given an integer $N_i \ge 1$ and a curve $C_{g_i}/\mathbb{Z}[1/N_i]$ of genus g_i which has big arithmetic monodromy. Suppose the generalized Sato-Tate conjecture holds. Then the double limit $\lim_{i\to\infty} \lim_{X\to\infty}$ of*

$$(1/\pi(X)) \sum_{p \le X \text{ prime to } N_i} \mathrm{discrep}(\mu(\vartheta(C_{g_i} \otimes \mathbb{F}_p/\mathbb{F}_p)), \mu(univ))$$

vanishes.

Question. Notations and hypotheses as in the corollary, suppose that all N_i have a common value N, cf. pages 12–13 of this Introduction for examples of such situations. What is the density of the set of primes p not dividing N for which

$$\lim_{g\to\infty \text{ in } \Gamma} \mathrm{discrep}(\mu(\vartheta(C_g \otimes \mathbb{F}_p/\mathbb{F}_p)), \mu(\mathrm{univ})) = 0?$$

Presumably this need not hold for **every** prime p not dividing N.

Now let us turn to the opposite extreme, cases in which either we can prove or we expect that the spacing measure is far from the GUE measure.

We first give, for every odd prime p, a sequence of curves over the prime field $\mathbb{F}_p$ whose genera go to infinity and whose spacing measures converge to the delta measure δ_0 supported at the origin. For each power $q = p^f$ of p, we consider the hyperelliptic curve $C_q/\mathbb{F}_p$ of equation

$$C_q : Y^2 = X^q - X.$$

This curve has genus g given by $2g = q - 1$. Over $\mathbb{F}_q$, this curve admits the Artin-Schreier action $X \mapsto X + \alpha, Y \mapsto Y$ of the additive group of $\mathbb{F}_q$. If we pick any prime $l \ne p$, and decompose the cohomology group $H^1(C_q \otimes \overline{\mathbb{F}_q}, \overline{\mathbb{Q}_l})$ under this action, each of the $q-1$ nontrivial additive characters ψ of $\mathbb{F}_q$ occurs with multiplicity one. On the corresponding one-dimensional eigenspace, the Frobenius with respect to $\mathbb{F}_q$, Frob_q, necessarily acts as a scalar, and that scalar is none other than minus the quadratic Gauss sum over $\mathbb{F}_q$:

$$-G_q(\psi, \chi_2) := - \sum_{x \text{ in } \mathbb{F}_q^\times} \psi(x)\chi_2(x),$$

where we have written χ_2 for the quadratic character of $\mathbb{F}_q^\times$. Now it is elementary that the quadratic Gauss sum $G_q(\psi, \chi_2)$ with any nontrivial ψ satisfies

$$(G_q(\psi, \chi_2))^2 = \chi_2(-1)q.$$

Moreover, both square roots occur as ψ varies. Thus Frob_q has precisely 2 distinct eigenvalues on H^1. To see what this means for Frob_p, write $q = p^f$. Then Frob_q is the f'th power of Frob_p, and hence Frob_p has at most $2f$ distinct eigenvalues on H^1. This means that among the $q - 1 = p^f - 1$ normalized spacings between the reciprocal zeroes of the zeta function of $C_q/\mathbb{F}_p$, all but at most $2f$ of the spacings are equal to zero. Since $2f/(p^f - 1) \to 0$ as $f \to \infty$, we see that the spacing measures $\mu(C_1/\mathbb{F}_p)$ approach δ_0 as $f \to \infty$. Since the GUE measure is absolutely continuous with respect to Lebesgue measure, it gives the origin mass zero. Hence we have

$$\mathrm{discrep}(\mu(C_q/\mathbb{F}_p), \mu(\mathrm{univ})) \geq 1 - 2f/(p^f - 1),$$

a crude quantification of the statement that $\mu(C_q/\mathbb{F}_p)$ is far from the GUE measure.

Here is another example, valid for any prime $p > 0$, of a sequence of curves over $\mathbb{F}_p$ whose genera go to infinity and whose spacing measures converge to δ_0. For each power $q := p^f$ of p, consider the degree $q + 1$ Fermat curve over $\mathbb{F}_p$, say, $\mathrm{Fermat}(q + 1)/\mathbb{F}_p$, of homogeneous equation

$$X^{q+1} + Y^{q+1} = Z^{q+1}.$$

This curve has genus g given by $2g = q(q - 1)$. It is elementary that over $\mathbb{F}_{q^2}$, this curve has $1 + q^3$ points. [Hint: for x in $\mathbb{F}_{q^2}$, x^{q+1} is its norm to $\mathbb{F}_q$, and the norm map is surjective.] But we readily compute that

$$1 + q^3 = 1 + q^2 + 2gq = 1 + q^2 - 2g(-q).$$

Thus the Weil bound is attained, and hence every eigenvalue of Frob_{q^2} on H^1 is $-q$. Therefore Frob_p has at most $2f$ distinct eigenvalues on H^1, while $\dim H^1 = p^f(p^f - 1)$, and we conclude as in the previous example that the spacing measure $\mu(\mathrm{Fermat}(q + 1)/\mathbb{F}_p)$ tends to δ_0 as $f \to \infty$, and that

$$\mathrm{discrep}(\mu(\mathrm{Fermat}(q + 1)/\mathbb{F}_p), \mu(\mathrm{univ})) \geq 1 - 2f/(p^f(p^f - 1)).$$

We now turn to a case in which we **expect** the spacing measure to be far from the GUE measure. For each prime l, consider the modular curve $X_0(l)/\mathbb{Z}[1/l]$, whose genus g_l is approximately $(l - 1)/12$. Choose any prime l'. When we decompose $H^1 := H^1(X_0(l) \otimes \overline{\mathbb{Q}}, \overline{\mathbb{Q}}_{l'})$ under the Hecke operators, we find a direct sum of g_l two-dimensional subspaces, corresponding to the g_l different weight two normalized ($a_1(f) = 1$) Hecke-eigenforms $f = \sum_{n\geq 1} a_n(f)q^n$ on $X_0(l)$. For each such eigenform f, and each prime p with $p \neq l$ and $p \neq l'$, the characteristic polynomial of Frob_p on the two-dimensional Hecke eigenspace in H^1 named by f is $X^2 - a_p(f)X + p$. We know by Deligne that $|a_p(f)| \leq 2\,\mathrm{Sqrt}(p)$, so the two eigenvalues of Frob_p here are $\mathrm{Sqrt}(p)e^{\pm i\vartheta_p(f)}$, where $\vartheta_p(f)$ is the unique angle in $[0, \pi]$ such that $a_p(f) = 2\,\mathrm{Sqrt}(p)\cos(\vartheta_p(f))$. We denote by ϑ_p in $[0, \pi]^{g_l}$ the g_l-tuple of angles $\vartheta_p(f)$ indexed by eigenforms f, and we view ϑ_p as a conjugacy class in the product group $USp(2)^{g_l} = SU(2)^{g_l}$.

This Hecke-eigenvalue decomposition of H^1 is respected by $\mathrm{Gal}(\overline{\mathbb{Q}}/\mathbb{Q})$, and forces the image of $\mathrm{Gal}(\overline{\mathbb{Q}}/\mathbb{Q})$ to land in the subgroup of the g_l-fold self product $GL(2, \overline{\mathbb{Q}}_{l'})^{g_l}$ of $GL(2, \overline{\mathbb{Q}}_{l'})$ with itself consisting of elements $(A_1, \ldots, A_{g_l})$ all of which have equal determinants. According to Ribet [**Rib**, 7.18], the image of $\mathrm{Gal}(\overline{\mathbb{Q}}/\mathbb{Q})$ is Zariski dense in this group. In view of Ribet's result, the natural Sato-Tate conjecture for $X_0(l)$ is the assertion that the conjugacy classes $\{\vartheta_p\}_{p\neq l}$ in the product group $SU(2)^{g_l}$ are equidistributed with respect to Haar measure in the sense that for any $\mathbb{C}$-valued continuous central function h on $SU(2)^{g_l}$, the integral

$$\int_{SU(2)^{g_l}} h(A)\,dA = \int_{[0,\pi]^{g_l}} h(x_1,\ldots,x_{g_l}) \prod_{i=1}^{g_l} (2/\pi)\sin^2(x_i)\,dx_i$$

can be computed as the limit

$$\lim_{X\to\infty} (1/\pi(X)) \sum_{p\le X, p\ne l} h(\vartheta_p).$$

Let us admit the truth of this Sato-Tate conjecture for $X_0(l)$. We denote by

$$F(x) := (2/\pi)\int_{[0,x]} \sin^2(t)\,dt = (2x - \sin(2x))/2\pi$$

the bijection from $[0,\pi]$ to $[0,1]$ which carries the measure $(2/\pi)\sin^2(t)\,dt = dF$ on $[0,\pi]$ to uniform measure dx on $[0,1]$. [We call F the straightening function for the measure in question.] Given an element ϑ in $[0,\pi]^{g_l}$, we denote by $F(\vartheta)$ in $[0,1]^{g_l}$ the result of applying F component by component. The Sato-Tate conjecture for $X_0(l)$ says precisely that the g_l-tuples $\{F(\vartheta_p)\}_{p\ne l}$ in $[0,1]^{g_l}$ are equidistributed in $[0,1]^{g_l}$ for uniform measure. Arrange the components of $F(\vartheta_p)$ in increasing numerical order, say

$$0 \le F(\vartheta_p)_1 \le F(\vartheta_p)_2 \le \cdots \le F(\vartheta_p)_{g_l} \le 1.$$

The F-straightened spacing measure $\mu_F(X_0(l)\otimes\mathbb{F}_p/\mathbb{F}_p)$ attached to $X_0(l)\otimes\mathbb{F}_p/\mathbb{F}_p$ is the measure of total mass $1 - 1/g_l$ on $\mathbb{R}$ which gives each of the $g_l - 1$ normalized spacings $s_i := g_l(F(\vartheta_p)_{i+1} - F(\vartheta_p)_i)$ the mass $1/g_l$. An elementary analysis of spacings between points in an interval which are randomly chosen for Lebesgue measure shows that in this kind of question, the limiting answer is not the rather exotic GUE, but rather the much more elementary and familiar exponential distribution $\mu(\text{Poisson})$, the measure on $\mathbb{R}$ supported in $\mathbb{R}_{\ge 0}$ and given there by $e^{-x}\,dx$.

Theorem. *Assume the Sato-Tate conjecture above for all the modular curves $X_0(l)/\mathbb{Z}[1/l], l$ any prime. Then the F-straightened spacing measures*

$$\mu_F(X_0(l)\otimes\mathbb{F}_p/\mathbb{F}_p)$$

are, for large l, very near the Poisson measure $\mu(\text{Poisson})$ for most primes $p\ne l$. More precisely, the double limit $\lim_{l\to\infty}\lim_{X\to\infty}$ of

$$(1/\pi(X)) \sum_{p\le X, p\ne l} \operatorname{discrep}(\mu_F(X_0(l)\otimes\mathbb{F}_p/\mathbb{F}_p), \mu(\text{Poisson}))$$

vanishes.

Question. Is it true that for each prime p, we have

$$\lim_{l\to\infty, l\ne p} \operatorname{discrep}(\mu_F(X_0(l)\otimes\mathbb{F}_p/\mathbb{F}_p), \mu(\text{Poisson})) = 0?$$

We now discuss another aspect of our work, the distribution in families of "low-lying zeroes". [This terminology "low-lying zeroes" is inspired by the number field picture, where we expect all the nontrivial zeroes to lie on a single vertical line, and we measure height from the real axis. In the finite field case, where the normalized zeroes lie on the unit circle, it would be more accurate to speak of "zeroes near 1".] For simplicity, we will discuss only the case of curves. Recall that the zeta function of a genus g curve $C/\mathbb{F}_q$ is of the form $P(T)/(1-T)(1-qT)$, for P a polynomial of degree $2g$ with the property that the auxiliary polynomial $P(T/\operatorname{Sqrt}(q))$ has all its $2g$ roots on the unit circle, and its $2g$ roots can be partitioned into g pairs of inverses $(\xi, 1/\xi)$. So we may write $P(T)$ as

$$P(T) = \prod_{j=1}^{g}(1-\alpha_j T)(1-\alpha_{-j}T),$$

with

$$\alpha_{-j} = \overline{\alpha}_j, \qquad \alpha_j\alpha_{-j} = q.$$

If we pick the α_j (rather than the α_{-j}) to lie in the upper half plane, we have

$$\alpha_j = \operatorname{Sqrt}(q)e^{i\varphi_j}, \qquad 0 \le \varphi_j \le \pi,$$

and with suitable renumbering the g angles φ_j in $[0,\pi]$ may be assumed to be in increasing order:

$$0 \le \varphi_1 \le \varphi_2 \le \cdots \le \varphi_g \le \pi.$$

With this numbering, we refer to $(g/\pi)\varphi_j$ as the j'th normalized angle attached to the curve $C/\mathbb{F}_q$, or, if we like, attached to the unitarized Frobenius conjugacy class $\vartheta(C/\mathbb{F}_q)$.

More generally, given any element A in $USp(2g)$, we have

$$\det(1-TA) = \prod_{j=1}^{g}(1-Te^{i\varphi_j})(1-Te^{-i\varphi_j})$$

for a unique sequence of angles $0 \le \varphi_1 \le \varphi_2 \le \cdots \le \varphi_g \le \pi$. For each integer $1 \le j \le g$, the function on $USp(2g)$ defined by

$$A \mapsto \varphi_j := \varphi_j(A)$$

is a continuous central function from $USp(2g)$ to $[0,\pi]$. We refer to $(g/\pi)\varphi_j(A)$ as the j'th normalized angle attached to the conjugacy class of the element A. The function

$$A \mapsto (g/\pi)\varphi_j(A)$$

is a continuous central function from $USp(2g)$ to $\mathbb{R}_{\ge 0}$.

If we take the direct image of normalized Haar measure μ_{Haar} on $USp(2g)$ by the map $A \mapsto (g/\pi)\varphi_j(A)$, we obtain a probability measure on $\mathbb{R}$ supported in $\mathbb{R}_{\ge 0}$, which we denote $\nu(j, USp(2g))$.

There are analogous constructions for the other classical groups, cf. 6.9 for the details, which give rise to probability measures $\nu(j, G(N))$ for $1 \le j \le N$ and $G(N)$ any of $U(N), USp(2N), SO(2N+1), SO(2N), O_-(2N+1), O_-(2N+2)$, all on the real line and all supported in $\mathbb{R}_{\ge 0}$. Now unlike the spacing measures, which were "universal" in the sense that the large N limit existed and was independent of which sequence of $G(N)$'s we ran through, these ν measures do depend on the sequence of $G(N)$'s chosen.

Theorem (7.5.5, 7.5.6). *For each integer $j \ge 1$, there exist*[2] *three probability measures $\nu(j), \nu(-,j)$, and $\nu(+,j)$ on $\mathbb{R}$, supported in $\mathbb{R}_{\ge 0}$ and having continuous*

[2]In fact, these measures all have densities, which for $j=1$ are shown in Appendix: Graphs.

CDF's, such that we have the following large N limit statements for convergence in the sense of uniform convergence of CDF's:

$$\lim_{N\to\infty} \nu(j, G(N)) = \begin{cases} \nu(j), & \textit{if } G(N) = U(N), \\ \nu(+,j), & \textit{if } G(N) = SO(2N) \textit{ or } O_-(2N+1), \\ \nu(-,j), & \textit{if } G(N) = USp(2N), SO(2N+1), O_-(2N+2). \end{cases}$$

Here is a convenient mnemonic to remember which sign in $\nu(\pm, j)$ is given by which orthogonal group series: the sign is the common sign of $\det(-A)$ for A in either $SO(2N)$ or $O_-(2N+1)$ or $SO(2N+1)$ or $O_-(2N+2)$. Experts will recognize this sign as being the **sign in the functional equation** of $P(T) := \det(1 - TA)$ under $T \mapsto 1/T$, which for orthogonal A is

$$T^{\deg(P)} P(1/T) = \det(-A) P(T).$$

To see what this means concretely for curves, fix an integer $N \geq 1$, and suppose for each genus $g \geq 1$ we are given a proper smooth family $\pi : \mathcal{C}_g \to S_g$ of genus g curves, parameterized by a scheme S_g which is smooth and surjective over $\mathrm{Spec}(\mathbb{Z}[1/N])$ with geometrically connected fibres. We further assume that for every prime number p which does not divide N, the geometric monodromy group of the family of curves

$$\pi \otimes \mathbb{F}_p : \mathcal{C}_g \otimes \mathbb{F}_p \to S_g \otimes \mathbb{F}_p$$

in characteristic p is the full symplectic group $Sp(2g)$. For example, we might take $N = 2$, and for $\mathcal{C}_g/S_g$ the universal family of hyperelliptic curves $Y^2 = f_{2g+1}(X)$ parameterized by the space $S_g := \mathcal{H}_{2g+1}$ of monic polynomials f_{2g+1} of degree $2g+1$ with invertible discriminant.

Let now $h(x)$ be any $\mathbb{C}$-valued continuous function on $\mathbb{R}$. By Deligne's equidistribution theorem, for each integer $j \geq 1$, and each genus $g \geq j$, we can compute the integral

$$\int_{\mathbb{R}} h \, d\nu(j, USp(2g)) := \int_{USp(2g)} h((g/\pi)\varphi_j(A)) \, dA$$

as the limit

$$\lim_{q\to\infty,\ \text{prime to } N} (1/|S_g(\mathbb{F}_q)|) \sum_{s \text{ in } S(\mathbb{F}_q)} h((g/\pi)\varphi_j(\vartheta(\mathcal{C}_{g,s}/\mathbb{F}_q))).$$

Using the theorem above, about the large N limit of the measures $\nu(j, USp(2g))$ being $\nu(-,j)$, we find that, for $h(x)$ any bounded $\mathbb{C}$-valued continuous function on $\mathbb{R}$, we can compute the integral

$$\int_{\mathbb{R}} h \, d(\nu)(-,j) = \lim_{g\to\infty} \int_{USp(2g)} h((g/\pi)\varphi_j(A)) \, dA$$

as the double limit

$$\lim_{g\to\infty} \lim_{q\to\infty,\ \text{prime to } N} (1/|S_g(\mathbb{F}_q)|) \sum_{s \text{ in } S(\mathbb{F}_q)} h((g/\pi)\varphi_j(\vartheta(\mathcal{C}_{g,s}/\mathbb{F}_q))).$$

If we look instead at universal families of hypersurfaces of even dimension, and average over those whose functional equation (for the factor of its zeta function corresponding to the primitive part of middle dimensional cohomology) has a fixed sign $\varepsilon = \pm 1$, we get double limit formulas for $\int_{\mathbb{R}} h \, d\nu(\varepsilon, j)$. [Universal families of odd dimensional hypersurfaces have monodromy group Sp, so will lead only to double limit formulas for $\int_{\mathbb{R}} h\, d\nu(-,j)$.]

We now leave the realm of what is proven, and discuss what might be true if, in the double limit formulas above, we were to omit the inner limit over q. Again to fix ideas, we return to the case of curves. Fix an integer $N \geq 1$, and an infinite subset Γ of $\mathbb{Z}_{\geq 1}$. Suppose for each genus g in Γ we are given a proper smooth family $\pi : \mathcal{C}_g \to S_g$ of genus g curves, parameterized by a scheme S_g which is smooth and surjective over $\mathrm{Spec}(\mathbb{Z}[1/N])$ with geometrically connected fibres. We further assume that for every prime number p which does not divide N, and for every g in Γ, the geometric monodromy group of the family of curves

$$\pi \otimes \mathbb{F}_p : \mathcal{C}_g \otimes \mathbb{F}_p \to S_g \otimes \mathbb{F}_p$$

in characteristic p is the full symplectic group $Sp(2g)$. To further simplify matters, we assume also that $S_g(\mathbb{F}_p)$ is nonempty for every prime p not dividing N, and for every genus g in Γ.

We say that this collection of families $\{\mathcal{C}_g/S_g\}_{g \text{ in } \Gamma}$ weakly calculates the measure $\nu(-,j)$ if, for every finite field $\mathbb{F}_q$ of characteristic prime to N, and for every bounded continuous function h on $\mathbb{R}$, we can calculate $\int_{\mathbb{R}} h\, d\nu(-,j)$ as

$$\lim_{n\to\infty} \left(1/\sum_{g\leq n \text{ in } \Gamma} |S_g(\mathbb{F}_q)|\right) \sum_{g\leq n \text{ in } \Gamma} \sum_{s \text{ in } S_g(\mathbb{F}_q)} h((g/\pi)\varphi_j(\vartheta(\mathcal{C}_{g,s}/\mathbb{F}_q))).$$

We say that this collection of families $\{\mathcal{C}_g, S_g\}_{g \text{ in } \Gamma}$ strongly calculates the measure $\nu(-,j)$ if, for every finite field $\mathbb{F}_q$ of characteristic prime to N, and for every bounded continuous function h on $\mathbb{R}$, we can calculate $\int_{\mathbb{R}} h\, d\nu(-,j)$ as

$$\lim_{g\to\infty \text{ in } \Gamma} (1/|S_g(\mathbb{F}_q)|) \sum_{s \text{ in } S_g(\mathbb{F}_q)} h((g/\pi)\varphi_j(\vartheta(\mathcal{C}_{g,s}/\mathbb{F}_q))).$$

It is elementary that if $\{\mathcal{C}_g/S_g\}_{g \text{ in } \Gamma}$ strongly calculates the measure $\nu(-,j)$, then $\{\mathcal{C}_g/S_g\}_{g \text{ in } \Gamma}$ weakly calculates it as well. If for every finite field $\mathbb{F}_q$ of characteristic prime to N, the cardinalities $|S_g(\mathbb{F}_q)|$ grow fast enough that the ratios

$$(1/|S_g(\mathbb{F}_q)|) \sum_{\gamma\leq g \text{ in } \Gamma} |S_\gamma(\mathbb{F}_q)|$$

stay bounded (as g varies over Γ, q fixed: the bound can vary with q), then the two notions, strong and weak calculation of $\nu(-,j)$, are equivalent.

Conjecture. *Fix an integer $N \geq 1$, and an infinite subset Γ of $\mathbb{Z}_{\geq 1}$. Suppose for each genus g in Γ we are given a proper smooth family $\pi : \mathcal{C}_g \to S_g$ of genus g curves, parameterized by a scheme S_g which is smooth and surjective over* $\mathrm{Spec}(\mathbb{Z}[1/N])$ *with geometrically connected fibres. Suppose that for every g in Γ and for every prime number p which does not divide N, the geometric monodromy group of the family of curves*

$$\pi \otimes \mathbb{F}_p : \mathcal{C}_g \otimes \mathbb{F}_p \to S_g \otimes \mathbb{F}_p$$

in characteristic p is the full symplectic group $Sp(2g)$. Suppose also that $S_g(\mathbb{F}_p)$ is nonempty for every prime p not dividing N, and for every genus g in Γ. Then for every integer $j \geq 1$, the collection of families $\{\mathcal{C}_g/S_g\}_{g \text{ in } \Gamma}$ weakly calculates the measure $\nu(-,j)$. Moreover, if in addition we have $\lim_{g\to\infty \text{ in } \Gamma} |S_g(\mathbb{F}_q)| = \infty$

for every finite field $\mathbb{F}_q$ of characteristic prime to N, then the collection of families $\{\mathcal{C}_g/S_g\}_{g \text{ in } \Gamma}$ strongly calculates the measure $\nu(-,j)$.[3]

Let us give some examples of situations $N, \Gamma, \{\mathcal{C}_g/S_g\}_{g \text{ in } \Gamma}$ which satisfy all of the hypotheses imposed in the statement of the conjecture.

1) $N = 1, \Gamma = \mathbb{Z}_{\geq 1}, \mathcal{M}_{g,3K}$, universal family of curves with $3K$ structure (cf. 10.6).

2) $N = 2, \Gamma = \mathbb{Z}_{\geq 1}, \mathcal{H}_{2g+1}$ or $\widetilde{\mathcal{H}}_{2g+1}$, family $Y^2 = f_{2g+1}(X)$ (cf. 10.1.18.3–4).

2bis) $N = 2, \Gamma = \mathbb{Z}_{\geq 1}, \mathcal{H}_{2g+2}$ or $\widetilde{\mathcal{H}}_{2g+2}$, family $Y^2 = f_{2g+2}(X)$ (cf. 10.1.18.4–5).

3) $N = 2, \Gamma = \mathbb{Z}_{\geq 1}$: write $2g$ to the base 2 as $\sum_{\text{some } \alpha \geq 1} 2^\alpha$, and define $F_{2,2g}(X)$ as the corresponding product of cyclotomic polynomials $\Phi_{2^{\alpha+1}}(X) = (X^{2^\alpha} + 1)$:

$$F_{2,2g}(X) := \prod_{\alpha \text{ in base 2 expansion of } 2g} (X^{2^\alpha} + 1).$$

Take for $\mathcal{C}_g/S_g$ the one-parameter ("T") family of hyperelliptic curves of equation $Y^2 = (X - T)F_{2,2g}(X)$.

4) $N = 2l$ for l a prime, $\Gamma =$ those integers $g \geq 1$ such that in the base l expression of $2g$, all the digits are either 0 or $l - 1$: write $2g$ to the base l as $\sum_{\text{some } \alpha \geq 0} (l-1)l^\alpha$, and define $F_{l,2g}(X)$ as the corresponding product of cyclotomic polynomials $\Phi_{l^{\alpha+1}}(X)$. Take for $\mathcal{C}_g/S_g$ the one-parameter ("T") family of hyperelliptic curves of equation $Y^2 = (X - T)F_{l,2g}(X)$. [Of course, if we take $l = 2$ in this example, we find example 3).]

5) $N = 2l$ for l a prime, $\Gamma =$ those $g \geq 1$ such that $2g = (l-1)l^\alpha$ for some integer $\alpha \geq 0$. Take for $\mathcal{C}_g/S_g$ the one-parameter ("T") family of hyperelliptic curves of equation $Y^2 = (X - T)\Phi_{l^{\alpha+1}}(X)$.

Notice that in examples 1), 2), and 2bis), we do have

$$\lim_{g \to \infty \text{ in } \Gamma} |S_g(\mathbb{F}_q)| = \infty.$$

But in examples 3) through 5), the parameter space S_g is always a Zariski open set in the affine line $\mathbb{A}^1$, so $|S_g(\mathbb{F}_q)| \leq q$ is uniformly bounded. The relevant sets $S_g(\mathbb{F}_q)$ are always nonempty, since both 0 and ± 1 are always allowed parameter values.

The conjecture in the examples 2) and 2bis) ($\widetilde{\mathcal{H}}$ version) can be viewed as a statement about the low-lying zeroes of the L-functions of all quadratic extensions of $\mathbb{F}_q(X)$. So seen, it has an analogue for Dirichlet L-series with quadratic character, which we will now formulate. Thus we take a quadratic extension $K/\mathbb{Q}$, of discriminant D_K, corresponding to the quadratic Dirichlet character χ_K. We assume that $L(s, \chi_K)$ satisfies the Riemann Hypothesis. We write the nontrivial zeroes of $L(s, \chi_K)$ (which by the (even!) functional equation occur in conjugate pairs) as $1/2 \pm i\gamma_{K,j}$ with $0 \leq \gamma_{K,1} \leq \gamma_{K,2} \leq \gamma_{K,3} \leq \cdots$.

Conjecture. *The low-lying zeroes of Dirichlet L-functions with quadratic character weakly calculate the measure $\nu(-,j)$, in the following sense. For any integer $j \geq 1$, and for any compactly supported continuous $\mathbb{C}$-valued function h on $\mathbb{R}$, we can calculate the integral $\int_{\mathbb{R}} h \, d\nu(-,j)$ as*

$$\lim_{X \to \infty} (1/|\{K \text{ with } D_K \leq X\}|) \sum_{K \text{ with } D_K \leq X} h(\gamma_{K,j} \log(D_K)/2\pi).$$

---

[3]This second part of the conjecture, about strong calculation, seems to us more speculative than the first part.

The measure $\nu(-,1)$ has a density which has the remarkable property of vanishing to second order at the origin, see Appendix: Graphs. Thus our conjecture implies that $L(s, \chi_K)$ rarely has a zero at or even near the point $1/2$. This sort of behaviour was observed at a crude level by Hazelgrove, who seems to have been the first to experiment numerically with zeroes of Dirichlet L-functions. The above conjecture offers a precise version, and hints at the existence, in the global case, of some remarkable analogue, yet to be discovered, of the symplectic monodromy which in the function field case binds together the L-functions of quadratic extensions of $\mathbb{F}_q(X)$.

For a global situation in which both measures $\nu(-,1)$ and $\nu(+,j)$ arise, one has elliptic curves over, say, $\mathbb{Q}$, where some analogue of orthogonal monodromy seems to enter. Let us grant that all elliptic curves over $\mathbb{Q}$ are modular, so that their L-functions are entire, and let us assume further that these L-functions have all their (nontrivial) zeroes on the line $\mathrm{Re}(s) = 1$. For each integer $n \geq 1$, denote by $\mathcal{E}_n$ the set of $\mathbb{Q}$-isogeny classes of elliptic curves over $\mathbb{Q}$ with conductor $N_E = n$, and by $\mathcal{E}_{n,+}$ and $\mathcal{E}_{n,-}$ the subsets of $\mathcal{E}_n$ consisting of those curves whose L-functions have an even or odd functional equation, respectively.

If $E/\mathbb{Q}$ has an even functional equation, the nontrivial zeroes of $L(s, E/\mathbb{Q})$ occur in conjugate pairs, and we write them as

$$1 \pm i\gamma_{K,j} \text{ with } 0 \leq \gamma_{E,1} \leq \gamma_{E,2} \leq \gamma_{E,3} \leq \cdots .$$

If $E/\mathbb{Q}$ has an odd functional equation, then 1 is a zero of $L(s, E/\mathbb{Q})$, and the remaining nontrivial zeroes of $L(s, E/\mathbb{Q})$ occur in conjugate pairs: we write the remaining zeroes as $1 \pm i\gamma_{E,j}$ with $0 \leq \gamma_{E,1} \leq \gamma_{E,2} \leq \gamma_{E,3} \leq \cdots$.

Conjecture. *The low-lying zeroes of L-functions of elliptic curves over $\mathbb{Q}$ weakly calculate the measure $\nu(\pm j)$, in the following sense. For any integer $j \geq 1$, and for any compactly supported continuous $\mathbb{C}$-valued function h on $\mathbb{R}$, we can calculate the integrals $\int_{\mathbb{R}} h\, d\nu(\pm, j)$ as follows:*

$$\int_{\mathbb{R}} h\, d\nu(-,j) = \lim_{X \to \infty} \left(1/\sum_{n \leq X} |\mathcal{E}_{n,-}| \right) \sum_{n \leq X, E \text{ in } \mathcal{E}_{n,-}} h(\gamma_{E,j} \log(N_E)/2\pi),$$

and

$$\int_{\mathbb{R}} h\, d(+,j) = \lim_{X \to \infty} \left(1/\sum_{n \leq X} |\mathcal{E}_{n,+}| \right) \sum_{n \leq X, E \text{ in } \mathcal{E}_{n,+}} h(\gamma_{E,j} \log(N_E)/2\pi).$$

As already remarked above, the measure $\nu(-,1)$ has a density which vanishes to second order at the origin. So the conjecture for $j = 1$ predicts that among L-functions of elliptic curves $E/\mathbb{Q}$ with odd functional equation, most should have only a simple zero at $s = 1$ and no zeroes very near to $s = 1$. In particular, just using the absolute continuity of $\nu(-,1)$ with respect to Lebesgue measure, the conjecture implies that

$$\lim_{X \to \infty} \left(1/\sum_{n \leq X} |\mathcal{E}_{n,-}| \right) \sum_{n \leq X} |\{E \text{ in } \mathcal{E}_{n,-} \text{ with } \gamma_{E,1} = 0\}| = 0.$$

The measure $\nu(+,1)$ also has a density. Its density, unlike that of $\nu(-,1)$, is nonzero at the origin. Nonetheless, $\nu(+,1)$ is absolutely continuous with respect to

Lebesgue measure on $\mathbb{R}$. So our conjecture for $j = 1$ implies that among $E/\mathbb{Q}$ with even functional equation, most should have no zero at $s = 1$, in the sense that

$$\lim_{X\to\infty}\left(1/\sum_{n\le X}|\mathcal{E}_{n,+}|\right)\sum_{n\le X}|\{E \text{ in } \mathcal{E}_{n,+} \text{ with } \gamma_{E,1}=0\}| = 0.$$

If we also admit the conjecture of Birch and Swinnerton-Dyer that for E/Q the rank of the Mordell-Weil group $E(\mathbb{Q})$ is equal to the order of vanishing of $L(s, E/\mathbb{Q})$ at $s = 1$, then the above consequences of the conjecture imply in turn that

$$\lim_{X\to\infty}\left(1/\sum_{n\le X}|\mathcal{E}_{n,-}|\right)\sum_{n\le X}|\{E \text{ in } \mathcal{E}_{n,-} \text{ with rank } (E(\mathbb{Q}))>1\}| = 0,$$

$$\lim_{X\to\infty}\left(1/\sum_{n\le X}|\mathcal{E}_{n,+}|\right)\sum_{n\le X}|\{E \text{ in } \mathcal{E}_{n,+} \text{ with rank } (E(\mathbb{Q}))>0\}| = 0.$$

These last two statements together imply in turn that

$$\lim_{X\to\infty}\left(1/\sum_{n\le X}|\mathcal{E}_{n}|\right)\sum_{n\le X}|\{E \text{ in } \mathcal{E}_{n} \text{ with rank } (E(\mathbb{Q}))\ge 2\}| = 0,$$

or, in words, zero percent of elliptic curves over $\mathbb{Q}$ have rank 2 or more. The truth of this has recently been called into question; cf. [**Kra-Zag**], [**Fer**], [**Sil**]. However, if our conjecture is correct, then the "contradictory" data is simply an artifact of too restricted a range of computation.

This ends our venture into speculation and conjecture. We now turn to a summary of the contents of this book. The book falls naturally into three parts. The first part, which consists of Chapters 1 through 8, is devoted to the theory of spacing measures on large classical groups. Chapter 1 is devoted to defining the spacing measures which are our main object of study, and to stating the main results about them. In Chapter 2, we define "naive" versions of the spacing measures which we find more amenable to combinatorial analysis. We then formulate versions of our main results for these "naive" spacing measures, and show that they imply the main results stated in the first chapter. The remainder of Chapter 2, along with all of Chapters 3 and 4, is devoted to successive reduction steps (2.9.1, 3.0.1, 3.1.9, 4.2.2–4) in proving the main results. By the end of Chapter 4 we are reduced to proving the three estimates of 4.2.2 and the "tail estimate" 3.1.9, iv). In Chapter 5, we first recall Weyl's explicit formulas for Haar measure on the classical groups. We then combine Weyl's formulas with a method of orthogonal polynomials (5.1.3) which goes back to Gaudin [**Gaudin**]. Thus armed, we establish (in 5.8.3, 5.10.3 and 5.11.2) the three estimates of 4.2.2. Chapter 6 is devoted to the proof of the tail estimate of 3.1.9, iv). To prove it, we introduce "eigenvalue location measures" in 6.9, and in 6.10.5 we give a tail estimate for the first of these measures. We then (6.11, 6.12) relate these measures to spacing measures, which allows us in 6.13 to prove the required tail estimate 3.1.9, iv). At this point, all (save 1.7.6) of the results announced in Chapters 1 and 2 have been proven. The remainder of Chapter 6 explores multi-variable versions of the eigenvalue location measures. In Chapter 7 we form generating series out of the limit spacing measures, we prove 1.7.6, and we relate these generating series, in the case of one variable, to certain infinite-dimensional Fredholm determinants, which are themselves large N limits of finite-dimensional Fredholm determinants. We use this theory to construct large N limits of the (one-variable) eigenvalue location measures, and establish the relations between the limit spacing measures and the limit eigenvalue location measures in one variable. Chapter 8 is devoted to a discussion of these same questions in several variables.

The second part of the book, which consists of Chapters 9 through 11, is devoted to algebro-geometric situations over finite fields which, by means of Deligne's equidistribution theorem and the determination of monodromy groups, provide us with "Frobenius conjugacy classes" in large compact classical groups which are suitably equidistributed for Haar measure. In Chapter 9, we give various "abstract" versions of Deligne's equidistribution theorem, in the language of pure lisse sheaves. Roughly speaking, these theorems assert an equidistribution (for Haar measure) of Frobenius conjugacy classes in the space of conjugacy classes of a maximal compact subgroup of the geometric monodromy group attached to the situation at hand. In Chapters 10 and 11, we prove various families (of curves, of abelian varieties, of hypersurfaces, of Kloosterman sums) to have geometric monodromy groups which are large classical groups.

The third part of the book, which consists of Chapters 12 and 13, applies the theory of the first part of the book to the families proven to have big monodromy in the second part of the book. Chapter 12 looks specifically at GUE discrepancies in these families, and Chapter 13 looks at the distribution of low-lying eigenvalues in these same families.

The book concludes with two appendices. The first appendix, Densities, develops an approach to eigenvalue location measures through densities, determines their large N limits, and presents a result of Harold Widom, that the large N limits of the densities, and hence of the eigenvalue location measures, for the groups $SU(N)$ exist and are equal to those for $U(N)$. The second appendix, Graphs, shows the densities of the GUE measure and of a few of the one-variable eigenvalue location measures.

CHAPTER 1

Statements of the Main Results

1.0. Measures attached to spacings of eigenvalues

1.0.1. Fix an integer $N \geq 1$. Given an element A in the unitary group $U(N)$, all of its eigenvalues lie on the unit circle, so there is a unique increasing sequence of angles in $[0, 2\pi)$,

$$0 \leq \varphi(1) \leq \varphi(2) \leq \cdots \leq \varphi(N) < 2\pi,$$

such that

$$\det(T - A) = \prod_j (T - \exp(i\varphi(j))).$$

The N nonnegative real numbers

$$\varphi(j+1) - \varphi(j), \text{ for } j = 1, \ldots, N-1,$$

and

$$2\pi + \varphi(1) - \varphi(N),$$

are called the "literal" spacings between the adjacent eigenvalues of A. Their sum is 2π, so their mean is $2\pi/N$. By the **normalized** spacings between the adjacent eigenvalues of A, we mean the N nonnegative real numbers $s_1, \ldots, s_N$, defined by

$$s_j := (N/2\pi)(\varphi(j+1) - \varphi(j)), \text{ for } j = 1, \ldots, N-1,$$

and

$$s_N := (N/2\pi)(2\pi + \varphi(1) - \varphi(N)).$$

Their sum is N, so their mean is 1.

1.0.2. There are more general sorts of spacing data it will be convenient to look at. To define these, we go back to the unique increasing sequence of angles in $[0, 2\pi)$,

$$0 \leq \varphi(1) \leq \varphi(2) \leq \cdots \leq \varphi(N) < 2\pi,$$

such that

$$\det(T - A) = \prod_j (T - \exp(i\varphi(j))).$$

We can uniquely prolong the sequence $i \mapsto \varphi(i)$ to all integers i by requiring that $\varphi(i+N) = \varphi(i) + 2\pi$. [This is just a convenient way to keep track of what happens as we wrap around the unit circle. In this numbering system, we recover the normalized spacings as

$$s_j := (N/2\pi)(\varphi(j+1) - \varphi(j)), \text{ for } j = 1, \ldots, N.]$$

1.0.3. Given an integer $r \geq 1$, and a strictly increasing sequence

$$c : 0 < c(1) < c(2) < \cdots < c(r)$$

of integers, which we will refer to as a vector c of "**offsets**", we start at one of the eigenvalue angles $\varphi(i)$, and then walk counterclockwise around the unit circle, stopping successively at the points $\varphi(i + c(1)), \varphi(i + c(2)), \ldots, \varphi(i + c(r))$. The distances we traverse in this r-stage journey form an r-vector,

$$(\varphi(i + c(1)) - \varphi(i), \varphi(i + c(2)) - \varphi(i + c(1)), \ldots, \varphi(i + c(r)) - \varphi(i + c(r - 1)),$$

called the literal spacing vector with offsets c starting at $\varphi(i)$. The sum, over $i = 1$ to N, of these literal spacing vectors is

$$2\pi(c(1), c(2) - c(1), c(3) - c(2), \ldots, c(r) - c(r - 1)).$$

1.0.3.1. We define the normalized spacing vectors with offsets c to be the N vectors, indexed by $i = 1$ to N,

$$s_i(\text{offsets } c) :=$$
$$(N/2\pi)(\varphi(i + c(1)) - \varphi(i), \varphi(i + c(2)) - \varphi(i + c(1)), \ldots, \varphi(i + c(r)) - \varphi(i + c(r - 1))).$$

Their mean is $(c(1), c(2) - c(1), c(3) - c(2), \ldots, c(r) - c(r - 1))$.

1.0.3.2. In the special case $r = 1, c = 1$, these normalized spacing vectors are just the normalized spacings between adjacent eigenvalues.

1.0.4. We now use the sequence of normalized spacing vectors with offsets c to define a probability measure $\mu(A, U(N), \text{offsets } c)$ on $\mathbb{R}^r$ by decreeing that each of the N spacing vectors $s_i(\text{offsets } c)$ has mass $1/N$. In other words,

$$\mu(A, U(N), \text{ offsets } c) := (1/N) \sum_j (\text{delta measure at } s_j(\text{offsets } c)),$$

i.e.,

$$\int f \, d\mu(A, U(N), \text{offsets } c) := (1/N) \sum_j f(s_j(\text{offsets } c))$$

for any $\mathbb{R}$-valued Borel measurable function f on $\mathbb{R}^r$, i.e.,

$$\mu(A, U(N), \text{ offsets } c)(E)$$
$$:= (1/N)(\text{number of indices } j \text{ such that } s_j(\text{offsets } c) \text{ lies in } E),$$

for any Borel set E in $\mathbb{R}^r$.

1.0.5. Given $r \geq 1$ and a vector $c = (c(1), \ldots, c(r))$ in $\mathbb{Z}^r$ of offsets,

$$0 < c(1) < c(2) < \cdots < c(r),$$

we attach to it two other vectors in $\mathbb{Z}^r$, the vector

$$b := (c(1), c(2) - c(1), c(3) - c(2), \ldots, c(r) - c(r - 1)),$$

which we call the vector of **steps**, and the vector

$$a := b - (1, 1, \ldots, 1),$$

which we call the vector of **separations**. [The names are chosen as follows. In the r-stage journey described by the vector of offsets, the i'th stage is to go from the eigenvalue where we are to the $b(i)$'th eigenvalue which comes after it, so if we think of ourselves as walking by stepping precisely on successive eigenvalues, then the i'th stage takes us $b(i)$ steps. The number $a(i)$ is the number of intermediate eigenvalues which separate the starting point from the finishing point of the i'th stage.] The step vector b has strictly positive components, the separation vector a has nonnegative components. Either b or a determines c, by the formulas

$$c(i) = b(1) + b(2) + \cdots + b(i) = i + a(1) + a(2) + \cdots + a(i).$$

1.0.6. It will sometimes be convenient to refer to the normalized spacings and to the measure defined by given offsets c in terms of the vector b of steps, or in terms of the vector a of separations. Thus we define

$$\begin{aligned}
&s_j(\text{steps } b) = s_j(\text{offsets } c),\\
&\mu(A, U(N), \text{ steps } b) := \mu(A, U(N), \text{ offsets } c),\\
&s_j(\text{separations } a) = s_j(\text{offsets } c),\\
&\mu(A, U(N), \text{ separations } a) := \mu(A, U(N), \text{ offsets } c).
\end{aligned}$$

1.0.7. In the above discussion, we chose the angles $\varphi(i)$ for $i = 1, \ldots, N$ in the particular fundamental domain $[0, 2\pi)$. Suppose instead we had fixed a real number α, and chosen the angles in $[\alpha, \alpha + 2\pi)$. Then the spacing vectors $s_j(\text{offsets } c)$ for A computed using $[\alpha, \alpha + 2\pi)$ would be some cyclic permutation of those computed using $[0, 2\pi)$. But the measure $\mu(A, U(N), \text{ offsets } c)$ would be the same.

1.0.8. We now discuss the sense in which the probability measure $\mu(A, U(N), \text{ offsets } c)$ depends continuously on A in $U(N)$.

Lemma 1.0.9. 1) *On $U(N)$, each of the angles $\varphi(i)$ (computed using $[0, 2\pi)$) is a bounded, Borel measurable function. Each is continuous on the open set*

$$U(N)[1/\det(1 - A)]$$

of $U(N)$ where 1 is not an eigenvalue. More generally, for each real α, the angles computed using $[\alpha, \alpha + 2\pi)$ are bounded, Borel measurable functions on $U(N)$, continuous on the open set $U(N)[1/\det(e^{i\alpha} - A)]$.

2) *For any $r \geq 1$, and any offset c in $\mathbb{Z}^r$, and any real number α, each of the N normalized spacing vectors $s_i(\text{offsets } c)$ computed using $[\alpha, \alpha + 2\pi)$ is a Borel measurable $\mathbb{R}^r$-valued function on $U(N)$, which is continuous on $U(N)[1/\det(e^{i\alpha} - A)]$.*

3) *Let f be an $\mathbb{R}$-valued continuous (respectively Borel measurable and bounded) function on $\mathbb{R}^r$. Then the $\mathbb{R}$-valued function on $U(N)$*

$$A \mapsto \int f \, d\mu(A, U(N), \text{ offsets } c)$$

is continuous (respectively Borel measurable and bounded) on $U(N)$.

PROOF. Assertion 1) is proven in the appendix, Corollary 1.8.5. Assertion 2) results immediately from 1), given the definition of the spacing vectors in terms of differences of the $\varphi(i)$'s. For 3), we write

$$\int f \, d\mu(A, U(N), \text{ offsets } c) := (1/N) \sum_j f(s_j(\text{offsets } c)).$$

If f is bounded and measurable on $\mathbb{R}^r$, each term $f(s_j(\text{offsets } c))$ is bounded and measurable on $U(N)$, hence also $\int f \, d\mu(A, U(N), \text{ offsets } c)$.

Suppose now that f is continuous and bounded on $\mathbb{R}^r$. If we compute angles using $[\alpha, \alpha + 2\pi)$, each term $f(s_j(\text{offsets } c))$ is continuous on $U(N)[1/\det(e^{i\alpha} - A)]$, so $A \mapsto \int f\,d\mu(A, U(N),$ offset $c)$ is continuous on $U(N)[1/\det(e^{i\alpha} - A)]$. These open sets cover $U(N)$ (any element of $U(N)$ has at most N distinct eigenvalues), hence the function $A \mapsto \int f\,d\mu(A, U(N),$ offsets $c)$ is continuous on $U(N)$. QED

1.0.10. Given a positive Borel measure ν on $\mathbb{R}^r$ which has finite total mass, its cumulative distribution function CDF_ν is the $\mathbb{R}_{\geq 0}$-valued function on $\mathbb{R}^r$ defined in terms of the ν-measure of standard r-dimensional rectangles by

$$\mathrm{CDF}_\nu(x) := \nu((-\infty, x]).$$

Given two such measures ν and μ on $\mathbb{R}^r$, we define the discrepancy between them, $\mathrm{discrep}(\nu, \mu)$ by

$$\mathrm{discrep}(\nu, \mu) := \mathrm{Sup}_{x \text{ in } \mathbb{R}^r} |\,\mathrm{CDF}_\nu(x) - \mathrm{CDF}_\mu(x)|.$$

If both μ and ν have total mass at most m, then $\mathrm{discrep}(\nu, \mu)$ lies in the closed interval $[0, m]$.

Lemma 1.0.11. *Fix an integer $r \geq 1$, and an offset c in $\mathbb{Z}^r$. Let ν be a probability measure ν on $\mathbb{R}^r$ whose cumulative distribution function CDF_ν is continuous. Then the function*

$$A \mapsto \mathrm{discrep}(\mu(A, U(N), \text{ offsets } c), \nu)$$

from $U(N)$ to $[0, 1]$ is continuous.

PROOF. It suffices to show that for every real number α, this function is continuous on the open set $U(N)[1/\det(e^{i\alpha} - A)]$. On this open set, the individual normalized spacing vectors s_i(offsets c) calculated using angles in $[\alpha, \alpha + 2\pi)$ are continuous functions. So the lemma results from the following lemma, which should be well-known, but for which we do not know a reference.

Lemma 1.0.12. *Fix $r \geq 1$, and a positive Borel measure ν on $\mathbb{R}^r$ of finite total mass whose cumulative distribution function is continuous. Fix an integer $N \geq 1$, and define for each N-tuple $P := x[1], \ldots, x[N]$ of points in $\mathbb{R}^r$ a measure $\mu(P)$ on $\mathbb{R}^r$ by*

$$\mu(P) := (1/N) \sum_j (\textit{delta measure at } x[j]).$$

The function

$$P \mapsto D(P) := \mathrm{discrep}(\mu(P), \nu)$$

from $\mathbb{R}^{rN}$ to $[0, 1]$ is continuous.

PROOF. We will show that $P \mapsto D(P)$ is continuous at any given point P in $\mathbb{R}^{rN}$. Thus let P_i be a sequence of points in $\mathbb{R}^{rN}$ which tends to P. We must show that

$$D(P) \geq \limsup D(P_i), \text{ and } \liminf D(P_i) \geq D(P).$$

Suppose not. Then there exists a real number $A > 0$ such that either

$$\limsup D(P_i) > D(P) + A, \text{ or } \liminf D(P_i) < D(P) - A.$$

Replacing the sequence P_i by a subsequence, we may further assume that either

$$D(P_i) \geq D(P) + A \text{ for every } i, \text{ or } D(P_i) \leq D(P) - A \text{ for every } i.$$

The space $\mathbb{R}^r$ is the increasing union of the countably many rectangles $[-L, L]^r$, $L = 1, 2, 3, \ldots$. Since ν is a measure, we have

$$\nu(\mathbb{R}^r) = \lim_{L \to \infty} \nu([-L, L]^r).$$

Therefore for large L we have

$$\nu([-L, L]^r) \geq \nu(\mathbb{R}^r) - A/100, \text{ and}$$
$$\text{each of the } N\ \mathbb{R}^r\text{-components of } P \text{ lies in } [1-L, L-1]^r.$$

We fix one such choice of L.

The cumulative distribution function of ν is continuous, by assumption, so its restriction to the compact set $[-L-1, L+1]^r$ is uniformly continuous. Therefore we may choose a real $\varepsilon > 0$ such that if x and y lie in $[-L-1, L+1]^r$ and are within ε of each other in the sense that $\sup_i |x_i - y_i| < \varepsilon$, then

$$|\operatorname{CDF}_\nu(x) - \operatorname{CDF}_\nu(y)| < A/100.$$

We may and will further suppose that $\varepsilon \leq 1$.

Choose i such that P_i is, coordinate by coordinate, within ε of P. Notice that, because P_i is within ε of P, both P and P_i lie in the rectangle $[-L, L]^r$. This rectangle contains all but at most $A/100$ of the mass of ν, and it contains all the mass of both $\mu(P)$ and $\mu(P_i)$.

Suppose first that

$$D(P_i) \geq D(P) + A.$$

In view of the definition of $D(P_i)$ as a sup, there exists some x in $\mathbb{R}^r$ such that

$$|\mu(P_i)((-\infty, x]) - \nu((-\infty, x])| > D(P_i) - A/100 \geq D(P) + (99/100)A.$$

We first remark that each coordinate x_i of x must be $\geq -L$. For if not, the rectangle $(-\infty, x]$ has empty intersection with $[-L, L]^r$, and hence has ν measure $\leq A/100$, and $\mu(P_i)$ measure zero, so

$$|\mu(P_i)((-\infty, x]) - \nu((-\infty, x])| = \nu((-\infty, x]) < A/100,$$

which is nonsense.

It may not be the case that $x_i \leq L$ for each coordinate x_i of x. We consider the auxiliary point y with coordinates $y_i = \min(x_i, L)$. The rectangle $(-\infty, y]$ is just the intersection of $(-\infty, x]$ with $(-\infty, L]^r$, and this last rectangle contains all but $A/100$ of the mass of ν, and all the mass of $\mu(P_i)$. Therefore

$$0 \leq \nu((-\infty, x)] - \nu((-\infty, y]) \leq A/100,$$
$$\mu(P_i)((-\infty, x]) = \mu(P_i)((-\infty, y]),$$

and hence

$$|\mu(P_i)((-\infty, y]) - \nu((-\infty, y])| > D(P) + (98/100)A.$$

What we have gained is that the point y lies in $[-L, L]^r$.

We now consider the sign of the difference

$$\mu(P_i)((-\infty, y]) - \nu((-\infty, y]).$$

If this is positive, consider the point $y + (\varepsilon, \varepsilon, \ldots, \varepsilon)$, which we write simply $y + \varepsilon$. Because P is ε close to P_i, $(-\infty, y+\varepsilon]$ contains at least as many P-points as $(-\infty, y]$

contains P_i points. Therefore

$$\begin{aligned}\mu(P)((-\infty, y+\varepsilon]) - \nu((-\infty, y+\varepsilon]) &\geq \mu(P_i)((-\infty, y]) - \nu((-\infty, y+\varepsilon]) \\ &\geq \mu(P_i)((-\infty, y]) - \nu((-\infty, y]) - (\nu((-\infty, y+\varepsilon]) - \nu((-\infty, y])) \\ &\geq D(P) + (98/100)A - A/100 > D(P),\end{aligned}$$

which contradicts the definition of $D(P)$ as a sup.

If the difference

$$\mu(P_i)((-\infty, y]) - \nu((-\infty, y])$$

is negative, we consider instead the point $y-\varepsilon$. Because P is ε close to P_i, $(-\infty, y-\varepsilon]$ contains at most as many P-points as $(-\infty, y]$ contains P_i points. Just as above, we find

$$\begin{aligned}\nu((-\infty, y-\varepsilon]) - \mu(P)((-\infty, y-\varepsilon]) &\geq \nu((-\infty, y-\varepsilon]) - \mu(P_i)((-\infty, y]) \\ &\geq \nu((-\infty, y]) - \mu(P_i)((-\infty, y]) - (\nu((-\infty, y]) - \nu((-\infty, y-\varepsilon])) \\ &\geq D(P) + (98/100)A - A/100 > D(P),\end{aligned}$$

again a contradiction.

It remains to deal with the case

$$D(P_i) \leq D(P) - A.$$

In view of the definition of $D(P)$ as a sup, there exists some x in $\mathbb{R}^r$ such that

$$|\mu(P)((-\infty, x]) - \nu((-\infty, x])| > D(P) - A/100.$$

Exactly as above, we show that $x_i \leq -L$ for every coordinate x_i of x, and that if we replace x by $y := \min(x, (L, \ldots, L))$, we have

$$|\mu(P)((-\infty, y]) - \nu((-\infty, y])| > D(P) - 2A/100.$$

What we have gained is that the point y lies in $[-L, L]^r$.

If $\mu(P)((-\infty, y]) - \nu((-\infty, y])$ is positive, then $(-\infty, y+\varepsilon]$ contains as least as many P_i-points as $(-\infty, y]$ contains P-points, and so we get

$$|\mu(P_i)((-\infty, y+\varepsilon]) - \nu((-\infty, y+\varepsilon])| > D(P) - 3A/100$$

(now use $D(P_i) \leq D(P) - A$)

$$\geq D(P_i) + A - 3A/100 > D(P_i),$$

again a contradiction of the definition of $D(P_i)$ as a sup.

If $\mu(P)((-\infty, y]) - \nu((-\infty, y])$ is negative, then $(-\infty, y-\varepsilon]$ contains at most as many P_i-points as $(-\infty, y]$ contains P-points, and this time we get

$$\begin{aligned}|\mu(P_i)((-\infty, y-\varepsilon]) - \nu((-\infty, y-\varepsilon])| &> D(P) - 3A/100 \\ &\geq D(P_i) + A - 3A/100 > D(P_i),\end{aligned}$$

again the same contradiction. QED

1.1. Expected values of spacing measures

1.1.1. Suppose now that we are given a compact group K, a continuous unitary representation

$$\rho : K \to U(N),$$

and an offset c in $\mathbb{Z}^r$ for some $r \geq 1$. Then for each element A in K, we have an element $\rho(A)$ in $U(N)$, and its associated spacing measure $\mu(\rho(A), U(N), \text{ offsets } c)$ on $\mathbb{R}^r$. When we wish to emphasize (K, ρ) rather than the dimension N of ρ, we will denote this spacing measure by

$$\mu(A, K, \rho, \text{ offsets } c) := \mu(\rho(A), U(N), \text{ offsets } c).$$

We denote by Haar_K the normalized (total mass 1) Haar measure on K.

1.1.2. We wish to define a probability measure $\mu(K, \rho, \text{ offsets } c)$ on $\mathbb{R}^r$ which is in a suitable sense the "expected value" (:= integral against Haar_K) of the function

$$A \mapsto \mu(\rho(A), U(N), \text{ offsets } c) = \mu(A, K, \rho, \text{ offsets } c)$$

from K to the space of probability measures on $\mathbb{R}^r$.

1.1.3. Here is an entirely elementary way to do this. Each of the functions $s_j(\text{offsets } c)$ is bounded and Borel measurable on $U(N)$, so each composite $s_j(\text{offsets } c) \circ \rho$ is bounded and Borel measurable on K. We define

$$\mu(K, \rho, \text{ offsets } c) := (1/N) \sum_j (s_j(\text{offsets } c) \circ \rho)_*(\text{Haar}_K),$$

which is visibly a probability measure on $\mathbb{R}^r$.

By the very definition of $\mu(K, \rho, \text{ offsets } c)$, for any Borel measurable function f on $\mathbb{R}^r$, we have the formula

$$\begin{aligned} \int_{\mathbb{R}^r} f \, d\mu(K, \rho, \text{ offsets } c) :=& (1/N) \sum_j \int_K f(s_j(\text{offsets } c)(\rho(A))) d\,\text{Haar}_K \\ =& \int_K \left((1/N) \sum_j f(s_j(\text{offsets } c)(\rho(A))) \right) d\,\text{Haar}_K \\ =& \int_K \left(\int_{\mathbb{R}^r} f \, d\mu(\rho(A), U(N), \text{ offsets } c) \right) d\,\text{Haar}_K . \end{aligned}$$

1.1.4. If the representation ρ is understood, we will omit it from the notation, and write $\mu(K, \rho, \text{ offsets } c)$ simply as $\mu(K, \text{ offsets } c)$.

1.1.5. In what follows, we will be primarily concerned with the case when K is one of compact classical groups $U(N)$ or $SU(N)$ or $SO(N)$ or $O(N)$ or, for N even, the compact form $USp(N)$ of $Sp(N)$, in its standard representation.

1.2. Existence, universality and discrepancy theorems for limits of expected values of spacing measures: the three main theorems

Theorem 1.2.1 (Main Theorem 0). *Fix an integer $r \geq 1$, and an offset vector c in $\mathbb{Z}^r$. Consider the expected value spacing measure $\mu(U(N)$, offsets $c)$ on $\mathbb{R}^r$ attached to $U(N)$. There exists a probability measure*

$$\mu(\textit{univ, offsets } c)$$

on $\mathbb{R}^r$ to which the measures $\mu(U(N)$, offsets $c)$ converge as $N \to \infty$, in the sense that for any $\mathbb{R}$-valued, bounded, Borel measurable function f of compact support on $\mathbb{R}^r$, we have

$$\int f\, d\mu(\textit{univ, offsets } c) = \lim_{N\to\infty} \int f\, d\mu(U(N), \textit{ offsets } c).$$

The probability measure μ(univ, offsets c) has a continuous cumulative distribution function.

1.2.2. The use of the notation "univ" (for "universal") is justified by the next theorem.

Theorem 1.2.3 (Main Theorem 1). *Fix an integer $r \geq 1$, and an offset vector c in $\mathbb{Z}^r$. Consider the expected value spacing measures $\mu(G(N)$, offsets $c)$ on $\mathbb{R}^r$ attached to each of the following sequences $G(N)$ of compact classical groups in their standard representations:*

$$U(N), SU(N), SO(2N+1), O(2N+1), USp(2N), SO(2N), O(2N).$$

In each of these sequences of classical groups, the expected value spacing measures $\mu(G(N)$, offsets $c)$ converge (in the sense above) to the ***same*** *limit μ(univ, offsets c) (which by Theorem 1.2.1 above is a probability measure with a continuous CDF).*

1.2.4. Fix an integer $r \geq 1$, and an offset vector c in $\mathbb{Z}^r$. According to Theorem 1.2.1, μ(univ, offsets c) has a continuous CDF. For each N, the $[0,1]$-valued function on $U(N)$

$$A \mapsto \text{discrep}(\mu(A, U(N), \text{ offsets } c), \mu(\text{univ, offsets } c))$$

is therefore continuous on $U(N)$, thanks to Lemma 1.0.11 above. So for any compact group K and any continuous unitary representation ρ of K, the $[0,1]$-valued function on K

$$A \mapsto \text{discrep}(\mu(A, K, \rho, \text{ offsets } c), \mu(\text{univ, offsets } c))$$

is continuous on K.

1.2.5. In particular, for $K = G(N)$ any of the compact classical groups in their standard representations:

$$U(N), SU(N), SO(2N+1), O(2+1), USp(2N), SO(2N), O(2N),$$

the function

$$A \mapsto \text{discrep}(\mu(A, G(N), \text{ offsets } c), \mu(\text{univ, offsets } c))$$

is a continuous function on $G(N)$, with values in $[0,1]$.

Theorem 1.2.6 (Main Theorem 2). *Fix an integer $r \geq 1$, an offset vector c in $\mathbb{Z}^r$ and a real number $\varepsilon > 0$. There exists an explicit constant $N(\varepsilon, r, c)$ with the following property: For $G(N)$ any of the compact classical groups in their standard representations,*

$$U(N), SU(N), SO(2N+1), O(2N+1), USp(2N), SO(2N), O(2N),$$

we have the inequality

$$\int_{G(N)} \mathrm{discrep}(\mu(A, G(N),\ \textit{offsets } c), \mu(\textit{univ},\ \textit{offsets } c)) d(\mathrm{Haar}_{G(N)})$$
$$\leq N^{\varepsilon - 1/(2r+4)},$$

provided that $N \geq N(\varepsilon, r, c)$.

1.3. Interlude: A functorial property of Haar measure on compact groups

Lemma 1.3.1. *Let H and K be compact groups, and $\pi : H \to K$ a continuous group homomorphism which is surjective. Denote by Haar_H and Haar_K the normalized (total mass 1) Haar measures on H and K respectively. Then $\pi_* \mathrm{Haar}_H = \mathrm{Haar}_K$ as Borel measures on K.*

PROOF. Since π is surjective, and Haar_H is translation invariant on H, its direct image $\pi_* \mathrm{Haar}_H$ is translation invariant on K. Thus $\pi_* \mathrm{Haar}_H$ is both translation invariant on K and of total mass 1, so it must coincide with Haar_K. QED

Corollary 1.3.2. *Let H be a compact group, Γ a compact normal subgroup of H, and K a closed subgroup of H such that $H = \Gamma K$. Then for any bounded measurable function f on H which is invariant by Γ, we have*

$$\int_H f d\,\mathrm{Haar}_H = \int_K f d\,\mathrm{Haar}_K .$$

PROOF. We consider the quotient $H/\Gamma \approx K/(\Gamma \cap K)$. By the lemma above, we have

$$\int_H f d\,\mathrm{Haar}_H = \int_{H/\Gamma} f d\,\mathrm{Haar}_{H/\Gamma} = \int_{K/(\Gamma\cap K)} f d\,\mathrm{Haar}_{K/(\Gamma \cap K)}$$
$$= \int_K f d\,\mathrm{Haar}_K . \text{ QED}$$

1.4. Application: Slight economies in proving Theorems 1.2.3 and 1.2.6

1.4.0. We apply this last result 1.3.2 as follows. The various spacing measures attached to an element A of the unitary group $U(N)$ are the same for A and for λA, for any unitary scalar λ in S^1, since multiplication by λ simply rotates the eigenvalues, but does not alter their spacings. So for each integer $r \geq 1$ and each offset vector c in $\mathbb{Z}^r$, the expected values of the "offsets c" spacing measures will coincide for $U(N)$ and for $SU(N)$. [Take Γ to be the subgroup of scalars in $H = U(N)$, and K the subgroup $SU(N)$.] Similarly, the expected values of the "offsets c" spacing measures wil coincide for $O(2N+1)$ and for $SO(2N+1)$. [Take Γ to be the subgroup $\{1, -1\}$ of scalars in $H = O(2N+1)$, and K the subgroup $SO(2N+1)$.] This means that the $SU(N)$ case of Theorem 1.2.3 is equivalent to the $U(N)$ case of it (i.e., to Theorem 1.2.1), and that the $SO(2N+1)$ case of Theorem

1.2.3 is equivalent to the $O(2N+1)$ case. Similarly, the discrepancy integrals in Theorem 1.2.6 coincide for $U(N)$ and $SU(N)$, and they coincide for $O(2N+1)$ and $SO(2N+1)$. This means that the $U(N)$ case of Theorem 1.2.6 is equivalent to the $SU(N)$ case of it, and that the $O(2N+1)$ case of Theorem 1.2.6 is equivalent to the $SO(2N+1)$ case of it.

1.5. Application: An extension of Theorem 1.2.6

Lemma 1.5.1. *Let N be an integer ≥ 1, $G(N) \subset H(N)$ compact groups in one of the following four cases:*
a) $G(N) = SU(N) \subset H(N) \subset$ *normalizer of* $G(N)$ *in* $U(N)$,
b) $G(N) = SO(2N+1) \subset H(N) \subset$ *normalizer of* $G(N)$ *in* $U(2N+1)$,
c) $G(N) = USp(2N) \subset H(N) \subset$ *normalizer of* $G(N)$ *in* $U(2N)$,
d) $G(N) = SO(2N) \subset H(N) \subset$ *normalizer of* $G(N)$ *in* $U(2N)$.
Let f be any bounded, Borel measurable function on the ambient group (i.e. on $U(N)$ in case a), *on $U(2N+1)$ in case* b), *on $U(2N)$ in cases* c) *and* d)) *which is invariant by the subgroup S^1 of unitary scalars. In cases* a), b), *and* c) *we have*

$$\int_{H(N)} f d\,\mathrm{Haar} = \int_{G(N)} f d\,\mathrm{Haar}.$$

In case d), *we have either*

$$\int_{H(N)} f d\,\mathrm{Haar} = \int_{SO(2N)} f d\,\mathrm{Haar}$$

or

$$\int_{H(N)} f d\,\mathrm{Haar} = \int_{O(2N)} f d\,\mathrm{Haar}$$

depending on whether or not every element of $H(N)$, acting by conjugation on $SO(2N)$, induces an inner automorphism of $SO(2N)$.

Proof. In cases a), b) and c), we claim that the normalizer of $G(N)$ in the ambient unitary group is $S^1G(N)$, while in case d) we claim this normalizer is $S^1O(2N)$. It is obvious that the named group normalizes $G(N)$, what must be shown is that the normalizer is no bigger. In other words, we must show that $H(N) \subset S^1G(N)$ in cases a), b), c), and we must show $H(N) \subset S^1O(2N)$ in case d).

In case a), $S^1G(N)$ is $U(N)$, so there is nothing to prove. In cases b) and c), use the fact that the Dynkin diagram has no nontrivial automorphisms, so every automorphism of $G(N)$ is inner. So any h in $H(N)$, acting by conjugation on $G(N)$, induces conjugation by some element g in $G(N)$. Then $h^{-1}g$ commutes with $G(N)$ in its standard representation. As this representation is irreducible, $h^{-1}g$ must be a scalar, and this scalar, being unitary, lies in S^1. Thus $H(N) \subset S^1G(N)$.

Case d) requires some additional attention. If N is 1, one argues by inspection. This case is left to the reader. If $N \geq 2$ and $N \neq 4$, there is one nontrivial automorphism of the Dynkin diagram, and it is induced by conjugation by any element of $O_-(2N)$. So any h in $H(N)$, acting by conjugation on $SO(2N)$, induces conjugation by some element g in $O(2N)$. Repeating the irreducibility argument given above in the b) and c) cases, we get $H(N) \subset S^1O(2N)$.

It remains to examine case d) with $N = 4$. Here the Dynkin diagram has three extreme points, and its automorphism group is the group Σ_3, acting by permutation of these three points. Think of these three extreme points as (the highest weights of) the three 8-dimensional irreducible representations of Lie($SO(8)$) (namely the "standard" one ρ_{std} and the two spin representations). The action of any h in $H(N)$ preserves the isomorphism class of ρ_{std}: indeed, for A in $SO(2N) \subset U(2N)$, and h in $U(2N)$ which normalizes $SO(2N)$,

$$\rho_{\text{std}}(hAh^{-1}) := hAh^{-1} = h\rho_{\text{std}}(A)h^{-1}.$$

Therefore conjugation by h is either inner, or it interchanges the two spin representations. In the first case, we have h in $S^1G(N)$. In the latter case, use the fact that any element γ_- of $O_-(2N)$ also interchanges the two spin representations. Then $h\gamma_-$ induces an inner automorphism, and hence $h\gamma_-$ is in $S^1G(N)$, whence h lies in $S^1O(2N)$, as required, even in the case $N = 4$.

In cases a), b) and c), the inclusions

$$G(N) \subset H(N) \subset S^1G(N)$$

make clear that we have an equality of groups

$$S^1H(N) = S^1G(N).$$

Using this equality, we apply Corollary 1.3.2 to the situations

$$H(N) \subset S^1H(N) \text{ and } G(N) \subset S^1G(N)$$

to infer that

$$\int_{H(N)} f d\,\text{Haar} = \int_{S^1H(N)} f d\,\text{Haar} = \int_{S^1G(N)} f d\,\text{Haar} = \int_{G(N)} f d\,\text{Haar},$$

as required.

In case d), we must distinguish two cases. If $H(N)$ acts on $SO(2N)$ through inner automorphisms, we have the inclusion

$$G(N) \subset H(N) \subset S^1G(N)$$

and we argue as above. If some h in $H(N)$ induces a noninner automorphism, then $H(N)$ contains an element of $S^1O_-(2N)$, in which case the inclusion $G(N) \subset H(N) \subset S^1O(2N)$ forces the equality

$$S^1H(N) = S^1O(2N),$$

and we apply Corollary 1.3.2 to the situations

$$H(N) \subset S^1H(N) \text{ and } O(2N) \subset S^1O(2N). \text{ QED}$$

Applying this Lemma 1.5.1 to f the discrepancy function, we get

Corollary 1.5.2. *Let N be an integer ≥ 1, $G(N) \subset H(N)$ compact groups in one of the following four cases*:

a) $G(N) = SU(N) \subset H(N) \subset$ *normalizer of* $G(N)$ *in* $U(N)$,

b) $G(N) = SO(2N+1) \subset H(N) \subset$ *normalizer of* $G(N)$ *in* $U(2N+1)$,

c) $G(N) = USp(2N) \subset H(N) \subset$ *normalizer of* $G(N)$ *in* $U(2N)$,

d) $G(N) = SO(2N) \subset H(N) \subset$ *normalizer of* $G(N)$ *in* $U(2N)$.

Let $r \geq 1$ be an integer, c in $\mathbb{Z}^r$ an offset vector. In cases a), b), c), *we have*

$$\int_{H(N)} \text{discrep}(\mu(A, H(N), \textit{ offsets } c), \mu(\textit{univ}, \textit{ offsets } c)) d\,\text{Haar}$$

$$= \int_{G(N)} \text{discrep}(\mu(A, G(N), \textit{ offsets } c), \mu(\textit{univ}, \textit{ offsets } c)) d\,\text{Haar}.$$

In case d), *we have*

$$\int_{H(N)} \operatorname{discrep}(\mu(A, H(N), \ \textit{offsets } c), \mu(\textit{univ, offsets } c)) d\,\mathrm{Haar}$$

$=$ *either*

$$\int_{SO(2N)} \operatorname{discrep}(\mu(A, SO(2N), \ \textit{offsets } c), \mu(\textit{univ, offsets } c)) d\,\mathrm{Haar}$$

or

$$\int_{O(2N)} \operatorname{discrep}(\mu(A, O(2N), \ \textit{offsets } c), \mu(\textit{univ, offsets } c)) d\,\mathrm{Haar},$$

depending on whether or not $H(N)$ acts by inner automorphism on $SO(2N)$.

Thanks to this result 1.5.2, Theorem 1.2.6 yields formally:

Theorem 1.5.3 (Theorem 1.2.6 extended). *Suppose Theorem* 1.2.6 *holds. Let N be an integer ≥ 1, $G(N) \subset H(N)$ compact groups in one of the following four cases*:

a) $G(N) = SU(N) \subset H(N) \subset$ *normalizer of* $G(N)$ *in* $U(N)$,
b) $G(N) = SO(2N+1) \subset H(N) \subset$ *normalizer of* $G(N)$ *in* $U(2N+1)$,
c) $G(N) = USp(2N) \subset H(N) \subset$ *normalizer of* $G(N)$ *in* $U(2N)$,
d) $G(N) = SO(2N) \subset H(N) \subset$ *normalizer of* $G(N)$ *in* $U(2N)$.

For any r, c, ε as in Theorem 1.2.6, *with explicit constant $N(\varepsilon, r, c)$, we have the equality*

$$\int_{H(N)} \operatorname{discrep}(\mu(A, H(N), \ \textit{offsets } c), \mu(\ \textit{univ, offsets } c)) d\,\mathrm{Haar}$$
$$\leq N^{\varepsilon - 1/(2r+4)}$$

provided that $N \geq N(\varepsilon, r, c)$.

1.6. Corollaries of Theorem 1.5.3

Corollary 1.6.1. *Fix an integer $r \geq 1$, an offset vector c in $\mathbb{Z}^r$ and a real number $0 < \varepsilon < 1/(2r+4)$. For any positive real numbers α and β with*

$$\alpha + \beta = 1/(2r+4) - \varepsilon,$$

and any $N \geq N(\varepsilon, r, c)$, then on $H(N)$ the inequality

$$\operatorname{discrep}(\mu(A, H(N), \ \textit{offsets } c), \mu(\textit{univ, offsets } c) \leq N^{-\alpha}$$

holds outside a set of measure $\leq N^{-\beta}$.

Proof. If discrep $> N^{-\alpha}$ on a set of measure $> N^{-\beta}$, then already integrating over this set gives a contribution $> N^{-\alpha-\beta}$ to the integral, which cannot be cancelled because discrep takes nonnegative values. QED

Corollary 1.6.2. *Fix an integer $r \geq 1$. Let N_n be any sequence of integers with $N_n \geq n^{2r+5}$. For each $N \geq 1$, pick one of the groups $H(N)$ listed in Theorem* 1.5.3, *and form the product space $\prod_n H(N_n)$ with its product Haar measure. There exists a set Z of measure zero in this product such that for every element $(A_n)_n$ not in Z, and for every offset vector c in $\mathbb{Z}^r$, we have the eventual inequality*

$$\operatorname{discrep}(\mu(A_n, H(N_n), \ \textit{offsets } c), \mu(\textit{univ, offsets } c)) \leq n^{-1/(2r+5)}$$
$$\textit{for all } n \gg 0.$$

PROOF. This follows from Corollary 1.6.1 and Borel-Cantelli. In Corollary 1.6.1, take

$$\begin{aligned}\varepsilon &= (2r+5)^{-2}(2r+4)^{-1} - (2r+5)^{-3} > 0,\\ \alpha &= (2r+5)^{-2},\\ \beta &= (2r+5)^{-1} + (2r+5)^{-3}.\end{aligned}$$

Then $\alpha + \beta + \varepsilon = (2r+4)^{-1}, (2r+5)\alpha = (2r+5)^{-1}$, and $(2r+5)\beta > 1$. So the series $\sum_n (N_n)^{-\beta}$ converges. Fix an offset vector c in $\mathbb{Z}^r$. By Borel-Cantelli and Corollary 1.6.1, in our product space, the sequences $(A_n)_n$ for which

$$\text{discrep}(\mu(A_n, H(N_n), \text{ offsets } c), \mu(\text{univ, offsets } c)) \leq N_n^{-\alpha}$$
$$\text{for all } n \gg 0$$

form a set of measure one, whose complement we denote Z_c. Since $(N_n)^\alpha \geq (n^{2r+5})^\alpha = n^{1/(2r+5)}$, we have $N_n^{-\alpha} \leq n^{-1/(2r+5)}$. Thus for $(A_n)_n$ outside the set $Z := \bigcup_c Z_c$ of measure zero, and for each offset vector c in $\mathbb{Z}^r$, we have

$$\text{discrep}(\mu(A_n, H(N_n), \text{offsets } c), \mu(\text{univ, offsets } c)) \leq N_n^{-\alpha}$$
$$\text{for all } n \gg 0. \text{ QED}$$

1.6.3. We next fix an $r \geq 1$ and an offset vector c in $\mathbb{Z}^r$, and ask how far the expected value measure $\mu(H(N), \text{offsets } c)$ is from the limit measure

$$\mu(\text{univ, offsets } c).$$

Corollary 1.6.4. *Let $r \geq 1$ be an integer, c in $\mathbb{Z}^r$ an offset vector, and $\varepsilon > 0$ a real number. For any integer $N \geq N(\varepsilon, r, c)$, and $H(N)$ any of the groups listed in Theorem 1.5.3, we have*

$$\text{discrep}(\mu(H(N), \textit{ offsets } c), \mu(\textit{univ, offsets } c)) \leq N^{\varepsilon - 1/(2r+4)}.$$

PROOF. For brevity, we denote

$$\begin{aligned}\mu &:= \mu(\text{univ, offsets } c),\\ \mu(N) &:= \mu(H(N), \text{ offsets } c),\\ \mu(A, N) &:= \mu(A, H(N), \text{ offsets } c).\end{aligned}$$

For any point x in $\mathbb{R}^r$, we denote by $R(x)$ the rectangle $(-\infty, x]$. We have

$$|\mu(N)(R(x)) - \mu(R(x))| = \left| \int_{H(N)} (\mu(A, N)(R(x)) - \mu(R(x))) d\,\text{Haar} \right|$$
$$\leq \int_{H(N)} |\mu(A, N)(R(x)) - \mu(R(x))| d\,\text{Haar}$$
$$\leq \int_{H(N)} (\text{Sup}_y |\mu(A, N)(R(y)) - \mu(R(y))|) d\,\text{Haar}$$
$$:= \int_{H(N)} |\,\text{discrep}(\mu(A, N), \mu)| d\,\text{Haar}$$
$$\leq N^{\varepsilon - 1/(2r+4)}.$$

Since this holds for all x, we have $\text{discrep}(\mu(N), \mu) \leq N^{\varepsilon - 1/(2r+4)}$. QED

1.7. Another generalization of Theorem 1.2.6

1.7.1. Before stating the generalization, we explain the motivation. Consider, for an A in $U(N)$, its N normalized eigenvalue spacings

$$s_j := (N/2\pi)(\varphi(j+1) - \varphi(j)).$$

If we are interested in the distribution of the sums of two adjacent spacings $s_j + s_{j+1}$, we look at the "offset 2" spacing measure. But suppose we are interested in the distribution of the differences of adjacent spacings, $s_{j+1} - s_j$, or more generally in the distribution of some real linear combination of two adjacent spacings, $\alpha s_j + \beta s_{j+1}$. Then what we are asking about is this: take the spacing measures on $\mathbb{R}^2$ given by the offset vector $(1,2)$, and take their direct images to $\mathbb{R}^1$ by the linear map $(x,y) \mapsto \alpha x + \beta y$. How do these measures on $\mathbb{R}^1$ behave? More generally, we might take the direct images of our measures on $\mathbb{R}^r$ by any surjective linear map to an $\mathbb{R}^k$.

1.7.2. Another problem is this. Because we began by considering spacings, our measures are measures on coordinatized Euclidean spaces $\mathbb{R}^r$. Our notion of CDF for measures on $\mathbb{R}^r$ depends entirely on these chosen coordinates. Consequently, the very notion of discrepancy depends on the choice of coordinates. The continuity property 1.0.11 of the discrepancy function depends on the fact that the measure "ν" has a continuous CDF, but this property itself depends on the coordinate system.

1.7.3. It is in order to deal with these two problems that we give the following generalization of Theorem 1.2.6. Before stating it, we make a somewhat ad hoc definition.

1.7.4. Given an integer $r \geq 1$, we say that a $\mathcal{C}^1$-diffeomorphism

$$\varphi : \mathbb{R}^r \cong \mathbb{R}^r,$$
$$x = (x_1, \ldots, x_r) \mapsto (\varphi_1(x), \ldots, \varphi_r(x)),$$

is of bounded distortion if it is "bi-bounded" in the sense that there exist strictly positive real constants κ and η such that

$$\eta \sum |x_i| \leq \sum |\varphi_i(x)| \leq \kappa \sum |x_i|.$$

The basic example we have in mind of such a φ is a linear automorphism of $\mathbb{R}^r$.

1.7.5. Given a second integer $k \geq 1$ with $k \leq r$, we say that a $\mathcal{C}^1$ map

$$\pi : \mathbb{R}^r \to \mathbb{R}^k$$

is a $\mathcal{C}^1$ partial coordinate system of bounded distortion if there exists a $\mathcal{C}^1$-diffeomorphism

$$\varphi : \mathbb{R}^r \cong \mathbb{R}^r$$

of bounded distortion such that, denoting by

$$\mathrm{pr}[\{1, 2, \ldots, k\}] : \mathbb{R}^r \to \mathbb{R}^k$$

the projection onto the first k coordinates, we have

$$\pi = \mathrm{pr}[\{1, 2, \ldots, k\}] \circ \varphi.$$

The basic example we have in mind of such a π is a surjective linear map.

Theorem 1.7.6. *Let $r \geq 1$ an integer, b in $\mathbb{Z}^r$ a step vector with corresponding separation vector a and offset vector c. Denote*

$$\mu := \mu(\textit{univ, offsets } c).$$

Suppose given an integer k with $1 \leq k \leq r$, and a surjective linear map

$$\pi : \mathbb{R}^r \to \mathbb{R}^k,$$

or, more generally, a partial $\mathcal{C}^1$ coordinate system of bounded distortion $\pi : \mathbb{R}^r \to \mathbb{R}^k$.

1) *The measure $\pi_*\mu$ on $\mathbb{R}^k$ is absolutely continuous with respect to Lebesgue measure, and (hence, by 2.11.18) has a continuous CDF.*

2) *Given any real $\varepsilon > 0$, there exists an explicit constant $N(\varepsilon, r, c, \pi)$ with the following property: For $G(N)$ any of the compact classical groups in their standard representations,*

$$U(N), SU(N), SO(2N+1), O(2N+1), USp(2N), SO(2N), O(2N),$$

and for

$$\mu(A, N) := \mu(A, G(N), \textit{ offsets } c), \textit{ for each } A \textit{ in } G(N),$$

we have the inequality

$$\int_{G(N)} \text{discrep}(\pi_*\mu(A, N), \pi_*\mu) d(\text{Haar}_{G(N)}) \leq N^{\varepsilon - 1/(2r+4)},$$

provided that $N \geq N(\varepsilon, r, c, \pi)$.

Theorem 1.7.6 will be proven in 7.1. It immediately implies:

Theorem 1.7.7. *Let $G(N) \subset H(N)$ be compact groups in one of the following four cases:*

a) *$G(N) = SU(N) \subset H(N) \subset$ normalizer of $G(N)$ in $U(N)$,*

b) *$G(N) = SO(2N+1) \subset H(N) \subset$ normalizer of $G(N)$ in $U(2N+1)$,*

c) *$G(N) = USp(2N) \subset H(N) \subset$ normalizer of $G(N)$ in $U(2N)$,*

d) *$G(N) = SO(2N) \subset H(N) \subset$ normalizer of $G(N)$ in $U(2N)$.*

For ε, r, c, π as in Theorem 1.7.6, with explicit constant $N(\varepsilon, r, c, \pi)$, we have the inequality

$$\int_{H(N)} \text{discrep}(\pi_*\mu(A, H(N), \textit{ offsets } c), \pi_*(\textit{univ, offsets } c)) d\,\text{Haar} \leq N^{\varepsilon - 1/(2r+4)}$$

provided that $N \geq N(\varepsilon, r, c, \pi)$.

PROOF. The proof is identical to that of Theorem 1.5.3. QED

1.7.8. Exactly as in 1.6, Theorem 1.7.7 has the following three corollaries.

Corollary 1.7.9. *Hypotheses and notations as in Theorem 1.7.7, for any real numbers α and β with*

$$\alpha + \beta = 1/(2r+4) - \varepsilon,$$

and any $N \geq N(\varepsilon, r, c, \pi)$, then on $H(N)$ any of the groups listed in Theorem 1.7.7, the inequality

$$\text{discrep}(\pi_*\mu(A, H(N), \textit{ offsets } c), \pi_*\mu(\textit{univ, offsets } c) \leq N^{-\alpha}$$

holds outside a set of measure $\leq N^{-\beta}$.

Corollary 1.7.10. *Fix an integer $r \geq 1$, an integer k with $1 \leq k \leq r$, and a surjective linear map $\pi : \mathbb{R}^r \to \mathbb{R}^k$. Let N_n be any sequence of integers with $N_n \geq n^{2r+5}$. For each $N \geq 1$, pick one of the groups $H(N)$ listed in Theorem 1.7.7, and form the product space $\prod_n H(N_n)$ with its product Haar measure. There exists a set Z of measure zero in this product such that for every element $(A_n)_n$ not in Z, and for every offset vector c in $\mathbb{Z}^r$, we have the eventual inequality*

$$\text{discrep}(\pi_*\mu(A_n, H(N_n),\ \textit{offsets } c), \pi_*\mu(\textit{univ},\ \textit{offsets } c)) \leq n^{-1/(2r+5)}$$
$$\textit{for all } n \gg 0.$$

Corollary 1.7.11. *Hypotheses and notations as in Theorem 1.7.7, for any integer $N \geq N(\varepsilon, r, c, \pi)$, and $H(N)$ any of the groups listed in Theorem 1.7.7, we have*

$$\text{discrep}(\pi_*\mu(H(N),\ \textit{offsets } c), \pi_*\mu(\textit{univ},\ \textit{offsets } c)) \leq N^{\varepsilon - 1/(2r+4)}.$$

1.8. Appendix: Continuity properties of "the i'th eigenvalue" as a function on $U(N)$

1.8.1. Given a nonempty set $Z \subset \mathbb{C}$, and an integer $N \geq 1$, denote by $\mathcal{P}_{N,Z}$ the set of those monic polynomials of degree N in $\mathbb{C}[T]$ all of whose roots lie in Z. Thus $\mathcal{P}_{N,\mathbb{C}}$ is the space of all monics of degree N in $\mathbb{C}[T]$, which we topologize by viewing it as the space $\mathbb{C}^N$ of coefficients, $\mathbb{C}$ with its usual topology. We topologize $\mathcal{P}_{N,Z}$ by viewing it as a subset of $\mathcal{P}_{N,\mathbb{C}}$.

Lemma 1.8.2. *If Z is bounded in $\mathbb{C}$ (resp. closed in $\mathbb{C}$, resp. compact), then $\mathcal{P}_{N,Z}$ is bounded in $\mathcal{P}_{N,\mathbb{C}}$ (resp. closed in $\mathcal{P}_{N,\mathbb{C}}$, resp. compact).*

PROOF. If Z is bounded, the set $\mathcal{P}_{N,Z}$ is bounded, because the coefficients of any element of $\mathcal{P}_{N,Z}$ are elementary symmetric functions of its N roots, all of which lie in the bounded set Z.

Suppose Z is closed. To show that $\mathcal{P}_{N,Z}$ is closed in $\mathcal{P}_{N,\mathbb{C}}$, we must show that if $\{f_n\}_n$ is a sequence of elements of $\mathcal{P}_{N,Z}$ which converges coefficientwise to some monic polynomial f, then f has all its roots in Z. Suppose not, and let α not in Z be a root of f. Then $f_n(\alpha)$ converges to $f(\alpha) = 0$. But since α is not in the closed set Z, its distance to Z, $d(\alpha, Z)$, is > 0. In particular, for any z in Z, we have

$$|\alpha - z| \geq d(\alpha, Z).$$

So for any g in $\mathcal{P}_{N,Z}$, say $g(T) = \prod(T - z_i)$, we have

$$|g(\alpha)| = \prod |\alpha - z_i| \geq \text{dist}(\alpha, Z)^N.$$

Applying this to $g = f_n$, we get $|f_n(\alpha)| \geq \text{dist}(\alpha, Z)^N$, so all the $|f_n(\alpha)|$ are uniformly bounded away from zero, and hence the $f_n(\alpha)$ do not converge to $f(\alpha) = 0$.

The compactness assertion results from the first two, since in both $\mathbb{C}$ and $\mathbb{C}^N$, compact is closed and bounded. QED

Lemma 1.8.3. *Let $a < b < a + 2\pi$ be real numbers, and denote by $Z[a, b]$ the closed interval in the unit circle which is the isomorphic image of the closed interval $[a, b]$ by the map $x \mapsto \exp(ix)$. Fix $N \geq 1$. Denote by $[a, b]^N$(ordered) the subset of $[a, b]^N$ consisting of those N-tuples $(\varphi_1, \ldots, \varphi_n)$ of elements of $[a, b]$ which satisfy*

$$a \leq \varphi(1) \leq \varphi(2) \leq \cdots \leq \varphi(N) \leq b.$$

Then the map

$$[a,b]^N(\textit{ordered}) \to \mathcal{P}_{N,Z[a,b]},$$
$$a \le \varphi(1) \le \varphi(2) \le \cdots \le \varphi(N) \le b \mapsto \prod_j (T - \exp(i\varphi(j)))$$

is bicontinuous. In particular, for $1 \le i \le N$, *the rule which to* f *in* $\mathcal{P}_{N,Z[a,b]}$ *attaches* $\varphi(i)$ *is a continuous function on* $\mathcal{P}_{N,Z[a,b]}$.

PROOF. The source $[a,b]^N$(ordered) is closed in the compact space $[a,b]^N$, being defined by $\le$ inequalities, hence the source is compact. Thus both source and target are compact. The map is obviously continuous, and obviously bijective. Therefore it is bicontinuous. QED

Lemma 1.8.4. *Fix a real number* α. *Consider the bijective map from* $\mathcal{P}_{N,S^1}$ *to the space* $[\alpha, \alpha+2\pi)^N$*(ordered) which to* f *in* $\mathcal{P}_{N,S^1}$ *attaches the unique increasing sequence of angles in* $[\alpha, \alpha+2\pi)$,

$$\alpha \le \varphi(1) \le \varphi(2) \le \cdots \le \varphi(N) < \alpha + 2\pi,$$

such that

$$f(T) = \prod_j (T - \exp(i\varphi(j))).$$

(1) *This map is continuous on the open set* $\mathcal{P}_{N,S^1-\{e^{i\alpha}\}}$ *of* $\mathcal{P}_{N,S^1}$ *consisting of those* f *with* $f(e^{i\alpha}) \ne 0$, *and induces a homeomorphism from* $\mathcal{P}_{N,S^1-\{e^{i\alpha}\}}$ *to* $(\alpha, \alpha+2\pi)^N$*(ordered).*

(2) *For each integer* $k \ge 1$, *denote by* W_k *the locally closed set in* $\mathcal{P}_{N,S^1}$ *consisting of those* f *which have a zero at* $T = e^{i\alpha}$ *of exact multiplicity* k. *This bijection is continuous on* W_k.

PROOF. It suffices to treat the case $\alpha = 0$.

(1) Any element f in $\mathcal{P}_{N,S^1-\{1\}}$ lies in $\mathcal{P}_{N,Z[\varepsilon,2\pi-\varepsilon]}$ for some $\varepsilon > 0$. If $f_n \to f$ is a Cauchy sequence in $\mathcal{P}_{N,S^1}$, we claim that each f_n itself lies in $\mathcal{P}_{N,Z[\varepsilon/2,2\pi-\varepsilon/2]}$ as soon as $n \gg 0$. To see this, take any point α in $Z[-\varepsilon/2, \varepsilon/2]$. Then we have

$$\operatorname{dist}(\alpha, Z[\varepsilon, 2\pi-\varepsilon]) \ge \varepsilon/\pi \quad \text{(the circular distance is } \ge \varepsilon/2),$$

and hence our f in $\mathcal{P}_{N,Z[\varepsilon,2\pi-\varepsilon]}$ satisfies

$$|f(\alpha)| \ge \operatorname{dist}(\alpha, Z[\varepsilon, 2\pi-\varepsilon])^N \ge (\varepsilon/\pi)^N.$$

But if f_n is within δ of f coefficient by coefficient, then for any α in S^1 we have $|f_n(\alpha) - f(\alpha)| \le N\delta$ (because $f_n - f$ has degree $\le N-1$, and all coefficients at most δ in absolute value). So if we choose δ so small that $N\delta < (\varepsilon/\pi)^N$, and $n \gg 0$ so that f_n is within δ of f, then $f_n(\alpha) \ne 0$ for any α in $Z[-\varepsilon/2, \varepsilon, 2]$. In particular, each such f_n lies in $\mathcal{P}_{N,Z[\varepsilon/2,2\pi-\varepsilon/2]}$.

Because any Cauchy sequence in $\mathcal{P}_{N,S^1}$ which converges to an element of $\mathcal{P}_{N,S^1-\{1\}}$ has all far out terms in a single closed set $\mathcal{P}_{N,Z[\varepsilon,2\pi-\varepsilon]}$, for some $\varepsilon > 0$, a function from $\mathcal{P}_{N,S^1-\{1\}}$ to a metric space is continuous if and only if its restriction to each closed set $\mathcal{P}_{N,Z[\varepsilon,2\pi-\varepsilon]}$, $\varepsilon > 0$, is continuous. Since the angles defined using $[0, 2\pi)$ are continuous on each $\mathcal{P}_{N,Z[\varepsilon,2\pi-\varepsilon]}$, $\varepsilon > 0$, they are continuous on $\mathcal{P}_{N,S^1-\{1\}}$, as asserted.

(2) On the one point set W_N, there is nothing to prove. On W_k, with $k < N$, $\varphi(1)$ through $\varphi(k)$ are identically zero. Division by $(T-1)^k$ is a continuous bijection

from W_k to $\mathcal{P}_{N-k,S^1-\{1\}}$. For each $i \geq 1$, the function $\varphi(i+k)$ on W_k is the composition $\varphi(i+k)(f) = \varphi(i)(f/(T-1)^k)$ of the continuous (thanks to part (1)) function $\varphi(i)$ on $\mathcal{P}_{N-k,S^1-\{1\}}$ with the division map. QED

Corollary 1.8.5. *On $U(N)$, the N angles $\varphi(i)$ defined using $[0, 2\pi)$ are continuous on the open set $U(N)_{[1/\det(1-A)]}$ of $U(N)$ where 1 is not an eigenvalue. They are also continuous on each of the N locally closed sets $\mathrm{Mult}_{=i}$ of $U(N)$ where 1 is a zero of the characteristic polynomial of fixed multiplicity $i \geq 1$. In particular, they are all Borel measurable functions on $U(N)$.*

CHAPTER 2

Reformulation of the Main Results

2.0. "Naive" versions of the spacing measures

2.0.1. Fix $r \geq 1$, and an offset vector $c = (c(1), \ldots, c(r))$ in $\mathbb{Z}^r$. We do not know how to do explicit calculations, say of expected values, with the spacing measures $\mu(A, G(N), \text{offsets}\, c)$ we have defined. The problem is caused by the fact that the eigenvalues of A lie on a circle, not on a line. For this reason, we will define, case by case for each $N > c(r)$, measures $\mu(\text{naive}, A, G(N), \text{offsets}\, c)$ on $\mathbb{R}^r$ which have total mass slightly (when $N \gg 0$) less than 1, but which are much better adapted to calculation because they only involve points on a line.

2.0.2. The $U(N)$ case. Here we assume $N > c(r)$. Attached to A in $U(N)$ are its N angles in $[0, 2\pi)$,

$$0 \leq \varphi(1) \leq \varphi(2) \leq \cdots \leq \varphi(N) < 2\pi.$$

Here we use only the first $N - c(r)$ of the normalized spacing vectors $s_i(\text{offsets}\, c)$, $i = 1, \ldots, N - c(r)$, i.e., precisely those which don't require wrapping around the unit circle.

We define

$$\mu(\text{naive}, A, U(N), \text{ offsets } c) := (1/N) \sum_{j=1}^{N-c(r)} (\text{delta measure at } s_j(\text{offsets } c)),$$

a measure of total mass $(N - c(r))/N$ on $\mathbb{R}^r$.

2.0.3. The $USp(2N)$ case. Here we assume $N > c(r)$. The eigenvalues of A in $USp(2N)$ occur in N complex conjugate pairs, which we may write uniquely as $e^{\pm i\varphi(j)}$ with N angles in $[0, \pi]$,

$$0 \leq \varphi(1) \leq \varphi(2) \leq \cdots \leq \varphi(N) \leq \pi.$$

Each of these angles $\varphi(i)$ is a continuous function of A in $USp(2N)$.

For $1 \leq i \leq N - c(r)$, we define the naive normalized spacing vectors $s_i(\text{naive, offsets } c)$ in terms of these angles in $[0, \pi]$ by

$$s_i(\text{naive, offsets } c) :=$$
$$(N/\pi)(\varphi(i+c(1)) - \varphi(i), \varphi(i+c(2)) - \varphi(i+c(1)), \ldots, \varphi(i+c(r)) - \varphi(i+c(r-1))).$$

We define

$$\mu(\text{naive}, A, USp(2N), \text{ offsets } c)$$
$$:= (1/N) \sum_{j=1}^{N-c(r)} (\text{delta measure at } s_j(\text{naive, offsets } c)),$$

a measure of total mass $(N - c(r))/N$ on $\mathbb{R}^r$.

2.0.4. The $SO(2N+1)$ case. Here we assume $N > c(r)$. In $SO(2N+1)$, 1 is an eigenvalue of every element A. The remaining eigenvalues of A occur in N complex conjugate pairs, which we may write uniquely as $e^{\pm i\varphi(j)}$ with N angles in $[0, \pi]$,

$$0 \le \varphi(1) \le \varphi(2) \le \cdots \le \varphi(N) \le \pi.$$

Each of these angles $\varphi(i)$ is a continuous function of A in $SO(2N+1)$.

For $1 \le i \le N - c(r)$, we define the naive normalized spacing vectors $s_i(\text{naive, offsets } c)$ in terms of these angles in $[0, \pi]$ by

$$s_i(\text{naive, offsets } c) := \\ ((N+1/2)/\pi)(\varphi(i+c(1))-\varphi(i), \varphi(i+c(2))-\varphi(i+c(1)), \ldots, \varphi(i+c(r))-\varphi(i+c(r-1))).$$

We define

$$\mu(\text{naive}, A, SO(2N+1), \text{ offsets } c) \\ := (1/(N+1/2)) \sum_{j=1}^{N-c(r)} (\text{delta measure at } s_j(\text{naive, offsets } c)),$$

a measure of total mass $(N - c(r))/(N + 1/2)$ on $\mathbb{R}^r$.

2.0.5. The $SO(2N)$ case. Here we assume $N > c(r)$. The eigenvalues of A in $SO(2N)$ occur in N complex conjugate pairs, which we may write uniquely as $e^{\pm i\varphi(j)}$ with N angles in $[0, \pi]$,

$$0 \le \varphi(1) \le \varphi(2) \le \cdots \le \varphi(N) \le \pi.$$

Each of these angles $\varphi(i)$ is a continuous function of A in $SO(2N)$.

For $1 \le i \le N - c(r)$, we define the naive normalized spacing vectors $s_i(\text{naive, offsets } c)$ in terms of these angles in $[0, \pi]$ by

$$s_i(\text{naive, offsets } c) := \\ (N/\pi)(\varphi(i+c(1))-\varphi(i), \varphi(i+c(2))-\varphi(i+c(1)), \ldots, \varphi(i+c(r))-\varphi(i+c(r-1))).$$

We define

$$\mu(\text{naive}, A, SO(2N), \text{ offsets } c) \\ := (1/N) \sum_{j=1}^{N-c(r)} (\text{delta measure at } s_j(\text{naive, offsets } c)),$$

a measure of total mass $(N - c(r))/N$ on $\mathbb{R}^r$.

2.0.6. The $O_-(2N+2)$ case. We assume $N > c(r)$. We denote by $O_-(2N+2)$ the set of elements in $O(2N+2)$ of determinant -1. Of course $O_-(2N+2)$ is not a group, but rather a principal homogeneous space under $SO(2N+2)$, and $O(2N+2)$ is the disjoint union

$$O(2N+2) = SO(2N+2) \cup O_-(2N+2).$$

We will denote by $\text{Haar}_{O_-(2N+2)}$ the restriction to $O_-(2N+2)$ of Haar measure on $O(2N+2)$, but normalized so that $O_-(2N+2)$ has measure 1.

Any element A in $O_-(2N+2)$ has both 1 and -1 as eigenvalues. The remaining eigenvalues of A occur in N complex conjugate pairs, which we may write uniquely as $e^{\pm i\varphi(j)}$ with N angles in $[0, \pi]$,

$$0 \le \varphi(1) \le \varphi(2) \le \cdots \le \varphi(N) \le \pi.$$

Each of these angles $\varphi(i)$ is a continuous function of A in $O_-(2N+2)$.

For $1 \leq i \leq N - c(r)$, we define the naive normalized spacing vectors s_i(naive, offsets c) in terms of these angles in $[0, \pi]$ by

$$s_i(\text{naive, offsets } c) :=$$
$$(N/\pi)(\varphi(i+c(1))-\varphi(i), \varphi(i+c(2))-\varphi(i+c(1)), \ldots, \varphi(i+c(r))-\varphi(i+c(r-1))).$$

We define

$$\mu(\text{naive}, A, O_-(2N+2), \text{ offsets } c)$$
$$:= (1/(N+1)) \sum_{j=1}^{N-c(r)} (\text{delta measure at } s_j(\text{naive, offsets } c)),$$

a measure of total mass $(N - c(r))/(N + 1)$ on $\mathbb{R}^r$.

2.0.7. This concludes the case by case definition of the naive spacing measure $\mu(\text{naive}, A, G(N), \text{ offsets } c)$ on $\mathbb{R}^r$, for c an offset vector in $\mathbb{Z}^r$, and $N > c(r)$. Denote by a and b in $\mathbb{Z}^r$ the separation and step vectors corresponding to the offset vector c. We will sometimes name this same measure by its a or b:

$$\mu(\text{naive}, A, G(N), \text{ separations } a) := \mu(\text{naive}, A, G(N), \text{ offsets } c),$$
$$\mu(\text{naive}, A, G(N), \text{ steps } b) := \mu(\text{naive}, A, G(N), \text{ offsets } c).$$

[In following these measures under direct image by certain linear maps, it is the "steps b" nomenclature which is most convenient. In expressing them as alternating sums of "correlations", it is the "separations a" nomenclature which is most convenient.]

2.1. Existence, universality and discrepancy theorems for limits of expected values of naive spacing measures: the main theorems bis

2.1.1. Fix an integer $r \geq 1$, and an offset vector c in $\mathbb{Z}^r$. For $N > c(r)$, and $G(N)$ any of $U(N), SO(2N+1), USp(2N), SO(2N), O_-(2N+2)$, we have defined a naive spacing measure

$$\mu(\text{naive}, A, G(N), \text{ offsets } c),$$

which is a positive Borel measure on $\mathbb{R}^r$ of total mass

$$(N - c(r))/N \quad \text{for } U(N), USp(2N), SO(2N),$$
$$(N - c(r))/(N + 1/2) \quad \text{for } SO(2N + 1),$$
$$(N - c(r))/(N + 1) \quad \text{for } O_-(2N + 2).$$

By integration over $G(N)$ with respect to normalized Haar measure, we define its expected value

$$\mu(\text{naive}, G(N), \text{ offsets } c),$$

which is a positive Borel measure on $\mathbb{R}^r$ of total mass given above.

Theorem 2.1.2 (Theorem 1.2.1 bis). *Fix an integer $r \geq 1$, and an offset vector c in $\mathbb{Z}^r$. For each $N > c(r)$, consider the expected value measure*

$$\mu(\textit{naive}, U(N), \textit{ offsets } c) \quad \textit{on } \mathbb{R}^r$$

attached to $U(N)$. As $N \to \infty$, these measures tend to a probability measure

$$\mu(\textit{naive, univ, offsets } c)$$

in the sense that for any $\mathbb{R}$-valued, bounded, Borel measurable function f of compact support on $\mathbb{R}^r$, we have

$$\int f\, d\mu(naive,\ univ,\ offsets\ c) = \lim_{N\to\infty} \int f\, d\mu(naive, U(N),\ offsets\ c).$$

The probability measure $\mu(naive,\ univ,\ offsets\ c)$ has a continuous CDF.

Theorem 2.1.3 (Theorem 1.2.3 bis). *Fix an integer $r \geq 1$, and an offset vector c in $\mathbb{Z}^r$. For each $N > c(r)$, consider the expected value spacing measures*

$$\mu(naive, G(N),\ offsets\ c)$$

on $\mathbb{R}^r$ attached to $G(N)$ one of

$$U(N),\ SO(2N+1),\ USp(2N),\ SO(2N),\ O_-(2N+2).$$

In each of these sequences, the expected value spacing measures

$$\mu(naive, G(N),\ offsets\ c)$$

converge (in the sense above) to the **same** *limit measure*

$$\mu(naive,\ univ,\ offsets\ c),$$

which is a probability measure with continuous CDF.

2.1.4. Fix an integer $r \geq 1$, and an offset vector c in $\mathbb{Z}^r$. For $N > c(r)$, and $G(N)$ one of $U(N)$, $SO(2N+1)$, $USp(2N)$, $SO(2N)$, $O_-(2N+2)$, the $[0,1]$-valued function on $G(N)$

$$A \mapsto \mathrm{discrep}(\mu(\mathrm{naive}, A, G(N),\ \mathrm{offsets}\ c), \mu(\mathrm{naive},\ \mathrm{univ},\ \mathrm{offsets}\ c))$$

is Borel measurable. For $G(N) = U(N)$, it is continuous on the open set of full measure $U(N)[1/\det(1-A)]$. For the other $G(N)$, this function is continuous on all of $G(N)$ (because in all the non-$U(N)$ cases, the individual angles $\varphi(i)$ in $[0,\pi]$ are each continuous functions on $G(N)$).

Theorem 2.1.5 (Theorem 1.2.6 bis). *Fix an integer $r \geq 1$, an offset vector c in $\mathbb{Z}^r$ and a real number $\varepsilon > 0$. There exists an explicit constant $N_1(\varepsilon, r, c)$ with the following property: for $G(N)$ one of*

$$U(N),\ SO(2N+1),\ USp(2N),\ SO(2N),\ O_-(2N+2),$$

and dA its total mass one Haar measure, we have the inequality

$$\int_{G(N)} \mathrm{discrep}(\mu(naive, A, G(N),\ offsets\ c), \mu(naive,\ univ,\ offsets\ c))\, dA$$
$$\leq N^{\varepsilon - 1/(2r+4)},$$

provided that $N \geq N_1(\varepsilon, r, c)$.

2.2. Deduction of Theorems 1.2.1, 1.2.3 and 1.2.6 from their bis versions

Lemma 2.2.1. *If Theorem 2.1.2 is true, then Theorem 1.2.1 is true, and the limit probability measure $\mu(univ,\ offsets\ c)$ of Theorem 1.2.1 is equal to the limit probability measure $\mu(naive,\ univ,\ offsets\ c)$ of Theorem 2.1.2.*

PROOF. The naive spacing measure $\mu(\text{naive}, A, U(N), \text{ offsets } c)$ differs from $\mu(A, U(N), \text{ offsets } c)$ in that it omits $c(r)$ of the spacing vectors, each of mass $1/N$. So for any Borel measurable function f on $\mathbb{R}^r$ of compact support whose sup norm is bounded by 1, we have the inequality

$$\left|\int f\, d\mu(A, U(N), \text{ offsets } c) - \int f\, d\mu(\text{naive}, A, U(N), \text{ offsets } c)\right| \leq c(r)/N.$$

Integrating this inequality over $U(N)$, we get the inequality

$$\left|\int f\, d\mu(U(N), \text{ offsets } c) - \int f\, d\mu(\text{naive}, U(N), \text{ offsets } c)\right| \leq c(r)/N.$$

Let $N \to \infty$. According to Theorem 2.1.2,

$$\int f\, d\mu(\text{naive}, U(N), \text{ offsets } c) \to \int f\, d\mu(\text{naive, univ, offsets } c).$$

Since $c(r)/N \to 0$ as $N \to \infty$, we deduce that

$$\int f\, d\mu(U(N), \text{ offsets } c) \to \int f\, d\mu(\text{naive, univ, offsets } c).$$

As this holds for all f as above, we see that Theorem 1.2.1 holds, with limit measure $\mu(\text{naive, univ, offsets } c)$. QED

2.2.2. Before we deduce Theorems 1.2.3 and 1.2.6 from their bis versions, we need to discuss a certain symmetry of $\mathbb{R}^r$, namely the "reversing" map

$$\text{rev} : \mathbb{R}^r \to \mathbb{R}^r, \quad \text{rev}(x_1, x_2, \ldots, x_r) := (x_r, x_{r-1}, \ldots, x_1).$$

For $r = 1$, rev is the identity; for $r \geq 2$, rev is an involution.

2.2.3. Given a measure ν on $\mathbb{R}^r$, we denote by $\text{rev}_* \nu$ the measure

$$\text{rev}_* \nu(E) = \nu(\text{rev}^{-1}(E)) = \nu(\text{rev}(E)),$$

the last equality because $\text{rev}^2 = \text{id}$.

Lemma 2.2.4. *Fix an integer $r \geq 1$, and an offset vector c in $\mathbb{Z}^r$. For every N, the expected value spacing measure $\mu(U(N), \text{ offsets } c)$ is invariant under reversal:*

$$\text{rev}_* \mu(U(N), \text{ offsets } c) = \mu(U(N), \text{ offsets } c).$$

PROOF. For A in $U(N)$, its complex conjugate $\overline{A}$ has the complex conjugate eigenvalues. So the N spacing vectors $s_i(\text{offsets } c)$ for $\overline{A}$ are just the reversals of those for A, listed in another order (the opposite order, if A has no eigenvalue 1). Therefore their spacing measures are related by

$$\mu(\overline{A}, U(N), \text{ offsets } c) = \text{rev}_* \mu(A, U(N), \text{ offsets } c).$$

Since $A \mapsto \overline{A}$ is an automorphism of $U(N)$, the normalized Haar measure on $U(N)$ is invariant under $A \mapsto \overline{A}$. Therefore the two measure-valued functions on $U(N)$,

$$A \mapsto \mu(A, U(N), \text{ offsets } c)$$

and

$$A \mapsto \mu(\overline{A}, U(N), \text{ offsets } c) = \text{rev}_* \mu(A, U(N), \text{ offsets } c),$$

have the same expected value over $U(N)$. QED

Corollary 2.2.5. *Suppose Theorem* 1.2.1 *holds. Then for every $r \geq 1$ and every offset vector c in $\mathbb{Z}^r$, the limit measure μ(univ, offsets c) is invariant under reversal*:

$$\mathrm{rev}_* \,\mu(\textit{univ, offsets } c) = \mu(\textit{univ, offsets } c).$$

PROOF. By Theorem 1.2.1, $\mu(U(N), \text{ offsets } c) \to \mu(\text{univ, offsets } c)$ as $N \to \infty$. Applying reversal, we get

$$\mathrm{rev}_* \,\mu(U(N), \text{ offsets } c) \to \mathrm{rev}_* \,\mu(\text{univ, offsets } c) \quad \text{as } N \to \infty.$$

But $\mathrm{rev}_* \,\mu(U(N), \text{ offsets } c) = \mu(U(N), \text{ offsets } c)$. QED

Corollary 2.2.6. *Suppose Theorem* 2.1.3 *holds. Then for every $r \geq 1$ and every offset vector c in $\mathbb{Z}^r$, the limit measure μ(naive, univ, offsets c) is invariant under reversal.*

PROOF. Theorem 2.1.3 implies Theorem 2.1.2, which in turn, by 2.2.1, implies that Theorem 1.2.1 holds and that $\mu(\text{naive, univ, offsets } c)$ is equal to $\mu(\text{univ, offsets } c)$. Now apply the above corollary 2.2.5. QED

Lemma 2.2.7. *Fix an integer $r \geq 1$, and an offset vector c in $\mathbb{Z}^r$. For $N > c(r)$ and $G(N)$ any of $SO(2N+1)$, $USp(2N)$, $SO(2N)$, $O_-(2N+2)$, the measures $\mu(A, G(N), \text{ offsets } c)$ and $\mu(\text{naive}, A, G(N), \text{ offsets } c)$ are related as follows*:

$$\begin{aligned}
&\mu(A, G(N), \textit{ offsets } c)\\
&\quad = 1/2(\mu(\textit{naive}, A, G(N), \textit{ offsets } c) + \mathrm{rev}_* \,\mu(\textit{naive}, A, G(N), \textit{ offsets } c))\\
&\qquad\quad + (\textit{a positive Borel measure of total mass } \leq (1 + c(r))/N).
\end{aligned}$$

PROOF. For A in $USp(2N)$ or $SO(2N)$, the $2N$ eigenvalues occur in conjugate pairs, N in the top half of the unit circle (i.e., angles in $[0, \pi]$) and their N mirror images in the bottom half of the unit circle. Our definition of the naive spacing measure made use of those in the top half. Its reversal is precisely what we would make using the spacings in the bottom half. In averaging them, we still omit the $2c(r)$ spacings which involve both top and bottom eigenvalues, each with mass $1/2N$.

For A in $SO(2N+1)$, 1 is an eigenvalue, and the other $2N$ eigenvalues fall into N pairs as above. This time the average of top and bottom omits $1 + 2c(r)$ spacings, each with mass $1/(2N+1)$.

For A in $O_-(2N+2)$, both ± 1 are eigenvalues, and the other $2N$ eigenvalues fall into N pairs as above. This time the average of top and bottom omits $2 + 2c(r)$ spacings, each with mass $1/(2N+2)$. QED

Corollary 2.2.8. *Fix an integer $r \geq 1$, and an offset vector c in $\mathbb{Z}^r$. For $N > c(r)$ and $G(N)$ any of $SO(2N+1)$, $USp(2N)$, $SO(2N)$, $O_-(2N+2)$, the expected values $\mu(G(N), \text{ offsets } c)$ and $\mu(\text{naive}, G(N), \text{ offsets } c)$ are related as follows*:

$$\begin{aligned}
&\mu(G(N), \textit{ offsets } c)\\
&\quad = 1/2(\mu(\textit{naive}, G(N), \textit{ offsets } c) + \mathrm{rev}_* \,\mu(\textit{naive}, G(N), \textit{ offsets } c))\\
&\qquad\quad + (\textit{a positive Borel measure of total mass } \leq (1 + c(r))/N).
\end{aligned}$$

PROOF. Integrate the previous Lemma 2.2.7 over $G(N)$. QED

Lemma 2.2.9. *If Theorem* 2.1.3 *holds, then Theorem* 1.2.3 *holds, and the limit probability measure* μ*(univ, offsets* c*) of Theorem* 1.2.3 *is equal to the limit probability measure* μ*(naive, univ, offsets* c*) of Theorem* 2.1.3.

PROOF. Suppose Theorem 2.1.3 holds. In view of Lemma 2.2.1 above and the discussion 1.4.0, it suffices to prove Theorem 1.2.3 for the cases $G(N) = SO(2N+1)$, $USp(2N)$, $SO(2N)$, and $O_-(2N)$ (these last two together trivially give the $O(2N)$ case).

By Theorem 2.1.3, we know that as $N \to \infty$, we have

$$\mu(\text{naive}, A, G(N),\ \text{offsets } c) \to \mu(\text{naive, univ, offsets } c).$$

Since μ(naive, univ, offsets c) is invariant under reversal, by 2.2.6, we get

$$\text{rev}_* \,\mu(\text{naive}, G(N),\ \text{offsets } c) \to \mu(\text{naive, univ, offsets } c).$$

The assertion now follows from the above Corollary 2.2.8. QED

Lemma 2.2.10. *Theorems* 2.1.3 *and* 2.1.5 *together imply* 1.2.6.

PROOF. By Theorem 2.1.3, there is no need to distinguish between the measures μ(naive, univ, offsets c) and μ(univ, offsets c): they are equal. So we may and will compute discrepancy from μ(univ, offsets c). We begin with the case of $U(N)$. As already remarked in the proof of Lemma 2.2.1, $\mu(\text{naive}, A, U(N),\ \text{offsets } c)$ differs from $\mu(A, U(N),\ \text{offsets } c)$ in that it omits $c(r)$ of the spacing vectors, each of mass $1/N$. In particular, the discrepancy between them is bounded by $c(r)/N$. So by the triangle inequality for discrepancy (sup norm for differences of CDF's), we have the inequality

$$\begin{aligned}&\text{discrep}(\mu(A, G(N),\ \text{offsets } c), \mu(\text{univ, offsets } c))\\ &\qquad \le c(r)/N + \text{discrep}(\mu(\text{naive}, A, U(N),\ \text{offsets } c), \mu(\text{univ, offsets } c)).\end{aligned}$$

Pick $\varepsilon > 0$. Integrating over $U(N)$, and using Theorem 2.1.5, we get

$$\begin{aligned}&\int_{G(N)} \text{discrep}(\mu(A, G(N),\ \text{offsets } c), \mu(\text{univ, offsets } c))d(\text{Haar}_{G(N)})\\ &\qquad \le c(r)/N + N^{\varepsilon/2-1/(2r+4)}\end{aligned}$$

provided that $N \ge N_1(\varepsilon/2, r, c)$. But clearly there exists an explicit constant $M(\varepsilon, r, c(r))$ such that if $N \ge M(\varepsilon, r, c(r))$, we have

$$(1 + c(r))/N + N^{\varepsilon/2-1/(2r+4)} \le N^{\varepsilon-1/(2r+4)}.$$

So Theorem 1.2.3 holds for $U(N)$, with

$$N_2(\varepsilon, r, c) := \text{Sup}(N_1(\varepsilon/2, r, c), M(\varepsilon, r, c(r))).$$

Clearly there exists an explicit constant $M_1(\varepsilon, r)$ such that for $N \ge M_1(\varepsilon, r)$, we have

$$(N-1)^{\varepsilon/2-1/(2r+4)} \le N^{\varepsilon-1/(2r+4)}.$$

We define

$$N(\varepsilon, r, c) := 1 + \text{Sup}(N_2(\varepsilon/2, r, c), M_1(\varepsilon, r)).$$

With this choice of $N(\varepsilon, r, c)$, one sees easily that Theorem 1.2.6 holds for the other $G(N)$ as well. The key in these cases is the relation 2.2.7:

$$\begin{aligned}
&\mu(A, G(N), \text{ offsets } c) \\
&\quad = 1/2(\mu(\text{naive}, A, G(N), \text{ offsets } c) + \text{rev}_* \, \mu(\text{naive}, A, G(N), \text{ offsets } c)) \\
&\qquad + (\text{a positive Borel measure of total mass } \le (1 + c(r))/N),
\end{aligned}$$

together with the reversal invariance of $\mu(\text{univ, offsets } c)$, thanks to which we have the equality of discrepancies

$$\begin{aligned}
&\text{discrep}(\mu(\text{naive}, A, G(N), \text{offsets } c), \mu(\text{univ, offsets } c)) \\
&\quad = \text{discrep}(\text{rev}_* \, \mu(\text{naive}, A, G(N), \text{ offsets } c), \mu(\text{univ, offsets } c)).
\end{aligned}$$

Using these together with the triangle inequality, we find that

$$\begin{aligned}
&\text{discrep}(\mu(A, U(N), \text{ offsets } c), \mu(\text{univ, offsets } c)) \\
&\quad \le (1 + c(r))/N + \text{discrep}(\mu(\text{naive}, A, G(N), \text{ offsets } c), \mu(\text{univ, offsets } c)),
\end{aligned}$$

and the argument proceeds exactly as in the $U(N)$ case. This proves Theorem 1.2.6 for $SO(2N+1)$, $USp(2N)$, $SO(2N)$ and $O_-(2N+2)$. [Indeed, in these cases $N_2(\varepsilon, r, c)$ would already work as our $N(\varepsilon, r, c)$. We only need the more conservative $N(\varepsilon, r, c)$ defined above to handle the $O_-(2N) = O(2(N-1)+2)$ case.] As already explained, the $SU(N)$ and $O(2N+1)$ cases result formally from the $U(N)$ and $SO(2N+1)$ cases. The $O(2N)$ case of Theorem 1.2.6 is just the average of the $SO(2N)$ and the $O_-(2N) = O(2(N-1)+2)$ cases. QED

2.3. The combinatorics of spacings of finitely many points on a line: first discussion

2.3.1. We give ourselves an integer $N \ge 2$ and a set X of N points on the real line, in increasing order,

$$X : x(1) \le x(2) \le \cdots \le x(N).$$

Formally, X is a nondecreasing map from the set $\{1, \dots, N\}$ to $\mathbb{R}$. But we will often speak of X as though it were a subset of $\mathbb{R}$ of cardinality N, which, strictly speaking, is only correct if all the $x(i)$ are distinct.

2.3.2. We fix a real number $s \ge 0$, and ask ourselves how many pairs of adjacent, or 0-separated, points in X are at distance s from each other, i.e., how many pairs of indices $1 \le i < j \le N$ are there for which

$$x(j) - x(i) = s \quad \text{and} \quad j - i = 1?$$

We call this number $\text{Sep}(0)(s)$.

2.3.3. We also ask how many pairs of points in X (not necessarily adjacent) are at distance s from each other, i.e., how many pairs of indices $1 \le i < j \le N$ are there for which

$$x(j) - x(i) = s?$$

We call this number $\text{Clump}(0)(s)$.

2.3.4. More generally, for each integer $a \geq 0$, we ask how many pairs of a-separated points in X are at distance s from each other, i.e., how many pairs of indices $1 \leq i < j \leq N$ are there for which

$$x(j) - x(i) = s \quad \text{and} \quad j - i = a + 1?$$

We call this number $\mathrm{Sep}(a)(s)$.

2.3.5. We also ask how many subsets of X of cardinality $a+2$ are there whose endpoints are at distance s from each other, i.e., how many $a+2$ tuples of indices $1 \leq i(0) < i(1) < \cdots < i(a+1) \leq N$ are there for which

$$x(i(a+1)) - x(i(0)) = s?$$

We call this number $\mathrm{Clump}(a)(s)$.

2.3.6. Obviously, both $\mathrm{Sep}(a)(s)$ and $\mathrm{Clump}(a)(s)$ vanish if $a \gg 0$ (in fact, if $a > N - 2$).

2.3.7. The equations whose solutions are counted by $\mathrm{Sep}(a)(s)$ and $\mathrm{Clump}(a)(s)$ make sense for all real s, but obviously have no solutions for $s < 0$. So we define

$$\mathrm{Sep}(a)(s) = \mathrm{Clump}(a)(s) = 0 \quad \text{if } s < 0.$$

Lemma 2.3.8. *For every integer $a \geq 0$, we have the identity*

$$\mathrm{Clump}(a)(s) = \sum_{b \geq a} \mathrm{Binom}(b, a)\,\mathrm{Sep}(b)(s).$$

PROOF. $\mathrm{Clump}(a)(s)$ is the number of $(a+2)$-tuples of indices

$$1 \leq i(0) < i(1) < \cdots < i(a+1) \leq N$$

for which

$$x(i(a+1)) - x(i(0)) = s.$$

Consider such an $(a+2)$-tuple. Its two endpoints are $b := i(a+1) - 1 - i(0)$ separated and at distance s, so these two endpoints are counted in $\mathrm{Sep}(b)(s)$. Between these endpoints we have chosen one of the $\mathrm{Binom}(b, a)$ subsets of cardinality a of the b intervening points. QED

Corollary 2.3.9. *We have the identities of generating polynomials in an indeterminate T,*

$$\sum_{a \geq 0} \mathrm{Clump}(a)(s) T^a = \sum_{b \geq 0} \mathrm{Sep}(b)(s)(1+T)^b,$$

$$\sum_{b \geq 0} \mathrm{Sep}(b)(s) T^b = \sum_{a \geq 0} \mathrm{Clump}(a)(s)(T-1)^a.$$

PROOF. The first identity is, coefficient by coefficient, the identity of Lemma 2.3.8. The second identity is obtained from the first by the change of variable $T \mapsto T - 1$. QED

Equating coefficients of like powers of T on both sides of the second identity, we obtain

Corollary 2.3.10. *For each integer $a \geq 0$, we have the identity*

$$\mathrm{Sep}(a)(s) = \sum_{n \geq a} (-1)^{n-a} \,\mathrm{Binom}(n,a)\,\mathrm{Clump}(n)(s).$$

Key Lemma 2.3.11. *For each integer $a \geq 0$ and each integer $m \geq a$, we have the inequalities*

$$\sum_{n=a}^{m} (-1)^{n-a} \,\mathrm{Binom}(n,a)\,\mathrm{Clump}(n)(s) \leq \mathrm{Sep}(a)(s) \quad \text{if } m-a \text{ is odd,}$$

$$\mathrm{Sep}(a)(s) \leq \sum_{n=a}^{m} (-1)^{n-a} \,\mathrm{Binom}(n,a)\,\mathrm{Clump}(n)(s) \quad \text{if } m-a \text{ is even.}$$

PROOF. Fix $m \geq a$. According to the previous corollary, we have

$$\begin{aligned}\mathrm{Sep}(a)(s) &= \sum_{n \geq a} (-1)^{n-a} \,\mathrm{Binom}(n,a)\,\mathrm{Clump}(n)(s) \\ &= \sum_{n=a}^{m} (-1)^{n-a} \,\mathrm{Binom}(n,a)\,\mathrm{Clump}(n)(s) \\ &\quad + \sum_{n \geq m+1} (-1)^{n-a} \,\mathrm{Binom}(n,a)\,\mathrm{Clump}(n)(s).\end{aligned}$$

So we must show that the tail has the correct sign, i.e., that

$$(-1)^{m+1-a} \sum_{n \geq m+1} (-1)^{n-a} \,\mathrm{Binom}(n,a)\,\mathrm{Clump}(n)(s) \geq 0,$$

i.e.,

$$\sum_{n \geq m+1} (-1)^{n-m-1} \,\mathrm{Binom}(n,a)\,\mathrm{Clump}(n)(s) \geq 0.$$

We use Lemma 2.3.8 to write Clump in terms of Sep, and this becomes

$$\sum_{n \geq m+1} (-1)^{n-m-1} \,\mathrm{Binom}(n,a) \sum_{b \geq n} \mathrm{Binom}(b,n)\,\mathrm{Sep}(b)(s) \geq 0,$$

i.e.,

$$\sum_{b \geq m+1} \mathrm{Sep}(b)(s) \sum_{n=m+1}^{b} (-1)^{n-m-1} \,\mathrm{Binom}(b,n)\,\mathrm{Binom}(n,a) \geq 0.$$

Since each term $\mathrm{Sep}(b)(s)$ is nonnegative, it suffices to show that for fixed $b \geq m+1 \geq a$, we have

$$\sum_{n=m+1}^{b} (-1)^{n-m-1} \,\mathrm{Binom}(b,n)\,\mathrm{Binom}(n,a) \geq 0.$$

At this point we make use of

Sublemma 2.3.12. *For $0 \leq a \leq n \leq b$, we have the identity*

$$\mathrm{Binom}(b,n)\,\mathrm{Binom}(n,a) = \mathrm{Binom}(b,a)\,\mathrm{Binom}(b-a,n-a).$$

PROOF. Write binomial coefficients as ratios of factorials. QED

Using the sublemma, this becomes

$$\sum_{n=m+1}^{b} (-1)^{n-m-1} \operatorname{Binom}(b,a) \operatorname{Binom}(b-a, n-a) \geq 0.$$

Factoring out $\operatorname{Binom}(b,a)$, this becomes

$$\sum_{n=m+1}^{b} (-1)^{n-m-1} \operatorname{Binom}(b-a, n-a) \geq 0.$$

In terms of the quantities $k := n-a$, $j := m+1-a$, and $l := b-a$, this becomes

$$\sum_{k=j}^{l} (-1)^{k-j} \operatorname{Binom}(l,k) \geq 0.$$

This positivity results from

Sublemma 2.3.13. *For integers $1 \leq j \leq l$, we have*

$$\sum_{k=j}^{l} (-1)^{k-j} \operatorname{Binom}(l,k) = \operatorname{Binom}(l-1, j-1).$$

PROOF. This is Pascal's triangle, read backwards. Explicitly, we have

$$\operatorname{Binom}(l,k) = \operatorname{Binom}(l-1,k) + \operatorname{Binom}(l-1,k-1).$$

So our sum telescopes:

$$\sum_{k=j}^{l} (-1)^{k-j} \operatorname{Binom}(l,k)$$
$$= \sum_{k=j}^{l} (-1)^{k-j} \operatorname{Binom}(l-1,k) + \sum_{k=j}^{l} (-1)^{k-j} \operatorname{Binom}(l-1,k-1).$$

In the first sum, the last $(k=l)$ term vanishes, and for $k < l$ the k'th term of the first sum cancels the $(k+1)$'st term of the second, leaving only the first $(k=j)$ term in the second sum. QED, for both 2.3.13 and 2.3.11.

2.4. The combinatorics of spacings of finitely many points on a line: second discussion

2.4.1. We continue with our given integer $N \geq 2$ and our given set X of N points on the real line, in increasing order,

$$X : x(1) \leq x(2) \leq \cdots \leq x(N).$$

2.4.2. Fix an integer $r \geq 1$, and a vector $s = (s(1), \ldots, s(r))$ in $\mathbb{R}^r$.

2.4.3. Suppose we are given an offset vector c,

$$c = (c(1), \dots, c(r)),$$

in $\mathbb{Z}^r$, and the corresponding separation vector a,

$$a = (a(1), \dots, a(r)),$$
$$a(1) = c(1) - 1, \qquad a(i) = c(i) - 1 - c(i-1) \quad \text{for } i \geq 2.$$

Using this data, we will now define two nonnegative integers

$$\operatorname{Sep}(a)(s) \quad \text{and} \quad \operatorname{Clump}(a)(s)$$

in such a way that for $r = 1$, these are exactly the quantities of the same name discussed in the previous section. It will be obvious from the definitions that

$$\operatorname{Sep}(a)(s) = \operatorname{Clump}(a)(s) = 0 \quad \text{unless } s \text{ has all } s(i) \geq 0.$$

2.4.4. $\operatorname{Clump}(a)(s)$ is the number of systems of indices

$$1 \leq i(0) < i(1) < \cdots < i(c(r)) \leq N$$

for which we have

$$x(i(c(1))) - x(i(0)) = s(1)$$

and

$$x(i(c(j))) - x(i(c(j-1))) = s(j) \quad \text{for } j = 2, \dots, r.$$

2.4.5. $\operatorname{Sep}(a)(s)$ is the number of systems as above which satisfy in addition the requirement that

$$i(j) - i(j-1) = 1 \quad \text{for } j = 1, \dots, c(r).$$

2.4.6. We can also describe $\operatorname{Sep}(a)(s)$ as the number of systems of indices

$$1 \leq k(0) < k(1) < \cdots < k(r) \leq N$$

such that

$$x(k(1)) - x(k(0)) = s(1)$$

and

$$x(k(j)) - x(k(j-1)) = s(j) \quad \text{for } j = 2, \dots, r,$$

and which satisfy

$$k(1) - k(0) = c(1),$$
$$k(j) - k(j-1) = c(j) \quad \text{for } j = 2, \dots, r.$$

[Given the indices $i(j)$ for $j = 0, \dots, c(r)$, define $k(0) = i(0)$, $k(j) = i(c(j))$ for $j = 1, \dots, r$. Given the indices $k(j)$ for $j = 0, \dots, r$, define $i(0) = k(0)$, $i(c(j)) = k(j)$ for $j = 1, \dots, r$; there is then a unique way to define the terms $i(l)$ for the remaining values of l so that $i(l) - i(l-1) = 1$ for $l = 1, \dots, c(r)$.]

2.4.7. In order to state the analogues in this more general context of the identities and inequalities given above for $r = 1$ in the "first discussion" 2.3, we will make use of the following notation.

2.4.8. Given $a = (a(1), \ldots, a(r))$ and $b = (b(1), \ldots, b(r))$ in $\mathbb{Z}^r$, we say that $a \geq 0$ if $a(i) \geq 0$ for each i. We say that $a \geq b$ if $a - b \geq 0$. If $a \geq 0$ and $b \geq 0$, we write

$$\mathrm{Binom}(b, a) := \prod_{i=1}^{r} \mathrm{Binom}(b(i), a(i)).$$

Given r indeterminates $T_1, \ldots, T_r$, and $a \geq 0$ in $\mathbb{Z}^r$, we write

$$T^a := \prod_{i=1}^{r} (T_i)^{a(i)}, \qquad (T \pm 1)^a := \prod_{i=1}^{r} (T_i \pm 1)^{a(i)}.$$

We define

$$\Sigma(a) := a(1) + a(2) + \cdots + a(r).$$

For x a scalar (i.e., an element of a commutative ring R) we define

$$x^a := x^{\Sigma(a)}$$

in R.

Lemma 2.4.9. *For every $a \geq 0$ in $\mathbb{Z}^r$, we have the identity*

$$\mathrm{Clump}(a)(s) = \sum_{b \geq a} \mathrm{Binom}(b, a)\, \mathrm{Sep}(b)(s).$$

PROOF. This is entirely analogous to the proof of Lemma 2.3.8. $\mathrm{Clump}(a)(s)$ is the number of systems of indices

$$1 \leq i(0) < i(1) < \cdots < i(c(r)) \leq N$$

for which we have

$$x(i(c(1))) - x(i(0)) = s(1),$$
$$x(i(c(j))) - x(i(c(j-1))) = s(j) \quad \text{for } j = 2, \ldots, r.$$

Given the indices $i(j)$ for $j = 0, \ldots, c(r)$, define $k(0) = i(0)$, $k(j) = i(c(j))$ for $j = 1, \ldots, r$. These indices are counted in $\mathrm{Sep}(b)(s)$ for b the separation vector of components

$$b(i) = k(i) - 1 - k(i-1), \quad \text{for } i = 1, \ldots, r.$$

Once the indices $k(j)$ for $j = 0, \ldots, r$ are marked, we recover our original point in $\mathrm{Clump}(a)(s)$ by picking, independently for $i = 1, \ldots, r$, one of the $\mathrm{Binom}(b(i), a(i))$ subsets of cardinality $a(i)$ of the $b(i)$ intervening points between $k(i)$ and $k(i-1)$. QED

Corollary 2.4.10. *We have the identities of generating polynomials in r indeterminates $T_1, \ldots, T_r$,*

$$\sum_{a \geq 0} \mathrm{Clump}(a)(s) T^a = \sum_{b \geq 0} \mathrm{Sep}(b)(s)(1 + T)^b,$$
$$\sum_{b \geq 0} \mathrm{Sep}(b)(s) T^b = \sum_{a \geq 0} \mathrm{Clump}(a)(s)(T - 1)^a.$$

PROOF. The first identity is, coefficient by coefficient, the identity of the lemma. The second identity is obtained from the first by the change of variable $T \mapsto T - 1$. QED

Equating coefficients of like powers of T on both sides of the second identity, we obtain

Corollary 2.4.11. *For each $a \geq 0$ in $\mathbb{Z}^r$, we have the identity*

$$\mathrm{Sep}(a)(s) = \sum_{n \geq a} (-1)^{n-a} \,\mathrm{Binom}(n,a)\,\mathrm{Clump}(n)(s).$$

Key Lemma 2.4.12. *For each $a \geq 0$ in $\mathbb{Z}^r$, and each integer $m \geq \Sigma(a)$, we have the inequalities*

$$\sum_{a \leq n, \Sigma(n) \leq m} (-1)^{n-a} \,\mathrm{Binom}(n,a)\,\mathrm{Clump}(n)(s) \leq \mathrm{Sep}(a)(s)$$

if $m - \Sigma(a)$ is odd, and

$$\mathrm{Sep}(a)(s) \leq \sum_{a \leq n, \Sigma(n) \leq m} (-1)^{n-a} \,\mathrm{Binom}(n,a)\,\mathrm{Clump}(n)(s)$$

if $m - \Sigma(a)$ is even.

Proof. Fix $m \geq \Sigma(a)$. According to the previous Corollary 2.4.11, we have

$$\begin{aligned}\mathrm{Sep}(a)(s) &= \sum_{n \geq a} (-1)^{n-a} \,\mathrm{Binom}(n,a)\,\mathrm{Clump}(n)(s) \\ &= \sum_{a \leq n, \Sigma(n) \leq m} (-1)^{n-a} \,\mathrm{Binom}(n,a)\,\mathrm{Clump}(n)(s) \\ &\quad + \sum_{a \leq n, \Sigma(n) \geq m+1} (-1)^{n-a} \,\mathrm{Binom}(n,a)\,\mathrm{Clump}(n)(s).\end{aligned}$$

So we must show that the tail has the correct sign, i.e., that

$$(-1)^{m+1-\Sigma(a)} \sum_{a \leq n, \Sigma(n) \geq m+1} (-1)^{n-a} \,\mathrm{Binom}(n,a)\,\mathrm{Clump}(n)(s) \geq 0,$$

i.e., we must show that

$$\sum_{a \leq n, \Sigma(n) \geq m+1} (-1)^{m+1+\Sigma(n)} \,\mathrm{Binom}(n,a)\,\mathrm{Clump}(n)(s) \geq 0.$$

We use Lemma 2.4.9 to write Clump in terms of Sep, and this reduces to showing that

$$\sum_{a \leq n, \Sigma(n) \geq m+1} (-1)^{m+1+\Sigma(n)} \,\mathrm{Binom}(n,a) \sum_{b \geq n} \mathrm{Binom}(b,n)\,\mathrm{Sep}(b)(s) \geq 0.$$

Using Sublemma 2.3.12, this becomes the inequality

$$\sum_{a \leq n, \Sigma(n) \geq m+1} (-1)^{m+1+\Sigma(n)} \sum_{b \geq n} \mathrm{Binom}(b,a)\,\mathrm{Binom}(b-a, n-a)\,\mathrm{Sep}(b)(s) \geq 0.$$

Since each term $\mathrm{Binom}(b,a)\,\mathrm{Sep}(b)(s)$ is nonnegative, it suffices to show that for fixed $b \geq a$, we have

$$\sum_{a \leq n \leq b, \Sigma(n) \geq m+1} (-1)^{m+1+\Sigma(n)} \,\mathrm{Binom}(b-a, n-a) \geq 0.$$

In terms of the vectors $k := n - a$ and $l := b - a$ in $\mathbb{Z}^r$, and the integer $j := m + 1 - \Sigma(a)$, this becomes the positivity statement

$$(-1)^j \sum_{0 \le k \le l, \Sigma(k) \ge j} (-1)^k \operatorname{Binom}(l, k) \ge 0.$$

This positivity results from

Sublemma 2.4.13. *For any integer $j \ge 1$, and any $l \ge 0$ in $\mathbb{Z}^r$ with $\Sigma(l) \ge 1$, we have*

$$(-1)^j \sum_{0 \le k \le l, \Sigma(k) \ge j} (-1)^k \operatorname{Binom}(l, k) = \operatorname{Binom}(\Sigma(l) - 1, j - 1).$$

PROOF. Consider the binomial expansion of

$$\begin{aligned}(1+T)^l &:= \prod_i (1 + T_i)^{l(i)} = \prod_i \left[\sum_{0 \le k(i) \le l(i)} \operatorname{Binom}(l(i), k(i))(T_i)^{k(i)} \right] \\ &= \sum_{0 \le k \le l} \operatorname{Binom}(l, k) T^k.\end{aligned}$$

Setting all T_i equal to a single indeterminate S, and equating coefficients of like powers of S, we find that for each integer $\alpha \ge 0$, we have the identity

$$\sum_{0 \le k \le l, \Sigma(k) = \alpha} \operatorname{Binom}(l, k) = \operatorname{Binom}(\Sigma(l), \alpha).$$

So the subject of our discussion is

$$\begin{aligned}&(-1)^j \sum_{0 \le k \le l, \Sigma(k) \ge j} (-1)^k \operatorname{Binom}(l, k) \\ &= (-1)^j \sum_{\alpha \ge j} \sum_{\substack{0 \le k \le l \\ \Sigma(k) = \alpha}} (-1)^k \operatorname{Binom}(l, k) \\ &= (-1)^j \sum_{\alpha \ge j} (-1)^\alpha \operatorname{Binom}(\Sigma(l), \alpha) \\ &= \sum_{\alpha \ge j} (-1)^{\alpha - j} \operatorname{Binom}(\Sigma(l), \alpha) \\ &= \operatorname{Binom}(\Sigma(l) - 1, j - 1),\end{aligned}$$

thanks to Sublemma 2.3.13. QED, for both 2.4.13 and 2.4.12.

2.5. The combinatorics of spacings of finitely many points on a line: third discussion: variations on Sep(a) and Clump(a)

2.5.1. We continue with our given integer $N \ge 2$ and our given set X of N points on the real line, in increasing order,

$$X : x(1) \le x(2) \le \cdots \le x(N).$$

2.5.2. Fix an integer $r \geq 1$, an offset vector

$$c = (c(1), \dots, c(r))$$

in $\mathbb{Z}^r$, and the corresponding separation vector a,

$$a = (a(1), \dots, a(r)),$$
$$a(1) = c(1) - 1, \qquad a(i) = c(i) - 1 - c(i-1) \quad \text{for } i \geq 2.$$

2.5.3. Out of this data, we define positive Borel measures (both will be finite sums of delta measures) $\mathrm{Sep}(a)$ and $\mathrm{Clump}(a)$ on $\mathbb{R}^r$ as follows. We have defined the nonnegative integers $\mathrm{Sep}(a)(s)$ and $\mathrm{Clump}(a)(s)$ for any s in $\mathbb{R}^r$. For any Borel set E in $\mathbb{R}^r$, we define

$$\mathrm{Sep}(a)(E) := \sum_{s \text{ in } E} \mathrm{Sep}(a)(s),$$
$$\mathrm{Clump}(a)(E) := \sum_{s \text{ in } E} \mathrm{Clump}(a)(s).$$

2.5.4. For f any $\mathbb{R}$-valued Borel measurable function on $\mathbb{R}^r$, we define

$$\mathrm{Sep}(a, f) := \int f \, d\,\mathrm{Sep}(a) := \sum_s f(s)\,\mathrm{Sep}(a)(s),$$
$$\mathrm{Clump}(a, f) := \int f \, d\,\mathrm{Clump}(a) := \sum_s f(s)\,\mathrm{Clump}(a)(s),$$

(in both cases the sum over all s in $\mathbb{R}^r$).

2.5.5. If we wish to emphasize the dependence of $\mathrm{Sep}(a, f)$ and $\mathrm{Clump}(a, f)$ on the initial choice of $N \geq 2$ and the increasing sequence of real numbers

$$X : x(1) \leq x(2) \leq \cdots \leq x(N),$$

we will write $\mathrm{Sep}(a, f)$ and $\mathrm{Clump}(a, f)$ as

$$\mathrm{Sep}(a, f, N, X) \text{ and } \mathrm{Clump}(a, f, N, X)$$

respectively.

2.5.6. Let us make explicit their dependence on X. We have

$$\mathrm{Clump}(a, f, N, X) =$$
$$\sum f(x(i(c(1))) - x(i(0)), x(i(c(2))) - x(i(c(1))), \dots, x(i(c(r))) - x(i(c(r-1)))),$$

the sum over all systems of indices

$$1 \leq i(0) < i(1) < \cdots < i(c(r)) \leq N.$$

$\mathrm{Sep}(a, f, N, X)$ is the same sum, but with the range of summation restricted to those systems of indices which in addition satisfy

$$i(j) - i(j-1) = 1 \quad \text{for } j = 1, \dots, c(r).$$

Lemma 2.5.7. *For any integer $N \geq 2$, denote by $\mathbb{R}^N$(ordered) the closed subset of $\mathbb{R}^N$ consisting of those points $(x(1), \dots, x(N))$ which satisfy*

$$x(1) \leq x(2) \leq \cdots \leq x(N).$$

The "order" map $\mathbb{R}^N \to \mathbb{R}^N(ordered) \subset \mathbb{R}^N$, "arrange the coordinates in increasing order",

$$X \mapsto \textit{order}\ (X),$$

is a continuous map of $\mathbb{R}^N$ to itself.

PROOF. To check that a map $X \to Y$ of topological spaces is continuous, it suffices to exhibit a finite covering of its source X by closed sets Z_i such that the map restricted to each Z_i is continuous. We apply this criterion, taking the closed sets to be indexed by the elements of the symmetric group S_N, the closed set Z_σ being the set of those points $(x(1), x(2), \ldots, x(N))$ which satisfy

$$x(\sigma(1)) \leq x(\sigma(2)) \leq \cdots \leq x(\sigma(N)).$$

Our map on Z_σ is certainly continuous, being the restriction to Z_σ of the linear automorphism of $\mathbb{R}^N$ given by σ. QED

Corollary 2.5.8. *Let $N \geq 2$. The "order" map $\mathbb{R}^N \to \mathbb{R}^N(ordered)$ makes $\mathbb{R}^N(ordered)$ the quotient space $\mathbb{R}^N/\Sigma_N$ in the category of topological spaces with continuous maps, and also in the category of topological spaces with Borel measurable maps. Concretely, for any topological space Y, to give a continuous (respectively Borel measurable) map $F\colon \mathbb{R}^N \to Y$ which is Σ_N-invariant is the same as to give a continuous (respectively Borel measurable) map $G\colon \mathbb{R}^N(ordered) \to Y$, with F and G determining each other by the rules*

$$G := F|\mathbb{R}^N(\textit{ordered}), \qquad F(X) = G(\textit{order}(X)).$$

PROOF. If we start with a Σ_N-invariant F, then $F(X) = F(\text{order}(X))$ by Σ_N-invariance. As the point order(X) lies in $\mathbb{R}^N$(ordered), we have

$$F(\text{order}(X)) = G(\text{order}(X)) \quad \text{for } G := F|\mathbb{R}^N(\text{ordered}).$$

If F is continuous (respectively Borel measurable) on $\mathbb{R}^N$, then so is G on $\mathbb{R}^N$(ordered). Conversely, given a continuous (respectively Borel measurable) map $G : \mathbb{R}^N(\text{ordered}) \to Y$, the composite $X \mapsto F(\text{order}(X))$ is Σ_N-invariant, and is continuous (resp. Borel measurable), because the "order" map is continuous. QED

Lemma 2.5.9. *Fix an integer $r \geq 1$ and a separation vector a in $\mathbb{Z}^r$. If f is a continuous (respectively Borel measurable) function on $\mathbb{R}^r$, then for each $N \geq 2$ the functions on $\mathbb{R}^N(ordered)$*

$$X \mapsto \text{Sep}(a, f, N, X) \quad \textit{and} \quad X \mapsto \text{Clump}(a, f, N, X)$$

are each continuous (respectively Borel measurable).

PROOF. This is obvious from the explicit formulas for these functions as finite sums of values of f at images of X under linear maps from $\mathbb{R}^N$ to $\mathbb{R}^r$. QED

2.5.10. We have defined $\text{Sep}(a, f, N, X)$ and $\text{Clump}(a, f, N, X)$ for X in $\mathbb{R}^N$(ordered). We now extend them to be Σ_N-invariant functions on X in $\mathbb{R}^N$, by defining

$$\text{Sep}(a, f, N, X) := \text{Sep}(a, f, N,\ \text{order}(X)),$$
$$\text{Clump}(a, f, N, X) := \text{Clump}(a, f, N,\ \text{order}(X)).$$

Lemma 2.5.11. *Fix an integer $r \geq 1$ and a separation vector a in $\mathbb{Z}^r$.*

1) *If f is a continuous (respectively Borel measurable) $\mathbb{R}$-valued function on $\mathbb{R}^r$, then for $N \geq 2$ the $\mathbb{R}$-valued functions on $\mathbb{R}^N$*

$$X \mapsto \mathrm{Sep}(a, f, N, X) \quad \textit{and} \quad X \mapsto \mathrm{Clump}(a, f, N, X)$$

are each continuous (respectively Borel measurable) and Σ_N-invariant.

2) *For t in $\mathbb{R}$, denote by $\Delta_N(t)$ the diagonal point $(t, t, \ldots, t)$ in $\mathbb{R}^N$. The functions*

$$X \mapsto \mathrm{Sep}(a, f, N, X) \quad \textit{and} \quad X \mapsto \mathrm{Clump}(a, f, N, X)$$

are each invariant under additive translations $X \mapsto X + \Delta_N(t)$, for all t in $\mathbb{R}$.

3) *Suppose f as a function on $\mathbb{R}^r$ has compact support, say supported in the set $\{s$ in $\mathbb{R}^r$ with $\sum_i |s(i)| \leq \alpha\}$. If $N = 1 + r + \Sigma(a)$, then the function*

$$F : X \mapsto \mathrm{Sep}(a, f, 1 + r + \Sigma(a), X) = \mathrm{Clump}(a, f, 1 + r + \Sigma(a), X)$$

vanishes unless X in $\mathbb{R}^{1+r+\Sigma(a)}$ lies within the tubular neighborhood

$$\left\{X \text{ in } \mathbb{R}^{1+r+\Sigma(a)} \text{ with } \max_{i,j} |x(i) - x(j)| \leq \alpha\right\}$$

of the diagonal. Moreover, if f is bounded on $\mathbb{R}^r$, then F is bounded on $\mathbb{R}^{1+r+\Sigma(a)}$, and $\|F\|_{\sup} \leq \|f\|_{\sup}$.

PROOF. Assertion 1) simply combines the previous two lemmas (2.5.7 and 2.5.9). Assertion 2) is an immediate consequence of the definitions. For assertion 3), by Σ_N-invariance, we may suppose X is ordered:

$$x(1) \leq x(2) \leq \cdots \leq x(N).$$

Consider the offset vector c in $\mathbb{Z}^r$ corresponding to a. Because

$$N = 1 + r + \Sigma(a) = 1 + c(r),$$

there is only one index system to be considered in the definition of

$$\begin{aligned} &\mathrm{Sep}(a, f, 1 + r + \Sigma(a), X) = \mathrm{Clump}(a, f, 1 + r + \Sigma(a), X) \\ &\quad := f(x(1 + c(1)) - x(1), x(c(2)) - x(c(1)), \ldots, x(c(r)) - x(c(r - 1))). \end{aligned}$$

For the unique point

$$s := (x(1 + c(1)) - x(1), x(c(2)) - x(c(1)), \ldots, x(c(r)) - x(c(r - 1)))$$

at which f in evaluated, we have, because X is ordered,

$$\sum_i |s(i)| = x(c(r)) - x(0) = \max_{i,j} |x(i) - x(j)|. \quad \text{QED}$$

Lemma 2.5.12. *Fix an integer $r \geq 1$, a separation vector $a \geq 0$ in $\mathbb{Z}^r$, a Borel measurable $\mathbb{R}$-valued function f on $\mathbb{R}^r$, an integer $N \geq 2$, and a point X in $\mathbb{R}^N$.*

1) *We have the identity*

$$\mathrm{Sep}(a, f, N, X) = \sum_{n \geq a} (-1)^{n-a} \,\mathrm{Binom}(n, a)\, \mathrm{Clump}(n, f, N, X).$$

2) *If $f \geq 0$ as function on $\mathbb{R}^r$, then for each integer $m \geq \Sigma(a)$, we have the inequalities*

$$\sum_{a \leq n, \Sigma(n) \leq m} (-1)^{n-a} \,\mathrm{Binom}(n, a)\, \mathrm{Clump}(n, f, N, X) \leq \mathrm{Sep}(a, f, N, X)$$

if $n - \Sigma(a)$ is odd, and

$$\mathrm{Sep}(a, f, N, X) \leq \sum_{a \leq n, \Sigma(n) \leq m} (-1)^{n-a} \,\mathrm{Binom}(n, a)\, \mathrm{Clump}(n, f, N, X)$$

if $m - \Sigma(a)$ is even.

PROOF. Assertion 1) is just a restatement of Corollary 2.4.11, and assertion 2) restates the Key Lemma 2.4.12. QED

2.5.13. Given an integer $N \geq 1$, an integer $n \geq 1$ and a subset T of $\{1, \dots, N\}$ with $\mathrm{Card}(T) = n$, written in increasing order

$$1 \leq t(1) < t(2) < \cdots < t(n) \leq N,$$

we denote by

$$\mathrm{pr}(T) : \mathbb{R}^N \to \mathbb{R}^n, \ X \mapsto \mathrm{pr}(T)(X),$$

the linear map

$$(x(1), x(2), \dots, x(N)) \mapsto (x(t(1)), x(t(2)), \dots, x(t(n))).$$

Lemma 2.5.14. *Let n and N be integers, both ≥ 1. Suppose we are given a continuous (respectively Borel measurable) $\mathbb{R}$-valued function F on $\mathbb{R}^n$ which is Σ_n-invariant. Then the $\mathbb{R}$-valued function $F[n, N]$ on $\mathbb{R}^N$ defined by*

$$F[n, N] : X \mapsto \sum_{\mathrm{Card}(T)=n} F(\mathrm{pr}(T)(X)),$$

the sum over all subsets T of $\{1, \dots, N\}$ with $\mathrm{Card}(T) = n$, is a continuous (respectively Borel measurable) $\mathbb{R}$-valued function on $\mathbb{R}^N$ which is Σ_N-invariant.

PROOF. If $n > N$, the sum is empty, and $F[n, N]$ vanishes. If $n \leq N$, each of the $\mathrm{Binom}(N, n)$ maps $\mathrm{pr}(T)$ is continuous, so $F[n, N]$ is continuous (respectively Borel measurable). For σ in Σ_N and X in $\mathbb{R}^N$, σX is the point with coordinates $(\sigma X)(i) := X(\sigma^{-1}(i))$. So for any subset T, $\mathrm{pr}(T)(\sigma(X))$ is a permutation of $\mathrm{pr}(\sigma^{-1}(T))(X)$. Since F is Σ_n-invariant, we have

$$\begin{aligned} F[n, N](\sigma X) &:= \sum_{\mathrm{Card}(T)=n} F(\mathrm{pr}(T)(\sigma X)) \\ &= \sum_{\mathrm{Card}(T)=n} F(\mathrm{pr}(\sigma^{-1}(T))(X)) = F[n, N](X). \quad \text{QED} \end{aligned}$$

Lemma 2.5.15. *Fix an integer $r \geq 1$, a separation vector $a \geq 0$ in $\mathbb{Z}^r$, the corresponding offset vector c in $\mathbb{Z}^r$, a Borel measurable $\mathbb{R}$-valued function f on $\mathbb{R}^r$, an integer $N \geq 2$, and a point X in $\mathbb{R}^N$. We have the identity*

$$\begin{aligned} \mathrm{Clump}(a, f, N, X) &= \sum_{\mathrm{Card}(T)=1+c(r)} \mathrm{Clump}(a, f, 1 + c(r), \mathrm{pr}(T)(X)) \\ &= \sum_{\mathrm{Card}(T)=1+r+\Sigma(a)} \mathrm{Clump}(a, f, 1 + r + \Sigma(a), \mathrm{pr}(T)(X)), \end{aligned}$$

the sum over all subsets T of $\{1, \dots, N\}$ of cardinality $1 + c(r) = 1 + r + \Sigma(a)$.

PROOF. Both sides are Σ_N-invariant, so it suffices to check when X lies in $\mathbb{R}^N$(ordered), in which case it is immediate from the explicit formula of 2.5.6. QED

2.6. The combinatorics of spacings of finitely many points of a line: fourth discussion: another variation on Clump(a)

2.6.1. We fix an integer $r \geq 1$, a separation vector a in $\mathbb{Z}^r$, an integer $k \geq \Sigma(a)$, an $\mathbb{R}$-valued Borel measurable function f on $\mathbb{R}^r$, an integer $N \geq 2$, and a point X in $\mathbb{R}^N$. We define a real number

$$\mathrm{TClump}(k, a, f, N, X) := \sum_{n \geq a, \Sigma(n)=k} \mathrm{Binom}(n, a)\,\mathrm{Clump}(n, f, N, X).$$

Lemma 2.6.2. *Fix an integer $r \geq 1$, a separation vector $a \geq 0$ in $\mathbb{Z}^r$, a Borel measurable $\mathbb{R}$-valued function f on $\mathbb{R}^r$, an integer $N \geq 2$, and a point X in $\mathbb{R}^N$.*

1) *We have the identity*

$$\mathrm{Sep}(a, f, N, X) = \sum_{k \geq \Sigma(a)} (-1)^{k-\Sigma(a)}\,\mathrm{TClump}(k, a, f, N, X).$$

2) *If $f \geq 0$ as function on $\mathbb{R}^r$, then for each integer $m \geq \Sigma(a)$, we have the inequalities*

$$\sum_{m \geq k \geq \Sigma(a)} (-1)^{k-\Sigma(a)}\,\mathrm{TClump}(k, a, f, N, X) \leq \mathrm{Sep}(a, f, N, X)$$

if $m - \Sigma(a)$ is odd, and

$$\mathrm{Sep}(a, f, N, X) \leq \sum_{m \geq k \geq \Sigma(a)} (-1)^{k-\Sigma(a)}\,\mathrm{TClump}(k, a, f, N, X)$$

if $m - \Sigma(a)$ is even.

PROOF. This is just Lemma 2.5.12 with terms collected according to the value k of $\Sigma(n)$. QED

Similarly, Lemma 2.5.15 gives

Lemma 2.6.3. *Fix an integer $r \geq 1$, a separation vector $a \geq 0$ in $\mathbb{Z}^r$, an integer $k \geq \Sigma(a)$, a Borel measurable $\mathbb{R}$-valued function f on $\mathbb{R}^r$, an integer $N \geq 2$, and a point X in $\mathbb{R}^N$. We have the identity*

$$\mathrm{TClump}(k, a, f, N, X) = \sum_{\mathrm{Card}(T)=1+r+k} \mathrm{TClump}(k, a, f, 1+r+k, \mathrm{pr}(T)(X)),$$

the sum over all subsets T of $\{1, \ldots, N\}$ of cardinality $1 + c(r) = 1 + r + k$.

2.7. Relation to naive spacing measures on $G(N)$: Int, Cor and TCor

2.7.1. We fix an integer $r \geq 1$, a separation vector a in $\mathbb{Z}^r$, an integer $k \geq \Sigma(a)$, and an $\mathbb{R}$-valued Borel measurable function f on $\mathbb{R}^r$ of compact support. For $G(N)$ any of $U(N), SO(2N+1), USp(2N), SO(2N), O_{-}(2N)$, and for A in $G(N)$, we will define real numbers

$$\mathrm{Int}(a, f, G(N), A), \qquad \mathrm{Cor}(a, f, G(N), A), \qquad \mathrm{TCor}(k, a, f, G(N), A).$$

The names "Int", "Cor", and "TCor" are intended to evoke "integral", "correlation" and "total correlation" respectively. We proceed case by case. We denote by c in $\mathbb{Z}^r$ the vector of offsets corresponding to the separation vector a.

2.7.2. The $U(N)$ case. Given A in $U(N)$, we define $X(A)$ in $\mathbb{R}^N$ to be the vector whose components $X(A)(i)$ are the N angles in $[0, 2\pi)$ of its eigenvalues,

$$0 \le \varphi(1) \le \varphi(2) \le \cdots \le \varphi(N) < 2\pi.$$

It is immediate from the definitions that we have

$$\int_{\mathbb{R}^r} d\,\mu(\text{naive}, A, U(N),\ \text{offsets } c) = (1/N)\,\text{Sep}(a, f, N, (N/2\pi)X(A)).$$

We define

$$\begin{aligned} \text{Int}(a, f, U(N), A) &:= \int_{\mathbb{R}^r} f\, d\mu(\text{naive}, A, U(N),\ \text{offsets } c) \\ &= (1/N)\,\text{Sep}(a, f, N, (N/2\pi)X(A)), \\ \text{Cor}(a, f, U(N), A) &:= (1/N)\,\text{Clump}(a, f, N, (N/2\pi)X(A)), \\ \text{TCor}(k, a, f, U(N), A) &:= (1/N)\,\text{TClump}(k, a, f, N, (N/2\pi)X(A)). \end{aligned}$$

2.7.3. The $USp(2N)$ and $SO(2N)$ cases. For $G(N)$ either $USp(2N)$ or $SO(2N)$, the eigenvalues of A in $G(N)$ occur in N complex conjugate pairs, which we may write uniquely as $e^{\pm i\varphi(j)}$ with N angles in $[0, \pi]$,

$$0 \le \varphi(1) \le \varphi(2) \le \cdots \le \varphi(N) \le \pi.$$

Each of these angles $\varphi(i)$ is a continuous function of A in $G(N)$. We denote by $X(A)$ in $\mathbb{R}^N$ the vector whose N components $X(A)(i)$ are these N angles $\varphi(i)$. It is immediate from the definitions that we have

$$\int_{\mathbb{R}^r} f\, d\mu(\text{naive}, A, G(N),\ \text{offsets } c) = (1/N)\,\text{Sep}(a, f, N, (N/\pi)X(A)),$$

for $G(N)$ either $USp(2N)$ or $SO(2N)$. We define

$$\begin{aligned} \text{Int}(a, f, G(N), A) &:= \int_{\mathbb{R}^r} f\, d\mu(\text{naive}, A, U(N),\ \text{offsets } c) \\ &= (1/N)\,\text{Sep}(a, f, N, (N/\pi)X(A)), \\ \text{Cor}(a, f, G(N), A) &:= (1/N)\,\text{Clump}(a, f, N, (N/\pi)X(A)), \\ \text{TCor}(k, a, f, G(N), A) &:= (1/N)\,\text{TClump}(k, a, f, N, (N/\pi)X(A)), \end{aligned}$$

for $G(N)$ either $USp(2N)$ or $SO(2N)$.

2.7.4. The $SO(2N+1)$ case. In $SO(2N+1)$, 1 is an eigenvalue of every element A. The remaining eigenvalues of A occur in N complex conjugate pairs, which we may write uniquely as $e^{\pm i\varphi(j)}$ with N angles in $[0, \pi]$,

$$0 \le \varphi(1) \le \varphi(2) \le \cdots \le \varphi(N) \le \pi.$$

Each of these angles $\varphi(i)$ is a continuous function of A in $SO(2N+1)$. We denote by $X(A)$ in $\mathbb{R}^N$ the vector whose N components $X(A)(i)$ are these N angles $\varphi(i)$. It is immediate from the definitions that for $G(N) = SO(2N+1)$ we have

$$\begin{aligned} &\int_{\mathbb{R}^r} f\, d\mu(\text{naive}, A, G(N),\ \text{offsets } c) \\ &\qquad = (1/(N+1/2))\,\text{Sep}(a, f, N, ((N+1/2)/\pi)X(A)). \end{aligned}$$

We define

$$\begin{aligned}\mathrm{Int}(a,f,G(N),A) &:= \int_{\mathbb{R}^r} f\,d\mu(\text{naive}, A, G(N),\ \text{offsets } c)\\ &= (1/(N+1/2))\,\mathrm{Sep}(a,f,N,((N+1/2)/\pi)X(A)),\end{aligned}$$

$$\mathrm{Cor}(a,f,G(N),A) := (1/(N+1/2))\,\mathrm{Clump}(a,f,N,((N+1/2)/\pi)X(A)),$$

$$\mathrm{TCor}(k,a,f,G(N),A) := (1/(N+1/2))\,\mathrm{TClump}(k,a,f,N,((N+1/2)/\pi)X(A)),$$

for $G(N) = SO(2N+1)$.

2.7.5. The $O_-(2N+2)$ case. Any element A in $O_-(2N+2)$ has both 1 and -1 as eigenvalues. The remaining eigenvalues of A occur in N complex conjugate pairs, which we may write uniquely as $e^{\pm i\varphi(j)}$ with N angles in $[0,\pi]$,

$$0 \le \varphi(1) \le \varphi(2) \le \cdots \le \varphi(N) \le \pi.$$

Each of these angles $\varphi(i)$ is a continuous function of A in $O_-(2N)$. We denote by $X(A)$ in $\mathbb{R}^N$ the vector whose N components $X(A)(i)$ are these N angles $\varphi(i)$. It is immediate from the definitions that for $G(N) = O_-(2N+2)$ we have

$$\begin{aligned}&\int_{\mathbb{R}^r} f\,d\mu(\text{naive}, A, G(N),\ \text{offsets } c)\\ &\qquad = (1/(N+1))\,\mathrm{Sep}(a,f,N,((N+1)/\pi)X(A)).\end{aligned}$$

We define

$$\begin{aligned}\mathrm{Int}(a,f,G(N),A) &:= \int_{\mathbb{R}^r} f\,d\mu(\text{naive}, A, G(N),\ \text{offsets } c)\\ &= (1/(N+1))\,\mathrm{Sep}(a,f,N,((N+1)/\pi)X(A)),\end{aligned}$$

$$\mathrm{Cor}(a,f,G(N),A) := (1/(N+1))\,\mathrm{Clump}(a,f,N,((N+1)/\pi)X(A)),$$

$$\mathrm{TCor}(k,a,f,G(N),A) := (1/(N+1))\,\mathrm{TClump}(k,a,f,N,((N+1)/\pi)X(A)),$$

for $G(N) = O_-(2N+2)$.

Lemma 2.7.6. *Fix an integer $r \ge 1$, a separation vector a in $\mathbb{Z}^r$, an $\mathbb{R}$-valued Borel measurable function f on $\mathbb{R}^r$ of compact support, $G(N)$ any of $U(N)$, $SO(2N+1), USp(2N), SO(2N), O_-(2N+2)$, and A an element of $G(N)$.*

0) *We have the identity*

$$\mathrm{TCor}(k,a,f,G(N),A) = \sum_{n\ge a,\Sigma(n)=k} \mathrm{Binom}(n,a)\,\mathrm{Cor}(n,f,N,A).$$

1) *We have the identity*

$$\mathrm{Int}(a,f,G(N),A) = \sum_{k\ge\Sigma(a)} (-1)^{k-\Sigma(a)}\,\mathrm{TCor}(k,a,f,G(N),A).$$

2) *If $f \ge 0$ as function on $\mathbb{R}^r$, then for each integer $m \ge \Sigma(a)$, we have the inequalities*

$$\sum_{m\ge k\ge\Sigma(a)} (-1)^{k-\Sigma(a)}\,\mathrm{TCor}(k,a,f,G(N),A) \le \mathrm{Int}(a,f,G(N),A)$$

if $m - \Sigma(a)$ *is odd, and*

$$\mathrm{Int}(a, f, G(N), A) \leq \sum_{m \geq k \geq \Sigma(a)} (-1)^{k-\Sigma(a)} \mathrm{TCor}(k, a, f, G(N), A)$$

if $m - \Sigma(a)$ *is even.*

PROOF. Assertion 0) is immediate from the definitions. The rest is just Lemma 2.5.12 in the new terminology. QED

2.8. Expected value measures via INT and COR and TCOR

2.8.1. We fix an integer $r \geq 1$, a separation vector a in $\mathbb{Z}^r$, an integer $k \geq \Sigma(a)$, and an $\mathbb{R}$-valued Borel measurable function f on $\mathbb{R}^r$ which is bounded and of compact support. For $G(N)$ any of $U(N), SO(2N+1), USp(2N), SO(2N)$, $O_-(2N)$, the three $\mathbb{R}$-valued functions on $G(N)$

$$A \mapsto \mathrm{Int}(a, f, G(N), A) \text{ or } \mathrm{Cor}(a, f, G(N), A) \text{ or } \mathrm{TCor}(k, a, f, G(N), A)$$

are Borel measurable and bounded, thanks to Lemma 2.5.9 and the definitions. The first two vanish for $N < 1 + r + \Sigma(a)$, the last for $N < 1 + r + k$. We define real numbers

$$\mathrm{INT}(a, f, G(N)), \qquad \mathrm{Cor}(a, f, G(N)), \qquad \mathrm{TCOR}(k, a, f, G(N))$$

by

$$\begin{aligned}\mathrm{INT}(a, f, G(N)) &:= \int_{G(N)} \mathrm{Int}(a, f, G(N), A) dA \\ &= \int_{G(N)} \int_{\mathbb{R}^r} f \, d\mu(\text{naive}, A, G(N), \text{ offsets } c) dA,\end{aligned}$$

$$\mathrm{COR}(a, f, G(N)) := \int_{G(N)} \mathrm{Cor}(a, f, G(N), A) dA,$$

$$\mathrm{TCOR}(k, a, f, G(N)) := \int_{G(N)} \mathrm{TCor}(k, a, f, G(N), A) dA.$$

Denote by c in $\mathbb{Z}^r$ the vector of offsets attached to the separation vector a. In terms of the expected value measure

$$\mu(\text{naive}, G(N), \text{ offsets } c) = \mu(\text{naive}, G(N), \text{ separations } a)$$

on $\mathbb{R}^r$, we have

$$\mathrm{INT}(a, f, G(N)) = \int_{\mathbb{R}^r} f \, d\mu(\text{naive}, G(N), \text{ offsets } c).$$

Lemma 2.8.2. *Fix an integer* $r \geq 1$, *a separation vector* a *in* $\mathbb{Z}^r$, *an* $\mathbb{R}$-*valued Borel measurable function* f *on* $\mathbb{R}^r$ *which is bounded and of compact support, and* $G(N)$ *any of* $U(N), SO(2N+1), USp(2N), SO(2N), O_-(2N+2)$.

0) *For each integer* $k \geq \Sigma(a)$, *we have the identity*

$$\mathrm{TCOR}(k, a, f, G(N)) = \sum_{n \geq a, \Sigma(n) = k} \mathrm{Binom}(n, a) \, \mathrm{COR}(n, f, G(N)).$$

1) *We have the identity*

$$\mathrm{INT}(a, f, G(N)) = \sum_{k \geq \Sigma(a)} (-1)^{k-\Sigma(a)} \mathrm{TCOR}(k, a, f, G(N)).$$

2) *If $f \geq 0$ as function on $\mathbb{R}^r$, then for each integer $m \geq \Sigma(a)$, we have the inequalities*

$$\sum_{m \geq k \geq \Sigma(a)} (-1)^{k-\Sigma(a)} \operatorname{TCOR}(k, a, f, G(N)) \leq \operatorname{INT}(a, f, G(N))$$

if $m - \Sigma(a)$ is odd, and

$$\operatorname{INT}(a, f, G(N)) \leq \sum_{m \geq k \geq \Sigma(a)} (-1)^{k-\Sigma(a)} \operatorname{TCOR}(k, a, f, G(N))$$

if $m - \Sigma(a)$ is even.

PROOF. Integrate Lemma 2.7.6 over $G(N)$. QED

2.9. The axiomatics of proving Theorem 2.1.3

Proposition 2.9.1. *Consider the following conditions* 1) *and* 2).

1) (existence of large N TCOR limits) *For every integer $r \geq 1$, every separation vector a in $\mathbb{Z}^r$, every integer $k \geq \Sigma(a)$, and every $\mathbb{R}$-valued Borel measurable function f on $\mathbb{R}^r$ which is bounded and of compact support, there exists a real number* $\operatorname{TCOR}(k, a, f, \textit{univ})$ *such that for $G(N)$ any of $U(N), SO(2N+1), USp(2N), SO(2N), O_-(2N+2)$, we have*

$$\lim_{N \to \infty} \operatorname{TCOR}(k, a, f, G(N)) = \operatorname{TCOR}(k, a, f, \textit{ univ}).$$

2) (convergence of large N TCOR limits) *For every integer $r \geq 1$, every separation vector a in $\mathbb{Z}^r$, and every $\mathbb{R}$-valued Borel measurable function f on $\mathbb{R}^r$ which is bounded and of compact support, the series*

$$\sum_{k \geq \Sigma(a)} |\operatorname{TCOR}(k, a, f, \textit{ univ})|$$

is convergent.

If both of these conditions hold, then Theorem 2.1.3 *holds. For every integer $r \geq 1$ and every separation vector a in $\mathbb{Z}^r$, the limit measure $\mu(\textit{naive, univ, sep.'s } a)$ on $\mathbb{R}^r$ is given by the explicit formula*

$$\int_{\mathbb{R}^r} f \, d\mu(\textit{naive, univ, sep's } a) = \sum_{k \geq \Sigma(a)} (-1)^{k-\Sigma(a)} \operatorname{TCOR}(k, a, f, \textit{ univ}),$$

valid for every $\mathbb{R}$-valued Borel measurable function f on $\mathbb{R}^r$ which is bounded and of compact support. We denote

$$\begin{aligned} \operatorname{INT}(a, f, \textit{ univ}) &:= \lim_{N \to \infty} \operatorname{INT}(a, f, G(N)) \\ &= \int_{\mathbb{R}^r} f \, d\mu(\textit{naive, univ, sep's } a). \end{aligned}$$

Moreover, if $f \geq 0$ as function on $\mathbb{R}^r$, then for each integer $m \geq \Sigma(a)$, we have the inequalities

$$\sum_{m \geq k \geq \Sigma(a)} (-1)^{k-\Sigma(a)} \operatorname{TCOR}(k, a, f, \textit{ univ}) \leq \operatorname{INT}(a, f, \textit{ univ})$$

if $m - \Sigma(a)$ is odd, and

$$\operatorname{INT}(a, f, \textit{ univ}) \leq \sum_{m \geq k \geq \Sigma(a)} (-1)^{k-\Sigma(a)} \operatorname{TCOR}(k, a, f, \textit{ univ})$$

if $m - \Sigma(a)$ is even.

PROOF. Fix the data r, a, f as above. Suppose that $f \geq 0$ on $\mathbb{R}^r$. For each N, and each $m \geq \Sigma(a)$ with $M \equiv 1 + \Sigma(a) \bmod 2$, we have the inequalities

$$\sum_{m \geq k \geq \Sigma(a)} (-1)^{k-\Sigma(a)} \operatorname{TCOR}(k, a, f, G(N)) \leq \operatorname{INT}(a, f, G(N))$$
$$\leq \sum_{m+1 \geq k \geq \Sigma(a)} (-1)^{k-\Sigma(a)} \operatorname{TCOR}(k, a, f, G(N)).$$

Fix m, and take the lim sup and the lim inf over N. By the existence of the large N TCOR limits, we get the inequalities

$$\sum_{m \geq k \geq \Sigma(a)} (-1)^{k-\Sigma(a)} \operatorname{TCOR}(k, a, f, \text{ univ}) \leq \liminf_N \operatorname{INT}(a, f, G(N))$$
$$\leq \limsup_N \operatorname{INT}(a, f, G(N)) \leq \sum_{m+1 \geq k \geq \Sigma(a)} (-1)^{k-\Sigma(a)} \operatorname{TCOR}(k, a, f, \text{ univ}).$$

Now take the limit over m of these inequalities. Since the series

$$\sum_{k \geq \Sigma(a)} (-1)^{k-\Sigma(a)} \operatorname{TCOR}(k, a, f, \text{ univ})$$

converges, we find that $\lim_{N \to \infty} \operatorname{INT}(a, f, G(N)) := \operatorname{INT}(a, f, \text{ univ})$ exists and is equal to the sum of this series. Once we know $\operatorname{INT}(a, f, \text{ univ})$ exists for $f \geq 0$, the inequalities above for finite m give the "moreover" conclusions upon passage to the limit with N.

To pass from $f \geq 0$ to the general case, write $f = f_+ - f_-$, with $f_+ := \operatorname{Sup}(f, 0)$ and $f_- := \operatorname{Sup}(-f, 0)$. Applying the above argument to $f_\pm$, one finds that $\lim_{N \to \infty} \operatorname{INT}(a, f, G(N))$ exists for every bounded, Borel measurable f of compact support on $\mathbb{R}^r$ (and is equal to the sum of the TCOR series). Therefore there exists a unique positive Borel measure $\mu(\text{naive, univ, sep.'s } a)$ on $\mathbb{R}^r$ such that for all such f we have

$$\int_{\mathbb{R}^r} f \, d\mu(\text{naive, univ, sep.'s } a) = \lim_{N \to \infty} \operatorname{INT}(a, f, G(N)),$$

cf. [**Fel**, page 243]. But this limit measure may be "defective", i.e., it may have total mass strictly less than 1. [Indeed, it may vanish, as happens when one takes a sequence of (delta measures concentrated at) points going off to infinity.]

2.9.2. Here is an argument to show that in our case, the limit measure, once it exists, must be a probability measure. We first treat the case $r = 1$, and then reduce the general case to the $r = 1$ case by a consideration of direct images.

Lemma 2.9.2.1 (Chebyshev). *Fix an integer $b \geq 1$ (to be thought of as a step vector in $\mathbb{Z}^r$ with $r = 1$). For any integer $N \geq 1$ and any A in $G(N)$, consider the measure $\mu(A, G(N), \text{ step } b)$ on $\mathbb{R}$. For any real $M > 0$, we have the inequality*

$$\mu(A, G(N), \textit{ step } b)(\{x \textit{ in } \mathbb{R} \textit{ with } |x| > MB\}) \leq 1/M.$$

PROOF. The spacings $s_j(\text{ step } b)$ are nonnegative, and they are normalized to have mean b. So $\mu(A, G(N), \text{ step } b)$ is supported in $\mathbb{R}_{\geq 0}$, and

$$\int x \, d\mu(A, G(N), \text{ step } b) = b.$$

We must show that

$$\mu(A, G(N),\ \text{step } b)(\{x > Mb\}) \leq 1/M.$$

We compute

$$b = \int x\, d\mu(A, G(N),\ \text{step } b) \geq \int_{\{x>Mb\}} x\, d\mu(A, G(N),\ \text{step } b)$$
$$\geq \int_{\{x>Mb\}} Mb\, d\mu(A, G(N),\ \text{step } b)$$
$$= Mb\mu(A, G(N),\ \text{step } b)(\{x > Mb\}).\quad \text{QED}$$

Corollary 2.9.2.2. *Fix an integer $b \geq 1$. For any integer $N \geq b$ and any A in $G(N)$, consider the measure $\mu(naive, A, G(N),\ step\ b)$ on $\mathbb{R}$. For any real $M > 0$, we have the inequality*

$$\mu(naive, A, G(N),\ step\ b)(\{x \text{ in } \mathbb{R} \text{ with } |x| > Mb\}) \leq 1/M.$$

PROOF. Indeed, for any (Borel measurable) set E, we have

$$\mu(\text{naive}, A, G(N),\ \text{step } b)(E) \leq \mu(A, G(N),\ \text{step } b)(E).\quad \text{QED}$$

Corollary 2.9.2.3. *Fix an integer $b \geq 1$. For any integer $N \geq b$ and any A in $G(N)$, consider the measure $\mu(naive, A, G(N),\ step\ b)$ on $\mathbb{R}$. For any real $M > 0$, we have the inequality*

$$\mu(naive, A, G(N),\ step\ b)(\{x \text{ in } \mathbb{R} \text{ with } |x| \leq Mb\}) \geq 1 - 1/N - b/N - 1/M.$$

PROOF. The measure $\mu(\text{naive}, A, G(N),\ \text{step b})$ has total mass

$$(N - b)/(N + \lambda) = 1 - (b + \lambda)/(N + \lambda) \geq 1 - 1/N - b/N,$$

λ being $0, 1/2$ or 1 depending on which $G(N)$ we are working on. QED

Lemma 2.9.2.4. *Fix an integer $b \geq 1$. For any integer $N \geq b$, consider the expected value measure $\mu(naive, G(N),\ step\ b)$ on $\mathbb{R}$. For any real $M > 0$, we have the inequality*

$$\mu(naive, G(N),\ step\ b)(\{x \text{ in } \mathbb{R} \text{ with } |x| \leq Mb\}) \geq 1 - 1/N - b/N - 1/M.$$

PROOF. Integrate the previous Corollary 2.9.2.3 over $G(N)$. QED

Lemma 2.9.2.5. *Fix an integer $b \geq 1$, and suppose that the limit measure $\mu(naive,\ univ,\ step\ b)$ on $\mathbb{R}$ exists. For any real $M > 0$, we have the inequality*

$$\mu(naive,\ univ,\ step\ b)(\{x \text{ in } \mathbb{R} \text{ with } |x| \leq Mb\}) \geq 1 - 1/M.$$

PROOF. Fix M, and take the limit as $N \to \infty$ in the previous result. QED

Corollary 2.9.2.6. *Fix an integer $b \geq 1$. If the limit measure*

$$\mu(naive,\ univ,\ step\ b)$$

on $\mathbb{R}$ exists, it is a probability measure.

PROOF. We already know it has total mass at most 1. By the above inequality, it has total mass $\geq 1 - 1/M$ for every $M > 0$. QED

2.9.2.7. We now explain how to deduce the general case from the $r = 1$ case.

Lemma 2.9.2.8. *Fix an integer $r \geq 1$ and a step vector b in $\mathbb{Z}^r$, with corresponding offset vector c. For any integer $N > \Sigma(b)$ and any A in $G(N)$, consider the measure $\mu(naive, A, G(N),\ step\ b)$ on $\mathbb{R}^r$. For any real $M > 0$, we have the inequality*

$$\mu(naive, A, G(N),\ steps\ b)(\{x \text{ in } \mathbb{R}^r \text{ with } |x(i)| > Mb(i) \text{ for some } i\}) \leq r/M.$$

PROOF. The set $\{x \text{ in } \mathbb{R}^r \text{ with } |x(i)| > Mb(i) \text{ for some } i\}$ is the union of the r sets $E_i := \{x \text{ in } \mathbb{R}^r \text{ with } |x(i)| > Mb(i)\}$. So it suffices to show that

$$\mu(\text{naive}, A, G(N),\ \text{steps } b)(E_i) \leq 1/M.$$

By definition of direct image, we have

$$\begin{aligned}&\mu(\text{naive}, A, G(N),\ \text{steps } b)(E_i)\\ &\quad = \text{pr}[i]_*\mu(\text{naive}, A, G(N),\ \text{steps b})(\{x \text{ in } \mathbb{R} \text{ with } |x| > Mb(i)\}).\end{aligned}$$

But we know (by 2.11.10, to follow) that

$$\begin{aligned}&\mu(\text{naive}, A, G(N),\ \text{step pr}[i](b) = b(i))\\ &\quad = \text{pr}[i]_*\mu(\text{naive}, A, G(N),\ \text{steps } b) + (\text{pos. meas. of total mass } \leq c(r)/N).\end{aligned}$$

So we have the inequality

$$\begin{aligned}&\text{pr}[i]_*\mu(\text{naive}, A, G(N),\ \text{steps } b)(\{x \text{ in } \mathbb{R} \text{ with } |x| > Mb(i)\})\\ &\quad \leq \mu(\text{naive}, A, G(N),\ \text{step } b(i))(\{x \text{ in } \mathbb{R} \text{ with } |x| > Mb(i)\})\\ &\quad \leq 1/M,\end{aligned}$$

the last step by Corollary 2.9.2.2. QED

Lemma 2.9.2.9. *Fix an integer $r \geq 1$ and a step vector b in $\mathbb{Z}^r$, with corresponding offset vector c. For any integer $N > \Sigma(b)$ and any A in $G(N)$, consider the measure $\mu(naive, A, G(N),\ step\ b)$ on $\mathbb{R}^r$. For any real $M > 0$, we have the inequality*

$$\begin{aligned}&\mu(naive, A, G(N),\ steps\ b)(\{x \text{ in } \mathbb{R}^r \text{ with } |x(i)| \leq Mb(i) \text{ for all } i\})\\ &\quad \geq 1 - 1/N - c(r)/N - r/M.\end{aligned}$$

PROOF. This is just the previous estimate applied to the complementary set, together with the fact that the measure $\mu(\text{naive}, A, G(N),\ \text{steps } b)$ has total mass $1 - 1/N - c(r)/N$.

Lemma 2.9.2.10. *Fix an integer $r \geq 1$ and a step vector b in $\mathbb{Z}^r$, with corresponding offset vector c. For $N > \Sigma(b)$, consider the expected value measure $\mu(naive, G(N),\ step\ b)$ on $\mathbb{R}^r$.*

1) *For any real $M > 0$, we have the inequality*

$$\begin{aligned}&\mu(naive, G(N).\ steps\ b)(\{x \text{ in } \mathbb{R}^r \text{ with } |x(i)| \leq Mb(i) \text{ for all }\})\\ &\quad \geq 1 - 1/N - c(r)/N - r/M.\end{aligned}$$

2) *If the limit measure $\mu(naive,\ univ,\ steps\ b)$ exists, then for real $M > 0$ we have the inequality*

$$\mu(naive,\ univ,\ steps\ b)(\{x \text{ in } \mathbb{R}^r \text{ with } |x(i)| \leq Mb(i) \text{ for all } i\}) \geq 1 - r/M,$$

and $\mu(naive,\ univ,\ steps\ b)$ is a probability measure.

PROOF. Assertion 1) is obtained by integrating the inequality of the previous lemma over $G(N)$. To obtain 2), we fix M and take the limit of 1) as $N \to \infty$. This inequality shows that $\mu(\text{naive, univ, steps } b)$ has total mass at least $1 - r/M$ for every M, and hence has total mass at least 1. But it a priori has total mass at most one. QED

2.9.3. These lemmas show that $\mu(\text{naive, univ, steps } b)$ is a probability measure. In order to complete the proof of Proposition 2.9.1, it remains only to show that $\mu(\text{naive, univ, steps } b)$ has a continuous CDF. As explained in the appendix 2.11.17, $\mu(\text{naive, univ, steps } b)$ has a continuous CDF if and only if for each $i = 1, \ldots, r$, its direct image by $\text{pr}[i] : \mathbb{R}^r \to \mathbb{R}$ has a continuous CDF. Thanks to Lemma 2.11.13, applicable because $\mu(\text{naive, univ, sep.'s } a)$ is a probability measure, we have

$$\text{pr}[i]_*\mu(\text{naive, univ, steps } b) = \mu(\text{naive, univ, step } b(i)).$$

Thus we are reduced to showing that each $\mu(\text{naive, univ, step } b(i))$ on $\mathbb{R}$ has a continuous CDF, i.e., to showing that this measure gives every point α in $\mathbb{R}$ measure zero. Think of this limit measure as being achieved through the sequence $G(N) = U(N)$. Since the characteristic function of a point is a bounded Borel measurable function of compact support, we have the limit formula

$$\mu(\text{naive, univ, step } b(i))(\{\alpha\}) = \lim_{N\to\infty} \mu(\text{naive}, U(N), \text{ step } b(i))(\{\alpha\}).$$

2.9.3.1. So it is sufficient to show that for every integer $b \geq 1$, every integer $N > b$, and every α in $\mathbb{R}$, we have

$$\mu(\text{naive}, U(N), \text{ step } b)(\{\alpha\}) = 0.$$

Lemma 2.9.3.2. *For every integer $b \geq 1$, every integer $N > b$, and every α in $\mathbb{R}$, we have*

$$\mu(\textit{naive}, U(N), \textit{ step } b)(\{\alpha\}) = 0,$$
$$\mu(U(N), \textit{ step } b)(\{\alpha\}) = 0.$$

The measures $\mu(\textit{naive}, U(N), \textit{ step } b)$ and $\mu(U(N), \textit{ step } b)$ on $\mathbb{R}$ have continuous CDF's.

PROOF. That the measures have continuous CDF's is equivalent to their giving each single point $\{\alpha\}$ measure zero. This is automatic if $\alpha < 0$, since spacings are nonnegative. So we may restrict attention to the case $\alpha \geq 0$.

The quantity $\mu(\text{naive}, U(N), \text{ step } b)(\{\alpha\})$ is, by definition, an integral over $U(N)$:

$$\mu(\text{naive}, U(N), \text{ step } b)(\{\alpha\}) := \int_{U(N)} \mu(\text{naive}, A, U(N), \text{ step } b)(\{\alpha\}) d\,\text{Haar}_{U(N)}.$$

Similarly,

$$\mu(U(N), \text{ step } b)(\{\alpha\}) := \int_{U(N)} \mu(A, U(N), \text{ step } b)(\{\alpha\}) d\,\text{Haar}_{U(N)}.$$

So it suffices to show that the integrand vanishes outside a set of measure zero in $U(N)$. Let us introduce the N angles

$$0 \le \varphi(1) \le \varphi(2) \le \cdots \le \varphi(N) < 2\pi$$

of the eigenvalues of A. Then $\mu(\text{naive}, A, U(N), \text{ step } b)(\{\alpha\})$ is $(1/N)$ times the number of indices $1 \le i \le N - b$ with

$$(N/2\pi)(\varphi(i+b) - \varphi(i)) = \alpha.$$

And $\mu(A, U(N), \text{ step } b)(\{\alpha\})$ is $(1/N)$ times the number of indices $1 \le i \le N$ with

$$(N/2\pi)(\varphi(i+b) - \varphi(i)) = \alpha$$

(with the wraparound convention $\varphi(i+N) := 2\pi + \varphi(i)$ of 1.0.2).

Consider the actual eigenvalues $\zeta(j) = e^{i\varphi(j)}$ of A, for $j = 1, \ldots, N$, extended to all j in $\mathbb{Z}$ by periodicity: $\zeta(j+N) := \zeta(j)$. Denote by β in S^1 the point

$$\beta := e^{2\pi i\alpha/N}.$$

Then $\mu(\text{naive}, A, U(N), \text{ step } b)(\{\alpha\})$ (resp. $\mu(A, U(N), \text{ step } b)(\{\alpha\})$) as function of A is supported in the set of those A in $U(N)$ for which there exists $1 \le i \le N-b$ (resp. $1 \le i \le N$) such that $\zeta(i+b)/\zeta(i) = \beta$. So both, as functions of A, are supported in the union, over $1 \le i \ne j \le N$, of the sets $Z_{i,j}$ of those A in $U(N)$ for which $\zeta(j)/\zeta(i) = \beta$. Let us call this finite union Z: thus $Z := \bigcup_{1 \le i \ne j \le N} Z_{i,j}$. We claim that Z has measure zero in $U(N)$. The set Z is defined by conditions on eigenvalues which are stable under the action of the symmetric group Σ_N (this group permutes the various $Z_{i,j}$), so Z is the inverse image of a set $\mathcal{Z}$ in the space of conjugacy classes of $U(N)$, and we can compute the measure of Z by using the Weyl integration formula. For $U(N)$, the space of conjugacy classes is $(S^1)^N/\Sigma_N$, and the direct image of Haar measure on $(S^1)^N/\Sigma_N$ is (the restriction to Σ_N-invariant functions of) a certain measure on $(S^1)^N$, the well-known formula for which will occupy us at great length later on. All that we need for now is that this Hermann Weyl measure on $(S^1)^N$ is absolutely continuous with respect to usual Lebesgue measure (:= the Haar measure on $(S^1)^N$ viewed as a compact group). The inverse image in $(S^1)^N$ of the set $\mathcal{Z}$ in $(S^1)^N/\Sigma_N$ is the union, over $1 \le i \ne j \le N$, of the sets

$$\{\zeta \text{ in } (S^1)^N \text{ such that } \zeta(j)/\zeta(i) = \beta\}.$$

This set obviously has Lebesgue measure zero, being a hyperplane, so by absolute continuity it has Hermann Weyl measure zero as well. Therefore Z has measure zero in $U(N)$. QED

This lemma in turn concludes the proof of Proposition 2.9.1. QED

2.10. Large N COR limits and formulas for limit measures

2.10.1. Since Proposition 2.9.1 is expressed entirely in terms of TCOR, the reader may wonder what happened to COR. The answer is that nothing happened, it's there also as a special case of TCOR.

From the definitions, we see that for an integer $r \ge 1$, a separation vector a in $\mathbb{Z}^r$, and an $\mathbb{R}$-valued Borel measurable function f on $\mathbb{R}^r$ which is bounded and of compact support, we have the identities

$$\begin{aligned}\mathrm{Clump}(a,f,N,X) &= \mathrm{TClump}(\Sigma(a),a,f,N,X) \quad \text{for } N\geq 2, X \text{ in } \mathbb{R}^N,\\ \mathrm{Cor}(a,f,G(N),A) &= \mathrm{TCor}(\Sigma(a),a,f,G(N),A) \quad \text{for } N\geq 2, A \text{ in } G(N),\\ \mathrm{COR}(a,f,G(N)) &= \mathrm{TCOR}(\Sigma(a),a,f,G(N)) \quad \text{for } N\geq 2, \text{ for any } G(N).\end{aligned}$$

Proposition 2.10.2. *Suppose that conditions* i) *and* ii) *of Proposition* 2.9.1 *hold. For every integer $r \geq 1$, every separation vector a in $\mathbb{Z}^r$, and every $\mathbb{R}$-valued Borel measurable function f on $\mathbb{R}^r$ which is bounded and of compact support, we have*:

1) *There exists a real number* $\mathrm{COR}(a,f,\ univ)$ *such that for $G(N)$ any of* $U(N), SO(2N+1), USp(2N), SO(2N), O_-(2N+2)$, *we have*

$$\lim_{N\to\infty} \mathrm{COR}(a,f,G(N)) = \mathrm{COR}(a,f,\ univ).$$

2) *For each integer $k \geq \Sigma(a)$, we have the identity*

$$\mathrm{TCOR}(k,a,f,\ univ) = \sum_{n\geq a, \Sigma(n)=k} \mathrm{Binom}(n,a)\,\mathrm{COR}(n,f,\ univ).$$

3) *The series*

$$\sum_{n\geq 0} \mathrm{Binom}(n,a)|\,\mathrm{COR}(n,f,\ univ)|$$

is convergent, and we have the identity

$$\begin{aligned}&\int_{\mathbb{R}^r} f\,d\mu(naive,\ univ,\ sep\text{'s } a)\\ &\quad= \sum_{k\geq\Sigma(a)} (-1)^{k-\Sigma(a)}\,\mathrm{TCOR}(k,a,f,\ univ)\\ &\quad= \sum_{n\geq 0} (-1)^{\Sigma(n-a)}\,\mathrm{Binom}(n,a)\,\mathrm{COR}(n,f,\ univ).\end{aligned}$$

PROOF. Assertion 1) is the special case $k=\Sigma(a)$ of condition i) of Proposition 2.9.1. Assertion 2) results from its finite N version (2.8.2, part 0), by passage to the limit over N. For assertion 3), remark that in the sum

$$\sum_{n\geq 0} \mathrm{Binom}(n,a)|\,\mathrm{COR}(n,f,\ \mathrm{univ})|,$$

only the terms with $n\geq a$ have $\mathrm{Binom}(n,a)$ nonzero, so we could as well write the sum as

$$\sum_{n\geq a} \mathrm{Binom}(n,a)|\,\mathrm{COR}(n,f,\ \mathrm{univ})|.$$

We write f as $f_+ - f_-$ to reduce to the case $f\geq 0$. In this case, each individual term $\mathrm{COR}(n,f,\ \mathrm{univ})\geq 0$, so the convergence is equivalent, thanks to 2), to the convergence of

$$\sum_{k\geq\Sigma(a)} |\,\mathrm{TCOR}(k,a,f,\ \mathrm{univ})|,$$

which holds by condition ii). Once we have the absolute convergence, 2) gives the equality

$$\sum_{k\geq\Sigma(a)}(-1)^{k-\Sigma(a)}\,\mathrm{TCOR}(k,a,f,\ \mathrm{univ})$$
$$=\sum_{n\geq 0}(-1)^{\Sigma(n-a)}\,\mathrm{Binom}(n,a)\,\mathrm{COR}(n,f,\ \mathrm{univ}),$$

and by Proposition 2.9.1 the first sum is the integral. QED

2.11. Appendix: Direct image properties of the spacing measures

2.11.0. Throughout this appendix, $G(N)$ is any of $U(N)$, $SO(2N+1)$, $USp(2N), SO(2N), O_{-}(2N+2)$.

Lemma 2.11.1. *Fix an integer $r\geq 1$, and a step vector b in $\mathbb{Z}^r$. For any integers $1\leq i\leq j\leq r$, denote by $\mathrm{pr}[i,j]:\mathbb{R}^r\to\mathbb{R}^{j+1-i}$ the projection*

$$\mathrm{pr}[i,j](x(1),\ldots,x(r)):=(x(i),x(i+1),\ldots,x(j)).$$

For any integer $N\geq 1$, we have an equality of measures on $\mathbb{R}^{j+1-i}$,

$$\mathrm{pr}[i,j]_{*}\mu(A,G(N),\ \textit{steps}\ b)=\mu(A,G(N),\ \textit{steps}\ \mathrm{pr}[i,j](b)).$$

PROOF. By definition, we have

$$\mu(A,G(N),\ \mathrm{steps}\ b):=\mu(\rho_{\mathrm{std}}(A),U(M),\ \mathrm{steps}\ b),$$

where $\rho_{\mathrm{std}}:G(N)\to U(M)$ is the standard representation of $G(N)$. So it suffices to treat universally the case when $G(N)=U(N)$.

In terms of the offset vector c corresponding to the step vector b, the measure $\mu(A,U(N),\ \mathrm{steps}\ b)$ is the average of the delta measures supported at the N normalized spacing vectors

$$s_k(\mathrm{steps}\ b):=s_k(\mathrm{offsets}\ c),\qquad k=1,\ldots,N.$$

The direct image measure is the average of the delta measures supported at the images of the points $s_k(\mathrm{steps}\ b)$ under $\mathrm{pr}[i,j]$. But the N points $\mathrm{pr}[i,j](s_k(\mathrm{steps}\ b))$ are a cyclic permutation of the N points $s_k(\mathrm{steps}\ \mathrm{pr}[i,j](b))$. QED

In the special case $i=j$, this gives

Corollary 2.11.2. *Fix an integer $r\geq 1$, and a step vector b in $\mathbb{Z}^r$. For any integers $1\leq i\leq r$, denote by $\mathrm{pr}[i]:\mathbb{R}^r\to\mathbb{R}$ the projection*

$$\mathrm{pr}[i](x(1),\ldots,x(r)):=x(i).$$

For any integer $N\geq 1$, we have an equality of measures on $\mathbb{R}$,

$$\mathrm{pr}[i]_{*}\mu(A,G(N),\ \textit{steps}\ b)=\mu(A,G(N),\ \textit{step}\ b(i)).$$

Lemma 2.11.3. *Fix an integer $r\geq 1$, and a step vector b in $\mathbb{Z}^r$, with corresponding offset vector c (thus $c(r)=\Sigma(b)$). Denote by*

$$\mathrm{Sum}[b]:\mathbb{R}^{c(r)}\to\mathbb{R}^r$$

the linear map

$$\mathrm{Sum}[b](s(1),\ldots,s(c(r))):=(x(1),\ldots,x(r)),$$

where

$$x(1) := \sum_{j=1}^{c(1)} s(j),$$

and

$$x(1) := \sum_{j=1+c(i-1)}^{c(i)} s(j) \quad \textit{for } 2 \le j \le r.$$

Then for any step vector B in $\mathbb{Z}^{c(r)}$ and for any integer $N \ge 1$, we have an equality of measures on $\mathbb{R}^r$,

$$\mathrm{Sum}[b]_*\mu(A, G(N),\ \textit{steps } B) = \mu(A, G(N),\ \textit{steps } \mathrm{Sum}[b](B)).$$

PROOF. Entirely analogous to the proof of the previous lemma. QED

In the special case $r = 1$, this becomes

Corollary 2.11.4. *Fix an integer $r \ge 1$, and a step vector b in $\mathbb{Z}^r$. Denote by $\mathrm{Sum} : \mathbb{R}^r \to \mathbb{R}$ the map $(x(1), \ldots, x(r)) \mapsto \sum_{j=1}^{r} x(j)$ For any integer $N \ge 1$, we have an equality of measures on $\mathbb{R}$,*

$$\mathrm{Sum}_*\, \mu(\textit{naive}, A, G(N),\ \textit{steps } b) = \mu(\textit{naive}, A, G(N),\ \textit{step } \mathrm{Sum}(b)).$$

In particular, taking $b = (1, 1, \ldots, 1)$ we get

$$\mathrm{Sum}_*\, \mu(A, G(N),\ \textit{steps } (1, 1, \ldots, 1)) = \mu(A, G(N),\ \textit{step } r).$$

Integrating the last two lemmas over $G(N)$, we get the analogous results for the expected value measures.

Lemma 2.11.5. *Fix an integer $r \ge 1$, a step vector b in $\mathbb{Z}^r$ with corresponding offset vector c, and a step vector B in $\mathbb{Z}^{c(r)}$. For any integers $1 \le i \le j \le r$, and for any integer $N \ge 1$, we have an equality of measures on $\mathbb{R}^{j+1-i}$,*

$$\mathrm{pr}[i, j]_*\mu(G(N),\ \textit{steps } b) = \mu(G(N),\ \textit{steps } \mathrm{pr}[i, j](b)),$$

and we have an equality of measures on $\mathbb{R}^r$,

$$\mathrm{Sum}[b]_*\mu(G(N),\ \textit{steps } B) = \mu(G(N),\ \textit{steps } \mathrm{Sum}[b](B)).$$

2.11.6. We now give the analogues of these results for the naive spacing measures.

Lemma 2.11.7. *Fix an integer $r \ge 1$, a step vector b in $\mathbb{Z}^r$, and c the corresponding offset vector. For any integers $1 \le i \le j \le r$, denote by*

$$\mathrm{pr}[i, j] : \mathbb{R}^r \to \mathbb{R}^{j+1-i}$$

the projection

$$\mathrm{pr}[i, j](x(1), \ldots, x(r)) := (x(i), x(i+1), \ldots, x(j)).$$

For any integer $N > c(r)$, we have an equality of measures on $\mathbb{R}^{j+1-i}$,

$$\begin{aligned} &\mu(\textit{naive}, A, G(N),\ \textit{steps } \mathrm{pr}[i, j](b)) \\ &\quad = \mathrm{pr}[i, j]_*\mu(\textit{naive}, A, G(N),\ \textit{steps } b) + (\textit{pos. meas. of total mass } \le c(r)/N). \end{aligned}$$

PROOF. The measure $\mu(A, G(N),\ \text{steps } b)$ is formed using $N - c(r)$ naive spacings $s_k(\text{naive, offsets } c) := s_k(\text{naive, steps } b)$, each with mass $1/(N+\lambda)$, where

$$\begin{aligned}\lambda &= 0 \quad \text{for } U(N), USp(2N), \text{ or } SO(2N),\\ \lambda &= \tfrac{1}{2} \quad \text{for } SO(2N+1),\\ \lambda &= 1 \quad \text{for } O_-(2N+2).\end{aligned}$$

The images under $\text{pr}[i,j]$ of these $N - c(r) = N - \Sigma(b)$ naive spacings are $N - c(r)$ out of the total number $N - \Sigma(\text{pr}[i,j](b))$ of naive spacings $s_k(\text{naive, steps } \text{pr}[i,j](b))$ defining $\mu(\text{naive}, A, G(N),\ \text{steps } \text{pr}[i,j](b))$. So we are "missing"

$$\Sigma(b) - \Sigma(\text{pr}[i,j](b)) \le \Sigma(b) = c(r)$$

naive spacings, each with mass $1/(N+\lambda) \le 1/N$. QED

Lemma 2.11.8. *Fix an integer $r \ge 1$, and a step vector b in $\mathbb{Z}^r$, with corresponding offset vector c (thus $c(r) = \Sigma(b)$). Denote by*

$$\text{Sum}[b] : \mathbb{R}^{c(r)} \to \mathbb{R}^r$$

the linear map

$$\text{Sum}[b](s(1), \ldots, s(c(r))) := (x(1), \ldots, x(r)),$$

where

$$x(1) := \sum_{j=1}^{c(1)} s(j),$$

and

$$x(i) := \sum_{j=1+c(i-1)}^{c(i)} s(j) \quad \text{for } 2 \le j \le r.$$

Then for any step vector B in $\mathbb{Z}^{c(r)}$ and for any integer $N \ge 1$, we have an equality of measures on $\mathbb{R}^r$,

$$\text{Sum}[b]_*\mu(\textit{naive}, A, G(N),\ \textit{steps } B) = \mu(\textit{naive}, A, G(N),\ \textit{steps } \text{Sum}[b](B)).$$

PROOF. Entirely analogous to the proof of the previous lemma, except this time there are no "missing" spacings, since $\Sigma(B) = \Sigma(\text{Sum}[b](B))$. QED

2.11.9. Integrating the last two lemmas over $G(N)$, we get the analogous results for the expected value measures.

Lemma 2.11.10. *Fix an integer $r \ge 1$, a step vector b in $\mathbb{Z}^r$, and c the corresponding offset vector. For any integers $1 \le i \le j \le r$, denote by*

$$\text{pr}[i,j] : \mathbb{R}^r \to \mathbb{R}^{j+1-i}$$

the projection

$$\text{pr}[i,j](x(1), \ldots, x(r)) := (x(i), x(i+1), \ldots, x(j)).$$

For any integer $N > c(r)$, we have an equality of measures on $\mathbb{R}^{j+1-i}$,

$$\begin{aligned}&\mu(\textit{naive}, G(N),\ \textit{steps } \text{pr}[i,j](b))\\ &\quad = \text{pr}[i,j]_*\mu(\textit{naive}, G(N),\ \textit{steps}\, b) + (\textit{pos. meas. of total mass} \le c(r)/N).\end{aligned}$$

Lemma 2.11.11. *Fix an integer $r \geq 1$, and a step vector b in $\mathbb{Z}^r$, with corresponding offset vector c (thus $c(r) = \Sigma(b)$). Denote by*

$$\mathrm{Sum}[b] : \mathbb{R}^{c(r)} \to \mathbb{R}^r$$

the linear map

$$\mathrm{Sum}[b](s(1), \ldots, s(c(r))) := (x(1), \ldots, x(r)),$$

where

$$x(1) := \sum_{j=1}^{c(1)} s(j),$$

and

$$x(i) := \sum_{j=1+c(i-1)}^{c(i)} s(j) \quad \text{for } 2 \leq j \leq r.$$

Then for any step vector B in $\mathbb{Z}^{c(r)}$ and for any integer $N \geq 1$, we have an equality of measures on $\mathbb{R}^r$,

$$\mathrm{Sum}[b]_* \mu(\textit{naive}, G(N), \ \textit{steps } B) = \mu(\textit{naive}, G(N), \ \textit{steps } \mathrm{Sum}[b](B)).$$

2.11.12. We now see what happens in the large N limit.

Lemma 2.11.13. *Fix an integer $r \geq 1$, a step vector b in $\mathbb{Z}^r$, and c the corresponding offset vector. For any integers $1 \leq i \leq j \leq r$, denote by*

$$\mathrm{pr}[i, j] : \mathbb{R}^r \to \mathbb{R}^{j+1-i}$$

the projection

$$\mathrm{pr}[i, j](x(1), \ldots, x(r)) := (x(i), x(i+1), \ldots, x(j)).$$

Suppose that both of the limit measures

$$\mu(\textit{naive, univ, steps } b) \quad \textit{and} \quad \mu(\textit{naive, univ, steps } \mathrm{pr}[i, j](b))$$

exist, and that the measure μ(naive, univ, steps b) is a probability measure. Then so is μ(naive, univ, steps $\mathrm{pr}[i, j](b)$), and we have

$$\mathrm{pr}[i, j]_* \mu(\textit{naive, univ, steps } b) = \mu(\textit{naive, univ, steps } \mathrm{pr}[i, j](b)).$$

PROOF. For finite N, the measure $\mu(\text{naive}, G(N), \text{ steps } b)$ has total mass $(N - \Sigma(b))/(N + \lambda)$, which tends to 1 as $N \to \infty$. Because of this, the standard argument [**Fel**, page 243] for proper convergence of probability measures [applied to the sequence of probability measures $((N - \Sigma(b))/(N + \lambda))^{-1} \mu(\text{naive}, G(N), \text{ steps } b)$, which converges properly to $\mu(\text{naive, univ, steps } b)$] shows that for any bounded continuous function on $\mathbb{R}^r$, we have

$$\int_{\mathbb{R}^r} f \, d\mu(\text{naive, univ, steps } b)$$
$$= \lim_{N \to \infty} \int_{\mathbb{R}^r} f \, d\mu(\text{naive}, G(N), \text{ steps } b).$$

Take now a continuous function g of compact support on $\mathbb{R}^{j+1-i}$, and take for f the function $g \circ \mathrm{pr}[i,j]$, which is a continuous bounded function on $\mathbb{R}^r$. Then the above limit formula gives

$$\int_{\mathbb{R}^{j+1-i}} g\, d\mathrm{pr}[i,j]_*\mu(\text{naive, univ, steps } b)$$
$$= \lim_{N\to\infty} \int_{\mathbb{R}^{j+1-i}} g\, d\mathrm{pr}[i,j]_*\mu(\text{naive}, G(N), \text{ steps } b).$$

By Lemma 2.11.10 above, we have

$$\mu(\text{naive}, G(N), \text{ steps } \mathrm{pr}[i,j](b))$$
$$= \mathrm{pr}[i,j]_*\mu(\text{naive}, G(N), \text{ steps } b) + (\text{pos. meas. of total mass} \le c(r)/N).$$

So we get

$$\int_{\mathbb{R}^{j+1-i}} g\, d\mathrm{pr}[i,j]_*\mu(\text{naive}, G(N), \text{ steps } b)$$
$$= \int_{\mathbb{R}^{j+1-i}} g\, d\mu(\text{naive}, G(N), \text{ steps } \mathrm{pr}[i,j](b)) + \mathrm{Error}(N),$$

with $|\mathrm{Error}(N)| \le \|g\|_{\sup}(c(r)/N)$. Therefore we get

$$\int_{\mathbb{R}^{j+1-i}} g\, d\mathrm{pr}[i,j]_*\mu(\text{naive, univ, steps } b)$$
$$= \lim_{N\to\infty} \int_{\mathbb{R}^{j+1-i}} g\, d\mu(\text{naive}, G(N), \text{ steps } \mathrm{pr}[i,j](b))$$
$$:= \int_{\mathbb{R}^{j+1-i}} g\, d\mu(\text{naive, univ, steps } \mathrm{pr}[i,j](b)).$$

Since this holds for every continuous g of compact support, we have the equality of measures

$$\mathrm{pr}[i,j]_*\mu(\text{naive, univ, steps } b) = \mu(\text{naive, univ, steps } \mathrm{pr}[i,j](b)).$$

From this it follows that $\mu(\text{naive, univ, steps } \mathrm{pr}[i,j](b))$ is itself a probability measure. QED

Lemma 2.11.14. *Fix an integer $r \ge 1$, and a step vector b in $\mathbb{Z}^r$, with corresponding offset vector c (thus $c(r) = \Sigma(b)$). Denote by*

$$\mathrm{Sum}[b] : \mathbb{R}^{c(r)} \to \mathbb{R}^r$$

the linear map

$$\mathrm{Sum}[b](s(1), \ldots, s(c(r))) := (x(1), \ldots, x(r)),$$

where

$$x(1) := \sum_{j=1}^{c(1)} s(j),$$

and

$$x(i) := \sum_{j=1+c(i-1)}^{c(i)} s(j) \quad \textit{for } 2 \le j \le r.$$

Fix a step vector B in $\mathbb{Z}^{c(r)}$. Suppose that both of the limit measures

$$\mu(\textit{naive, univ, steps } B) \quad \textit{and} \quad \mu(\textit{naive, univ, steps } \mathrm{Sum}[b(B)])$$

exist, and that the measure $\mu(\textit{naive, univ, steps } B)$ is a probability measure. Then so is $\mu(\textit{naive, univ, steps } \mathrm{Sum}[b](B))$, and we have

$$\mathrm{Sum}[b]_*\mu(\textit{naive, univ, steps } B) = \mu(\textit{naive, univ, steps } \mathrm{Sum}[b](B)).$$

PROOF. Entirely analogous to the previous proof, except easier, since this time thanks to Lemma 2.11.1 there is no error term. QED

2.11.15. To conclude this appendix, we give the following lemma and corollaries, which are surely well-known, but for which we do not know an explicit reference.

Lemma 2.11.16. *Let $r \geq 1$ be an integer. A positive Borel measure μ of $\mathbb{R}^r$ of finite total mass has a continuous CDF if and only for each $i = 1, \dots, r$, and for each real α, the translated coordinate hyperplane $x(i) = \alpha$ has μ-measure zero.*

Before giving the entirely elementary proof, we state the main applications we have in mind (cf. the discussion 2.9.3).

Corollary 2.11.17. *Let $r \geq 1$ be an integer. A positive Borel measure μ on $\mathbb{R}^r$ of finite total mass has a continuous CDF if and only if for each $i = 1, \dots, r$, the direct image measure $\mathrm{pr}[i]_*\mu$ on $\mathbb{R}$ has a continuous CDF.*

PROOF OF COROLLARY 2.11.17. Using the criterion of the lemma for having a continuous CDF, this equivalence is immediate from the tautogous identity

$$\mu(\{x(i) = \alpha\}) = (\mathrm{pr}[i]_*\mu)(\{\alpha\}). \quad \text{QED}$$

Corollary 2.11.18. *Let $r \geq 1$ be an integer. A positive Borel measure μ on $\mathbb{R}^r$ of finite total mass which is absolutely continuous with respect to Lebesgue measure on $\mathbb{R}^r$ has a continuous CDF.*

PROOF OF COROLLARY 2.11.18. Absolute continuity forces each translated coordinate hyperplane $x(i) = \alpha$ to have μ-measure zero. QED

We now turn to the proof of Lemma 2.11.16.

PROOF OF LEMMA 2.11.16. Given $\varepsilon > 0$ real, we write $\Delta(\varepsilon)$ for the point $(\varepsilon, \varepsilon, \dots, \varepsilon)$ in $\mathbb{R}^r$. If no confusion seems likely, we will write $x + \varepsilon$ to mean $x + \Delta(\varepsilon)$. We denote by $F(x) := \mu((-\infty, x])$ the CDF of the measure μ.

Suppose first that $\mu(\{x(i) = \alpha\}) = 0$ for all i and all real α. Fix a point X in $\mathbb{R}^r$, and let $\{X_n\}_n$ be a sequence of points in $\mathbb{R}^r$ converging to X. We must show that $F(X_n) \to F(X)$. Because $X_n \to X$, there is a sequence of strictly positive real numbers $\varepsilon_n \to 0$ such that X_n lies in the open interval $(X - \varepsilon_n, X + \varepsilon_n)$. Thus we have

$$X - \varepsilon_n \leq X_n \leq X + \varepsilon_n \quad \text{and} \quad X - \varepsilon_n \leq X \leq X + \varepsilon_n.$$

Because F is a CDF, it has the monotonicity property

$$F(x) \leq F(y) \quad \text{if } x \leq y.$$

So we have

$$F(X - \varepsilon_n) \leq F(X_n) \leq F(X + \varepsilon_n) \quad \text{and} \quad F(X - \varepsilon_n) \leq F(X) \leq F(X + \varepsilon_n).$$

Therefore it suffices to show that $F(X+\varepsilon_n) \to F(X)$ and $F(X-\varepsilon_n) \to F(X)$, for any sequence of strictly positive real numbers $\varepsilon_n \to 0$.

Again using the monotonicity property, it suffices to show that

$$F(X+1/n) \to F(X) \quad \text{and} \quad F(X-1/n) \to F(X).$$

The set $(-\infty, X]$ is the intersection of the decreasing sets $(-\infty, X+1/n]$, so $F(X+1/n) \to F(X)$ for F the CDF of any positive measure of finite total mass.

Because $\mu(\{x(i)=\alpha\}) = 0$ for all i and all α, for any point s in $\mathbb{R}^r$ we have $\mu((-\infty, s)) = \mu((-\infty, s])$ (since the difference is contained in the union of the hyperplanes $\{x \text{ in } \mathbb{R}^r \text{ with } x(i) = s(i)\}$, each of which has μ-measure zero). In particular, $F(X) = \mu((-\infty, X))$, and $F(x-1/n) = \mu((-\infty, X-1/n))$ for each n. The set $(-\infty, X)$ is the increasing union of the sets $(-\infty, X-1/n)$, and μ is a measure, so $F(X-1/n) \to F(X)$.

Suppose now that μ has continuous CDF. We must show that $\mu(\{x(i)=\alpha\}) = 0$ for all i and all α. By symmetry, it suffices to treat the case $i=1$. If $r=1$, then

$$\mu(\{\alpha\}) = F(\alpha) - \mu((-\infty, \alpha)) = F(\alpha) - \lim_{n\to\infty} F(\alpha - 1/n) = 0.$$

If $r > 1$, fix a large real number M, and consider the point

$$X = (\alpha, M, M, \ldots, M)$$

and the approaching sequence

$$X_n = (\alpha - 1/n, M, M, \ldots, M).$$

Because μ is a measure, $\lim_{n\to\infty} F(X_n)$ is the measure of

$$\{Y \text{ in } \mathbb{R}^r \text{ with } Y(1) < \alpha \text{ and } Y(i) \leq M \text{ for } i = 2, \ldots, r\}.$$

But $F(X) = \mu((-\infty, X])$, so

$$\begin{aligned} &F(X) - \lim_{n\to\infty} F(X_n) \\ &\quad = \mu(\{Y \text{ in } \mathbb{R}^r \text{ with } Y(1) = \alpha \text{ and } Y(i) \leq M \text{ for } i = 2, \ldots, r\}). \end{aligned}$$

Thus if F is continuous, we have

$$\mu(\{Y \text{ in } \mathbb{R}^r \text{ with } Y(1) = \alpha \text{ and } Y(i) \leq M \text{ for } i = 2, \ldots, r\}) = 0$$

for every M, and hence $\mu(\{Y \text{ in } \mathbb{R}^r \text{ with } Y(1) = \alpha\}) = 0$. QED

CHAPTER 3

Reduction Steps in Proving the Main Theorems

3.0. The axiomatics of proving Theorems 2.1.3 and 2.1.5

Proposition 3.0.1. *Suppose that both of the following conditions* i) *and* ii) *hold.*

i) (Convergence rate for TCOR's) *For every integer $r \geq 1$, every separation vector a in $\mathbb{Z}^r$, every integer $k \geq \Sigma(a)$, and every $\mathbb{R}$-valued Borel measurable function f on $\mathbb{R}^r$ which is bounded and supported in $\sum_i |s(i)| \leq \alpha$, there exists a real number*

$$\mathrm{TCOR}(k, a, f, \ \mathit{univ})$$

such that for every $G(N)$ as in 2.1.3 *above, we have the estimate*

$$\begin{aligned} &|\,\mathrm{TCOR}(k, a, f, G(N)) - \mathrm{TCOR}(k, a, f, \ \mathit{univ})| \\ &\quad \leq \mathrm{Binom}(r + k - 1, r + \Sigma(a) - 1)\|f\|_{\sup}(8\alpha)^{k+r}((\pi\alpha)^2 + \alpha + 1 + 10\log(N))/N. \end{aligned}$$

ii) (Bound for finite N TCOR's) *For every integer $r \geq 1$, every separation vector a in $\mathbb{Z}^r$, every integer $k \geq \Sigma(a)$, and every $\mathbb{R}$-valued Borel measurable function f on $\mathbb{R}^r$ which is bounded, supported in $\sum_i |s(i)| \leq \alpha$, and every $G(N)$ as above, we have the estimate (independent of N)*

$$\begin{aligned} &|\,\mathrm{TCOR}(k, a, f, G(N))| \\ &\quad \leq \mathrm{Binom}(r + k - 1, r + \Sigma(a) - 1)\|f\|_{\sup}2(2\alpha)^{k+r}/(k + r)!. \end{aligned}$$

Then conditions 1) *and* 2) *of Proposition* 2.9.1 *are satisfied, and hence Theorem* 2.1.3 *holds.*

PROOF. Assertion i) gives the existence of large N TCor limits (as well as an estimate for the rate of convergence). Since the large N limit exists, assertion ii) gives the estimate

$$\begin{aligned} &|\,\mathrm{TCOR}(k, a, f, \ \mathrm{univ})| \\ &\quad \leq \mathrm{Binom}(r + k - 1, r + \Sigma(A) - 1)\|f\|_{\sup}2(2\alpha)^{k+r}/(k + r)!. \end{aligned}$$

The convergence of $\sum_{k \geq \Sigma(a)} |\,\mathrm{TCOR}(k, a, f, \ \mathrm{univ})|$ follows easily. Indeed, since r and $\Sigma(a)$ are fixed, if we put $d := r + \Sigma(a) - 1$, we get

$$\mathrm{Binom}(r + k - 1, r + \Sigma(a) - 1) = \mathrm{Binom}(r + k - 1, d) \leq \mathrm{Binom}(r + k, d),$$

whence

$$\sum_{k\geq\Sigma(a)} |\operatorname{TCOR}(k,a,f,\ \text{univ})|$$
$$\leq 2\|f\|_{\sup} \sum_{k\geq\Sigma(a)} \operatorname{Binom}(r+k,d)(2\alpha)^{k+r}/(k+r)!$$
$$\leq 2\|f\|_{\sup} \sum_{n\geq 0} \operatorname{Binom}(n,d)(2\alpha)^{n}/n!$$
$$= 2\|f\|_{\sup} \times \text{ the value at } z=2\alpha \text{ of the entire function}$$
$$(z^d(d/dz)^d/d!)(e^z) = z^d e^z/d!. \quad \text{QED}$$

3.1. A mild generalization of Theorem 2.1.5: the φ-version

3.1.1. Fix an integer $r \geq 1$, and a step vector b in $\mathbb{R}^r$, with separation vector a and offset vector c. Denote by μ the probability measure on $\mathbb{R}^r$

$$\mu := \mu(\text{naive, univ, steps } b) = \mu(\text{univ, steps } b)$$

given by Theorem 2.1.3. For each integer i with $1 \leq i \leq r$, we denote by μ_i the probability measure on $\mathbb{R}$ which is its i'th projection (cf. 2.11.13):

$$\mu_i := \operatorname{pr}[i]_*\mu = \mu(\text{naive, univ, step } b(i)) = \mu(\text{univ, step } b(i)).$$

The measure μ is supported in $(\mathbb{R}_{\geq 0})^r$, and each μ_i is supported in $\mathbb{R}_{\geq 0}$. By Theorem 2.1.3, we know that each μ_i has a continuous CDF.

3.1.2. We wish to understand the role played by the choice of coordinates on $\mathbb{R}^r$. Thus we fix

$$\varphi : \mathbb{R}^r \to \mathbb{R}^r \text{ a homeomorphism of } \mathbb{R}^r,$$
$$x = (x_1, \ldots, x_r) \mapsto (\varphi_1(x), \ldots, \varphi_r(x)),$$

which is "bi-bounded" in the sense that there exist strictly positive real constants κ and η such that

$$\eta \sum |x_i| \leq \sum |\varphi_i(x)| \leq \kappa \sum |x_i|.$$

The basic example we have in mind of such a φ is a linear automorphism of $\mathbb{R}^r$.

3.1.3. For each integer i with $1 \leq i \leq r$, we denote by $\mu_{i,\varphi}$ the probability measure on $\mathbb{R}$ which is the i'th projection of μ in φ-coordinates:

$$\mu_{i,\varphi} := (\varphi_i)_*\mu.$$

We make the following **assumption**[1] 3.1.4:

3.1.4. For each i, the measure $\mu_{i,\varphi}$ on $\mathbb{R}$ has a continuous CDF.

3.1.5. Attached to each point x in $(\mathbb{R} \cup \{\pm\infty\})^r$ is the set (a semi-infinite rectangle in φ-coordinates)

$$R(x,\varphi) := \{z \text{ in } \mathbb{R}^r \text{ with } \varphi_i(z) \leq x(i) \text{ for all } i\}.$$

We will use the expression "semi-infinite φ-rectangle" to mean a set of this form. If any coordinate of x is $-\infty$, the set $R(x,\varphi)$ is empty. If x is a finite point, i.e., a point in $\mathbb{R}^r$, we call $R(x,\varphi)$ a "semi-finite φ-rectangle".

---

[1] We will see in 7.0.13 that 3.1.4 automatically holds for φ any C^1 diffeomorphism, in particular for φ any linear automorphism of $\mathbb{R}^r$.

3.1.5.1. For ν_1 and ν_2 any two measures on $\mathbb{R}^r$ of finite total mass, we define their **φ-discrepancy**

$$\begin{aligned}\varphi\text{-discrep}(\nu_1, \nu_2) &:= \operatorname*{Sup}_{x \text{ in } (\mathbb{R}\cup\{+\infty\})^r} |\nu_1(R(x,\varphi)) - \nu_2(R(x,\varphi))| \\ &= \operatorname*{Sup}_{x \text{ in } \mathbb{R}^r} |\nu_1(R(x,\varphi)) - \nu_2(R(x,\varphi))|.\end{aligned}$$

[A priori, the first Sup might exceed the second, since it takes more sets into account. In fact, the two Sup's are equal, because for x in $(\mathbb{R}\cup\{+\infty\})^r$ with some coordinates $+\infty$, $R(x,\varphi)$ is the increasing union of φ-rectangles $R(x_n,\varphi)$, for x_n the point in $\mathbb{R}^r$ defined by replacing each $+\infty$ coordinate in x by n. By the countable additivity of measures, $\nu_1(R(x,\varphi))$ is the limit of $\nu_1(R(x_n,\varphi))$, and similarly for ν_2. Therefore for this fixed x, we have

$$|\nu_1(R(x,\varphi)) - \nu_2(R(x,\varphi))| = \lim_{n\to\infty} |\nu_1(R(x_n,\varphi)) - \nu_2(R(x_n,\varphi))|,$$

and each term $|\nu_1(R(x_n,\varphi)) - \nu_2(R(x_n,\varphi))|$ is bounded by the second, "smaller" Sup.]

Theorem 3.1.6 (φ-version of 2.1.5). *Fix an integer $r \geq 1$, a step vector b in $\mathbb{Z}^r$, and strictly positive constants ε, η, and κ. There exists an explicit constant $N_1(\varepsilon, r, c, \eta, \kappa)$ with the following property:*

For any homeomorphism $\varphi : \mathbb{R}^r \to \mathbb{R}^r$,

$$x = (x_1, \ldots, x_r) \mapsto (\varphi_1(x), \ldots, \varphi_r(x)),$$

which is "(η,κ) bi-bounded" in the sense that

$$\eta \sum |x_i| \leq \sum |\varphi_i(x)| \leq \kappa \sum |x_i|,$$

and which satisfies the property

(3.1.4) *for each i, $\mu_{i,\varphi} := (\varphi_i)_*\mu$ on $\mathbb{R}$ has a continuous CDF,*

for any $N \geq N_1(\varepsilon, r, c, \eta, \kappa)$, for $G(N)$ any of $U(N)$, $USp(2N)$, $SO(2N+1)$, $SO(2N), O_-(2N+2)$, for μ the probability measure on $\mathbb{R}^r$

$$\mu := \mu(\textit{naive, univ, steps } b) = \mu(\textit{univ, steps } b),$$

and for

$$\mu(A, N) := \mu(\textit{naive}, A, G(N), \textit{ steps } b),$$

we have

$$\int_{G(N)} \varphi\text{-discrep}(\mu, \mu(A,N))dA \leq N^{\varepsilon - 1/(2r+4)}.$$

Corollary 3.1.7. *Hypotheses and notations as in 3.1.6, let $J \subset \{1, 2, \ldots, r\}$ be any nonempty subset of the indices; say J is $j_1 < j_2 < \cdots < j_k$. Consider the map*

$$\varphi(J) : \mathbb{R}^r \to \mathbb{R}^k, x \mapsto (\varphi_{j_1}(x), \varphi_{j_2}(x), \ldots, \varphi_{j_k}(x)).$$

Then for $N \geq N_1(\varepsilon, r, c, \eta, \kappa)$, $G(N)$ any of $U(N), USp(2N), SO(2N+1), SO(2N)$, $O_-(2N+2)$, we have

$$\int_{G(N)} \text{discrep}(\varphi(J)_*\mu, \varphi(J)_*\mu(A,N))dA \leq N^{\varepsilon - 1/(2r+4)}.$$

PROOF OF COROLLARY 3.1.7. We claim that we have the inequality

$$\text{discrep}(\varphi(J)_*\mu, \varphi(J)_*\mu(A,N)) \leq \varphi\text{-discrep}(\mu, \mu(A,N)),$$

which when integrated over $G(N)$ gives the corollary. This inequality holds for any two measures of finite total mass ν_1 and ν_2, i.e., we have

$$\text{discrep}(\varphi(J)_*\nu_1, \varphi(J)_*\nu_2) \leq \varphi\text{-discrep}(\nu_1, \nu_2).$$

To see this, notice that for any x in $\mathbb{R}^k$,

$$\varphi(J)^{-1}(\text{the "usual" rectangle } R(x) \text{ in } \mathbb{R}^k) = R(y, \varphi)$$

for y in $(\mathbb{R} \cup \{+\infty\})^r$ the point with coordinates $y_{j_i} = x_i$, and $y_\lambda = +\infty$ for λ not in J. Thus we get

$$\begin{aligned}|(\varphi(J)_*\nu_1)(R(x)) - (\varphi(J)_*\nu_2)(R(x))| &= |\nu_1(R(y,\varphi)) - \nu_2(R(y,\varphi))| \\ &\leq \varphi\text{-discrep}(\nu_1, \nu_2). \quad \text{QED}\end{aligned}$$

3.1.8. The main task of this chapter is to prove the following proposition.

Proposition 3.1.9. *Suppose that in addition to conditions* i) *and* ii) *of the previous Proposition* 3.0.1, *the following conditions* iii) *and* iv) *hold*:

iii) (Bound for variance of TCor as function of A in $G(N)$) *For every integer $r \geq 1$, every separation vector a in $\mathbb{Z}^r$, every integer $k \geq \Sigma(a)$, and every $\mathbb{R}$-valued Borel measurable function f on $\mathbb{R}^r$ which is bounded, supported in $\sum_i |s(i)| \leq \alpha$, and every $G(N)$ as above, we have the estimate*

$$\begin{aligned}&\text{Var}(A \mapsto \text{TCor}(k, a, f, G(N), A) \textit{ on } G(N)) \\ &\qquad \leq [\text{Binom}(r+k-1, r+\Sigma(a)-1)\|f\|_{\text{sup}}]^2(3(8\alpha)^{k+r} + 65(8\alpha)^{2k+2r})/N.\end{aligned}$$

iv) (Estimate for the tail of the most classical spacing measure) *There exist explicit real constants $A > 0$ and $B > 0$ such that the limit measure*

$$\mu(\textit{naive, univ, step } 1)$$

on $\mathbb{R}$ satisfies

$$\mu(\textit{naive, univ, step } 1)(\{|x| > s\}) \leq Ae^{-Bs^2} \textit{ for every real } s \geq 0.$$[2]

Then Theorem 3.1.6, *the φ-version of Theorem* 2.1.5, *holds.*

PROOF. Because Theorem 2.1.3 holds (thanks to the previous proposition), there is no need to distinguish between the limit measures μ(naive, univ, steps b) and μ(univ, steps b). Our first task is to deduce from iv) a tail estimate for the most general limit measure μ(naive, univ, steps b).

Lemma 3.1.10. *Suppose that Theorem* 2.1.3 *holds. Fix an integer $r \geq 1$, and a step vector b in $\mathbb{Z}^r$. For any real $s \geq 0$, we have the inequality*

$$\begin{aligned}&\mu(\textit{univ, steps } b)(\{x \textit{ in } \mathbb{R}^r \textit{ with } |x(i)| > sb(i) \textit{ for some } i\}) \\ &\qquad \leq \Sigma(b) \times \mu(\textit{univ, step } 1)(\{|x| > s\}).\end{aligned}$$

In particular, if iv) holds, then for any real $s \geq 0$, we have

$$\mu(\textit{univ, steps } b)(\{x \textit{ in } \mathbb{R}^r \textit{ with } |x(i)| > sb(i) \textit{ for some } i\}) \leq \Sigma(b) \times Ae^{-Bs^2}.$$

[2] We will show in 6.13.4 that we can take $A = 4/3$ and $B = 1/8$.

PROOF. The set $\{x \text{ in } \mathbb{R}^r \text{ with } |x(i)| > sb(i) \text{ for some } i\}$ is the union of the r sets $E_i := \{x \text{ in } \mathbb{R}^r \text{ with } |x(i)| > sb(i)\}$. So we have

$$\mu(\text{univ, steps } b)\left(\left\{\sum |x(i)| > s\Sigma(b)\right\}\right) \leq \sum_i \mu(\text{univ, steps } b)(E_i).$$

By definition of direct image, we have

$$\begin{aligned}\mu(\text{univ, steps } b)(E_i) &= (\text{pr}[i]_*\mu(\text{univ, steps } b))(\{|x| > sb(i)\}) \\ &= \mu(\text{univ, step } b(i))(\{|x| > sb(i)\}),\end{aligned}$$

the last equality by Lemma 2.11.13.

For any $n \geq 1$, denote by $\mathbb{1}_n$ the step vector $(1, 1, 1, \ldots, 1)$ in $\mathbb{R}^n$. By Lemma 2.11.14, under the "sum of the coordinates" map $\text{Sum} : \mathbb{R}^n \to \mathbb{R}$, we have

$$\text{Sum}_* \, \mu(\text{univ, steps } \mathbb{1}_n) = \mu(\text{univ, step } n).$$

Applying this with $n = b(i)$, we get

$$\mu(\text{univ, step } b(i))(\{|x| > sb(i)\}) = \mu(\text{univ, steps } \mathbb{1}_{b(i)})\left(\left\{\left|\sum_j x(j)\right| > sb(i)\right\}\right).$$

For each i, the set in $\mathbb{R}^{b(i)}$ where $|\sum_j x(j)| > sb(i)$ lies in the union of the $b(i)$ sets $F_j := \{x \text{ in } \mathbb{R}^{b(i)} \text{ with } |x(j)| > s\}$. So again by definition of direct image, we have

$$\begin{aligned}&\mu(\text{univ, steps } \mathbb{1}_{b(i)})\left(\left\{\left|\sum_j x(j)\right| > sb(i)\right\}\right) \\ &\leq \sum_j \mu(\text{univ, steps } \mathbb{1}_{b(i)})(\{|x(j)| > s\}) \\ &= \sum_j (\text{pr}[j]_*\mu(\text{univ, steps } \mathbb{1}_{b(i)}))(\{|x| > s\}) \\ &= \sum_j \mu(\text{univ, step } 1)(\{|x| > s\}) \\ &= b(i)\mu(\text{univ, step } 1)(\{|x| > s\}). \quad \text{QED}\end{aligned}$$

3.2. M-grid discrepancy, L cutoff and dependence on the choice of coordinates

3.2.1. Throughout this section, we will assume that conditions i), ii) and iii) of Propositions 3.0.1 and 3.1.9 hold, but we will not assume iv). The main result of this section is Corollary 3.2.29.

3.2.2. We first explain our general strategy for dealing with discrepancy. Fix the following data:

$r \geq 1$ an integer,

b in $\mathbb{R}^r$ a step vector, with separation vector a and offset vector c,

$M \geq 2$ a (large) integer, the "grid size",

$L \geq \Sigma(a)$ a (large) integer, the "cutoff",

$N \geq 1$ a (large) integer, the "group size",

$G(N)$, one of $U(N), USp(2N), SO(2N+1), SO(2N), O_-(2N+2)$.

Denote by μ the probability measure on $\mathbb{R}^r$

$$\mu := \mu(\text{naive, univ, steps } b) = \mu(\text{univ, steps } b)$$

given by Theorem 2.1.3. We further fix

$$\varphi : \mathbb{R}^r \to \mathbb{R}^r \text{ a homeomorphism of } \mathbb{R}^r,$$
$$x = (x_1, \ldots, x_r) \mapsto (\varphi_1(x), \ldots, \varphi_r(x)),$$

which is "bi-bounded" in the sense that there exist strictly positive real constants κ and η such that

$$\eta \sum |x_i| \leq \sum |\varphi_i(x)| \leq \kappa \sum |x_i|.$$

For each integer i with $1 \leq i \leq r$, we denote by $\mu_{i,\varphi}$ the probability measure on $\mathbb{R}$ which is the i'th projection of μ in φ-coordinates:

$$\mu_{i,\varphi} := (\varphi_i)_* \mu.$$

We make **the assumption 3.1.4**:

3.2.3 = 3.1.4. For each i, the measure $\mu_{i,\varphi}$ on $\mathbb{R}$ has a continuous CDF.

3.2.4. We denote by $G_{i,\varphi}$ the CDF of $\mu_{i,\varphi}$. Because $G_{i,\varphi}$ is continuous and nondecreasing, with $\lim_{x \to -\infty} G_{i,\varphi}(x) = 0$ and $\lim_{x \to \infty} G_{i,\varphi}(x) = 1$, there exists a set S_i of $M + 1$ distinct points in $\mathbb{R} \cup \{\pm\infty\}$,

$$-\infty = s(0,i) < s(1,i) < s(2,i) < \cdots < s(M-1,i) < s(M,i) = +\infty$$

such that

$$G_{i,\varphi}(s(j,i)) = j/M, \text{ for } j = 0, \ldots, M.$$

We fix, for each $1 \leq i \leq r$, a choice of such a set S_i. The positive real number

$$\beta := \sum_i \text{Max}(|s(1,i)|, |s(M-1,i)|) \tag{3.2.4.1}$$

is called the M-φ-**grid diameter**.

3.2.4.2. We view the set

$$S := S_1 \times S_2 \times \cdots \times S_r \subset (\mathbb{R} \cup \{\pm\infty\})^r$$

as a set of $(M+1)^r$ grid points in $(\mathbb{R} \cup \{\pm\infty\})^r$, and call it **the** set of grid points, or the set of M-φ-grid points relative to μ if confusion is possible. We say that a grid point is **finite** if none of its coordinates is $\pm\infty$, i.e., if the point lies in $\mathbb{R}^r$.

3.2.5. Attached to each point x in $(\mathbb{R} \cup \{\pm\infty\})^r$ is the set (a semi-infinite rectangle in φ-coordinates)

$$R(x, \varphi) := \{z \text{ in } \mathbb{R}^r \text{ with } \varphi_i(z) \leq x(i) \text{ for all } i\}.$$

We will use the expression "semi-infinite φ-rectangle" to mean a set of this form. If any coordinate of x is $-\infty$, the set $R(x, \varphi)$ is empty. If x is a finite point, i.e., a point in $\mathbb{R}^r$, we call $R(x, \varphi)$ a "semi-finite φ-rectangle". The **marked φ-rectangles** are those attached to grid points. The semi-finite marked φ-rectangles are those attached to the $(M-1)^r$ finite grid points.

3.2.6. Recall (3.1.5.1) that for ν_1 and ν_2 any two measures on $\mathbb{R}^r$ of finite total mass, we defined their **φ-discrepancy**

$$\varphi\text{-discrep}(\nu_1, \nu_2) := \operatorname*{Sup}_{x \text{ in } (\mathbb{R}\cup\{\pm\infty\})^r} |\nu_1(R(x,\varphi)) - \nu_2(R(x,\varphi))|$$
$$= \operatorname*{Sup}_{x \text{ in } \mathbb{R}^r} |\nu_1(R(x,\varphi)) - \nu_2(R(x,\varphi))|.$$

3.2.7. We denote

$$R_{\max} := \{z \text{ in } \mathbb{R}^r \text{ with } s(1,i) < \varphi_i(z) \le s(M-1,i) \text{ for all } i\}.$$

Lemma 3.2.8. *We have the following inequalities:*
1) $\mu(\mathbb{R}^r - R_{\max}) \le 2r/M$.
2) *For any Borel measurable set E in $\mathbb{R}^r$, and any Borel measure ν on $\mathbb{R}^r$ of total mass ≤ 1, we have*

$$|\mu(E) - \nu(E)|$$
$$\le 4r/M + |\mu(E \cap R_{\max}) - \nu(E \cap R_{\max})| + |\mu(R_{\max}) - \nu(R_{\max})|.$$

PROOF. For 1), notice that $\mathbb{R}^r - R_{\max}$ is contained in the union of the $2r$ sets

$$\{z \text{ in } \mathbb{R}^r \text{ with } \varphi_i(z) \le s(1,i)\}, \quad i = 1, \dots, r,$$

and

$$\{z \text{ in } \mathbb{R}^r \text{ with } \varphi_i(z) > s(M-1,i)\}, \quad i = 1, \dots, r.$$

By construction of our M-φ-grid, each of these sets has μ-measure $1/M$. The point is that, by definition of direct image,

$$\mu(\{z \text{ in } \mathbb{R}^r \text{ with } \varphi_i(z) \le s(1,i)\}) = \mu_{i,\varphi}((-\infty, s(1,i))) = G_{i,\varphi}(s(1,i)) = 1/M.$$

Now that we have proven 1), 2) is the special case $X = \mathbb{R}^r$, $K = R_{\max}$, $\varepsilon = 2r/M$, of the following standard lemma, which is implicit in [**Fel**, pages 243–244].

Lemma 3.2.9. *Let (X, μ) be a probability space, $\varepsilon > 0$ real, $K \subset X$ a measurable set with $\mu(X - K) \le \varepsilon$. For any measure ν on X of total mass at most one, and any measurable set $E \subset X$, we have*

$$|\mu(E) - \nu(E)|$$
$$\le 2\varepsilon + |\mu(E \cap K) - \nu(E \cap K)| + |\mu(K) - \nu(K)|.$$

PROOF. Write E as the disjoint union

$$E = (E \cap K) \bigsqcup (E - E \cap K),$$

so

$$\mu(E) - \nu(E) = \mu(E \cap K) - \nu(E \cap K) + \mu(E - E \cap K) - \nu(E - E \cap K).$$

Therefore we have

$$|\mu(E) - \nu(E)| \le |\mu(E \cap K) - \nu(E \cap K)| + \mu(E - E \cap K) + \nu(E - E \cap K).$$

But we have

$$\begin{aligned}
&\mu(E - E\cap K) + \nu(E - E\cap K)\\
&\quad\le \mu(X-K) + \nu(X-K)\\
&\quad= \mu(X-K) + \nu(X) - \nu(K)\\
&\quad\le \mu(X-K) + 1 - \nu(K)\\
&\quad= \mu(X-K) + \mu(X-K) + \mu(K) - \nu(K)\\
&\quad\le 2\varepsilon + \mu(K) - \nu(K)\\
&\quad\le 2\varepsilon + |\mu(K) - \nu(K)|. \quad \text{QED}
\end{aligned}$$

3.2.10. For ν_1 and ν_2 any two Borel measures on $\mathbb{R}^r$ of finite total mass, we define their "**M-φ-grid discrepancy**"

$$\begin{aligned}
&M\text{-}\varphi\text{-discrep}(\nu_1,\nu_2)\\
&\quad := \mathrm{Sup}_{\text{semi-finite marked } \varphi\text{-rectangles } R}\, |\nu_1(R\cap R_{\max}) - \nu_2(R\cap R_{\max})|.
\end{aligned}$$

The M-φ-grid discrepancy is "easy" to calculate, in that it requires looking only at the $(M-2)^r$ semi-finite marked φ-rectangles none of whose coordinates is either $s(0,i) = -\infty$ (such rectangles being empty) or $s(1,i)$ (such rectangles having empty intersection with $R_{\max}$), whereas the φ-discrepancy (cf. 3.1.5.1) takes the sup over all φ-rectangles.

Lemma 3.2.11. *Suppose given a semi-finite φ-rectangle $R = R(x,\varphi)$, x in $\mathbb{R}^r$.*

1) *Among all marked φ-rectangles contained in R, there is a maximal one, say R_1. Among all marked φ-rectangles containing R, there is a minimal one, say R_2. We have*

$$R_1 \subset R \subset R_2,\ R_1 \text{ is semi-finite or empty, and}$$
$$\mu(R_2) - \mu(R_1) \le r/M.$$

2) $\mu(R) - \mu(R\cap R_{\max}) \le 2r/M$.

Proof. For each $1 \le i \le r$, see where $x(i)$ sits among the points of S_i: there is a largest index $j(i)$ such that

$$s(j(i),i) \le x(i)$$

and a smallest index $k(i)$ such that

$$x(i) \le s(k(i),i).$$

It is clear that $k(i) \ge j(i)$, and that $k(i)$ is either $j(i)$ or $1 + j(i)$, depending on whether or not $s(j(i),i) = x(i)$. We take for R_1 (resp. R_2) the φ-rectangle attached to the point s_1 (resp. s_2) whose i'th coordinate is $s(j(i),i)$ (resp. $s(k(i),i)$) for each i. These marked φ-rectangles obviously have the asserted maximality and minimality properties, R_1 is semi-finite if no $j(i)$ vanishes, and R_1 is empty otherwise, $R_1 \subset R \subset R_2$, and $R_2 - R_1$ is contained in the union of the r sets $E_i := \{z \text{ in } \mathbb{R}^r \text{ with } s(j(i),i) < \varphi_i(z) \le s(k(i),i)\}$, for $i = 1,\ldots,r$. By the definition of $\mu_{i,\varphi}$ as direct image,

$$\begin{aligned}
\mu(E_i) &= \mu_{i,\varphi}((s(j(i),i), s(k(i),i)])\\
&= G_i(s(k(i),i)) - G_i(s(j(i),i)) = k(i)/M - j(i)/M \le 1/M,
\end{aligned}$$

whence $\mu(R_2) - \mu(R_1) \leq r/M$, as required for 1). Statement 2) is immediate from part 1) of the previous lemma 3.2.8. QED

Lemma 3.2.12. *For ν any measure on $\mathbb{R}^r$ of total mass ≤ 1, we have the inequalities*

$$\varphi\text{-discrep}(\mu, \nu) \leq 5r/M + |\mu(R_{\max}) - \nu(R_{\max})| + M\text{-}\varphi\text{-discrep}(\mu, \nu),$$

$$|\mu(R_{\max}) - \nu(R_{\max})| \leq M\text{-}\varphi\text{-discrep}(\mu, \nu),$$

and

$$\varphi\text{-discrep}(\mu, \nu) \leq 5r/M + 2(M\text{-}\varphi\text{-discrep}(\mu, \nu)).$$

PROOF. Given a semi-finite φ-rectangle $R = R(x, \varphi)$, x in $\mathbb{R}^r$, we find marked φ-rectangles $R_1 \subset R \subset R_2$ with $\mu(R_2) - \mu(R_1) \leq r/M$, thanks to the previous lemma. So we have the inequalities

$$\mu(R_1) \leq \mu(R) \leq \mu(R_2) \text{ and } \nu(R_1) \leq \nu(R) \leq \nu(R_2),$$

from which we infer

$$|\mu(R) - \nu(R)| \leq \mathrm{Max}(|\mu(R_2) - \nu(R_1)|, |\nu(R_2) - \mu(R_1)|).$$

From the inequality $\mu(R_2) - \mu(R_1) \leq r/M$, we get

$$|\mu(R) - \nu(R)| \leq r/M + \mathrm{Max}(|\mu(R_1) - \nu(R_1)|, |\nu(R_2) - \mu(R_2)|).$$

For $i = 1$ or 2, we apply 3.2.9, to bound

$$\begin{aligned}&|\mu(R_i) - \nu(R_i)| \\ &\qquad \leq 4r/M + |\mu(R_i \cap R_{\max}) - \nu(R_i \cap R_{\max})| + |\mu(R_{\max}) - \nu(R_{\max})|.\end{aligned}$$

Thus we get

$$\begin{aligned}|\mu(R) - \nu(R)| &\leq r/M + \mathrm{Max}(|\mu(R_1) - \nu(R_1)|, |\nu(R_2) - \mu(R_2)|) \\ &\leq 5r/M + |\mu(R_{\max}) - \nu(R_{\max})| \\ &\quad + \mathrm{Max}(|\mu(R_1 \cap R_{\max}) - \nu(R_1 \cap R_{\max})|, |\nu(R_2 \cap R_{\max}) - \mu(R_2 \cap R_{\max})|).\end{aligned}$$

Although R_2 need not be semi-finite, its intersection with $R_{\max}$ is equal to $R_3 \cap R_{\max}$, for a semi-finite marked R_3. [Indeed, if R_2 is $R(x)$, we can take R_3 to be $R(y)$ for y the finite grid point with coordinates $y(i) = x(i)$ if $x(i) < +\infty$, $y(i) = s(M-1, i)$ if $x(i) = +\infty$.] Therefore the inequality above gives us

$$|\mu(R) - \nu(R)| \leq 5r/M + |\mu(R_{\max}) - \nu(R_{\max})| + M\text{-}\varphi\text{-discrep}(\mu, \nu).$$

Since this holds for all semi-finite R, we have

$$\varphi\text{-discrep}(\mu, \varphi) \leq 5r/M + |\mu(R_{\max}) - \nu(R_{\max})| + M\text{-}\varphi\text{-discrep}(\mu, \nu).$$

We note that $R_{\max}$ is itself of the form $R_4 \cap R_{\max}$, for R_4 the semi-finite marked φ-rectangle $R(x, \varphi)$ with $x(i) = s(M-1, i)$ for all i. Therefore we have

$$|\mu(R_{\max}) - \nu(R_{\max})| \leq M\text{-}\varphi\text{-discrep}(\mu, \nu).$$

Combining these last two inequalities, we get

$$\varphi\text{-discrep}(\mu, \varphi) \leq 5r/M + 2(M\text{-}\varphi\text{-discrep}(\mu, \nu)). \quad \text{QED}$$

3.2.13. For each element A in $G(N)$, we denote by $\mu(A, N)$ the measure of total mass $(N - c(r))/(N + \lambda)$ on $\mathbb{R}^r$,

$$\mu(A, N) := \mu(\text{naive}, A, G(N), \text{ steps } b).$$

Applying the previous result, we get

Corollary 3.2.14. *For any A in $G(N)$, we have*

$$\varphi\text{-discrep}(\mu, \mu(A, N)) \leq 5r/M + 2(M\text{-}\varphi\text{-discrep}(\mu, \mu(A, N))).$$

3.2.15. At this point, we recall (cf. 2.7.6, 2.8.2, 2.9.1) the combinatorial formulas which express the measures

$$\begin{aligned}
\mu(A, N) &:= \mu(\text{naive}, A, G(N), \text{ separations } a),\\
\mu(N) &:= \mu(\text{naive}, G(N), \text{ separations } a),\\
\mu &:= \mu(\text{naive, univ, separations } a)
\end{aligned}$$

as an alternating sum of TCor's and TCOR's. For f an $\mathbb{R}$-valued Borel measurable function on $\mathbb{R}^r$ of compact support, and A in $G(N)$, we have

$$\begin{aligned}
\text{Int}(a, f, G(N), A) &:= \int_{\mathbb{R}^r} f\, d\mu(A, N),\\
\text{INT}(a, f, G(N)) &:= \int_{\mathbb{R}^r} f\, d\mu(N),\\
\text{INT}(a, f, \text{ univ}) &:= \int_{\mathbb{R}^r} f\, d\mu,
\end{aligned}$$

$$\begin{aligned}
\text{Int}(a, f, G(N), A) &= \sum_{k \geq \Sigma(a)} (-1)^{k - \Sigma(a)}\, \text{TCor}(k, a, f, G(N), A),\\
\text{INT}(a, f, G(N)) &= \sum_{k \geq \Sigma(a)} (-1)^{k - \Sigma(a)}\, \text{TCOR}(k, a, f, G(N)),\\
\text{INT}(a, f, \text{ univ}) &= \sum_{k \geq \Sigma(a)} (-1)^{k - \Sigma(a)}\, \text{TCOR}(k, a, f, \text{ univ}).
\end{aligned}$$

If in addition $f \geq 0$ as function on $\mathbb{R}^r$, each of the terms

$$\text{TCor}(k, a, f, G(N), A), \quad \text{TCOR}(k, a, f, G(N)), \quad \text{TCOR}(k, a, f, \text{ univ})$$

is nonnegative, and for each integer $m \geq \Sigma(a)$, we have the inequalities

$$\begin{aligned}
\sum_{m \geq k \geq \Sigma(a)} (-1)^{k - \Sigma(a)}\, \text{TCor}(k, a, f, G(N), A) &\leq \text{Int}(a, f, G(N), A),\\
\sum_{m \geq k \geq \Sigma(a)} (-1)^{k - \Sigma(a)}\, \text{TCOR}(k, a, f, G(N)) &\leq \text{INT}(a, f, G(N)),\\
\sum_{m \geq k \geq \Sigma(a)} (-1)^{k - \Sigma(a)}\, \text{TCOR}(k, a, f, \text{ univ}) &\leq \text{INT}(a, f, \text{univ}),
\end{aligned}$$

if $m - \Sigma(a)$ is odd, and we have the inequalities

$$\begin{aligned}
\mathrm{Int}(a, f, G(N), A) &\leq \sum_{m \geq k \geq \Sigma(a)} (-1)^{k-\Sigma(a)} \,\mathrm{TCor}(k, a, f, G(N), A), \\
\mathrm{INT}(a, f, G(N)) &\leq \sum_{m \geq k \geq \Sigma(a)} (-1)^{k-\Sigma(a)} \,\mathrm{TCOR}(k, a, f, G(N)), \\
\mathrm{INT}(a, f, \text{ univ}) &\leq \sum_{m \geq k \geq \Sigma(a)} (-1)^{k-\Sigma(a)} \,\mathrm{TCOR}(k, a, f, \text{ univ}),
\end{aligned}$$

if $m - \Sigma(a)$ is even.

Lemma 3.2.16. *For $f \geq 0$ a nonnegative, bounded, Borel measurable function of compact support on $\mathbb{R}^r$, and $L \geq \Sigma(a)$ a cutoff, we have the inequalities*

$$\begin{aligned}
&|\,\mathrm{INT}(a, f, \textit{ univ}) - \mathrm{Int}(a, f, G(N), A)| \\
&\quad \leq \sum_{L \geq k \geq \Sigma(a)} |\,\mathrm{TCOR}(k, a, f, \textit{ univ}) - \mathrm{TCor}(k, a, f, G(N), A)| \\
&\quad\quad + \mathrm{TCOR}(L, a, f, \textit{ univ}) + \mathrm{TCOR}(L+1, a, f, \textit{ univ}) \\
&\quad \leq \sum_{L \geq k \geq \Sigma(a)} |\,\mathrm{TCOR}(k, a, f, G(N)) - \mathrm{TCor}(k, a, f, G(N), A)| \\
&\quad\quad + \sum_{L \geq k \geq \Sigma(a)} |\,\mathrm{TCOR}(k, a, f, G(N)) - \mathrm{TCOR}(k, a, f, \textit{ univ})| \\
&\quad\quad + \mathrm{TCOR}(L, a, f, \textit{ univ}) + \mathrm{TCOR}(L+1, a, f, \textit{ univ}).
\end{aligned}$$

PROOF. The second inequality is immediate from the first, by the triangle inequality. For the first, we argue as follows. Fix A in $G(N)$. Let $m \geq \Sigma(a)$ be a cutoff having the same parity as $\Sigma(a)$. Suppose first that

$$\mathrm{INT}(a, f, \text{ univ}) \geq \mathrm{Int}(a, f, G(N), A).$$

Then we have

$$\begin{aligned}
0 &\leq \mathrm{INT}(a, f, \text{ univ}) - \mathrm{Int}(a, f, G(N), A) \\
&\leq \sum_{m \geq k \geq \Sigma(a)} (-1)^{k-\Sigma(a)} \,\mathrm{TCOR}(k, a, f, \text{ univ}) \\
&\quad\quad - \sum_{m-1 \geq k \geq \Sigma(a)} (-1)^{k-\Sigma(a)} \,\mathrm{TCor}(k, a, f, G(N), A) \\
&\leq \mathrm{TCOR}(m, a, f, \text{ univ}) \\
&\quad\quad + \sum_{m-1 \geq k \geq \Sigma(a)} |\,\mathrm{TCOR}(k, a, f, \text{ univ}) - \mathrm{TCor}(k, a, f, G(N), A)|.
\end{aligned}$$

If L is m, this inequality trivially implies the asserted one. If L is $m+1$, this inequality for $m+2$ trivially implies the asserted one.

Now suppose that

$$\mathrm{Int}(a, f, G(N), A) \geq \mathrm{INT}(a, f, \text{ univ}).$$

Then we have

$$\begin{aligned}
0 &\le \mathrm{Int}(a,f,G(N),A) - \mathrm{INT}(a,f,\ \mathrm{univ}) \\
&\le \sum_{m\ge k\ge \Sigma(a)} (-1)^{k-\Sigma(a)}\,\mathrm{TCor}(k,a,f,G(N),A) \\
&\qquad - \sum_{m+1\ge k\ge \Sigma(a)} (-1)^{k-\Sigma(a)}\,\mathrm{TCOR}(k,a,f,\ \mathrm{univ}) \\
&\le \mathrm{TCOR}(m+1,a,f,G(N)) \\
&\qquad + \sum_{m\ge k\ge \Sigma(a)} |\,\mathrm{TCOR}(k,a,f,\ \mathrm{univ}) - \mathrm{TCor}(k,a,f,G(N),A)|.
\end{aligned}$$

If L is either m or $m+1$, this inequality trivially implies the asserted one. QED

3.2.17. It is the second inequality of 3.2.16 which will be useful to us. We will use our estimate for the rate of convergence of large N limits of TCOR's to estimate the terms of the second line, and our bounds on TCOR's to estimate the terms of the last line. Let us recall a crude form of these bounds.

Lemma 3.2.18. *Suppose that conditions* i), ii) *and* iii) *of Propositions* 3.0.1 *and* 3.1.9 *hold. For $f \ge 0$ a bounded, Borel measurable function of compact support on $\mathbb{R}^r$ with $\|f\|_{\sup} \le 1$, supported in $\sum_i |x(i)| \le \alpha$, we have the following estimates.*

i) $$\begin{aligned}&|\,\mathrm{TCOR}(k,a,f,G(N)) - \mathrm{TCOR}(k,a,f,\ \mathit{univ})| \\ &\qquad\le (16\alpha)^{k+r}((\pi\alpha)^2+\alpha+1+10\log(N))/N.\end{aligned}$$

ii) $$|\,\mathrm{TCOR}(k,a,f,G(N))| \le (4\alpha)^{k+r}/(k+r)!,$$

ii bis) $$|\,\mathrm{TCOR}(k,a,f,\ \mathit{univ})| \le (4\alpha)^{k+r}/(k+r)!,$$

iii) $$\begin{aligned}&\mathrm{Sqrt}(\mathrm{Var}(A \mapsto \mathrm{TCor}(k,a,f,G(N),A)\ \mathit{on}\ G(N))) \\ &\qquad\le (3(32\alpha)^{k+r}+1)/\,\mathrm{Sqrt}(N).\end{aligned}$$

PROOF. Use the trivial bound $\mathrm{Binom}(r+k-1, r+\Sigma(a)-1) \le 2^{r+k-1}$ to get i), ii) and iii) from their cited counterparts, and let $N \to \infty$ to get ii bis) from ii). QED

3.2.19. We now use these bounds for f the characteristic function of a set $R \cap R_{\max}$, for R a finite marked φ-rectangle. Such an f is certainly supported in

$$R_{\max} := \{z \text{ in } \mathbb{R}^r \text{ with } s(1,i) < \varphi_i(z) \le s(M-1,i) \text{ for all } i\}.$$

This set lies in the set

$$\{z \text{ in } \mathbb{R}^r \text{ with } |\varphi_i(z)| \le \mathrm{Max}(|s(M-1,i)|, |s(1,i)|) \text{ for all } i\},$$

which set in turn lies in the set

$$\left\{z \text{ in } \mathbb{R}^r \text{ with } \sum_i |\varphi_i(z)| \le \sum_i \mathrm{Max}(|s(M-1,i)|, |s(1,i)|) := \beta\right\}.$$

[Recall that

$$\beta := \sum_i \mathrm{Max}(|s(1,i)|, |s(M-1,i)|)$$

is the M-φ-**grid diameter**, cf. 3.2.4.1.] In view of the dilational assumption on φ, namely

$$\eta \sum |x_i| \leq \sum |\varphi_i(x)| \leq \kappa \sum |x_i|,$$

this last set, and hence $R_{\max}$, is contained in the set

$$\left\{ x \text{ in } \mathbb{R}^r \text{ with } \sum_i |x_i| \leq \beta/\eta \right\}.$$

For east of later reference, we record this fact.

Lemma 3.2.20. *We have the inclusion*

$$R_{\max} \subset \left\{ x \text{ in } \mathbb{R}^r \text{ with } \sum_i |x_i| \leq \beta/\eta \right\}.$$

It will be convenient to know that β/η is not too small.

Lemma 3.2.21. *If $M \geq 4r$, then $\beta/\eta > 1/10$.*

PROOF. Let $\alpha > 0$. For f the characteristic function of any Borel set E contained in the region $\sum |x_i| \leq \alpha$, and a in $\mathbb{Z}^r$ the separation vector such that μ is μ(univ, sep.'s a), we have, as recalled above,

$$\begin{aligned}\mu(E) &= \int_{\mathbb{R}^r} f\, d\mu = \mathrm{INT}(a, f, \text{ univ}) \\ &= \sum_{k \geq \Sigma(a)} (-1)^{k-\Sigma(a)}\, \mathrm{TCOR}(k, a, f, \text{ univ}),\end{aligned}$$

with the estimate

$$|\,\mathrm{TCOR}(k, a, f, \text{ univ})| \leq (4\alpha)^{k+r}/(k+r)!.$$

Therefore we have the inequality

$$\begin{aligned}\mu(E) &\leq \sum_{k \geq \Sigma(a)} (4\alpha)^{k+r}/(k+r)! \leq \sum_{k \geq 0} (4\alpha)^{k+r}/(k+r)! \\ &= \sum_{k \geq r} (4\alpha)^k/k! \leq \sum_{k \geq 1} (4\alpha)^k/k! = e^{4\alpha} - 1.\end{aligned}$$

Take $\alpha = \beta/\eta$, and $E = R_{\max}$. We know by 3.2.8, part 1), that

$$\mu(R_{\max}) \geq 1 - 2r/M,$$

so we obtain

$$1 - 2r/M \leq \mu(R_{\max}) \leq e^{4\beta/\eta} - 1,$$

i.e.,

$$e^{4\beta/\eta} \geq 2 - 2r/M \geq 1.5.$$

Taking logs, we find $4\beta/\eta \geq \log(1.5) > 0.405$, whence $\beta/\eta > 1/10$. QED

Lemma 3.2.22. *For any Borel set R contained in $R_{\max}$, and for $L \geq \Sigma(a)$ a cutoff, we have the inequality*

$$\begin{aligned}
&|\mu(R) - \mu(A,N)(R)| \\
&\quad \leq \sum_{L \geq k \geq \Sigma(a)} |\,\mathrm{TCOR}(k,a,f,G(N)) - \mathrm{TCor}(k,a,f,G(N),A)| \\
&\quad + 9(16\beta/\eta)^{L+r+2}(1 + 2\log(N))/N \\
&\quad + (4\beta/\eta)^{L+r}/(L+r)! + (4\beta/\eta)^{L+1+r}/(L+1+r)!.
\end{aligned}$$

PROOF. For f the characteristic function of R, Lemma 3.2.16 gives

$$\begin{aligned}
|\mu(R) - \mu(A,N)(R)| &= |\,\mathrm{INT}(a,f,\ \mathrm{univ}) - \mathrm{Int}(a,f,G(N),A)| \\
&\leq \sum_{L \geq k \geq \Sigma(a)} |\,\mathrm{TCOR}(k,a,f,G(N)) - \mathrm{TCor}(k,a,f,G(N),A)| \\
&\quad + \sum_{L \geq k \geq \Sigma(a)} |\,\mathrm{TCOR}(k,a,f,G(N)) - \mathrm{TCOR}(k,a,f,\ \mathrm{univ})| \\
&\quad + \mathrm{TCOR}(L,a,f,\ \mathrm{univ}) + \mathrm{TCOR}(L+1,a,f,\ \mathrm{univ})
\end{aligned}$$

(using the bounds of 3.2.18 and 3.2.20)

$$\begin{aligned}
&\leq \sum_{L \geq k \geq \Sigma(a)} |\,\mathrm{TCOR}(k,a,f,G(N)) - \mathrm{TCor}(k,a,f,G(N),A)| \\
&\quad + \sum_{L \geq k \geq \Sigma(a)} (16\beta/\eta)^{k+r}((\pi\beta/\eta)^2 + \beta/\eta + 1 + 10\log(N))/N \\
&\quad + (4\beta/\eta)^{L+r}/(L+r)! + (4\beta/\eta)^{L+1+r}/(L+1+r)!.
\end{aligned}$$

Recall that $\beta/\eta > 1/10$, so $16\beta/\eta > 1.6 > 3/2$. For a finite geometric series $\Sigma_{L \geq k \geq \Sigma(a)} \gamma^{k+r}$ with $\gamma > 3/2$, we have

$$\begin{aligned}
\sum_{L \geq k \geq \Sigma(a)} \gamma^{k+r} &\leq \sum_{L \geq k \geq 0} \gamma^{k+r} = (\gamma^{r+1+L} - 1)/(\gamma - 1) \\
&\leq \gamma^{r+1+L}/(\gamma - 1) \leq (\gamma/(\gamma-1))\gamma^{r+L} = (1 + 1/(\gamma-1))\gamma^{r+L} \leq 3\gamma^{r+L}.
\end{aligned}$$

Thus we have

$$\begin{aligned}
&\sum_{L \geq k \geq \Sigma(a)} (16\beta/\eta)^{k+r}((\pi\beta/\eta)^2 + \beta/\eta + 1 + 10\log(N)) \\
&\quad \leq 3(16\beta/\eta)^{L+r}((\pi\beta/\eta)^2 + \beta/\eta + 1 + 10\log(N)) \\
&\quad \leq 3(16\beta/\eta)^{L+r}(3(16\beta/\eta)^2 + 10\log(N)) \\
&\quad \leq 3(16\beta/\eta)^{L+r}(3(16\beta/\eta)^2(1 + 2\log(N))) \\
&\quad \leq 9(16\beta/\eta)^{L+r+2}(1 + 2\log(N)). \quad \text{QED}
\end{aligned}$$

Corollary 3.2.23. *For each semi-finite marked φ-rectangle R, denote by χ_R the characteristic function of $R \cap R_{\max}$. Then we have the inequality*

$$\begin{aligned}
&M\text{-}\varphi\text{-discrep}(\mu, \mu(A,N)) \\
&\quad := \mathrm{Sup}_{\textit{semi-finite marked } R} |\mu(R \cap R_{\max}) - \mu(A,N)(R \cap R_{\max})| \\
&\quad \leq 9(16\beta/\eta)^{L+r+2}(1 + 2\log(N))/N \\
&\qquad + (4\beta/\eta)^{L+r}/(L+r)! + (4\beta/\eta)^{L+1+r}/(L+1+r)! \\
&\qquad + \sum_R \sum_{L \geq k \geq \Sigma(a)} |\,\mathrm{TCOR}(k, a, \chi_R, G(N)) - \mathrm{TCor}(k, a, \chi_R, G(N), A)|,
\end{aligned}$$

the sum extended over the $(M-2)^r$ finite marked φ-rectangles none of whose coordinates is $s(0,i) = -\infty$ or $s(1,i)$.

PROOF. For each of the $(M-2)^r$ semi-finite marked φ-rectangles R for which $R \cap R_{\max}$ is nonempty, the previous lemma gives an upper bound for the quantity $|\mu(R \cap R_{\max}) - \mu(A,N)(R \cap R_{\max})|$ as a sum of two positive terms $A + B(R)$, with A constant and $B(R)$ depending on R. Each of these is bounded by $A + \sum_R B(R)$. QED

Integrating over $G(N)$, we obtain

Lemma 3.2.24. *We have the estimate*

$$\begin{aligned}
&\int_{G(N)} M\text{-}\varphi\text{-discrep}(\mu, \mu(A,N))dA \\
&\quad \leq 9(16\beta/\eta)^{L+r+2}(1 + 2\log(N))/N \\
&\qquad + (4\beta/\eta)^{L+r}/(L+r)! + (4\beta/\eta)^{L+1+r}/(L+1+r)! \\
&\qquad + \sum_R \sum_{L \geq k \geq \Sigma(a)} \int_{G(N)} |\,\mathrm{TCOR}(k, a, \chi_R, G(N)) - \mathrm{TCor}(k, a, \chi_R, G(N), A)|dA.
\end{aligned}$$

3.2.25. We now use Cauchy-Schwarz to estimate

$$\begin{aligned}
&\int_{G(N)} |\,\mathrm{TCOR}(k, a, \chi_R, G(N)) - \mathrm{TCor}(k, a, \chi_R, G(N), A)|\, dA \\
&\quad \leq \mathrm{Sqrt}\left(\int_{G(N)} |\,\mathrm{TCOR}(k, a, \chi_R, G(N)) - \mathrm{TCor}(k, a, \chi_R, G(N), A)|^2\, dA\right) \\
&\quad := \mathrm{Sqrt}(\mathrm{Var}(A \mapsto \mathrm{TCOR}(k, a, \chi_R, G(N)) \text{ on } G(N))).
\end{aligned}$$

So from Lemma 3.2.24 we get

Lemma 3.2.26. *We have the estimate*

$$\begin{aligned}
&\int_{G(N)} M\text{-}\varphi\text{-discrep}(\mu, \mu(A,N))\, dA \\
&\quad \leq 9(16\beta/\eta)^{L+r+2}(1 + 2\log(N))/N \\
&\qquad + (4\beta/\eta)^{L+r}/(L+r)! + (4\beta/\eta)^{L+1+r}/(L+1+r)! \\
&\qquad + \sum_R \sum_{L \geq k \geq \Sigma(a)} \mathrm{Sqrt}(\mathrm{Var}(A \mapsto \mathrm{TCOR}(k, a, \chi_R, G(N)) \textit{ on } G(N))).
\end{aligned}$$

Now plugging in the estimates for variance and TCOR's recalled above, with $f = \chi_R$ and $\alpha = \beta/\eta$, we find

Corollary 3.2.27. *We have the estimate*

$$\begin{aligned}\int_{G(N)} & M\text{-}\varphi\text{-discrep}(\mu, \mu(A, N))\, dA \\ & \leq 27(16\beta/\eta)^{L+r+2}/\operatorname{Sqrt}(N) \\ & \quad + (4\beta/\eta)^{L+r}/(L+r)! + (4\beta/\eta)^{L+1+r}/(L+1+r)! \\ & \quad + (M-2)^r 12(32\beta/\eta)^{L+r}/\operatorname{Sqrt}(N).\end{aligned}$$

PROOF. We have, from the previous result,

$$\begin{aligned}\int_{G(N)} & M\text{-}\varphi\text{-discrep}(\mu, \mu(A, N))\, dA \\ & \leq 9(16\beta/\eta)^{L+r+2}(1 + 2\log(N))/N \\ & \quad + (4\beta/\eta)^{L+r}/(L+r)! + (4\beta/\eta)^{L+1+r}/(L+1+r)! \\ & \quad + \sum_R \sum_{L \geq k \geq \Sigma(a)} \operatorname{Sqrt}(\operatorname{Var}(A \mapsto \operatorname{TCOR}(k, a, \chi_R, G(N)) \text{ on } G(N)))\end{aligned}$$

(using 3.2.18 and 3.2.20)

$$\begin{aligned} & \leq 9(16\beta/\eta)^{L+r+2}(1 + 2\log(N))/N \\ & \quad + (4\beta/\eta)^{L+r}/(L+r)! + (4\beta/\eta)^{L+1+r}/(L+1+r)! \\ & \quad + \sum_R \sum_{L \geq k \geq 0} (3(32\beta/\eta)^{k+r} + 1)/\operatorname{Sqrt}(N)\end{aligned}$$

(using $32\beta/\eta \geq 1$)

$$\begin{aligned} & \leq 9(16\beta/\eta)^{L+r+2}(1 + 2\log(N))/N \\ & \quad + (4\beta/\eta)^{L+r}/(L+r)! + (4\beta/\eta)^{L+1+r}/(L+1+r)! \\ & \quad + (M-2)^r \sum_{L \geq k \geq 0} 4(32\beta/\eta)^{k+r}/\operatorname{Sqrt}(N)\end{aligned}$$

(using $32\beta/\eta \geq 3/2$)

$$\begin{aligned} & \leq 9(16\beta/\eta)^{L+r+2}(1 + 2\log(N))/N \\ & \quad + (4\beta/\eta)^{L+r}/(L+r)! + (4\beta/\eta)^{L+1+r}/(L+1+r)! \\ & \quad + (M-2)^r 12(32\beta/\eta)^{L+r}/\operatorname{Sqrt}(N).\end{aligned}$$

Now use the fact that for any $N \geq 1$, we have

$$\begin{aligned} 1/N &\leq 1/\operatorname{Sqrt}(N), \\ \log(N)/N &\leq 1/\operatorname{Sqrt}(N)\end{aligned}$$

to bound $(1 + 2\log(N))/N$ by $3/\operatorname{Sqrt}(N)$. QED

Corollary 3.2.28. *We have the estimate*

$$\begin{aligned}\int_{G(N)} &\varphi\text{-discrep}(\mu, \mu(A, N))\, dA \\ &\le 5r/M \\ &\quad + 2(27)(16\beta/\eta)^{L+r+2}/\operatorname{Sqrt}(N) \\ &\quad + 2((4\beta/\eta)^{L+r}/(L+r)! + (4\beta/\eta)^{L+1+r}/(L+1+r)!) \\ &\quad + 2(M-2)^r 12(32\beta/\eta)^{L+r}/\operatorname{Sqrt}(N).\end{aligned}$$

PROOF. We have the inequality (3.2.12, third inequality)

$$\varphi\text{-discrep}(\mu, \nu) \le 5r/M + 2(M\text{-}\varphi\text{-discrep}(\mu, \nu)).$$

Integrate it over $G(N)$ and use the previous result. QED

For ease of later computation, we record a very crude form of the previous result.

Corollary 3.2.29. *We have the estimate*

$$\begin{aligned}\int_{G(N)} &\varphi\text{-discrep}(\mu, \mu(A, N))\, dA \\ &\le 5r/M + 27(32\beta/\eta)^{L+r+2}(M^r/\operatorname{Sqrt}(N) + 1/(L+r)!).\end{aligned}$$

PROOF. We bound 2 by 2^r, so

$$2(27)(16\beta/\eta)^{L+r+2} \le 27(32\beta/\eta)^{L+r+2}.$$

We bound $(M-2)^r$ by M^r, so

$$2(M-2)^r 12(32\beta)^{L+r} \le 24M^r(32\beta)^{L+r+2}.$$

We bound

$$\begin{aligned}&2((4\beta\eta)^{L+r}/(L+r)! + (4\beta/\eta)^{L+1+r}/(L+1+r)!) \\ &\quad \le 2^r((4\beta/\eta)^{L+r} + (4\beta/\eta)^{L+1+r})/(L+r)! \le (32\beta/\eta)^{L+r+2}/(L+r)!.\end{aligned}$$

QED

3.3. A weak form of Theorem 3.1.6

Proposition 3.3.1. *Suppose that conditions* i), ii) *and* iii) *of Propositions* 3.0.1 *and* 3.1.9 *hold. Then the following weak form of Theorem* 3.1.6 *holds:*

Theorem 3.3.2. *Fix an integer $r \ge 1$, an offset vector c in $\mathbb{Z}^r$, with corresponding separation vector a and step vector b, and a real number $\varepsilon > 0$. For any φ as in* 3.1.20 *and satisfying* 3.1.4, *there exists an explicit constant $N_3(\varepsilon, r, c, \varphi)$ with the following property: for $G(N)$ any of*

$$U(N), SO(2N+1), USp(2N), SO(2N), O_-(2N+2),$$

and for

$$\mu := \mu(\textit{naive, univ, offsets } c),$$
$$\mu(A, N) := \mu(\textit{naive}, A, G(N), \textit{ offsets } c)$$

we have the inequality

$$\int_{G(N)} \varphi\text{-discrep}(\mu, \mu(A, N))\, dA \le \varepsilon,$$

provided that $N \geq N_3(\varepsilon, r, c, \varphi)$.

PROOF. For any grid size $M \geq 4r$, with M-φ-grid diameter denoted β, and any cutoff $L \geq \Sigma(a)$, we have, by Corollary 3.2.29 above,

$$\int_{G(N)} \varphi\text{-discrep}(\mu, \mu(A, N))\, dA$$
$$\leq 5r/M + 27(32\beta/\eta)^{L+r+2}(M^r / \operatorname{Sqrt}(N) + 1/(L+r)!).$$

We first choose M large enough that $M \geq 4r$ and $5r/M \leq \varepsilon/3$. This choice of M gives us a β, namely the M-φ-grid diameter. The power series for $27x^2e^x$,

$$27 \sum_{n \geq 0} x^{n+2}/n!,$$

is everywhere convergent, in particular at $x = 32\beta/\eta$. So the sequence, indexed by L,

$$27(32\beta/\eta)^{L+r+2}/(L+r)!$$

tends to zero as $L \to \infty$. We choose L large enough that

$$27(32\beta/\eta)^{L+r+2}/(L+r)! \leq \varepsilon/3.$$

Finally, having chosen M and L, we need only take N so large that

$$27(32\beta/\eta)^{L+r+2}M^r / \operatorname{Sqrt}(N) \leq \varepsilon/3. \quad \text{QED}$$

3.4. Conclusion of the axiomatic proof of Theorem 3.1.6

3.4.1. We now make use of condition iv) of Proposition 3.1.9, the tail estimate for the "classical" one variable spacing measure. We have restricted the use of the tail estimate to this section in order to clarify what we can get without it, namely the weak version 3.3.2 of Theorem 3.1.6 given above. Another reason for isolating it is that, although we give a proof of it (6.13.4) at the level of the Weyl integration formula, it seems to use to be somewhat deeper than our other three axiomatic inputs. [We should mention here that the asymptotic behavior of the tail for large s, not just an upper bound for it, was given by [**Widom**]. Presumably one could make effective his result and so get from it the kind of tail estimate, valid for all $s > 0$, that we require.]

3.4.2. Thus there exist explicit real constants $A > 0$ and $B > 0$ such that the limit measure $\mu(\text{univ, sep. } 0) = \mu(\text{univ, step } 1)$ on $\mathbb{R}$ satisfies

$$\mu(\text{univ, step } 1)(\{|x| > s\}) \leq Ae^{-Bs^2} \text{ for every real } s \geq 0.$$

We may and will assume that the constant A is ≥ 1, and that the constant B is ≤ 1, cf. 6.13.4, where we show that we may take $A = 4/3$ and $B = 1/8$.

Lemma 3.4.3. *Suppose* $M \geq 3$. *Then for* C *the constant*

$$C := (1/\eta)\kappa r^2 \Sigma(b) \operatorname{Sqrt}((1 + \log(rA\Sigma(b)))/B),$$

the M-φ-*grid diameter* β *satisfies*

$$\beta/\eta \leq C \times \operatorname{Sqrt}(\log(M)).$$

PROOF. Recall (3.2.4.1) that the M-φ-grid diameter β is defined as

$$\beta := \sum_i \mathrm{Max}(|s(1,i)|, |s(M-1,i)|),$$

where, for each i, $s(1,i) < s(M-1,i)$ are real numbers chosen so that

$$\mu(\{x \text{ in } \mathbb{R}^r \text{ with } \varphi_i(x) \le s(1,i)\}) = 1/M,$$
$$\mu(\{x \text{ in } \mathbb{R}^r \text{ with } \varphi_i(x) > s(M-1,i)\}) = 1/M.$$

It will be convenient to introduce the quantities, for $1 < i < r$,

$$\beta_i := \mathrm{Max}(|s(1,i)|, |s(M-1,i)|).$$

Our strategy to estimate $\beta = \sum \beta_i$ is to estimate separately each β_i.

We claim that for each i we have the inequality

$$\mu(\{x \text{ in } \mathbb{R}^r \text{ with } |\varphi_i(x)| > \beta_i\}) \ge 1/M.$$

Indeed, if $\beta_i = |s(M-1,i)| > |s(1,i)|$, then we must have $s(M-1,i) > 0$ (for if $s(M-1,i) \le 0$ then $s(1,i) < s(M-1,i)$ implies $|s(1,i)| > |s(M-1,i)|$) and hence

$$\{x \text{ in } \mathbb{R}^r \text{ with } |\varphi_i(x)| > \beta_i\} \supset \{x \text{ in } \mathbb{R}^r \text{ with } \varphi_i(x) > s(M-1,i)\}.$$

This latter set has μ-measure $1/M$ by construction.

If $\beta_i = |s(1,i)| \ge |s(M-1,i)|$, then $s(1,i) < 0$ (because if $s(1,i) \ge 0$, then $s(1,i) < s(M-1,i)$ implies $|s(M-1,i)| > |s(1,i)|$), and hence

$$\{x \text{ in } \mathbb{R}^r \text{ with } |\varphi_i(x)| > \beta_i\} \supset \{x \text{ in } \mathbb{R}^r \text{ with } \varphi_i(x) < s(1,i)\}.$$

This latter set has μ-measure $1/M$, because by construction the set

$$\{x \text{ in } \mathbb{R}^r \text{ with } \varphi_i(x) \le s(1,i)\}$$

has μ-measure $1/M$, and by the hypothesis 3.1.4 that $\mu_{i,\varphi}$ has a continuous CDF, the set $\{z \text{ in } \mathbb{R}^r \text{ with } \varphi_i(z) = s(1,i)\}$ has μ-measure zero.

We next use the inequality

$$|\varphi_i(x)| \le \sum_j |\varphi_j(x)| \le \kappa \sum_j |x_j|,$$

which gives an inclusion

$$\{x \text{ in } \mathbb{R}^r \text{ with } |\varphi_i(x)| > \beta_i\} \subset \{x \text{ in } \mathbb{R}^r \text{ with } \kappa \sum_j |x_j| > \beta_i\}$$
$$\subset \bigcup_j \{x \text{ in } \mathbb{R}^r \text{ with } |x_j| > \beta_i/\kappa r\}.$$

Therefore this union $\bigcup_j \{x \text{ in } \mathbb{R}^r \text{ with } |x_j| > \beta_i/\kappa r\}$ has measure $\ge 1/M$, and hence at least one of the unionees has measure $\ge 1/rM$. This says that for each i there exists an index j such that

$$\mu(\{x \text{ in } \mathbb{R}^r \text{ with } |x_j| > \beta_i/\kappa r\}) \ge 1/rM.$$

Now recall that the direct image measure $\mu_j := \mathrm{pr}[j]_*\mu$ defined in terms of the standard coordinates $x_1, \ldots, x_r$ on $\mathbb{R}^r$ of our spacing measure $\mu = \mu(\text{univ, steps } b)$ is the measure $\mu(\text{univ, step } b(j))$. According to 3.1.10, $\mu_j = \mu(\text{univ, step } b(j))$ has the tail estimate

$$\mu_j(\{|x| > sb(j)\}) \le b(j)Ae^{-Bs^2},$$

which we rewrite as

$$\mu(\{x \text{ in } \mathbb{R}^r \text{ with } |x_j| > sb(j)\}) \le b(j)Ae^{-Bs^2}.$$

Taking s to be $\beta_i/\kappa r b(j)$, we have the inequality

$$\mu(\{x \text{ in } \mathbb{R}^r \text{ with } |x_j| > \beta_i/\kappa r\}) \le b(j)A\exp(-B(\beta_i/\kappa r b(j))^2)$$
$$\le \Sigma(b)A\exp(-B(\beta_i/\kappa r \sum(b))^2).$$

But recall that for each i there exists j such that

$$1/rM \le \mu(\{x \text{ in } \mathbb{R}^r \text{ with } |x_j| > \beta_i/\kappa r\}).$$

Thus we get, for each i, the inequality

$$1/rM \le \Sigma(b)A\exp(-B(\beta_i/\kappa r\Sigma(b))^2),$$

i.e.,

$$\exp(B(\beta_i/\kappa r\Sigma(b))^2) \le MrA\Sigma(b),$$

or, taking logs and remembering that $M \ge 3 > e$ and $A \ge 1$,

$$B(\beta_i/\kappa r\Sigma(b))^2 \le \log(M) + \log(rA\Sigma(b)) \le \log(M)(1 + \log(rA\Sigma(b))),$$

i.e.

$$(\beta_i)^2 \le \log(M)(\kappa r\Sigma(b))^2(1 + \log(rA\Sigma(b)))/B,$$

i.e.,

$$\beta_i \le \mathrm{Sqrt}(\log(M))(kr\Sigma(b))\,\mathrm{Sqrt}((1 + \log(rA\Sigma(b)))/B).$$

Since $\beta = \sum_i \beta_i$ is the sum of r such terms, we get

$$\beta \le \kappa r^2\Sigma(b)\,\mathrm{Sqrt}((1 + \log(rA\Sigma(b)))/B) \times \mathrm{Sqrt}(\log(M)) = \eta C \times \mathrm{Sqrt}(\log(M)),$$

for C the constant

$$C := (1/\eta)\kappa r^2\Sigma(b)\,\mathrm{Sqrt}((1 + \log(rA\Sigma(b)))/B). \quad \text{QED}$$

3.4.4. We now explain the **idea** of the argument, to show how the exponent $1/(2r+4)$ arises. Using 3.2.29 and the above estimate for β/η, we get

$$\int_{G(N)} \varphi\text{-discrep}(\mu, \mu(A, N))\, dA$$
$$\le 5r/M + 27(32C)^{L+r+2}\,\mathrm{Sqrt}(\log(M))^{L+r+2}(M^r/\,\mathrm{Sqrt}(N) + 1/(L+r)!).$$

We wish to pick M and L as functions of N so as to exploit this. Our rough idea, which we will make precise in a moment, is to take

$$M = N^\alpha, \qquad (L+r)! = N^\gamma$$

for positive real α and γ to be determined. With such a choice, we have

$$\log(M) = \alpha\log(N) = (\alpha/\gamma)\log((L+r)!).$$

Before proceeding, let us explicate some standard inequalities.

Lemma 3.4.5. *Let $\varepsilon > 0$.*

1) *Given real $K > 0$, there is an explicit constant $N_4(\varepsilon, K)$ such that*

$$K^{x+1} \leq \Gamma(x)^{\varepsilon} \text{ for real } x \geq N_4(\varepsilon, K).$$

We can take

$$N_4(\varepsilon, K) = 2 \text{ if } K \leq 1,$$
$$N_4(\varepsilon, K) = eK^{2/\varepsilon} \text{ if } K > 1.$$

2) *There is an explicit constant $N_5(\varepsilon)$ such that*

$$(\log(\Gamma(x)))^{x+1} \leq \Gamma(x)^{1+\varepsilon} \text{ for real } x \geq N_5(\varepsilon).$$

For $\varepsilon \leq 1/2$, we can take $N_5(\varepsilon)$ to be the unique real number $s > e^e$ for which $\log\log(s)/\log(s) = \varepsilon/2$. For $\varepsilon > 1/2$, we can take $N_5(\varepsilon)$ to be $N_5(1/2)$.

PROOF. Both result from Stirling's formula, in the form [**W-W**, page 253] that for $x > 0$ real,

$$\log(\Gamma(x)) = (x - 1/2)\log(x) - x + \log(\mathrm{Sqrt}(2\pi)) + \varphi(x),$$
$$\text{with } 0 < \varphi(x) < 1/12x.$$

We will use the two following consequences of Stirling's formula. For $x > 0$,

$$\begin{aligned}\log(\Gamma(x)) &> (x - 1/2)\log(x) - x + 1/2 + \log(\mathrm{Sqrt}(2\pi/e))\\ &> (x - 1/2)(\log(x) - 1),\end{aligned}$$

and for $x \geq 1$,

$$\begin{aligned}\log(\Gamma(x)) &< (x - 1/2)\log(x) - x + \log(\mathrm{Sqrt}(2\pi)) + 1/12x\\ &< (x - 1/2)\log(x) - x + 1/12x\\ &< (x - 1/2)\log(x).\end{aligned}$$

To prove 1), note first that if $K \leq 1$, then $N_4(\varepsilon, K) = 2$ works, because $\Gamma(2) = 1$ and $\Gamma(x)$ is increasing in $x \geq 2$. If $K > 1$, we claim that

$$N_4(\varepsilon, K) = eK^{2/\varepsilon}$$

does the job. Suppose that $x \geq eK^{2/\varepsilon}$. Then $x \geq e$. By Stirling,

$$\log(\Gamma(x)) > (x - 1/2)(\log(x) - 1),$$

so it suffices to show that

$$(x + 1)\log(K) \leq \varepsilon(x - 1/2)(\log(x) - 1),$$

i.e.,

$$\begin{aligned}(1/\varepsilon)\log(K) &\leq ((x - 1/2)/(x + 1))(\log(x) - 1)\\ &= (1 - (3/2(x + 1)))(\log(x) - 1).\end{aligned}$$

Since $x \geq e > 2, (1 - (3/2(x + 1))) > 1/2$, and $\log(x) - 1 \geq 0$, and so it suffices if

$$(1/\varepsilon)\log(K) \leq (1/2)(\log(x) - 1),$$

which is precisely the condition $x \geq eK^{2/\varepsilon}$.

To prove 2), we may assume $\varepsilon \leq 1/2$, since $N_5(1/2)$ will work as $N_5(\varepsilon)$ for any larger ε. We now use also the inequality

$$\log(\Gamma(x)) < (x - 1/2)\log(x) \text{ for } x \geq 1,$$

so for $x \geq e^e$ we have

$$1 \leq \log\log(\Gamma(x)) < \log(x - 1/2) + \log\log(x).$$

Thus to have $(\log(\Gamma(x)))^{x+1} \leq \Gamma(x)^{1+\varepsilon}$, it suffices to have

$$(x+1)(\log(x-1/2) + \log\log(x)) \leq (1+\varepsilon)(x-1/2)(\log(x) - 1),$$

i.e.,

$$\log(x-1/2) + \log\log(x) \leq (1+\varepsilon)((x-1/2)/(x+1))(\log(x) - 1).$$

So it suffices if

$$\log(x) + \log\log(x) \leq (1+\varepsilon)(1 - 1/(2x+2))(\log(x) - 1),$$

i.e.,

$$1 + \log\log(x)/\log(x) \leq (1+\varepsilon)(1 - 1/(2x+2))(1 - 1/\log(x)),$$

and for this it suffices if

$$1 + \log\log(x)/\log(x) \leq (1+\varepsilon)(1 - 1/(2x+2) - 1/\log(x)).$$

For $0 < \varepsilon \leq 1/2$, we have $\varepsilon^2 = \varepsilon\varepsilon \leq \varepsilon/2$, so we have

$$1 + \varepsilon/2 = 1 + 2\varepsilon/3 - \varepsilon/6 \leq 1 + 2\varepsilon/3 - \varepsilon^2/3 = (1+\varepsilon)(1 - \varepsilon/3).$$

So we need simply choose $x \geq e^e$ large enough that

$$\log\log(x)/\log(x) \leq \varepsilon/2 \quad \text{and} \quad 1/(2x+2) + 1/\log(x) \leq \varepsilon/3.$$

For this, it is enough if $x \geq e^e$ satisfies

$$\log\log(x)/\log(x) \leq \varepsilon/2,$$
$$1/\log(x) \leq \varepsilon/4,$$
$$1/(2x+2) \leq \varepsilon/12.$$

Introduce the quantity $t := \log\log(x)$. Thus $t \geq 1$. Then

$$\log\log(x)/\log(x) = te^{-t} \quad \text{and} \quad 1/\log(x) = e^{-t}.$$

The function $x \mapsto \log\log(x)$ is an order preserving bijection from (e^e, ∞) to $(1, \infty)$. The function $t \mapsto te^{-t}$ is strictly decreasing for $t \geq 1$ (its derivative, $e^{-t} - te^{-t} = e^{-t}(1-t)$ is < 0 for $t > 1$), so it defines an order reversing bijection of $(1, \infty)$ with $(0, 1/e)$. Thus the function

$$x \mapsto \log\log(x)/\log(x)$$

defines an order reversing bijection of (e^e, ∞) with $(0, 1/e)$.

Because $\varepsilon/2 \leq 1/4 < 1/e$, there is a unique $s > e^e$ with $\log\log(s)/\log(s) = \varepsilon/2$. For any $x \geq s$, we have $\log\log(x)/\log(x) \leq \varepsilon/2$. We define $N_5(\varepsilon) := s$, and define $\xi := \log\log(s)$ (so $\xi e^{-\xi} = \varepsilon/2$).

We must show that if $x \geq s$, then we also have

$$1/\log(x) \leq \varepsilon/4 \quad \text{and} \quad 1/(2x+2) \leq \varepsilon/12.$$

To show that $1/\log(x) \leq \varepsilon/4$ if $x \geq s$, it suffices to show that $1/\log(s) \leq \varepsilon/4$, i.e., that $e^{-\xi} \leq \varepsilon/4$. Suppose not. Then $e^{-\xi} > \varepsilon/4$, and so

$$\varepsilon/2 = \xi e^{-\xi} > \xi\varepsilon/4,$$

which implies $2 > \xi$, in which case we must have

$$2e^{-2} < \xi e^{-\xi} = \varepsilon/2 \leq 1/4,$$

which implies $8 < e^2$, which is false (e^2 is $7.389\ldots$). This contradiction shows that if $x \geq s$, then $1/\log(x) \leq \varepsilon/4$.

We now show that if $x \geq s$, then $1/(2x+2) \leq \varepsilon/12$. Again, it suffices to show that $1/(2s+2) \leq \varepsilon/12$. Since we just showed that $1/\log(s) \leq \varepsilon/4$, it suffices to show that $1/(2s+2) \leq 1/3\log(s)$, which is equivalent to $2s+2 \geq 3\log(s)$. The function

$$x \mapsto 2x + 2 - 3\log(x)$$

is strictly increasing for $x > 3/2$, and already at $x = 2$ it takes the value $6 - 3\log(2) > 6 - 3 > 0$. As $s > e^e > 2$, we must have $2s + 2 - 3\log(s) > 0$. QED

3.4.6. With $\varepsilon > 0$ fixed, apply the lemma with $x = L + r + 1$ and $K = 32C\,\mathrm{Sqrt}(\alpha/\gamma)$. For $L = r + 1 \geq \mathrm{Sup}(N_4(\varepsilon, K), N_5(\varepsilon))$, which in turn forces $N = ((L+r)!)^{1+\gamma}$ to be impressively large, we get

$$\begin{aligned}
&(32C)^{L+r+2}\,\mathrm{Sqrt}(\log(M))^{L+r+2}\\
&\quad = (32C \times \mathrm{Sqrt}(\alpha/\gamma))^{L+r+2}\,\mathrm{Sqrt}(\log((L+r)!))^{L+r+2}\\
&\quad \leq ((L+r)!)^{\varepsilon}((L+r)!)^{(1+\varepsilon)/2} = N^{\gamma/2+3\gamma\varepsilon/2},
\end{aligned}$$

and thus

$$\begin{aligned}
&\int_{G(N)} \varphi\text{-discrep}(\mu, \mu(A, N))dA\\
&\quad \leq 5r/M + 27(32C)^{L+r+2}\,\mathrm{Sqrt}(\log(M))^{L+r+2}(M^r/\mathrm{Sqrt}(N) + 1/(L+r)!)\\
&\quad \leq 5rN^{-\alpha} + 27N^{\gamma/2+3\gamma\varepsilon/2}(N^{r\alpha-1/2} + N^{-\gamma})\\
&\quad = 5rN^{-\alpha} + 27N^{\gamma/2+3\gamma\varepsilon/2+r\alpha-1/2} + 27N^{-\gamma/2+3\gamma\varepsilon/2}.
\end{aligned}$$

Let us now equate the three different exponents to which N occurs:

$$-\alpha = (\gamma/2)(1+3\varepsilon) + r\alpha - 1/2 = (\gamma/2)(1-3\varepsilon).$$

Equating the last two and then doubling gives

$$2\gamma + 2r\alpha = 1.$$

Equating the first and last and then doubling gives $2\alpha = \gamma(1-3\varepsilon)$, so we find

$$2\gamma + r\gamma(1-3\varepsilon) = 1,$$

i.e.,

$$\begin{aligned}
\gamma &= 1/(2 + r - 3r\varepsilon),\\
\alpha &= \gamma(1-3\varepsilon)/2 = (1-3\varepsilon)/(4 + 2r - 6r\varepsilon).
\end{aligned}$$

Thus we find

$$\begin{aligned}
&\int_{G(N)} \varphi\text{-discrep}(\mu, \mu(A, N))\,dA\\
&\quad \leq 5rN^{-\alpha} + 27N^{\gamma/2+3\gamma\varepsilon/2+r\alpha-1/2} + 27N^{-\gamma/2+3\gamma\varepsilon/2}\\
&\quad \leq (5r+54)N^{-\alpha}
\end{aligned}$$

with

$$\alpha = (1 - 3\varepsilon)/(4 + 2r - 6r\varepsilon),$$

provided that N is sufficiently large.

3.4.7. We now make precise this argument. Given $\varepsilon > 0$, we will produce an explicit constant $N_1(\varepsilon, r, c, \eta, \kappa)$ such that for $N \geq N_1(\varepsilon, r, c, \eta, \kappa)$, we have

$$\int_{G(N)} \varphi\text{-discrep}(\mu, \mu(A, N))\, dA \leq N^{\varepsilon - 1/(2r+4)}.$$

We may and will suppose that $\varepsilon < 1/6$, since otherwise the statement is trivially true, and holds for all $N \geq 1$.

3.4.8. We define strictly positive real numbers α and γ by

$$\alpha = (1 - 3\varepsilon)/(4 + 2r - 6r\varepsilon),$$
$$\gamma = 1/(2 + r - 3r\varepsilon).$$

3.4.9. Now suppose that $N^\alpha \geq 4r$ and that $N^\gamma > (\Sigma(a) + r)!$. We define the grid size $M \geq 4r$ to be the integral part of N^α. Thus

$$(1/2)N^\alpha \leq N^\alpha - 1 < M \leq N^\alpha.$$

We define the cutoff $L \geq 1 + \Sigma(a)$ to be the largest positive integer L such that $N^\gamma > (L + r - 1)!$. Since $n \mapsto n!$ is strictly increasing on integers $n \geq 1$, we have

$$(L + r)! \geq N^\gamma > (L + r - 1)! = (L + r)!/(L + r) > N^\gamma/(L + r).$$

Using these inequalities, we infer that

$$\log(M) \leq \log(N^\alpha) = (\alpha/\gamma)\log(N^\gamma) \leq (\alpha/\gamma)\log((L + r)!),$$

and hence

$$\begin{aligned}&(32C)^{L+r+2}\,\mathrm{Sqrt}(\log(M))^{L+r+2}\\ &\quad \leq (32C \times \mathrm{Sqrt}(\alpha/\gamma))^{L+r+2}\,\mathrm{Sqrt}(\log((L + r)!))^{L+r+2}.\end{aligned}$$

Suppose $L + r + 1 \geq \mathrm{Sup}(N_4(\varepsilon, K), N_5(\varepsilon))$, for $K = 32C\,\mathrm{Sqrt}(\alpha/\gamma)$. Then we can continue this chain of inequalities:

$$\begin{aligned}&\leq ((L + r)!)^\varepsilon((L + r)!)^{(1+\varepsilon)/2} = ((L + r)!)^{(1+3\varepsilon)/2}\\ &= (L + r)^{(1+3\varepsilon)/2}((L + r - 1)!)^{(1+3\varepsilon)/2}\\ &\leq (L + r)^{(1+3\varepsilon)/2}N^{\gamma(1+3\varepsilon)/2}.\end{aligned}$$

We also record for use below the inequalities

$$1/M \leq 2N^{-\alpha}, \qquad 1/(L + r)! \leq N^{-\gamma}.$$

3.4.10. Thus from 3.2.29 we get

$$\begin{aligned}
&\int_{G(N)} \varphi\text{-discrep}(\mu, \mu(A, N))\, dA \\
&\quad \leq 5r/M + 27(32C)^{L+r+2} \operatorname{Sqrt}(\log(M))^{L+r+2}(M^r / \operatorname{Sqrt}(N) + 1/(L+r)!) \\
&\quad \leq 10rN^{-\alpha} + 27(L+r)^{(1+3\varepsilon)/2} N^{\gamma(1+3\varepsilon)/2}(N^{r\alpha-1/2} + N^{-\gamma}) \\
&\quad = 10rN^{-\alpha} + 27(L+r)^{(1+3\varepsilon)/2}(N^{-\alpha} + N^{-\alpha}),
\end{aligned}$$

the last equality by our calculated choice of α and γ.

Thus we have

$$\begin{aligned}
&\int_{G(N)} \varphi\text{-discrep}(\mu, \mu(A, N))\, dA \\
&\quad \leq (10r + 54(L+r)^{(1+3\varepsilon)/2})N^{-\alpha} \\
&\quad \leq (10r + 54(L+r))N^{-\alpha} \\
&\quad = (64r + 54L)N^{-\alpha},
\end{aligned}$$

provided that $N^\alpha \geq 2$ and that $N^\gamma > (\Sigma(a) + r)!$.

3.4.11. Suppose in addition that $L + r \geq e^2$. We use

$$N^\gamma > (L + r - 1)! = \Gamma(L + r)$$

and Stirling's formula to get

$$\begin{aligned}
\gamma \log(N) &> (L + r - 1/2)\log(L + r) - L - r + \log(\operatorname{Sqrt}(2\pi)) \\
&= (L + r - 1/2)(\log(L + r) - 1) + \log(\operatorname{Sqrt}(2\pi/e)) \\
&> (L + r - 1/2)(\log(L + r) - 1) \\
&\geq L + r - 1/2 > L,
\end{aligned}$$

and thus we have

$$\int_{G(N)} \varphi\text{-discrep}(\mu, \mu(A, N))\, dA \leq (64r + 54\gamma \log(N))N^{-\alpha}.$$

Finally, we choose N sufficiently large that

$$64r + 54\gamma \log(N) \leq N^{\varepsilon/2}.$$

Then we have

$$\int_{G(N)} \varphi\text{-discrep}(\mu, \mu(A, N))\, dA \leq N^{\varepsilon/2-\alpha}.$$

Returning to the definition of α, we see that

$$\begin{aligned}
\alpha - \varepsilon/2 &= (1 - 3\varepsilon)/(4 + 2r - 6r\varepsilon) - \varepsilon/2 \\
&> (1 - 3\varepsilon)/(4 + 2r) - \varepsilon/2 \\
&= 1/(4 + 2r) - 3\varepsilon/(4 + 2r) - \varepsilon/2 \\
&\geq 1/(4 + 2r) - 3\varepsilon/6 - \varepsilon/2 = 1/(4 + 2r) - \varepsilon,
\end{aligned}$$

and hence

$$\int_{G(N)} \varphi\text{-discrep}(\mu, \mu(A, N))\, dA \leq N^{\varepsilon-1/(2r+4)},$$

provided that N is sufficiently large.

3.5. Making explicit the constants

3.5.1. Let us now make explicit **how** large we must take N, i.e., let us calculate the constant $N_1(\varepsilon, r, c, \eta, \kappa)$. We first review the various constants $A, B, \Sigma(b)$, $\eta, \kappa, C, \varepsilon, \alpha, \gamma, K, N_4(\varepsilon, K), N_5(\varepsilon)$ which have arisen, and the constraints placed upon N.

3.5.2. $\mu(\text{univ, step } 1)(\{|x| > s\}) \leq Ae^{-Bs^2}$ for real $s \geq 0$, with $A \geq 1$, $0 < B \leq 1$ (in fact, we can take $A = 4/3, B = 1/8$).

3.5.3. μ is $\mu(\text{univ, steps } b)$ on $\mathbb{R}^r$, separation vector a and offset vector c.

3.5.4. $\varphi : \mathbb{R}^r \to \mathbb{R}^r$ is a homeomorphism of $\mathbb{R}^r$,

$$x = (x_1, \ldots, x_r) \mapsto (\varphi_1(x), \ldots, \varphi_r(x)),$$

which is "bi-bounded" in the sense that there exist strictly positive real constants κ and η such that $\eta \sum |x_i| \leq \sum |\varphi_i(x)| \leq \kappa \sum |x_i|$. We may, at the cost of increasing κ, assume that $\kappa/\eta \geq 1$. We assume that each $(\varphi_i)_*\mu$ has a continuous CDF.

3.5.5. $C := (1/\eta)\kappa r^2 \Sigma(b) \operatorname{Sqrt}((1 + \log(rA\Sigma(b)))/B)$, so $C \geq 1$.

3.5.6. $0 < \varepsilon < 1/6$.

3.5.7. $\alpha = (1 - 3\varepsilon)/(4 + 2r - 6r\varepsilon)$.

3.5.8. $\gamma = 1/(2 + r - 3r\varepsilon)$.

3.5.9. $\alpha/\gamma = (1 - 3\varepsilon)/2$ lies in $(1/4, 1/2)$.

3.5.10. $K = 32C \operatorname{Sqrt}(\alpha/\gamma)$; visibly $K > 1$.

3.5.11. $K^{x+1} \leq \Gamma(x)^\varepsilon$ for real $x \geq N_4(\varepsilon, K) = eK^{2/\varepsilon}$.

3.5.12. $(\log(\Gamma(x)))^{x+1} \leq \Gamma(x)^{1+\varepsilon}$ for real $x \geq N_5(\varepsilon)$.

3.5.13. $N_5(\varepsilon) =$ unique real $s > e^e$ with $\log\log(s)/\log(s) = \varepsilon/2$.

3.5.14. $N^\alpha \geq 2$ and $N^\gamma > (\Sigma(a) + r)! = \Sigma(b)!$.

3.5.15. $(1/2)N^\alpha \leq N^\alpha - 1 < M \leq N^\alpha$; this defines M.

3.5.16. $(L + r)! \geq N^\gamma > (L + r - 1)!$; this defines L.

3.5.17. $L + r + 1 \geq \operatorname{Sup}(N_4(\varepsilon, K), N_5(\varepsilon))$; this is equivalent to

$$\Gamma(L + r + 1) \geq \Gamma(\operatorname{Sup}(N_4(\varepsilon, K), N_5(\varepsilon)))$$

and is implied by

$$N^\gamma \geq \Gamma(\operatorname{Sup}(N_4(\varepsilon, K), N_5(\varepsilon))).$$

3.5.18. $L + r \geq e^2$; equivalent to

$$\Gamma(L + r + 1) \geq \Gamma(e^2 + 1),$$

implied by

$$N^\gamma \geq \Gamma(e^2 + 1).$$

3.5.19. $64r + 54\gamma \log(N) \leq N^{\varepsilon/2}$.

[In the discussion of $L + r + 1$ and of $L + r$, we use the well known fact that $\Gamma(s)$ is strictly increasing on $[2, \infty)$, and so defines an order preserving bijection of $[2, \infty)$ with $[1, \infty)$. This is equivalent to the fact that $\log \Gamma$ is strictly increasing on $[2, \infty)$. It holds because the derivative of $\log \Gamma$ is > 0 on $[2, \infty)$. One sees this positivity from Gauss's integral formula [**W-W**, page 247] for

$$\Psi(s) := (\log \Gamma)'(s) = \int_{(0,\infty)} ((e^{-t}/t) - (e^{-ts}/(1 - e^{-t})))\, dt,$$

in which the integrand is > 0 for $s \geq 2$. The positivity of the integrand is the assertion that for $s \geq 2$ and $t > 0$, we have

$$1 - e^{-t} > te^{-t(s-1)},$$

or equivalently that for $u \geq 1$ and $t > 0$ we have

$$1 - e^{-t} - te^{-tu} > 0.$$

View $u \geq 1$ as fixed. The function $f(t) := 1 - e^{-t} - te^{-tu}$ vanishes at $t = 0$, and is strictly increasing for $t \geq 0$, as its derivative at $t > 0$ is

$$e^{-t} - e^{-tu} + tue^{-tu} \geq tue^{-tu} > 0.$$

Therefore $f(t) > 0$ for $t > 0$, as required.]

3.5.20. We now look more closely at the last condition,

$$64r + 54\gamma \log(N) \leq N^{\varepsilon/2}.$$

Put $x := \log(N^{\varepsilon/2})$ and write this condition as

$$128r\varepsilon + 216\gamma x \leq 2\varepsilon e^{x},$$

which is implied by

$$128r \leq e^{x} \text{ and } 216\gamma x \leq \varepsilon e^{x}.$$

Since $x \leq e^{x/2}$ for $x \geq \log(4)$, these will both be satisfied if $N^{\varepsilon/2} \geq 4$ and if

$$128r \leq N^{\varepsilon/2} \text{ and } 216\gamma/\varepsilon \leq N^{\varepsilon/4}.$$

3.5.21. So the estimate

$$\int_{G(N)} \varphi\text{-discrep}(\mu, \mu(A, N))\, dA \leq N^{\varepsilon - 1/(2r+4)},$$

holds provided N is strictly larger than each of the following six quantities:

$$2^{1/\alpha},$$
$$(\Sigma(b)!)^{1/\gamma},$$
$$(\Gamma(\mathrm{Sup}(N_4(\varepsilon, K), N_5(\varepsilon))))^{1/\gamma},$$
$$(\Gamma(e^2 + 1))^{1/\gamma},$$
$$(128r)^{2/\varepsilon},$$
$$(216\gamma/\varepsilon)^{4/\varepsilon},$$

and we may take $N_1(\varepsilon, r, c, \eta, \kappa)$ in Theorem 3.1.6 to be the sup of these six explicit though gigantic quantities.

CHAPTER 4

Test Functions

4.0. The classes $\mathcal{T}(n)$ and $\mathcal{T}_0(n)$ of test functions

4.0.1. For each integer $n \geq 2$, we denote by $\mathcal{T}(n)$ the $\mathbb{R}$-vector space consisting of all bounded, Borel measurable, $\mathbb{R}$-valued functions F on $\mathbb{R}^n$ which satisfy the following two conditions:

1) F is Σ_n-invariant,

2) F is invariant under additive translation by the diagonal vector $\Delta_n(t) := (t, t, \ldots, t)$, for any t in $\mathbb{R}$.

4.0.2. We denote by $\mathcal{T}_0(n)$ the subspace of $\mathcal{T}(n)$ consisting of those functions F in $\mathcal{T}(n)$ which satisfy the additional condition:

3) F has "compact support modulo the diagonal" in the sense that there exists a real $\alpha \geq 0$ such that $F(X) = 0$ if $\mathrm{Sup}_{i,j} |x(i) - x(j)| > \alpha$. [If F satisfies this condition with a given α, we write $\mathrm{supp}(F) \leq \alpha$.]

4.0.3. Given F in $\mathcal{T}(n)$, we denote by $\|F\|_{\mathrm{sup}}$ its sup norm as function on $\mathbb{R}^n$. Under pointwise multiplication of functions, $\mathcal{T}(n)$ is a ring. For each real $\alpha \geq 0$, $\{F \text{ in } \mathcal{T}_0(n) \text{ with } \mathrm{supp}(F) \leq \alpha\}$ is an ideal, say $I(\alpha)$, in $\mathcal{T}(n)$. For $\alpha \leq \beta$, we have $I(\alpha) \subset I(\beta)$, and $\mathcal{T}_0(n) = \bigcup_{\alpha \geq 0} I(\alpha)$. We also have $I(\alpha) = \bigcap_{\beta > \alpha} I(\beta)$.

4.0.4. The motivation for introducing the class $\mathcal{T}_0(n)$ is given by the following lemma.

Lemma 4.0.5. *Fix an integer $r \geq 1$, a separation vector $a \geq 0$ in $\mathbb{Z}^r$, and an integer $k \geq \Sigma(a)$. Let f be a bounded, Borel measurable, $\mathbb{R}$-valued function on $\mathbb{R}^r$ [respectively, which in addition is of compact support, and $\alpha \geq 0$ a real number such that f is supported in the compact set $\{s \text{ in } \mathbb{R}^r \text{ with } \sum_i |s(i)| \leq \alpha\}$]. Then*

1) The function F on $\mathbb{R}^{r+\Sigma(a)+1}$ defined by

$$F(X) := \mathrm{Clump}(a, f, r + \Sigma(a) + 1, X)$$

lies in $\mathcal{T}(r + \Sigma(a) + 1)$ [resp. in $\mathcal{T}_0(r + \Sigma(a) + 1)$], and satisfies

$$\|F\|_{\mathrm{sup}} \leq \|f\|_{\mathrm{sup}}$$

$$[\textit{resp., and } \mathrm{supp}(F) \leq \alpha].$$

2) The function G on $\mathbb{R}^{r+k+1}$ defined by

$$G(X) := \mathrm{TClump}(k, a, f, r + k + 1, X)$$

lies in $\mathcal{T}(r + k + 1)$ [resp. in $\mathcal{T}_0(r + k + 1)$], and satisfies

$$\|G\|_{\mathrm{sup}} \leq \mathrm{Binom}(r + k - 1, r + \Sigma(a) - 1)\|f\|_{\mathrm{sup}}$$

$$[\textit{resp., and } \mathrm{supp}(G) \leq \alpha].$$

PROOF. Since $\mathrm{Clump}(a, f, r + \Sigma(a) + 1, X)$ is symmetric in X by definition, assertion 1) is just a restatement of Lemma 2.5.11. For assertion 2), we use the definition

$$\begin{aligned}&\mathrm{TClump}(k, a, f, r + k + 1, X)\\&\quad := \sum_{n \geq a, \Sigma(n) = k} \mathrm{Binom}(n, a)\, \mathrm{Clump}(n, f, r + k + 1, X).\end{aligned}$$

By 1) applied to each term $\mathrm{Clump}(n, f, r + k + 1, X)$, each Clump term lies in $\mathcal{T}(r + k + 1)$, has sup norm $\leq \|f\|_{\mathrm{sup}}$ [resp. and has supp $\leq \alpha$]. So it remains only to prove

Sublemma 4.0.6. *Fix an integer $r \geq 1$, a separation vector $a \geq 0$ in $\mathbb{Z}^r$, and an integer $k \geq \Sigma(a)$. Then we have*

$$\sum_{n \geq a, \Sigma(n) = k} \mathrm{Binom}(n, a) = \mathrm{Binom}(r + k - 1, r + \Sigma(a) - 1).$$

PROOF. For $r = 1$, there is nothing to prove. For $r \geq 2$, we argue as follows. Recall first that for a fixed integer $l \geq 1$, the series $(1 - T)^{-l}$ is the generating series for the number of monomials of degree d in l variables. Thus

$$(-1)^d \,\mathrm{Binom}(-l, d) = \mathrm{Binom}(l - 1 + d, l - 1)$$

is the number of monomials of degree d in l variables. We restate this as saying that, for l, d integers ≥ 0, $\mathrm{Binom}(l + d, l)$ is the number of monomials of degree d in $l + 1$ variables.

Now consider r sets of distinct independent variables, the i'th set consisting of $a(i) + 1$ variables. We will count the number of monomials in these $\Sigma(a) + r$ variables which are homogeneous of degree D. Any such monomial is multi-homogeneous, i.e., it is homogeneous of some degree $\delta(i)$ in the variables from the i'th set, and the degrees $\delta(i)$ are subject only to the condition that each $\delta(i)$ is a nonnegative integer, and that $\sum_i \delta(i) = D$. Thus $\delta := (\delta(1), \ldots, \delta(r))$ is a separation vector, and $\Sigma(\delta) = D$. Now how many monomials in all the variables are multi-homogeneous of multi-degree δ? This number is

$$\begin{aligned}&\prod_i (\text{number of monomials of degree } \delta(i) \text{ in } a(i) + 1 \text{ variables})\\&\quad = \prod_i \mathrm{Binom}(a(i) + \delta(i), a(i)) := \mathrm{Binom}(a + \delta, a).\end{aligned}$$

So if we break up the monomials of degree D in $\Sigma(a) + r$ variables according to their multi-degrees, we get

$$\begin{aligned}&\mathrm{Binom}(D + \Sigma(a) + r - 1, \Sigma(a) + r - 1)\\&\quad = \text{number of monomials of degree } D \text{ in } \Sigma(a) + r \text{ variables}\\&\quad = \sum_{\delta \geq 0 \text{ with } \Sigma(\delta) = D} \mathrm{Binom}(a + \delta, a).\end{aligned}$$

Taking $D = k - \Sigma(a)$, $n = a + \delta$ gives the asserted identity. QED

4.1. The random variable $Z[n, F, G(N)]$ on $G(N)$ attached to a function F in $\mathcal{T}(n)$

4.1.1. Given an integer $n \geq 2$, a function F in $\mathcal{T}(n)$, and an integer $N \geq 1$, recall (2.5.14) that we denote by $F[n, N]$ the function on $\mathbb{R}^N$ defined by

$$F[n, N] : X \mapsto \sum_{\mathrm{Card}(T)=n} F(\mathrm{pr}(T)(X)),$$

the sum over all subsets T of $\{1, \ldots, N\}$ with $\mathrm{Card}(T) = n$. Thus $F[n, N]$ vanishes for $N < n$, and $F[n, n]$ is F itself.

4.1.2. It is immediate from the definitions that if F lies in $\mathcal{T}(n)$, then $F[n, N]$ lies in $\mathcal{T}(N)$. However, even if F lies in $\mathcal{T}_0(n)$, $F[n, N]$ does not, in general, lie in $\mathcal{T}_0(N)$.

4.1.3. Given F in $\mathcal{T}(n)$ as above, and an integer $N \geq 1$, for $G(N)$ any of $U(N), SO(2N+1), USp(2N), SO(2N), O_-(2N+2)$, and A in $G(N)$, we will define a function $Z[n, F, G(N)]$ on $G(N)$ as follows, cf. the definitions of Int and Cor in 2.7.2–5. Given A in $G(N)$, we denote by $X(A)$ in $\mathbb{R}^N$ the N-tuple of its angles of eigenvalues, cf. 2.7.2–5, and we define:
if $G(N) = U(N)$,

$$\begin{aligned} Z[n, F, G(N)](A) &:= (1/N) F[n, N]((N/2\pi) X(A)) \\ &= (1/N) \sum_{\mathrm{Card}(T)=n} F((N/2\pi)\, \mathrm{pr}(T)(X(A))), \end{aligned}$$

if $G(N) = USp(2N)$ or $SO(2N)$,

$$\begin{aligned} Z[n, F, G(N)](A) &:= (1/N) F[n, N]((N/\pi) X(A)) \\ &= (1/N) \sum_{\mathrm{Card}(T)=n} F((N/\pi)\, \mathrm{pr}(T)(X(A))), \end{aligned}$$

if $G(N) = SO(2N+1)$,

$$\begin{aligned} Z[n, F, G(N)](A) &:= (1/(N+1/2)) F[n, N](((N+1/2)/\pi) X(A)) \\ &:= (1/(N+1/2)) \sum_{\mathrm{Card}(T)=n} F(((N+1/2)/\pi)\, \mathrm{pr}(T)(X(A))), \end{aligned}$$

if $G(N) = O_-(2N+2)$,

$$\begin{aligned} Z[n, F, G(N)](A) &:= (1/(N+1)) F[n, N](((N+1)/\pi) X(A)) \\ &:= (1/(N+1)) \sum_{\mathrm{Card}(T)=n} F(((N+1)/\pi)\, \mathrm{pr}(T)(X(A))). \end{aligned}$$

4.1.4. It is obvious from these definitions and from Lemma 2.5.14 that $Z[n, F, G(N)]$ is a bounded, Borel measurable function on $G(N)$. For the reader's convenience, we record the relation of the random variable $Z[n, F, G(N)]$ to the Cor and TCor functions of 2.7.

Lemma 4.1.5. *Fix an integer $r \geq 1$, a separation vector a in $\mathbb{Z}^r$, an integer $k \geq \Sigma(a)$, and a bounded $\mathbb{R}$-valued Borel measurable function f on $\mathbb{R}^r$.*

1) *Denote by H in $\mathcal{T}(r + \Sigma(a) + 1)$ the function*

$$H(X) := \mathrm{Clump}(a, f, r + \Sigma(a) + 1, X).$$

Then for every $N \geq 2$ and every X in $\mathbb{R}^N$ we have

$$H[r + \Sigma(a) + 1, N](X) = \mathrm{Clump}(a, f, N, X),$$

and for every A in $G(N)$ we have

$$Z[r + \Sigma(a) + 1, H, G(N)](A) = \mathrm{Cor}(a, f, G(N), A).$$

2) *Denote by F in $\mathcal{T}(r + k + 1)$ the function*

$$F(X) := \mathrm{TClump}(k, a, f, r + k + 1, X).$$

Then for every $N \geq 2$ and every X in $\mathbb{R}^N$ we have

$$F[r + k + 1, N](X) = \mathrm{TClump}(k, a, f, N, X),$$

and for every A in $G(N)$ we have

$$Z[r + k + 1, F, G(N)](A) = \mathrm{TCor}(k, a, f, G(N), A).$$

PROOF. Immediate from 2.5.15, 2.6.3, and the definitions. QED

Remark 4.1.6. Although we have defined $Z[n, F, G(N)]$ for F in $\mathcal{T}(n)$, it is only for F in $\mathcal{T}_0(n)$ that we will be able to say much of interest.

4.2. Estimates for the expectation $E(Z[n, F, G(N)])$ and variance $\mathrm{Var}(Z[n, F, G(N)])$ of $Z[n, F, G(N)]$ on $G(N)$

4.2.1. For any bounded Borel measurable $\mathbb{R}$-valued function f on $G(N)$, $G(N)$ any of $U(N), SO(2N+1), USp(2N), SO(2N), O_-(2N+2)$, we denote

$$E(f) := \int_{G(N)} f \, d\,\mathrm{Haar}, \qquad \mathrm{Var}(f) := \int_{G(N)} f^2 \, d\,\mathrm{Haar} - \left(\int_{G(N)} f \, d\,\mathrm{Haar} \right)^2.$$

The key estimate we need is the following:

Theorem 4.2.2. *Let $n \geq 2$ be an integer, F in $\mathcal{T}_0(n)$, $\alpha \geq 0$ a real number such that $\mathrm{supp}(F) \leq \alpha$.*

i) *There exists a real number $E(n, F, \text{ univ})$ such that for any $N \geq 2$ and any $G(N)$, we have*

$$|E(Z[n, F, G(N)]) - E(n, F, \text{ univ})| \\ \leq \|F\|_{\mathrm{sup}} (8\alpha)^{n-1} ((\pi\alpha)^2 + \alpha + 1 + 10 \log(N))/N.$$

ii) *For any $N \geq 2$ and any $G(N)$ we have*

$$|E(Z[n, F, G(N)])| \leq \|F\|_{\mathrm{sup}} 2(2\alpha)^{n-1}/(n-1)!.$$

iii) *For any $N \geq 2$ and any $G(N)$ we have*

$$|\mathrm{Var}(Z[n, F, G(N)])| \leq (3(8\alpha)^{n-1} + 65(8\alpha)^{2n-2})(\|F\|_{\mathrm{sup}})^2/N.$$

Proposition 4.2.3. *Suppose that Theorem 4.2.2 holds. Then conditions* i) *and* ii) *of Proposition 3.0.1 hold, and condition* iii) *of Proposition 3.1.9 holds. In particular, parts* i) *and* ii) *of Theorem 4.2.2 imply Theorem 2.1.3.*

PROOF. In order to get the desired conclusions for the data k, a, f, and $G(N)$, simply apply Theorem 4.2.2 to $n = r + k + 1$ and the function

$$F(X) := \mathrm{TClump}(k, a, f, r + k + 1, X)$$

in $\mathcal{T}_0(n)$, making use of the estimates 4.0.5 for $\|F\|_{\mathrm{sup}}$ and $\mathrm{supp}(F)$, and of the compatibility 4.1.5. QED

Corollary 4.2.4. *Suppose that Theorem* 4.2.2 *holds, and that the following condition* iv) *holds.*

iv) (*estimate for the tail of the most classical spacing measure*) *There exist explicit real constants* $A > 0$ *and* $B > 0$ *such that the limit measure*

$$\mu(\textit{naive, univ, sep. } 0)$$

on $\mathbb{R}$ *satisfies*

$$\mu(\textit{naive, univ, sep. } 0)(\{|x| > s\}) \leq Ae^{-Bs^2} \text{ for every real } s \geq 0.$$

Then Theorem 3.1.6, *the* φ*-version of* 2.1.5, *holds.*

Proof. In view of the previous result, this is just a restatement of 3.1.9. QED

CHAPTER 5

Haar Measure

5.0. The Weyl integration formula for the various $G(N)$

5.0.1. We give first a case by case account of the explicit shape the Weyl integration formula takes for each of the $G(N)$. The version of the formula which we need, especially in the non-$U(N)$ case, is precisely the one given in [**Weyl**, pages 197 (7.4B), 218 (7.8B), 224 (9.7) and 226 (9.15)]. In all but the $O_-(2N+2)$ case, this formula is, to a modern reader, a straightforward deciphering of the "intrinsic" one given in [**Bour-L9**, §6, No. 2, Cor. 1]. The $O_-(2N+2)$ formula seems to have been all but forgotten in modern times.

5.0.2. For A in $G(N)$, we denote by $X(A)$ in $\mathbb{R}^N$ its vector of eigenvalue angles, cf. 2.7.2–5. Thus for A in $U(N)$, $X(A)$ lies in $[0, 2\pi)^N$, while in the other cases, $X(A)$ lies in $[0, \pi]^N$.

5.0.3. The $U(N)$ case [**Weyl**, p. 197 (7.4B)]. An element A in $U(N)$ is determined up to conjugacy by its vector of angles $X(A)$. Bounded, Borel measurable $\mathbb{R}$-valued central functions g on $U(N)$ are in one-one correspondence with bounded, Borel measurable $\mathbb{R}$-valued functions $\tilde{g}$ on $[0, 2\pi)^N$ which are Σ_N-invariant, via $g(A) = \tilde{g}(X(A))$. We denote by $\mu(U(N))$ the measure on $[0, 2\pi)^N$ (with coordinates $x(i)$, $i = 1, \ldots, N$) given by

$$\mu(U(N)) := (1/N!) \left(\prod_{j<k} |e^{ix(j)} - e^{ix(k)}|^2 \right) \prod_i d(x(i)/2\pi).$$

The Weyl integration formula asserts that for g a central function on $U(N)$, corresponding to a Σ_N-invariant $\tilde{g}$ on $[0, 2\pi)^N$, we have

$$\int_{U(N)} g \, d\,\mathrm{Haar} = \int_{[0,2\pi)^N} \tilde{g} \, d\mu(U(N)).$$

5.0.4. The $USp(2N)$ case [**Weyl**, p. 218 (7.8B)]. An element A in $USp(2N)$ is determined up to conjugacy by its vector of angles $X(A)$. Bounded, Borel measurable $\mathbb{R}$-valued central functions g on $USp(2N)$ are in one-one correspondence with bounded, Borel measurable $\mathbb{R}$-valued functions $\tilde{g}$ on $[0, \pi]^N$ which are Σ_N-invariant, via $g(A) = \tilde{g}(X(A))$. We denote by $\mu(USp(2N))$ the measure on $[0, \pi]^N$ (with coordinates $x(i)$, $i = 1, \ldots, N$) given by

$$\mu(USp(2N))$$
$$:= (1/N!) \left(\prod_{i<j} (2\cos(x(i)) - 2\cos(x(j))) \right)^2 \prod_i ((2/\pi) \sin^2(x(i)) \, dx(i)).$$

The Weyl integration formula asserts that for g a central function on $USp(2N)$, corresponding to a Σ_N-invariant $\tilde{g}$ on $[0,\pi]^N$, we have

$$\int_{USp(2N)} g\, d\,\text{Haar} = \int_{[0,\pi]^N} \tilde{g}\, d\mu(USp(2N)).$$

5.0.5. The $SO(2N+1)$ case [Weyl, p. 224 (9.7)]. An element A in $SO(2N+1)$ is determined up to conjugacy by its vector of angles $X(A)$. Bounded, Borel measurable $\mathbb{R}$-valued central functions g on $SO(2N+1)$ are in one-one correspondence with bounded, Borel measurable $\mathbb{R}$-valued functions $\tilde{g}$ on $[0,\pi]^N$ which are Σ_N-invariant, via $g(A) = \tilde{g}(X(A))$. We denote by $\mu(SO(2N+1))$ the measure on $[0,\pi]^N$ (with coordinates $x(i)$, $i = 1, \ldots, N$) given by

$$\mu(SO(2N+1))$$
$$:= (1/N!)\left(\prod_{i<j}([1+2\cos(x(i))]-[1+2\cos(x(j))])\right)^2 \prod_i((2/\pi)\sin^2(x(i)/2)\,dx(i)).$$

The Weyl integration formula asserts that for g a central function on $SO(2N+1)$, corresponding to a Σ_N-invariant $\tilde{g}$ on $[0,\pi]^N$, we have

$$\int_{SO(2N+1)} g\, d\,\text{Haar} = \int_{[0,\pi]^N} \tilde{g}\, d\mu(SO(2N+1)).$$

5.0.6. The $SO(2N)$ case [Weyl, p. 228 (9.15)]. An element A in $SO(2N)$ is determined up to conjugation by elements in the ambient group $O(2N)$ by its vector of angles $X(A)$. Bounded, Borel measurable $\mathbb{R}$-valued $O(2N)$-central (i.e., invariant by $O(2N)$ conjugation) functions g on $SO(2N)$ are in one-one correspondence with bounded, Borel measurable $\mathbb{R}$-valued functions $\tilde{g}$ on $[0,\pi]^N$ which are Σ_N-invariant, via $g(A) = \tilde{g}(X(A))$. We denote by $\mu(SO(2N))$ the measure on $[0,\pi]^N$ (with coordinates $x(i)$, $i = 1, \ldots, N$) given by

$$\mu(SO(2N)) := (2/N!)\left(\prod_{i<j}(2\cos(x(i)) - 2\cos(x(j)))\right)^2 \prod_i((1/2\pi)\,dx(i)).$$

The Weyl integration formula asserts that for g an $O(2N)$-central function on $SO(2N)$, corresponding to a Σ_N-invariant $\tilde{g}$ on $[0,\pi]^N$, we have

$$\int_{SO(2N)} g\, d\,\text{Haar} = \int_{[0,\pi]^N} \tilde{g}\, d\mu(SO(2N)).$$

5.0.7. The $O_-(2N+2)$ case [Weyl, p. 228 (9.15)]. An element A in $O_-(2N+2)$ is determined up to $O(2N+2)$-conjugation by its vector of angles $X(A)$. Bounded, Borel measurable $\mathbb{R}$-valued $O(2N+2)$-central functions g on $O_-(2N+2)$ are in one-one correspondence with bounded, Borel measurable $\mathbb{R}$-valued functions $\tilde{g}$ on $[0,\pi]^N$ which are Σ_N-invariant, via $g(A) = \tilde{g}(X(A))$. We denote by $\mu(O_-(2N+2))$ the measure on $[0,\pi]^N$ (with coordinates $x(i)$, $i = 1, \ldots, N$) given by

$$\mu(O_-(2N+2)) := \mu(USp(2N))$$
$$:= (1/N!)\left(\prod_{i<j}(2\cos(x(i)) - 2\cos(x(j)))\right)^2 \prod_i((2/\pi)\sin^2(x(i))\,dx(i)).$$

The Weyl integration formula asserts that for g an $O(2N+2)$-central function on $O_-(2N+2)$, corresponding to a Σ_N-invariant $\tilde{g}$ on $[0,\pi]^N$, we have

$$\int_{O_-(2N+2)} g\, d\,\mathrm{Haar} = \int_{[0,\pi]^N} \tilde{g}\, d\mu(O_-(2N+2)).$$

Remarks 5.0.8. 1) In all the cases, the measure $\mu(G(N))$ is visibly Σ_N-invariant. So for any (bounded, Borel measurable) function f on $[0,2\pi)^N$ in the $U(N)$ case or on $[0,\pi]^N$ in the non-$U(N)$ cases, both f and its Σ_N-symmetrization have the same integral against $\mu(U(N))$.

2) In the case of either $SO(2N)$ or $O_-(2N)$, the normalized Haar measure is, by its uniqueness, necessarily invariant under $O(2N)$-conjugation. Therefore any (bounded, Borel measurable) function f on $SO(2N)$ or $O_-(2N)$ has the same integral against Haar measure as its $O(2N)$-centralization, the function

$$x \mapsto \int_{O(2N)} f(gxg^{-1})\, d\,\mathrm{Haar}(g).$$

5.1. The $K_N(x,y)$ version of the Weyl integration formula

5.1.1. In this section, we give another expression for the measure $\mu(G(N))$, for $G(N)$ each of $U(N), USp(2N), SO(2N+1)$, and $SO(2N)$, which shows how it is built up from the $N=1$ case. [Since $\mu(O_-(2N+2)) = \mu(USp(2N))$, we do not discuss the $O_-(2N+2)$ case separately.] This version of the Weyl integration formula, which we learned from [**Mehta**] in the $U(N)$ case, is what allows us to do effective calculations of expectation and variance for the functions $Z[n,F,G(N)]$.

5.1.2. We first recall the Vandermonde determinant. Given an integer $N \geq 1$, and N elements $f(1),\ldots,f(N)$ in a commutative ring R, the Vandermonde determinant Vandermonde$(f(1),\ldots,f(N))$ is the $N \times N$ determinant whose (i,j) entry is $f(i)^{j-1}$, $1 \leq i,j \leq N$. One has

$$\mathrm{Vandermonde}(f(1),\ldots,f(N)) = \prod_{i<j}(f(j)-f(i)).$$

Key Lemma 5.1.3 (compare [**Mehta**, 5.2.1]). *Suppose that (T,μ) is a measure space with a positive measure μ of finite total mass. Let f be a bounded measurable $\mathbb{C}$-valued function on T with*

$$\int_T f\, d\mu = 0, \qquad \int_T |f|^2\, d\mu = 1.$$

Suppose that for every integer $n \geq 1$, there exists a monic polynomial in one variable $P_n(X)$ in $\mathbb{C}[X]$ of degree n such that the sequence of functions $\{\varphi_n\}_{n\geq 0}$ on T defined by

$$\varphi_0 := 1/\mathrm{Sqrt}(\mu(T)),$$
$$\varphi_n := P_n(f) \quad \text{for } n \geq 1,$$

is an orthonormal sequence:

$$\int_T \varphi_i \overline{\varphi}_j\, d\mu = \delta_{i,j} \quad \text{for all } i,j \geq 0.$$

For any integer $N \geq 1$, consider the N-fold product T^N. For each $i=1,\ldots,N$, denote by

$$\mathrm{pr}[i] : T^N \to T, \qquad t \mapsto t(i),$$

the i'th projection. Denote by $K_N(x,y)$ *the function on* T^2

$$K_N(x,y) := \sum_{n=0}^{N-1} \varphi_n(x)\overline{\varphi}_n(y).$$

For each integer $n \geq 1$*, we define*

$$D(n,N) := \det_{n\times n}(K_N(t(i),t(j))),$$

a function on T^n*. We define*

$$D(0,N) := 1,$$

viewed as function on the one point space T^0*.*

1) *We have the identity of functions on* T^N

$$(1/\operatorname{Sqrt}(\mu(T)))\operatorname{Vandermonde}(f(t(1)),\ldots,f(t(N))) = \det_{N\times N}(\varphi_{i-1}(t(j))).$$

2) *We have the identity of functions on* T^N

$$(1/\mu(T))|\operatorname{Vandermonde}(f(t(1)),\ldots,f(t(N)))|^2 = D(N,N).$$

3) *For* $1 \leq n \leq N$*, we have the identity of functions on* T^{n-1}

$$\int_T D(n,N)(t(1),\ldots,t(n))\,d\mu(t(n)) = (N+1-n)D(n-1,N).$$

4) *For* $1 \leq n \leq N$*, the function* $D(n,N)$ *on* T^n *is* $\mathbb{R}$*-valued, nonnegative, and symmetric, i.e.,* Σ_n*-invariant. For* $n > N$*, the function* $D(n,N)$ *vanishes identically.*

5) *For* $n \geq 1$*, let* F *be a (bounded, measurable,* $\mathbb{R}$*-valued) function on* T^n*, and for any* $N \geq 1$ *denote by* $F[n,N]$ *the function on* T^N *defined by*

$$F[n,N](t(1),\ldots,t(N)) := \sum_{1\leq i(1)<i(2)<\cdots<i(n)\leq N} F(t(i(1)),\ldots,t(i(n))).$$

Denote by $\mu(n,N)$ *the measure on* T^n *defined by*

$$\mu(n,N) := (1/n!)D(n,N)\,d\mu_1\cdots d\mu_n.$$

The measure $\mu(n,N)$ *is invariant by* Σ_n*, and we have the integration formula*

$$\int_{T^N} F[n,N]\,d\mu(N,N) = \int_{T^n} F\,d\mu(n,N).$$

PROOF. 1) For $n \geq 1$, $\varphi_n := P_n(f) = f^n +$ lower terms. This means that we can pass from $\det_{N\times N}(\varphi_{i-1}(t(j)))$ to the $N{\times}N$ determinant whose i'th row for $i \geq 2$ is $(f(t(j))^{i-1})_j$ and whose first row is $(\varphi_0(t(j)))_j = (1/\operatorname{Sqrt}(\mu(T)))(1,1,\ldots,1)$, by elementary row operations. Equating determinants, we get the assertion.

2) Taking the square absolute value of 1), we get

$$\begin{aligned}
&(1/\mu(T))|\operatorname{Vandermonde}(f(t(1)),\ldots,f(t(N)))|^2 \\
&\quad = |\det_{N\times N}(\varphi_{i-1}(t(j)))|^2 = \det_{N\times N}(\varphi_{i-1}(t(j))) \times \det_{N\times N}(\overline{\varphi}_{i-1}(t(j))) \\
&\quad = \det_{N\times N}((i,j)\mapsto \varphi_{j-1}(t(i))) \times \det_{N\times N}(\overline{\varphi}_{i-1}(t(j))) \\
&\quad = \det_{N\times N}((i,j)\mapsto \sum_k \varphi_{k-1}(t(i))\overline{\varphi}_{k-1}(t(j))) := D(N,N).
\end{aligned}$$

3) From the orthonormality of the φ_n's and the definition of K_N, we get the integration formulas

$$\int_T K_N(t,t)\,d\mu(t) = N,$$

$$\int_T K_N(x,t)K_N(t,y)\,d\mu(t) = K_N(x,y).$$

The first of these is precisely the $n=1$ case of the assertion.

For $n>1$, we expand $D(n,N)$ by its n'th column:

$$D(n,N) = \sum_{k=1}^{n}(-1)^{k+n}K_N(t(k),t(n))\,\mathrm{Cofactor}(k,n).$$

The term with $k = n$ is $K_N(t(n),t(n))D(n-1,N)$, which integrates to give $N \times D(n-1,N)$.

It remains to see that for each of the $n-1$ values of k from 1 to $n-1$, the term $(-1)^{k+n}K_N(t(k),t(n)) \times \mathrm{Cofactor}(k,n)$ integrates to give $-D(n-1,N)$. For each such k, we expand $\mathrm{Cofactor}(k,n)$ by its n'th row:

$$\mathrm{Cofactor}(k,n) = \sum_{l=1}^{n-1}(-1)^{n-1+l}K_N(t(n),t(l)) \times \mathrm{Cofactor}(\{n,k\},\{l,n\}),$$

where

$$\mathrm{Cofactor}(\{n,k\},\{l,n\}) := \text{ the } (n,l)\text{-cofactor of } \mathrm{Cofactor}(k,n)$$

is the $n-2 \times n-2$ matrix obtained by removing the indicated rows and columns. So we obtain

$$\begin{aligned}&(-1)^{k+n}K_N(t(k),t(n)) \times \mathrm{Cofactor}(k,n)\\ &\quad= \sum_{l=1}^{n-1}(-1)^{k-1+l}K_N(t(n),t(l))K_N(t(k),t(n))\,\mathrm{Cofactor}(\{n,k\},\{l,n\}).\end{aligned}$$

The term $\mathrm{Cofactor}(\{n,k\},\{l,n\})$ is just the (k,l) cofactor of $D(n-1,N)$ (itself the (n,n)-cofactor of $D(n,N)$), and $K_N(t(n),t(l))K_N(t(k),t(n))$ integrates to give $K_N(t(k),t(l))$. So after integration, we get

$$= \sum_{l=1}^{n-1}(-1)^{k-1+l}K_N(t(k),t(l)) \times (\text{the } (k,l) \text{ cofactor of } D(n-1,N)),$$

which is precisely (-1) times (the expansion by minors along the k'th row of) $D(n-1,N)$.

4) In view of the integration formula 3), it suffices, to treat the case $n \le N$, to show that $D(N,N)$ is real, nonnegative, and symmetric in its N variables. This is obvious from 2), since the Vandermonde determinant transforms under Σ_N by the sign character, and hence its square absolute value is symmetric, as well as real and nonnegative.

To show that $D(n,N)$ vanishes for $n > N$, think of $D(n,N)$ as the $n \times n$ determinant made from $K_N(x,y)$. Introduce functions ψ_i on T for $i = 0,\dots,n-1$ by defining

$$\psi_i := \varphi_i \quad \text{for } 0 \le i \le N-1, \quad \text{and}$$
$$\psi_i := 0 \quad \text{for } N \le i \le n-1.$$

Then it is trivially true that

$$K_N(x,y) = \sum_{i=0}^{n-1} \psi_i(x)\overline{\psi}_i(y),$$

and hence (cf. the proof of 2)) that

$$\det_{n\times n}(K_N(x(i),x(j))) = |\det_{n\times n}(\psi_{i-1}(x(j)))|^2.$$

But as $n > N$, the last row of this last determinant is identically zero.

5) The Σ_n-invariance of $\mu(n,N)$ is obvious from 4). If $n > N$, both $F[n,N]$ and $\mu(n,N)$ vanish identically, so the integration formula is true but nugatory. For $n = N$, there is nothing to prove. Suppose now that $1 \le n < N$. Consider the integral

$$\begin{aligned} &\int_{T^N} F[n,N] \times D(N,N)\,d\mu_1 \cdots d\mu_N \\ &\quad = \sum_{1\le i(1)<i(2)<\cdots<i(n)\le N} \int_{T^N} F(t(i(1)),\ldots,t(i(n)))D(N,N)\,d\mu_1\cdots d\mu_N. \end{aligned}$$

By symmetry of $D(N,N)$ under Σ_N, each summand is equal to

$$\int_{T^N} F(t(1),\ldots,t(n))D(N,N)\,d\mu_1\cdots\,d\mu_N.$$

Using 3) successively to integrate out the variables $t(N), t(N-1),\ldots,t(n+1)$, we get

$$\begin{aligned} &\int_{T^N} F(t(1),\ldots,t(n))D(N,N)\,d\mu_1\cdots d\mu_N \\ &\quad = (1)(2)\cdots(N-n)\int_{T^n} F(t(1),\ldots,t(n))D(n,N)\,d\mu_1\cdots d\mu_n \\ &\quad = (N-n)!\int_{T^n} F\times D(n,N)\,d\mu_1\cdots d\mu_n. \end{aligned}$$

Since there are $\mathrm{Binom}(N,n)$ summands, we get

$$\begin{aligned} &(1/N!)\int_{T^N} F[n,N]\times D(N,N)\,d\mu_1\cdots d\mu_N \\ &\quad = (1/N!)\,\mathrm{Binom}(N,n)(N-n)!\int_{T^n} F\times D(n,N)\,d\mu_1\cdots d\mu_n \\ &\quad = (1/n!)\int_{T^n} F\times D(n,N)\,d\mu_1\cdots d\mu_n. \quad \text{QED} \end{aligned}$$

Remark 5.1.3.1. In part 5), if F is symmetric on T^n, then $F[n,N]$ is symmetric on T^N. In nearly all applications, the input function F will in fact be symmetric, but this symmetry is not needed for the validity of part 5).

5.1.4. We now explain how to apply this lemma to rewrite the Weyl measure $\mu(G(N))$ as being the measure $\mu(N,N)$ on T^N for a suitable situation of the type (T,μ,f) considered in the lemma. We proceed case by case.

5.1.5. The $U(N)$ case. The group $G(1) = U(1)$ is abelian, so $U(1)$ is its own space of conjugacy classes. We take for T this space of conjugacy classes, viewed not so canonically as being $[0, 2\pi)$, endowed with the normalized Haar measure $\mu := dx/2\pi$. For the function f, we take the function $f(x) := e^{ix}$, the character of the standard representation of $U(1)$. The powers f^n, n in $\mathbb{Z}$, are orthonormal, so we may take $P_n(X)$ to be X^n. Since μ has total mass 1, we have $\varphi_n = f^n$ for all $n \geq 0$. For $N \geq 1$, the function $K_N(x,y)$ is

$$K_N(x,y) := \sum_{n=0}^{N-1} e^{in(x-y)}.$$

The measure $\mu(U(N))$ on $T^N = [0, 2\pi)^N$ is equal to

$$\begin{aligned}\mu(U(N)) &:= (1/N!)\left(\prod_{j<k} |e^{ix(j)} - e^{ix(k)}|^2\right)\prod_i d(x(i)/2\pi)\\ &= (1/N!)|\operatorname{Vandermonde}(f(x(1)), \ldots, f(x(N)))|^2 \prod_i d(x(i)/2\pi)\\ &= (1/N!)\det_{N\times N}(K_N(x(i), x(j)))\prod_i d(x(i)/2\pi),\end{aligned}$$

which is the measure $\mu(N,N)$ attached to the data

$$(T = [0, 2\pi), \mu = dx/2\pi, f(x) = e^{ix}).$$

The measure $\mu(n,N)$ attached to this data is

$$\mu(n,N) = (1/n!)\det_{n\times n}(K_N(x(i), x(j)))\prod_i d(x(i)/2\pi).$$

5.1.6. The $USp(2N)$ case. The group $G(1) = USp(2) = SU(2)$ has as its space of conjugacy classes the space $T := [0, \pi]$, and the Weyl measure on T is $(2/\pi)\sin^2(x)\,dx$. We take for f the function

$$f(x) := 2\cos(x) = \sin(2x)/\sin(x),$$

which is the character of the standard 2-dimensional representation of $SU(2)$. The group $SU(2)$ has a single irreducible representation of each degree $n \geq 1$ (namely Sym^{n-1} of the standard representation), whose character is

$$\sin(nx)/\sin(x) = \delta_{n,\text{ odd}} + \sum_{\substack{1\leq k\leq n-1\\ k\equiv n-1 \bmod 2}} (e^{ikx} + e^{-ikx}).$$

This formula makes it obvious that $\sin(nx)/\sin(x)$ is a monic $\mathbb{Z}$-polynomial of degree $n-1$ in the quantity $e^{ix} + e^{-ix} = 2\cos(x)$. By the Peter-Weyl theorem (or by trigonometry), we know that the functions $\sin(nx)/\sin(x)$ are orthonormal on $[0, \pi]$ for the measure $\mu = (2/\pi)\sin^2(x)\,dx$. So we have

$$\varphi_n = \sin((n+1)x)/\sin(x) \quad \text{for all } n \geq 0.$$

For $N \geq 1$, the function $K_N(x,y)$ is therefore

$$K_N(x,y) := (1/\sin(x)\sin(y))\sum_{n=1}^{N} \sin(nx)\sin(ny).$$

The measure $\mu(USp(2N))$ on $T^N = [0,\pi]^N$ is equal to

$$
\begin{aligned}
&:= (1/N!)\left(\prod_{i<j}(2\cos(x(i)) - 2\cos(x(j)))\right)^2 \prod_i((2/\pi)\sin^2(x(i))\,dx(i)) \\
&= (1/N!)|\,\mathrm{Vandermonde}(f(x(1)),\ldots,f(x(N)))|^2 \prod_i((2/\pi)\sin^2(x(i))\,dx(i)) \\
&= (1/N!)\det_{N\times N}(K_N(x(i),x(j)))\prod_i((2/\pi)\sin^2(x(i))\,dx(i)),
\end{aligned}
$$

which is the measure $\mu(N,N)$ attached to the data

$$(T = [0,\pi], \mu = (2/\pi)\sin^2(x)\,dx, f(x) = 2\cos(x)).$$

The measure $\mu(n,N)$ attached to this data is

$$\mu(n,N) = (1/n!)\det_{n\times n}(K_N(x(i),x(j)))\prod_i((2/\pi)\sin^2(x(i))\,dx(i)).$$

5.1.7. The $SO(2N+1)$ case. The group $G(1) = SO(3)$ has as its space of conjugacy classes the space $T := [0,\pi]$, and the Weyl measure on T is

$$(2/\pi)\sin^2(x/2)\,dx.$$

We take for f the function

$$f(x) := 1 + 2\cos(x) = \sin(3x/2)/\sin(x/2),$$

which is the character of the standard 3-dimensional representation of $SO(3)$. The group $SO(3)$ has a single irreducible representation of each odd degree $2n+1$ (namely Sym^{2n} of the standard representation of $SU(2)$, viewed as a representation of $SU(2)/(\pm 1) = SO(3)$), whose character is

$$\sin((2n+1)x/2)/\sin(x/2) = 1 + \sum_{1\le k\le n}(e^{ikx} + e^{-ikx}).$$

This formula makes it obvious that $\sin((2n+1)x/2)/\sin(x/2)$ is a monic $\mathbb{Z}$-polynomial of degree n in the quantity

$$1 + e^{ix} + e^{-ix} = 1 + 2\cos(x) = \sin(3x/2)/\sin(x/2).$$

By the Peter-Weyl theorem (or by trigonometry), we know that the functions $\{\sin((2n+1)x/2)/\sin(x/2)\}_{n\ge 0}$ are orthonormal on $[0,\pi]$ for the measure $\mu = (2/\pi)\sin^2(x/2)\,dx$. So we have

$$\varphi_n = \sin((2n+1)x/2)/\sin(x/2) \quad \text{for all } n \ge 0.$$

For $N \ge 1$, the function $K_N(x,y)$ is therefore

$$K_N(x,y) = (1/\sin(x/2)\sin(y/2))\sum_{n=0}^{N-1}\sin((2n+1)x/2)\sin((2n+1)y/2).$$

The measure $\mu(SO(2N+1))$ on $T^N = [0,\pi]^N$ is equal to

$$\begin{aligned}
&(1/N!)\left(\prod_{i<j}([1+2\cos(x(i))]-[1+2\cos(x(j))])\right)^2 \prod_i((2/\pi)\sin^2(x(i)/2)\,dx(i))\\
&= (1/N!)|\,\mathrm{Vandermonde}(f(x(1)),\ldots,f(x(N)))|^2\prod_i((2/\pi)\sin^2(x(i)/2)\,dx(i))\\
&= (1/N!)\det_{N\times N}(K_N(x(i),x(j)))\prod_i((2/\pi)\sin^2(x(i)/2)\,dx(i)),
\end{aligned}$$

which is the measure $\mu(N,N)$ attached to the data

$$(T=[0,\pi], \mu = (2/\pi)\sin^2(x/2)\,dx, f(x) = 1+2\cos(x)).$$

The measure $\mu(n,N)$ attached to this data is

$$\mu(n,N) = (1/n!)\det_{n\times n}(K_N(x(i),x(j)))\prod_i((2/\pi)\sin^2(x(i)/2)\,dx(i)).$$

5.1.8. The $SO(2N)$ case. The group $G(1) = SO(2)$ is the abelian group $U(1)$. The conjugation action of $O(2)/SO(2) = \pm 1$ on $SO(2)$ is inversion. The quotient of $SO(2)$ by $O(2)$-conjugation is the space $T := [0,\pi]$. We take for μ the measure $dx/2\pi$ on T, which has total mass $1/2$, and is **half** of the direct image of normalized Haar measure from $SO(2)$. We take for f the function

$$f(x) := 2\cos(x) = e^{ix}+e^{-ix},$$

which is the character of the standard 2-dimensional representation of $SO(2)$. For every $n \geq 1$, we define

$$f_n(x) := 2\cos(nx) = e^{inx}+e^{-inx}.$$

Each f_n is the character of a representation V_n of $SO(2)$ which is the sum of two inequivalent irreducible representations, and for $n \neq m$ the representations V_n and V_m have no common constituent. Intrinsically, the V_n for $n \geq 1$ are precisely the restrictions to $SO(2)$ of the nontrivial irreducible representations of $O(2)$.

It is obvious that $f_n(x)$ is a monic $\mathbb{Z}$-polynomial of degree n in the quantity

$$f(x) = e^{ix}+e^{-ix} = 2\cos(x).$$

By the Peter-Weyl theorem (or by trigonometry), we know that the functions $\{2\cos(nx)\}_{n\geq 1}$ are orthonormal on $[0,\pi]$ for the measure $\mu = dx/2\pi$. [It was to insure this ortho**normality** that we chose μ as we did.] So we have

$$\varphi_n = 2\cos(nx) \quad \text{for all } n \geq 1.$$

But precisely because the measure μ has total mass $1/2$, we have

$$\varphi_0(x) = \mathrm{Sqrt}(2).$$

For $N \geq 1$, the function $K_N(x,y)$ is therefore

$$K_N(x,y) = 2 + 4\sum_{n=1}^{N-1}\cos(nx)\cos(ny).$$

The measure $\mu(SO(2N))$ on $T^N = [0, \pi]^N$ is equal to

$$\begin{aligned}(2/N!)&\left(\prod_{i<j}(2\cos(x(i)) - 2\cos(x(j)))\right)^2 \prod_i((1/2\pi)\,dx(i))\\ &= (2/N!)|\operatorname{Vandermonde}(f(x(1)), \ldots, f(x(N)))|^2 \prod_i((1/2\pi)\,dx(i))\\ &= (1/N!)\det_{N\times N}(K_N(x(i), x(j))) \prod_i((1/2\pi)\,dx(i)),\end{aligned}$$

which is the measure $\mu(N, N)$ attached to the data

$$(T = [0, \pi], \mu = dx/2\pi, f(x) = 2\cos(x)).$$

The measure $\mu(n, N)$ attached to this data is

$$\mu(n, N) = (1/n!)\det_{n\times n}(K_N(x(i), x(j))) \prod_i((1/2\pi)\,dx(i)).$$

5.2. The $L_N(x, y)$ rewriting of the Weyl integration formula

5.2.1. In this section, we record the "euclidean" version of the formulas of the last section, i.e., we write the measure $\mu(n, N)$ on $[0, 2\pi)^n$ or $[0, \pi]^n$ as an explicit "density", given by a determinant, times normalized (total mass 1) Lebesgue measure.

5.2.2. The $U(N)$ case. In this case, there is nothing to change. We define

$$L_N(x, y) := K_N(x, y) = \sum_{n=0}^{N-1} e^{in(x-y)}.$$

The measure $\mu(n, N)$ on $[0, 2\pi)^n$ is

$$\mu(n, N) = (1/n!)\det_{n\times n}(L_N(x(i), x(j))) \prod_i(dx(i)/2\pi).$$

5.2.3. The $USp(2N)$ (and $O_-(2N+2)$) case. In this case, we define

$$\begin{aligned}L_N(x, y) &:= 2\sin(x)\sin(y)K_N(x, y)\\ &= \sum_{n=1}^{N} 2\sin(nx)\sin(ny).\end{aligned}$$

The measure $\mu(n, N)$ on $[0, \pi]^n$ is

$$\mu(n, N) = (1/n!)\det_{n\times n}(L_N(x(i), x(j))) \prod_i(dx(i)/\pi).$$

5.2.4. The $SO(2N+1)$ case. In this case, we define

$$\begin{aligned}L_N(x, y) &:= 2\sin(x/2)\sin(y/2)K_N(x, y)\\ &= \sum_{n=0}^{N-1} 2\sin((2n+1)x/2)\sin((2n+1)y/2).\end{aligned}$$

The measure $\mu(n, N)$ on $[0, \pi]^n$ is

$$\mu(n, N) = (1/n!)\det_{n\times n}(L_N(x(i), x(j))) \prod_i(dx(i)/\pi).$$

5.2.5. The $SO(2N)$ case. In this case, we define

$$L_N(x,y) := (1/2)K_N(x,y)$$
$$= 1 + \sum_{n=1}^{N-1} 2\cos(nx)\cos(ny).$$

The measure $\mu(n,N)$ on $[0,\pi]^n$ is

$$\mu(n,N) = (1/n!)\det_{n\times n}(L_N(x(i),x(j)))\prod_i (dx(i)/\pi).$$

5.3. Estimates for $L_N(x,y)$

Lemma 5.3.1. *For every (x,y) in $\mathbb{R}^2$, and every $N \geq 1$, we have the estimate*

$|L_N(x,y)| \leq N$ *in the $U(N)$ case,*

$|L_N(x,y)| \leq 2N$ *in the $USp(2N)$, $SO(2N+1)$ and $SO(2N)$ cases.*

PROOF. Obvious from the expression of L_N as a sum of trig functions. QED

Lemma 5.3.2. *For every $N \geq 1$, $L_N(x,y)$ is a periodic of period 4π in each variable separately, and viewed as a function on the probability space $[0,4\pi]^2$ with normalized Lebesgue measure $dx\,dy/16\pi^2$ it has L_2 norm $=$ Sqrt(N).*

PROOF. The periodicity is obvious from the expression of L_N as a sum of trig functions (we need 4π for the $SO(2N+1)$ case), as is the L_2 estimate. QED

Lemma 5.3.3. *For every $1 \leq n \leq N$ and every $(x(1),\ldots,x(n))$ in $\mathbb{R}^n$ we have the estimate*

$|\det_{n\times n}(L_N(x(i),x(j)))| \leq N^n$ *in the $U(N)$ case,*

$|\det_{n\times n}(L_N(x(i),x(j)))| \leq (2N)^n$ *in the $USp(2N)$, $SO(2N+1)$ and $SO(2N)$ cases.*

PROOF. Interpret $L_n(x(i),x(j))$ as the standard Hermitian inner product $\langle v(i),v(j)\rangle$ of vectors in $\mathbb{C}^N$, where $v(j)$ in $\mathbb{C}^N$ is

for $U(N)$: $(1, e^{ix(j)}, e^{2ix(j)}, \ldots, e^{i(N-1)x(j)})$,

for $USp(2N)$: $\sqrt{2}(\sin(x(j)), \sin(2x(j)), \ldots, \sin(Nx(j)))$,

for $SO(2N+1)$: $\sqrt{2}(\sin(x(j)/2), \sin(3x(j)/2), \ldots, \sin((2N-1)x(j)/2))$,

for $SO(2N)$: $(1, \sqrt{2}\cos(x(j)), \sqrt{2}\cos(2x(j)), \ldots, \sqrt{2}\cos((N-1)x(j)))$.

Since each vector $v(j)$ has

$\|v(j)\|^2 \leq N$ in the $U(N)$ case,

$\|v(j)\|^2 \leq 2N$ in the $USp(2N)$, $SO(2N+1)$ and $SO(2N)$ cases,

our assertion amounts to the well-known Hadamard inequality:

Lemma 5.3.4. *Given $n \geq 1$ vectors $v(1),\ldots,v(n)$ in a Hilbert space, we have the inequality*

$$|\det_{n\times n}(\langle v(i),v(j)\rangle)| \leq \prod_i \|v(i)\|^2.$$

PROOF. The truth of the asserted inequality is invariant under scaling the vectors $v(i)$ by strictly positive real constants β_i. Indeed, under such scaling the (i, j) entry is multiplied by $\beta_i\beta_j$, so each term in the full Σ_n expansion of the determinant is multiplied by $\prod_i \beta_i\beta_{\sigma(i)} = (\prod_i \beta_i)^2$. This allows us to reduce to the case where each vector $v(i)$ is either zero or has $\|v(i)\| = 1$. If any $v(i)$ is zero, the assertion is obvious, since both sides vanish. So we may assume that each $\|v(i)\| = 1$, and we must prove that

$$|\det_{n\times n}(\langle v(i), v(j)\rangle)| \leq 1.$$

If the vectors $v(i)$ are linearly dependent, say $\sum_i \alpha_i v(i) = 0$, then for every $j, \sum_i \alpha_i\langle v(i), v(j)\rangle = 0$, so the determinant vanishes, and there is nothing to prove. If the vectors $v(i)$ are linearly independent, then the $n\times n$ matrix $A := (\langle v(i), v(j)\rangle)$ is the matrix of a positive definite Hermitian form (namely the Hilbert space inner product) on the n-dimensional space spanned by the $v(i)$, expressed in that basis. Therefore the n eigenvalues $\lambda_1, \ldots, \lambda_n$ of A are real and positive. Therefore $|\det A| = \prod_i \lambda_i$. By the inequality between the geometric and arithmetic mean, we have

$$|\det_{n\times n}(\langle v(i), v(j)\rangle)|^{1/n} = |\det A|^{1/n} = \left(\prod_i \lambda_i\right)^{1/n} \leq (1/n)\sum_i \lambda_i$$
$$= (1/n)\operatorname{Trace}(A) = (1/n)\sum_i \langle v(i), v(i)\rangle = (1/n)\sum_i \|v(i)\|^2 = 1. \quad \text{QED}$$

5.4. The $L_N(x, y)$ determinants in terms of the sine ratios $S_N(x)$

5.4.1. It will be convenient to adapt the following notation. For x real and $N \geq 1$ an integer, we define

$$S_N(x) := \sin(Nx/2)/\sin(x/2), \tag{5.4.2}$$

a Laurent polynomial in the quantity $e^{ix/2}$ with $\mathbb{Z}$ coefficients.

5.4.3. The $U(N)$ case. For each $n \geq 1$, we have

$$\det_{n\times n}(L_N(x(i), x(j))) = \det_{n\times n}(S_N(x(i) - x(j))).$$

To see this, recall that

$$L_N(x, y) := \sum_{n=0}^{N-1} e^{in(x-y)}$$

is the value at $z = x - y$ of

$$(e^{iNz} - 1)/(e^{iz} - 1) = [e^{iNz/2}\sin(Nz/2)]/[e^{iz/2}\sin(z/2)]$$
$$= e^{i(N-1)z/2}S_N(z).$$

Therefore the $n \times n$ matrix made from $L_N(x(i), x(j))$ is conjugate to that made from $S_N(x(i) - x(j))$ by the diagonal matrix whose j'th entry is $e^{i(N-1)x(j)/2}$.

Remark 5.4.3.1. This formula makes visible the fact that

$$\det_{n\times n}(L_N(x(i), x(j)))$$

is real, a fact we know a priori because the matrix is hermitian.

5.4.4. The $USp(2N)$ (and $O_-(2N+2)$) case. In this case, we have

$$L_N(x, y) = (1/2)(S_{2N+1}(x-y) - S_{2N+1}(x+y)).$$

To see this, recall that

$$L_N(x, y) := 2\sin(x)\sin(y)K_N(x, y) = \sum_{n=1}^{N} 2\sin(nx)\sin(ny).$$

Subtract the cosine addition identities

$$\cos(x+y) = \cos(x)\cos(y) - \sin(x)\sin(y),$$
$$\cos(x-y) = \cos(x)\cos(y) + \sin(x)\sin(y),$$

to get

$$2\sin(x)\sin(y) = \cos(x-y) - \cos(x+y).$$

Taking nx and ny in the above and summing over n, we get

$$L_N(x, y) = \sum_{n=1}^{N} \cos(n(x-y)) - \cos(n(x+y)).$$

The identity

$$\sin((2N+1)x)/\sin(x) = 1 + 2\sum_{n=1}^{N} \cos(2nx)$$

gives

$$(1/2)S_{2N+1}(x) = 1/2 + \sum_{n=1}^{N} \cos(nx),$$

so we get

$$L_N(x, y) = (1/2)(S_{2N+1}(x-y) - S_{2N+1}(x+y)),$$

as asserted.

5.4.5. The $SO(2N+1)$ case. In this case, we have

$$L_N(x, y) = (1/2)(S_{2N}(x-y) - S_{2N}(x+y)).$$

To see this, recall that

$$\begin{aligned} L_N(x, y) &:= 2\sin(x/2)\sin(y/2)K_N(x, y) \\ &= \sum_{n=0}^{N-1} 2\sin((2n+1)x/2)\sin((2n+1)y/2). \end{aligned}$$

Again using the cosine addition formula, we rewrite this as

$$= \sum_{n=0}^{N-1} \cos((2n+1)(x-y)/2) - \cos((2n+1)(x+y)/2).$$

The identity

$$\sin(2Nx)/\sin(x) = 2\sum_{n=0}^{N-1} \cos((2n+1)x)$$

gives

$$(1/2)S_{2N}(x) = \sum_{n=0}^{N-1} \cos((2n+1)x/2),$$

so we get

$$L_N(x,y) = (1/2)(S_{2N}(x-y) - S_{2N}(x+y)),$$

as asserted.

5.4.6. The $SO(2N)$ case. In this case, we have

$$L_N(x,y) = (1/2)(S_{2N-1}(x-y) + S_{2N+1}(x+y)).$$

To see this, recall that

$$L_N(x,y) := (1/2)K_N(x,y) = 1 + \sum_{n=1}^{N-1} 2\cos(nx)\cos(ny).$$

Using the cosine addition formula, we rewrite this as

$$= 1 + \sum_{n=1}^{N-1} \cos(n(x-y)) + \cos(n(x+y)).$$

Now use the identity (cf. the $USp(2N)$ case)

$$(1/2)S_{2N-1}(x) = 1/2 + \sum_{n=1}^{N} \cos(nx),$$

to get the asserted identity.

5.5. Case by case summary of explicit Weyl measure formulas via S_N

5.5.1. The $U(N)$ case. The measure $\mu(n,N)$ on $[0,2\pi)^n$ is

$$\mu(n,N) = (1/n!)\det_{n\times n}(S_N(x(i)-x(j)))\prod_i (dx(i)/2\pi).$$

5.5.2. The $USp(2N)$ and $O_-(2N+2)$ cases. The measure $\mu(n,N)$ on $[0,\pi]^n$ is

$$\mu(n,N) = (1/n!)\det_{n\times n}(L_N(x(i),x(j)))\prod_i (dx(i)/\pi),$$

where

$$L_N(x,y) = (1/2)(S_{2N+1}(x-y) - S_{2N+1}(x+y)).$$

5.5.3. The $SO(2N+1)$ case. The measure $\mu(n,N)$ on $[0,\pi]^n$ is

$$\mu(n,N) = (1/n!)\det_{n\times n}(L_N(x(i),x(j)))\prod_i (dx(i)/\pi),$$

where

$$L_N(x,y) = (1/2)(S_{2N}(x-y) - S_{2N}(x+y)).$$

5.5.4. The $SO(2N)$ case. The measure $\mu(n, N)$ on $[0, \pi]^n$ is

$$\mu(n, N) = (1/n!)\det_{n\times n}(L_N(x(i), x(j)))\prod_i (dx(i)/\pi),$$

where

$$L_N(x, y) = (1/2)(S_{2N-1}(x-y) + S_{2N-1}(x+y)).$$

Remark 5.5.5. We have proven (Lemma 5.1.3, part 5) that the measures $\mu(n, N)$ are Σ_n-invariant. On the other hand, this invariance is obvious from the explicit formulas, since for any $f(x, y)$, the determinant $\det_{n\times n}(f(x(i), x(j)))$ is Σ_n-invariant.

5.6. Unified summary of explicit Weyl measure formulas via S_N

5.6.1. In order to unify these formulas, we introduce the quantities $\lambda, \sigma, \rho, \tau$ and ε according to the following table:

$G(N)$	λ	σ	ρ	τ	ε
$U(N)$	0	2	1	0	0
$USp(2N)$	0	1	2	1	-1
$SO(2N+1)$	$\frac{1}{2}$	1	2	0	-1
$SO(2N)$	0	1	2	-1	1
$O_-(2N+2)$	1	1	2	1	-1

The measure $\mu(n, N)$ on $[0, \sigma\pi]^n$ is

$$(1/n!)\det_{n\times n}(L_N(x(i), x(j)))\prod_i (dx(i)/\sigma\pi)$$

with

$$L_N(x, y) = (\sigma/2)[S_{\rho N+\tau}(x-y) + \varepsilon S_{\rho N+\tau}(x+y)].$$

Remark 5.6.1.1. The attentive reader will have noticed that, in the $U(N)$ case, we have replaced $[0, 2\pi)^n$ by $[0, 2\pi]^n$. This change is harmless because the measure

$$(1/n!)\det_{n\times n}(L_N(x(i), x(j)))\prod_i (dx(i)/\sigma\pi)$$

on $[0, 2\pi]^n$ is absolutely continuous with respect to Lebesgue measure, so it gives the entire boundary measure zero.

5.6.2. With an eye to what will be useful later, we define

$$L_{N,-}(x, y) := (\sigma/2)S_{\rho N+\tau}(x-y) = (1/\rho)S_{\rho N+\tau}(x-y),$$
$$L_{N,+}(x, y) := (\varepsilon\sigma/2)S_{\rho N+\tau}(x+y).$$

Lemma 5.6.3. 1) *For any (x, y) in $\mathbb{R}^2$, and any integer $n \geq 1$, we have*

$$|L_{N,-}(x, y)| \leq (2N+\tau)/2 = N + \tau/2 \leq N+1,$$
$$|L_{N,+}(x, y)| \leq |\varepsilon|(2N+\tau)/2 = |\varepsilon|(N+\tau/2) \leq |\varepsilon|(N+1).$$

2) *For any x in $\mathbb{R}^n$, we have*

$$|\det_{n\times n}(L_N(x(i), x(j)))| \leq (\rho N)^n,$$
$$|\det_{n\times n}(L_{N,-}(x(i), x(j)))| \leq ((2N+\tau)/2)^n.$$

PROOF. Assertion 1) is just the fact, obvious from its expression as a trigonometric polynomial, that $|S_N(x)| \leq N$ for all real x. The first statement of 2) just repeats Lemma 5.3.1. The second statement of 2) for $U(N)$ is the same as the first, and the second statement for other $G(N)$ is the same as the first for $U(2N+\tau)$. QED

5.6.4. We denote by $\mu_-(n,N)$ the measure on $[0,\sigma\pi]^n$ given by

$$\mu_-(n,N) := (1/n!)\det_{n\times n}(L_{N,-}(x(i),x(j)))\prod_i(dx(i)/\sigma\pi)$$
$$= (1/n!)\det_{n\times n}(S_{\rho N+\tau}(x(i)-x(j)))\prod_i(dx(i)/2\pi).$$

This last expression shows that $\mu_-(n,N)$ for $G(N)$ is indeed a measure; namely, it is the restriction to $[0,\sigma\pi]^n$ of the measure $\mu(n,2N+\tau)$ from $U(2N+\tau)$ on $[0,2\pi]^n$. It also reminds us that in the $U(N)$ case, we have $\mu_-(n,N)=\mu(n,N)$.

5.7. Formulas for the expectation $E(Z[n,F,G(N)])$

5.7.1. Given an integer $n\geq 2$, a function F in $\mathcal{T}(n)$, and an integer $N\geq 1$, recall (from 2.5.14) that we denote by $F[n,N]$ the function on $\mathbb{R}^N$ defined by

$$F[n,N] : X \mapsto \sum_{\mathrm{Card}(T)=n} F(\mathrm{pr}(T)(X)),$$

the sum over all subsets T of $\{1,\ldots,N\}$ with $\mathrm{Card}(T)=n$. Recall (from 4.1.3) that $Z[n,F,G(N)]$ is the function on $G(N)$ defined by

$$Z[n,FG(N)](A) := (1/(N+\lambda))F[n,N](((N+\lambda)/\sigma\pi)X(A))$$
$$:= (1/(N+\lambda))\sum_{\mathrm{Card}(T)=n} F(((N+\lambda)/\sigma\pi)\,\mathrm{pr}(T)(X(A))).$$

Lemma 5.7.2. *Given an integer $n\geq 2$, a function F in $\mathcal{T}(n)$, and an integer $N\geq 1$, the expectation $E(Z[n,F,G(N)])$ is given by the integral*

$$E(Z[n,F,G(N)]) = (1/(N+\lambda))\int_{[0,\sigma\pi]^n} F((N+\lambda)x/\sigma\pi)\,d\mu(n,N).$$

PROOF. The function $A\mapsto Z[n,F,G(N)](A)$ is a symmetric function of $X(A)$, corresponding to the symmetric function

$$X \mapsto (1/(N+\lambda))\sum_{\mathrm{Card}(T)=n} F(((N+\lambda)/\sigma\pi)\,\mathrm{pr}(T)(X)),$$

so we may apply the Weyl integration formula to express it as the integral against $\mu(N,N)$ of this function. But this function is of the form $G[n,N]$, for G the symmetric function

$$X \mapsto (1/(N+\lambda))F(((N+\lambda)/\sigma\pi)X).$$

We then apply part 5) of the Key Lemma 5.1.3 to write the integral of $G[n,N]$ against $\mu(N,N)$ as the integral of G against $\mu(n,N)$. The "unified formula" is just a rewriting of this, except that in the $U(N)$ case the domain of integration is $[0,2\pi]^n$ instead of $[0,2\pi)^n$. But as already remarked in 5.6.1.1 above, the measure $\mu(n,N)$ is absolutely continuous with respect to Lebesgue measure, so it is the same to integrate over $[0,2\pi]^n$ as over $[0,2\pi)^n$. QED

5.8. Upper bound for $E(Z[n, F, G(N)])$

5.8.1. We now show that part ii) of Theorem 4.2.2 holds.

Proposition 5.8.2. *Let $n \geq 2$ be an integer, F in $\mathcal{T}_0(n)$, $\alpha \geq 0$ a real number such that* $\mathrm{supp}(F) \leq \alpha$. *For any $N \geq 2$ and any $G(N)$ we have*

$$|E(Z[n, F, G(N)])| \leq \|F\|_{\sup}(2/\sigma)(2\alpha/\sigma)^{n-1}/(n-1)!$$
$$\leq \|F\|_{\sup} 2(2\alpha)^{n-1}/(n-1)!.$$

PROOF. We may assume $n \leq N$, since if not the function $Z[n, F, G(N)]$ vanishes. By the previous lemma, we have

$$E(Z[n, F, G(N)]) = (1/(N+\lambda)) \int_{[0,\sigma\pi]^n} F((N+\lambda)x/\sigma\pi)\, d\mu(n, N).$$

By scaling, we may assume $\|F\|_{\sup} = 1$. Among all such F with $\mathrm{supp}(F) \leq \alpha$, the expectation is largest when F is the characteristic function of the "α-neighborhood of the diagonal" set

$$\Delta(n, \alpha) := \{x \text{ in } \mathbb{R}^n \text{ with } \operatorname*{Sup}_{i,j} |x(i) - x(j)| \leq \alpha\}.$$

Then $F((N+\lambda)x/\sigma\pi)$ is the characteristic function of the set $\Delta(n, \alpha\sigma\pi/(N+\lambda))$.

Thanks to 5.2 and 5.3.3, the measure $\mu(n, N)$ is dominated by $(1/n!)(2N/\sigma)^n$ times normalized Lebesgue measure on $[0, \sigma\pi]^n$. So we have

$$|E(Z[n, F, G(N)])| = \left|(1/(N+\lambda)) \int_{[0,\sigma\pi]^n} F((N+\lambda)x/\sigma\pi)\, d\mu(n, N)\right|$$
$$\leq (1/n!)(2N/\sigma)^n(1/(N+\lambda)) \int_{[0,\sigma\pi]^n} F((N+\lambda)x/\sigma\pi) \prod_i (dx(i)/\sigma\pi)$$
$$= (1/n!)(2N/\sigma)^n(1/(N+\lambda)) \int_{[0,1]^n} F((N+\lambda)x) \prod_i dx(i)$$
$$= (1/n!)(2N/\sigma)^n(1/(N+\lambda))\, \mathrm{Vol}(\Delta(n, \alpha/(N+\lambda)) \cap [0,1]^n)$$
$$\leq (1/n!)(2N/\sigma)^n(1/N)\, \mathrm{Vol}(\Delta(n, \alpha/N) \cap [0,1]^n)$$
$$= (1/n!)(2/\sigma)^n(N)^{n-1}\, \mathrm{Vol}(\Delta(n, \alpha/N) \cap [0,1]^n).$$

At this point we need the following

Lemma 5.8.3. *For $n \geq 2$ and any real $\alpha \geq 0$, we have*

$$(1/n!)\, \mathrm{Vol}(\Delta(n, \alpha) \cap [0,1]^n) \leq \alpha^{n-1}/(n-1)!.$$

PROOF. The region $\Delta(n, \alpha) \cap [0,1]^n$ is stable by the symmetric group Σ_n, so $(1/n!)\, \mathrm{Vol}(\Delta(n, \alpha) \cap [0,1]^n)$ is the volume of the region in $\mathbb{R}^n$ defined by the inequalities

$$0 \leq x(1) \leq x(2) \leq \cdots \leq x(n) \leq 1,$$
$$x(n) - x(1) \leq \alpha.$$

This region lies in the region

$$0 \leq x(1) \leq 1,$$
$$0 \leq x(1) \leq x(2) \leq \cdots \leq x(n),$$
$$x(n) - x(1) \leq \alpha.$$

By the unimodular change of coordinates

$$y(0) = x(1),$$
$$y(i) = x(i+1) - x(1) \quad \text{for } i = 1, \ldots, n-1,$$

this last region is the product

$$[0,1] \times \{y \text{ in } \mathbb{R}^{n-1} \text{ with } 0 \le y(1) \le \cdots \le y(n-1) \le \alpha\},$$

whose volume in $\mathbb{R}^n$ is that of the region

$$\{y \text{ in } \mathbb{R}^{n-1} \text{ with } 0 \le y(1) \le \cdots \le y(n-1) \le \alpha\}$$

in $\mathbb{R}^{n-1}$. This last region is a fundamental domain for the action of Σ_{n-1} on the α-cube $[0,\alpha]^{n-1}$, so has volume $\alpha^{n-1}/(n-1)!$. QED

5.8.4. Using Lemma 5.8.3, applied with α/N, we get

$$(1/n!)(2/\sigma)^n(N)^{n-1}\operatorname{Vol}(\Delta(n,\alpha/N)\cap[0,1]^n)$$
$$\le (2/\sigma)^n(N)^{n-1}(\alpha/N)^{n-1}/(n-1)! = (2/\sigma)^n\alpha^{n-1}/(n-1)!,$$

which completes the proof of Proposition 5.8.2. QED

5.9. Interlude: The $\sin(\pi x)/\pi x$ kernel and its approximations

5.9.1. For an integer $n \ge 2$, and x in $\mathbb{R}^n$, we define

$$W(n)(x(1),\ldots,x(n)) := \det_{n\times n}(\sin(\pi(x(i)-x(j)))/\pi(x(i)-x(j))).$$

Lemma 5.9.2. 1) *For x in $\mathbb{R}$, we have* $|\sin(\pi x)/\pi x| \le 1$.
2) *For $n \ge 2$ and x in $\mathbb{R}^n$, we have* $|W(n)(x(1),\ldots,x(n))| \le 1$.

PROOF. For any fixed x in $\mathbb{R}$, we have the limit formula

$$\sin(\pi x)/\pi x = \lim_{N\to\infty} \sin(\pi x)/N\sin(\pi x/N)$$
$$= \lim_{N\to\infty} (1/N)S_N(2\pi x/N).$$

So we have the limit formula

$$W(n)(x(1),\ldots,x(n)) = \lim_{N\to\infty} \det_{n\times n}((1/N)S_N((2\pi/N)(x(i)-x(j)))).$$

So 1) results from the estimate $|S_N(x)| \le N$, and 2) results from the $U(N)$ case of 5.3.3 (via 5.4.3), by passage to the limit. QED

Lemma 5.9.3. *For x real with $|x| \le 1$, we have*

$$|(\sin(x)/x) - 1| \le x^2/5.$$

PROOF. Expanding $\sin(x)/x$ in power series, we get

$$|(\sin(x)/x) - 1| = \left|\sum_{n\ge1}(-1)^n x^{2n}/(2n+1)!\right| \le \sum_{n\ge1} x^{2n}/(2n+1)!.$$

For $|x| \le 1$, each term x^{2n} is bounded by x^2, so

$$|(\sin(x)/x) - 1| \le x^2\left(\sum_{n\ge1} 1/(2n+1)!\right),$$

and

$$\begin{aligned}\sum_{n\geq 1} 1/(2n+1)! &= e - 1 - 1 - \sum_{n\geq 1} 1/(2n)! \leq e - 2 - 1/2! - 1/4! \\ &\leq 2.72 - 2 - 1/2 - 1/24 \leq 1/5. \quad \text{QED}\end{aligned}$$

Lemma 5.9.3.1. *For x real and $M > 0$ real, we have*

$$|1 - (x/M\sin(x/M))| \leq (1/4)(x/M)^2 \quad \textit{if } |x| \leq M.$$

PROOF. Changing variable, this becomes

$$|1 - (x/\sin(x))| \leq x^2/4 \quad \text{if } |x| \leq 1.$$

By 5.9.3, we know that

$$|(\sin(x)/x) - 1| \leq x^2/5 \quad \text{if } |x| \leq 1.$$

Fix x, $|x| \leq 1$, and put $A := \sin(x)/x$. Then $|A - 1| \leq 1/5$, so $|A| \geq 4/5$, and

$$|1 - A^{-1}| = |(A-1)|/|A| \leq |A-1|/(4/5) \leq (x^2/5)/(4/5) = x^2/4. \quad \text{QED}$$

Lemma 5.9.4. *For x real and $M > 0$ real, we have*

$$|\sin(x)/x - \sin(x)/M\sin(x/M)| \leq (x/M)^2/4 \quad \textit{if } |x| \leq M.$$

PROOF. We have $|\sin(x)| \leq |x|$ for all real x, and

$$|\sin(x)/x - \sin(x)/M\sin(x/M)| = |\sin(x)/x|\,|1 - (x/M\sin(x/M))|.$$

Now apply the previous Lemma 5.9.3.1. QED

Lemma 5.9.5. *For x real, δ real and $M > 0$ real, if $|x| \leq M$ we have*

$$|\sin(x)/x - \sin((1+\delta)x)/M\sin(x/M)| \leq |\delta| + (1+|\delta|)(x/M)^2/4.$$

PROOF. We suppose $|x| \leq M$. We write the quantity to be estimated as

$$\begin{aligned}&|\sin(x)/x - \sin(x)/M\sin(x/M) + (\sin(x) - \sin((1+\delta)x))/M\sin(x/M)| \\ &\quad \leq |\sin(x)/x - \sin(x)/M\sin(x/M)| + |(\sin(x) - \sin((1+\delta)x))/M\sin(x/M)|.\end{aligned}$$

The first term is bounded by $(x/M)^2/4$, by 5.9.4 above. We use the mean value theorem and then 5.9.3 to estimate the second:

$$\begin{aligned}|(\sin(x) - \sin((1+\delta)x))/M\sin(x/M)| &\leq |\delta x/M\sin(x/M)| \\ &= |\delta + \delta(x/M\sin(x/M) - 1)| \leq |\delta| + |\delta|(x/M)^2/4. \quad \text{QED}\end{aligned}$$

Corollary 5.9.6. *For any integer $N \geq 1$ and any $G(N)$, we have:*
1) *For real x with $|\pi x| \leq N$,*

$$\begin{aligned}&|\sin(\pi x)/\pi x - (N+\lambda)^{-1}(\sigma/2)S_{\rho N+\tau}((N+\lambda)^{-1}\sigma\pi x)| \\ &\quad \leq 1/2N + (1+1/2N)(\pi x/2N)^2.\end{aligned}$$

2) *For all real x,*

$$|\sin(\pi x)/\pi x - (N+\lambda)^{-1}(\sigma/2)S_{\rho N+\tau}((N+\lambda)^{-1}\sigma\pi x)| \leq 1/2N + 2(\pi x/N)^2.$$

PROOF. Let us simplify

$$\begin{aligned}(N+\lambda)^{-1}&(\sigma/2)S_{\rho N+\tau}((N+\lambda)^{-1}\sigma\pi x)\\ &= (\rho(N+\lambda))^{-1}S_{\rho N+\tau}((\rho(N+\lambda))^{-1}2\pi x)\\ &= \sin((\rho N+\tau)\pi x/(\rho(N+\lambda)))/(\rho(N+\lambda))\sin((\rho(N+\lambda))^{-1}\pi x).\end{aligned}$$

Suppose first that $|\pi x/N| \le 1$. The first assertion is the special case

$$\delta = (\rho N+\tau)/\rho(N+\lambda) - 1 = (\tau - \rho\lambda)/\rho(N+\lambda), \qquad M = \rho N+\tau$$

of the previous lemma, since in all cases $|\delta| \le 1/2N$. It trivially implies the second.

If $|\pi x/N| > 1$, the second assertion holds trivially, since the terms being subtracted have absolute values at most 1 and $(\rho N+\tau)/(\rho N+\rho\lambda) \le 1 + 1/2N$ respectively. QED

5.9.7. For each $G(N)$ and for each integer $n \ge 2$, we define a function $W(n,N)$ on $\mathbb{R}^n$ by

$$\begin{aligned}W(n,N) &:= \det_{n\times n}((N+\lambda)^{-1}L_{N,-}((N+\lambda)^{-1}\sigma\pi x(i),(N+\lambda)^{-1}\sigma\pi x(j))),\\ &= \det_{n\times n}((N+\lambda)^{-1}(\sigma/2)S_{\rho N+\tau}((N+\lambda)^{-1}\sigma\pi(x(i)-x(j)))).\end{aligned}$$

By Lemma 5.6.3, part 2), we have

Lemma 5.9.8. *For any $G(N)$, any $n \ge 2$ and any x in $\mathbb{R}^n$, we have*

$$|W(n,N)(x)| \le ((2N+\tau)/(2N+2\lambda))^n \le (1+1/2N)^n \le 2^n.$$

5.9.9. The next lemma shows that for fixed n, $W(n,N)(x)$ is a good approximation to $W(n)(x)$, provided N is very large and x is near the diagonal.

Lemma 5.9.10. *Fix integers $n \ge 2$ and $N \ge 1$, and a choice of $G(N)$. Let $\alpha \ge 0$ be real, and x in $\mathbb{R}^n$ a point with $\operatorname{Sup}_{i,j}|x(i)-x(j)| \le \alpha$. Then we have the inequality*

$$|W(n)(x) - W(n,N)(x)| \le n! \times n \times 2^{n-1} \times (1/2N + 2(\pi\alpha/N)^2).$$

PROOF. We are comparing the determinants of two $n\times n$ matrices, say A and B, whose individual entries $a(i,j)$ and $b(i,j)$ are bounded respectively by 1 and by $(\rho N+\tau)/(\rho N+\rho\lambda) \le 1+1/2N \le 2$ in absolute value, and whose differences $|a(i,j)-b(i,j)|$ are bounded by $1/2N+2(\pi\alpha/N)^2$, thanks to the previous Corollary 5.9.6. So the result follows from (the $t=2, s=1/2N+2(\pi\alpha/N)^2$ case of) the following crude lemma.

Lemma 5.9.11. *Let $n \ge 2$ be an integer, and $s \ge 0$ and $t \ge 0$ real. Let $A = (a(i,j))$ and $B = (b(i,j))$ in $M_n(\mathbb{C})$ be $n\times n$ matrices with*

$$\operatorname*{Sup}_{i,j}|a(i,j)| \le t,$$

$$\operatorname*{Sup}_{i,j}|b(i,j)| \le t,$$

$$\operatorname*{Sup}_{i,j}|a(i,j)-b(i,j)| \le s.$$

Then $|\det(A) - \det(B)| \le (n!)nt^{n-1}s$.

PROOF. Expand out $\det(A)$ and $\det(B)$ and separately compare each of the $n!$ terms. For a fixed φ in Σ_n, let $a(i) := a(i, \varphi(i)), b(i) := b(i, \varphi(i))$. The telescoping sum

$$\sum_{j=1}^{n} \left(\prod_{i<j} b(i)\right) (b(j) - a(j)) \left(\prod_{i>j} a(i)\right) = \prod_i b(i) - \prod_i a(i)$$

is the sum of n terms, each bounded in absolute value by $t^{n-1}s$. QED

5.10. Large N limit of $E(Z[n, F, G(N)])$ via the $\sin(\pi x)/\pi x$ kernel

Definition 5.10.1. Let $n \geq 2$ be an integer, F in $\mathcal{T}_0(n)$, $\alpha \geq 0$ a real number such that $\operatorname{supp}(F) \leq \alpha$. Define the real number $E(n, F, \text{ univ})$ by

$$E(n, F, \text{ univ}) = \int_{[0,\alpha]^{n-1}(\text{order})} F(0, z) W(n)(0, z) \prod_i dz(i).$$

The integral is visibly independent of the auxiliary choice of $\alpha \geq 0$ such that $\operatorname{supp}(F) \leq \alpha$.

5.10.2. We now show that part i) of Theorem 4.2.2 holds, with the above explicit formula for the large N limit $E(n, F, \text{ univ})$.

Proposition 5.10.3. *Let $n \geq 2$ be an integer, F in $\mathcal{T}_0(n)$, $\alpha \geq 0$ a real number such that* $\operatorname{supp}(F) \leq \alpha$. *For any integer $N \geq 2$ and any $G(N)$, we have*

$$\begin{aligned} &|E(Z[n, F, G(N)]) - E(n, F, \textit{ univ})| \\ &\quad \leq \|F\|_{\sup}(8\alpha)^{n-1}((\pi\alpha)^2 + \alpha + 1 + 10\log(N))/N. \end{aligned}$$

PROOF. We will prove this in a series of three lemmas.

By Lemma 5.7.2, the expectation $E(Z[n, F, G(N)])$ is given by

$$E(Z[n, F, G(N)]) = (1/(N+\lambda)) \int_{[0,\sigma\pi]^n} F((N+\lambda)x/\sigma\pi)\, d\mu(n, N)$$

$$= (1/n!(N+\lambda)) \int_{[0,\sigma\pi]^n} F((N+\lambda)x/\sigma\pi) \det_{n\times n}(L_N(x(i), x(j))) \prod_i (dx(i)/\sigma\pi).$$

We denote by $E_-(Z[n, F, G(N)])$ the integral

(5.10.3.1)

$$E_-(Z[n, F, G(N)]) = (1/(N+\lambda)) \int_{[0,\sigma\pi]^n} F((N+\lambda)x/\pi)\, d\mu_-(n, N)$$

$$= (1/n!(N+\lambda)) \int_{[0,\sigma\pi]^n} F((N+\lambda)x/\sigma\pi) \det_{n\times n}(L_{N,-}(x(i), x(j))) \prod_i (dx(i)/\sigma\pi).$$

Lemma 5.10.4. *Let $n \geq 2$ be an integer, F in $\mathcal{T}_0(n)$, $\alpha \geq 0$ a real number such that* $\operatorname{supp}(F) \leq \alpha$. *For $N \geq 2$ and any $G(N)$, we have the estimate*

$$\begin{aligned} &|E(Z[n, F, G(N)]) - E_-(Z[n, F, G(N)])| \\ &\quad \leq \|F\|_{\sup}(8\alpha)^{n-1} 10\log(N)/N. \end{aligned}$$

PROOF. If $G(N)$ is $U(N)$, then $\mu(n, N) = \mu_-(n, N)$, and there is nothing to prove. In the other cases, we have $\sigma = 1, \rho = 2$.

We expand the $n \times n$ determinant of the $L_N(x,y)$ into $n!$ terms of type $\mathrm{sgn}(\varphi)\prod_i L_N(x(i), x(\varphi(i)))$ for φ in Σ_n. Writing

$$L_N(x,y) = L_{N,-}(x,y) + L_{N,+}(x,y),$$

we expand each n-fold product $\prod_i L_N(x(i), x(\varphi(i)))$ into 2^n terms, corresponding to which factors we replace by $L_{N,-}(x,y)$ and which by $L_{N,+}(x,y)$. The choice of all "$-$" gives the $n \times n$ determinant of the $L_{N,-}(x,y)$; the remaining $(2^n-1)n!$ terms are individually to be regarded as error terms. Thus the difference of the two integrals is a sum, with signs, of $(2^n-1)n!$ integrals of the form

$$(1/(N+\lambda))(1/n!)\int_{[0,\pi]^n} F((N+\lambda)x/\pi)\left(\prod_i L_{N,\pm}(x(i), x(\varphi(i)))\right)\prod_i (dx(i)/\pi),$$

where in the product $\prod_i L_{N,\pm}(x(i), x(\varphi(i)))$, at least one $\pm$ is $+$. We choose one particular term $L_{N,+}(x(i_0), x(\varphi(i_0)))$ which has the $+$, and use the trivial estimate

$$|L_{N,\pm}(x,y)| \le (1/2)(2N+\tau) \le (1/2)(2N+1) \le N+1$$

to deal with the $n-1$ other terms. We also use the trivial estimate $1/(N+\lambda) \le 1/N$. Thus each of $(2^n-1)n!$ integrals is bounded in absolute value by one of the form

$$(1/n!\,N)(N+1)^{n-1}\int_{[0,\pi]^n} |F((N+\lambda)x/\pi)|\,|L_{N,+}(x(i_0), x(j_0))|\prod_i (dx(i)/\pi).$$

Because $\mathrm{supp}(F) \le \alpha$, $|F((N+\lambda)x/\pi)|$ is supported in the region $\Delta(n, \pi\alpha/N)$ defined by

$$\mathrm{Sup}_{i,j} |x(i) - x(j)| \le \pi\alpha/(N+\lambda) \le \pi\alpha/N.$$

So the above integral is bounded by

$$(1/n!\,N)(N+1)^{n-1}\|F\|_{\sup}\int_{[0,\pi]^n\cap\Delta(n,\pi\alpha/N)} |L_{N,+}(x(i_0), x(j_0))|\prod_i (dx(i)/\pi).$$

By renumbering, we may assume that $i_0 = 1$ and $j_0 = 1$ or 2. We pass to the coordinates

$$x(1), \quad \delta(j) := x(j) - x(1) \text{ for } j = 2, \dots, n.$$

In this coordinate system, each $|\delta(j)| \le \pi\alpha/N$, and so the domain of integration $[0,\pi]^n \cap \Delta(n, \pi\alpha/N)$ lies in the product region

$$[0,\pi] \times [-\pi\alpha/N, \pi\alpha/N]^{n-1}.$$

Thus

$$\begin{aligned}
&\int_{[0,\pi]^n\cap\Delta(n,\pi\alpha/N)} |L_{N,+}(x(1), x(j_0))|\prod_i (dx(i)/\pi) \\
&\quad\le \int_{[0,\pi]\times[-\pi\alpha/N,\pi\alpha/N]^{n-1}} |L_{N,+}(x(1), x(j_0))| d(x(1)/\pi)\prod_{i\ge 2}(d\delta(i)/\pi) \\
&\quad= \int_{[0,\pi]\times[-\pi\alpha/N,\pi\alpha/N]^{n-1}} |S_{2N+\tau}(2x(1)+\delta(j_0))| d(x(1)/2\pi)\prod_{i\ge 2}(d\delta(i)/\pi).
\end{aligned}$$

If $j_0 = 1$, the integrand is a function of $x(1)$ alone, and this integral is

$$= (2\alpha/N)^{n-1} \int_{[0,\pi]} |S_{2N+\tau}(2x(1))|\, dx(1)/2\pi$$
$$= (2\alpha/N)^{n-1} \int_{[0,\pi]} |\sin((2N+\tau)x)/\sin(x)|\, dx/2\pi.$$

If $j_0 = 2$, the integrand is a function of $x(1)$ and $\delta(2)$ alone, and the integral is

$$(2\alpha/N)^{n-2} \int_{[-\pi\alpha/N, \pi\alpha/N]} \left(\int_{[0,\pi]} |S_{2N+\tau}(2x+\delta)|\, dx/2\pi \right) d\delta/\pi$$
$$\le (2\alpha/N)^{n-1} \operatorname*{Sup}_{|\delta| \le \alpha\pi/N} \int_{[0,\pi]} |S_{2N+\tau}(2x+\delta)|\, dx/2\pi$$
$$\le (2\alpha/N)^{n-1} \operatorname*{Sup}_{|\delta| \le \alpha\pi/N} \int_{[0,\pi]} |\sin((2N+\tau)(x+\delta/2))/\sin(x+\delta/2)|\, dx/2\pi.$$

Let us admit temporarily the truth of the following sublemma.

Sublemma 5.10.5. *For any integer $N \ge 2$, and any real y, we have*

$$\int_{[0,\pi]} |\sin(N(x+y))/\sin(x+y)|\, dx/2\pi \le 2 + \log(N-1).$$

By the sublemma, we have

$$\int_{[0,\pi]^n \cap \Delta(n, \pi\alpha/N)} |L_{N,+}(x(i_0), x(j_0))| \prod_i (dx(i)/\pi)$$
$$\le (2\alpha/N)^{n-1}(2 + \log(2N+\tau-1))$$
$$\le (2\alpha/N)^{n-1}(2 + \log(2N))$$
$$\le (2\alpha/N)^{n-1}(5\log(N)) \qquad \text{(since } N \ge 2\text{)}.$$

Hence, retracing our steps, we find that

$$|E(Z[n,F,G(N)]) - E_-(Z[n,F,G(N)])|$$
$$\le (2^n - 1)n!\,(1/n!\,N)(N+1)^{n-1}\|F\|_{\sup}(2\alpha/N)^{n-1}(5\log(N))$$
$$= \|F\|_{\sup}(2^n - 1)(2\alpha)^{n-1}((N+1)/N)^{n-1} 5\log(N)/N$$
$$\le \|F\|_{\sup} 2^n (4\alpha)^{n-1} 5\log(N)/N = \|F\|_{\sup}(8\alpha)^{n-1} 10\log(N)/N.$$

QED for Lemma 5.10.4, modulo Sublemma 5.10.5.

Proof of Sublemma 5.10.5. The function $\sin(Nx)/\sin(x)$ is periodic of period 2π, so for any y we have

$$\int_{[0,\pi]} |\sin(N(x+y))/\sin(x+y)|\, dx/2\pi$$
$$\le \int_{[-\pi,\pi]} |\sin(N(x+y))/\sin(x+y)|\, dx/2\pi$$
$$= \int_{[-\pi,\pi]} |\sin(Nx)/\sin(x)|\, dx/2\pi \qquad \text{(by periodicity)}$$
$$= \int_{[0,\pi]} |\sin(Nx)/\sin(x)|\, dx/\pi \qquad \text{(by evenness)}.$$

Break up the interval $[0, \pi]$ into the $2N$ subintervals

$$I_j := [j\pi/2N, (j+1)\pi/2N] \quad \text{for } j = 0, \ldots, 2N-1.$$

We get

$$\int_{[0,\pi]} |\sin(Nx)/\sin(x)|\, dx/\pi = \sum_{j=0}^{2N-1} \int_{I_j} |\sin(Nx)/\sin(x)|\, dx/\pi.$$

We use the estimate $|\sin(Nx)/\sin(x)| \le N$ for the $j = 0$ and $j = 2N-1$ terms. We use the estimate $|\sin(Nx)/\sin(x)| \le |1/\sin(x)|$ for the others. As each interval I_j has dx/π length $1/2N$, this gives

$$\int_{[0,\pi]} |\sin(Nx)/\sin(x)|\, dx/\pi \le 1 + \sum_{j=1}^{2N-2} \int_{I_j} |1/\sin(x)|\, dx/\pi.$$

Because $\sin(x)$ in $[0, \pi]$ is symmetric about the midpoint $\pi/2$, the terms $\int_{I_j} |1/\sin(x)|\, dx/\pi$ match in pairs (j and $2N-1-j$), so we get

$$\int_{[0,\pi]} |\sin(Nx)/\sin(x)|\, dx/\pi \le 1 + 2\sum_{j=1}^{N-1} \int_{I_j} |1/\sin(x)|\, dx/\pi.$$

For x in $[0, \pi/2]$, we have $x \ge \sin(x) \ge 2x/\pi$, so on I_j above we have

$$1/\sin(x) \le \pi/2x \le \pi/(2j\pi/2N) = N/j,$$

so

$$\int_{I_j} |1/\sin(x)|\, dx/\pi \le (N/j)\int_{I_j} dx/\pi = 1/2j,$$

and thus we get

$$\int_{[0,\pi]} |\sin(Nx)/\sin(x)|\, dx/\pi \le 1 + \sum_{j=1}^{N-1} 1/j$$

$$\le 2 + \sum_{j=2}^{N-1} 1/j \le 2 + \log(N-1). \quad \text{QED for 5.10.5.}$$

Lemma 5.10.6. *Let $n \ge 2$ be an integer, F in $\mathcal{T}_0(n)$, $\alpha \ge 0$ a real number such that* $\mathrm{supp}(F) \le \alpha$. *For $N \ge 2$ and any $G(N)$, if we define*

$$E_-(n, F, N, G(N)) := \int_{[0,\alpha]^{n-1}(order)} F(0, z) W(n, N)(0, z) \prod_i dz(i),$$

we have the estimate

$$|E_-(n, F, N, G(N)) - E_-(Z[n, F, G(N)])| \le 2(2\alpha)^n \|F\|_{\sup}/(n-1)!\, N.$$

Proof. Recall that $E_-(Z[n, F, G(N)])$ is the integral

$$(1/n!\,(N+\lambda)) \int_{[0,\sigma\pi]^n} F((N+\lambda)x/\sigma\pi) \det_{n\times n}(L_{N,-}(x(i), x(j))) \prod_i (dx(i)/\sigma\pi).$$

Making the change of variable

$$y = (N+\lambda)x/\sigma\pi,$$

we may rewrite $E_-(Z[n, F, G(N)])$ as

$$(1/n!\,(N+\lambda)) \int_{[0,N+\lambda]^n} F(y)W(n,N)(y) \prod_i dy(i).$$

A key point is that $W(n, N)$ is itself a function in the class $\mathcal{T}(n)$, i.e., it is both Σ_n-invariant [being of the form $\det_{n\times n}(f(y(i), y(j)))$] and invariant by additive translations by all $\Delta_n(t) := (t, \ldots, t)$ [because $f(x, y)$ is of the form $g(x - y)$].

By the Σ_n-invariance, we may rewrite $E_-(Z[n, F, G(N)])$ as

$$(1/(N+\lambda)) \int_{[0,N+\lambda]^n(\text{order})} F(y)W(n,N)(y) \prod_i dy(i),$$

where $[0, N + \lambda]^n(\text{order})$ is the region

$$\{y \text{ in } \mathbb{R}^n \text{ with } 0 \le y(1) \le y(2) \le \cdots \le y(n) \le N + \lambda\}.$$

By the invariance of both $F(y)$ and $W(n, N)(y)$ by additive translation by diagonal vectors $\Delta_n(t)$, we have

$$\begin{aligned} F(y) &= F(0, y(2) - y(1), \ldots, y(n) - y(1)), \\ W(n,N)(y) &= W(n,N)(0, y(2) - y(1), \ldots, y(n) - y(1)). \end{aligned}$$

Make the further change of variable

$$t = y(1), \qquad z(i) = y(i) - y(1) \quad \text{for } i = 2, \ldots, n.$$

In the coordinates (t, z) our integral $E_-(Z[n, F, G(N)])$ is

$$\begin{aligned} &(1/(N+\lambda)) \int_{[0,N+\lambda]} \left(\int_{[0,N+\lambda-t]^{n-1}(\text{order})} F(0,z)W(n,N)(0,z) \prod_i dz(i) \right) dt \\ &\quad = (1/(N+\lambda)) \int_{[0,N+\lambda]} g(t)\, dt \end{aligned}$$

with

$$g(t) := \int_{[0,N+\lambda-t]^{n-1}(\text{order})} F(0,z)W(n,N)(0,z) \prod_i dz(i).$$

The function F has $\text{supp}(F) \le \alpha$, so for z in $[0, N + \lambda - t]^{n-1}(\text{order})$ the function $F(0, z)$ vanishes unless $z(n) \le \alpha$. Thus we may rewrite the inner integral $g(t)$ as

$$g(t) = \int_{[0,\min(\alpha,N+\lambda-t)]^{n-1}(\text{order})} F(0,z)W(n,N)(0,z) \prod_i dz(i).$$

For any t in $[0, N + \lambda]$, we have the bound

$$\begin{aligned} |g(t)| &\le \int_{[0,\alpha]^{n-1}(\text{order})} |F(0,z)W(n,N)(0,z)| \prod_i dz(i) \\ &\le (\alpha^{n-1}/(n-1)!)\|F\|_{\text{sup}} \|W(n,N)\|_{\text{sup}} \le 2^n(\alpha^{n-1}/(n-1)!)\|F\|_{\text{sup}} \end{aligned}$$

by 5.9.8, which also establishes that

$$|E_-(n, F, N, G(N))| \le 2^n(\alpha^{n-1}/(n-1)!)\|F\|_{\text{sup}}.$$

Suppose first that $N+\lambda > \alpha$. Then we have

$$\begin{aligned}(1/(N+\lambda))&\int_{[0,N+\lambda]} g(t)\,dt\\ &= (1/(N+\lambda))\int_{[0,N+\lambda-\alpha]} g(t)\,dt + (1/(N+\lambda))\int_{[N+\lambda-\alpha,N+\lambda]} g(t)\,dt.\end{aligned}$$

For t in $[0, N+\lambda-\alpha]$, the inner integral $g(t)$ is

$$\int_{[0,\alpha]^{n-1}(\text{order})} F(0,z)W(n,N)(0,z)\prod_i dz(i),$$

independent of t, and so the first term is

$$\begin{aligned}(1/(N+\lambda))&\int_{[0,N+\lambda-\alpha]} g(t)\,dt\\ &= ((N+\lambda-\alpha)/(N+\lambda))\int_{[0,\alpha]^{n-1}(\text{order})} F(0,z)W(n,N)(0,z)\prod_i dz(i)\\ &= (1-(\alpha/(N+\lambda)))E_-(n,F,N,G(N)).\end{aligned}$$

We may estimate the second term by

$$\begin{aligned}&\left|(1/(N+\lambda))\int_{[N+\lambda-\alpha,N+\lambda]} g(t)\,dt\right|\\ &\qquad\le (\alpha/(N+\lambda))2^n(\alpha^{n-1}/(n-1)!)\|F\|_{\text{sup}} \le (2\alpha)^n\|F\|_{\text{sup}}/(n-1)!\,N.\end{aligned}$$

Thus we have

$$|E_-(n,F,N,G(N)) - E_-(Z[n,F,G(N)])| \le 2(2\alpha)^n\|F\|_{\text{sup}}/(n-1)!\,N,$$

provided $N+\lambda>\alpha$.

Suppose now that $\alpha \ge N+\lambda$. Then the above estimate holds trivially. Indeed, for any α, we have

$$\begin{aligned}&|E_-(Z[n,F,G(N)])|\\ &\qquad= \left|(1/(N+\lambda))\int_{[0,N+\lambda]} g(t)\,dt\right| \le 2^n(\alpha^{n-1}/(n-1)!)\|F\|_{\text{sup}},\end{aligned}$$

and we have already noted that

$$|E_-(n,F,N,G(N))| \le 2^n(\alpha^{n-1}/(n-1)!)\|F\|_{\text{sup}}.$$

Thus we find that

$$|E_-(n,F,N,G(N)) - E_-(Z[n,F,G(N)])| \le 2^{n+1}\alpha^{n-1}\|F\|_{\text{sup}}/(n-1)!.$$

Because $\alpha \ge N+\lambda \ge N$, we have $1 \le \alpha/N$, so

$$2^{n+1}\alpha^{n-1}\|F\|_{\text{sup}}/(n-1)! \le 2(2\alpha)^n\|F\|_{\text{sup}}/(n-1)!\,N. \quad \text{QED}$$

Lemma 5.10.7. *Let $n \ge 2$ be an integer, F in $\mathcal{T}_0(n)$, $\alpha \ge 0$ a real number such that* $\text{supp}(F) \le \alpha$. *For $N \ge 2$ and any $G(N)$, we have the estimate*

$$\begin{aligned}&|E_-(n,F,N,G(N)) - E(n,F,\text{ univ})|\\ &\qquad\le n^2(2\alpha)^{n-1}(1/2N + 2(\pi\alpha/N)^2)\|F\|_{\text{sup}}.\end{aligned}$$

PROOF. The difference is the absolute value of

$$\int_{[0,\alpha]^{n-1}(\text{order})} F(0,z)(W(n,N)(0,z) - W(n)(0,z)) \prod_i dz(i).$$

The domain of integration has area $\alpha^{n-1}/(n-1)!$. In it, we have, by Lemma 5.9.10,

$$|W(n)(0,z) - W(n,N)(0,z)| \leq n! \times n \times 2^{n-1} \times (1/2N + 2(\pi\alpha/N)^2),$$

and so the assertion is obvious. QED

5.10.8. We can now finish the proof of Proposition 5.10.3, by combining the three lemmas. Taken together, they give an error of

$$\begin{aligned} &\|F\|_{\sup}(8\alpha)^{n-1} 10\log(N)/N \\ &\quad + \|F\|_{\sup} 2(2\alpha)^n/(n-1)!\,N \\ &\quad + \|F\|_{\sup} n^2(2\alpha)^{n-1}(1/2N + 2(\pi\alpha/N)^2). \end{aligned}$$

We leave the first alone, replace the second by the cruder

$$\|F\|_{\sup}(8\alpha)^{n-1}\alpha/N,$$

and use $n^2 \leq 4^{n-1}$ and $N \geq 2$ to replace the third by

$$\begin{aligned} &\|F\|_{\sup}(8\alpha)^{n-1}(1/2N + 2(\pi\alpha/N)^2) \\ &\quad \leq \|F\|_{\sup}(8\alpha)^{n-1}(1/N + (\pi\alpha)^2/N) \\ &\quad = \|F\|_{\sup}(8\alpha)^{n-1}(1 + (\pi\alpha)^2)/N. \quad \text{QED} \end{aligned}$$

5.11. Upper bound for the variance

5.11.1. In this section, we show that part iii) of Theorem 4.2.2 holds.

Proposition 5.11.2. *Let $n \geq 2$ be an integer, F in $\mathcal{T}_0(n)$, $\alpha \geq 0$ a real number such that* $\text{supp}(F) \leq \alpha$. *For any $N \geq 2$ and any $G(N)$ we have*

$$|\operatorname{Var}(Z[n,F,G(N)])| \leq (3(8\alpha)^{n-1} + 65(8\alpha)^{2n-2})(\|F\|_{\sup})^2/N.$$

PROOF. For A in $G(N)$, with angles $X(A)$ in $[0,\sigma\pi]^N$, $Z[n,F,G(N)](A)$ was defined (in 4.1.3) by

$$\begin{aligned} Z[n,FG(N)](A) &:= (1/(N+\lambda))F[n,N](((N+\lambda)/\sigma\pi)X(A)) \\ &:= (1/(N+\lambda)) \sum_{\text{Card}(T)=n} F(((N+\lambda)/\sigma\pi)\,\text{pr}(T)(X(A))). \end{aligned}$$

For any function W on $G(N)$, its variance $\operatorname{Var}(W)$ is defined as

$$\operatorname{Var}(W) := \int_{G(N)} W^2\, d\,\text{Haar} - \left(\int_{G(N)} W\, d\,\text{Haar}\right)^2.$$

So our first task is to square $Z[n,F,G(N)](A)$.

Let us fix $F, N, G(N)$, and A in $G(N)$, and denote

$$Z := Z[n,F,G(N)](A).$$

For a subset T of $\{1,2,\ldots,N\}$ of cardinality n, we define

$$f(T) := F(((N+\lambda)/\sigma\pi)\,\text{pr}(T)(X(A))).$$

Thus we have

$$(N+\lambda)Z = \sum_{\#T=n} f(T),$$

$$(N+\lambda)^2 Z^2 = \sum_{\#T=n,\#S=n} f(T)f(S).$$

We now break up this sum according to the value of $C := T \cup S$:

$$Z^2 = (N+\lambda)^{-2} \sum_{n \le l \le 2n} \sum_{\#C=l} \sum_{\substack{C=T\cup S\\ \#T=n,\#S=n}} f(T)f(S).$$

For each subset C of $\{1,2,\ldots,N\}$ of cardinality l, we define

$$h(C) := \sum_{\substack{C=T\cup S\\ \#T=n,\#S=n}} f(T)f(S).$$

Thus we have

$$Z^2 = (N+\lambda)^{-1} \sum_{n \le l \le 2n} (N+\lambda)^{-1} \sum_{\#C=l} h(C).$$

For each l with $n \le l \le 2n$, it is tautological that the inner summand

$$(N+\lambda)^{-1} \sum_{\#C=l} h(C)$$

is itself of the form $Z[l, H_l, G(N)](A)$ for H_l the function on $\mathbb{R}^l$ defined by

$$H_l(X) := \sum_{\substack{\{1,2,\ldots,l\}=T\cup S\\ \#T=n,\#S=n}} F(\mathrm{pr}(T)(X))F(\mathrm{pr}(S)(X)).$$

Thus we have

$$Z[n,F,G(N)]^2 = (N+\lambda)^{-1} \sum_{n\le l\le 2n} Z[l,H_l,G(N)]. \tag{5.11.2.1}$$

Integrating over $G(N)$, we get

$$E(Z[n,F,G(N)]^2) = \sum_{n\le l\le 2n} (N+\lambda)^{-1} E(Z[l,H_l,G(N)]). \tag{5.11.2.2}$$

5.11.3. Our first task is to show that the terms with $l < 2n$ are negligible.

Lemma 5.11.4. *The function H_l lies in $\mathcal{T}(l)$, and*

$$\|H_l\|_{\sup} \le \mathrm{Binom}(l,n)\,\mathrm{Binom}(n,l-n)(\|F\|_{\sup})^2.$$

If $l < 2n$, H_l lies in $\mathcal{T}_0(l)$, and $\mathrm{supp}(H_l) \le 2\alpha$.

Proof. The key is that F lies in $\mathcal{T}_0(n)$, with $\mathrm{supp}(F) \le \alpha$. Because F lies in $\mathcal{T}(n)$, i.e., is Σ_n-invariant and invariant by diagonal translation in $\mathbb{R}^n$, we see easily that H_l is Σ_l-invariant and invariant by diagonal translation in $\mathbb{R}^l$, i.e., H_l lies in $\mathcal{T}(l)$.

To get the asserted bound on $\|H_l\|_{\sup}$, look at the formula for H_l in terms of F, each term of which is bounded by $(\|F\|_{\sup})^2$. This formula for H_l in terms of F has $\mathrm{Binom}(l,n)\,\mathrm{Binom}(n,l-n)$ terms. Pick first T, which can be any of the $\mathrm{Binom}(l,n)$ possible subsets of $\{1,\ldots,l\}$ of cardinality n. Having picked T, S may be any set of cardinality n which contains $C(T) := \{1,\ldots,l\} - T$, a set of cardinality $\#C(T) = l - n$. So picking S amounts to picking $S - C(T)$, which may be any subset of cardinality $n - (l - n) = 2n - l$ of T. Hence, given T, there are $\mathrm{Binom}(n, 2n - l) = \mathrm{Binom}(n, l - n)$ choices for S.

Suppose now that $l < 2n$. We claim that $\mathrm{supp}(H_l) \leq 2\alpha$. In each term $F(\mathrm{pr}(T)(X))F(\mathrm{pr}(S)(X))$, the two sets S and T cannot be disjoint (because $S \cup T$ is $\{1, \ldots, l\}$, and $l < 2n$). So there is an index i_0 which lies in both S and T. Any index j with $1 \leq j \leq l$ lies in either S or T. If j lies in S, then $F(\mathrm{pr}(S)(X))$ vanishes if $|x(i_0) - x(j)| > \alpha$, because $\mathrm{supp}(F) \leq \alpha$. If j is in T, then $F(\mathrm{pr}(T)(X))$ vanishes if $|x(i_0) - x(j)| > \alpha$. So the product $F(\mathrm{pr}(T)(X))F(\mathrm{pr}(S)(X))$ vanishes if $|x(i_0) - x(j)| > \alpha$ for any j. By the triangle inequality, $F(\mathrm{pr}(T)(X))F(\mathrm{pr}(S)(X))$, and hence $H_l(X)$ itself, vanishes if there are any two indices j and k such that $|x(k) - x(j)| > 2\alpha$. This means precisely that $\mathrm{supp}(H_l) \leq 2\alpha$. QED

Corollary 5.11.5. *For $l < 2n$, we have the estimate*

$$
\begin{aligned}
&|(N+\lambda)^{-1}E(Z[l, H_l, G(N)])| \\
&\quad \leq (\|F\|_{\mathrm{sup}})^2 \,\mathrm{Binom}(l, n)\,\mathrm{Binom}(n, l-n) 2(2\alpha)^{l-1}/(l-1)!\,N.
\end{aligned}
$$

PROOF. Simply combine Proposition 5.8.2 with the above Lemma 5.11.4. QED

5.11.6. We now turn to a detailed look at the $l = 2n$ term

$$(N+\lambda)^{-1}E(Z[2n, H_{2n}, G(N)]).$$

Lemma 5.11.7. *We have the estimate*

$$
\begin{aligned}
&|(N+\lambda)^{-1}E(Z[2n, H_{2n}, G(N)]) - (E(Z[n, F, G(N)]))^2| \\
&\quad \leq (\|F\|_{\mathrm{sup}})^2 \,\mathrm{Binom}(2n, n) 4^2 (4\alpha)^{2n-2}/N.
\end{aligned}
$$

PROOF. We apply the general formula (5.7.2),

$$E(Z[n, F, G(N)]) = (1/(N+\lambda)) \int_{[0,\sigma\pi]^n} F((N+\lambda)x/\sigma\pi)\, d\mu(n, N),$$

to find

$$
\begin{aligned}
&(N+\lambda)^{-1}E(Z[2n, H_{2n}, G(N)]) \\
&\quad = (N+\lambda)^{-2} \int_{[0,\sigma\pi]^{2n}} H_{2n}((N+\lambda)x/\sigma\pi)\, d\mu(2n, N) \\
&\quad = (N+\lambda)^{-2} \sum_{\substack{S \coprod T = \{1,\ldots,2n\} \\ \#S = \#T = n}} \mathrm{Integral}(S, T),
\end{aligned}
$$

where $\mathrm{Integral}(S, T)$ is the integral

$$\int_{[0,\sigma\pi]^{2n}} F((N+\lambda)\,\mathrm{pr}(T)(x)/\sigma\pi) F((N+\lambda)\,\mathrm{pr}(S)(x)/\sigma\pi)\, d\mu(2n, N).$$

For brevity of notation, let us denote

$$
\begin{aligned}
x(T) &:= \mathrm{pr}(T)(x), \\
f(x) &:= F((N+\lambda)x/\sigma\pi).
\end{aligned}
$$

Then we may rewrite $\mathrm{Integral}(S, T)$ as

$$\mathrm{Integral}(S, T) = \int_{[0,\sigma\pi]^{2n}} f(x(T)) f(x(S))\, d\mu(2n, N).$$

Because the measure $\mu(2n, N)$ is Σ_{2n}-invariant, and Σ_{2n} acts transitively on the set of all partitions (S, T) of $\{1, \ldots, 2n\}$ into two disjoint subsets of cardinality n, we see that Integral(S, T) is independent of (S, T). There are Binom$(2n, n)$ such partitions. Fixing one such (S, T), say $(\{1, 2, \ldots, n\}, \{n+1, n+2, \ldots, 2n\})$, we have

$$\begin{aligned}(N+\lambda)^{-1}&E(Z[2n, H_{2n}, G(N)])\\ &= \mathrm{Binom}(2n, n)(N+\lambda)^{-2}\,\mathrm{Integral}(S, T).\end{aligned}$$

The measure $\mu(2n, N)$ is of the form (5.6.1)

$$(1/(2n)!)\det_{2n\times 2n}(L_N(x(i), x(j)))\prod_i (dx(i)/\sigma\pi).$$

Let us denote

$$D(2n, N)(x) := \det_{2n\times 2n}(L_N(x(i), x(j))).$$

Then we have

$$\begin{aligned}&(N+\lambda)^{-1}E(Z[2n, H_{2n}, G(N)])\\ &\quad= \mathrm{Binom}(2n, n)(N+\lambda)^{-2}\int_{[0,\sigma\pi]^{2n}} f(x(T))f(x(S))\,d\mu(2n, N)\\ &\quad= (n!\,(N+\lambda))^{-2}\int_{[0,\sigma\pi]^{2n}} f(x(T))f(x(S))D(2n, N)(x)\prod_i (dx(i)/\sigma\pi).\end{aligned}$$

Expand $D(2n, N)(x) := \det_{2n\times 2n}(L_N(x(i), x(j)))$ as the sum of $(2n)!$ terms

$$\mathrm{sgn}(\varphi)\prod_i L_N(x(i), x(\varphi(i)))$$

indexed by φ in Σ_{2n}. We must distinguish two sorts of elements φ, those which respect the chosen partition (S, T), and those which do not. We group the corresponding terms,

$$D_{\mathrm{resp}}(2n, N)(x) := \sum_{\varphi \text{ respects } S\coprod T} \mathrm{sgn}(\varphi)\prod_i L_N(x(i), x(\varphi(i))),$$

and

$$D_{\mathrm{nonresp}}(2n, N)(x) := \sum_{\varphi \text{ does not respect } S\coprod T} \mathrm{sgn}(\varphi)\prod_i L_N(x(i), x(\varphi(i))).$$

Thus we have

$$\begin{aligned}&(N+\lambda)^{-1}E(Z[2n, H_{2n}, G(N)])\\ &\quad= (n!\,(N+\lambda))^{-2}\int_{[0,\sigma\pi]^{2n}} f(x(T))f(x(S))D_{\mathrm{resp}}(2n, N)(x)\prod_i (dx(i)/\sigma\pi)\\ &\qquad+ (n!\,(N+\lambda))^{-2}\int_{[0,\sigma\pi]^{2n}} f(x(T))f(x(S))D_{\mathrm{nonresp}}(2n, N)(x)\prod_i (dx(i)/\sigma\pi).\end{aligned}$$

It is tautological that we have the product decomposition

$$D_{\mathrm{resp}}(2n, N)(x) = D(n, N)(x(S))D(n, N)(x(T)).$$

From this, we see that

$$(n!\,(N+\lambda))^{-2}\int_{[0,\sigma\pi]^{2n}} f(x(T))f(x(S))D_{\mathrm{resp}}(2n, N)(x)\prod_i (dx(i)/\sigma\pi)$$

is the **square** of

$$(n!\,(N+\lambda))^{-1}\int_{[0,\sigma\pi]^n} f(x)D(n,N)(x)\prod_i (dx(i)/\sigma\pi) = E(Z[n,F,G(N)]).$$

Thus we have

$$\begin{aligned}&(N+\lambda)^{-1}E(Z[2n,H_{2n},G(N)]) - (E(Z[n,F,G(N)]))^2\\ &\quad = (n!\,(N+\lambda))^{-2}\int_{[0,\sigma\pi]^{2n}} f(x(T))f(x(S))D_{\text{nonresp}}(2n,N)(x)\prod_i (dx(i)/\sigma\pi).\end{aligned}$$

5.11.8. It remains only to bound the "remainder" term

$$(n!(N+\lambda))^{-2}\int_{[0,\sigma\pi]^{2n}} f(x(T))f(x(S))D_{\text{nonresp}}(2n,N)(x)\prod_i (dx(i)/\sigma\pi).$$

Lemma 5.11.7 will follow from

Lemma 5.11.9. *We have the estimate*

$$\begin{aligned}&\left|(n!\,(N+\lambda))^{-2}\int_{[0,\sigma\pi]^{2n}} f(x(T))f(x(S))D_{nonresp}(2n,N)(x)\prod_i (dx(i)/\sigma\pi)\right|\\ &\quad\le (\|F\|_{\sup})^2\,\text{Binom}(2n,n)4^2(4\alpha)^{2n-2}/N.\end{aligned}$$

Proof. It suffices to show that for each φ in $\sum_{2n}$ which does not respect (S,T), of which there are at most $(2n)!$, we have

$$\begin{aligned}&\left|(n!\,(N+\lambda))^{-2}\int_{[0,\sigma\pi]^{2n}} f(x(T))f(x(S))\prod_i L_N(x(i),x(\varphi(i)))\prod_i (dx(i)/\sigma\pi)\right|\\ &\quad\le (\|F\|_{\sup})^2 4^2(4\alpha)^{2n-2}/(n!)^2N.\end{aligned}$$

Look at the cycle decomposition of φ. At least one of its cycles contains elements of both S and T. Renumbering, we may suppose that (S,T) is

$$(\{1,2,\ldots,n\},\{n+1,n+2,\ldots,2n\}),$$

that both 1 and $n+1$ are in the same φ-orbit, and that $\varphi(1)=n+1$. Because the forward φ-orbit of $n+1$ does not stay in T, there will be some $n+a>n$ and some $b\le n$ for which $\varphi(n+a)=b$. The idea is to pay special attention to the factor

$$L_N(x(1),x(n+1))L_N(x(n+a),x(b)),$$

and to use the trivial bound $|L_N(x(i),x(\varphi(i)))|\le 2N$, cf. 5.3.1, on the other $2n-2$ terms in the product. So we have

$$\begin{aligned}&\left|(n!\,(N+\lambda))^{-2}\int_{[0,\sigma\pi]^{2n}} f(x(T))f(x(S))\prod_i L_N(x(i),x(\varphi(i)))\prod_i (dx(i)/\sigma\pi)\right|\\ &\le (2N)^{2n-2}(n!\,(N+\lambda))^{-2}\\ &\times\int_{[0,\sigma\pi]^{2n}} |f(x(T))f(x(S))|\,|L_N(x(1),x(n+1))L_N(x(n+a),x(b))|\prod_i (dx(i)/\sigma\pi).\end{aligned}$$

Make the change of coordinates

$$\begin{aligned}s &= x(1), \quad \delta(i)=x(i)-x(1) \text{ for } i=2,\ldots,n,\\ t &= x(n+1), \quad \varepsilon(i)=x(n+i)-x(n+1) \text{ for } i=2,\ldots,n,\end{aligned}$$

and put

$$\delta(1)=\varepsilon(1)=0.$$

Because $\text{supp}(F) \le \alpha$, $f(x(S)) := F((N+\lambda)x(S)/\sigma\pi)$ is supported in the region $\text{Sup}_i |\delta(i)| \le \sigma\pi\alpha/(N+\lambda) \le \sigma\pi\alpha/N$, and similarly $f(x(T))$ is supported in the region $\text{Sup}_i |\varepsilon(i)| \le \sigma\pi\alpha/N$. So we have

$$\begin{aligned}\int_{[0,\sigma\pi]^{2n}} &|f(x(T))f(x(S))|\,|L_n(x(1),x(n+1))L_N(x(n+a),x(b))| \prod_i (dx(i)/\sigma\pi) \\ &\le (\|F\|_{\text{sup}})^2 \int_{[-\sigma\pi\alpha/N,\sigma\pi\alpha/N]^{2n-2}} \text{Int}_2(\varepsilon,\delta) \prod_i (d\delta(i)/\sigma\pi) \prod_i (d\varepsilon(i)/\sigma\pi)\end{aligned}$$

where

$$\text{Int}_2(\varepsilon,\delta) := \int_{[0,\sigma\pi]^2} |L_N(s,t)L_N(t+\varepsilon(a), s+\delta(b))| d(s/\sigma\pi)d(t/\sigma\pi).$$

Thus we have the inequality

$$\begin{aligned}\int_{[0,\sigma\pi]^{2n}} &|f(x(T))f(x(S))|\,|L_N(x(1),x(n+1))L_N(x(n+a),x(b))| \prod_i (dx(i)/\sigma\pi) \\ &\le (\|F\|_{\text{sup}})^2 (2\alpha/N)^{2n-2} \sup_{\varepsilon,\delta} |\text{Int}_2(\varepsilon,\delta)|.\end{aligned}$$

But for any values of the ε's and the δ's, we have, by enlarging the domain of integration from $[0,\sigma\pi]^2$ to $[0,4\pi]^2$, the inequality

$$|\text{Int}_2(\varepsilon,\delta)| \le 16 \int_{[0,4\pi]^2} |L_N(s,t)L_N(t+\varepsilon(a), s+\delta(b))| d(s/4\pi)d(t/4\pi).$$

By the Cauchy-Schwarz inequality on the probability space $[0,4\pi]^2$, $d(s/4\pi)d(t/4\pi)$, we get

$$|\text{Int}_2(\varepsilon,\delta)| \le 16\|L_N(s,t)\|_{L_2}\|L_N(t+\varepsilon(a), s+\delta(b))\|_{L_2}.$$

The function $L_N(s,t)$ is periodic of period 4π in each variable separately, and has L_2 norm $\text{Sqrt}(N)$, by 5.3.2. By the translation invariance and symmetry of Lebesgue measure, we have

$$\|L_N(s,t)\|_{L_2} = \|L_N(t+\varepsilon(a), s+\delta(b))\|_{L_2}.$$

So we get

$$|\text{Int}_2(\varepsilon,\delta)| \le 16N.$$

Tracing our way back, we get the required estimate, and so Lemma 5.11.9, and hence Lemma 5.11.7. QED

5.11.10. We may now conclude the proof of Proposition 5.11.2. By combining 5.11.4, 5.11.5, and 5.11.7, we find that $|\text{Var}(Z[n,F,G(N)])|$ is bounded by $(\|F\|_{\text{sup}})^2/N$ times

$$\sum_{l=n}^{2n-1} \text{Binom}(l,n)\,\text{Binom}(n,l-n) 2(2\alpha)^{l-1}/(l-1)! \\ + \text{Binom}(2n,n)4^2(4\alpha)^{2n-2}.$$

Let us admit temporarily that

$$\text{Binom}(l,n)\,\text{Binom}(n,l-n)/(l-1)! \le 3.$$

Then our bounding factor is

$$\leq 6 \sum_{l=n}^{2n-1} (2\alpha)^{l-1} + \text{Binom}(2n, n) 4^2 (4\alpha)^{2n-2}.$$

In the partial geometric series, either the first term or the last term is largest, depending on the size of 2α, so each term is bounded by the sum of the first and last terms. We bound $\text{Binom}(2n, n)$ by 2^{2n} (expand $(1+1)^{2n}$). Thus our bounding factor is

$$\begin{aligned}
&\leq 6n((2\alpha)^{n-1} + (2\alpha)^{2n-2}) + 2^{2n} 4^2 (4\alpha)^{2n-2} \\
&= (6n/4^{n-1})(8\alpha)^{n-1} + (6n/4^{2n-2})(8\alpha)^{2n-2} + 4^3 (8\alpha)^{2n-2} \\
&\leq 3(8\alpha)^{n-1} + 65(8\alpha)^{2n-2},
\end{aligned}$$

which proves the proposition. It remains to explain why

$$\text{Binom}(l, n)\,\text{Binom}(n, l-n)/(l-1)! \leq 3,$$

which in fact holds for all $0 \leq n \leq l \leq 2n$. Expanding out in terms of factorials, it amounts to

$$l \leq 3(l-n)!\,(l-n)!\,(2n-l)!,$$

which is the case $r = 3$ of the statement that, given $r \geq 1$ nonnegative integers $k(i)$, we have

$$(1/r) \sum_i k(i) \leq \prod_i (k(i)!).$$

To see this, proceed by induction on r, the case $r = 1$ being clear. If any $k(i)$ vanishes, the r case is trivially implied by the $r-1$ case. If all $k(i)$ are ≥ 1, the assertion results from the stronger assertion

$$(1/r) \sum_i k(i) \leq \prod_i k(i) \quad \text{if all } k(i) \geq 1,$$

which is obvious from writing $k(i)$ as $1 + x(i)$ and expanding out both sides. QED

CHAPTER 6

Tail Estimates

6.0. Review: Operators of finite rank and their (reversed) characteristic polynomials

6.0.1. The following lemma is well known, we give it for ease of reference.

Lemma 6.0.2. *Let k be a field, V a k-vector space, and $L : V \to V$ a k-linear endomorphism which is of finite rank in the sense that the image space $L(V)$ is finite-dimensional. Then V has a unique direct sum decomposition*

$$V = V_{\text{nilp}} \oplus V_{\text{inv}}$$

into L-stable subspaces such that L is invertible on V_{inv} and such that L is nilpotent on V_{nilp}. The subspace V_{inv} is finite-dimensional.

PROOF. Since the image space $L(V)$ is finite-dimensional, the decreasing sequence of subspaces $L(V) \supset L^2(V) \supset L^3(V) \supset \cdots$ must stabilize, say

$$L^N(V) = L^{N+1}(V) = L^{N+k}(V) \text{ for all } k \geq 1.$$

Then $L^N(V)$ is finite-dimensional, and L maps it onto itself, so L is invertible on $L^N(V)$. We will take V_{inv} to be $L^N(V)$, and we will take V_{nilp} to be $\text{Ker}(L^N)$. These two subspaces have zero intersection in V, since their intersection is the kernel of L^N in $L^N(V)$, while L is invertible on $L^N(V)$. To see that $\text{Ker}(L^N) \oplus L^N(V)$ maps onto V, take any v in V. Then $L^N(v)$ lies in $L^N(V) = L^{2N}(V)$, say $L^N(v) = L^{2N}(w)$ for some w in V. Thus L^N kills $v - L^N(w)$, so $v = (v - L^N(w)) + L^N(w)$, as required.

Suppose we have some other L-stable direct sum decomposition $V = A \oplus B$ with L invertible on B and nilpotent on A. It suffices to show that $A \subset \text{Ker}(L^N)$ and that $B \subset L^N(V)$. Now $L^N(A)$ lies in $L^N(V)$, but is killed by some power of L, so $L^N(A) = 0$, and thus A lies in $\text{Ker}(L^N)$. Since L is invertible on B, we have $B = L^N(B)$, so $B \subset L^N(V)$. QED

6.0.3. Given an operator L on V of finite rank, we define its (reversed) characteristic polynomial in $1 + Tk[T]$ by

$$\det(1 - TL|V) := \det(1 - TL|V_{\text{inv}}).$$

6.0.4. For any L-stable subspace V_0 of V with $V_{\text{inv}} \subset V_0$, e.g. for $V_0 = L(V)$, we have

$$\det(1 - TL|V) = \det(1 - TL|V_0).$$

6.1. Integral operators of finite rank: a basic compatibility between spectral and Fredholm determinants

6.1.1. Suppose we are given a measure space X, a measure μ on X of finite total mass, and a bounded measurable function $K(x, y)$ on $X \times X$. We denote by K the integral operator $f \mapsto \int K(x,y)f(y)\,d\mu(y)$ on $L_2(X,\mu)$. Its Fredholm determinant $\det(1+TK)$ is the formal series in $1+T\mathbb{C}[[T]]$ defined by

$$\det(1+TK) = 1 + \sum_{n\geq 1}(T^n/n!)\int_{X^n}\det_{n\times n}(K(x(i),x(j)))\prod_i d\mu(i).$$

6.1.2. Suppose now that the kernel $K(x,y)$ can be written as a finite sum $\sum_{i=1}^n f_i(x)g_i(y)$ with f_i and g_i bounded measurable functions on X. Then the operator K is visibly of finite rank: its image lies in the span of the f_i. Therefore we may also speak of the determinant $\det(1+TK)$ in the sense of operators of finite rank, which is a priori an element of $1+T\mathbb{C}[T] \subset 1+T\mathbb{C}[[T]]$. Let us refer to this determinant as the spectral determinant.

6.1.3. The following lemma is well known. We give it for ease of reference.

Lemma 6.1.4. *In the situation of* 6.1.2, *the Fredholm determinant is equal to the spectral determinant.*

PROOF. If $K=0$, both determinants are the constant function 1. If not, pick a basis $\varphi_1,\ldots,\varphi_d$ of the $\mathbb{C}$-span of the f_i's. Expressing the f_i in terms of these, we get an expression of $K(x,y)$ as $\sum\varphi_i(x)\gamma_i(y)$. So it suffices to treat universally the case in which the f_i are linearly independent. Then the image $K(L_2)$ lies in the $\mathbb{C}$-span of the f_i, a K-stable subspace which has basis the f_i. On this basis, K acts as

$$f_i \mapsto K(f_i) = \int_X \Big(\sum f_j(x)g_j(y)\Big) f_i(y)\,d\mu(y) = \sum_j a_{j,i}f_j,$$

where the coefficients $a_{j,i}$ are given by

$$a_{j,i} = \int g_j f_i\,d\mu.$$

So if we denote by A the $n\times n$ matrix $(a_{i,j})$, the spectral determinant $\det(1+TK)$ is equal to $\det(1+TA)$, which is a polynomial of degree at most n.

Let us first check that the Fredholm determinant is also a polynomial of degree at most n. This amounts to showing that

$$\int_{X^d}\det_{d\times d}(K(x(i),x(j)))\prod_i d\mu(i) = 0 \quad \text{for } d>n.$$

In fact $\det_{d\times d}(K(x(i),x(j)))$ vanishes identically for $d>n$. To see this, notice that for any $d\geq 1$, the $d\times d$ matrix of functions $K(x(i),x(j))$ is a matrix product

$$\begin{pmatrix} f_1(x(1)),\ldots,f_n(x(1)) \\ f_1(x(2)),\ldots,f_n(x(2)) \\ \vdots \\ f_1(x(d)),\ldots,f_n(x(d)) \end{pmatrix} \quad \text{times} \quad \begin{pmatrix} g_1(x(1)),\ldots,g_1(x(d)) \\ g_2(x(1)),\ldots,g_2(x(d)) \\ \vdots \\ g_n(x(1)),\ldots,g_n(x(d)) \end{pmatrix}$$

of the shape

$$(d \text{ rows by } n \text{ columns}) \times (n \text{ rows by } d \text{ columns}),$$

intrinsically the matrix of an endomorphism of a d-dimensional space which factors through an n-dimensional space, and hence has rank at most n. So if $d>n$, the $d\times d$ determinant vanishes identically, as asserted.

For each $d = 1, \ldots, n$, we must now compare the coefficient of $T^d/d!$ in the two sorts of $\det(1 + TK)$. On the Fredholm side, this coefficient is

$$\begin{aligned}
&\int_{X^d} \det_{d\times d}(K(x(i), x(j))) \prod_i d\mu(i) \\
&\quad = \int_{X^d} \prod_i d\mu(i) \sum_{\sigma \text{ in } S_d} \operatorname{sgn}(\sigma) \prod_{i=1}^{d} K(x(i), x(\sigma(i))) \\
&\quad = \int_{X^d} \prod_i d\mu(i) \sum_{\sigma \text{ in } S_d} \operatorname{sgn}(\sigma) \prod_{i=1}^{d} \sum_{j=1}^{n} f_j(x(i)) g_j(x(\sigma(i))).
\end{aligned}$$

Expanding out the product, we find

$$= \sum_{J \text{ in } [1,n]^d} \int_{X^d} \prod_i d\mu(i) \sum_{\sigma \text{ in } S_d} \operatorname{sgn}(\sigma) \prod_{i=1}^{d} f_{J(i)}(x(i)) g_{J(i)}(x(\sigma(i))).$$

Putting like variables together, we rewrite this as

$$\begin{aligned}
&\sum_{J \text{ in } [1,n]^d} \int_{X^d} \prod_i d\mu(i) \sum_{\sigma \text{ in } S_d} \operatorname{sgn}(\sigma) \prod_{i=1}^{d} f_{J(i)}(x(i)) G_{J(\sigma^{-1}i)}(x(i)) \\
&\quad = \sum_{J \text{ in } [1,n]^d} \sum_{\sigma \text{ in } S_d} \operatorname{sgn}(\sigma) \prod_{i=1}^{d} \left(\int_X f_{J(i)} g_{J(\sigma^{-1}(i))}\, d\mu \right) \\
&\quad = \sum_{J \text{ in } [1,n]^d} \sum_{\sigma \text{ in } S_d} \operatorname{sgn}(\sigma) \prod_{i=1}^{d} a_{J(i), J(\sigma^{-1}(i))} \\
&\quad = \sum_{J \text{ in } [1,n]^d} \det_{d\times d}(a_{J(i),J(j)}).
\end{aligned}$$

In this last expression, the terms indexed by a J such that the set $\{J(1), \ldots, J(d)\}$ consists of less than d distinct elements all vanish. Indeed, if, say, $J(1) = J(2)$, the $d \times d$ matrix in question has the same first and second rows.

Now consider those J for which the set $\{J(1), J(2), \ldots, J(d)\}$ is a given set $S = \{s_1, s_2, \ldots, s_d\}$ consisting of d distinct elements of $\{1, \ldots, n\}$. For a given S, there are precisely $d!$ distinct J giving rise to it, and for each the contribution is the same, namely it is the $S \times S$ minor of the matrix A. Thus the coefficient of $T^d/d!$ is equal to

$$d! \sum_{S \subset \{1,\ldots,n\}, \operatorname{Card}(S)=d} (\text{the } S \times S \text{ minor of } A) = d! \operatorname{Trace}(\Lambda^d(A)),$$

and this is precisely the coefficient of $T^d/d!$ in $\det(1 + TA)$. QED

6.2. An integration formula

6.2.1. Let us return to the situation of 5.1.3. Thus we are given a measure space (X, μ) with μ a positive measure of finite total mass, and a bounded, Borel measurable $\mathbb{C}$-valued function f on X with

$$\int_X f\, d\mu = 0, \qquad \int_X |f|^2\, d\mu = 1.$$

We suppose that for every integer $n \geq 1$, there exists a monic polynomial in one variable $P_n(T)$ in $\mathbb{C}[T]$ of degree n such that the sequence of functions $\{\varphi_n\}_{n\geq 0}$ on X defined by

$$\varphi_0 := 1/\operatorname{Sqrt}(\mu(X)),$$
$$\varphi_n := P_n(f) \quad \text{for } n \geq 1,$$

is an orthonormal sequence:

$$\int_X \varphi_i \overline{\varphi}_j \, d\mu = \delta_{i,j} \quad \text{for all } i, j \geq 0.$$

For any integer $N \geq 1$, consider the N-fold product X^N. For each $i = 1, \ldots, N$, denote by

$$\operatorname{pr}[i] : X^N \to X, x \mapsto x(i)$$

the i'th projection. Denote by $K_N(x, y)$ the function on X^2

$$K_N(x, y) := \sum_{n=0}^{N-1} \varphi_n(x)\overline{\varphi}_n(y).$$

For each integer $n \geq 1$, we define

$$D(n, N) := \det_{n\times n}(K_N(x(i), x(j))),$$

a function on X^n. We denote by $\mu(n, N)$ the measure on X^n defined by

$$\mu(n, N) := (1/n!)D(n, N)\, d\mu_1 \cdots d\mu_n.$$

Lemma 6.2.2. *In the above situation* 6.2.1, *let* $F : X \to \mathbb{C}$ *be a* $\mathbb{C}$*-valued bounded measurable function on* X*. Fix an integer* $N \geq 1$*. Consider the integral operator of finite rank on* $L_2(X, \mu)$ *with kernel* $K_N(x, y)F(y)$*. Consider also the matrix-valued function*

$$F_{\text{mat}} : X^N \to N \times N \textit{ matrices},$$
$$F_{\text{mat}}(x) := \operatorname{Diag}(f(x(1)), f(x(2)), \ldots, f(x(N))).$$

1) *We have the identity of polynomials in* $\mathbb{C}[T]$

$$\int_{X^N} \det(1 + TF_{\text{mat}}(x))\mu(N, N) = \det(1 + TK_N(x, y)F(y)|L_2(X, \mu)).$$

2) *More generally, given finitely many* $\mathbb{C}$*-valued bounded measurable functions* $F_1, \ldots, F_n$ *on* X*, we have the identity*

$$\int_{X^N} \det\left(1 + \sum_i T_i F_{i,\text{mat}}(x)\right)\mu(N, N) = \det\left(1 + K_N(x, y)\sum_i T_i F_i(y)|L_2(X, \mu)\right)$$

of polynomials in $\mathbb{C}[T_1, \ldots, T_n]$.

Proof. For any t in $\mathbb{C}^n$, the integral operator $K_N(x, y)\sum_i t_i F_i(y)$ has image contained in the $\mathbb{C}$-span of the functions $\varphi_0, \ldots, \varphi_{N-1}$, and the matrix coefficients of its restriction to that space are polynomials (in fact linear forms) in t. So both sides of the identity asserted in 2) are in fact polynomials in T. To see that they coincide as polynomials, it suffices to show that they coincide as functions on $\mathbb{C}^n$. To show that they coincide at a given t in $\mathbb{C}^n$, it suffices to prove 1) for the function $\sum_i t_i F_i$, and then evaluate at $T = 1$.

Assertion 1) is nothing other than Key Lemma 5.1.3, part 5), together with the explicit formula for a Fredholm determinant. Indeed, if we expand out the integrand in terms of the elementary symmetric functions σ_n of N variables, we get

$$\begin{aligned}
&\int_{X^N} \det(1 + TF_{\mathrm{mat}}(x))\mu(N,N) \\
&\quad = \int_{X^N} \prod_i (1 + TF(x(i)))\mu(N,N) \\
&\quad = 1 + \sum_{n\geq 1} T^n \int_{X^N} \sigma_n(F(x(1)), F(x(2)), \ldots, F(x(N)))\mu(N,N).
\end{aligned}$$

The function $\sigma_n(F(x(1)), F(x(2)), \ldots, F(x(N)))$ on X^N is of the form $G_n[n,N]$ for the symmetric function G_n on X^n defined by

$$x \text{ in } X^n \mapsto \prod_i F(x(i)).$$

So by 5.1.3, part 5), we have

$$\begin{aligned}
&\int_{X^N} \sigma_n(F(x(1)), F(x(2)), \ldots, F(x(N)))\mu(N,N) \\
&\quad = \int_{X^N} G_n[n,N]\mu(N,N) = \int_{X^n} G_n\mu(n,N) \\
&\quad = (1/n!) \int_{X^n} \left(\prod_{i=1}^n F(x(i))\right) \det_{n\times n}(K_N(x(i), x(j)))\, d\mu_1 \cdots d\mu_n \\
&\quad = (1/n!) \int_{X^n} \det_{n\times n}(K_N(x(i), x(j))F(x(j)))\, d\mu_1 \cdots d\mu_n,
\end{aligned}$$

and this last expression is precisely the coefficient of T^n in the Fredholm determinant $\det(1 + TK_N(x,y)F(y)|L_2(X,\mu))$. QED

6.3. Integrals of determinants over $G(N)$ as Fredholm determinants

6.3.1. In this section, we give a very simple integral formula, which for a step function reduces to a formula of Tracy-Widom [**T-W**, Thm. 6], and which for a characteristic function goes back to [**Gaudin**] and [**Mehta**, A.7.21]. The only novelty here is in the formulation of the result: once formulated, it very nearly "proves itself".

6.3.2. We begin with an elementary instance of "functional calculus". Suppose we are given a $\mathbb{C}$-valued function $f : S^1 \to \mathbb{C}$ on the unit circle. For each $N \geq 1$, we can uniquely extend f to a function

$$F : U(N) \to N \times N \text{ matrices over } \mathbb{C}$$

which satisfies

$$\begin{aligned}
F(\mathrm{Diag}(\alpha_1, \ldots, \alpha_N)) &= \mathrm{Diag}(f(\alpha_1), \ldots, f(\alpha_N)), \\
F(ABA^{-1}) &= AF(B)A^{-1} \quad \text{for all } A, B \text{ in } U(N).
\end{aligned}$$

To see this, think of $U(N)$ as $\mathrm{Aut}(V, \langle,\rangle)$ for $(V, \langle,\rangle)$ an N-dimensional Hilbert space. Then we define

$$F : \mathrm{Aut}(V, \langle,\rangle) \to \mathrm{End}(V)$$

as follows. Any A in $U(N)$ is semi-simple, so we have a direct sum decomposition of V as $\bigoplus_{\lambda \text{ in } S^1} \operatorname{Ker}(A-\lambda)$. We define $F(A)$ to be the operator

$$\bigoplus_{\alpha \text{ in } S^1} (\text{scalar multiplication by } f(\alpha) \text{ on } \operatorname{Ker}(A-\alpha)).$$

Lemma 6.3.3. *A $\mathbb{C}$-valued function f on S^1 is continuous (resp. bounded and Borel measurable), if and only if the extended function*

$$F : \operatorname{Aut}(V, \langle,\rangle) \to \operatorname{End}(V)$$

is continuous (resp. bounded for the operator norm, and Borel measurable) on $\operatorname{Aut}(V, \langle,\rangle)$.

PROOF. To show the "if", denote by $\mathbb{1}$ the identity element in $\operatorname{Aut}(V, \langle,\rangle)$. Then we may recover f on S^1 from its attached F on $\operatorname{Aut}(V, \langle,\rangle)$ as the composite

$$S^1 \xrightarrow{z \mapsto z\mathbb{1}} \operatorname{Aut}(V, \langle,\rangle) \xrightarrow{F} \operatorname{End}(V) \xrightarrow{(1/N)\operatorname{Trace}} \mathbb{C}.$$

This makes clear that f has whatever good properties F does.

To show the "only if", begin with f on S^1. For f bounded on S^1, and any A in $\operatorname{Aut}(V, \langle,\rangle)$, the explicit recipe for $f(A)$ on the eigenspaces of A shows that the operator norm of $f(A)$ is equal to the sup norm of f on S^1:

$$\|f(A)\| = \|f\|_{\text{sup}}.$$

Any continuous f on $S^1 = \{z \text{ in } \mathbb{C}, |z| = 1\}$ is the uniform limit of trigonometric polynomials, i.e. of Laurent polynomials $f_n(z)$. But for any such Laurent polynomial f_n, $A \mapsto f_n(A)$ is visibly a continuous function of A in $\operatorname{Aut}(V, \langle,\rangle)$. By the above norm equality, the sequence of functions $A \mapsto f_n(A)$ on $\operatorname{Aut}(V, \langle,\rangle)$ converges uniformly to the function $A \mapsto f(A)$, which is therefore continuous.

Here is an argument to show that if f is Borel measurable on S^1, then its F is Borel measurable on $\operatorname{Aut}(V, \langle,\rangle)$. Inside the $\mathbb{C}$-vector space of all $\mathbb{C}$-valued functions on S^1, consider the $\mathbb{C}$-vector subspace, call it $\mathcal{L}$, consisting of those f's whose F is Borel measurable. It is trivial from the definitions that if a sequence of $\mathbb{C}$-valued functions g_n on S^1 converges pointwise on S^1 to a $\mathbb{C}$-valued function g on S^1, then the sequence of associated functions G_n converges pointwise on $\operatorname{Aut}(V, \langle,\rangle)$ to the function G. Since a pointwise limit of Borel measurable functions on $\operatorname{Aut}(V, \langle,\rangle)$ is again Borel measurable, the vector space $\mathcal{L}$ is closed under the operation of taking pointwise limits: if each g_n is in $\mathcal{L}$, so is g. We proved above that $\mathcal{L}$ contains all the continuous functions. Therefore $\mathcal{L}$ contains all "Baire class 1" functions (functions which are pointwise limits of continuous functions), all pointwise limits of Baire class 1 functions, et cetera. Thus $\mathcal{L}$ contains all functions in all the successive Baire classes, i.e., $\mathcal{L}$ contains all the Borel measurable functions. QED

6.3.4. With these preliminaries out of the way, we can state and prove our integration formula. We begin with the unitary group, where the formulation is the most natural.

Theorem 6.3.5. *Let $N \geq 1$. Denote by $K_N(x,y)$ the kernel attached to $U(N)$, viewed as a function on $S^1 \times S^1$. For any bounded, Borel measurable function $f : S^1 \to \mathbb{C}$, consider the integral operator on $L_2(S^1, \mu)$, μ the normalized Haar measure on S^1, with kernel $K_N(x,y)f(y)$. Then we have the polynomial identity*

$$\int_{U(N)} \det(1 + TF(A))\, dA = \det(1 + TK_N(x,y)f(y) | L_2(S^1, \mu)),$$

where we have written dA for the normalized Haar measure on $U(N)$.

More generally, given finitely many $\mathbb{C}$-valued, bounded, Borel measurable functions $f_1, \ldots, f_n$ on S^1, we have the identity

$$\int_{U(N)} \det\left(1 + \sum_i T_i F_i(A)\right) dA = \det\left(1 + K_N(x,y) \sum_i T_i f_i(y) | L_2(S^1, \mu)\right)$$

of polynomials in $\mathbb{C}[T_1, \ldots, T_n]$.

PROOF. This results from the Weyl integration formula on $U(N)$ in its $K_N(x,y)$ form, together with 6.2.2. Just as in 6.2.2, it suffices to prove the first version. We have

$$\begin{aligned}
&\int_{U(N)} \det(1 + TF(A))\, dA \\
&\quad = \int_{(S^1)^N} \det(1 + TF(\mathrm{Diag}(x(i), \ldots, x(N))))\mu(N,N) \\
&\quad = \int_{(S^1)^N} \left(\prod_i (1 + Tf(x(i)))\right) \mu(N,N) \\
&\quad = \int_{(S^1)^N} \det(1 + Tf_{\mathrm{mat}}(x))\mu(N,N) \\
&\quad = \det(1 + TK_N(x,y)f(y)|L_2(S^1, \mu)). \quad \text{QED}
\end{aligned}$$

Variant 6.3.6. *Hypotheses and notations as in Theorem 6.3.5, suppose that $J \subset S^1$ is a Borel measurable set outside of which all the functions f_i vanish. Then we have the identities*

$$\int_{U(N)} \det(1 + TF(A))\, dA = \det(1 + TK_N(x,y)Tf(y)|L_2(J, \mu|J)),$$

$$\int_{U(N)} \det\left(1 + \sum_i T_i F_i(A)\right) dA = \det\left(1 + K_N(x,y) \sum_i T_i f_i(y) | L_2(J, \mu|J)\right).$$

PROOF. Again it suffices to prove the one-variable version, so what must be shown is that if f vanishes outside J, then we have

$$\begin{aligned}
&\det(1 + TK_N(x,y)Tf(y)|L_2(S^1, \mu)) \\
&\quad = \det(1 + TK_N(x,y)f(y)|L_2(J, \mu|J)).
\end{aligned}$$

This is obvious coefficient by coefficient from the explicit integral formula for the coefficients of a Fredholm determinant. QED

Corollary 6.3.7 (Compare [**Gaudin**], [**Mehta**, A.7.27], and [**T-W**, discussion preceding Theorem 6]).

1) *Let $J \subset S^1$ be a Borel measurable set. We have the identity*

$$\det(1 + TK_N(x,y)|L_2(J, \mu|J)) = \sum_{n \geq 0} (1+T)^n \,\mathrm{eigen}(n, J, U(N)),$$

where $\mathrm{eigen}(n, J, U(N))$ *is the normalized Haar measure of the set*

$$\mathrm{Eigen}(n, J, U(N)) \subset U(N)$$

consisting of those elements A exactly n of whose eigenvalues lie in J.

2) *More generally, given a collection $\mathcal{J}$ of $r \geq 1$ pairwise disjoint Borel measurable subsets $J_1, \ldots, J_r$ of S^1, with characteristic functions χ_{J_i}, we have the identity*

$$\det\left(1 + TK_N(x,y)\sum_i T_i\chi_{J_i}(y)|L_2\left(\bigcup_i J_i, \mu|\bigcup_i J_i\right)\right)$$
$$= \sum_{n\geq 0 \text{ in } \mathbb{Z}^r} (1+T)^n \operatorname{eigen}(n, \mathcal{J}, U(N)),$$

where for each $n \geq 0$ in $\mathbb{Z}^r$, $\operatorname{eigen}(n, \mathcal{J}, U(N))$ is the normalized Haar measure of the set $\operatorname{Eigen}(n, J, U(N)) \subset U(N)$ consisting of those elements A exactly n_i of whose eigenvalues lie in J_i for each i.

PROOF. To prove the single J case, apply the previous result to f the characteristic function of J. Then the integrand $\det(1 + TF(A))$ is equal to $(1+T)^n$ precisely on the set $\operatorname{Eigen}(n, J, U(N))$, so we have

$$\int_{U(N)} \det(1 + TF(A))\,dA = \sum_{n\geq 0}(1+T)^n \operatorname{eigen}(n, J, U(N)).$$

On the other side, we have

$$\det(1 + TK_N(x,y)f(y)|L_2(J,\mu|J)) = \det(1 + TK_N(x,y)|L_2(J,\mu|J)),$$

exactly because f is the characteristic function of J. To do the $\mathcal{J}$ case, apply the previous result to the functions $f_i := \chi_{J_i}$. Because the J_i's are disjoint, the integrand $\det(1 + \sum_i T_iF_i(A))$ takes the value $(1+T)^n$ precisely on the set $\operatorname{Eigen}(n, J, U(N))$. QED

6.3.8. We now turn to the other $G(N)$, where the formulation is more cumbersome. Thus we fix an integer $N \geq 1$, and take $G(N)$ to be one of $SO(2N+1)$, $USp(2N), O_-(2N+2), SO(2N)$. In the discussion 5.1.5–8 of the Weyl integration formula for these $G(N)$, the measure space (T,μ) which underlay the theory had $T = [0,\pi]$, and the measure μ was given by the following table:

$G(N)$	measure μ on $[0,\pi]$
$USp(2N)$	$(2/\pi)\sin^2(x)\,dx$
$SO(2N+1)$	$(2/\pi)\sin^2(x/2)\,dx$
$SO(2N)$	$dx/2\pi$
$O_-(2N+2)$	$(2/\pi)\sin^2(x)\,dx$

We denote by $K_N(x,y)$ the K_N-kernel on $[0,\pi]$ attached to $G(N)$.

Theorem 6.3.9. *Fix an integer $N \geq 1$, and take $G(N)$ to be one of $SO(2N+1), USp(2N), O_-(2N+2), SO(2N)$. Let $f : [0,\pi] \to \mathbb{C}$ be a bounded, Borel measurable function on $[0,\pi]$ with $f(0) = f(\pi) = 0$. We extend f by zero to a function F on $[0,2\pi) \cong S^1$. Viewing each $G(N)$ inside its ambient unitary group $U(2N+2\lambda)$ of the same size, we consider the composite map*

$$G(N) \subset U(2N+2\lambda) \xrightarrow{F} \{(2N+2\lambda)\times(2N+2\lambda) \text{ matrices}\},$$

which we continue to denote $A \mapsto F(A)$. Then we have the integration formula

$$\int_{G(N)} \det(1 + TF(A))\,dA = \det(1 + TK_N(x,y)f(y)|L_2(T,\mu)),$$

where we write dA for the normalized Haar measure on $G(N)$.

More generally, given finitely many $\mathbb{C}$-valued bounded measurable functions $f_1, \ldots, f_n$ on $[0, \pi]$, each of which vanishes at the endpoints, we have the identity

$$\int_{G(N)} \det\left(1 + \sum_i T_i F_i(A)\right) dA = \det\left(1 + K_N(x, y) \sum_i T_i f_i(y) | L_2(T, \mu)\right)$$

of polynomials in $\mathbb{C}[T_1, \ldots, T_n]$.

PROOF. As in the proof of the $U(N)$ case, the several variable version results from the one-variable version. Given an element A in $G(N)$, denote by $\varphi(A)$ in $[0, \pi]^N$ its "vector of eigenvalue angles",

$$0 \leq \varphi(1)(A) \leq \varphi(2)(A) \leq \cdots \leq \varphi(N)(A) \leq \pi$$

as defined in 2.0.3–6. Then the eigenvalues of A are the $2N + 2\lambda$ numbers

$$e^{\pm i\varphi(j)(A)}, \qquad j = 1, \ldots, N,$$

together with

$$\{1 \text{ if } G(N) \text{ is } SO(2N+1)\}, \qquad \{\pm 1 \text{ if } G(N) \text{ is } O_-(2N+2)\}.$$

Because $f(0) = f(\pi) = 0$, we have $F(1) = F(-1) = 0$, so the only eigenvalues of A at which F is possibly nonzero are the $e^{i\varphi(j)(A)}$, $j = 1, \ldots, N$. Thus $F(A)$ has as its possibly nonzero eigenvalues precisely the N quantities $F(e^{i\varphi(j)(A)})$, $j = 1, \ldots, N$.

But tautologically we have $f(x) = F(e^{ix})$ for any x in $[0, \pi]$, so we have

$$\det(1 + TF(A)) = \prod_j (1 + Tf(\varphi(j)(A))) = \det(1 + Tf_{\text{mat}}(\varphi(A))).$$

Thus we have

$$\int_{G(N)} \det(1 + TF(A))\, dA = \int_{G(N)} \det(1 + Tf_{\text{mat}}(\varphi(A)))\, dA$$

$$= \int_{T^N} \det(1 + Tf_{\text{mat}}(x))\mu(N, N) = \det(1 + TK_N(x, y)f(y)|L_2(T, \mu)). \quad \text{QED}$$

Variant 6.3.10. *Hypotheses and notations as in 6.3.9, denote by $L_N(x, y)$ the kernel attached to $G(N)$ in 5.2.3–5. Then*

$$\int_{G(N)} \det(1 + TF(A))\, dA = \det(1 + TL_N(x, y)f(y)|L_2([0, \pi], dx/\pi)),$$

and

$$\int_{G(N)} \det\left(1 + \sum_i T_i F_i(A)\right) dA = \det\left(1 + L_N(x, y) \sum_i T_i f_i(y) | L_2([0, \pi], dx/\pi)\right).$$

PROOF. Again it suffices to prove the one-variable statement, i.e., to show that

$$\det(1 + TL_N(x, y)f(y)|L_2([0, \pi], dx/\pi))$$
$$= \det(1 + TK_N(x, y)f(y)|L_2([0, \pi], \mu)).$$

But the measure μ is of the form $D(x)\, dx/\pi$, and $L_N(x, y)$ is defined to be $D(x)D(y)K_N(x, y)$, so the identity is obvious coefficient by coefficient. More intrinsically, the map "multiplication by $D(x)$" is a unitary isomorphism from $L_2([0, \pi], \mu)$ to $L_2([0, \pi], dx/\pi)$ which carries the integral operator on $L_2([0, \pi], \mu)$ with kernel $K_N(x, y)f(y)$ to that on $L_2([0, \pi], dx/\pi)$ with kernel $L_N(x, y)f(y)$. QED

Variant 6.3.11. *Hypotheses and notations as in* 6.3.9, *suppose that* $J \subset [0,\pi]$ *is a Borel measurable set outside of which all the functions* f_i *vanish. Then we have the identities*

$$\int_{G(N)} \det(1+TF(A))\,dA = \det(1+TK_N(x,y)f(y)|L_2(J,\mu))$$
$$= \det(1+TL_N(x,y)f(y)|L_2(J,dx/\pi)),$$

and

$$\int_{G(N)} \det\left(1+\sum_i T_iF_i(A)\right) dA = \det\left(1+K_N(x,y)\sum_i T_if_i(y)|L_2(J,\mu)\right)$$
$$= \det\left(1+L_N(x,y)\sum_i T_if_i(y)|L_2(J,dx/\pi)\right).$$

6.3.12. Here is another version of the result for $G(N)$ which does not impose the "vanishing at the endpoints" condition. Its only disadvantage is that the integrand is specific to the particular $G(N)$ in question, rather than being the restriction to $G(N)$ of a single integrand on the ambient unitary group.

Proposition 6.3.13. *Suppose* f *is a bounded, Borel measurable function on* $[0,\pi]$. *Given an element* A *in one of* $SO(2N+1), USp(2N), O_-(2N+2), SO(2N)$, *with "angles"*

$$0 \le \varphi(1)(A) \le \varphi(2)(A) \le \cdots \le \varphi(N)(A) \le \pi,$$

form the expression

$$\prod_j (1+Tf(\varphi(j)(A))),$$

which for variable A *is a bounded, Borel measurable function on* $G(N)$ (*the individual* $\varphi(i)(A)$ *are continuous functions of* A *for these* $G(N)$). *Let* J *be a measurable subset of* $[0,\pi]$ *outside of which* f *vanishes. Then we have the integration formulas*

$$\int_{G(N)} \left(\prod_j (1+Tf(\varphi(j)(A)))\right) dA = \det(1+TK_N(x,y)f(y)|L_2(J,\mu)),$$
$$= \det(1+TL_N(x,y)f(y)|L_2(J,dx/\pi)).$$

Similarly for several f_i, *all of which vanish outside of* J:

$$\int_{G(N)} \left(\prod_j \left(1+\sum_i T_if_i(\varphi(j)(A))\right)\right) dA = \det\left(1+K_N(x,y)\sum_i T_if_i(y)|L_2(J,\mu)\right)$$
$$= \det\left(1+L_N(x,y)\sum_i T_if_i(y)|L_2(J,dx/\pi)\right).$$

PROOF. Immediate from 6.2.2 and the Weyl integration formula. QED

6.3.14. Exactly as in the $U(N)$ case above, we find

Corollary 6.3.15 (Compare [**Gaudin**], [**Mehta**, A.7.27], and [**T-W**, discussion preceding Theorem 6]). *Let* $G(N)$ *be one of* $SO(2N+1)$, $USp(2N)$, $O_-(2N+2)$, *or* $SO(2N)$.

1) *Let $J \subset [0, \pi]$ be a Borel measurable set. We have the identity*

$$\det(1 + TL_N(x, y)|L_2(J, dx/\pi)) = \sum_{n\geq 0}(1+T)^n \operatorname{eigen}(n, J, G(N)),$$

where $\operatorname{eigen}(n, J, G(N))$ *is the normalized ($G(N)$ gets total mass one) Haar measure of the set* $\operatorname{Eigen}(n, J, G(N)) \subset G(N)$ *consisting of those elements A exactly n of whose angles $\{\varphi(i)(A)\}_{i=1}^N$ lie in J.*

2) *More generally, given a collection $\mathcal{J}$ of $r \geq 1$ pairwise disjoint Borel measurable subsets $J_1, \ldots, J_r$ of $[0, \pi]$, with characteristic functions χ_{J_i}, we have the identity*

$$\det\left(1 + L_N(x, y)\sum_i T_i\chi_{J_i}(y)|L_2\left(\bigcup_i J_i, dx/\pi\right)\right)$$
$$= \sum_{n\geq 0 \text{ in } \mathbb{Z}^r}(1+T)^n \operatorname{eigen}(n, \mathcal{J}, G(N)),$$

where for each $n \geq 0$ in $\mathbb{Z}^r$, $\operatorname{eigen}(n, \mathcal{J}, G(N))$ *is the normalized Haar measure of the set* $\operatorname{Eigen}(n, J, G(N)) \subset G(N)$ *consisting of those elements A exactly n_i of whose angles $\{\varphi(i)(A)\}_{i=1}^N$ lie in J_i for each $i = 1, \ldots, r$.*

6.4. A new special case: $O_-(2N+1)$

6.4.1. In previous discussions, we never considered separately the case of $O_-(2N+1) = (-1)SO(2N+1)$, because our main interest was in spacings, and these are insensitive to replacing A by $-A$. Moreover, the Haar measure on $O_-(2N+1)$ (i.e., the restriction from $O(2N+1)$ of its Haar measure, but normalized to give $O_-(2N+1)$ total mass one) is, via the bijection $A \mapsto -A$ of $SO(2N+1)$ with $O_-(2N+1)$, just the (direct image of) Haar measure on $SO(2N+1)$. However, when we come to questions of **location** of eigenvalues, the situation is no longer the same for $SO(2N+1)$ as for $O_-(2N+1)$.

6.4.2. Let us be more precise. Given an element B in $O_-(2N+1)$, there is a unique sequence of N angles

$$0 \leq \varphi(1)(B) \leq \varphi(2)(B) \leq \cdots \leq \varphi(N)(B) \leq \pi$$

such that the $2N+1$ eigenvalues of B are $\{-1\}$ together with the $2N$ quantities $e^{\pm i\varphi(j)(B)}$, $j = 1, \ldots, N$.

6.4.3. On the other hand, the element $A := -B$ lies in $SO(2N+1)$, and its eigenvalues are thus $\{1\}$ together with the $2N$ quantities

$$-e^{\pm i\varphi(j)(B)} = e^{i\pi}e^{\pm i\varphi(j)(B)} = e^{\pm i(\pi - \varphi(j)(B))}, \qquad j = 1, \ldots, N.$$

So the angles of A are related to those of $B := -A$ by

$$\varphi(j)(A) = \pi - \varphi(N+1-j)(B).$$

In other words, the involution $x \mapsto \pi - x$ interchanges the angles of A with those of B.

Lemma 6.4.4. *Let f be a $\mathbb{C}$-valued, bounded, Borel measurable function on $[0, \pi]$, and denote by g the function on $[0, \pi]$ defined by*

$$g(x) = f(\pi - x).$$

Then we have the identity

$$\int_{O_-(2N+1)} \left(\prod_j (1 + Tf(\varphi(j)(B))) \right) dB$$
$$= \int_{SO(2N+1)} \left(\prod_j (1 + Tg(\varphi(j)(A))) \right) dA,$$

and similarly for several f_i's:

$$\int_{O(2N+1)} \left(\prod_j \left(1 + \sum_i T_i f_i(\varphi(j)(B)) \right) \right) dB$$
$$= \int_{SO(2N+1)} \left(\prod_j \left(1 + \sum_i T_i g_i(\varphi(j)(A)) \right) \right) dA.$$

6.4.5. Combining 6.4.4 with our integration formula 6.3.15 for $SO(2N+1)$, we obtain

Corollary 6.4.6. *Denote by $L_N(x,y)$ the L_N kernel for $SO(2N+1)$. Denote by $G(N)$ the kernel*

$$G_N(x,y) := L_N(\pi - x, \pi - y).$$

1) *For f any $\mathbb{C}$-valued, bounded, Borel measurable function on $[0,\pi]$, and any Borel measurable subset $J \subset [0,\pi]$ outside of which f vanishes, we have the identity*

$$\int_{O_-(2N+1)} \left(\prod_j (1 + Tf(\varphi(j)(B))) \right) dB$$
$$= \det(1 + TG_N(x,y)f(y)|L_2(J, dx/\pi)).$$

In particular, taking for f the characteristic function of J, we have the identity

$$\sum_{n \geq 0} (1+T)^n \operatorname{eigen}(n, J, O_-(2N+1)) = \det(1 + TG_N(x,y)|L_2(J, dx/\pi)).$$

2) *Similarly, for several f_i, all of which vanish outside of J, we have*

$$\int_{O_-(2N+1)} \left(\prod_j \left(1 + \sum_i T_i f_i(\varphi(j)(B)) \right) \right) dB$$
$$= \det \left(1 + G_N(x,y) \sum_i T_i f_i(y) | L_2(J, dx/\pi) \right).$$

In particular, given a collection $\mathcal{J}$ of $r \geq 1$ pairwise disjoint Borel measurable subsets $J_1, \ldots, J_r$ of $[0,\pi]$, and taking for the f_i their characteristic functions χ_{J_i}, we have the identity

$$\det \left(1 + G_N(x,y) \sum_i T_i \chi_{J_i}(y) | L_2 \left(\bigcup_i J_i, dx/\pi \right) \right)$$
$$= \sum_{n \geq 0 \text{ in } \mathbb{Z}^r} (1+T)^n \operatorname{eigen}(n, \mathcal{J}, O_-(2N+1)),$$

where for each $n \geq 0$ in $\mathbb{Z}^r$, $\mathrm{eigen}(n, \mathcal{J}, O_-(2N+1))$ *is the normalized Haar measure of the set* $\mathrm{Eigen}(n, J, O_-(2N+1)) \subset O_-(2N+1)$ *consisting of those elements B exactly n_i of whose angles $\{\varphi(i)(B)\}_{i=1}^N$ lie in J_i for each $i = 1, \ldots, r$.*

6.4.7. We will refer to $G_N(x,y)$ as the L_N kernel for $O_-(2N+1)$. It differs by a quite interesting sign from the L_N kernel for $SO(2N+1)$, as we will now show. Recall (5.4.5) that the L_N kernel for $SO(2N+1)$ is given in terms of the function $S_N(x) := \sin(Nx/2)/\sin(x/2)$ by the formula

$$L_N(x,y) = (1/2)(S_{2N}(x-y) - S_{2N}(x+y)).$$

Lemma 6.4.8. *The L_N kernel for $O_-(2N+1)$ is given by*

$$G_N(x,y) = (1/2)(S_{2N}(x-y) + S_{2N}(x+y)).$$

PROOF. We readily calculate

$$\begin{aligned} G_N(x,y) &:= L_N(\pi - x, \pi - y) \\ &= (1/2)(S_{2N}((\pi - x) - (\pi - y)) - S_{2N}((\pi - x) + (\pi - y))) \\ &= (1/2)(S_{2N}(y - x) - S_{2N}(2\pi - (x+y))). \end{aligned}$$

So our assertion comes down to the two identities

$$S_{2N}(x) = S_{2N}(-x), \qquad S_{2N}(2\pi - x) = -S_{2N}(x).$$

But for any integer j, $S_j(x)$ is an even function of x, and satisfies

$$S_j(2\pi - x) = (-1)^{j+1} S_j(x).$$

Let us deduce this last identity from the representation theory of $SU(2)$. Remember that $T_j(x) := S_j(2x)$ is the character (trace of $\mathrm{Diag}(e^{ix}, e^{-ix})$) of the j-dimensional irreducible representation ρ_j of $SU(2)$, which is

$$\mathrm{Symm}^{j-1}(\mathrm{std}_2).$$

Since the element -1 acts as the scalar -1 in std_2, it acts as the scalar $(-1)^{j-1}$ in $\mathrm{Symm}^{j-1}(\mathrm{std}_2) = \rho_j$. Therefore for any A in $SU(2)$, we have

$$\rho_j(-A) = (-1)^{j-1}\rho_j(A),$$

and taking traces gives

$$\mathrm{Trace}(\rho_j(-A)) = (-1)^{j-1}\,\mathrm{Trace}(\rho_j(A)).$$

If $A = \mathrm{Diag}(e^{ix}, e^{-ix})$, $-A$ is $SU(2)$-conjugate to $\mathrm{Diag}(e^{i(\pi - x)}, e^{-i(\pi - x)})$, whence $T_j(\pi - x) = (-1)^{j-1} T_j(x)$, as required. QED

6.4.9. For ease of later reference, we record here the fact that the L_N kernel for $O_-(2N+1)$ does indeed provide for $O_-(2N+1)$ the analogue of the integration formulas of 5.1.3, part 5), and 5.5.

Lemma 6.4.10. *Let $1 \leq n \leq N$ be integers. Let F be a $\mathbb{C}$-valued, bounded, Borel measurable function on $[0,\pi]^n$ which is Σ_n-invariant, and denote by $F[n,N]$ the function on $[0,\pi]^N$ defined by*

$$F[n,N](t(1),\ldots,t(N)) := \sum_{1 \leq i(1) < \cdots < i(n) \leq N} F(t(i(1)),\ldots,t(i(n))).$$

For B in $O_-(2N+1)$, denote by $X(B)$ in $[0,\pi]^N$ its vector of angles of eigenvalues. Denote by dB the total mass one Haar measure on $O_-(2N+1)$. Denote by $G_N(x,y)$ the kernel

$$G_N(x,y) = (1/2)(S_{2N}(x-y) + S_{2N}(x+y)).$$

Denote by $\mu(n,N)$ the measure on $[0,\pi]^n$ defined by

$$\mu(n,N) = (1/n!)\det_{n\times n}(G_N(x(i),x(j)))\prod_i (dx(i)/\pi).$$

Then we have the integration formula

$$\int_{O_-(2N+1)} F[n,N](X(B))\,dB = \int_{[0,\pi]^n} F\,d\mu(n,N).$$

PROOF. This results immediately from the corresponding integration formula on $SO(2N+1)$, applied to the function $H(X) := F(\pi\mathbb{1}_n - X)$ on $[0,\pi]^n$. We have, by 5.1.3, part 5), 5.5.3, and 6.4.8,

$$\int_{SO(2N+1)} H[n,N](X(A))\,dA$$
$$= (1/n!)\int_{[0,\pi]^n} H(x(1),\ldots,x(n))\det_{n\times n}(L_N(x(i),x(j)))\prod_i (dx(i)/\pi),$$

for $L_N(x,y)$ the kernel $(1/2)(S_{2N}(x-y) - S_{2N}(x+y)) = G_N(\pi - x, \pi - y)$.

Making the change of variable $B := -A$, we have

$$H[n,N](X(A)) = H[n,N](\pi\mathbb{1}_n - X(B)) = F[n,N](X(B)),$$

and the first integral becomes

$$\int_{SO(2N+1)} H[n,N](X(A))\,dA = \int_{O_-(2N+1)} F[n,N](X(B))\,dB.$$

Make the change of variable $x(i) \mapsto \pi - x(i)$, and recall that

$$G_N(x,y) := L_N(\pi - x, \pi - y).$$

The second integral becomes

$$(1/n!)\int_{[0,\pi]^n} H(x(1),\ldots,x(n))\det_{n\times n}(L_N(x(i),x(j)))\prod_i (dx(i)/\pi)$$
$$= (1/n!)\int_{[0,\pi]^n} F(x(1),\ldots,x(n))\det_{n\times n}(G_N(x(i),x(j)))\prod_i (dx(i)/\pi).$$

Thus we find the asserted integration formula. QED

6.5. Interlude: A determinant-trace inequality

Lemma 6.5.1. *For real x in $[0,1]$, we have the inequality*

$$1 - x \le e^{-x}.$$

PROOF. For $x = 1$, the assertion is obvious. For x in $[0,1)$, the assertion is equivalent to

$$e^x \le 1/(1-x)$$

which is obvious from the power series expansions. QED

Corollary 6.5.2. *Suppose that* $\lambda_1 \geq \lambda_2 \geq \lambda_3 \geq \cdots$ *is a decreasing sequence of real numbers in* $[0,1]$, *such that* $\sum_i \lambda_i$ *converges, say to* S. *Denote by* $E(T)$ *the entire function* $\prod_i (1 + \lambda_i T)$, *and consider its power series expansion around the point* $T = -1$, *say*

$$E(T) = \sum_{n \geq 0} (1+T)^n E_n.$$

Then the expansion coefficients E_n *all lie in* $[0,1]$, $\sum_n E_n = 1$, *and we have the (crude) estimate*

$$0 \leq E_n \leq 2^n e^{-S/2}.$$

We have the more precise estimates

$$0 \leq E_0 \leq e^{-S},$$

and

$$0 \leq E_n \leq (e/n)^n S^n e^{-S} \quad \textit{for each } n \textit{ satisfying } 1 \leq n \leq S.$$

Proof. Write $E(T)$ as $\prod_i ((1-\lambda_i) + \lambda_i(1+T))$ to see that each E_n is ≥ 0. If we evaluate $E(T) = \prod_i (1+\lambda_i T) = \sum_{n \geq 0} (1+T)^n E_n$ at $T = 0$, we get $\sum_n E_n = 1$. Thus each E_n lies in $[0,1]$.

To estimate the E_n, we argue as follows. We have

$$E_0 = E(-1) = \prod_i (1-\lambda_i) \leq \prod_i e^{-\lambda_i} = e^{-S}.$$

To estimate E_n for $n \geq 1$, we use the calculus of residues. By the Cauchy formula, for any real $r > 0$ we have

$$\begin{aligned} E_n &= (1/2\pi i) \int_{|1+T|=r} (E(T)/(1+T)^n)\, d\log(1+T) \\ &= (1/2\pi) \int_{[0,2\pi]} (E(re^{i\vartheta} - 1)/(re^{i\vartheta})^n)\, d\vartheta, \end{aligned}$$

so we have

$$\begin{aligned} |E_n| &\leq (1/2\pi) \int_{[0,2\pi]} |E(re^{i\vartheta} - 1)/(re^{i\vartheta})^n|\, d\vartheta \\ &\leq (r^{-n}) \operatorname*{Sup}_{\vartheta} |E(re^{i\vartheta} - 1)|. \end{aligned}$$

Take $r \leq 1$. Then for any real ϑ we have

$$\begin{aligned} |E(re^{i\vartheta} - 1)| &= \prod_i |(1-\lambda_i) + \lambda_i(re^{i\vartheta})| \leq \prod_i ((1-\lambda_i) + r\lambda_i) \\ &= \prod_i (1 - (1-r)\lambda_i) \leq \prod_i e^{-(1-r)\lambda_i} = e^{-(1-r)S} = e^{-S} e^{rS}. \end{aligned}$$

Thus we find that for each $n \geq 1$ and each r in $(0,1]$, we have an estimate

$$0 \leq E_n \leq e^{-S} e^{rS} / r^n.$$

If we take $r = 1/2$, we get the asserted estimate $E_n \leq 2^n e^{-S/2}$.

The attentive reader will have noticed that for $n \geq S$, this estimate for E_n is worse than the trivial estimate $E_n \leq 1$. The most extreme instance of this failure of our method is when $S = 0$: then all $\lambda_i = 0$, so $E(T)$ is the constant function 1, $E_0 = 1$, and $E_n = 0$ for $n \geq 1$.

To analyse the general case, we fix $S > 0$ and $n \geq 1$, and we view $e^{-S}e^{rS}/r^n$ as a function of the strictly positive real variable r. This function is strictly decreasing in $(0, n/S)$, attains its minimum at $r = n/S$, and is then strictly increasing in $(n/S, \infty)$. So if $S \geq n$ we may take $r = n/S$ (remember we need $r \leq 1$). But if $S \leq n$, our function is decreasing on $(0, 1)$, so the "best" r is $r = 1$, giving only the trivial estimate $E_n \leq 1$. QED

6.5.3. If we take the λ_i to be the eigenvalues of an operator A of finite rank, then $E(T)$ becomes $\det(1 + TA)$, S becomes $\mathrm{Trace}(A)$, and we get

Corollary 6.5.4 (determinant-trace inequality). *Let A be an endomorphism of finite rank of a $\mathbb{C}$-vector space, all of whose eigenvalues are real numbers in $[0, 1]$. Put $S := \mathrm{Trace}(A)$. Denote by $E(T)$ the characteristic polynomial*

$$E(T) := \det(1 + TA),$$

and consider its expansion around $T = -1$,

$$E(T) = \sum_{n \geq 0} (1 + T)^n E_n.$$

Then the expansion coefficients E_n all lie in $[0, 1]$, $\sum_n E_n = 1$, and we have the (crude) estimate

$$0 \leq E_n \leq 2^n e^{-S/2}.$$

We have the more precise estimates

$$0 \leq E_0 \leq e^{-S}$$

and

$$0 \leq E_n \leq (e/n)^n S^n e^{-S} \quad \textit{for each } n \textit{ satisfying } 1 \leq n \leq S.$$

6.6. First application of the determinant-trace inequality

6.6.1. Recall that for each integer $N \geq 1$, $S_N(x)$ is the function

$$\sin(Nx/2)/\sin(x/2) = \sum_{j=0}^{N-1} e^{i(N-1-2j)x/2}.$$

We defined kernels $S_N(x, y)$ and $S_{\pm,N}(x, y)$ as follows:

$$S_N(x, y) := S_N(x - y),$$
$$S_{\pm,N}(x, y) := S_N(x, y) \pm S_N(-x, y).$$

For real α and real $s \geq 0$, we define integral operators

$K_{N,s,\alpha}$:= the integral operator with kernel $S_N(x, y)$ on $L_2([\alpha, \alpha + s], dx/2\pi)$,
$K_{\pm,N,s}$:= the integral operator with kernel $S_{\pm,N}(x, y)$ on $L_2([0, s], dx/2\pi)$
= the integral operator with kernel
$(1/2)S_{\pm,N}(x, y) = (1/2)(S_N(x, y) \pm S_N(-x, y))$
on $L_2([0, s], dx/\pi)$.

Lemma 6.6.2. *For real s in $[0, 2\pi]$, and any real α, the integral operator $K_{N,s,\alpha}$ on $L_2([\alpha, \alpha + s], dx/2\pi)$, whose unitary isomorphism class is independent of α, is positive (and hence self-adjoint) of finite rank, and all its eigenvalues lie in $[0, 1]$. Its Fredholm determinant $E(N, T, s)$ is equal to its spectral determinant.*

PROOF. The integral operator $K_{N,s,\alpha}$ is of finite rank, because its kernel is a finite sum of terms $f(x)g(y)$. So its spectral determinant is equal to its Fredholm determinant. To show that all its eigenvalues lie in $[0,1]$, it suffices to show that this operator is positive, and of operator norm ≤ 1. We have an orthogonal direct sum decomposition

$$\begin{aligned}&L_2([\alpha,\alpha+2\pi],dx/2\pi)\\&\quad=L_2([\alpha,\alpha+s],dx/2\pi)\oplus L_2([\alpha+s,\alpha+2\pi],dx/2\pi).\end{aligned}$$

We denote by $P(s)$ the orthogonal projection of $L_2([\alpha,\alpha+2\pi],dx/2\pi)$ onto $L_2([\alpha,\alpha+s],dx/2\pi)$, and by $K_{N,\alpha}$ the integral operator on $L_2([\alpha,\alpha+2\pi],dx/2\pi)$ with kernel $S_N(x,y)$. Then we have the direct sum decomposition

$$P(s)\circ K_{N,\alpha}\circ P(s)=K_{N,s,\alpha}\oplus 0.$$

To show that $K_{N,s,\alpha}$ is positive and of operator norm ≤ 1, it suffices, using the direct sum decomposition, to show that $P(s)\circ K_{N,\alpha}\circ P(s)$ is positive and of operator norm ≤ 1. To show this, it suffices in turn to show that $K_{N,\alpha}$ is positive and of operator norm ≤ 1. But $K_{N,\alpha}$ is unitarily conjugate (via multiplication by the function $e^{i(N-1)x/2}$) to the operator given by the kernel

$$\sum_{0\leq j\leq N-1} e^{-ij(x-y)}.$$

So it suffices to show that this operator is positive and of operator norm ≤ 1. But this last operator is an orthogonal projection (onto the span of the N functions e^{-ijx}, $0\leq j\leq N-1$), so is certainly a positive operator of norm ≤ 1. QED

6.6.3. Since the function $S_N(x)$ is even, the kernel K_N satisfies

$$K_N(x,-y)=K_N(-x,y),$$

and hence the operator $K_{N,2s,\alpha=-s}$ on $L_2([-s,s],dx/2\pi)$ respects the subspaces of odd and even functions.

6.6.4. More precisely, we have the orthogonal decomposition

$$\begin{aligned}L_2([-s,s],dx/2\pi)&=L_2([-s,s],dx/2\pi)_{\text{even}}\oplus L_2([-s,s],dx/2\pi)_{\text{odd}},\\ f&=f_++f_-,\end{aligned}$$

with

$$f_\pm(x):=(1/2)(f(x)\pm f(-x)).$$

6.6.5. Via the isometric isomorphisms

$$\begin{aligned}\text{Restriction}:L_2([-s,s],dx/2\pi)_{\text{even}}&\cong L_2([0,s],dx/\pi),\\ \text{Restriction}:L_2([-s,s],dx/2\pi)_{\text{odd}}&\cong L_2([0,s],dx/\pi),\end{aligned}$$

we get an orthogonal direct sum decomposition

$$L_2([-s,s],dx/2\pi)=L_2([0,s],dx/\pi)\oplus L_2([0,s],dx/\pi), \tag{6.6.6}$$

of spaces and of operators

$$K_{N,2s,\alpha=-s}=K_{+,N,s}\oplus K_{-,N,s}. \tag{6.6.7}$$

From this direct sum decomposition, we obtain

Corollary 6.6.8. *For real s in $[0,\pi]$, and each choice of $\pm$, the integral operator $K_{\pm,N,s}$ on $L_2([0,s], dx/\pi)$ with kernel*

$$(1/2)(S_N(x,y) \pm S_N(-s,y))$$

is of finite rank, positive and self-adjoint, and all its eigenvalues lie in $[0,1]$. Its Fredholm determinant $E_\pm(N,T,s)$ is equal to its spectral determinant. We have the product formula

$$E(N,T,2s) = E_+(N,T,s)E_-(N,T,s).$$

6.6.9. Our next task is to compute the traces of these operators.

Lemma 6.6.10. 1) *For any real s in $[0,\pi]$ and any integer $N \geq 1$, we have the exact formula*

$$\mathrm{Trace}(K_{N,2s}) = Ns/\pi.$$

2) *For any real s in $(0,\pi/2]$ and any integer $N \geq 1$, we have the estimates*

$$|\,\mathrm{Trace}(K_{\pm,N,s}) - Ns/2\pi| \leq (1/2\pi)(1 + (\pi/2)|\log(sN)|).$$

Proof. By the compatibility of spectral and Fredholm determinants for an integral operator K on a finite measure space (X,μ) with a bounded kernel $K(x,y) = \sum_{i=1}^n f_i(x)g_i(y)$ of finite rank, its trace is the integral of $K(x,y)$ over the diagonal:

$$\mathrm{Trace}(K) = \int_X K(x,x)\,d\mu.$$

Since the function $S_N(x) := \sin(Nx/2)/\sin(x/2)$ takes the value N at $x = 0$, the kernel $S_N(x,y) := S_N(x-y)$ when restricted to the diagonal is the constant function N, and hence we get

$$\mathrm{Trace}(K_{N,2s}) = \int_{[-s,s]} S_N(x,x)\,dx/2\pi = \int_{[-s,s]} N\,dx/2\pi = Ns/\pi.$$

Similarly, we get

$$\begin{aligned}
&\mathrm{Trace}(K_{\pm,N,s})\\
&= (1/2)\int_{[0,s]} S_N(x,x)\,dx/\pi \pm (1/2)\int_{[0,s]} S_N(-x,x)\,dx/\pi\\
&= (1/2)\int_{[0,s]} N\,dx/\pi \pm (1/2)\int_{[0,s]} S_N(-2x)\,dx/\pi\\
&= Ns/2\pi \pm (1/2\pi)\int_{[0,s]} (\sin(Nx)/\sin(x))\,dx.
\end{aligned}$$

It now suffices to prove the estimate

$$\int_{[0,s]} |\sin(Nx)/\sin(x)|\,dx \leq 1 + (\pi/2)|\log(Ns)|,$$

provided that s lies in $(0,\pi/2]$ and $N \geq 1$.

The integrand $|\sin(Nx)/\sin(x)|$ is bounded by N (think of $\sin(Nx)/\sin(x)$ as the character of the N-dimensional irreducible representation of $SU(2)$), so we have the trivial estimate

$$\int_{[0,s]} |\sin(Nx)/\sin(x)|\,dx \leq Ns.$$

For $s \leq 1/N$, this gives what we need. For $s > 1/N$, break the interval $[0, s]$ into $[0, 1/N]$ and $[1/N, s]$, and use the inequality above to bound the integral over $[0, 1/N]$ by 1. It remains to check that

$$\int_{[1/N,s]} |\sin(Nx)/\sin(x)|\, dx \leq (\pi/2)\log(Ns)$$

for s in $[1/N, \pi/2]$.

For x in $[0, \pi/2]$, we have $\sin(x) \geq 2x/\pi$. Since $|\sin(Nx)| \leq 1$ for all real x, we have

$$\int_{[1/N,s]} |\sin(Nx)/\sin(x)|\, dx \leq \int_{[1/N,s]} (\pi/2x)\, dx = (\pi/2)\log(Ns),$$

as required. QED

6.6.11. We can now apply the determinant-trace inequality 6.5.4 to estimate the expansion coefficients $E_n(N, s)$ and $E_{\pm,n}(N, s)$ of the Fredholm determinants of these operators around the point -1. Thus we define

$$\det(1 + TK_{N,s}) = E(N, T, s) := \sum_n E_n(N, s)(1 + T)^n, \tag{6.6.12}$$

$$\det(1 + TK_{\pm,N,s}) = E_\pm(N, T, s) := \sum_n E_{\pm,n}(N, s)(1 + T)^n. \tag{6.6.13}$$

Proposition 6.6.14. 1) *For any real s in $[0, \pi]$, any integer $N \geq 1$, and any integer $n \geq 0$, we have the estimate*

$$0 \leq E_n(N, 2s) \leq 2^n \exp(-Ns/2\pi).$$

2) *For any real s in $(0, \pi/2]$, any integer $N \geq 2$ with $(N-1)s \geq 12$, and any integer $n \geq 0$, we have the estimate*

$$0 \leq E_{\pm,n}(N, s) \leq 2^n \exp(-(N+1)s/8\pi).$$

PROOF. For 1), we have $\mathrm{Trace}(K_{N,2s}) = Ns/\pi$, and we use the crude form of the determinant-trace inequality. For 2), we use the inequality

$$|\,\mathrm{Trace}(K_{\pm,N,s}) - Ns/2\pi| \leq (1/2\pi)(1 + (\pi/2)|\log(sN)|)$$

to infer that

$$\begin{aligned}
\mathrm{Trace}&(K_{\pm,N,s}) \geq Ns/2\pi - (1/2\pi)(1 + (\pi/2)|\log(sN)|)\\
&= (N+1)s/4\pi + \{(N-1)s/4\pi - (1/2\pi)(1 + (\pi/2)|\log(sN)|)\}\\
&= (N+1)s/4\pi + (1/4\pi)\{(N-1)s - 2 - \pi|\log(sN)|\}\\
&= (N+1)s/4\pi + (1/4\pi)\{(N-1)s - 2 - \pi|\log(s(N-1))| - \pi\log(N/(N-1))\}\\
&\geq (N+1)s/4\pi + (1/4\pi)\{(N-1)s - 2 - \pi|\log(s(N-1))| - \pi\log(2)\}.
\end{aligned}$$

For real $x \geq 12$, we have $x - 2 - \pi\log(x) - \pi\log(2) > 0$ (check numerically at $x = 12$ and observe that this function is increasing in $x > \pi$). So for $(N-1)s \geq 12$ we have $\mathrm{Trace}(K_{\pm,N,s}) \geq (N+1)s/4\pi$, and we use the crude form of the determinant-trace inequality. QED

6.7. Application: Estimates for the numbers eigen$(n, s, G(N))$

6.7.1. Fix an integer $N \geq 1$, $G(N)$ any of $U(N), SO(2N+1), USp(2N)$, $SO(2N), O_-(2N+2), O_-(2N+1)$. Recall the table of auxiliary constants (with a new entry for $O_-(2N+1)$ in accordance with 6.4.8):

$G(N)$	λ	σ	ρ	τ	ε
$U(N)$	0	2	1	0	0
$USp(2N)$	0	1	2	1	-1
$SO(2N+1)$	$\frac{1}{2}$	1	2	0	-1
$SO(2N)$	0	1	2	-1	1
$O_-(2N+2)$	1	1	2	1	-1
$O_-(2N+1)$	$\frac{1}{2}$	1	2	0	1

6.7.2. Let us recall that kernel $L_N(x,y)$ attached to $G(N)$ in 5.5 and 6.4.8. It is given by the following table:

$G(N)$	$L_N(x,y)$
$U(N)$	$\sum_{n=0}^{N-1} e^{in(x-y)} = S_N(x,y)e^{i(N-1)(x-1)/2}$
other $G(N)$, i.e.,	$(\sigma/2)[S_{\rho N+\tau}(x-y) + \varepsilon S_{\rho N+\tau}(x+y)]$, i.e.,
$SO(2N+1)$	$(1/2)(S_{2N}(x-y) - S_{2N}(x+y))$
$USp(2N)$ or $O_-(2N+2)$	$(1/2)(S_{2N+1}(x-y) - S_{2N+1}(x+y))$
$SO(2N)$	$(1/2)(S_{2N+1}(x-y) + S_{2N-1}(x+y))$
$O_-(2N+1)$	$(1/2)(S_{2N}(x-y) + S_{2N}(x+y))$.

6.7.3. Given an integer $n \geq 0$, and a Borel measurable set $\mathcal{J}$ in $[0, \sigma\pi]$, we defined the subset

$$\text{Eigen}(n, \mathcal{J}, G(N)) \subset G(N)$$

to consist of those elements A in $G(N)$ exactly n of whose angles

$$\begin{aligned} &0 \leq \varphi(1) \leq \varphi(2) \leq \cdots \leq \varphi(N) < 2\pi \quad \text{if } G(N) = U(N), \\ &0 \leq \varphi(1) \leq \varphi(2) \leq \cdots \leq \varphi(N) \leq \pi \quad \text{for the other } G(N), \end{aligned}$$

lie in $\mathcal{J}$. We defined

$$\text{eigen}(n, \mathcal{J}, G(N)) := \text{ Haar measure of Eigen}(n, \mathcal{J}, G(N)),$$

the measure normalized to give $G(N)$ total mass one.

6.7.4. When $\mathcal{J}$ is the closed interval $[0, s]$, for some real s in $[0, \sigma\pi]$, we write

$$\begin{aligned} \text{Eigen}(n, s, G(N)) &:= \text{Eigen}(n, [0, s], G(N)), \\ \text{eigen}(n, s, G(N)) &:= \text{eigen}(n, [0, s], G(N)). \end{aligned}$$

6.7.5. Thus Eigen$(0, s, G(N))$ is the locus defined by

$$\varphi(1) > s.$$

For $1 \leq n \leq N-1$, Eigen$(n, s, G(N))$ is the locus defined by

$$\varphi(n) \leq s \text{ and } \varphi(n+1) > s.$$

For $n = N$, Eigen$(n, s, G(N))$ is the locus defined by

$$\varphi(N) \leq s,$$

and for $n > N$, Eigen$(n, s, G(N))$ is empty.

Combining the tables above with 6.3.7, 6.3.15, and 6.4.6, we get the following.

Proposition 6.7.6. *Fix an integer $N \geq 1$, $G(N)$ any of $U(N), SO(2N+1), USp(2N), SO(2N), O_-(2N+2), O_-(2N+1)$. For s real in $[0, \sigma\pi]$, the polynomial*

$$\sum_{n \geq 0} (1+T)^n \operatorname{eigen}(n, s, G(N))$$

is equal to the following Fredholm determinant:

$G(N)$	*Fredholm* det *giving* $\sum_{n \geq 0}(1+T)^n \operatorname{eigen}(n, s, G(N))$
$U(N)$	$E(N, T, s)$
$SO(2N+1)$	$E_-(2N, T, s)$
$USp(2N)$ *or* $O_-(2N+2)$	$E_-(2N+1, T, s)$
$SO(2N)$	$E_+(2N-1, T, s)$
$O_-(2N+1)$	$E_+(2N, T, s)$

More explicitly, we have the following table:

$G(N)$	eigen$(n, s, G(N))$
$U(N)$	$E_n(N, s)$
$SO(2N+1)$	$E_{-,n}(2N, s)$
$USp(2N)$ *or* $O_-(2N+2)$	$E_{-,n}(2N+1, s)$
$SO(2N)$	$E_{+,n}(2N-1, s)$
$O_-(2N+1)$	$E_{+,n}(2N, s)$

6.7.7. Putting this together with the estimates 6.6.14, we get the following estimates.

Proposition 6.7.8: Estimates for eigen$(n, S, G(N))$. 1) *For any real s in $(0, \pi]$, any integer $N \geq 1$ and any integer $n \geq 0, n \leq N$, we have*

$$\operatorname{eigen}(n, 2s, U(N)) \leq 2^n \exp(-Ns/2\pi).$$

2) *For any s in $(0, \pi/2]$, any integer $N \geq 2$ with $(2N-2)s \geq 12$, and any integer $n \geq 0, n \leq N$, we have*

$$\begin{aligned}
\operatorname{eigen}(n, s, SO(2N+1)) &\leq 2^n \exp(-(2N+1)s/8\pi),\\
\operatorname{eigen}(n, s, USp(2N)) &\leq 2^n \exp(-(2N+2)s/8\pi),\\
\operatorname{eigen}(n, s, O_-(2N+2)) &\leq 2^n \exp(-(2N+2)s/8\pi),\\
\operatorname{eigen}(n, s, SO(2N)) &\leq 2^n \exp(-2Ns/8\pi),\\
\operatorname{eigen}(n, s, O_-(2N+1)) &\leq 2^n \exp(-(2N+1)s/8\pi).
\end{aligned}$$

6.7.9. In terms of the auxiliary quantities σ and λ,

$G(N)$	λ	σ
$U(N)$	0	2
$USp(2N)$	0	1
$SO(2N+1)$	$\frac{1}{2}$	1
$O_-(2N+2)$	1	1
$SO(2N)$	0	1
$O_-(2N+1)$	$\frac{1}{2}$	1

we can rewrite the previous result more compactly as

Lemma 6.7.10. 1) *For any real s in $(0, \pi]$, any integer $N \geq 1$ and any integer $n \geq 0, n \leq N$, we have*

$$\text{eigen}(n, \sigma s, U(N)) \leq 2^n \exp(-\sigma N s/4\pi).$$

2) *For any s in $(0, \pi/2]$, any integer $N \geq 2$ with $(2N-2)s \geq 12$, and any integer $n \geq 0, n \leq N$, we have*

$$\text{eigen}(n, \sigma s, G(N)) \leq 2^n \exp(-\sigma(N+\lambda)s/4\pi)$$

for $G(N)$ each of $SO(2N+1), USp(2N), O_-(2N+2), SO(2N), O_-(2N+1)$.

6.7.11. After the change of variable from s to $s\pi/(N+\lambda)$, we find the following slightly cruder but more useful form.

Lemma 6.7.12. *For any integer $N \geq 2$, any integer n in $[0, N]$, and any real s in $[8/\pi, N/2]$, we have the estimate*

$$\text{eigen}(n, s\sigma\pi/(N+\lambda), G(N)) \leq 2^n e^{-\sigma s/4}$$

for $G(N)$ any of $U(N), SO(2N+1), USp(2N), O_-(2N+2), SO(2N), O_-(2N+1)$.

6.8. Some curious identities among various eigen$(n, s, G(N))$

Lemma 6.8.1. *Fix a real number s in $(0, \pi]$. The polynomials $E(N, T, s)$ and $E_\pm(N, T, s)$ have the following degrees in T:*

$$\begin{aligned} \deg_T E(N, T, 2s) &= N, \\ \deg_T E_-(2N+1, T, s) &= N, \\ \deg_T E_+(2N-1, T, s) &= N, \\ \deg_T E_-(2N, T, s) &= N, \\ \deg_T E_+(2N, T, s) &= N. \end{aligned}$$

PROOF. By 6.7.6, these are the polynomials

$$\sum_{n \geq 0} (1+T)^n \,\text{eigen}(n, \sigma s, G(N))$$

for $G(N)$ respectively $U(N), USp(2N), SO(2N), SO(2N+1), O_-(2N+1)$. The sets Eigen$(n, \sigma s, G(N))$ are empty for $n > N$. For $n = N$ and $s > 0$, the set Eigen$(N, \sigma s, G(N))$ has nonzero Haar measure, because it contains an open neighborhood of $\mathbb{1}$ for the four groups, and an open neighborhood of $-\mathbb{1}$ for $O_-(2N+1)$. QED

6.8.2. The curious identities are these.

Lemma 6.8.3. 1) *For each integer $N \geq 1$, and each real s in $[0, \pi]$, we have the following measure identities:*

$$\begin{aligned} \text{eigen}(0, 2s, U(2N-1)) &= \text{eigen}(0, s, SO(2N)) \times \text{eigen}(0, s, USp(2N-2)) \\ &= \text{eigen}(0, s, SO(2N)) \times \text{eigen}(0, s, O_-(2N)), \end{aligned}$$

and

$$\text{eigen}(0, 2s, U(2N)) = \text{eigen}(0, s, SO(2N+1)) \times \text{eigen}(0, s, O_-(2N+1)).$$

2) *More generally, for each integer* $n \geq 0$, *we have the identities*

$$\begin{aligned}
&\mathrm{eigen}(n, 2s, U(2N-1)) \\
&\quad = \sum_{a+b=n} \mathrm{eigen}(a, s, SO(2N)) \times \mathrm{eigen}(b, s, USp(2N-2)) \\
&\quad = \sum_{a+b=n} \mathrm{eigen}(a, s, SO(2N)) \times \mathrm{eigen}(b, s, O_{-}(2N)),
\end{aligned}$$

and

$$\begin{aligned}
&\mathrm{eigen}(n, 2s, U(2N)) \\
&\quad = \sum_{a+b=n} \mathrm{eigen}(a, s, SO(2N+1)) \times \mathrm{eigen}(b, s, O_{-}(2N+1)).
\end{aligned}$$

PROOF. Expand out in powers of $1+T$ the identities

$$\begin{aligned}
E(2N-1, T, 2s) &= E_{+}(2N-1, T, s)E_{-}(2N-1, T, s), \\
E(2N, T, 2s) &= E_{+}(2N, T, s)E_{-}(2N, T, s). \quad \text{QED}
\end{aligned}$$

Question 6.8.4. Is there an intrinsic proof of these rather mysterious measure identities? The $U(2N-1)$ identity is reminiscent of the equality of dimensions

$$\dim(U(2N-1)) = \dim(SO(2N) \times USp(2N-2)),$$

i.e.,

$$(2N-1)^2 = (2N)(2N-1)/2 + (2N-2)(2N-1)/2.$$

On the other hand, the $U(2N)$ identity does not seem to have a dimensional counterpart. And the equality of dimensions,

$$\dim(U(2N)) = \dim(SO(2N) \times USp(2N)),$$

which is the dimensional consequence of the decomposition

$$V^{\otimes 2} = \Lambda^2(V) \oplus \mathrm{Sym}^2(V)$$

for a $2N$-dimensional vector space V, does not seem to have a measure formula to go along with it.

6.9. Normalized "n'th eigenvalue" measures attached to $G(N)$

6.9.1. For $G(N)$ any of $U(N), SO(2N+1), USp(2N)SO(2N), O_{-}(2N+2)$, $O_{-}(2N+1)$, and A in $G(N)$, we have its sequence of angles

$$\begin{aligned}
&0 \leq \varphi(1) \leq \varphi(2) \leq \cdots \leq \varphi(N) < 2\pi \quad \text{if } G(N) = U(N), \\
&0 \leq \varphi(1) \leq \varphi(2) \leq \cdots \leq \varphi(N) \leq \pi \quad \text{for the other } G(N).
\end{aligned}$$

In terms of the auxiliary constants λ and σ we define its sequence of **normalized angles**

$$0 \leq \vartheta(1) \leq \vartheta(2) \leq \cdots \leq \vartheta(N) \leq N + \lambda$$

by

$$\vartheta(n) := (N+\lambda)\varphi(n)/\sigma\pi.$$

Concretely,

$$\begin{aligned}
\vartheta(n) &:= N\varphi(n)/2\pi && \text{for } U(N),\\
\vartheta(n) &:= (N+1/2)\varphi(n)/\pi = (2N+1)\varphi(n)/2\pi && \text{for } SO(2N+1),\\
\vartheta(n) &:= N\varphi(n)/\pi = 2N\varphi(n)/2\pi && \text{for } USp(2N) \text{ or } SO(2N),\\
\vartheta(n) &:= (N+1)\varphi(n)/\pi = (2N+2)\varphi(n)/2\pi && \text{for } O_-(2N+2),\\
\vartheta(n) &:= (N+1/2)\varphi(n)/\pi = (2N+1)\varphi(n)/2\pi && \text{for } O_-(2N+1).
\end{aligned}$$

We will view the $\vartheta(n)$ as functions from $G(N)$ to $\mathbb{R}$. For $U(N)$, each $\vartheta(n)$ is Borel measurable, and its restriction to the A with $\det(1-A)$ nonzero is continuous (by 1.8.5). On the other $G(N)$, each $\vartheta(n)$ is continuous.

6.9.2. It is worth pointing out that, unlike the normalized spacings (1.0.1), the normalized angles of an element A of $G(N)$ depend on A as an element of $G(N)$, not just on A as an element of the ambient unitary group. For example, in $SO(2N+1)$, every element A has 1 as an eigenvalue, and so there is a shift in numbering:

$$\vartheta(n)(A \text{ in } SO(2N+1)) = \vartheta(n+1)(A \text{ in } U(2N+1))$$

for $1 \le n \le N$. Similarly, in $O_-(2N+2)$, every element A has both ± 1 as eigenvalues, and again there is a shift in numbering:

$$\vartheta(n)(A \text{ in } O_-(2N+2)) = \vartheta(n+1)(A \text{ in } U(2N+2))$$

for $1 \le n \le N$. In the case of A in $USp(2N)$ or $SO(2N)$, the eigenvalue 1 occurs with even multiplicity $2k$, and there is a shift in numbering which depends upon this multiplicity:

$$\vartheta(n)(A \text{ in } SO(2N) \text{ or } USp(2N)) = \vartheta(n+k)(A \text{ in } U(2N))$$

for $1 \le n \le N$. The only slightly redeeming feature is that for $G(N)$ one of $U(N), USp(2N), O_-(2N+1)$ or $SO(2N)$, the set $G(N)[1/\det(1-A)]$ of elements for which 1 is not an eigenvalue has full measure one, and for $G(N)$ one of $SO(2N+1)$ or $O_-(2N+2)$, the set $G(N)[1/\det'(1-A)]$ of elements A for which 1 is an eigenvalue with multiplicity one has full measure one.

6.9.3. For $G(N)$ any of $U(N), SO(2N+1), USp(2N), SO(2N), O_-(2N+2)$, $O_-(2N+1)$ and any integer n with $1 \le n \le N$, we define a probability measure

$$\nu(n, G(N))$$

on $\mathbb{R}$, supported in $[0, N+\lambda]$, by

$$\nu(n, G(N)) := \vartheta(n)_* \operatorname{Haar}_{G(N)}. \tag{6.9.4}$$

Thus $\nu(n, G(N))$ is the probability measure on $\mathbb{R}$ whose cumulative distribution function $\operatorname{CDF}_{\nu(n,G(N))}$ is

$$\operatorname{CDF}_{\nu(n,G(N))}(s) = \text{ Haar measure of } \{A \text{ in } G(N) \text{ with } \vartheta(n) \le s\}. \tag{6.9.5}$$

It is convenient to work also with the tails of some of these measures. Recall that for any probability measure μ on $\mathbb{R}$, its tail, Tail_μ, is the function on $\mathbb{R}$ with values in $[0,1]$ defined by

$$\operatorname{Tail}_\mu(s) := 1 - \operatorname{CDF}_\mu(s) = \mu((s, \infty)).$$

Thus we have

$$\text{Tail}_{\nu(n,G(N))}(s) = \text{Haar measure of } \{A \text{ in } G(N) \text{ with } \vartheta(n) > s\}. \tag{6.9.6}$$

Lemma 6.9.7. *For $N \geq 2$, $G(N)$ any of $U(N), SO(2N+1), USp(2N), SO(2N), O_-(2N+2), O_-(2N+1)$, and real s in $(0, N+\lambda)$, we have the identities*

1)
$$\begin{aligned}\text{eigen}(0, s\sigma\pi/(N+\lambda), G(N)) &= \text{Tail}_{\nu(1,G(N))}(s)\\ &= 1 - \text{CDF}_{\nu(1,G(N))}(s),\end{aligned}$$

2) *for* $1 \leq n \leq N-1$,

$$\begin{aligned}&\text{eigen}(n, s\sigma\pi/(N+\lambda), G(N))\\ &\quad= \text{CDF}_{\nu(n,G(N))}(s) - \text{CDF}_{\nu(n+1,G(N))}(s)\\ &\quad= \text{Tail}_{\nu(n+1,G(N))}(s) - \text{Tail}_{\nu(n,G(N))}(s),\end{aligned}$$

3)
$$\text{eigen}(N, s\sigma\pi/(N+\lambda), G(N)) = \text{CDF}_{\nu(N,G(N))}(s).$$

Equivalently, we have the identities, for $1 \leq n \leq N$,

4)
$$\text{Tail}_{\nu(n,G(N))}(s) = \sum_{j=0}^{n-1} \text{eigen}(j, s\sigma\pi/(N+\lambda), G(N)).$$

PROOF. The set $\{A \text{ in } G(N) \text{ with } \vartheta(n) > s\}$ is the set

$$\begin{aligned}&\{A \text{ in } G(N) \text{ with } \leq n-1 \text{ normalized angles in } [0,s]\}\\ &= \coprod_{j=0}^{n-1} \{A \text{ in } G(N) \text{ with exactly } j \text{ normalized angles in } [0,s]\}\\ &= \coprod_{j=0}^{n-1} \text{Eigen}(j, s\sigma\pi/(N+\lambda), G(N)),\end{aligned}$$

and its complement $\{A \text{ in } G(N) \text{ with } \vartheta(n) \leq s\}$ is the set

$$\begin{aligned}&\{A \text{ in } G(N) \text{ with } \geq n \text{ normalized angles in } [0,s]\}\\ &= \coprod_{j \geq n} \{A \text{ in } G(N) \text{ with exactly } j \text{ normalized angles in } [0,s]\}\\ &= \coprod_{j \geq n} \text{Eigen}(j, s\sigma\pi/(N+\lambda), G(N)).\end{aligned}$$

Taking the Haar measures of these sets, and remembering that for $j > N$ the set $\text{Eigen}(j, s\sigma\pi/(M+\lambda), G(N))$ is empty, gives the assertions. QED

Proposition 6.9.8: Tail Estimates for $\nu(n, G(N))$. *For any integer $N \geq 2$, any integer n in $[0, N]$, and any real s in $[8/\pi, N/2]$, we have the estimate*

$$\text{Tail}_{\nu(n,G(N))}(s) \leq 2^n e^{-\sigma s/4}.$$

PROOF. This is immediate from 6.7.12, thanks to the formula 6.9.7, part 4),

$$\text{Tail}_{\nu(n,G(N))}(s) = \sum_{j=0}^{n-1} \text{eigen}(j, s\sigma\pi/(N+\lambda), G(N)). \quad \text{QED}$$

6.10. Interlude: Sharper upper bounds for eigen$(0, s, SO(2N))$, for eigen$(0, s, O_-(2N+1))$, and for eigen$(0, s, U(N))$

Proposition 6.10.1 (compare [**Mehta**, 6.8]). *For each $N \geq 1$, and each real s in $[0, \pi]$, we have the inequality*

$$\mathrm{eigen}(0, s, SO(2N)) \leq (\cos^2(s/2))^{N^2 - N/2} \leq (1 - (s/\pi)^2)^{N^2 - N/2}.$$

PROOF. It is tautological that the measure of Eigen$(0, s, SO(2N))$ is the integral

$$\int_{(s,\pi]^N} d\mu(N, N)$$

for the measure $\mu(N, N)$ attached to $SO(2N)$. Rather than expand this out in terms of the L_N kernel, we will go back to the Weyl integration formula on $SO(2N)$ in its original form as found in [**Weyl**, page 228, (9.15)] and recalled in 5.0.6. Thus the measure of Eigen$(0, s, SO(2N))$ is

$$(2/N!) \int_{(s,\pi]^N} \left(\prod_{i<j} (2\cos(x(i)) - 2\cos(x(j))) \right)^2 \prod_i (dx(i)/2\pi).$$

In terms of the quantities $\vartheta(i) := x(i)/2$, this becomes

$$(2/N!) \int_{(s/2,\pi/2]^N} \left(\prod_{i<j} (2\cos(2\vartheta(i)) - 2\cos(2\vartheta(j))) \right)^2 \prod_i (d\vartheta(i)/\pi).$$

Using the identity

$$(2\cos(\vartheta))^2 = 2\cos(2\vartheta) + 2,$$

we may rewrite this as

$$(2/N!) \int_{(s/2,\pi/2]^N} \left(\prod_{i<j} (4\cos^2(\vartheta(i)) - 4\cos^2(\vartheta(j))) \right)^2 \prod_i (d\vartheta(i)/\pi).$$

We now express this integral in terms of $w_i := \cos^2(\vartheta(i))$. The map

$$\vartheta \mapsto \cos^2(\vartheta) := w$$

is an orientation-reversing bijection of $[0, \pi/2]$ with $[0, 1]$, and

$$dw = -2\cos(\vartheta)\sin(\vartheta)\, d\vartheta = -2\,\mathrm{Sqrt}(w(1-w))\, d\vartheta.$$

Thus our integral becomes

$$(2/N!) \int_{[0,\cos^2(s/2))^N} \left(\prod_{i<j} (4w_i - 4w_j) \right)^2 \prod_i (dw_i/2\pi\,\mathrm{Sqrt}(w_i(1-w_i))).$$

To clarify what is going on, let us introduce, for any real λ in $[0, 1]$, the integral $I(\lambda)$ defined as

$$I(\lambda) := f(N) \int_{[0,\lambda)^N} \left(\prod_{i<j} (4w_i - 4w_j) \right)^2 \prod_i (dw_i/\,\mathrm{Sqrt}(w_i(1-w_i))),$$

where

$$f(N) := (2/(2\pi)^N N!).$$

Thus $I(0) = 0$, and $\mathrm{Eigen}(0, s, SO(2N)) = I(\cos^2(s/2))$, for s in $[0, \pi]$. In particular, taking $s = 0$, we see that $I(1) = 1$. We claim that

$$I(\lambda) \le \lambda^{N^2 - N/2} I(1)$$

for λ in $[0, 1]$. This is clear for $\lambda = 0$. For $\lambda > 0$, write w_i as λz_i for z_i in $[0, 1]$. Making the change of variable, we get a factor of λ^2 from each of the $N(N-1)/2$ terms $(4\lambda z_i - 4\lambda z_j)$, and a factor of $\lambda^{1/2}$ from each of the N terms $d(\lambda z_i)/\mathrm{Sqrt}(\lambda z_i)$:

$$I(\lambda) = f(N) \int_{[0,1)^N} \left(\prod_{i<j} (4\lambda z_i - 4\lambda z_j) \right)^2 \prod_i (d(\lambda z_i)/\mathrm{Sqrt}(\lambda z_i(1 - \lambda z_i)))$$

$$= [\lambda^{N(N-1/2)}] f(N) \int_{[0,1)^N} \left(\prod_{i<j} (4z_i - 4z_j) \right)^2 \prod_i (dz_i/\mathrm{Sqrt}(z_i(1 - \lambda z_i)))$$

$$\le [\lambda^{N(N-1/2)}] f(N) \int_{[0,1)^N} \left(\prod_{i<j} (4z_i - 4z_j) \right)^2 \prod_i (dz_i/\mathrm{Sqrt}(z_i(1 - z_j))),$$

the last inequality because $1/\mathrm{Sqrt}(1 - \lambda z_i) \le 1/\mathrm{Sqrt}(1 - z_i)$. Taking λ to be $\cos^2(s/2)$, we get the asserted estimate

$$\mathrm{eigen}(0, s, SO(2N)) \le (\cos^2(s/2))^{N^2 - N/2}.$$

To get the final inequality, recall that for x in $[0, \pi/2]$, we have the inequality

$$2x/\pi \le \sin(x) \le x.$$

Thus for s in $[0, \pi]$ we have

$$\cos^2(s/2) = 1 - \sin^2(s/2) \le 1 - (s/\pi)^2. \quad \text{QED}$$

6.10.2. A similar argument leads to the next estimate.

Proposition 6.10.3. *For each $N \ge 1$, and each real s in $[0, \pi]$, we have the inequality*

$$\mathrm{eigen}(0, s, O_-(2N+1)) \le (\cos^2(s/2))^{N^2 + N/2} \le (1 - (s/\pi)^2)^{N^2 + N/2}.$$

PROOF. By the $\vartheta \mapsto \pi - \vartheta$ symmetry relating the angles of an element B in $O_-(2N+1)$ to those of the corresponding element $-B$ in $SO(2N+1)$, we have

$$\mathrm{eigen}(0, s, O_-(2N+1)) := \mathrm{eigen}(0, [0, s], O_-(2N+1))$$
$$= \mathrm{eigen}(0, [\pi - s, \pi], SO(2N+1)).$$

To calculate $\mathrm{eigen}(0, [\pi - s, \pi], SO(2N+1))$, we return to Weyl's form of the Weyl integration for $SO(2N+1)$, as given in [**Weyl**, page 224, (9.7)] and recalled in 5.0.5. We have

$$\mathrm{eigen}(0, [\pi - s, \pi], SO(2N+1))$$

$$= (1/N!) \int_{[0,\pi-s)^N} \left(\prod_{i<j} (2\cos(x(i)) - 2\cos(x(j))) \right)^2 \prod_i ((2/\pi)\sin^2(x(i)/2)\, dx(i)).$$

In terms of the quantities $\vartheta_i := x(i)/2$, this becomes

$$(1/N!)\int_{[0,\pi/2-s/2)^N}\left(\prod_{i<j}(2\cos(2\vartheta_i)-2\cos(2\vartheta_j))\right)^2\prod_i((4/\pi)\sin^2(\vartheta_i)\,d\vartheta_i).$$

Again using the identity

$$(2\cos(\vartheta))^2 = 2\cos(2\vartheta)+2,$$

we may rewrite this as

$$(1/N!)\int_{[0,\pi/2-s/2)^N}\left(\prod_{i<j}(4\cos^2(\vartheta_i)-4\cos^2(\vartheta_j))\right)^2\prod_i((4/\pi)\sin^2(\vartheta_i)\,d\vartheta_i).$$

Now make the change of variable $w_i := \sin^2(\vartheta_i)$. We have

$$dw_i = 2\sin(\vartheta_i)\cos(\vartheta_i)\,d\vartheta_i = 2\,\mathrm{Sqrt}(w_i(1-w_i))\,d\vartheta_i.$$

Thus we have

$$\begin{aligned}(4/\pi)\sin^2(\vartheta_i)\,d\vartheta_i &= (4/\pi)w_i\,dw_i/2\,\mathrm{Sqrt}(w_i(1-w_i))\\ &= (2/\pi)\,\mathrm{Sqrt}(w_i/(1-w_i))\,dw_i,\end{aligned}$$

and our integral becomes

$$g(N)\int_{[0,\sin^2(\pi/2-s/2))^N}\left(\prod_{i<j}(4w_i-4w_j)\right)^2\prod_i(\mathrm{Sqrt}(w_i/(1-w_i))\,dw_i),$$

with

$$g(N) := 2^N/N!\pi^N.$$

For real λ in $[0,1]$, we denote by $I(\lambda)$ the integral defined by

$$I(\lambda) := g(N)\int_{[0,\lambda)^N}\left(\prod_{i<j}(4w_i-4w_j)\right)^2\prod_i(\mathrm{Sqrt}(w_i/(1-w_i))\,dw_i).$$

Thus $I(0)=0$, and

$$\mathrm{Eigen}(0,s,SO_-(2N+1)) = I(\sin^2(\pi/2-s/2)),\quad \text{for } s \text{ in } [0,\pi].$$

In particular, taking $s=0$, we see that $I(1)=1$. We claim that

$$I(\lambda)\le\lambda^{N^2+N/2}I(1).$$

Making the substitution $w_i=\lambda z_i$, we get a factor of λ^2 from each of the $N(N-1)/2$ terms $(4w_i-4w_j)^2$, and a factor $\lambda^{3/2}$ from each of the N factors $\mathrm{Sqrt}(w_i)\,dw_i$. And as before we have $\mathrm{Sqrt}(1/(1-\lambda z_i))\le\mathrm{Sqrt}(1/(1-z_i))$. Taking λ to be $\sin^2(\pi/2-s/2)$, we get the inequality

$$\mathrm{eigen}(0,s,O_-(2N+1))\le(\sin^2(\pi/2-s/2))^{N^2+N/2}.$$

But $\sin(\pi/2-s/2)=\cos(s/2)$, and we conclude as in the proof of 6.10.1 above. QED

Proposition 6.10.4. *For s real in $[0,\pi]$, and any integer $N\ge1$, we have the estimate*

$$\mathrm{eigen}(0,2s,U(N))\le(1-(s/\pi)^2)^{N(N+1)/4}.$$

PROOF. From 6.8.3, we have

$$\text{eigen}(0, 2s, U(2N-1)) = \text{eigen}(0, s, SO(2N)) \times \text{eigen}(0, s, USp(2N-2))$$

for $N \geq 1$. But eigen$(0, s, USp(2N-2))$ lies in $[0, 1]$, because it is the Haar measure of a subset of $USp(2N-2)$. So we have

$$\text{eigen}(0, 2s, U(2N-1)) \leq \text{eigen}(0, s, SO(2N)) \leq (1-(s/\pi)^2)^{N^2-N/2},$$

the last inequality by 6.10.1.

Similarly, we have

$$\text{eigen}(0, 2s, U(2N)) = \text{eigen}(0, s, SO(2N+1)) \times \text{eigen}(0, s, O_-(2N+1)),$$

which gives

$$\text{eigen}(0, 2s, U(2N)) \leq \text{eigen}(0, s, O_-(2N+1)) \leq (1-(s/\pi)^2)^{N^2+N/2},$$

this time using 6.10.3. QED

Corollary 6.10.5 (tail estimate for $\nu(1, U(N))$). *For any integer $N \geq 1$ and any real $s > 0$, we have*

$$\text{Tail}_{\nu(1,U(N))}(s) \leq e^{-s^2/4}.$$

PROOF. The measure $\nu(1, U(N))$ is supported in $[0, N]$, so its tail vanishes for $s \geq N$. Thus the assertion is trivially correct for $s \geq N$. By definition, we have

$$\text{Tail}_{\nu(1,U(N))}(s) = \text{eigen}(0, 2\pi s/N, U(N)).$$

By the above estimate we have

$$\text{eigen}(0, 2\pi s/N, U(N)) \leq (1-(s/N)^2)^{N(N+1)/4}.$$

For s in $[0, N]$, $(s/N)^2$ lies in $[0, 1]$, so by the inequality $1 - x \leq e^{-x}$ for x in $[0, 1]$ (cf. 6.5.1), we find

$$\begin{aligned}(1-(s/N)^2)^{N(N+1)/4} &\leq \exp(-(s/N)^2)^{N(N+1)/4} \\ &\leq \exp(-(s/N)^2)^{N^2/4} = e^{-s^2/4}. \quad \text{QED}\end{aligned}$$

6.11. A more symmetric construction of the "n'th eigenvalue" measures $\nu(n, U(N))$

6.11.1. Consider the product group $U(1) \times U(N)$, and on it consider, for each integer n with $1 \leq n \leq N$, the function

$$F_n : U(1) \times U(N) \to \mathbb{R}_{\geq 0} \subset \mathbb{R}$$

defined by

$F_n(e^{i\varphi}, A) :=$ normalized distance from $e^{i\varphi}$ to the n'th eigenvalue of A which one encounters starting at $e^{i\varphi}$ and walking counterclockwise around the unit circle, measuring distances so that the unit circle has circumference N.

Lemma 6.11.2. *The direct image of normalized Haar measure on $U(1) \times U(N)$ by the map F_n is the measure $\nu(n, U(N))$ on $\mathbb{R}$.*

PROOF. In terms of the n'th normalized angle $\vartheta(n)(A)$ attached to an element A in $U(N)$, we have

$$F_n(e^{i\varphi}, A) = \vartheta(n)(e^{-i\varphi}A).$$

So if we denote by

$$\pi : U(1) \times U(N) \to U(N)$$

the surjective homomorphism

$$\pi(e^{i\varphi}, A) := e^{-i\varphi}A,$$

we have a commutative diagram

$$\begin{array}{ccc} U(1) \times U(N) & \xrightarrow{F_n} & \mathbb{R} \\ \downarrow \pi & \nearrow \vartheta(n) & \\ U(N). & & \end{array}$$

Since π_* Haar $=$ Haar for any surjective homomorphism of compact groups (by 1.3.1), the assertion is obvious from the transitivity of direct image:

$$\begin{aligned} F_{n*} \operatorname{Haar}_{U(1)\times U(N)} &= \vartheta(n)_* \pi_* \operatorname{Haar}_{U(1)\times U(N)} \\ &= \vartheta(n)_* \operatorname{Haar}_{U(N)} := \nu(n, U(N)). \quad \text{QED} \end{aligned}$$

6.12. Relation between the "n'th eigenvalue" measures $\nu(n, U(N))$ and the expected value spacing measures $\mu(U(N)$, sep. $k)$ on a fixed $U(N)$

6.12.1. Before proceeding, it will be useful to have the following elementary but useful version of integration by parts.

Lemma 6.12.2 (Tail integration lemma). *Let μ be a probability measure on $\mathbb{R}_{\geq 0}$. For any positive, increasing $\mathcal{C}^\infty$ function $f : \mathbb{R}_{\geq 0} \to \mathbb{R}_{\geq 0}$ with $f(0) = 0$, e.g. $f(x) := x^n$, any $n \geq 1$, we have the integration formula*

$$\int_{[0,\infty)} f'(s) \operatorname{Tail}_\mu(s)\, ds = \int_{[0,\infty)} f(x)\, d\mu(x).$$

More generally, let $g : \mathbb{R}_{\geq 0} \to \mathbb{R}_{\geq 0}$ be any positive Borel measurable function, with indefinite integral $G(x) := \int_{[0,x]} g(s)\, ds$. Then we have the integration formula

$$\int_{[0,\infty)} g(s) \operatorname{Tail}_\mu(s)\, ds = \int_{[0,\infty)} G(x)\, d\mu(x).$$

PROOF. The second statement for $g := f'$ implies the first for f, so it suffices to prove the second. The functions $g(s)$ and $\operatorname{Tail}_\mu(x)$ are both positive and Borel measurable, so both sides make sense as elements of $\mathbb{R}_{\geq 0} \cup \{+\infty\}$. It is there that the asserted equality holds. We have

$$\int_{[0,\infty)} g(s)\,\mathrm{Tail}_\mu(s)\,ds := \int_{[0,\infty)} g(s)\left(\int_{(s,\infty)} d\mu(x)\right) ds$$

$$= \iint_{0\le s<x<\infty} g(s)\,ds\,d\mu(x) = \int_{(0,\infty)}\left(\int_{[0,x)} g(s)\,ds\right) d\mu(x)$$

$$= \int_{(0,\infty)} G(x)\,d\mu(x) = \int_{[0,\infty)} G(x)\,d\mu(x),$$

the last equality because $G(0)=0$. QED

6.12.3. We now turn to the subject proper of this section. We begin with the simplest, and in many ways the most important, case.

Proposition 6.12.4 (compare [**Mehta**, 5.1.16a,b,c, A.8]). *For any integer $N \ge 1$, we have an equality of measures on $\mathbb{R}_{\ge 0}$*

$$\nu(1, U(N)) = \mathrm{Tail}_{\mu(U(N),\ step\ 1)}(s)\,ds.$$

PROOF. Both measures are supported in $[0, N]$, so it suffices to show that they agree on all Borel measurable subsets of $[0, N]$. For this, it suffices to show that for every positive, bounded, Borel measurable function $g : \mathbb{R}_{\ge 0} \to \mathbb{R}_{\ge 0}$ of compact support, we have

$$\int_{[0,\infty)} g\,d\nu(1, U(N)) = \int_{[0,\infty)} g(s)\,\mathrm{Tail}_{\mu(U(N),\ \mathrm{step}\ 1)}(s)\,ds.$$

Let us denote by G the indefinite integral $G(x) := \int_{[0,x]} g(s)\,ds$, so that by the tail integration Lemma 6.12.2 above we have

$$\int_{[0,\infty)} g(s)\,\mathrm{Tail}_{\mu(U(N),\ \mathrm{step}\ 1)}(s)\,ds = \int_{[0,\infty)} G(x)\,d\mu(x).$$

Thus we are reduced to showing that

$$\int_{[0,\infty)} g\,d\nu(1, U(N)) = \int_{[0,\infty)} G\,d\mu(U(N)),\ \mathrm{step}\ 1).$$

The idea is to express both integrals as integrals over $U(N)$ against Haar measure, and then to show that the integrands coincide outside a set of Haar measure zero. In the case at hand, we will show the integrands agree on the set $U(N)^{\mathrm{reg}}$ of "regular elements" in $U(N)$, i.e., those with N distinct eigenvalues.

The description 6.11.2 of $\nu(1, U(N))$ as a direct image from $U(1) \times U(N)$ gives

$$\int_{[0,\infty)} g\,d\nu(1, U(N)) = \int_{U(N)} \int_{[0,2\pi)} g(\vartheta(1)(e^{-i\varphi}A))d(\varphi/2\pi)\,dA.$$

The definition 1.1.3 of $\mu(U(N),\ \mathrm{step}\ 1)$ as an expected value over $U(N)$ of the measures $\mu(A, U(N),\ \mathrm{step}\ 1)$ gives

$$\int_{[0,\infty)} G\,d\mu(U(N),\ \mathrm{step}\ 1) = \int_{U(N)}\left(\int_{[0,\infty)} G\,d\mu(A, U(N),\ \mathrm{step}\ 1)\right) dA.$$

We will now show that for each A in $U(N)^{\mathrm{reg}}$, we have

$$\int_{[0,2\pi)} g(\vartheta(1)(e^{-i\varphi}A))\,d\varphi/2\pi = \int_{[0,\infty)} G\,d\mu(A, U(N),\ \mathrm{step}\ 1).$$

Let us denote by

$$0 \leq \varphi(1) < \varphi(2) < \cdots < \varphi(N) < 2\pi$$

the (nonnormalized) angles of A, and put $\varphi(N+1) = 2\pi + \varphi(1)$. For $k = 1, \ldots, N$, let

$$s_k = (N/2\pi)(\varphi(k+1) - \varphi(k))$$

be the normalized spacings of A. Let us denote by

$$S_k := (\varphi(k), \varphi(k+1)] \subset U(1)$$

the half open interval between the k'th and $(k+1)$'st eigenvalues of A (counting as we start counterclockwise from the origin in $U(1)$).

By definition of $\mu(A, U(N), \text{ step } 1)$, we have

$$\int_{[0,\infty)} G\, d\mu(A, U(N), \text{ step } 1) = (1/N) \sum_k G(s_k).$$

The integral over S^1 is

$$\int_{[0,2\pi)} g(\vartheta(1)(e^{-i\varphi}A))\, d\varphi/2\pi = \sum_k \int_{S_k} g(\vartheta(1)(e^{-i\varphi}A))\, d\varphi/2\pi.$$

We will show that for each k we have

$$\int_{S_k} g(\vartheta(1)(e^{-i\varphi}A))\, d\varphi/2\pi = (1/N)G(s_k).$$

The key point is that when the variable φ is in the interval S_k, then the first eigenvalue past φ has angle $\varphi(k+1)$, and the normalized distance from φ to it is $(N/2\pi)(\varphi(k+1) - \varphi)$. So for each k we have

$$\begin{aligned}
\int_{S_k} g(\vartheta(1)(e^{-i\varphi}A))\, d\varphi/2\pi &= \int_{(\varphi(k),\varphi(k+1)]} g((N/2\pi)(\varphi(k+1) - \varphi))\, d\varphi/2\pi \\
&= \int_{(0,\varphi(k+1)-\varphi(k)]} g((N/2\pi)(\varphi(k+1) - \varphi(k) - \varphi))\, d\varphi/2\pi \\
&= \int_{(0,\varphi(k+1)-\varphi(k)]} g(s_k - N\varphi/2\pi)\, d\varphi/2\pi \\
&= \int_{(0,s_k]} g(s_k - x)\, dx/N = \int_{[0,s_k)} g(x)\, dx/N = (1/N)G(s_k). \quad \text{QED}
\end{aligned}$$

6.12.5. A similar argument gives

Proposition 6.12.6 (compare [**Mehta**, 5.1.16a,b,c, A.8]). *For any integer $N \geq 1$, and any integer r with $1 \leq r \leq N$, we have an equality of measures on $\mathbb{R}_{\geq 0}$*

$$\sum_{j=1}^{r} \nu(j, U(N)) = \mathrm{Tail}_{\mu(U(N),\ step\ r)}(s)\, ds.$$

PROOF. We adopt the notations of the previous proof. Exactly as above, we reduce to showing that for each regular element A in $U(N)$, we have

$$\sum_{j=1}^{r} \int_{[0,2\pi)} g(\vartheta(j)(e^{-i\varphi}A))\, d\varphi/2\pi = \int_{[0,\infty)} G\, d\mu(A, U(N), \text{ step } r).$$

The second integral is given by

$$\int_{[0,\infty)} G\, d\mu(A, U(N), \text{ step } r) := (1/N)\sum_{k=1}^{N} G(s_k + s_{k+1} + \cdots + s_{k+r-1}).$$

The first expression we rewrite as a sum over the intervals S_k:

$$\sum_{j=1}^{r}\int_{[0,2\pi)} g(\vartheta(j)(e^{-i\varphi}A))\, d\varphi/2\pi = \sum_{j=1}^{r}\sum_{k=1}^{N}\int_{S_k} g(\vartheta(j)(e^{-i\varphi}A))\, d\varphi/2\pi.$$

For φ in the interval S_k, the j'th eigenvalue after φ is $\varphi(k+j)$, and the normalized distance from φ to it is

$$\begin{aligned}(N/2\pi)(\varphi(k+j) - \varphi) &= (N/2\pi)(\varphi(k+j) - \varphi(k+1)) + (N/2\pi)(\varphi(k+1) - \varphi)\\ &= \sum_{l=1}^{j-1} s_{k+l} \;+\; (N/2\pi)(\varphi(k+1) - \varphi).\end{aligned}$$

So exactly as in the previous proof we find

$$\begin{aligned}&\int_{S_k} g(\vartheta(j)(e^{-i\varphi}A))\, d\varphi/2\pi\\ &\quad = \int_{[\sum_{l=1}^{j-1} s_{k+l}, \sum_{l=0}^{j-1} s_{k+l})} g(x)\, dx/N\\ &\quad = (1/N)\left(G\left(\sum_{l=0}^{j-1} s_{k+l}\right) - G\left(\sum_{l=1}^{j-1} s_{k+l}\right)\right).\end{aligned}$$

Summing over k and j, we get the required telescoping. QED

Corollary 6.12.7. *For $N \geq k \geq 2$, the measure $\nu(k, U(N))$ on $\mathbb{R}_{\geq 0}$ is given by*

$$\nu(k, U(N)) = (\mathrm{Tail}_{\mu(U(N),\ step\ k)}(t) - \mathrm{Tail}_{\mu(U(N),\ step\ k-1)}(t))\, dt.$$

6.12.8. We now return to the measures $\nu(k, U(N))$ and $\mu(U(N), \text{step } k)$.

Corollary 6.12.9. *For any integer $n \geq 1$, and any integer $N > n$, we have the integration formula*

$$\int_{[0,\infty)} \mathrm{Tail}_{\mu(U(N),\ step\ n)}(t)\, dt = n.$$

First Proof. Each measure $\nu(k, U(N))$ with $1 \leq k < N$ is a probability measure on $\mathbb{R}$, supported in $\mathbb{R}_{\geq 0}$.

Second Proof. By the tail integration lemma above, we have

$$\int_{(0,\infty)} \mathrm{Tail}_{\mu(U(N),\text{ step } n)}(t)\, dt = \int_{(0,\infty)} x\, d\mu(U(N), \text{ step } n)(x).$$

The measure $\mu(U(N), \text{step } n)$ on $\mathbb{R}$ is supported in $[0, N]$, because it is the integral over $U(N)$ of the function $A \mapsto \mu(A, U(N), \text{step } n)$, and each $\mu(A, U(N), \text{step } n)$ is tautologically supported in $[0, N]$. So we may rewrite the integral in terms of the characteristic function $\chi_{(0,N]}$ of the interval $(0, N]$ as

$$\int_{(0,\infty)} x\, d\mu(U(N), \text{ step } n)(x) = \int_{\mathbb{R}_{\geq 0}} x\chi_{(0,N]}(x)\, d\mu(U(N), \text{ step } n)(x).$$

Here the integrand $x\chi_{(0,N]}(x)$ is a bounded Borel function, so by the definition of $\mu(U(N), \text{step } n)$ as an integral over $U(N)$, this integral is

$$= \int_{U(N)} \left(\int_{\mathbb{R}_{\geq 0}} x\chi_{(0,N]}(x)\, d\mu(A, U(N), \text{ step } n)(x) \right) dA$$
$$= \int_{U(N)} \left(\int_{\mathbb{R}_{\geq 0}} x\, d\mu(A, U(N), \text{ step } n)(x) \right) dA,$$

the last equality because each measure $\mu(A, U(N), \text{ step } n)$ has support in $[0, N]$. The innermost integral is equal to n, because for each A the mean of its normalized n-step spacings is tautologically equal to n (cf. 1.0.3). QED

6.13. Tail estimate for $\mu(U(N), \text{sep. } 0)$ and $\mu(\text{univ}, \text{sep. } 0)$

Proposition 6.13.1. *For any real $s \geq 0$, and any integer $N \geq 1$, we have the estimate*

$$\text{Tail}_{\mu(U(N),\ sep.\ 0)}(s+1) \leq e^{-s^2/4}.$$

PROOF. The function $\text{Tail}_{\mu(U(N),\ \text{sep. } 0)}(s)$ is decreasing, so we have

$$\begin{aligned} \text{Tail}_{\mu(U(N),\ \text{sep. } 0)}(s+1) &\leq \int_{[s,s+1]} \text{Tail}_{\mu(U(N),\ \text{sep. } 0)}(x)\, dx \\ &= \int_{[s,s+1]} \nu(1, U(N)) \quad \text{(by 6.12.4)} \\ &\leq \int_{[s,\infty]} \nu(1, U(N)) := \text{Tail}_{\nu(1,U(N))}(s) \\ &\leq e^{-s^2/4}, \end{aligned}$$

the last inequality by 6.10.5. QED

Corollary 6.13.2 (Tail Estimate). *For any real $s \geq 0$, and any integer $N \geq 1$, we have the estimate*

$$\text{Tail}_{\mu(U(N),\ sep.\ 0)}(s) \leq e^{1/4}e^{-s^2/8} \leq (4/3)e^{-s^2/8}.$$

PROOF. For $0 \leq s \leq 1$, we have $e^{1/4-s^2/8} \geq e^{1/8} > 1$, but Tail takes values in $[0, 1]$. For $s \geq 1$, we know that $\text{Tail}(s) \leq e^{-(s-1)^2/4}$, so we need only check that

$$e^{-(s-1)^2/4} \leq e^{1/4-s^2/8} \quad \text{for real } s \geq 1,$$

i.e.,

$$1 \leq e^{1/4+(s-1)^2/4-s^2/8} \quad \text{for real } s \geq 1.$$

But this holds for all real s, since

$$1/4 + (s-1)^2/4 - s^2/8 = (1/8)(2 + 2(s-1)^2 - s^2) = (1/8)(s-2)^2 \geq 0. \quad \text{QED}$$

6.13.3. We now pass to the large N limit.

Proposition 6.13.4 (Tail estimate for μ(univ, sep. 0)). *For real $s \geq 0$, the measure μ(univ, sep. 0) has the tail estimate*

$$\mathrm{Tail}_{\mu(univ,\ sep.\ 0)}(s) \leq (4/3)e^{-s^2/8}.$$

PROOF. For fixed $s \geq 0$, we have

$$\begin{aligned}
\mathrm{Tail}_{\mu(\text{univ, sep. } 0)}(s) &:= 1 - \mu(\text{univ, sep. } 0)([0,s]) \\
&= 1 - \lim_{N\to\infty} \mu(U(N), \text{ sep. } 0)([0,s]) \\
&= \lim_{N\to\infty} (1 - \mu(U(N), \text{ sep. } 0)([0,s])) \\
&= \lim_{N\to\infty} \mathrm{Tail}_{\mu(U(N), \text{ sep. } 0)}(s),
\end{aligned}$$

so the result follows from the finite N tail estimate above. QED

Corollary 6.13.5. *For any integer $r \geq 1$, and for any step vector b in $\mathbb{Z}^r$, the measures μ(univ, steps b) and $\mu(U(N)$, steps b), $N \geq 2$, satisfy the tail estimate*

$$\mu(\{x \text{ in } \mathbb{R}^r \text{ with } |x(i)| > sb(i) \text{ for some } i\}) \leq \Sigma(b)(4/3)e^{-s^2/8},$$

for every real $s \geq 0$.

PROOF. This follows from the tail estimate for the case $r = 1 = b$ proven above by the argument used in the proof of 3.1.10. QED

In particular, in one variable we have

Corollary 6.13.6. *For any integer $b \geq 1$, the measures μ(univ, step b) and $\mu(U(N)$, step b), $N \geq 2$, on $\mathbb{R}$ satisfy the tail estimates*

$$\mathrm{Tail}_{\mu(univ,\ step\ b)}(s) \leq b(4/3)e^{-s^2/8b^2},$$
$$\mathrm{Tail}_{\mu(U(N),\ step\ b)}(s) \leq b(4/3)e^{-s^2/8b^2}.$$

6.14. Multi-eigenvalue location measures, static spacing measures and expected values of several variable spacing measures on $U(N)$

6.14.1. Given an element A in $U(N)$, we have defined in 1.0.1 its sequence of angles

$$0 \leq \varphi(1) \leq \varphi(2) \leq \cdots \leq \varphi(N) < 2\pi,$$

and then extended the definition of $\varphi(i)$ to all integers i by requiring

$$\varphi(i + N) = \varphi(i) + 2\pi.$$

Up to now, we have taken the point of view that we start at the origin in S^1 and walk counterclockwise around the unit circle without looking back, noting the eigenvalues of A as we pass them. Now we allow ourselves to look both ways from the origin, i.e., we consider the angles $\varphi(i)$ with i negative as well as positive. For $1 - N \leq i \leq N$, these angles satisfy

$$-2\pi \leq \varphi(1-N) \leq \cdots \leq \varphi(-1) \leq \varphi(0) < 0 \leq \varphi(1) \leq \varphi(2) \leq \cdots \leq \varphi(N) < 2\pi.$$

6.14.2. In terms of the complex conjugate $\overline{A}$ of A in $U(N)$, we have

$$\begin{aligned}\varphi(0)(A) &= -\varphi(1)(\overline{A}),\\ \varphi(-1)(A) &= -\varphi(2)(\overline{A}),\\ \varphi(1-k)(A) &= -\varphi(k)(\overline{A}),\end{aligned}$$

for every integer k.

6.14.3. For every integer n, we define the n'th normalized angle $\vartheta(n)$ of A in $U(N)$ by

$$\vartheta(n) := (N/2\pi)\varphi(n),$$

and we denote by $F_n : U(1) \times U(N) \to \mathbb{R}$ the map defined by

$$F_n(e^{i\varphi}, A) := \vartheta(n)(e^{-i\varphi}A).$$

6.14.4. We first define multi-eigenvalue location measures. Given an integer $r \geq 1$ and a vector c in $\mathbb{Z}^r$ with

$$c(1) < c(2) < \cdots < c(r),$$

we denote by $\nu(c, U(N))$ the probability measure on $\mathbb{R}^r$ which is the direct image of Haar measure by the map $U(N) \to \mathbb{R}^r$ defined by the normalized angles

$$A \mapsto (\vartheta(c(1)), \ldots, \vartheta(c(r))).$$

Thus

$$\nu(c, U(N)) := (\vartheta(c(1)), \ldots, \vartheta(c(r)))_* \operatorname{Haar}_{U(N)}.$$

We call $\nu(c, U(N))$ the **multi-eigenvalue location measure** for the eigenvalues named by the vector c.

Lemma 6.14.5. *Suppose given an integer $r \geq 1$, and a vector c in $\mathbb{Z}^r$ with*

$$c(1) < c(2) < \cdots < c(r).$$

The measure $\nu(c, U(N))$ is supported in $(\mathbb{R}_{\geq 0})^r(\text{order})$. For any nonempty subset

$$S = \{1 \leq s_1 < \cdots < s_k \leq r\}$$

of the index set $\{1, \ldots, r\}$, we denote by

$$\operatorname{pr}[S] : \mathbb{R}^r \to \mathbb{R}^{\operatorname{Card}(S)}$$

the projection onto the named coordinates. We have an equality

$$\operatorname{pr}[S]_* \nu(c, U(N)) = \nu(\operatorname{pr}[S](c), U(N))$$

of measures on $\mathbb{R}^{\operatorname{Card}(S)}$.

Proof. A tautology. QED

6.14.6. Exactly as in 6.11.2, we have a more symmetric description of the ν's in terms of the F_n's.

Lemma 6.14.7. *Suppose given an integer $r \geq 1$, and a vector c in $\mathbb{Z}^r$ with*

$$c(1) < c(2) < \cdots < c(r).$$

In terms of the map

$$(F_{c(1)}, \ldots, F_{c(r)}) : U(1) \times U(N) \to \mathbb{R}^r,$$

we have

$$\nu(c, U(N)) = (F_{c(1)}, \ldots, F_{c(r)})_* \operatorname{Haar}_{U(1)\times U(N)}.$$

6.14.8. We next define "static spacing measures". Given an integer $r \geq 2$ and a vector c in $\mathbb{Z}^r$ with

$$c(1) < c(2) < \cdots < c(r),$$

consider the map $U(N) \to \mathbb{R}^{r-1}$ given by

$$(\vartheta(c(2)) - \vartheta(c(1)), \vartheta(c(3)) - \vartheta(c(2)), \ldots, \vartheta(c(r)) - \vartheta(c(r-1))).$$

We denote by $\xi(c, U(N))$ the probability measure on $\mathbb{R}^{r-1}$ given by

$$(\vartheta(c(2)) - \vartheta(c(1)), \vartheta(c(3)) - \vartheta(c(2)), \ldots, \vartheta(c(r)) - \vartheta(c(r)) - \vartheta(c(r-1)))_* \operatorname{Haar}_{U(N)}.$$

We call $\xi(c, U(N))$ the **static spacing measure** for the eigenvalues named by the vector c.

Lemma 6.14.9. *Under the "successive subtraction" map*

$$\operatorname{SuccSub} : \mathbb{R}^r \to \mathbb{R}^{r-1},$$
$$(x_1, \ldots, x_r) \mapsto (x_2 - x_1, \ldots, x_r - x_{r-1}),$$

we have

$$\operatorname{SuccSub}_* \nu(c, U(N)) = \xi(c, U(N)).$$

PROOF. A tautology. QED

Corollary 6.14.10. *In terms of the map*

$$(F_{c(2)} - F_{c(1)}, \ldots, F_{c(r)} - F_{c(r-1)}) : U(1) \times U(N) \to \mathbb{R}^{r-1},$$

we have

$$\xi(c, U(N)) = (F_{c(2)} - F_{c(1)}, \ldots, F_{c(r)} - F_{c(r-1)})_* \operatorname{Haar}_{U(1)\times U(N)}.$$

PROOF. Immediate from the last two lemmas, by the transitivity of direct images. QED

6.14.11. We now turn to the relations between the multi-eigenvalue location measures, the static spacing measures, and the expected value spacing measures.

Proposition 6.14.12. *Fix an integer $N \geq 2$. Let $a \geq 0$ and $b \geq 0$ be integers. Consider the integer vector*

$$[-a, b+1] := (-a, 1-a, 2-a, \ldots, 0, 1, \ldots, b+1) \text{ in } \mathbb{Z}^{a+b+2},$$

and the associated multi-eigenvalue location measure

$$\nu([-a, b+1], U(N)) \quad \text{on } \mathbb{R}^{a+b+2}.$$

This measure is related to the expected value spacing measure

$$\mu(U(N),\ \textit{sep. } 0_{a+b+1}) \quad \textit{on } \mathbb{R}^{a+b+1}$$

as follows. Denote by $L : \mathbb{R}^{a+b+1} \to \mathbb{R}^{a+b+2}, x \mapsto L(x)$, *the linear map defined as follows: given a vector* $x := (x(-a), \ldots, x(b))$, *its image* $L(x)$ *is the vector whose components are given by*

$$\begin{aligned}(L(x))(-j) &= -\sum_{l=1}^{j} x(-l), \quad \textit{for } j = 1, \ldots, a,\\ (L(x))(0) &= 0,\\ (L(x))(j) &= \sum_{l=0}^{j-1} x(l), \quad \textit{for } j = 1, \ldots, b+1.\end{aligned}$$

[Strictly speaking, we should denote this map $L_{a,b}$, *since it depends on both* a *and* b, *not just on the integer* $a+b+1$.*]*

Given a nonnegative Borel measurable function $f \geq 0$ *on* $\mathbb{R}^{a+b+2}$, *denote by* F *the nonnegative Borel measurable function on* $\mathbb{R}^{a+b+1}$ *whose value at* $x := (x(-a), \ldots, x(b))$ *is given by the Lebesgue integral*

$$F(x) := \int_{[0,x(0)]} f(L(x) - t\mathbb{1})\, dt := |x(0)| \int_{[0,1]} f(L(x) - tx(0)\mathbb{1})\, dt.$$

Then we have the identity

$$\int_{\mathbb{R}^{a+b+2}} f\, d\nu([-a, b+1], U(N)) = \int_{\mathbb{R}^{a+b+2}} F\, d\mu(U(N),\ \textit{sep. } 0_{a+b+1}).$$

PROOF. The idea of the proof is that already used in proving 6.12.4 and 6.12.6, namely to express both integrals as integrals over $U(N)$ against Haar measure, and then to show that the integrands coincide on the set $U(N)^{\text{reg}}$ of "regular elements" in $U(N)$, i.e., those with N distinct eigenvalues.

The definition 6.14.7 of $\nu([-a, b+1], U(N))$ as a direct image from $U(1) \times U(N)$ gives

$$\begin{aligned}&\int f\, d\nu([-a, b+1], U(N))\\ &\quad = \int_{U(N)} \int_{[0,2\pi)} f(\vartheta(-a)(e^{-i\varphi}A), \ldots, \vartheta(b+1)(e^{-i\varphi}A))\, d(\varphi/2\pi)\, dA.\end{aligned}$$

The definition 1.1.3 of $\mu(U(N), \text{ sep. } 0_{a+b+2})$ as the expected value over $U(N)$ of the measures $\mu(A, U(N), \text{ sep. } 0_{a+b+2})$ gives

$$\begin{aligned}&\int_{\mathbb{R}^{a+b+2}} F\, d\mu(U(N),\ \text{sep. } 0_{a+b+1})\\ &\quad = \int_{U(N)} \left(\int_{\mathbb{R}^{a+b+2}} F\, d\mu(A, U(N),\ \text{sep. } 0_{a+b+1}) \right) dA.\end{aligned}$$

We will show that for A with N distinct eigenvalues, we have

$$\begin{aligned}&\int_{[0,2\pi)} f(\vartheta(-a)(e^{-i\varphi}A), \ldots, \vartheta(b+1)(e^{-i\varphi}A)) d(\varphi/2\pi)\\ &\quad = \int_{\mathbb{R}^{a+b+1}} F\, d\mu(A, U(N),\ \text{sep. } 0_{a+b+1}).\end{aligned}$$

Let us denote by $\varphi(i) := \varphi(i)(A)$ the (nonnormalized) angles of A, defined for all i in $\mathbb{Z}$. For each i, let

$$s_i = (N/2\pi)(\varphi(i+1) - \varphi(i))$$

be the i'th normalized spacing of A. Let us denote by

$$S_i := (\varphi(i), \varphi(i+1)] \subset U(1)$$

the half open interval between $\varphi(i)$ and $\varphi(i+1)$.

By definition of $\mu(A, U(N), \text{ sep. } 0_{a+b+1})$, we have

$$\int_{\mathbb{R}^{a+b+1}} F\, d\mu(A, U(N), \text{ sep. } 0_{a+b+1}) = (1/N) \sum_{l \bmod N} F(s_l, s_{l+1}, \ldots, s_{l+a+b}),$$

which we rewrite as

$$= (1/N) \sum_{l \bmod N} F(s_{l-a}, s_{l+1-a}, \ldots, s_{l+b}).$$

We claim that for each integer l, we have

$$\begin{aligned} &\int_{S_l} f(\vartheta(-a)(e^{-i\varphi}A), \ldots, \vartheta(b+1)(e^{-i\varphi}A))d(N\varphi/2\pi) \\ &\qquad = F(s_{l-a}, s_{l+1-a}, \ldots, s_{l+b}). \end{aligned}$$

For φ in the interval S_l, and $j \geq 1$, the j'th eigenvalue after φ is $\varphi(l+j)$, and the normalized distance from φ to it is

$$\begin{aligned} &(N/2\pi)(\varphi(l+j) - \varphi) \\ &\quad = (N/2\pi)(\varphi(l+1) - \varphi) + (N/2\pi)(\varphi(l+j) - \varphi(l+1)) \\ &\quad = (N/2\pi)(\varphi(l+1) - \varphi) + \sum_{k=1}^{j-1} s_{k+l} \\ &\quad = \left(\sum_{k=0}^{j-1} s_{k+l}\right) - (N/2\pi)(\varphi - \varphi(l)). \end{aligned}$$

For $j \geq 0$, the $j+1$st eigenvalue before φ is $\varphi(l-j)$, and **minus** the normalized distance from φ to it is

$$-(N/2\pi)(\varphi - \varphi(l)) - \sum_{k=1}^{j} s_{l-k}.$$

The point is that for φ in S_l, the vector

$$(\vartheta(-a)(e^{-i\varphi}A), \ldots, \vartheta(b+1)(e^{-i\varphi}A))$$

is equal to

$$L(s_{l-a}, s_{l+1-a}, \ldots, s_{l+b}) - (N/2\pi)(\varphi - \varphi(l))\mathbb{1}.$$

As φ runs over S_l, the parameter $(N/2\pi)(\varphi - \varphi(l))$ runs from 0 to s_l. Thus we get

$$\begin{aligned} &\int_{S_l} f(\vartheta(-a)(e^{-i\varphi}A), \ldots, \vartheta(b+1)(e^{-i\varphi}A))\, d(N\varphi/2\pi) \\ &\qquad = \int_{[0,s_l]} f(L(s_{l-a}, s_{l+1-a}, \ldots, s_{l+b}) - t\mathbb{1})\, dt, \end{aligned}$$

which, by the very definition of F, is equal to $F(s_{l-a}, s_{l+1-a}, \ldots, s_{l+b})$. QED

Corollary 6.14.13. *Hypotheses and notations as in 6.14.12, let Y be a topological space, and let $\pi : \mathbb{R}^{a+b+2} \to Y$ be a Borel measurable map such that $\pi(x - t\mathbb{1}) = \pi(x)$ for all $x := (x(-a), \ldots, x(b+1))$ in $\mathbb{R}^{a+b+2}$ and all t in $\mathbb{R}$. Then we have equalities of measures on Y,*

$$\begin{aligned}\pi_*\nu([-a,b+1],U(N)) &= (\pi\circ L)_*(|x(0)|\mu(U(N),\ \textit{sep. } 0_{a+b+1}))\\ &= (\pi\circ L)_*(x(0)\mu(U(N),\ \textit{sep. } 0_{a+b+1})).\end{aligned}$$

Proof. For g a positive Borel measurable function on Y, and f the function $x \mapsto g(\pi(x))$ on $\mathbb{R}^{a+b+2}$, the invariance of π under $t\mathbb{1}$ translation gives

$$\begin{aligned}F(x) &:= \int_{[0,x(0)]} f(L(x)-t\mathbb{1})\,dt := \int_{[0,x(0)]} g(\pi(L(x)-t\mathbb{1}))\,dt\\ &= \int_{[0,x(0)]} g(\pi(L(x)))\,dt = |x(0)|g(\pi(L(x))).\end{aligned}$$

Using Proposition 6.14.12, we get

$$\begin{aligned}&\int_Y g\,d\pi_*\nu([-a,b+1],U(N))\\ &:= \int_{\mathbb{R}^{a+b+2}} f\,d\nu([-a,b+1],U(N)) = \int_{\mathbb{R}^{a+b+1}} F\,d\mu(U(N),\ \text{sep. } 0_{a+b+1})\\ &= \int_{\mathbb{R}^{a+b+1}} |x(0)|g(\pi(L(x)))\,d\mu(U(N),\ \text{sep. } 0_{a+b+1})\\ &= \int_Y g\,d(\pi\circ L)_*(|x(0)|\mu(U(N),\ \text{sep. } 0_{a+b+1})).\end{aligned}$$

This shows the first equality. But the measure $\mu(U(N),\ \text{sep. } 0_{a+b+1})$ is supported in $(\mathbb{R}_{\geq 0})^{a+b+1}$, so we have the trivial equality

$$|x(0)|\mu(U(N),\ \text{sep. } 0_{a+b+1}) = x(0)\mu(U(N),\ \text{sep. } 0_{a+b+1})$$

of measures on $\mathbb{R}^{a+b+1}$. QED

Corollary 6.14.14. *Let $r \geq 1$ be an integer. On $\mathbb{R}^r$ with coordinates $x(1), \ldots, x(r)$, for each integer $i = 1, \ldots, r$, the measure $x(i)\mu(U(N),\ \textit{sep. } 0_r)$ on $\mathbb{R}^r$ is equal to the static spacing measure $\xi([1-i, r+1-i], U(N))$.*

Proof. For each $a \geq 0, b \geq 0$ with $a + b = r - 1$, consider the successive subtraction map

$$\pi := \text{SuccSub} : \mathbb{R}^{a+b+2} \to \mathbb{R}^{a+b+1} = \mathbb{R}^r,$$

which is visibly invariant under $t\mathbb{1}$ translation. In coordinates $X(-a), \ldots, X(b+1)$ on $\mathbb{R}^{a+b+2}$, the previous corollary gives

$$\begin{aligned}&\text{SuccSub}_*\,\nu([-a,b+1],U(N))\\ &\quad= (\text{SuccSub}\circ L)_*(X(0)\mu(U(N),\ \text{sep. } 0_{a+b+1})).\end{aligned}$$

But the composite map

$$\text{SuccSub}\circ L : \mathbb{R}^{a+b+1} \to \mathbb{R}^{a+b+1}$$

is the identity, so we get

$$\text{SuccSub}_*\,\nu([-a,b+1],U(N)) = X(0)\mu(U(N),\ \text{sep. } 0_{a+b+1}).$$

The coordinates $x(1), \ldots, x(r)$ on $\mathbb{R}^r = \mathbb{R}^{a+b+1}$ are related to the $X(i)$ by

$$x(i) = X(i-1-a),$$

so we get

$$\text{SuccSub}_* \, \nu([-a, b+1], U(N)) = x(a+1)\mu(U(N), \text{ sep. } 0_{a+b+1}),$$

i.e.,

$$\xi([-a, b+1], U(N)) = x(a+1)\mu(U(N), \text{ sep. } 0_r).$$

Taking successively $a = 0, 1, \ldots, r-1$ gives the assertion. QED

6.14.15. Taking $r = 1$, we get perhaps the most striking corollary.

Corollary 6.14.16. *The measure $x\mu(U(N), \text{sep. } 0)$ on $\mathbb{R}$ is equal to the static spacing measure $\xi((0,1), U(N))$ on $\mathbb{R}$, i.e., the direct image of Haar measure on $U(N)$ by the map $A \mapsto$ the normalized spacing s_0 of A which contains the point 1 on the unit circle.*

6.14.17. For higher r, we get a description of $x\mu(U(N), \text{ step } r)$ as a sum of all the r-fold static spacing measures corresponding to the r different r-fold spacings which contain the point 1 on the unit circle.

Corollary 6.14.18. *The measure $x\mu(U(N), \text{ step } r)$ is the sum of the static spacing measures $\xi((a-r, a), U(N))$, $a = 1, \ldots, r$.*

PROOF. We know (by 6.14.14) that for $i = 1, \ldots, r$ we have

$$x(i)\mu(U(N), \text{ sep. } 0_r) = \xi([1-i, r+1-i], U(N)).$$

Summing over i, we get

$$\left(\sum_{i=1}^{r} x(i)\right) \mu(U(N), \text{ sep. } 0_r) = \sum_{i=1}^{r} \xi([1-i, r+1-i], U(N)),$$

an equality of measures on $\mathbb{R}^r$. Take the direct image to $\mathbb{R}$ by the map $\text{Sum} : \mathbb{R}^r \to \mathbb{R}$, $(x(1), \ldots, x(r)) \mapsto \sum_{i=1}^{r} x(i)$. On the left, we have

$$\begin{aligned} \text{Sum}_* &\left(\left(\sum_{i=1}^{r} x(i)\right) \mu(U(N), \text{ sep. } 0_r)\right) \\ &= \text{Sum}_*(\text{Sum}^*(x)\mu(U(N), \text{ sep. } 0_r)) \\ &= x \, \text{Sum}_* \, \mu(u(N), \text{ sep. } 0_r) \\ &= x\mu(U(N), \text{ step } r). \end{aligned}$$

On the right, we have the tautological relation

$$\text{Sum}_* \, \xi([1-i, r+1-i], U(N)) = \xi((1-i, r+1-i), U(N)).$$

Reindexing by taking a to be $r+1-i$, we get the assertion. QED

Remark 6.14.19. This expression of $x\mu(U(N), \text{ step } r)$ as the sum of r probability measures gives yet "another" proof that

$$\int_{\mathbb{R}} x\mu(U(N), \text{ step } r) = r,$$

or equivalently that

$$\int_{[0,\infty)} \mathrm{Tail}_{\mu(U(N),\ \mathrm{step}\ r)}(t)\, dt = r,$$

cf. the proof of 6.12.9.

6.14.20. Here is the several variable version of the above result.

Proposition 6.14.21. *Fix an integer $N \geq 2$. Let $r \geq 1$ be an integer, and c in $\mathbb{Z}^r$ an offset vector,*

$$0 < c(1) < c(2) < \cdots < c(r).$$

Denote by $0 \oplus c$ in $\mathbb{Z}^{r+1}$ the vector $(0, c(1), \ldots, c(r))$, and define

$$c(0) := 0.$$

For each $i = 1, \ldots, r$, define

$$d(i) := c(i) - c(i-1).$$

Then for each $i = 1, \ldots, r$, the measure $x(i)\mu(U(N),$ offset $c)$ on $\mathbb{R}^r$ is the sum of $d(i)$ static spacing measures

$$x(i)\mu(U(N),\ \textit{offset}\ c) = \sum_{j=c(i-1)}^{c(i)-1} \xi(0 \oplus c\ - j\mathbb{1}_{r+1}, U(N)),$$

corresponding to the $d(i)$ different spacing vectors of offset c whose i'th constituent spacing contains the point 1 on the unit circle.

PROOF. By the result 6.14.14 in $c(r)$ variables, in $\mathbb{R}^{c(r)}$ for every integer a with $1 \leq a \leq c(r)$ we have

$$x(a)\mu(U(N),\ \mathrm{sep.}\ 0_{c(r)}) = \xi([1-a, c(r)+1-a], U(N)).$$

Take the sum of these equalities as a runs from $c(i-1)+1$ to $c(i)$:

$$\sum_{a=1+c(i-1)}^{c(i)} x(a)\mu(U(N),\ \mathrm{sep.}\ 0_{c(r)})$$

$$= \sum_{a=1+c(i-1)}^{c(i)} \xi([1-a, c(r)+1-a], U(N)).$$

Now take the direct image to $\mathbb{R}^r$ of this equality by the map "successive partial sums": $\mathrm{SPS}_c : \mathbb{R}^{c(r)} \to \mathbb{R}^r$ given by

$$(x(a))_{a=1,\ldots,c(r)} \mapsto \left(\sum_{a=1+c(i-1)}^{c(i)} x(a) \right)_{i=1,\ldots,r}$$

to obtain the asserted identity. QED

It is perhaps worth pointing out the following special case.

Corollary 6.14.22. *Notations and hypotheses as in 6.4.21, suppose in addition that $c(1) = 1$. Then we have*

$$x(1)\mu(U(N),\ \textit{offset}\ c) = \xi(0 \oplus c, U(N)).$$

6.15. A failure of symmetry

6.15.1. One might think that "by symmetry", all the static spacing measures of given offset coincide. This is **not** the case. We will illustrate this in the simplest case, that of the nearest neighbor static spacing measures $\xi((i, i+1), U(N))$ for various values of $i \bmod N$, any $N \geq 2$. Given A in $U(N)$, we denote by $s_i(A)$, $i \bmod N$, its N normalized spacings

$$s_i(A) := (N/2\pi)(\varphi(i+1)(A) - \varphi(i)(A)).$$

So viewing $A \mapsto s_i(A)$ as a function on $U(N)$, we have

$$(s_i)_* \operatorname{Haar}_{U(N)} = \xi((i, i+1), U(N)).$$

We do have the relations

$$s_i(\overline{A}) = s_{-i}(A)$$

for each $i \bmod N$. Because Haar measure is invariant under the automorphism $A \mapsto \overline{A}$ of $U(N)$, these relations imply that we have equalities of measures

$$\xi((i, i+1), U(N)) = \xi((-i, 1-i), U(N))$$

for every $i \bmod N$.

Question 6.15.2. Are these the only $\mathbb{R}$-linear relations among the N measures $\xi((i, i+1), U(N))$, i.e., are the measures $\xi((i, i+1), U(N))$ with $0 \leq i \leq [N/2]$ $\mathbb{R}$-linearly independent?

Proposition 6.15.3. *Fix an integer $N \geq 2$. The N measures*

$$\xi((i, i+1), U(N)), \quad i \bmod N,$$

are not all equal.

PROOF. We argue by contradiction. The sum of these measures is

$$N\mu(U(N), \text{ sep. } 0);$$

indeed this was how we saw back in 1.1.3 that the expected value measures $\mu(U(N), \text{ sep. } 0)$ made good sense. So if all these measures coincide, each is equal to $\mu(U(N), \text{ sep. } 0)$. But the N'th one is $x\mu(U(N), \text{ sep. } 0)$. Therefore we find an equality of measures

$$\mu(U(N), \text{ sep. } 0) = x\mu(U(N), \text{ sep. } 0).$$

Multiplying both sides by x^k for any integer $k \geq 0$, we find

$$x^k\mu(U(N), \text{ sep. } 0) = x^{k+1}\mu(u(N), \text{ sep. } 0).$$

Therefore for all $k \geq 0$ we have

$$x^k\mu(U(N), \text{ sep. } 0) = \mu(U(N), \text{ sep. } 0).$$

Integrating both sides over $\mathbb{R}$, we find that $\mu(U(N), \text{ sep. } 0)$ has all its moments 1. Since these moments grow slowly, $\mu(U(N), \text{ sep. } 0)$ is the unique measure with these moments, and hence $\mu(U(N), \text{ sep. } 0)$ must be equal to δ_1, the delta measure at the point 1. Therefore each measure $\xi((i, i+1), U(N))$ is δ_1. This means that outside a set of measure zero in $U(N)$, every element of $U(N)$ has all of its normalized spacings equal to 1, in other words is conjugate to an element of the form

$$(\text{a unitary scalar}) \times (\text{the diagonal matrix } \operatorname{Diag}(e^{2\pi i j/N}, j = 1, \ldots, N)).$$

Taking traces, we conclude that outside a set of measure zero in $U(N)$, the trace vanishes (remember $N \geq 2$, so $\sum_j e^{2\pi ij/N} = 0$). But the trace is nonzero at the identity, so must be nonzero in a nonempty open set. But any nonempty open set has nonzero Haar measure. Contradiction. QED

6.15.4. Let us treat the $U(2)$ case explicitly, since here the failure of the symmetry intuition is perhaps the most shocking. The Weyl integration formula in this case asserts that if we view the space of conjugacy classes in $U(2)$ as the space of pairs of angles in order

$$0 \leq \varphi(1) \leq \varphi(2) < 2\pi,$$

then the (direct image from $U(2)$ of) Haar measure is the measure

$$(1/2\pi)^2 |e^{i\varphi(2)} - e^{i\varphi(1)}|^2 \, d\varphi(1)\, d\varphi(2).$$

We reparameterize this space by passing to coordinates $\varphi(1)$ and $x = \varphi(2) - \varphi(1)$. In these coordinates the space of conjugacy classes is

$$\varphi(1) \geq 0, \quad x \geq 0, \quad \varphi(1) + x < 2\pi,$$

and the measure is

$$(1/2\pi)^2 |e^{ix} - 1|^2 \, d\varphi(1)\, dx.$$

The normalized spacing $\vartheta(2) - \vartheta(1)$ is $(N/2\pi)x = x/\pi$. Thus for f a function on $\mathbb{R}$, we have

$$\begin{aligned}
&\int_{\mathbb{R}} f(x)\, d(\vartheta(2) - \vartheta(1))_* \operatorname{Haar}_{U(2)} \\
&\quad = (1/2\pi)^2 \iint_{\varphi(1)\geq 0, x \geq 0, \varphi(1)+x<2\pi} f(x/\pi) |e^{ix} - 1|^2 \, d\varphi(1)\, dx \\
&\quad = (1/2\pi)^2 \int_{[0,2\pi)} \left(\int_{[0,2\pi - x)} d\varphi(1) \right) f(x/\pi) |e^{ix} - 1|^2 \, dx \\
&\quad = (1/2\pi)^2 \int_{[0,2\pi)} (2\pi - x) f(x/\pi) |e^{ix} - 1|^2 \, dx \\
&\quad = (1/2)^2 \int_{[0,2\pi)} (2 - x/\pi) f(x/\pi) |e^{ix} - 1|^2 d(x/\pi) \\
&\quad = (1/2)^2 \int_{[0,2)} (2 - t) f(t) |e^{i\pi t} - 1|^2 \, dt \\
&\quad = (1/2)^2 \int_{[0,2)} (2 - t) f(t) (2 - 2\cos(\pi t))\, dt \\
&\quad = \int_{[0,2)} (2 - t) f(t) \sin^2(\pi t/2)\, dt.
\end{aligned}$$

Thus we have

$$(\vartheta(2) - \vartheta(1))_* \operatorname{Haar}_{U(2)} = I_{[0,2)}(t)(2 - t)\sin^2(\pi t/2)\, dt.$$

In the same $\varphi(1), x$ coordinates, the normalized spacing $\vartheta(3) - \vartheta(2)$ is equal to $2 - x/\pi$, since the sum of the two normalized spacings is 2. Thus we have

$$\begin{aligned}
&\int_{\mathbb{R}} f(x) d(\vartheta(3) - \vartheta(2))_* \operatorname{Haar}_{U(2)} \\
&= (1/2\pi)^2 \iint_{\varphi(1)\geq 0, x\geq 0, \varphi(1)+x<2\pi} f(2 - x/\pi)|e^{ix} - 1|^2 \, d\varphi(1) \, dx \\
&= (1/2\pi)^2 \int_{[0,2\pi)} \left(\int_{[0,2\pi - x)} d\varphi(1) \right) f(2 - x/\pi)|e^{ix} - 1|^2 \, dx \\
&= (1/2\pi)^2 \int_{[0,2\pi)} (2\pi - x) f(2 - x/\pi)|e^{ix} - 1|^2 \, dx \\
&= \int_{[0,2)} (2 - t) f(2 - t) \sin^2(\pi t/2) \, dt \\
&= \int_{(0,2]} t f(t) \sin^2(\pi - \pi t/2) \, dt \\
&= \int_{(0,2]} t f(t) \sin^2(\pi t/2) \, dt,
\end{aligned}$$

and thus

$$(\vartheta(3) - \vartheta(2))_* \operatorname{Haar}_{U(2)} = I_{(0,2]}(t) t \sin^2(\pi t/2) \, dt.$$

Thus we see explicitly that

$$(\vartheta(3) - \vartheta(2))_* \operatorname{Haar}_{U(2)} \neq (\vartheta(2) - \vartheta(1))_* \operatorname{Haar}_{U(2)}.$$

Adding these two, we get $2\mu(U(2), \text{ sep. } 0)$, so we must have

$$\mu(U(2), \text{ sep. } 0) = I_{(0,2]}(t) \sin^2(\pi t/2) \, dt.$$

6.16. Offset spacing measures and their relation to multi-eigenvalue location measures on $U(N)$

6.16.1. Given an integer $r \geq 1$, we denote by

$$\operatorname{Off} : \mathbb{R}^r \to \mathbb{R}^r$$

the linear automorphism of $\mathbb{R}^r$ defined on $x = (x(1), \dots, x(r))$ in $\mathbb{R}^r$ by

$$\operatorname{Off}(x) = (x(1), x(1) + x(2), x(1) + x(2) + x(3), \dots),$$

i.e., $\operatorname{Off}(x)$ is the vector whose j'th component is $\sum_{1\leq i\leq j} x(i)$. The inverse automorphism is $y \mapsto \operatorname{Diff}(y)$, defined by

$$\operatorname{Diff}(y)(i) = \begin{cases} y(1), & \text{for } i = 1, \\ y(i) - y(i-1), & \text{for } i = 2, \dots, r. \end{cases}$$

6.16.2. Given a Borel probability measure μ on $\mathbb{R}^r$, we denote by

$$\operatorname{Off} \mu := \operatorname{Off}_* \mu$$

the direct image of μ by the automorphism $\operatorname{Off} : \mathbb{R}^r \to \mathbb{R}^r$. For any bounded, Borel measurable function $f(x)$ on $\mathbb{R}^r$, we have the tautological integration formula

$$\int_{\mathbb{R}^r} f(x) d(\operatorname{Off} \mu) = \int_{\mathbb{R}^r} f(\operatorname{Off}(x)) \, d\mu.$$

6.16.3. We now fix a separation vector a in $\mathbb{Z}^r$, with corresponding step vector $b := a + \mathbb{1}_r$ and corresponding offset vector $c := \mathrm{Off}(b)$. For any $N \geq 2$, we have the spacing measure $\mu(U(N), \text{offsets } c)$ on $\mathbb{R}^r$. In this section, we will be concerned with the "offset spacing measure"

$$\mathrm{Off}\,\mu(U(N), \text{ offsets } c) := \mathrm{Off}_*\,\mu(U(N), \text{ offsets } c).$$

Just as $\mu(U(N), \text{offsets } c)$ was defined as the expected value over $U(N)$ of the measures $\mu(A, U(N), \text{offsets } c)$ attached to elements A of $U(N)$, so the measure $\mathrm{Off}\,\mu(U(N), \text{ offsets } c)$ is the expected value over $U(N)$ of the measures $\mathrm{Off}\,\mu(A, U(N), \text{offsets } c)$.

Lemma 6.16.4 (compare Lemma 6.14.5). 1) *For any nonempty subset S of $\{1, 2, \ldots, r\}$, denote by*

$$\mathrm{pr}[S] : \mathbb{R}^r \to \mathbb{R}^S$$

the projection onto the named variables. Then we have

$$\mathrm{pr}[S]_*(\mathrm{Off}\,\mu(A, U(N), \text{ offsets } c)) = \mathrm{Off}\,\mu(A, U(N), \text{ offsets } \mathrm{pr}[S](c))$$

for every A in $U(N)$, and

$$\mathrm{pr}[S]_*(\mathrm{Off}\,\mu(U(N), \text{ offsets } c)) = \mathrm{Off}\,\mu(U(N), \text{ offsets } \mathrm{pr}[S](c)).$$

2) *The measures $\mathrm{Off}\,\mu(A, U(N), \text{ offsets } c)$ and $\mathrm{Off}\,\mu(U(N), \text{ offsets } c)$ on $\mathbb{R}^r$ are supported in $(\mathbb{R}_{\geq 0})^r(\text{order})$.*

PROOF. For both 1) and 2), the assertion for $\mathrm{Off}\,\mu(A, U(N) \text{ offsets } c)$ is tautologous, and the assertion for $\mathrm{Off}\,\mu(U(N), \text{offsets } c)$ results from this one by integration over $U(N)$. QED

6.16.5. The main result of this section is the relation of the offset spacing measure to the multi-eigenvalue location measure $\nu(c, U(N))$. In the case $r = 1$, where $\mathrm{Off} : \mathbb{R} \to \mathbb{R}$ is the identity, this relation is 6.12.6. Here is the general case.

Proposition 6.16.6. *Fix integers $r \geq 1$ and $N \geq 2$, and denote*

$$\mathbb{1} := \mathbb{1}_r := (1, 1, \ldots, 1) \text{ in } \mathbb{R}^r.$$

For any nonnegative Borel measurable function $g \geq 0$ on $\mathbb{R}^r$, denote by G the nonnegative Borel measurable function $G \geq 0$ defined by the Lebesgue integral

$$G(x) := \int_{[0,x(1)]} g(x - t\mathbb{1})\, dt := |x(1)| \int_{[0,1]} g(x - tx(1)\mathbb{1})\, dt.$$

Fix an offset vector c in $\mathbb{Z}^r$:

$$1 \leq c(1) < c(2) < \cdots < c(r).$$

For each integer k with $0 \leq k \leq c(1) - 1$, $c - k\mathbb{1}$ is again an offset vector, and we have the identity

$$\int_{\mathbb{R}^r} G\, d\,\mathrm{Off}\,\mu(U(N), \text{ offsets } c) = \sum_{0 \leq k \leq c(1)-1} \int_{\mathbb{R}^r} g\, d\nu(c - k\mathbb{1}, U(N)).$$

PROOF. The idea of the proof is that already used in proving 6.12.4, 6.12.6, and 6.14.12, namely to express the integrals involved as integrals over $U(N)$ against Haar measure, and then to show that the integrands coincide on the set $U(N)^{\mathrm{reg}}$ of elements with N distinct eigenvalues.

The definition of $\nu(c, U(N))$ as a direct image from $U(N) \times U(1)$ gives

$$\int_{\mathbb{R}^r} g\, d\nu(c, U(N))$$
$$= \int_{U(N)} \int_{[0,2\pi)} f(\vartheta(c(1))(e^{-i\varphi}A), \dots, \vartheta(c(r))(e^{-i\varphi}A))d(\varphi/2\pi)\, dA$$

for any offset vector c. The definition of $\operatorname{Off}\mu(U(N), \text{ offsets } c)$ as the expected value over $U(N)$ of the measures $\operatorname{Off}\mu(A, U(N), \text{ offsets } c)$ gives

$$\int_{\mathbb{R}^r} Gd \operatorname{Off}\mu(U(N), \text{ offsets } c)$$
$$= \int_{U(N)} \left(\int_{\mathbb{R}^r} Gd \operatorname{Off}\mu(A, U(N), \text{ offsets } c) \right) dA.$$

We will show that for each A in $U(N)$ with N distinct eigenvalues, we have

$$\int_{\mathbb{R}^r} G\, d \operatorname{Off}\mu(A, U(N), \text{ offsets } c)$$
$$= \sum_{0 \le k \le c(1)-1} \int_{[0,2\pi)} f(\vartheta(c(1)-k)(e^{-i\varphi}A), \dots, \vartheta(c(r)-k)(e^{-i\varphi}A))d(\varphi/2\pi).$$

To show this, we proceed as follows. Denote by $\varphi(i) := \varphi(i)(A)$ the (nonnormalized) angles of A, defined for all i in $\mathbb{Z}$. For each i, let

$$s_i := (N/2\pi)(\varphi(i+1) - \varphi(i))$$

be the i'th normalized spacing of A, and let

$$S_i := (\varphi(i), \varphi(i+1)] \subset U(1)$$

be the half open interval between $\varphi(i)$ and $\varphi(i+1)$. By definition of $\operatorname{Off}\mu(A, U(N), \text{ offsets } c)$, we have

$$N \int_{\mathbb{R}^r} G\, d \operatorname{Off}\mu(A, U(N), \text{ offsets } c)$$
$$= \sum_{l \bmod N} G(s_{l+1} + s_{l+2} + \dots + s_{l+c(1)}, \dots, s_{l+1} + s_{l+2} + \dots + s_{l+c(r)}).$$

Let us introduce the scalars

$$s_{l,a,b} := \sum_{a \le i \le b} s_{l+i}, \quad \text{if } a \le b,$$
$$:= 0, \quad \text{if } a > b.$$

Then

$$N \int_{\mathbb{R}^r} G\, d \operatorname{Off}\mu(A, U(N), \text{ offsets } c)$$
$$= \sum_{l \bmod N} G(s_{l,1,c(1)}, s_{l,1,c(2)}, \dots, s_{l,1,c(r)})$$
$$= \sum_{l \bmod N} G(s_{l,\mathbb{1},c}),$$

where we denote by $s_{l,\mathbb{1},c}$ the vector $(s_{l,1,c(1)}, s_{l,1,c(2)}, \dots, s_{l,1,c(r)})$.

Now recall the definition of G in terms of g, to see that

$$G(s_{l,\mathbb{1},c}) = \int_{[0,s_{l,1,c(1)}]} g(s_{l,\mathbb{1},c} - t\mathbb{1})\,dt$$
$$= \int_{(0,s_{l,\mathbb{1},c(1)}]} g(s_{l,\mathbb{1},c} - t\mathbb{1})\,dt.$$

We break the interval $(0, s_{l,1,c(1)}]$ into $c(1)$ disjoint intervals

$$(0, s_{l,1,c(1)}] = (0, \sum_{1\le i\le c(1)} s_{l+i}] = \bigsqcup_{0\le k\le c(1)-1} (s_{l,1,k},\ s_{l,1,k+1}].$$

Thus we get

$$G(s_{l,\mathbb{1},c}) = \sum_{0\le k\le c(1)-1} \int_{(s_{l,1,k},\ s_{l,1,k+1}]} g(s_{l,\mathbb{1},c} - t\mathbb{1})\,dt$$
$$= \sum_{0\le k\le c(1)-1} \int_{(0,s_{l,k,k+1}]} g(s_{l,\mathbb{1},c} - s_{l,1,k}\mathbb{1} - t\mathbb{1})\,dt.$$

At this point, we observe that we have the relations

$$s_{l,\mathbb{1},c} - s_{l,1,k}\mathbb{1} = s_{l+k,\mathbb{1},c-k\mathbb{1}}, \qquad s_{l,k,k+1} = s_{l+k,0,1} = s_{l+k}.$$

So the previous identity becomes

$$G(s_{l,\mathbb{1},c}) = \sum_{0\le k\le c(1)-1} \int_{[0,s_{l+k}]} g(s_{l+k,\mathbb{1},c-k\mathbb{1}} - t\mathbb{1})\,dt.$$

Summing over l and shifting l by k, we obtain

$$N\int_{\mathbb{R}^r} G\,d\,\mathrm{Off}\,\mu(A, U(N),\ \text{offsets } c)$$
$$= \sum_{0\le k\le c(1)-1} \ \sum_{l \bmod N} \int_{[0,s_l]} g(s_{l,\mathbb{1},c-k\mathbb{1}} - t\mathbb{1})\,dt.$$

So we are reduced to showing that for each k with $0 \le k \le c(1) - 1$, we have

$$(1/N) \sum_{l \bmod N} \int_{[0,s_l]} g(s_{l,\mathbb{1},c-k\mathbb{1}} - t\mathbb{1})\,dt$$
$$= \int_{[0,2\pi)} f(\vartheta(c(1) - k)(e^{-i\varphi}A), \ldots, \vartheta(c(r) - k)(e^{-i\varphi}A))\,d(\varphi/2\pi).$$

This is a statement about the offset vector $c - k\mathbb{1}$, so it suffices to treat universally the case when $k = 0$, i.e., to show that for any offset vector c in $\mathbb{Z}^r$ we have

$$(1/N) \sum_{l \bmod N} \int_{[0,s_l]} g(s_{l,\mathbb{1},c} - t\mathbb{1})\,dt$$
$$= \int_{[0,2\pi)} f(\vartheta(c(1))(e^{-i\varphi}A), \ldots, \vartheta(c(r))(e^{-i\varphi}A))\,d(\varphi/2\pi).$$

To show this, it suffices to show that for each l we have

$$\int_{S_l} g(\vartheta(c(1))(e^{-i\varphi}A), \ldots, \vartheta(c(r))(e^{-i\varphi}A))d(\varphi/2\pi)$$
$$= (1/N) \int_{[0,s_l]} g(s_{l,\mathbb{1},c} - t\mathbb{1})\, dt.$$

But this is a tautology: as φ runs in $(\varphi(l), \varphi(l+1)]$, $\vartheta(c)(e^{-i\varphi}A)$ runs from $s_{l,\mathbb{1},c}$ to $s_{l,\mathbb{1},c} - s_l\mathbb{1}$. QED

Corollary 6.16.7. *The relation*

$$\int_{\mathbb{R}^r} G\, d\,\mathrm{Off}\, \mu(U(N),\ \textit{offsets}\ c) = \sum_{0 \le k \le c(1)-1} \int_{\mathbb{R}^r} g\, d\nu(c - k\mathbb{1}, U(N))$$

for $G(x)$ defined as the Lebesgue integral $G(x) := \int_{[0,x(1)]} g(x - t\mathbb{1})\, dt$, remains valid if we allow g to be a Borel measurable function of polynomial growth (or indeed any Borel measurable function which is bounded on compact sets).

PROOF. The measures involved are all probability measures of compact support. Because we have measures of finite mass, we may extend the proposition from nonnegative Borel functions g to bounded Borel functions f, and using the compact support we may extend to any Borel functions g whose restrictions to all compact sets are bounded, in particular to Borel functions of polynomial growth. QED

6.17. Interlude: "Tails" of measures on $\mathbb{R}^r$

6.17.1. Let $r \ge 1$ be an integer, μ a Borel probability measure on $\mathbb{R}^r$. For s in $\mathbb{R}^r$, we defined the rectangle $R(s) \subset \mathbb{R}^r$ as

$$R(s) := \{x \text{ in } \mathbb{R}^r \text{ such that } x(i) \le s(i) \text{ for } i = 1, \ldots, r\},$$

and we defined the cumulative distribution function (CDF) of μ by

$$\mathrm{CDF}_\mu(s) := \mu(R(s)).$$

6.17.2. We now define, for s in $\mathbb{R}^r$, a "tail rectangle" $T(s)$ in $\mathbb{R}^r$ by

$$T(s) := \{x \text{ in } \mathbb{R}^r \text{ such that } x(i) > s(i) \text{ for } i = 1, \ldots, r\}.$$

For $r = 1$, $T(s)$ is just the complement of $R(s)$, whence the terminology "tail". For $r = 2$, $R(s)$ is a closed third quadrant and $T(s)$ is the "opposite" open first quadrant. Here is the $r = 2$ picture:

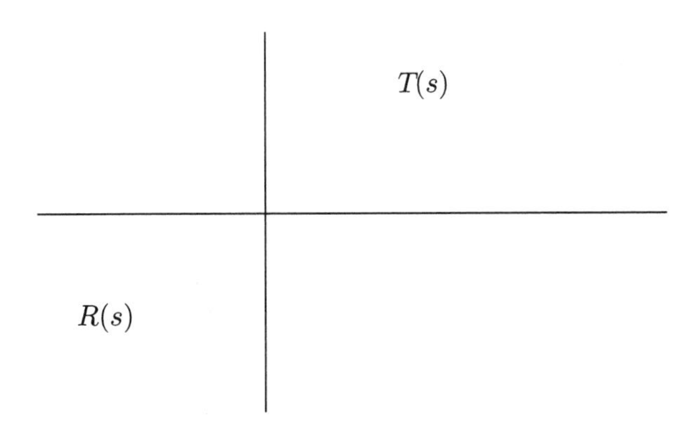

6.17.3. We define the tail function of μ, Tail_μ, by

$$\mathrm{Tail}_\mu(s) := \mu(T(s)).$$

Lemma 6.17.4. *Let $r \geq 1$ be an integer, μ a Borel probability measure on $\mathbb{R}^r$. Denote by $[-1] : \mathbb{R}^r \to \mathbb{R}^r$ the map $x \mapsto -x$.*

1) *We have*

$$\mathrm{CDF}_{[-1]_*\mu}(-s) = \lim_{n\to\infty} \mathrm{Tail}_\mu(s - (1/n)\mathbb{1}),$$
$$\mathrm{Tail}_\mu(s) = \lim_{n\to\infty} \mathrm{CDF}_{[-1]_*\mu}(-s - (1/n)\mathbb{1}).$$

2) *The measure μ is determined by its tail function.*
3) *The following conditions are equivalent:*
 3a) Tail_μ *is continuous,*
 3b) $\mathrm{CDF}_{[-1]_*\mu}$ *is continuous,*
 3c) CDF_μ *is continuous.*

Moreover, if these equivalent conditions hold, then for every s in $\mathbb{R}^r$ we have

$$\mathrm{Tail}_\mu(s) = \mathrm{CDF}_{[-1]_*\mu}(-s) = \mu(\{x \text{ in } \mathbb{R}^r \text{ with } x(i) \geq s(i) \text{ for all } i\}).$$

PROOF. We first prove 1). It is tautologous that

$$\begin{aligned}
\mathrm{CDF}_{[-1]_*\mu}(-s) &= \mu(\{x \text{ in } \mathbb{R}^r \text{ such that } -x(i) \leq -s(i) \text{ for all } i\}) \\
&= \mu(\{x \text{ in } \mathbb{R}^r \text{ such that } x(i) \geq s(i) \text{ for all } i\}) \\
&= \mu\left(\bigcap_{n\geq 1} \{x \text{ in } \mathbb{R}^r \text{ such that } x(i) > s(i) - 1/n \text{ for all } i\}\right) \\
&= \mu\left(\bigcap_{n\geq 1} T(s - (1/n)\mathbb{1})\right).
\end{aligned}$$

Because μ has finite mass, the measure of this decreasing intersection is

$$\lim_{n\to\infty} \mathrm{Tail}_\mu(s - (1/n)\mathbb{1}).$$

Similarly,

$$\begin{aligned}
\mathrm{Tail}_\mu(s) &= \mu(\{x \text{ in } \mathbb{R}^r \text{ such that } x(i) > s(i) \text{ for all } i\}) \\
&= \mu(\{x \text{ in } \mathbb{R}^r \text{ such that } -x(i) < -s(i) \text{ for all } i\}) \\
&= \mu\left(\bigcup_{n\geq 1} \{x \text{ in } \mathbb{R}^r \text{ such that } -x(i) \leq -s(i) - 1/n \text{ for all } i\}\right) \\
&= \lim_{n\to\infty} \mu(\{x \text{ in } \mathbb{R}^r \text{ such that } -x(i) \leq -s(i) - 1/n \text{ for all } i\}) \\
&= \lim_{n\to\infty} \mathrm{CDF}_{[-1]_*\mu}(-s - (1/n)\mathbb{1}).
\end{aligned}$$

Assertion 2) follows from the first formula of 1), which shows that Tail_μ determines $\mathrm{CDF}_{[-1]_*\mu}$, which in turn determines $[-1]_*\mu$ and then μ itself. To prove 3), we argue as follows. If Tail_μ is continuous, then by 1) we have

$$\mathrm{CDF}_{[-1]_*\mu}(-s) = \mathrm{Tail}_\mu(s),$$

which shows that $\mathrm{CDF}_{[-1]_*\mu}$ is continuous. $\mathrm{CDF}_{[-1]_*\mu}$ is continuous if and only if each hyperplane $x(i) = \alpha$ has $[-1]_*\mu$-measure zero, if and only if each hyperplane $x(i) = \alpha$ has μ-measure zero, if and only if CDF_μ is continuous. Finally, if $\mathrm{CDF}_{[-1]_*\mu}$ is continuous, then by 1) we have $\mathrm{Tail}_\mu(s) = \mathrm{CDF}_{[-1]_*\mu}(-s)$, whence Tail_μ is continuous. The final assertion results from the equality

$$\mathrm{Tail}_\mu(s) = \mathrm{CDF}_{[-1]_*\mu}(-s)$$

proven above, along with the explicitation of $\mathrm{CDF}_{[-1]_*\mu}(-s)$. QED

Lemma 6.17.5. *A real-valued continuous function $F(s)$ on $\mathbb{R}^r$ with values in $[0, 1]$ is Tail_μ for some Borel probability measure μ on $\mathbb{R}^r$ with a continuous* Tail *if and only if F satisfies the following three conditions:*

1) $\lim_{n\to\infty} F(n\mathbb{1}) = 0$.

2) $\lim_{n\to\infty} F(-n\mathbb{1}) = 1$.

3) *F satisfies the Lebesgue-Stieltjes positivity condition (which expresses that the rectangle $(s, s + t]$ is to have nonnegative measure): For every s in $\mathbb{R}^r$, and for every t in $(\mathbb{R}_{>0})^r$, we have*

$$\sum_{\text{subsets } S \text{ of } \{1,2,\ldots,r\}} (-1)^{\mathrm{Card}(S)} F\left(s + \sum_{i \text{ in } S} t(i)e(i)\right) \geq 0.$$

PROOF. It is elementary that if μ is a Borel probability measure with Tail_μ continuous, then $F(s) := \mathrm{Tail}_\mu(s)$ satisfies these conditions. Conversely, given a continuous F which satisfies these conditions, we define $G(s) := F(-s)$. Then G satisfies the usual conditions to be the CDF of a Borel probability measure, say ν, with a continuous CDF, and if we take $\mu := [-1]_*\nu$, then μ has tail F. QED

Corollary 6.17.6 (Limit Lemma). *Let $r \geq 1$ be an integer, $\{\mu_n\}_{n\geq 1}$ a sequence of Borel probability measures on $\mathbb{R}^r$, each of which has a continuous tail function. Suppose that the sequence of functions $\{\mathrm{Tail}_{\mu_n}\}_n$ converges uniformly on $\mathbb{R}^r$ to some function F. Then F is continuous, and there exists a unique Borel probability measure μ on $\mathbb{R}^r$ whose tail function is F. For any bounded continuous function g on $\mathbb{R}^r$, we have the limit formula*

$$\int_{\mathbb{R}^r} g\, d\mu = \lim_{n\to\infty} \int_{\mathbb{R}^r} g\, d\mu_n.$$

PROOF. Let us write $F_n := \mathrm{Tail}_{\mu_n}$. The uniform convergence of F_n to F shows that F is continuous, takes values in $[0, 1]$, and satisfies the Lebesgue-Stieltjes positivity condition 3) of the previous lemma. To construct μ with $\mathrm{Tail}_\mu = F$, it suffices to show that F satisfies conditions 1) and 2) of the previous lemma as well. For each n, the function of one real variable $x \mapsto F_n(x\mathbb{1}) := \mu_n(T(x\mathbb{1}))$ is a decreasing function of x, which is 1 at $-\infty$ and 0 at $+\infty$. By the uniformity of the convergence of F_n to F, $x \mapsto F(x\mathbb{1})$ has these properties as well, and so satisfies 1) and 2) of the previous lemma.

Once we know that F is Tail_μ for some Borel probability measure μ with a continuous tail, we use the relations

$$\mathrm{CDF}_{[-1]_*\mu_n}(s) = \mathrm{Tail}_{\mu_n}(-s), \qquad \mathrm{CDF}_{[-1]_*\mu}(s) = \mathrm{Tail}_\mu(-s),$$

to see that the measures $[-1]_*\mu_n$ converge properly to the probability measure μ, from which it follows [**Fel**, page 243] that

$$\int_{\mathbb{R}^r} g\, d([-1]_*\mu) = \lim_{n\to\infty} \int_{\mathbb{R}^r} g\, d([-1]_*\mu_n)$$

for every bounded continuous g. Applying this to the function $g(-x)$ gives the asserted relation

$$\int_{\mathbb{R}^r} g\, d\mu = \lim_{n\to\infty} \int_{\mathbb{R}^r} g\, d\mu_n. \quad \text{QED}$$

6.17.7. There is a simple Moebius inversion relation between CDF's and Tail's.

Lemma 6.17.8. *Let $r \geq 1$ be an integer, μ a Borel probability measure on $\mathbb{R}^r$. For each subset S of $\{1, 2, \dots, r\}$, denote by $\mathrm{pr}[S] : \mathbb{R}^r \to \mathbb{R}^S$ the projection onto the named coordinates, with the convention that $\mathbb{R}^{\varnothing} = \{0\}$ for $S = \varnothing$. Then for s in $\mathbb{R}^r$ we have the inversion formulas*

$$\mathrm{Tail}_\mu(s) = \sum_{S\subset\{1,2,\dots,r\}} (-1)^{\#S}\, \mathrm{CDF}_{\mathrm{pr}[S]_*\mu}(\mathrm{pr}[S](s)),$$

$$\mathrm{CDF}_\mu(s) = \sum_{S\subset\{1,2,\dots,r\}} (-1)^{\#S}\, \mathrm{Tail}_{\mathrm{pr}[S]_*\mu}(\mathrm{pr}[S](s)).$$

PROOF. Fix s in $\mathbb{R}^r$. For each $j = 1, \dots, r$, denote by χ_j and ρ_j the characteristic functions of the half-planes $x(j) \leq s(j)$ and $x(j) > s(j)$ respectively. Then $\chi_j + \rho_j = 1$, the characteristic function of $R(s)$ is $\prod_j \chi_j = \prod_j (1 - \rho_j)$, and the characteristic function of $T(s)$ is $\prod_j \rho_j = \prod_j (1 - \chi_j)$. Similarly, for each $S \subset \{1, 2, \dots, r\}$, the characteristic function of $\mathrm{pr}[S]^{-1}(R(\mathrm{pr}[S](s)))$ is $\prod_{j \text{ in } S} \chi_j$, and the characteristic function of $\mathrm{pr}[S]^{-1}(T(\mathrm{pr}[S](s)))$ is $\prod_{j \text{ in } S} \rho_j$. The asserted inversion formulas are obtained by expanding

$$\chi_{R(s)} = \prod_j \chi_j = \prod_j (1 - \rho_j)$$

and

$$\chi_{T(s)} = \prod_j \rho_j = \prod_j (1 - \chi_j)$$

respectively by the binomial theorem, and integrating against μ. QED

Corollary 6.17.9. *Let $r \geq 1$ be an integer, $\{\mu_n\}_{n\geq 1}$ and μ Borel probability measures on $\mathbb{R}^r$. The following conditions are equivalent.*

1) *For every subset S of $\{1, 2, \dots, r\}$, the sequence of functions $\{\mathrm{Tail}_{\mathrm{pr}[S]_*\mu_n}\}_n$ converges pointwise (respectively uniformly) to $\mathrm{Tail}_{\mathrm{pr}[S]_*\mu}$.*

2) *For every subset S of $\{1, 2, \dots, r\}$, the sequence of functions $\{\mathrm{CDF}_{\mathrm{pr}[S]_*\mu_n}\}_n$ converges pointwise (respectively uniformly) to $\mathrm{CDF}_{\mathrm{pr}[S]_*\mu}$.*

6.18. Tails of offset spacing measures and tails of multi-eigenvalue location measures on $U(N)$

Proposition 6.18.1. *Fix an integer $r \geq 1$ and an offset vector c in $\mathbb{Z}^r$,*

$$1 \leq c(1) < c(2) < \cdots < c(r).$$

For each integer $N \geq 2$, and each s in $\mathbb{R}^r$ with $s(1) \geq 0$, we have the relation

$$\sum_{0 \leq k \leq c(1)-1} \mathrm{Tail}_{\nu(c-k\mathbb{1}, U(N))}(s) = \int_{[0,\infty]} \mathrm{Tail}_{\mathrm{Off}\,\mu(U(N),\ \textit{offsets}\ c)}(s + t\mathbb{1})\, dt.$$

PROOF. Fix s in $\mathbb{R}^r$ with $s(1) \geq 0$. We apply 6.16.6 with the nonnegative Borel function g taken to be the characteristic function of the tail rectangle $T(s)$. Then for t in $\mathbb{R}$, the function $x \mapsto g(x - t\mathbb{1})$ is the characteristic function of the rectangle $T(s + t\mathbb{1})$. The general relation

$$\sum_{0 \leq k \leq c(1)-1} \int_{\mathbb{R}^r} g\, d\nu(c - k\mathbb{1}, U(N)) = \int_{\mathbb{R}^r} G\, d\,\mathrm{Off}\,\mu(U(N),\ \text{offsets}\ c)$$

becomes, for this choice of g, the relation

$$\sum_{0 \leq k \leq c(1)-1} \mathrm{Tail}_{\nu(c-k\mathbb{1}, U(N))}(s)$$
$$= \int_{\mathbb{R}^r} \left(\int_{[0,x(1)]} g(x - t\mathbb{1})\, dt \right) d\,\mathrm{Off}\,\mu(U(N),\ \text{offsets}\ c).$$

Because the measure $\mathrm{Off}\,\mu(U(N),\ \text{offsets}\ c)$ is supported in $(\mathbb{R}_{\geq 0})^r(\text{order})$, in particular it is supported in $x(1) \geq 0$, so we may rewrite this integral as

$$= \int_{\mathbb{R}^r, x(1) \geq 0} \left(\int_{[0,x(1)]} g(x - t\mathbb{1})\, dt \right) d\,\mathrm{Off}\,\mu(U(N),\ \text{offsets}\ c)$$
$$= \int_{\mathbb{R}^r \times \mathbb{R}, x(1) \geq t \geq 0, x > s + t\mathbb{1}} d\,\mathrm{Off}\,\mu(U(N),\ \text{offsets}\ c)\, dt.$$

Because $s(1) \geq 0$, the condition $x \geq s + t\mathbb{1}$ implies $x(1) \geq t$, so we may rewrite the integral as

$$= \int_{\mathbb{R}^r \times \mathbb{R}, t \geq 0, x > s + t\mathbb{1}} d\,\mathrm{Off}\,\mu(U(N),\ \text{offsets}\ c)\, dt$$
$$= \int_{[0,\infty]} \left(\int_{\mathbb{R}^r, x > s + t\mathbb{1}} d\,\mathrm{Off}\,\mu(U(N),\ \text{offsets}\ c) \right) dt$$

and the inner integral is $\mathrm{Tail}_{\mathrm{Off}\,\mu(U(N),\ \text{offsets}\ c)}(s + t\mathbb{1})$. QED

Corollary 6.18.2. *Hypotheses and notations as in 6.18.1 above, for s in $\mathbb{R}^r$ with $s(1) > 0$, we have*

$$\mathrm{Tail}_{\mathrm{Off}\,\mu(U(N),\ \textit{offsets}\ c)}(s)$$
$$= -\left(\sum_i \partial / \partial s(i) \right) \sum_{0 \leq k \leq c(1)-1} \mathrm{Tail}_{\nu(c-k\mathbb{1}, U(N))}(s).$$

PROOF. Indeed, for $\varepsilon > 0$ in $\mathbb{R}$ we have

$$\sum_{0 \le k \le c(1)-1} \mathrm{Tail}_{\nu(c-k\mathbb{1},U(N))}(s+\varepsilon\mathbb{1})$$

$$= \int_{[0,\infty]} \mathrm{Tail}_{\mathrm{Off}\,\mu(U(N),\ \text{offsets } c)}(s+\varepsilon\mathbb{1}+t\mathbb{1})\,dt$$

$$= \int_{[\varepsilon,\infty]} \mathrm{Tail}_{\mathrm{Off}\,\mu(U(N),\ \text{offsets } c)}(s+t\mathbb{1})\,dt.$$

So we readily compute the difference

$$\sum_{0 \le k \le c(1)-1} (\mathrm{Tail}_{\nu(c-k\mathbb{1},U(N))}(s+\varepsilon\mathbb{1}) - \mathrm{Tail}_{\nu(c-k\mathbb{1},U(N))}(s))$$

$$= -\int_{[0,\varepsilon]} \mathrm{Tail}_{\mathrm{Off}\,\mu(U(N),\ \text{offsets } c)}(s+t\mathbb{1})\,dt.$$

Dividing by ε and letting $\varepsilon \to 0$ gives the assertion for the derivative from above. Similarly, for small ε, the vector $s - \varepsilon\mathbb{1}$ still has first coordinate ≥ 0, so we have

$$\sum_{0 \le k \le c(1)-1} \mathrm{Tail}_{\nu(c-k\mathbb{1},U(N))}(s-\varepsilon\mathbb{1})$$

$$= \int_{[0,\infty]} \mathrm{Tail}_{\mathrm{Off}\,\mu(U(N),\ \text{offsets } c)}(s-\varepsilon\mathbb{1}+t\mathbb{1})\,dt$$

$$= \int_{[-\varepsilon,\infty]} \mathrm{Tail}_{\mathrm{Off}\,\mu(U(N),\ \text{offsets } c)}(s+t\mathbb{1})\,dt.$$

So we readily compute the difference

$$\sum_{0 \le k \le c(1)-1} (\mathrm{Tail}_{\nu(c-k\mathbb{1},U(N))}(s-\varepsilon\mathbb{1}) - \mathrm{Tail}_{\nu(c-k\mathbb{1},U(N))}(s))$$

$$= \int_{[-\varepsilon,0]} \mathrm{Tail}_{\mathrm{Off}\,\mu(U(N),\ \text{offsets } c)}(s+t\mathbb{1})\,dt.$$

Dividing by $-\varepsilon$ and letting $\varepsilon \to 0$ gives the assertion for the derivative from below. QED

6.19. Moments of offset spacing measures and of multi-eigenvalue location measures on $U(N)$

Proposition 6.19.1. *Fix an integer $r \ge 1$ and an offset vector c in $\mathbb{Z}^r$,*

$$1 \le c(1) < c(2) < \cdots < c(r).$$

For each integer $N \ge 2$, we have the following relations among moments: for any polynomial function $H(x)$ on $\mathbb{R}^r$ which is divisible by $x(1)$ as a polynomial, denote by $h(x)$ the polynomial

$$h := \left(\sum_i \partial/\partial x(i)\right) H.$$

Then for such an H we have the relation

$$\sum_{0 \le k \le c(1)-1} \int_{\mathbb{R}^r} h\,d\nu(c-k\mathbb{1},U(N)) = \int_{\mathbb{R}^r} H\,d\,\mathrm{Off}\,\mu(U(N),\ \textit{offsets } c).$$

PROOF. This is seen most easily by writing H as an $\mathbb{R}$-linear sum of monomials

$$x(1)^{a(1)}\prod_{i=2}^{r}(x(i)-x(i-1))^{a(i)}$$

with a in $\mathbb{Z}^r$ having $a(1)\geq 1$ and all $a(i)\geq 0$. For H an individual such monomial, we have

$$h:=\left(\sum_i \partial/\partial x(i)\right)H=a(1)x(1)^{a(1)-1}\prod_{i=2}^{r}(x(i)-x(i-1))^{a(i)}.$$

For $x(1)\geq 0$, we recover

$$H(x):=\int_{[0,x(1)]}h(x-t\mathbb{1})\,dt.$$

Since the measure Off $\mu(U(N)$, offsets $c)$ is supported in $x(1)\geq 0$, the assertion results immediately from 6.16.7, applied to the polynomial function h. QED

6.20. Multi-eigenvalue location measures for the other $G(N)$

6.20.1. For $G(N)$ one of $SO(2N+1), USp(2N), SO(2N), O_-(2N+2)$, or $O_-(2N+1)$, and A in $G(N)$, we have defined its angles

$$0\leq\varphi(1)(A)\leq\varphi(2)(A)\leq\cdots\leq\varphi(N)(A)\leq\pi,$$

and its normalized angles

$$\vartheta(i)(A):=(N+\lambda)\varphi(i)(A)/\pi,\qquad i=1,\ldots,N.$$

Given an integer $r\geq 1$, an offset vector c in $\mathbb{Z}^r$,

$$0<c(1)<c(2)<\cdots<c(r),$$

and an integer $N\geq c(r)$, we denote by $\nu(c,G(N))$ the probability measure on $\mathbb{R}^r$ which is the direct image of total mass one Haar measure on $G(N)$ by the map $G(N)\to\mathbb{R}^r$ defined by the normalized angles

$$A\mapsto(\vartheta(c(1)),\ldots,\vartheta(c(r))).$$

Thus

$$\nu(c,G(N)):=(\vartheta(c(1)),\ldots,\vartheta(c(r)))_*\operatorname{Haar}_{G(N)}.$$

6.20.2. Exactly as in the case of $U(N)$, cf. 6.14.5, we have

Lemma 6.20.3. *Suppose given an integer $r\geq 1$, and an offset vector c in $\mathbb{Z}^r$. For any nonempty subset*

$$S=\{1\leq s_1<\cdots<s_k\leq r\}$$

of the index set $\{1,\ldots,r\}$, we denote by

$$\operatorname{pr}[S]:\mathbb{R}^r\to\mathbb{R}^{\operatorname{Card}(S)}$$

the projection onto the named coordinates. We have an equality

$$\operatorname{pr}[S]_*\nu(c,G(N))=\nu(\operatorname{pr}[S](c),G(N))$$

of measures on $\mathbb{R}^{\operatorname{Card}(S)}$.

Proposition 6.20.4. *Given an integer $r \geq 1$, an offset vector c in $\mathbb{Z}^r$,*

$$0 < c(1) < c(2) < \cdots < c(r),$$

and an integer $N \geq c(r)$, for $G(N)$ any of $U(N), SO(2N+1), USp(2N), SO(2N)$, $O_-(2N+2), O_-(2N+1)$, the measure $\nu(c, G(N))$ on $\mathbb{R}^r$ is absolutely continuous with respect to Lebesgue measure.

PROOF. Denote by S the subset $\{c(1), \ldots, c(r)\}$ of $\{1, 2, \ldots, N\}$. Then

$$\nu(c, G(N)) = \mathrm{pr}[S]_* \nu((1, 2, \ldots, N), G(N)).$$

Now $\mathrm{pr}[S]_*$(Lebesgue measure on $\mathbb{R}^N$) is absolutely continuous with respect to Lebesgue measure on $\mathbb{R}^r$, so it suffices to show that $\nu((1, 2, \ldots, N), G(N))$ is absolutely continuous with respect to Lebesgue measure on $\mathbb{R}^N$. This is obvious from the fact that the Hermann Weyl measure on T/W, viewed as a W-invariant measure on T, is absolutely continuous with respect to Haar measure on T. QED

Corollary 6.20.5. *Hypotheses and notations as in* 6.20.4 *above, the measure $\nu(c, G(N))$ has a continuous CDF and hence (by 6.17.4) has a continuous tail.*

PROOF. In view of 2.11.17, it suffices to show that each one-variable projection $\mathrm{pr}[i]_* \nu(c, G(N)) = \nu(c(i), G(N))$ on $\mathbb{R}$ has no point masses, and this is guaranteed by the absolute continuity of $\nu(c(i), G(N))$ with respect to Lebesgue measure. QED

CHAPTER 7

Large N Limits and Fredholm Determinants

7.0. Generating series for the limit measures μ(univ, sep.'s a) in several variables: absolute continuity of these measures

7.0.1. Let us fix an integer $r \geq 1$, and a bounded, Borel measurable $\mathbb{R}$-valued function f on $\mathbb{R}^r$ of compact support. For any separation vector a in $\mathbb{Z}^r$, with corresponding offset vector c, the measure

$$\mu(\text{univ, sep.'s } a) = \mu(\text{naive, univ, sep.'s } a)$$

is given (2.10.1) by

$$\int_{\mathbb{R}^r} f\, d\mu(\text{univ, sep.'s } a) \\ = \sum_{n \geq 0} (-1)^{\Sigma(n-a)} \operatorname{Binom}(n, a) \operatorname{COR}(n, f, \text{univ}).$$

Lemma 7.0.2. *Suppose that f on $\mathbb{R}^r$ is supported in $\sum |s(i)| \leq \alpha$. For any separation vector n in $\mathbb{Z}^r$, we have the estimate*

$$|\operatorname{COR}(n, f, \textit{ univ})| \leq \|f\|_{\sup} (\alpha)^{r+\Sigma(n)}/(r + \Sigma(n))!.$$

If n corresponds to the offset vector c, so $c(r) = r + \Sigma(n)$, we have the formula

$$\operatorname{COR}(n, f, \textit{ univ}) = \int_{[0,\alpha]^{r+\Sigma(n)}(\textit{order})} H(0, z) W(r + \Sigma(n) + 1)(0, z) \prod_i dz(i),$$

where

$$H(x(0), x(1), \ldots, x(c(r))) \\ = f(x(c(1)) - x(0), x(c(2)) - x(c(1)), \ldots, x(c(r)) - x(c(r-1))).$$

PROOF. Because f has compact support, the function

$$H(X) := \operatorname{Clump}(n, f, r + \Sigma(n) + 1, X) \\ := f(x(c(1)) - x(0), x(c(2)) - x(c(1)), \ldots, x(c(r)) - x(c(r-1)))$$

lies in $\mathcal{T}_0(r + \Sigma(n) + 1)$, cf. 4.0.5. According to 4.1.5, for every A in $G(N)$ we have

$$\operatorname{Cor}(n, f, G(N), A) = Z[r + \Sigma(n) + 1, H, G(N)](A).$$

Integrating over $G(N)$, we get

$$\operatorname{COR}(n, f, G(N)) = E(Z[r + \Sigma(n) + 1, H, G(N)]).$$

If we take $G(N)$ to be $U(N)$, then 5.8.2 gives (because $\sigma = 2$ for $U(N)$) the estimate

$$|E(Z[r + \Sigma(n) + 1, H, G(N)])| \leq \|H\|_{\sup} (\alpha)^{r+\Sigma(n)}/(r + \Sigma(n))! \\ \leq \|f\|_{\sup} (\alpha)^{r+\Sigma(n)}/(r + \Sigma(n))!.$$

Taking the limit over N and using 5.10.3, we get the asserted estimate for $|\,\mathrm{COR}(n,f,\ \mathrm{univ})|$, and the asserted formula

$$\begin{aligned}\mathrm{COR}(n,f,\ \mathrm{univ}) &= E(r+\Sigma(n)+1, H,\ \mathrm{univ})\\ &= \int_{[0,\alpha]^{r+\Sigma(n)}(\mathrm{order})} H(0,z)W(r+\Sigma(n)+1)(0,z)\prod_i dz(i). \quad \mathrm{QED}\end{aligned}$$

Proposition 7.0.3. *Fix an integer $r \geq 1$. For any separation vector a in $\mathbb{Z}^r$, the measure*

$$\mu(\mathit{univ},\ \mathit{sep.'s}\ a) = \mu(\mathit{naive},\ \mathit{univ},\ \mathit{sep.'s}\ a)$$

on $\mathbb{R}^r$ is absolutely continuous with respect to Lebesgue measure.

PROOF. It suffices to show that any bounded set $E \subset \mathbb{R}^r$ of Lebesgue measure zero has μ(univ, sep.'s a)-measure zero. Denote by f the characteristic function of E. In view of the fundamental identity 2.10.2,

$$\begin{aligned}&\int_{\mathbb{R}^r} f\, d\mu(\mathrm{univ},\ \mathrm{sep.'s}\ a)\\ &\quad= \sum_{n\geq 0} (-1)^{\Sigma(n-a)} \operatorname{Binom}(n,a)\,\mathrm{COR}(n,f,\ \mathrm{univ}),\end{aligned}$$

it suffices to show that $\mathrm{COR}(n,f,\ \mathrm{univ}) = 0$ for $n \geq 0$ in $\mathbb{Z}^r$. For this, we use the identity of the previous lemma,

$$\mathrm{COR}(n,f,\ \mathrm{univ}) = \int_{[0,\alpha]^{r+\Sigma(n)}(\mathrm{order})} H(0,z)W(r+\Sigma(n)+1)(0,z)\prod_i dz(i),$$

where c is the offset vector corresponding to n, and

$$\begin{aligned}&H(x(0),x(1),\ldots,x(c(r)))\\ &\quad= f(x(c(1))-x(0), x(c(2))-x(c(1)),\ldots,x(c(r))-x(c(r-1))).\end{aligned}$$

So it suffices if $H(0,z)$ as function on $\mathbb{R}^{c(r)}$ is the characteristic function of a set of Lebesgue measure zero in $\mathbb{R}^{c(r)}$. To see this, view $\mathbb{R}^{c(r)}$ as $\mathbb{R}^r \times \mathbb{R}^{c(r)-r}$ via coordinates (the $z(c(i))$'s, the other $z(j)$'s), and recall the inverse linear automorphisms Off and Diff of $\mathbb{R}^r$ (cf. 6.16.1). Then $H(0,z)$ is the characteristic function of the product set $\mathrm{Diff}^{-1}(E) \times \mathbb{R}^{c(r)-r}$. QED

7.0.4. Let us now consider the formal power series in r variables

$$G_r(T)(f) = G_r(T_1,\ldots,T_r)(f)$$

defined by

$$G_r(T_1,\ldots,T_r)(f) := \sum_{n\geq 0} \mathrm{COR}(n,f,\ \mathrm{univ})T^n.$$

Lemma 7.0.5. *For any bounded, Borel measurable $\mathbb{R}$-valued function f on $\mathbb{R}^r$ of compact support, the formal power series $G_r(T)(f)$ is everywhere convergent.*

PROOF. If f is supported in $\sum |s(i)| \le \alpha$, then by the previous lemma $G_r(T)(f)$ is term by term majorized (denoted temporarily as $\lll$) by

$$\sum_{n\ge 0} \alpha^{r+\Sigma(n)}T^n/(r+\Sigma(n))! \lll \alpha^r \sum_{n\ge 0} \alpha^{\Sigma(n)}T^n/(r+\Sigma(n))!$$

$$\lll \alpha^r \sum_{n\ge 0} \alpha^{\Sigma(n)}T^n/\Sigma(n)! \lll \alpha^r \sum_{n\ge 0} \alpha^{\Sigma(n)}T^n \Big/ \left(\prod_i (n(i)!)\right)$$

$$= \alpha^r \exp\left(\alpha \sum_i T_i\right). \quad \text{QED}$$

Lemma 7.0.6. *For any bounded, Borel measurable $\mathbb{R}$-valued function f on $\mathbb{R}^r$ of compact support, and any separation vector a in $\mathbb{Z}^r$, we have*

$$\int_{\mathbb{R}^r} f\, d\mu(\textit{univ, sep.'s } a)$$
$$= \prod_i ((d/dT_i)^{a(i)}/a(i)!) G_r(T_1, \ldots, T_r)(f)|_{\textit{all } T_i=-1}.$$

Equivalently, we have the identity of entire functions of r variables

$$G_r(T)(f) = \sum_{n\ge 0 \textit{ in } \mathbb{Z}^r} (1+T)^n \int_{\mathbb{R}^r} f\, d\mu(\textit{univ, sep.'s } n),$$

with the usual notational convention $(1+T)^n := \prod_i (1+T_i)^{n(i)}$.

PROOF. In view of the definition of $G_r(T)$, this is just a compact restatement of 2.10.2. QED

7.0.7. Given a point s in $(\mathbb{R}_{\ge 0})^r$, we denote by $R(s)$ the rectangle $[0,s]$ in $(\mathbb{R}_{\ge 0})^r$, and by $I_{R(s)}$ its characteristic function. Given an offset vector c in $\mathbb{Z}^r$, we define

$$\text{Domain}(c,s) \subset (\mathbb{R}_{\ge 0})^{c(r)}(\text{order})$$

to be the set of those points $0 \le x(1) \le x(2) \le \cdots \le x(c(r))$ which satisfy

$$x(c(1)) \le s(1),$$

and

$$x(c(i)) - x(c(i-1)) \le s(i) \quad \text{for } i = 2, \ldots, r.$$

Lemma 7.0.8. *For each separation vector n in $\mathbb{Z}^r$, with corresponding offset vector c, and for each s in $(\mathbb{R}_{\ge 0})^r$, we have the identity*

$$\text{COR}(n, I_{R(s)}, \textit{ univ}) = \int_{\text{Domain}(c,s)} W(1+c(r))(0,z) \prod_i dz(i),$$

and the estimate

$$|\,\text{COR}(n, I_{R(s)}, \textit{ univ})| \le (\Sigma(s))^{r+\Sigma(n)}/(r+\Sigma(n))!.$$

PROOF. This is just 7.0.2, applied to the function $f = I_{R(s)}$, whose α is $\Sigma(s)$ and whose $H(0,z)$ is the characteristic function of Domain(c,s). QED

Lemma 7.0.9. *For each separation vector n in $\mathbb{Z}^r$, with corresponding step vector b and offset vector c, the function*

$$s \mapsto \mathrm{COR}(n, I_{R(s)},\ univ)$$

is the restriction to $(\mathbb{R}_{\geq 0})^r$ of an entire function of s which is divisible by $\prod_i s(i)^{b(i)}$ (and hence in particular divisible by $\prod_i s(i)$). For s in $\mathbb{C}^r$, we have

$$|\,\mathrm{COR}(n, I_{R(s)},\ univ)|$$
$$\leq \left(\prod_i (|s(i)|^{b(i)}/b(i)!)\right) \mathrm{Sqrt}(1+c(r))^{1+c(r)} \exp((1+c(r))\sum_i |s(i)|\pi).$$

Proof. We first prove everything except the estimate for complex s. That it is entire will result from the formula

$$\mathrm{COR}(n, I_{R(s)},\ \mathrm{univ}) = \int_{\mathrm{Domain}(c,s)} W(1+c(r))(0,z) \prod_j dz(j),$$

and the fact that $W(1+c(r))(0,z)$ is the restriction to $\mathbb{R}^{c(r)}$ of an entire function of z. Let us begin with an s in $(\mathbb{R}_{\geq 0})^r$. If any of the coordinates $s(i)$ of s vanishes, $\mathrm{Domain}(c,s)$ has Lebesgue measure zero, and $\mathrm{COR}(n, I_{R(s)},\ \mathrm{univ}) = 0$. Thus it suffices to study

$$s \mapsto \mathrm{COR}(n, I_{R(s)},\ \mathrm{univ})$$

for s in $(\mathbb{R}_{>0})^r$. To do this, we introduce the difference variables

$$x(1) = z(1),$$
$$x(j) = z(j) - z(j-1) \quad \text{for } 2 \leq j \leq c(r).$$

Seen in the x variables, $\mathrm{Domain}(c,s)$ is the region $\mathrm{Region}(c,s)$ defined by the inequalities

$$\text{all } x(j) \geq 0, \text{ and } \sum_{j=1+c(i-1)}^{c(i)} x(j) \leq s(i) \quad \text{for } i = 1, \ldots, r,$$

with the convention that $c(0) = 0$. We define an entire function of x's by

$$V(c(r))(x) := W(1+c(r))(0,z).$$

We define scalars $t(j)$, $j = 1, \ldots, c(r)$, by

$$t(j) := s(i) \quad \text{if } 1 + c(i-1) \leq j \leq c(i),$$

and we make the further change of variables

$$x(j) = t(j)u(j), \quad j = 1, \ldots, c(r).$$

In the u coordinates, we have

$$\mathrm{COR}(n, I_{R(s)},\ \mathrm{univ}) = \int_{\mathrm{Region}(c,1)} V(c(r))(tu) \prod_j (t(j)\, du(j))$$
$$= \left(\prod_j t(j)\right) \int_{\mathrm{Region}(c,1)} V(c(r))(tu) \prod_j du(j)$$
$$= \left(\prod_i s(i)^{b(i)}\right) \int_{\mathrm{Region}(c,1)} V(c(r))(tu) \prod_j du(j),$$

where Region$(c, 1)$ is the region of u space defined by the inequalities

$$\text{all } u(j) \geq 0, \text{ and } \sum_{j=1+c(i-1)}^{c(i)} u(j) \leq 1 \quad \text{for } i = 1, \dots, r.$$

Thus it suffices to show that

$$\int_{\text{Region}(c,1)} V(c(r))(tu) \prod_j du(j)$$

is an entire function of t. To see this, expand the entire function $V(c(r))(x)$ in an everywhere convergent power series, say $\sum_w A_w x^w$. For fixed t, we may apply dominated convergence to integrate term by term over the compact Region$(c, 1)$:

$$\begin{aligned}
&\int_{\text{Region}(c,1)} V(c(r))(tu) \prod_i du(i) \\
&\quad = \int_{\text{Region}(c,1)} \sum_w A_w (tu)^w \prod_i du(i) \\
&\quad = \sum_w A_w (t)^w \int_{\text{Region}(c,1)} u^w \prod_i du(i).
\end{aligned}$$

To see that this last series in t is everywhere convergent, we argue as follows. The region Region$(c, 1)$ is contained in the unit cube $[0,1]^{c(r)}$, so $|u^w| \leq 1$ in the integral. Moreover, Region$(c, 1)$ has Euclidean volume $1/\prod_i b(i)!$; this amounts to the statement that for any integer $n \geq 1$, the region of $\mathbb{R}^n$ defined by

$$\text{all } x(i) \geq 0, \quad \sum_i x(i) \leq 1$$

has volume $1/n!$. To see this, use the variables $z(i) = \sum_{j \leq i} x(j)$, in which this is the region $[0,1]^n$(order),

$$0 \leq z(1) \leq z(2) \leq \cdots \leq z(n) \leq 1,$$

which has volume $1/n!$, being, up to a set of measure zero, a fundamental domain for the action of Σ_n on $[0,1]^n$. Thus we may estimate

$$\left| \int_{\text{Region}(c,1)} u^w \prod_i du(i) \right| \leq \int_{\text{Region}(c,1)} \prod_i du(i) = 1/\prod_i b(i)! \leq 1,$$

and so the series $\sum_w A_w(t)^w \int_{\text{Region}(c,1)} u^w \prod_i du(i)$ is dominated term by term by the entire function $\sum_w A_w(t)^w = V(c(r))(t)$.

It remains to prove the estimate asserted for complex s. We use the formula

$$\text{COR}(n, I_{R(s)}, \text{univ}) = \left(\prod_i s(i)^{b(i)} \right) \int_{\text{Region}(c,1)} V(c(r))(tu) \prod_j du(j),$$

and remind the reader that t in $\mathbb{C}^{c(r)}$ is the vector

$$(s(1) \text{ repeated } b(1) \text{ times}, \dots, s(r) \text{ repeated } b(r) \text{ times}).$$

From this, we get

$$\begin{aligned}
&|\operatorname{COR}(n, I_{R(s)}, \text{univ})| \\
&\leq \left(\prod_i |s(i)|^{b(i)}\right) \int_{\text{Region}(c,1)} |V(c(r))(tu)| \prod_j du(j) \\
&\leq \left(\prod_i |s(i)|^{b(i)}\right) \operatorname*{Sup}_{u \text{ in Region}(c,1)} |V(c(r))(tu)| \int_{\text{Region}(c,1)} \prod_j du(j), \\
&= \left(\prod_i (|s(i)|^{b(i)}/b(i)!)\right) \operatorname*{Sup}_{u \text{ in Region}(c,1)} |V(c(r))(tu)|.
\end{aligned}$$

Thus it suffices to establish the inequality

$$|V(c(r))(ut)| \leq \operatorname{Sqrt}(1+c(r))^{1+c(r)} \exp\left((1+c(r))\sum_i |s(i)|\pi\right)$$

for t in $\mathbb{C}^{c(r)}$ and u in Region$(c,1)$.

In the $c(r)$ variables $z(j) := \sum_{i \leq j} u(i)t(i)$, $1 \leq j \leq c(r)$, and with $z(0) := 0$, we have

$$\begin{gathered}
V(c(r))(ut) = W(1+c(r))(0,z) = \det(A), \\
A := (f(z(i)-z(j)))_{0 \leq i,j \leq c(r)}, \qquad f(x) = \sin(\pi x)/\pi x.
\end{gathered}$$

Thus each entry of the matrix A is of the form

$$f(\pm(\text{a partial sum of } u(j)t(j)\text{'s})).$$

From the power series expansion of $f(x) = \sin(\pi x)/\pi x$, we see that for complex x we have

$$\begin{aligned}
|f(x)| = |\sin(\pi x)/\pi x| &= \left|\sum_{n \geq 0} (-1)^n (\pi x)^{2n}/(2n+1)!\right| \\
&\leq \sum_{n \geq 0} |\pi x|^{2n}/(2n)! \leq \sum_{n \geq 0} |\pi x|^n/n! = \exp(\pi|x|).
\end{aligned}$$

Since $\exp(x)$ is increasing for real x, we have

$$|f(\pm(\text{a partial sum of } u(j)t(j)\text{'s}))| \leq \exp\left(\sum_j |u(j)t(j)|\pi\right).$$

But recall that for those j with $1+c(i-1) \leq j \leq c(i)$, we have $t(j) = s(i)$. Moreover, because u lies in Region$(c,1)$, we have all $u(j) \geq 0$, and $\sum_{1+c(i-1)}^{c(i)} u(j) \leq 1$. Thus we have

$$\sum_j |u(j)t(j)|\pi = \sum_{i=1}^{r} \sum_j u(j)|s(i)|\pi \leq \sum_i |s(i)|\pi.$$

Thus $V(c(r))(t)$ is a determinant of size $1+c(r)$, each of whose entries is bounded in absolute value by $\exp(\sum_i |s(i)|\pi)$. The Hadamard determinant inequality

$$|\text{an } n \times n \text{ determinant } (a_{i,j})| \leq (n^{1/2} \operatorname*{Max}_{i,j} |a_{i,j}|)^n$$

gives the asserted estimate. QED

Proposition 7.0.10. *For each integer $r \geq 1$, the series*

$$G_r(T, s) := G_r(T)(I_{R(s)}) := \sum_{n \geq 0} T^n \operatorname{Cor}(n, I_{R(s)},\ univ),$$

which for each s in $(\mathbb{R}_{\geq 0})^r$ is an entire function of T in $\mathbb{C}^r$, is the restriction to $\mathbb{C}^r \times (\mathbb{R}_{\geq 0})^r$ of an entire function on $\mathbb{C}^{2r}$ which is divisible by $\prod_i s(i)$.

PROOF. We know that each coefficient $\operatorname{COR}(n, I_{R(s)},$ univ$)$ is an entire function of s which is divisible by $\prod_i s(i)$, and hence each finite sum

$$\sum_{l \geq n \geq 0} T^n \operatorname{Cor}(n, I_{R(s)}, \text{ univ})$$

is an entire function of (T, s) which vanishes when any $s(i) = 0$. Since a uniform limit of holomorphic functions is holomorphic, it suffices to check that the series

$$\sum_{n \geq 0} T^n \operatorname{Cor}(n, I_{R(s)}, \text{ univ})$$

converges uniformly on compact sets in $\mathbb{C}^{2r}$. For then the limit series will be entire, and it will vanish when any $s(i) = 0$, so will be divisible by $\prod_i s(i)$.

To do this, pick a real number $M > 1$, and suppose all $|T_i| \leq M$ and all $|s(i)| \leq M$. Then

$$\begin{aligned}
&|\operatorname{COR}(n, I_{R(s)}, \text{ univ})| \\
&\leq \left(\prod_i (M^{b(i)}/b(i)!)\right) \operatorname{Sqrt}(1 + c(r))^{1+c(r)} \exp((1 + c(r))rM\pi) \\
&= \left(\Sigma(b)!/\prod_i b(i)!\right) (M^{\Sigma(b)}/\Sigma(b)!) \operatorname{Sqrt}(1 + c(r))^{1+c(r)} \exp((1 + c(r))rM\pi) \\
&\leq r^{\Sigma(b)}(M^{\Sigma(b)}/\Sigma(b)!) \operatorname{Sqrt}(1 + c(r))^{1+c(r)} \exp((1 + c(r))rM\pi).
\end{aligned}$$

Recalling that $c(r) = \Sigma(b) \geq r \geq 1$, we have $1 + c(r) \leq 2c(r)$, so we may continue

$$\begin{aligned}
&\leq r^{\Sigma(b)}(M^{\Sigma(b)}/\Sigma(b)!) \operatorname{Sqrt}(1 + c(r))^{1+c(r)} \exp(2c(r)rM\pi) \\
&= ((\exp(2rM\pi)rM)^{\Sigma(b)}/\Sigma(b)!) \operatorname{Sqrt}(1 + \Sigma(b))^{1+\Sigma(b)}.
\end{aligned}$$

The number of step vectors b with a given value of $\Sigma(b)$ is trivially bounded by $(\Sigma(b))^{r-1}$, since each $b(i)$ is an integer in $[1, \Sigma(b)]$, and the last one $b(r)$ is determined by the first $r - 1$. Thus if all $|T_i| \leq M$ and all $|s(i)| \leq M$, we have, summing over possible values k of $\Sigma(b)$,

$$\begin{aligned}
&\sum_{n \geq 0} |T^n \operatorname{Cor}(n, I_{R(s)}, \text{ univ})| \\
&\qquad \leq \sum_{k \geq r} k^{r-1} M^{k-r} (\exp(2rM\pi)rM)^k (\operatorname{Sqrt}(1 + k)^{1+k}/k!) \\
&\qquad \leq \sum_{k \geq r} k^{r-1} M^k (\exp(2rM\pi)rM)^k (\operatorname{Sqrt}(1 + k)^{1+k}/k!) \\
&\qquad \leq \sum_{k \geq 0} k^{r-1} (\exp(2rM\pi)rM^2)^k (\operatorname{Sqrt}(1 + k)^{1+k}/k!).
\end{aligned}$$

So it suffices that the function of one variable

$$\sum_{k\geq 0} k^{r-1}X^k(\mathrm{Sqrt}(1+k)^{1+k}/k!)$$

be entire. This function is the result of applying $(Xd/dX)^{r-1}$ to

$$\sum_{k\geq 0} X^k(\mathrm{Sqrt}(1+k)^{1+k}/k!),$$

so it suffices that this last function be entire. This is immediate from Stirling's inequality (cf. the proof of 3.4.5)

$$\log(\Gamma(x)) > (x-1/2)(\log(x)-1) \quad \text{for real } x>0. \quad \text{QED}$$

Corollary 7.0.11. *For each integer $r \geq 1$, and each separation vector n in $\mathbb{Z}^r$, the cumulative distribution function (CDF) of the measure $\mu(univ, sep.\text{'s } n)$,*

$$s \text{ in } (\mathbb{R}_{\geq 0})^r \mapsto \mu(univ, sep.\text{'s } n)(R(s)),$$

is (the restriction to $(\mathbb{R}_{\geq 0})^r$ of) an entire function on $\mathbb{C}^r$ which is divisible by $\prod_i s(i)$.

PROOF. Obvious from 7.0.6 and the fact that $G_r(T,s)$ is entire and divisible by $\prod_i s(i)$. QED

Proposition 7.0.12. *For each integer $r \geq 1$, and each separation vector n in $\mathbb{Z}^r$, consider the entire function $p(n;s)$ of $(s(1),\ldots,s(r))$ defined by*

$$p(n,s) = \left(\prod_i (d/ds(i))\right)(\mu(univ, sep.\text{'s } n)(R(s))).$$

The measure $\mu(univ, sep.\text{'s } n)$ on $\mathbb{R}^r$ is the direct image (i.e., extension by zero) from $(\mathbb{R}_{\geq 0})^r$ of the measure on $(\mathbb{R}_{\geq 0})^r$ given by

$$p(n;s)\prod_i ds(i).$$

PROOF. Let us denote temporarily by $f(n,s)$ the entire function

$$f(n,s) := \mu(\text{univ, sep.'s } n)(R(s)).$$

For s in $(\mathbb{R}_{\geq 0})^r$, $p(n,s)$ is real and ≥ 0, because it is the limit, as $\varepsilon > 0$ goes to zero, of $\mu(\text{univ, sep.'s } n)(\prod_i [s(i), s(i)+\varepsilon])/\varepsilon^r$. So $p(n;s)\prod_i ds(i)$ is a positive measure on $(\mathbb{R}_{\geq 0})^r$.

We know that $\mu(\text{univ, sep.'s } n)$ is supported in $(\mathbb{R}_{\geq 0})^r$, so it suffices to show that it coincides with $p(n;s)\prod_i ds(i)$ on $(\mathbb{R}_{\geq 0})^r$. Since $f(n,s)$ vanishes whenever any $s(i) = 0$, the fundamental theorem of calculus together with the previous corollary gives, for any s in $(\mathbb{R}_{\geq 0})^r$,

$$f(n,s) = \int_{\prod_i [0,s(i)]} p(n;x)\prod_i dx(i),$$

i.e.,

$$\int_{R(s)} \mu(\text{univ, sep.'s } n) = \int_{R(s)} p(n;x)\prod_i dx(i)$$

for every rectangle $R(s)$ in $(\mathbb{R}_{\geq 0})^r$. Thus our two Borel measures on $(\mathbb{R}_{\geq 0})^r$ agree on all rectangles $R(s)$, hence must be equal as Borel measures on $(\mathbb{R}_{\geq 0})^r$. QED

Corollary 7.0.13. 1) *For each integer $r \geq 1$, and each separation vector n in $\mathbb{Z}^r$, the measure μ(univ, sep.'s n) on $\mathbb{R}^r$ is absolutely continuous with respect to Lebesgue measure.*

2) *For any C^1 diffeomorphism of $\mathbb{R}^r$,*

$$\varphi : \mathbb{R}^r \to \mathbb{R}^r, \quad \varphi(x) = (\varphi_1(x), \ldots, \varphi_r(x)),$$

we have:

2a) *The direct image measure $\varphi_*\mu$(univ, sep.'s n) is absolutely continuous with respect to Lebesgue measure.*

2b) *Each of the direct image measures $(\varphi_i)_*\mu$(univ, sep.'s n) on $\mathbb{R}$ is absolutely continuous with respect to Lebesgue measure, and so in particular has a continuous CDF.*

PROOF. 1) The absolute continuity was proven above in 7.0.3. It is also obvious from the explicit formula for μ(univ, sep.'s n) as the extension by zero of $p(n,x)\prod_i dx(i)$.

For 2a), we argue as follows. Since φ is a C^1 diffeomorphism, the Jacobian formula

$$(\varphi^{-1})_*(\text{Lebesgue measure} \prod_i dx(i)) = \det_{r\times r}(\partial\varphi_i/\partial x(j)) \prod_i dx(i)$$

shows that $(\varphi^{-1})_*$(Lebesgue measure) is absolutely continuous with respect to Lebesgue measure. But $(\varphi^{-1})_*\mu$(univ, sep.'s n) is absolutely continuous with respect to $(\varphi^{-1})_*$(Lebesgue measure), just by transport of structure from 1), so $(\varphi^{-1})_*\mu$(univ, sep.'s n) is absolutely continuous with respect to Lebesgue measure on $\mathbb{R}^r$. Replacing φ by φ^{-1}, we find that $\varphi_*\mu$(univ, sep.'s n) is absolutely continuous with respect to Lebesgue measure.

For 2b), consider the i'th projection $\mathrm{pr}[i] : \mathbb{R}^r \to \mathbb{R}$. For any Borel set E in $\mathbb{R}$ of Lebesgue measure zero, its inverse image $\mathrm{pr}[i]^{-1}(E)$ in $\mathbb{R}^r$ certainly has Lebesgue measure zero. Therefore if ν is any Borel measure on $\mathbb{R}^r$ which is absolutely continuous with respect to Lebesgue measure, its direct image $\mathrm{pr}[i]_*\nu$ on $\mathbb{R}$ is absolutely continuous with respect to Lebesgue measure on $\mathbb{R}$. Applying this to the measure $\nu = \varphi_*\mu$(univ, sep.'s n), we find that

$$\mathrm{pr}[i]_*\nu = \mathrm{pr}[i]_*\varphi_*\mu(\text{univ, sep.'s } n) = (\varphi_i)_*\mu(\text{univ, sep.'s } n)$$

is absolutely continuous with respect to Lebesgue measure. QED

7.1. Interlude: Proof of Theorem 1.7.6

Let us recall the statement.

Theorem 1.7.6. *Let $r \geq 1$ be an integer, b in $\mathbb{Z}^r$ a step vector with corresponding separation vector a and offset vector c. Denote*

$$\mu := \mu(\textit{univ, offsets } c).$$

Suppose given an integer k with $1 \leq k \leq r$, and a surjective linear map

$$\pi : \mathbb{R}^r \to \mathbb{R}^k,$$

or, more generally, a partial C^1 coordinate system of bounded distortion $\pi : \mathbb{R}^r \to \mathbb{R}^k$.

1) *The measure $\pi_*\mu$ on $\mathbb{R}^k$ is absolutely continuous with respect to Lebesgue measure, and (consequently) has a continuous CDF.*

2) *Given any real* $\varepsilon > 0$, *there exists an explicit constant* $N(\varepsilon, r, c, \pi)$ *with the following property*: *For* $G(N)$ *any of the compact classical groups in their standard representations,*

$$U(N), SU(N), SO(2N+1), O(2N+1), USp(2N), SO(2N), O(2N),$$

and for

$$\mu(A, N) := \mu(A, G(N), \textit{ offsets } c), \textit{ for each } A \textit{ in } G(N),$$

we have the inequality

$$\int_{G(N)} \text{discrep}(\pi_*\mu(A, N), \pi_*\mu)\, dA \leq N^{\varepsilon - 1/(2r+4)},$$

provided that $N \geq N(\varepsilon, r, c, \pi)$.

PROOF. We first "complete" $\pi : \mathbb{R}^r \to \mathbb{R}^k$ to a $\mathcal{C}^1$-diffeomorphism of bounded distortion $\varphi : \mathbb{R}^r \to \mathbb{R}^r$ in such a way that for $J := \{1, 2, \ldots, k\}$, we have

$$\pi = \text{pr}[J] \circ \varphi.$$

To prove assertion 1), we use the fact (7.0.13) that μ on $\mathbb{R}^r$ is absolutely continuous with respect to Lebesgue measure ν_r on $\mathbb{R}^r$. It is then tautological that $\pi_*\mu$ is absolutely continuous with respect to $\pi_*\nu_r$. We claim that $\pi_*\nu_r$ is absolutely continuous with respect to Lebesgue measure ν_k on $\mathbb{R}^k$. To see this, use the fact that $\pi_*\nu_r = \text{pr}[J]_*\varphi_*\nu_r$. Because φ is a $\mathcal{C}^1$-diffeomorphism, the Jacobian rule gives $\varphi_*\nu_r =$ (the Jacobian of $\varphi^{-1})\nu_r$, and hence $\varphi_*\nu_r$ is absolutely continuous with respect to ν_r. Taking direct image by $\text{pr}[J]$, we get that $\pi_*\nu_r$ is absolutely continuous with respect to $\text{pr}[J]_*\nu_r$ on $\mathbb{R}^k$. But $\text{pr}[J]_*\nu_r$ on $\mathbb{R}^k$ is absolutely continuous with respect to Lebesgue measure ν_k. Indeed, ν_r is a product: $\nu_r = \nu_k \times \nu_{r-k}$, so $\text{pr}[J]_*\nu_r$ is a product

$$\text{pr}[J]_*\nu_r = \nu_k \times (\text{direct image from } \mathbb{R}^{r-k} \text{ to } \mathbb{R}^0 \text{ of } \nu_{k-r}).$$

More explicitly, if $k = r$, then $\text{pr}[J]_*\nu_r = \nu_k$. If $k < r$, then $\text{pr}[J]_*\nu_r$ is the measure on $\mathbb{R}^k$ defined on Lebesgue measurable sets E by

$$\begin{aligned}(\text{pr}[J]_*\nu_r)(E) = \nu_r(E \times \mathbb{R}^{r-k}) &= 0 \text{ if } \nu_k(E) = 0,\\ &= \infty \text{ if } \nu_k(E) > 0.\end{aligned}$$

To recapitulate, $\pi_*\nu_r$ is absolutely continuous with respect to $\text{pr}[J]_*\nu_r$, and $\text{pr}[J]_*\nu_r$ on $\mathbb{R}^k$ is absolutely continuous with respect to Lebesgue measure ν_k. Hence $\pi_*\mu$ is absolutely continuous with respect to Lebesgue measure. The continuity of its CDF then results from 2.11.18.

To prove assertion 2), we first reduce to the case $k = r$ by using $\pi = \text{pr}[J] \circ \varphi$. We then use the trivial inequality that for any two Lebesgue measures ν_1 and ν_2 of finite total mass on $\mathbb{R}^r$, we have

$$\text{discrep}(\text{pr}[J]_*\nu_1, \text{pr}[J]_*\nu_2) \leq \text{discrep}(\nu_1, \nu_2),$$

cf. 3.1.7.

Applying this inequality to ν_1 and ν_2 the measures $\varphi_*\mu(A,N)$ and $\varphi_*\mu$, we find that

$$\begin{aligned}\int_{G(N)} &\operatorname{discrep}(\pi_*\mu(A,N), \pi_*\mu)\, dA \\ &= \int_{G(N)} \operatorname{discrep}(\operatorname{pr}[J]_*\varphi_*\mu(A,N), \operatorname{pr}[J]_*\varphi_*\mu)\, dA \\ &\leq \int_{G(N)} \operatorname{discrep}(\varphi_*\mu(A,N), \varphi_*\mu)\, dA.\end{aligned}$$

We now wish to apply Theorem 3.1.6, which says that

$$\int_{G(N)} \operatorname{discrep}(\varphi_*\mu(\text{naive}, A, N), \varphi_*\mu)\, dA \leq N^{\varepsilon - 1/(2r+4)},$$

provided that $N > N_1(\varepsilon, r, c, \eta, \kappa)$, where κ is the L_1 norm of φ, and η is the L_1 norm of φ^{-1}. The hypothesis 3.1.4 of 3.1.6 holds by part 1), according to which $\varphi_*\mu$ has a continuous CDF, or, what is the same (by 2.11.17), each $\operatorname{pr}[i]_*\varphi_*\mu$ has a continuous CDF.

We also apply the same result with φ replaced by $\varphi \circ \operatorname{rev}$, where $\operatorname{rev} : \mathbb{R}^k \to \mathbb{R}^k$ is the "reverse the coordinates" automorphism of $\mathbb{R}^k$. This gives us the inequality

$$\begin{aligned}\int_{G(N)} &\operatorname{discrep}(\varphi_* \operatorname{rev}_* \mu(\text{naive}, A, N), \varphi_* \operatorname{rev}_* \mu)\, dA \\ &\leq N^{\varepsilon - 1/(2r+4)},\end{aligned}$$

provided that $N > N_1(\varepsilon, r, c, \eta, \kappa)$. As already noted in 2.2.8 and 2.2.6, we have

$$\begin{aligned}\operatorname{rev}_* \mu &= \mu, \\ \mu(A,N) &= (1/2)(\operatorname{rev}_* \mu(\text{naive}, A, N) + \mu(\text{naive}, A, N)) \\ &\quad + \text{ a positive Borel measurable of total mass } \leq (1 + c(r))/(N-1).\end{aligned}$$

Apply φ_* to these equalities. We get

$$\begin{aligned}\varphi_* \operatorname{rev}_* \mu &= \varphi_*\mu, \\ \varphi_*\mu(A,N) &= (1/2)(\varphi_* \operatorname{rev}_* \mu(\text{naive}, A, N) + \varphi_*\mu(\text{naive}, A, N)) \\ &\quad + \text{ a positive Borel measure of total mass } \leq (1 + c(r))/(N-1).\end{aligned}$$

By the triangle inequality for discrepancy (L_∞ norm of differences of CDF's), we get

$$\begin{aligned}\operatorname{discrep}&(\varphi_*\mu(A,N), \varphi_*\mu) \\ &\leq (1 + c(r))/(N-1) \\ &\quad + (1/2) \operatorname{discrep}(\varphi_* \operatorname{rev}_*(\text{naive}, A, N), \varphi_* \operatorname{rev}_* \mu) \\ &\quad + (1/2) \operatorname{discrep}(\varphi_*\mu(\text{naive}, A, N), \varphi_*\mu).\end{aligned}$$

Integrating this inequality over $G(N)$, we get

$$\begin{aligned}\int_{G(N)} &\operatorname{discrep}(\varphi_*\mu(A,N), \varphi_*\mu)\, dA \\ &\leq (1 + c(r))/(N-1) + N^{\varepsilon - 1/(2r+4)}\end{aligned}$$

provided $N > N_1(\varepsilon, r, c, \eta, \kappa)$, Thus if we take N so large that all three of the following conditions hold,

$$N > N_1(\varepsilon/2, r, c, \eta, \kappa),$$
$$N^{\varepsilon/2-1/(2r+4)} \leq (1/2)N^{\varepsilon-1/(2r+4)},$$
$$(1+c(r))/(N-1) \leq N^{\varepsilon/2-1/(2r+4)},$$

we get

$$\int_{G(N)} \operatorname{discrep}(\varphi_*\mu(A,N), \varphi_*\mu)\, dA \leq N^{\varepsilon-1/(2r+4)}.$$

So we may take

$$N(\varepsilon, r, c, \pi) = \operatorname{Max}(N_1(\varepsilon/2, r, c, \eta, \kappa), 2^{2/\varepsilon}, 1 + (1+c(r))^{6/5}). \quad \text{QED}$$

7.2. Generating series in the case $r = 1$: relation to a Fredholm determinant

7.2.1. We specialize now to the case $r = 1$, and discuss in detail the measures on $\mathbb{R}$

(7.2.1.1) $$\mu_a := \mu(\text{univ, sep. } a), \qquad a = 0, 1, \ldots,$$

the classical spacing measures. In this case, the general formulas of 7.0.1 and 7.0.2 specialize to

(7.2.1.2) $$\int_{\mathbb{R}} f\, d\mu_a = \sum_{n\geq 0} (-1)^n \operatorname{Binom}(n,a) \operatorname{COR}(n, f, \text{ univ}),$$

and

(7.2.1.3)
$$\operatorname{COR}(n, f, \text{ univ}) = \int_{0\leq z(1)\leq\cdots\leq z(n+1)} f(z(n+1)) W(n+2)(0,z) \prod_i dz(i),$$

for any bounded measurable f of compact support on $\mathbb{R}$. The series $G_1(T,s)$ is given by

$$G_1(T,s) := \sum_{n\geq 0} T^n \operatorname{Cor}(n, I_{R(s)}, \text{ univ})$$

(7.2.1.4) $$= \sum_{n\geq 0} T^n \int_{0\leq z(1)\leq\cdots\leq z(n+1)\leq s} W(n+2)(0,z) \prod_i dz(i)$$

$$= \sum_{n\geq 0} (T^n/(n+1)!) \int_{[0,s]^{n+1}} W(n+2)(0,z) \prod_i dz(i).$$

Thus for each integer $a \geq 0$ and each real $s \geq 0$ we have the identity

(7.2.1.5) $$\operatorname{CDF}_{\mu_a}(s) := \int_{[0,s]} d\mu(\text{univ, sep. } a) = ((d/dT)^a/a!) G_1(T,s)|_{T=-1}.$$

Equivalently, we have the series expansion

(7.2.1.6) $$G_1(T,s) = \sum_{n\geq 0} (1+T)^n \operatorname{CDF}_{\mu_n}(s),$$

for (T,s) in $\mathbb{C} \times \mathbb{R}_{\geq 0}$.

Lemma 7.2.2. *For each integer $n \geq 1$, consider the function $e_n(s)$ on $\mathbb{R}_{\geq 0}$ defined by*

$$e_n(s) := \int_{[0,s]^n} W(n)(x) \prod_i dx(i).$$

1) *The function e_n is the restriction to $\mathbb{R}_{\geq 0}$ of an entire function of s, which is divisible by s^n and which satisfies the estimate*

$$|e_n(s)| \leq |s|^n \exp(n|s|\pi) \operatorname{Sqrt}(n)^n \quad \textit{for all } s \textit{ in } \mathbb{C}.$$

2) *For $n = 1$, we have $e_1(s) = s$.*
3) *For $n \geq 0$, we have*

$$\begin{aligned}(d/ds)(e_{n+2}(s)/(n+2)!) &= \mathrm{COR}(n, I_{R(s)},\ \textit{univ}) \\ &= (1/(n+1)!) \int_{[0,s]^{n+1}} W(n+2)(0,z) \prod_i dz(i).\end{aligned}$$

PROOF. To prove 1), notice that $e_n(0) = 0$. For $s > 0$, we make the change of variable $sy(i) = x(i)$, so that

$$e_n(s) := \int_{[0,s]^n} W(n)(x) \prod_i dx(i) = s^n \int_{[0,1]^n} W(n)(sy) \prod_i dy(i).$$

The rest of the proof of 1) is entirely similar to the proof of 7.0.9, and is left to the reader. Assertion 2) is obvious, since $W(1)(x) = \det_{1\times 1}(f(x-x)) = f(0)$, with $f(x) = \sin(\pi x)/\pi x$, and $f(0) = 1$. To prove 3), which is an identity between entire functions, it suffices to check for s in $\mathbb{R}_{\geq 0}$. We view $e_n(s)$ as

$$\int_{[0,s]^n} W(n)(x) \prod_i dx(i),$$

and expand to second order: we introduce ε with $\varepsilon^2 = 0$, and compute

$$\begin{aligned}&e_n(s+\varepsilon) - e_n(s) \\ &\quad = \int_{[0,s+\varepsilon]^n} W(n)(x) \prod_i dx(i) - \int_{[0,s]^n} W(n)(x) \prod_i dx(i) \\ &\quad = \sum_i \int_{[0,s]^{i-1}\times[s,s+\varepsilon]\times[0,s]^{n-i}} W(n)(x) \prod_i dx(i) \\ &\quad = \varepsilon \sum_i \int_{[0,s]^{n-1}} W(n)(x(1),\ldots,x(i-1),s,x(i+1),\ldots,x(n)) \prod_i dx(i) \\ &\quad = n\varepsilon \int_{[0,s]^{n-1}} W(n)(s,x(2),\ldots,x(n)) \prod_i dx(i),\end{aligned}$$

the last step using the fact that $W(n)$ is Σ_n-invariant. To conclude, we use the fact that $W(n)(x)$ is invariant under both

$$x \mapsto -x$$

and

$$x \mapsto x + \Delta_n(t) := x + (t,t,\ldots,t)$$

to make the change of variable $x \mapsto s - x$, which carries

$$W(n)(s,x(2),\ldots,x(n)) \mapsto W(n)(0,s-x(2),\ldots,s-x(n)),$$

and preserves the domain of integration. Thus we get

$$\begin{aligned}(d/ds)(e_n(s)) &= n\int_{[0,s]^{n-1}} W(n)(0, s-x(2), \ldots, s-x(n)) \prod_i dx(i) \\ &= n\int_{[0,s]^{n-1}} W(n)(0, x(2), \ldots, x(n)) \prod_i dx(i) \\ &= n!\ \mathrm{COR}(n-2, f,\ \mathrm{univ}). \quad \text{QED}\end{aligned}$$

Proposition 7.2.3. *The series*

$$E(T,s) := 1 + \sum_{n\geq 1} T^n e_n(s)/n!$$

is the restriction to $\mathbb{C}\times\mathbb{R}_{\geq 0}$ *of an entire function of* (T,s)*, and*

$$(d/ds)E(T,s) = T + T^2 G_1(T,s).$$

PROOF. The proof that $E(T,s)$ is entire is entirely similar to the proof of 7.0.10 that $G_r(T,s)$ is entire, and is left to the reader. The asserted formula for $(d/ds)E(T,s)$ results from parts 2) and 3) of 7.2.2. QED

7.2.4. By definition, we have

$$\begin{aligned}E(T,s) &:= 1 + \sum_{n\geq 1} T^n(1/n!)\int_{[0,s]^n} W(n)(x)\prod_i dx(i) \\ &= 1 + \sum_{n\geq 1} T^n(1/n!)\int_{[0,s]^n} \det_{n\times n}(K(x(i),x(j)))\prod_i dx(i)\end{aligned} \tag{7.2.4.1}$$

for $K(x,y)$ the kernel

$$K(x,y) := \sin(\pi(x-y))/\pi(x-y). \tag{7.2.4.2}$$

Because $K(x,y) = K(x+\alpha, y+\alpha)$ for any α, for each n we have

$$\begin{aligned}&\int_{[0,s]^n} \det_{n\times n}(K(x(i),x(j)))\prod_i dx(i) \\ &= \int_{[\alpha,\alpha+s]^n} \det_{n\times n}(K(x(i),x(j)))\prod_i dx(i).\end{aligned} \tag{7.2.4.3}$$

Thus we find [**W-W**, page 215, 11.21] the apparently miraculous

Identity 7.2.5. *For any real* α*, and any real* $s \geq 0$*, consider the integral operator* $K_{s,\alpha}$ *with kernel* $K(x,y) := \sin(\pi(x-y))/\pi(x-y)$ *acting on the space* $L_2([\alpha,\alpha+s],dx)$:

$$(K_{s,\alpha}f)(x) := \int_{[\alpha,\alpha+s]} K(x,y)f(y)\,dy.$$

This operator has a well defined Fredholm determinant, and we have the identity

$$E(T,s) = \det(1 + TK_{s,\alpha}).$$

[For $s=0$*,* $L_2([\alpha,\alpha+s],dx)$ *is the zero space, and this identity says* $1=1$*.]*

Remark 7.2.6. For fixed s and variable α, all the operators $K_{s,\alpha}$ are isometrically isomorphic, via the translation isomorphisms from $L_2([\alpha_1,\alpha_1+s],dx)$ to $L_2([\alpha_2,\alpha_2+s],dx)$. We put the "$\alpha$" in the notation because we will have occasion to consider also kernels which are not translation-invariant.

7.3. The Fredholm determinants $E(T,s)$ and $E_\pm(T,s)$

7.3.1. In the preceding section, we considered the kernel

$$K(x,y) := \sin(\pi(x-y))/\pi(x-y).$$

In this section, we will consider also the kernels

$$K_+(x,y) := K(x,y) + K(-x,y),$$
$$K_-(x,y) := K(x,y) - K(-x,y).$$

In terms of these kernels, we define integral operators K_{2s}, $K_{+,s}$ and $K_{-,s}$ as follows, for each real $s \geq 0$:

$$K_{2s} := \text{the integral operator with kernel } K(x,y) \text{ on } L_2([-s,s],\, dx),$$
$$K_{\pm,s} := \text{the integral operator with kernel } K_\pm(x,y) \text{ on } L_2([0,s],dx).$$

7.3.2. It is proven in 7.10.1 that the above integral operators K_{2s} and $K_{\pm,s}$ are positive operators of trace class, whose Fredholm determinants agree with their spectral determinants.

7.3.3. We have an orthogonal decomposition

$$L_2([-s,s],dx) = L_2([-s,s],dx)_{\text{even}} \oplus L_2([-s,s],dx)_{\text{odd}},$$
$$f = f_+ + f_-,$$

with

$$f_\pm(x) := (1/2)(f(x) \pm f(-x)).$$

Both of the subspaces are stable by the integral operator K_{2s}, because, $\sin(x)/x$ being an even function, we have

$$K(x,-y) = K(-x,y).$$

7.3.4. Via the isometric isomorphisms

$$(1/\operatorname{Sqrt}(2))\,\text{Restriction} : L_2([-s,s],dx)_{\text{even}} \cong L_2([0,s],dx),$$
$$(1/\operatorname{Sqrt}(2))\,\text{Restriction} : L_2([-s,s],dx)_{\text{odd}} \cong L_2([0,s],dx),$$

we get the isometric isomorphisms

$$K_{2s}|L_2([-s,s]dx)_{\text{even}} \cong K_{+,s}|L_2([0,s],dx),$$
$$K_{2s}|L_2([-s,s]dx)_{\text{odd}} \cong K_{-,s}|L_2([0,s],dx).$$

7.3.5. We define the Fredholm determinant

$$E_\pm(T,s) := 1 + \sum_{n\geq 1} T^n(1/n!) \int_{[0,s]^n} \det_{n\times n}(K_\pm(x(i),x(j))) \prod_i dx(i).$$

By 7.10.1, $E_\pm(T,s)$ is also the spectral determinant

$$\begin{aligned} E_\pm(T,s) &= \det(1 + TK_{\pm,s}|L_2([0,s],dx)) \\ &= \det(1 + TK_{2s}|L_2([-s,s],dx)_{\text{parity } \pm}). \end{aligned} \tag{7.3.5.1}$$

Exactly as in proving 7.0.10, one proves that $E_\pm(T,s)$ is (the restriction to $\mathbb{C}\times\mathbb{R}_{\geq 0}$ of) an entire function of (T,s). Explicitly, writing

$$e_{\pm,n}(s) := \int_{[0,s]^n} \det_{n\times n}(K_\pm(x(i),x(j))) \prod_i dx(i), \tag{7.3.5.2}$$

we have the estimate (compare 7.2.2; here each matrix entry is at worst twice as large)

$$|e_{\pm,n}(s)| \leq |s|^n \exp(2n|s|\pi)\,\mathrm{Sqrt}(n)^n \quad \text{for all } s \text{ in } \mathbb{C}. \tag{7.3.5.3}$$

In view of the decomposition

$$L_2([-s,s],dx) = L_2([-s,s],dx)_{\text{even}} \oplus L_2([-s,s],dx)_{\text{odd}} \tag{7.3.5.4}$$

and the spectral interpretation of the Fredholm determinants in question, we have (by 7.10.1, part 4) the identity

$$E(T,2s) = E_+(T,s)E_-(T,s). \tag{7.3.5.5}$$

Notice that at $s = 0$, all three spaces in the decomposition above are the zero space, so we have

$$1 = E(T,0) = E_+(T,0) = E_-(T,0), \tag{7.3.5.6}$$

and the above identity reduces to $1 = 1 \times 1$.

7.3.6. It will be convenient to name the coefficients of the series expansions of these functions around the point $T = -1$. Thus we define entire functions of s, $E_n(s)$ and $E_{\pm,n}(s)$, for each integer $n \geq 0$ by

$$E(T,s) = \sum_{n\geq 0}(1+T)^n E_n(s),$$
$$E_\pm(T,s) = \sum_{n\geq 0}(1+T)^n E_{\pm,n}(s),$$

or equivalently,

$$E_n(s) := ((d/dT)^n/n!)E(T,s)|_{T=-1},$$
$$E_{\pm,n}(s) := ((d/dT)^n/n!)E_\pm(T,s)|_{T=-1}.$$

7.3.7. We will see shortly, in 7.5.3, that the functions $E_n(s)$ and $E_{\pm,n}(s)$ have exponential decay for large real s. Much stronger results are available, cf. [**Widom**], [**T-W**] and [**B-T-W**], who give detailed asymptotics of these functions for large real s.

7.4. Interpretation of $E(T,s)$ and $E_\pm(T,s)$ as large N scaling limits of $E(N,T,s)$ and $E_\pm(N,T,s)$

7.4.1. Recall from 5.4.2 that for each integer $N \geq 1$, $S_N(x)$ is the function $\sin(Nx/2)/\sin(x/2) = \sum_{j=0}^{N-1} e^{i(N-1-2j)x/2}$. We defined kernels $S_N(x,y)$ and $S_{\pm,N}(x,y)$ as follows:

$$S_N(x,y) := S_N(x-y),$$
$$S_{\pm,N}(x,y) := S_N(x,y) \pm S_N(-x,y).$$

7.4.2. For real α and real $s \geq 0$, we defined integral operators

$$\begin{aligned} K_{N,s,\alpha} &:= \text{the integral operator with kernel } S_N(x,y) \text{ on } L_2([\alpha,\alpha+s], dx/2\pi), \\ K_{\pm,N,s} &:= \text{the integral operator with kernel } S_{\pm,N}(x,y) \text{ on } L_2([0,s], dx/2\pi) \\ &= \text{the integral operator with kernel} \\ &\qquad (1/2)S_{\pm,N}(x,y) = (1/2)(S_N(x,y) \pm S_N(-x,y)) \\ &\qquad \text{on } L_2([0,s], dx/\pi). \end{aligned}$$

[For $s = 0$, the spaces $L_2([\alpha,\alpha+s], dx/2\pi)$ and $L_2([0,s], dx/\pi)$ are the zero spaces.] For fixed $s \geq 0$ and variable α, all the operators $K_{N,s,\alpha}$ are isometrically equivalent.

7.4.3. We defined (variants of) characteristic polynomials

$$E(N,T,s) := \det(1 + TK_{N,s,\alpha}) \quad \text{(independent of } \alpha\text{)},$$
$$E_{\pm}(N,T,s) := \det(1 + TK_{\pm,N,s}).$$

7.4.4. Explicitly, we have the formulas

$$E(N,T,s) = \sum_{k \geq 0} (T^k/k!) \int_{[0,s]^k} \det_{k\times k}(S_N(x(i),x(j))) \prod_i (dx(i)/2\pi),$$

$$E_{\pm}(N,T,s) = \sum_{k \geq 0} (T^k/k!) \int_{[0,s]^k} \det_{k\times k}(S_{\pm,N}(x(i),x(j))) \prod_i (dx(i)/2\pi),$$

with the convention that a 0×0 determinant is 1. These are polynomials in T of degree at most N. [For $s = 0$, these polynomials are identically 1.] Their coefficients are controlled by the following lemma.

Lemma 7.4.5. *Given integers $N \geq 1$ and $k \geq 1$, the functions $A_{k,N}(s)$ and $A_{\pm,k,N}(s)$ defined for real $s \geq 0$ by the integrals*

$$\int_{[0,s]^k} \det_{k\times k}(S_N(x(i),x(j))) \prod_i (dx(i)/2\pi)$$

and

$$\int_{[0,s]^k} \det_{k\times k}(S_{\pm,N}(x(i),x(j))) \prod_i (dx(i)/2\pi)$$

respectively are the restriction to $\mathbb{R}_{>0}$ of entire functions of s, which are both bounded in absolute value by

$$|s/2\pi|^k \operatorname{Sqrt}(k)^k (2N)^k \exp(kN|s|/2)$$

for all s in $\mathbb{C}$.

PROOF. At $s = 0$, all these integrals vanish. If $s > 0$, then by the change of variable $x(i) = sy(i)$, our integrals become

$$(s/2\pi)^k \int_{[0,1]^k} \det_{k\times k}(S_N(sy(i),sy(j))) \prod_i dy(i)$$

and

$$(s/2\pi)^k \int_{[0,1]^k} \det_{k\times k}(S_{\pm,N}(sy(i),sy(j))) \prod_i dy(i)$$

respectively, which makes clear that the functions are entire in s. The function $S_N(x) = \sum_{j=0}^{N-1} e^{i(N-1-2j)x/2}$ obviously satisfies $|S_N(x)| \leq N\exp(N|x|/2)$ for all

x in $\mathbb{C}$. So for complex s, but x and y in $[0,1]^k$, the entries $S_N(sy(i), sy(j))$ and $S_{\pm,N}(sy(i), sy(j))$ of the determinants are bounded by

$$|S_N(sy(i), sy(j))| \leq N \exp(N|s|/2),$$
$$|S_{\pm,N}(sy(i), sy(j))| \leq 2N \exp(N|s|/2).$$

The Hadamard determinant inequality now gives the asserted bounds, cf. the proof of 7.0.9. QED

Proposition 7.4.6. *For any fixed real number α, the three sequences of entire functions of (T, s)*

$$E(N, T, 2\pi s/(N+\alpha)), \quad E_+(N, T, 2\pi s/(N+\alpha)), \quad E_-(N, T, 2\pi s/(N+\alpha))$$

indexed by integers $N > \mathrm{Max}(2, -\alpha)$ converge to the entire functions $E(T, s)$, $E_+(T, s)$, and $E_-(T, s)$ respectively, uniformly on compact subsets of $\mathbb{C}^2$.

PROOF. For N large, the ratio $N/(N+\alpha) \leq 2$. For any such N, the coefficient of $T^k/k!$ in any of the N'th terms is bounded on complex s by

$$(4|s|\,\mathrm{Sqrt}(k)e^{2\pi|s|})^k,$$

as is immediate from 7.4.5. The same bound holds for the coefficient of $T^k/k!$ in $E(T, s)$ or $E_\pm(T, s)$ by 7.2.2 and 7.3.5.3.

Using these estimates, we see that given $M > 0$ and $\varepsilon > 0$, there exists an integer L such that in the region $(|T| \leq M, |s| \leq M)$, each of the functions in question, namely either $E(T, s)$ or $E_\pm(T, s)$ or one of its finite N alleged approximants with N large enough that $N/(N+\alpha) < 2$, is approximated within ε by the sum of its first L terms as a series in T with coefficients functions of s. So we need only prove that the coefficients of individual powers of T converge uniformly on compact subsets of the complex s plane. Let us do this explicitly for, say, $E_\pm(T, s)$, the $E(T, s)$ case being entirely similar. Fix an integer $k \geq 1$. The coefficient $e_{\pm,k}(s)$ of $T^k/k!$ in $E_\pm(T, s)$ is

$$\int_{[0,s]^k} \det_{k\times k}(K_\pm(x(i), x(j))) \prod_i dx(i)$$
$$= \int_{[0,1]^k} \det_{k\times k}(sK_\pm(sy(i), sy(j))) \prod_i dy(i).$$

The coefficient of $T^k/k!$ in $E_\pm(N, T, 2\pi s/(N+\alpha))$ is

$$\int_{[0,2\pi s/(N+\alpha)]^k} \det_{k\times k}(S_{\pm,N}(x(i), x(j))) \prod_i (dx(i)/2\pi),$$

which by the change of variable $x(i) = 2\pi sy(i)/(N+\alpha)$ becomes

$$= \int_{[0,1]^k} \det_{k\times k}((s/(N+\alpha))S_{\pm,N}(2\pi sy(i)/(N+\alpha), 2\pi sy(j)/(N+\alpha))) \prod_i dy(i).$$

The determinant is integrated over a compact region, so it suffices that as $N \to \infty$, the determinants converge uniformly, for y in $[0,1]^k$ and s in a compact set of $\mathbb{C}$, to

$$\det_{k\times k}(sK_\pm(sy(i), sy(j))).$$

So it suffices for each of the individual entries

$$(s/(N+\alpha))S_{\pm,N}(2\pi sy(i)/(N+\alpha), 2\pi sy(j)/(N+\alpha))$$

to so converge to $sK_{\pm}(sy(i), sy(j))$. This in turn reduces to the standard fact that for s in a compact subset of $\mathbb{C}$ we have uniform convergence of

$$(1/(N+\alpha))S_N(2\pi s/(N+\alpha)) = \sin(\pi s N/(N+\alpha))/(N+\alpha)\sin(\pi s/(N+\alpha))$$

to $\sin(\pi s)/\pi s$. QED

7.5. Large N limits of the measures $\nu(n, G(N))$: the measures $\nu(n)$ and $\nu(\pm, n)$

Proposition 7.5.1. *For $s \geq 0$ real, $n \geq 0$ an integer, and $G(N)$ any of $U(N)$, $SO(2N+1), USp(2N), SO(2N), O_-(2N+2), O_-(2N+1)$, the limit*

$$\lim_{N\to\infty} \text{eigen}(n, \sigma\pi s/(N+\lambda), G(N))$$

exists. In terms of the expansion coefficients $E_{\text{sign}(\varepsilon),n}(s)$ defined by

$$E_{\text{sign}(\varepsilon)}(T, s) = \sum_{n\geq 0}(1+T)^n E_{\text{sign}(\varepsilon),n}(s),$$

i.e.,

$$E_{\text{sign}(\varepsilon),n}(s) := ((d/dT)^n/n!)E_{\text{sign}(\varepsilon)}(T, s)|_{T=-1},$$

we have the limit formula

$$E_{\text{sign}(\varepsilon),n}(s) = \lim_{N\to\infty} \text{eigen}(n, \sigma\pi s/(N+\lambda), G(N)).$$

Explicitly, we have the limit formulas

1) $$E_n(s) = \lim_{N\to\infty} \text{eigen}(n, 2\pi s/N, U(N)).$$

2) $$E_{-,n}(s) = \lim_{N\to\infty} \text{eigen}(n, 2\pi s/(N+1/2), SO(2N+1)).$$

3) $$E_{-,n}(s) = \lim_{N\to\infty} \text{eigen}(n, \pi s/N, USp(2N)) = \lim_{N\to\infty} \text{eigen}(n, \pi s/(N+1), O_-(2N+2)).$$

4) $$E_{+,n}(s) = \lim_{N\to\infty} \text{eigen}(n, \pi s/N, SO(2N)) = \lim_{N\to\infty} \text{eigen}(n, \pi s/(N+1/2), O_-(2N+1)).$$

Moreover, the convergence is uniform on compact subsets of $\mathbb{R}_{\geq 0}$.

PROOF. This is a simple application of Proposition 7.4.6, together with the fact that if one has convergence, uniformly on compact sets, of a sequence of entire functions $f_n \to f$ of several complex variables, then for any analytic differential operator D with entire coefficients, the sequence $D(f_n)$ converges, uniformly on compact sets, to $D(f)$.

According to 6.7.6, for $0 \leq s \leq \sigma\pi$, we have

$$\text{eigen}(n, s, G(N)) = ((d/dT)^n/n!)E_{\text{sign}(\varepsilon)}(\rho N+\tau, T, s)|_{T=-1}.$$

Rescaling, we find that for $N > s \geq 0$ we have

$$\begin{aligned}\text{eigen}(n, \sigma\pi s/(N+\lambda), G(N)) &= ((d/dT)^n/n!)E_{\text{sign}(\varepsilon)}(\rho N+\tau, T, \sigma\pi s/(N+\lambda))|_{T=-1}\\ &= ((d/dT)^n/n!)E_{\text{sign}(\varepsilon)}(\rho N+\tau, T, 2\pi s/\rho(N+\lambda))|_{T=-1}.\end{aligned}$$

Thanks to 7.4.6, we have convergence

$$E_{\text{sign}(\varepsilon)}(\rho N+\tau, T, 2\pi s/\rho(N+\lambda)) \to E_{\text{sign}(\varepsilon)}(T, s),$$

uniformly on compact sets of $\mathbb{C}^2$. Applying $((d/dT)^n/n!)$ and evaluating at $T=-1$ gives the asserted limit formulas, with uniform convergence on compact subsets of $\mathbb{R}_{\geq 0}$. QED

Remark 7.5.2. For any $G(N)$, and any real s with $0 \leq s \leq \sigma\pi$, we have the relation

$$1 = \sum_{n\geq 0} \operatorname{eigen}(n, s, G(N)),$$

simply because the sets $\operatorname{Eigen}(n, s, G(N))$ are a partition of $G(N)$. In view of 6.7.6, this amounts to the statement that

$$1 = E_{\operatorname{sign}(\varepsilon)}(\rho N + \tau, T, s)|_{T=0},$$

which is obvious from the definition of $E_{\operatorname{sign}(\varepsilon)}(\rho N+\tau, T, s)$ as a modified characteristic polynomial $\det(1 + T(\text{something}))$. Similarly, the series $E(T,s)$ and $E_{\pm}(T,s)$, being modified Fredholm determinants, satisfy

$$1 = E(0,s) = E_{\pm}(0,s),$$

or equivalently,

$$1 = \sum_{n\geq 0} E_n(s) = \sum_{n\geq 0} E_{\pm,n}(s).$$

So we get the not entirely obvious relation

$$1 = \sum_{n\geq 0} \lim_{N\to\infty} \operatorname{eigen}(n, \sigma\pi s/(N+\lambda), G(N)), \tag{7.5.2.1}$$

for every real $s \geq 0$.

Corollary 7.5.3. *For any real $s > 0$, and any integer $n \geq 0$, we have the estimates*

$$0 \leq E_n(s) \leq 2^n \exp(-s/2),$$
$$0 \leq E_{\pm,n}(s) \leq 2^n \exp(-s/4).$$

PROOF. Combine the estimates of 6.7.8 for $\operatorname{eigen}(n, s, G(N))$ with the above limit formulas. QED

Corollary 7.5.4. *For $s \geq 0$ real, $n \geq 1$ an integer, and $G(N)$ any of $U(N)$, $SO(2N+1), USp(2N), SO(2N), O_-(2N+2), O_-(2N+1)$, the limits*

$$\lim_{N\to\infty} \operatorname{Tail}_{\nu(n,G(N))}(s), \qquad \lim_{N\to\infty} \operatorname{CDF}_{\nu(n,G(N))}(s)$$

exist, and we have the following limit formulas:

1) $$\lim_{N\to\infty} \operatorname{Tail}_{\nu(n,G(N))}(s) = \sum_{j=0}^{n-1} E_{\operatorname{sign}(\varepsilon),j}(s),$$

2) $$\lim_{N\to\infty} \operatorname{CDF}_{\nu(n,G(N))}(s) = \sum_{j\geq n} E_{\operatorname{sign}(\varepsilon),j}(s),$$

3) $$\lim_{N\to\infty} \operatorname{CDF}_{\nu(n,G(N))}(s) = 1 - \sum_{j=0}^{n-1} E_{\operatorname{sign}(\varepsilon),j}(s).$$

Moreover, the convergence is uniform on $\mathbb{R}_{\geq 0}$.

PROOF. Fix $n \geq 1$. The limit formula for $\mathrm{Tail}_{\nu(n,G(N))}(s)$ results from its finite analogue 6.9.7, part 4),

$$\mathrm{Tail}_{\nu(n,G(N))}(s) = \sum_{j=0}^{n-1} \mathrm{eigen}(j, s\sigma\pi/(N+\lambda), G(N)),$$

valid for $N \geq n$, using the limit formulas of 7.5.1,

$$E_{\mathrm{sign}(\varepsilon),j}(s) = \lim_{N\to\infty} \mathrm{eigen}(j, \sigma\pi s/(N+\lambda), G(N)),$$

for $j = 0, 1, \ldots, n-1$. The uniformity on compact subsets of $\mathbb{R}_{\geq 0}$ of the convergence for each of these n limits gives the same uniformity for the Tail limit formula. But since we have the tail estimate 6.9.8,

$$0 \leq \mathrm{Tail}_{\nu(n,G(N))}(s) \leq 2^n e^{-\sigma s/4},$$

the convergence is necessarily uniform on $\mathbb{R}_{\geq 0}$: given $\varepsilon > 0$, choose S large enough that $2^n e^{-S/4} < \varepsilon$, and then take N_0 large enough that, for all $N \geq N_0$, $\mathrm{Tail}_{\nu(n,G(N))}(s)$ is within ε of the limit for s in $[0, S]$.

The limit formulas 2) and 3) with uniformity are equivalent to 1), because

$$\mathrm{CDF}_{\nu(n,G(N))}(s) = 1 - \mathrm{Tail}_{\nu(n,G(N))}(s),$$

and because (by 7.5.2) we have

$$1 = \sum_{n\geq 0} E_{\mathrm{sign}(\varepsilon),n}(s). \quad \text{QED}$$

Proposition 7.5.5. *For every integer $n \geq 1$, there exist positive Borel probability measures on $\mathbb{R}$,*

$$\nu(n) \text{ and } \nu(\pm, n),$$

supported in $\mathbb{R}_{\geq 0}$, and having continuous CDF's whose restrictions to $\mathbb{R}_{\geq 0}$ are entire, such that

1) $\lim_{N\to\infty} \nu(n, U(N)) = \nu(n)$,
2) $\lim_{N\to\infty} \nu(n, SO(2N+1)) = \nu(-, n)$,
3) $\lim_{N\to\infty} \nu(n, USp(2N)) = \nu(-, n)$,
4) $\lim_{N\to\infty} \nu(n, SO(2N)) = \nu(+, n)$,
5) $\lim_{N\to\infty} \nu(n, O_-(2N+2)) = \nu(-, n)$,
6) $\lim_{N\to\infty} \nu(n, O_-(2N+1)) = \nu(+, n)$,

in the sense of convergence of cumulative distribution functions which is uniform on $\mathbb{R}_{\geq 0}$. The cumulative distribution funcitions of these measures are given by the explicit formulas

$$\mathrm{CDF}_{\nu(n)}(s) = 1 - \sum_{j=0}^{n-1} E_j(s),$$

$$\mathrm{CDF}_{\nu(\pm,n)}(s) = 1 - \sum_{j=0}^{n-1} E_{\pm,j}(s).$$

The tails of these measures satisfy the estimates

$$0 \leq \mathrm{Tail}_{\nu(n)}(s) \leq 2^n e^{-s/2},$$

$$0 \leq \mathrm{Tail}_{\nu(\pm,n)}(s) \leq 2^n e^{-s/4}.$$

PROOF. In the previous result, we proved that

$$\lim_{N\to\infty} \mathrm{CDF}_{\nu(n,G(N))}(s) = 1 - \sum_{j=0}^{n-1} E_{\mathrm{sign}(\varepsilon),j}(s)$$

with convergence uniform on $\mathbb{R}_{\geq 0}$. Consider the limit function

$$s \mapsto \lim_{N\to\infty} \mathrm{CDF}_{\nu(n,G(N))}(s) = 1 - \sum_{j=0}^{n-1} E_{\mathrm{sign}(\varepsilon),j}(s),$$

say $f(s)$. This f is a nondecreasing function on $\mathbb{R}_{\geq 0}$ with values in $[0,1]$, because it is the pointwise limit of such functions. We claim that $f(0) = 0$. This follows from the fact 7.3.5.6 that $E_{\mathrm{sign}(\varepsilon)}(T,0) = 1$, so its expansion coefficients around $T = -1$ are given by $E_{\mathrm{sign}(\varepsilon),j}(0) = \delta_{j,0}$, and hence we find

$$f(0) = 1 - \sum_{j=0}^{n-1} E_{\mathrm{sign}(\varepsilon),j}(0) = 1 - \left(1 + \sum_{j=1}^{n-1} 0\right) = 0.$$

Moreover, f is the restriction to $\mathbb{R}_{\geq 0}$ of an entire function, so in particular it is continuous. As it vanishes at $s = 0$, we may extend f to all of $\mathbb{R}$ as a continuous nondecreasing function from $\mathbb{R}$ to $[0,1]$ by decreeing that it vanish for $s < 0$. From the tail estimate 6.9.8, we see that the function $1 - f(s)$ satisfies

$$0 \leq 1 - f(s) \leq 2^n \exp(-\sigma s/4)$$

for $s \geq 0$. Therefore f is the CDF of a positive Borel probability measure [namely the Lebesgue-Stieltjes measure for which every interval $[a,b]$ or $(a,b]$ or (a,b) gets measure $f(b) - f(a)$.] Its tail is the function $1 - f(s)$, which, as noted just above, satisfies the asserted tail estimate. QED

Proposition 7.5.6. *For any integer $n \geq 1$, and $G(N)$ any of $U(N)$, $SO(2N+1), USp(2N), SO(2N), O_-(2N+2), O_-(2N+1)$, we have:*

1) *The measures $\nu(n,G(N)), \nu(n)$ and $\nu(\pm,n)$ have moments of all orders, and each of these measures is uniquely determined by its moments.*

2) *For any continuous function $f(x)$ of polynomial growth, we have the limit formula*

$$\lim_{N\to\infty} \int f(x)\, d\nu(n,G(N)) = \int f(x)\, d\nu(\mathrm{sign}(\varepsilon),n),$$

i.e.,

$$\lim_{N\to\infty} \int f(x)\, d\nu(n,U(N)) = \int f(x)\, d\nu(n),$$

$$\lim_{N\to\infty} \int f(x)\, d\nu(n,SO(2N+1)) = \int f(x)\, d\nu(-,n),$$

$$\lim_{N\to\infty} \int f(x)\, d\nu(n,USp(2N)) = \int f(x)\, d\nu(-,n),$$

$$\lim_{N\to\infty} \int f(x)\, d\nu(n,SO(2N)) = \int f(x)\, d\nu(+,n),$$

$$\lim_{N\to\infty} \int f(x)\, d\nu(n,O_-(2N+2)) = \int f(x)\, d\nu(-,n),$$

$$\lim_{N\to\infty} \int f(x)\, d\nu(n, O_-(2N+1)) = \int f(x)\, d\nu(+, n).$$

PROOF. We have uniform tail estimates for the measures $\nu(n, G(N))$ and $\nu(\mathrm{sign}(\varepsilon), n)$, and uniform convergence of their CDF's. So the proposition results from the case $\lambda = 1$ of the next three standard lemmas, which we give for ease of reference.

Lemma 7.5.7. *Let η be a positive Borel measure on $\mathbb{R}$ under which finite intervals have finite measure. Suppose that there exist strictly positive real constants A, B, and λ such that for every real $s \geq 0$, we have the tail estimate*

$$\eta(\{x \text{ in } \mathbb{R} \text{ with } |x| > s\}) \leq Ae^{-Bs^\lambda}.$$

Then η has moments of all orders, i.e., the functions $|x|^n$ lie in $L_1(\mathbb{R}, \eta)$ for all integers $n \geq 0$. If in addition $\lambda \geq 1$, then η is uniquely determined by its moments $m_n := \int_\mathbb{R} x^n \, d\eta$, $n = 0, 1, \ldots$.

PROOF. If $n = 0$, then

$$\int_\mathbb{R} d\eta = \eta(\mathbb{R}) = \eta(\{0\}) + \eta(\{x \text{ in } \mathbb{R} \text{ with } |x| > 0\}) \leq \eta(\{0\}) + A.$$

If $n \geq 1$, then $|x|^n$ vanishes at 0, and we readily calculate

$$\begin{aligned}
\int_\mathbb{R} |x|^n \, d\eta &= \int_{\{0\}} |x|^n \, d\eta + \sum_{k\geq 0} \int_{k<|x|\leq k+1} |x|^n \, d\eta \\
&= \sum_{k\geq 0} \int_{k<|x|\leq k+1} |x|^n \, d\eta \\
&\leq \sum_{k\geq 0} \int_{k<|x|\leq k+1} (k+1)^n \, d\eta \\
&\leq \sum_{k\geq 0} (k+1)^n \int_{k<|x|} d\eta \\
&\leq \sum_{k\geq 0} (k+1)^n Ae^{-Bk^\lambda} \\
&= A + \sum_{k\geq 1} (k+1)^n Ae^{-Bk^\lambda} \\
&\leq A + A\sum_{k\geq 1} \int_{[k-1,k]} (x+2)^n e^{-Bx^\lambda} \, dx \\
&= A + A\int_{[0,\infty)} (x+2)^n e^{-Bx^\lambda} \, dx \\
&= A + A\int_{[0,1)} (x+2)^n e^{-Bx^\lambda} \, dx + A\int_{[1,\infty)} (x+2)^n e^{-Bx^\lambda} \, dx \\
&\leq A + 3^n A + 3^n A \int_{[1,\infty)} x^n e^{-Bx^\lambda} \, dx \\
&\leq A + 3^n A + 3^n A \int_{[1,\infty)} x^{n+1} e^{-Bx^\lambda} \, dx/x
\end{aligned}$$

(now set $y := Bx^{\lambda}$)

$$
\begin{aligned}
&= A + 3^n A + 3^n A(1/\lambda)\int_{[B,\infty)} (y/B)^{(n+1)/\lambda} e^{-y}\, dy/y \\
&\leq A + 3^n A + 3^n A(1/\lambda) B^{-(n+1)/\lambda} \int_{(0,\infty)} y^{(n+1)/\lambda} e^{-y}\, dy/y \\
&= A + 3^n A + 3^n A(1/\lambda) B^{-(n+1)/\lambda}\Gamma((n+1)/\lambda).
\end{aligned}
$$

This shows that all moments exist. Now it is known [**Fel**, page 487] that a sufficient condition for η to be determined by its moments m_n is that the series

$$\sum_n M_n x^n/n!$$

built from its absolute moments

$$M_n := \int_{\mathbb{R}} |x|^n\, d\eta$$

have a nonzero radius of convergence. If $\lambda \geq 1$, then our estimate, together with Stirling's formula, shows that this condition is satisfied, and that for $\lambda > 1$ this series is entire. QED

7.5.8. For ease of later reference, let us record a several variable version of this result.

Lemma 7.5.9. *Let $r \geq 1$ be an integer, and η a positive Borel measure on $\mathbb{R}^r$ which gives finite measure to compact sets. Suppose that there exist strictly positive real constants A, B, and λ such that for every real $s \geq 0$, and every $i = 1, \ldots, r$, we have the tail estimate*

$$\mathrm{pr}[i]_*\eta(\{x \text{ in } \mathbb{R} \text{ with } |x| > s\}) := \eta(\{x \text{ in } \mathbb{R}^r \text{ with } |x(i)| > s\}) \leq Ae^{-Bs^{\lambda}}.$$

Then η has moments of all orders, i.e., for all $n \geq 0$ in $\mathbb{Z}^r$, the functions $|x|^n := \prod_i |x(i)|^{n(i)}$ lie in $L_1(\mathbb{R}^r, \eta)$. If in addition $\lambda \geq 1$, then η is uniquely determined by its moments $m_n := \int_{\mathbb{R}^r} x^n\, d\eta$, $n \geq 0$ in $\mathbb{Z}^r$.

PROOF. For real $s \geq 0$, let us denote by $E(s) \subset \mathbb{R}^r$ the closed set

$$E(s) := \{x \text{ in } \mathbb{R}^r \text{ with } \operatorname{Max}_i |x(i)| \leq s\}.$$

For each $i = 1, \ldots, r$, we denote by $E_i(s) \subset \mathbb{R}^r$ the closed set

$$E_i(s) := \{x \text{ in } \mathbb{R}^r \text{ with } |x(i)| \leq s\}.$$

Our hypothesis is that for $s \geq 0$ real, for each $i = 1, \ldots, r$ we have

$$\eta(\mathbb{R}^r - E_i(s)) \leq Ae^{-Bs^{\lambda}}.$$

Now $E(s)$ is the intersection of the $E_i(s)$, so $\mathbb{R}^r - E(s)$ is the union of the $\mathbb{R}^r - E_i(s)$, and hence we have

$$\eta(\mathbb{R}^r - E(s)) \leq rAe^{-Bs^{\lambda}}.$$

For $n = 0$, we use the fact that $E(0) = \{0\}$ to write

$$\int_{\mathbb{R}^r} |x|^n\, d\eta = \eta(\mathbb{R}^r) = \eta(E(0)) + \eta(\mathbb{R}^r - E(0)) \leq \eta(\{0\}) + rA.$$

For $n \geq 0$ and $\neq 0$, $|x|^n$ vanishes at 0, and we readily calculate

$$\begin{aligned}\int_{\mathbb{R}^r} |x|^n \, d\eta &= \int_{E(0)} |x|^n \, d\eta + \sum_{k\geq 0} \int_{E(k+1)-E(k)} |x|^n \, d\eta \\ &= \sum_{k\geq 0} \int_{E(k+1)-E(k)} |x|^n \, d\eta \\ &\leq \sum_{k\geq 0} \int_{E(k+1)-E(k)} (k+1)^{\Sigma(n)} \, d\eta \\ &\leq \sum_{k\geq 0} (k+1)^{\Sigma(n)} \int_{\mathbb{R}^r - E(k)} d\eta \\ &\leq \sum_{k\geq 0} (k+1)^{\Sigma(n)} rAe^{-Bk^\lambda} \\ &\leq r[A + 3^{\Sigma(n)}A + 3^{\Sigma(n)}A(1/\lambda)B^{-(\Sigma(n)+1)/\lambda}\Gamma((\Sigma(n)+1)/\lambda)],\end{aligned}$$

the last inequality obtained just as in the proof of the previous lemma. This shows that all moments exist. A sufficient condition [**Fel**, page 463; the same "analyticity of the Fourier transform in a neighborhood of $\mathbb{R}^r$ in $\mathbb{C}^r$" works for any r] for η to be uniquely determined by its moments is that the series in r variables

$$\sum_{n\geq 0 \text{ in } \mathbb{Z}^r} M_n x^n / n!$$

made from the absolute moments $M_n := \int_{\mathbb{R}^r} |x|^n \, d\eta$ converge in a nonempty open polydisc around the origin. If $\lambda \geq 1$, then our estimate for M_n, together with Stirling's formula, shows that this condition is satisfied, and that if $\lambda > 1$ this series is entire. QED

7.5.10. Given (A, B, λ) in $(\mathbb{R}_{>0})^3$, we say that a positive Borel measure η on $\mathbb{R}^r$ which is finite on compact sets has exponential decay of type (A, B, λ) if it satisfies the hypotheses of the previous lemma for this choice of (A, B, λ).

Lemma 7.5.11. *Let $r \geq 1$ be an integer, (A, B, λ) in $(\mathbb{R}_{>0})^3$, and η_k, $k \geq 1$, a sequence of positive Borel measures on $\mathbb{R}^r$, each of which has exponential decay of type (A, B, λ). Suppose that*

$$\lim_{k\to\infty} \eta_k(\mathbb{R}^r) = 1.$$

Suppose further that the sequence of cumulative distribution functions $\mathrm{CDF}_{\eta_k}(x)$ converges, uniformly on compact subsets of $\mathbb{R}^r$, to some function $F(x)$ on $\mathbb{R}^r$. Then $F(x)$ is the CDF of a positive Borel probability measure η_∞ on $\mathbb{R}^r$ which has exponential decay of type (A, B, λ). For any $\mathbb{R}$-valued continuous function f on $\mathbb{R}^r$ which has polynomial growth, i.e. which satisfies

$$|f(x)| = O((1 + \|x\|)^d), \qquad \|x\| := \sum_i |x(i)|,$$

for some integer $d \geq 0$, we have

$$\lim_{k\to\infty} \int_{\mathbb{R}^r} f \, d\eta_k = \int_{\mathbb{R}^r} f \, d\eta_\infty.$$

PROOF. First, we should remark that each η_k has finite total mass (it has moments of all orders, in particular a zeroth moment), so each CDF_{η_k} is well-defined as an $\mathbb{R}_{\geq 0}$-valued function on $\mathbb{R}^r$. To show that $F(x)$ is the CDF of a positive Borel measure, it suffices, by the Lebesgue-Stieltjes construction, to check that $F(x)$ satisfies the Lebesgue-Stieltjes positivity condition, and that $F(x)$ is continuous from above in the sense that $F(x + \varepsilon \mathbb{1}_r) \to F(x)$ as $\varepsilon \to 0+$. The positivity results from the fact that F is the pointwise limit of functions with this positivity. The continuity from above of F at any given point x in $\mathbb{R}^r$ results from the fact that F is, in a neighborhood of x, the uniform limit of functions which are continuous from above.

Once η_∞ exists, then for any rectangle $R(x) = (-\infty, x]$ in $\mathbb{R}^r$, we have

$$\eta_\infty(R(x)) = \lim_{k\to\infty} \eta_k(R(x)) = \lim_{k\to\infty} [\eta_k(\mathbb{R}^r) - \eta_k(\mathbb{R}^r - R(x))].$$

Taking x to be $s\mathbb{1}_r$ with $s \geq 0$, we have $E(s) \subset R(s\mathbb{1}_r)$, so

$$0 \leq \eta_k(\mathbb{R}^r - R(s\mathbb{1}_r)) \leq \eta_k(\mathbb{R}^r - E(s)) \leq rAe^{-Bs^\lambda},$$

so we get

$$\eta_\infty(R(s\mathbb{1}_r)) \geq \left[\lim_{k\to\infty} \eta_k(\mathbb{R}^r)\right] - rAe^{-Bs^\lambda} = 1 - rAe^{-Bs^\lambda}.$$

Taking s large, we see that η_∞ has total mass at least one. To show that η_∞ has mass at most one, we argue as follows. Because $\mathbb{R}^r$ is the increasing union of the $R(s\mathbb{1}_r)$, $\eta_\infty(\mathbb{R}^r) = \lim_{n\to\infty} \eta_\infty(R(n\mathbb{1}_r))$. So it suffices to show that for fixed n, $\eta_\infty(R(n\mathbb{1}_r)) \leq 1$. But

$$\eta_\infty(R(n\mathbb{1}_r)) = \lim_{k\to\infty} \eta_k(R(n\mathbb{1}_r)) \leq \lim_{k\to\infty} \eta_k(\mathbb{R}^r) = 1.$$

Therefore η_∞ is a probability measure.

We must now show that

$$\eta_\infty(\mathbb{R}^r - E_i(s)) \leq Ae^{-Bs^\lambda}.$$

View $E_i(s)$ as the increasing union of $E_i(s) \cap E(n)$, as $n \to \infty$. For fixed n, think of $\mathbb{R}^r$ as the disjoint union of $E(n)$ and $\mathbb{R}^r - E(n)$. We have

$$\begin{aligned}
&\eta_\infty(\mathbb{R}^r - E_i(s)) \\
&\quad = \eta_\infty(E(n) - E_i(s) \cap E(n)) + \eta_\infty((\mathbb{R}^r - E_i(s)) \cap (\mathbb{R}^r - E(n))) \\
&\quad \leq \eta_\infty(E(n) - E_i(s) \cap E(n)) + \eta_\infty(\mathbb{R}^r - E(n)) \\
&\quad = \eta_\infty(E(n) - E_i(s) \cap E(n)) + 1 - \eta_\infty(E(n)) \\
&\quad = \lim_{k\to\infty} [\eta_k(E(n) - E_i(s) \cap E(n)) + 1 - \eta_k(E(n))] \\
&\quad = \lim_{k\to\infty} [\eta_k(E(n) - E_i(s) \cap E(n)) + \eta_k(\mathbb{R}^r) - \eta_k(E(n))] \\
&\quad = \lim_{k\to\infty} [\eta_k(E(n) - E_i(s) \cap E(n)) + \eta_k(\mathbb{R}^r - E(n))] \\
&\quad \leq rAe^{-Bn^\lambda} + \lim_{k\to\infty} \eta_k(E(n) - E_i(s) \cap E(n)) \\
&\quad \leq rAe^{-Bn^\lambda} + \lim_{k\to\infty} \eta_k(\mathbb{R}^r - E_i(s)) \\
&\quad \leq rAe^{-Bn^\lambda} + Ae^{-Bs^\lambda}.
\end{aligned}$$

Letting n get large, we get the asserted inequality.

Let us say that a real $s > 0$ is a continuity point for η_∞ if the nondecreasing function $s \mapsto \eta_\infty(E(s))$ from $\mathbb{R}_{\geq 0}$ to $[0,1]$ is continuous at s. There are at most countably many points at which this function is discontinuous (because for each $k \geq 1$ there are at most k points where it jumps by more than $1/k$). Denote by $E(s-)$ the half open rectangle $(-s\mathbb{1}_r, s\mathbb{1}_r]$; thus $\text{Interior}(E(s)) \subset E(s-) \subset E(s)$. For s a point of continuity of η_∞, all three sets have the same η_∞-measure. Similarly, each η_k, having finite mass, has at most countably many points of discontinuity. Thus every s outside a countable set is a point of continuity for every η_k and for η_∞.

Suppose now that f is a continuous function on $\mathbb{R}^r$ with polynomial growth. For any continuous f, and any real $s \geq 0$, f restricted to $E(s)$ is uniformly continuous. For all rectangles $R = (a, b]$ contained in $E(s)$, the sequence $\eta_k(R)$ converges to $\eta_\infty(R)$. The usual argument [**Fel**, pages 243, 244] of partitioning $E(s-)$ into finitely many small half open rectangles $(a, b]$ on each of which f is very nearly constant shows that for any s which is a point of continuity of all η_n and of η_∞, we have

$$\lim_{k \to \infty} \int_{E(s)} f\, d\eta_k = \lim_{k \to \infty} \int_{E(s-)} f\, d\eta_k = \int_{E(s-)} f\, d\eta_\infty = \int_{E(s)} f\, d\eta_\infty.$$

For any measure η with exponential decay of type (A, B, λ), and any function f with $|f(x)| \leq C(1 + \|x\|)^d$, we have

$$\begin{aligned}
\left| \int_{\mathbb{R}^r - E(s)} f\, d\eta \right| &= \left| \sum_{k \geq 0} \int_{E(s+k+1)-E(s+k)} f\, d\eta \right| \\
&\leq \sum_{k \geq 0} \int_{E(s+k+1)-E(s+k)} |f|\, d\eta \\
&\leq \sum_{k \geq 0} C \int_{E(s+k+1)-E(s+k)} \left(1 + \sum_i |x(i)|\right)^d d\eta \\
&\leq \sum_{k \geq 0} C(1 + r(s+k+1))^d \int_{E(s+k+1)-E(s+k)} d\eta \\
&\leq \sum_{k \geq 0} C(1 + r(s+k+1))^d \int_{\mathbb{R}^r - E(s+k)} d\eta \\
&\leq \sum_{k \geq 0} C(1 + r(s+k+1))^d A e^{-B(s+k)^\lambda} \\
&\leq \sum_{k \geq 0} C(r(s+k+2))^d A e^{-B(s+k)^\lambda}.
\end{aligned}$$

Let us denote by $[s]$ the integer part of s, so $[s] \leq s < [s] + 1$. Then we have

$$\left| \int_{\mathbb{R}^r - E(s)} f\, d\eta \right| \leq ACr^d \sum_{k \geq [s]} (k+3)^d e^{-Bk^\lambda}.$$

Since the series

$$ACr^d \sum_{k \geq 0} (k+3)^d e^{-Bk^\lambda}$$

is convergent, its tails go to zero, so given $\varepsilon > 0$ there exists n such that for any $s \geq n$,

$$\left| \int_{\mathbb{R}^r - E(s)} f\, d\eta \right| \leq ACr^d \sum_{k \geq n} (k+3)^d e^{-Bk^\lambda} \leq \varepsilon$$

for all η of exponential decay (A, B, λ) and all f with $|f(x)| \leq C(1 + \|x\|)^d$. Fix a choice of $s > n$ which is a point of continuity for all the η_k and for η_∞. We can choose k_0 such that

$$\left| \int_{E(s)} f\, d\eta_k - \int_{E(s)} f\, d\eta_\infty \right| \leq \varepsilon$$

for $k \geq k_0$. Then we have

$$\left| \int_{\mathbb{R}^r} f\, d\eta_k - \int_{\mathbb{R}^r} f\, d\eta_\infty \right| \leq 3\varepsilon$$

for all $k \geq k_0$. Since $\varepsilon > 0$ was arbitrary, we get

$$\lim_{k \to \infty} \int_{\mathbb{R}^r} f\, d\eta_k = \int_{\mathbb{R}^r} f\, d\eta_\infty. \quad \text{QED}$$

Corollary 7.5.12. *Hypotheses and notations as in Lemma* 7.5.11 *above, suppose we are given in addition an integer* $n \geq 1$ *and a continuous map* $\varphi : \mathbb{R}^r \to \mathbb{R}^n$ *which is of polynomial growth of degree d in the sense that for some real* $C > 0$, $\|\varphi(x)\| \leq C(1 + \|x\|)^d$. *Let*

$$D := \operatorname*{Sup}_k(\eta_k(\mathbb{R}^r)).$$

Then we have:

1) *The direct image measures* $\varphi_*\eta_k$ *and* $\varphi_*\eta_\infty$ *are of exponential decay* $(A', B', \lambda/d)$ *with*

$$B' = B/(2^\lambda r^\lambda C^{\lambda/d}) \quad \text{and} \quad A' = \mathrm{Max}(rA, De^{B'(C2^d)^{\lambda/d}}).$$

2) *For any continuous function* f *of polynomial growth on* $\mathbb{R}^n$, *we have*

$$\lim_{k \to \infty} \int_{\mathbb{R}^n} f\, d(\varphi_*\eta_k) = \int_{\mathbb{R}^n} f\, d(\varphi_*\eta_\infty).$$

PROOF. Assertion 2) is obvious from the final assertion of the lemma, since for η any of η_k or η_∞, we have

$$\int_{\mathbb{R}^n} f\, d(\varphi_*\eta) := \int_{\mathbb{R}^r} \varphi^*(f)\, d\eta,$$

and the function $\varphi^*(f) := f \circ \varphi$ is of polynomial growth on $\mathbb{R}^r$. To check assertion 1), write $\varphi(x) = (\varphi_1(x), \ldots, \varphi_n(x))$. For any measure η on $\mathbb{R}^r$ of exponential decay (A, B, λ) and total mass $\leq D$, we have

$$\begin{aligned}(\mathrm{pr}[i]_*\varphi_*\eta)(\{x \text{ in } \mathbb{R} \text{ with } |x| > C(1+s)^d\}) \\ := \eta(\{x \text{ in } \mathbb{R}^r \text{ with } |\varphi_i(x)| > C(1+s)^d\}).\end{aligned}$$

But we have

$$|\varphi_i(x)| \leq \sum_j |\varphi_j(x)| = \|\varphi(x)\| \leq C(1 + \|x\|)^d,$$

so if x in $\mathbb{R}^r$ has $|\varphi_i(x)| > C(1+s)^d$, then $\|x\| > s$, so x lies in $\mathbb{R}^r - E(s/r)$. Thus we have

$$\{x \text{ in } \mathbb{R}^r \text{ with } |\varphi_i(x)| > C(1+s)^d\} \subset \mathbb{R}^r - E(s/r),$$

and hence

$$(\mathrm{pr}[i]_*\varphi_*\eta)(\{x \text{ in } \mathbb{R} \text{ with } |x| > C(1+s)^d\}) \leq \eta(\mathbb{R}^r - E(s/r)) \leq rAe^{-B(s/r)^\lambda}.$$

For $s \geq 1$, put $t = C(1+s)^d$, so $s = (t/C)^{1/d} - 1$. Because $s \geq 1$, we have $(t/C)^{1/d} \geq 2$, and hence $(t/C)^{1/d} - 1 \geq (1/2)(t/C)^{1/d}$. Thus for $t \geq C2^d$, we have

$$\begin{aligned}(\mathrm{pr}[i]_*\varphi_*\eta)(\{x \text{ in } \mathbb{R} \text{ with } |x| > t\}) &\leq rAe^{-B(((t/C)^{1/d}-1)/r)^\lambda} \\ &\leq rAe^{-B((1/2)(t/C)^{1/d}/r)^\lambda} = rAe^{-B't^{\lambda/d}}, \quad \text{if } t \geq C2^d.\end{aligned}$$

For $0 \leq t < C2^d$, we have the trivial estimate

$$(\mathrm{pr}[i]_*\varphi_*\eta)(\{x \text{ in } \mathbb{R} \text{ with } |x| > t\}) \leq D \leq A'e^{-B'(C2^d)^{\lambda/d}} \leq A'e^{-B't^{\lambda/d}},$$

as required. QED

7.5.13. Here is a second application of these general results on convergence of moments.

Proposition 7.5.14. *Let $r \geq 1$ be an integer, and b in $\mathbb{Z}^r$ a step vector.*

1) The measures $\mu(univ$, steps $b)$ and $\mu(U(N)$, steps $b)$, $N \geq 2$, on $\mathbb{R}^r$ have moments of all orders, and each of these measures is uniquely determined by its moments.

2) For any continuous function $f(x)$ on $\mathbb{R}^r$ of polynomial growth, we have the limit formula

$$\lim_{N\to\infty} \int f(x)\, d\mu(U(N), \textit{steps } b) = \int f(x)\, d\mu(\textit{univ}, \textit{steps } b).$$

PROOF. Thanks to 6.13.5, these measures are all of exponential decay

$$(A, B, \lambda) = ((4/3)\Sigma(b), (1/8)\operatorname{Sup}_i(b(i)), 2),$$

so the first assertion follows from 7.5.9. According to 1.6.4, the CDF's of the measures $\mu(U(N)$, steps $b)$ converge uniformly to the CDF of μ(univ, steps b), so the second assertion results from 7.5.11. QED

Corollary 7.5.15. *For each $i = 1, \ldots, r$, the first moments of these measures are given by*

$$\int x(i)\, d\mu(U(N), \textit{steps } b) = \int x(i)\, d\mu(\textit{univ}, \textit{steps } b) = b(i).$$

PROOF. For finite N, this reduces by direct image to the $r = 1$ case, in which case it is proven in 6.12.9. Now take the large N limit, using 2) above. QED

7.6. Relations among the measures μ_n and the measures $\nu(n)$

7.6.1. In order to manipulate these objects more conveniently, we introduce three entire functions on $\mathbb{C}^2$:

$$F(T,s) := (1+T)(E(T,s)-1)/T,$$
$$F_\pm(T,s) := (1+T)(E_\pm(T,s)-1)/T.$$

These functions are entire because $E(0,s) = E_\pm(0,s) = 1$, and they visibly vanish at $T = -1$. Their higher derivatives at $T = -1$ are given by

Lemma 7.6.2. *For each real $s \geq 0$, we have the identities*

$$F(T,s) = \sum_{n\geq 1}(1+T)^n \operatorname{CDF}_{\nu(n)}(s),$$

$$F_\pm(T,s) = \sum_{n\geq 1}(1+T)^n \operatorname{CDF}_{\nu(\pm,n)}(s).$$

PROOF. We will write out the proof for $F(T,s)$. The $F_\pm$ cases are exactly the same and are left to the reader. Since $F(T,s)$ is, for fixed s, entire in T, it suffices to show that the asserted series expansion is valid in a neighborhood of $T = -1$. We begin by writing the power series expansion of $E(T,s)$ around $T = -1$:

$$E(T,s) = \sum_{n\geq 0}(1+T)^n E_n(s).$$

Multiplying by $1+T$, we get

$$(1+T)E(T,s) = \sum_{n\geq 1}(1+T)^n E_{n-1}(s).$$

Multiplying by

$$1/T = -1/(1-(1+T)) = -\sum_{n\geq 0}(1+T)^n,$$

we find

$$(1+T)E(T,s)/T = -\sum_{n\geq 1}(1+T)^n \sum_{0\leq j\leq n-1} E_j(s).$$

By 7.5.5, we may rewrite this as

$$\begin{aligned}(1+T)E(T,s)/T &= \sum_{n\geq 1}(1+T)^n(\operatorname{CDF}_{\nu(n)}(s)-1)\\ &= \Big\{\sum_{n\geq 1}(1+T)^n \operatorname{CDF}_{\nu(n)}(s)\Big\} - \sum_{n\geq 1}(1+T)^n\\ &= \Big\{\sum_{n\geq 1}(1+T)^n \operatorname{CDF}_{\nu(n)}(s)\Big\} - (1+T)/(1-(1+T))\\ &= \Big\{\sum_{n\geq 1}(1+T)^n \operatorname{CDF}_{\nu(n)}(s)\Big\} + (1+T)/T,\end{aligned}$$

so we get

$$(1+T)E(T,s)/T - (1+T)/T = \sum_{n\geq 1}(1+T)^n \operatorname{CDF}_{\nu(n)}(s). \quad \text{QED}$$

Lemma 7.6.3. *We have the identity*

$$(d/ds)F(T,s) = (1+T) - (1+T)G_1(T,s) + (1+T)^2 G_1(T,s)$$

of entire functions on $\mathbb{C}^2$.

PROOF. This results from the fundamental relation 7.2.3

$$(d/ds)E(T,s) = T + T^2 G_1(T,s)$$

and the definition of $F(T,s)$ as

$$F(T,s) := (1+T)(E(T,s)-1)/T.$$

We readily calculate

$$\begin{aligned}(d/ds)F(T,s) &= (1+T)(d/ds)E(T,s)/T \\ &= (1+T)(T+T^2G_1(T,s))/T \\ &= (1+T)(1+TG_1(T,s)) \\ &= (1+T)(1-G_1(T,s)+(1+T)G_1(T,s)). \quad \text{QED}\end{aligned}$$

Corollary 7.6.4. *The restrictions to $\mathbb{R}_{\geq 0}$ of the CDF's of the measures $\nu(n)$ for $n \geq 1$ and of the measures*

$$\mu_n := \mu(\mathit{univ},\ \mathit{sep.}\ n)$$

for $n \geq 0$ are related by

1) $$(d/ds)\,\mathrm{CDF}_{\nu(1)} = 1 - \mathrm{CDF}_{\mu_0},$$

and for each $n \geq 2$,

2) $$(d/ds)\,\mathrm{CDF}_{\nu(n)} = \mathrm{CDF}_{\mu_{n-2}} - \mathrm{CDF}_{\mu_{n-1}}.$$

Equivalently, on $\mathbb{R}_{\geq 0}$ we have the formulas

3) $$\mathrm{Tail}_{\mu_n} = \sum_{0\leq j\leq n} (d/ds)\,\mathrm{CDF}_{\nu(j+1)} \quad \text{for } n \geq 0,$$

or

4) $$\mathrm{Tail}_{\mu_n}(s) = -\sum_{0\leq j\leq n} (n+1-j)(d/ds)E_j(s) \quad \text{for } n \geq 0.$$

We have the identities of measures on $\mathbb{R}_{\geq 0}$

5) $$\sum_{j=0}^{n} \nu(j+1) = \mathrm{Tail}_{\mu_n}(s)\,ds.$$

In statements 1) through 4), derivatives at $s=0$ are taken from above.

PROOF. The first set of formulas 1) and 2) are just the spelling out of the identity

$$(d/ds)F(T,s) = (1+T) - (1+T)G_1(T,s) + (1+T)^2 G_1(T,s),$$

together with the expansions

$$F(T,s) = \sum_{n\geq 1} (1+T)^n\,\mathrm{CDF}_{\nu(n)}(s),$$

and

$$G_1(T,s) = \sum_{n\geq 0} (1+T)^n\,\mathrm{CDF}_{\mu_n}(s).$$

Because each μ_n is a probability measure, $\mathrm{Tail}_{\mu_n} = 1 - \mathrm{CDF}_{\mu_n}$, so 3) results from 1) and 2) by partial summation, and implies them by successive subtraction. The next one 4) is equivalent to 3), as one sees by applying d/ds to the relations

$$\mathrm{CDF}_{\nu(n)}(s) = 1 - \sum_{j=0}^{n-1} E_j(s) \quad \text{for } n \geq 1.$$

The final one 5) is just the integrated form of 3). QED

Remarks 7.6.5. Since $\nu(n)$ is a positive measure, its CDF is a nondecreasing function, so $(d/ds)\,\mathrm{CDF}_{\nu(n)} \geq 0$. So we learn that $1 - \mathrm{CDF}_{\mu_0} \geq 0$, which we already knew, and we learn that for each $n \geq 0$,

$$\mathrm{CDF}_{\mu_n} \geq \mathrm{CDF}_{\mu_{n+1}}.$$

This we also see directly, as follows. Already at the level of individual elements A of $U(N)$ with $N > n+1$, we claim that

$$\mathrm{CDF}_{\mu(A,U(N),\text{ sep. } n)} \geq \mathrm{CDF}_{\mu(A,U(N),\text{ sep. } n+1)}.$$

Indeed, in terms of the sequence of angles of A,

$$0 \leq \varphi(1) \leq \varphi(2) \leq \cdots \leq \varphi(N) < 2\pi,$$

extended to all integers by $\varphi(j+N) = \varphi(j) + 2\pi$, we have the tautological formula

$$\begin{aligned} &N \times \mathrm{CDF}_{\mu(A,U(N),\text{ sep. } n)}(s) \\ &\quad = \mathrm{Card}\{\text{indices } i \text{ with } 1 \leq i \leq N \text{ and } \varphi(i+n+1) - \varphi(i) \leq 2\pi s/N\}, \end{aligned}$$

which makes obvious that for fixed s, $\mathrm{CDF}_{\mu(A,U(N),\text{ sep. } n)}(s)$ is a decreasing function of n.

7.7. Recapitulation, and concordance with the formulas in [Mehta]

7.7.1. For real $s \geq 0$, with derivatives at $s = 0$ taken from above, we have the following summarizing list of formulas:

$$E(T,s) = \det(1 + TK_{s,\alpha}), \tag{7.7.1.1}$$

$$E(T,0) = 1, \tag{7.7.1.2}$$

$$E(T,s) = \sum_{n \geq 0} (1+T)^n E_n(s), \tag{7.7.1.3}$$

$$E_n(s) = \lim_{N \to \infty} \mathrm{eigen}(n, 2\pi s/N, U(N)) \quad \text{for } n \geq 0, \tag{7.7.1.4}$$

$$F(T,s) := (1+T)(E(T,s) - 1)/T, \tag{7.7.1.5}$$

$$F(T,s) = \sum_{n \geq 1} (1+T)^n \,\mathrm{CDF}_{\nu(n)}(s), \tag{7.7.1.6}$$

$$\nu(n) := \lim_{N \to \infty} \vartheta(n)_*(\text{Haar measure on } U(N)) \quad \text{for } n \geq 1, \tag{7.7.1.7}$$

$$\mathrm{CDF}_{\nu(n)}(s) = 1 - \sum_{j=0}^{n-1} E_j(s) \quad \text{for } n \geq 1, \tag{7.7.1.8}$$

$$\mathrm{Tail}_{\nu(n)}(s) = \sum_{j=0}^{n-1} E_j(s) \quad \text{for } n \geq 1, \tag{7.7.1.9}$$

$$E_0(s) = 1 - \mathrm{CDF}_{\nu(1)}(s), \tag{7.7.1.10}$$

$$(7.7.1.11) \qquad E_n(s) = \mathrm{CDF}_{\nu(n)}(s) - \mathrm{CDF}_{\nu(n+1)}(s) \quad \text{for } n \geq 1,$$

$$(7.7.1.12) \qquad G_1(T, s) = \sum_{n \geq 0} (1+T)^n \,\mathrm{CDF}_{\mu_n}(s),$$

$$(7.7.1.13) \qquad (d/ds)E(T, s) = T + T^2 G_1(T, s),$$

$$(7.7.1.14) \qquad (d/ds)F(T, s) = (1+T) - (1+T)G_1(T, s) + (1+T)^2 G_1(T, s),$$

$$(7.7.1.15) \qquad \mu_n := \mu(\text{univ, sep. } n),$$

$$(7.7.1.16) \qquad (d/ds)\,\mathrm{CDF}_{\nu(1)} = 1 - \mathrm{CDF}_{\mu_0},$$

$$(7.7.1.17) \qquad (d/ds)\,\mathrm{CDF}_{\nu(n)} = \mathrm{CDF}_{\mu_{n-2}} - \mathrm{CDF}_{\mu_{n-1}} \quad \text{for } n \geq 2,$$

$$(7.7.1.18) \qquad \begin{aligned} \mathrm{Tail}_{\mu_n} &= \sum_{1 \leq j \leq n+1} (d/ds)\,\mathrm{CDF}_{\nu(j)} = -\sum_{1 \leq j \leq n+1} (d/ds)\,\mathrm{Tail}_{\nu(j)} \\ &= -\sum_{0 \leq j \leq n} (n+1-j)(d/ds)E_j(s) \quad \text{for } n \geq 0, \end{aligned}$$

$$(7.7.1.19) \qquad \mathrm{Tail}_{\mu_n}(s)\,ds = \sum_{1 \leq j \leq n+1} \nu(j) \quad \text{as measures on } \mathbb{R}_{\geq 0}.$$

7.7.2. We see from formulas 7.7.1.1–2 and 7.7.1.13 above that knowing $E(T, s)$ is equivalent to knowing $G_1(T, s)$, and from 7.7.1.12 that knowing $G_1(T, s)$ is equivalent to knowing all the measures μ_a. If we compare these equations with those in [**Mehta**, 5.1.16–18 and A.7.27], we find the following concordance between Mehta's objects and ours:

range of validity	Mehta's	ours
$n \geq 0$	$E_2(n, s)$	$E_n(s)$
$n \geq 0$	$F_2(n, s)$	$(d/ds)\,\mathrm{CDF}_{\nu(n+1)}(s)$
$n \geq 0$	$F_2(n, s)\,ds$	$\nu(n+1)$
$n \geq 0$	$p_2(n, s)$	$(d/ds)\,\mathrm{CDF}_{\mu_n}(s)$
$n \geq 0$	$p_2(n, s)\,ds$	μ_n

It is interesting to note that Mehta [**Mehta**, discussion page 84, discussion page 88, appendix A8] defines his $p_2(0, s)$ as a conditional probability. He then speaks of $p_2(0, s)\,ds$ [**Mehta**, 5.1.38] as though it were equal to what we have defined as the spacing measure μ_0. At finite level N, the justification for doing so is provided by 6.12.4.

7.8. Supplement: Fredholm determinants and spectral determinants, with applications to $E(T, s)$ and $E_{\pm}(T, s)$

7.8.1. In this supplement, we give some basic compatibilities between Fredholm determinants and spectral determinants of integral operators.

7.8.2. Recall from 7.3 that

$$E(T,s) = \sum_{n\geq 0}(1+T)^n E_n(s),$$
$$E_\pm(T,s) = \sum_{n\geq 0}(1+T)^n E_{\pm,n}(s),$$

where $E(T,2s)$ and $E_\pm(T,s)$ are the Fredholm determinants

$$E(T,2s) := \det(1+TK_{2s}|L_2([-s,s],dx)),$$
$$E_\pm(T,s) := \det(1+TK_{\pm,s}|L_2([0,s],dx)),$$

for K_{2s} the integral operator on $L_2([-s,s],dx)$ with kernel

$$K(x,y) := \sin(\pi(x-y))/\pi(x-y),$$

and for $K_{\pm,s}$ the integral operator on $L_2([0,s],dx)$ with kernel

$$K_\pm(x,y) := K(x,y) \pm K(-x,y).$$

Because $\sin(\pi x)/\pi x$ is an even function, the integral operator K_{2s} preserves the subspaces of odd and of even functions. We have an isometric isomorphism

$$K_{\pm,s}|L_2([0,s],dx) = K_{2s}|L_2([-s,s],dx)_{\text{parity }\pm}.$$

7.8.3. It is technically convenient to interpret all of our operators as acting on a single space, which here we will take to be $L_2(\mathbb{R},dx)$. For each real $s>0$, we have an orthogonal direct sum decomposition

$$L_2(\mathbb{R},dx) = L_2([-s,s],dx) \oplus L_2(\mathbb{R}-[-s,s],dx).$$

7.8.4. We denote by $P(s)$ the orthogonal projection of $L_2(\mathbb{R},dx)$ onto $L_2([-s,s],dx)$. Concretely, if we denote by I_s the characteristic function of the interval $[-s,s]$, then for f in $L_2(\mathbb{R},dx)$ we have

$$P(s)(f) = I_s f.$$

7.8.5. We also have an orthogonal direct sum decomposition into odd and even functions,

$$L_2(\mathbb{R},dx) := L_2(\mathbb{R},dx)_{\text{even}} \oplus L_2(\mathbb{R},dx)_{\text{odd}}.$$

The corresponding orthogonal projections are denoted $P(\pm)$. These projections both commute with $P(s)$. We denote by

$$P(\pm,s) := P(\pm)P(s) = P(s)P(\pm)$$

the orthogonal projection of $L_2(\mathbb{R},dx)$ onto $L_2([-s,s],dx)_{\text{parity }\pm}$.

7.8.6. We denote by K the integral operator on $L_2(\mathbb{R},dx)$ given by the kernel

$$K(x,y) := \sin(\pi(x-y))/\pi(x-y).$$

7.8.7. For real $s>0$, we define operators $K(2s)$ and $K(\pm,s)$ (sic) on $L_2(\mathbb{R},dx)$ by

$$\begin{aligned} K(2s) &= P(s)\circ K\circ P(s),\\ K(\pm,s) &= P(\pm)\circ K(2s)\circ P(\pm)\\ &= P(\pm,s)\circ K\circ P(\pm,s) = P(s)\circ P(\pm)\circ K\circ P(\pm)\circ P(s). \end{aligned}$$

7.8.8. In terms of the decomposition

$$L_2(\mathbb{R}, dx) = L_2([-s, s], dx) \oplus L_2(\mathbb{R} - [-s, s], dx),$$

we have

$$K(2s) = K_{2s}|L_2([-s, s], dx) \oplus 0.$$

7.8.9. In terms of the further decomposition

$$L_2(\mathbb{R}, dx) = L_2([-s, s], dx)_{\text{even}} \oplus L_2([-s, s], dx)_{\text{odd}} \oplus L_2(\mathbb{R} - [-s, s], dx),$$

we have

$$K(+, s) = K_{2s}|L_2([-s, s], dx)_{\text{even}} \oplus 0 \oplus 0,$$
$$K(-, s) = 0 \oplus K_{2s}|L_2([-s, s], dx)_{\text{odd}} \oplus 0.$$

Thus the operators $K(2s)$ and $K(\pm, s)$ are compact for all real $s > 0$.

Lemma 7.8.10. *The operator K on $L_2(\mathbb{R}, dx)$ is the Fourier transform of the orthogonal projection $P(\pi)$. In particular, K is itself an orthogonal projection, with both kernel and range of infinite dimension.*

PROOF. The operator K is convolution with the function $k(x) := \sin(\pi x)/\pi x$. The Fourier transform on $L_2(\mathbb{R}, dx)$, defined by

$$FT(f)(y) := (2\pi)^{-1/2} \int_{\mathbb{R}} f(x)e^{ixy}\,dx,$$

is an isometry, whose square is $f \mapsto f_- :=$ the function $x \mapsto f(-x)$. Because of the normalizing factor, the relation of FT to convolution is

$$FT(f * g) = (2\pi)^{1/2} FT(f)FT(g) = FT(f)FT((2\pi)^{1/2} g).$$

One readily calculates that for the function $I_\pi :=$ the characteristic function of $[-\pi, \pi]$, we have $FT(I_\pi) = (2\pi)^{1/2}k$. By inversion, we have $FT((2\pi)^{1/2}k) = I_\pi$. So taking $g := k$, we get

$$FT(K(f)) = FT(f * k) = FT(f)FT((2\pi)^{1/2}k) = I_\pi FT(f) = P(\pi)FT(f).$$

In other words, we have $FT \circ K = P(\pi) \circ FT$, or $K = FT^{-1} \circ P(\pi) \circ FT$. QED

Corollary 7.8.10.1. *The operator K is self-adjoint, positive, and of operator norm 1. For real $s > 0$, each of the compact operators $K(2s)$ and $K(\pm, s)$ is self-adjoint, positive, and of operator norm ≤ 1.*

PROOF. We know that K is a nonzero orthogonal projection, and any such is self-adjoint, positive, and of operator norm 1. For any second orthogonal projection P, the operator PKP remains self-adjoint and positive, and of operator norm ≤ 1. Apply this with P either $P(s)$ or $P(\pm, s)$. QED

Corollary 7.8.10.2. *The operators $K(\pm) := P(\pm) \circ K \circ P(\pm)$ are the Fourier transforms of the orthogonal projections $P(\pm, \pi)$, so in particular they are orthogonal projections with kernels and ranges of infinite dimension.*

PROOF. The projections $P(\pm)$ commute with Fourier transform. QED

Lemma 7.8.11. *In the strong topology on operators on $L_2(\mathbb{R}, dx)$, we have*

$$\lim_{s \to +\infty} P(s) = 1, \text{ the identity operator } f \mapsto f.$$

PROOF. This is the statement that for any $f(x)$ in $L_2(\mathbb{R}, dx)$, the truncated functions $I_s(x)f(x)$ tend to $f(x)$ in $L_2(\mathbb{R}, dx)$, i.e., that for f in $L_2(\mathbb{R}, dx)$ we have

$$\lim_{s\to+\infty} \int_{|x|>s} |f(x)|^2\, dx = 0.$$

The integral in question is a decreasing function of $s > 0$, so it suffices to check for integer values of s. But the series

$$\sum_{n\geq 0} \int_{n+1\geq|x|>n} |f(x)|^2\, dx$$

is convergent (to $\int_{\mathbb{R}} |f(x)|^2\, dx$), so its tails go to zero. QED

Corollary 7.8.12. *In the strong topology on operators on* $L_2(\mathbb{R}, dx)$, *we have*

$$\lim_{s\to+\infty} K(2s) = K,$$
$$\lim_{s\to+\infty} K(\pm, s) = K(\pm).$$

PROOF. This is obvious from $\lim_{s\to+\infty} P(s) = 1$ and the formulas

$$K(2s) = P(s)\circ K\circ P(s),$$
$$K(\pm, s) = P(s)\circ K(\pm)\circ P(s).$$

Indeed, for any bounded operator L, we have

$$L = \lim_{s\to+\infty} P(s)\circ L\circ P(s).$$

The point is that all the operator norms $\|P(s)\circ L\|$ are uniformly bounded (by $\|L\|$, as the $P(s)$ are projections), so the assertion is obvious from the identity

$$L - P(s)\circ L\circ P(s) = (1-P(s))\circ L + (P(s)\circ L)\circ(1-P(s)),$$

cf. [**Riesz-Sz.-Nagy**, §84, top of page 201]. QED

7.9. Interlude: Generalities on Fredholm determinants and spectral determinants

7.9.1. For any positive, self-adjoint compact operator L on a separable Hilbert space $\mathcal{H}$, one knows that its spectrum $\sigma(L)$ is a countable subset of the closed real interval $[0, \|L\|]$ which contains $\|L\|$ and which has no nonzero limit points. One knows further that for every nonzero λ in $\sigma(L)$, the subspace

$$\bigcup_n \mathrm{Ker}(L-\lambda)^n$$

has finite dimension $m(\lambda) \geq 1$, called the multiplicity of λ, cf. [**Reed-Simon**, VI.15] and [**Riesz-Sz.-Nagy**, §§93–95]. If we write down each nonzero λ in the spectrum as many times as its multiplicity $m(\lambda)$, proceeding by decreasing size of λ, we get a list which starts with $\|L\|$ and which is either finite or countable:

$$\|L\| = \lambda_1 \geq \lambda_2 \geq \cdots.$$

We adopt the convention (which is reasonable only in the infinite-dimensional context) that if the list is finite, say of length N, then

$$\lambda_i := 0 \quad \text{for } i > N.$$

Thus we may speak unambiguously of the n'th eigenvalue λ_n of L, for every integer $n \geq 1$. If $\mathcal{H}$ is infinite dimensional, then by the Hilbert-Schmidt theorem [**Reed-Simon**, VI.16] there is an orthonormal basis $\{e_n\}_{n\geq 1}$, such that $Le_n = \lambda_n e_n$, and the λ_n tend to zero as $n \to \infty$.

7.9.2. In this generality, there is nothing one can say about the rate of decay of the λ_n: any real sequence $\lambda_1 \geq \lambda_2 \geq \cdots$ with limit 0 is the sequence of eigenvalues of some compact, positive, self-adjoint L. Indeed, if we choose an orthonormal basis $\{e_n\}_{n\geq 1}$, the operator $v \mapsto \sum_n \lambda_n(e_n, v)e_n$ does the job, cf. [**Reed-Simon**, VI, Problem 45 (a)] for the compactness. In particular, the series $\sum \lambda_n$ may diverge, so in general the formal expression $\prod_{n\geq 1}(1+\lambda_n T)$ fails to make sense, even as a formal power series in $1 + T\mathbb{R}[[T]]$. Notice, however, that if $\sum \lambda_n$ converges (and hence converges absolutely, as all $\lambda_n \geq 0$), then the formal expression $\prod_{n\geq 1}(1+\lambda_n T)$ is an entire function, in the sense that the sequence of polynomials $\{\prod_{1\leq n\leq k}(1+\lambda_n T)\}_k$ converges uniformly on compact subsets of $\mathbb{C}$, cf. [**W-W**, 3.341].

7.9.3. Under what circumstances can we be sure that $\sum \lambda_n$ converges, so that $\prod_{n\geq 1}(1 + \lambda_n T)$ is an entire function of T? A trivial case is when there are only finitely many nonzero λ_n's. The next simplest case, and one that will be adequate for our purposes, is this.

Lemma 7.9.4. *On $\mathcal{H} := L_2([a, b], dx)$ for a compact interval $[a, b]$, let L be a positive self-adjoint compact operator given by a self-adjoint kernel $L(x, y)$ on $[a, b] \times [a, b]$ which is continuous. Denote by $\{\lambda_n\}_n$ its sequence of eigenvalues. Then $\sum \lambda_n$ converges, and the entire function $\prod_{n\geq 1}(1 + \lambda_n T)$ is equal to the Fredholm determinant*

$$\det(1 + TL|\mathcal{H}) := 1 + \sum_{n\geq 1} T^n(1/n!) \int_{[a,b]^n} \det_{n\times n}(L(x(i), x(j))) \prod_i dx(i).$$

PROOF. Denote by $\{\varphi_n(x)\}_n$ an orthonormal set of eigenfunctions for the nonzero λ_n's. Then by Mercer's theorem [**Riesz-Sz.-Nagy**, §98] each φ_n is a continuous function, and the series development

$$L(x, y) = \sum_n \lambda_n \varphi_n(x)\overline{\varphi}_n(y)$$

is uniformly convergent. In particular, the series

$$L(x, x) = \sum_n \lambda_n \varphi_n(x)\overline{\varphi}_n(x)$$

is uniformly convergent, so may be integrated term by term to give

$$\int_{[a,b]} L(x, x)\, dx = \sum_n \lambda_n.$$

This shows that $\sum \lambda_n$ converges, and hence that $\prod_{n\geq 1}(1 + \lambda_n T)$ is an entire function of T, to which the partial products $\prod_{1\leq n\leq k}(1 + \lambda_n T)$ converge uniformly on compacta.

Now consider the Fredholm determinant

$$\det(1 + TL|\mathcal{H}) := 1 + \sum_{n\geq 1} T^n(1/n!) \int_{[a,b]^n} \det_{n\times n}(L(x(i), x(j))) \prod_i dx(i),$$

which we know [**W-W**, 11.21] to be an entire function. We claim that

$$\det(1+TL|\mathcal{H}) = \prod_{n\geq 1}(1+\lambda_n T).$$

If L has only finitely many nonzero λ_n's, the series $\sum_n \lambda_n\varphi_n(x)\overline{\varphi}_n(y)$ has finitely many terms, and the asserted identity is 6.1.4. In the general case, for each $k\geq 1$ define

$$L_k(x,y) := \sum_{n\leq k}\lambda_n\varphi_n(x)\overline{\varphi}_n(y).$$

Then L_k converges uniformly to L as $k\to\infty$, by Mercer's theorem. Looking at the formulas for the individual coefficients of a Fredholm determinant, we see that

$$\det(1+TL_k|\mathcal{H}) \to \det(1+TL|\mathcal{H})$$

in the sense of coefficient by coefficient convergence. But for each finite k we have

$$\det(1+TL_k|\mathcal{H}) = \prod_{1\leq n\leq k}(1+\lambda_n T),$$

so we have

$$\prod_{1\leq n\leq k}(1+\lambda_n T) \to \det(1+TL|\mathcal{H}),$$

coefficient by coefficient. Therefore the power series around $T=0$ of the two entire functions $\prod_{n\geq 1}(1+\lambda_n T)$ and $\det(1+TL|\mathcal{H})$ are identical, and hence these functions are equal. QED

7.9.5. This result 7.9.4 motivates the following definition. Suppose we are given a bounded operator on a separable Hilbert space $\mathcal{H}$. We say that L is "positive of trace class", or PTC, if both the following conditions are satisfied:

1) L is positive, self-adjoint, and compact.

2) $\sum\lambda_n$ converges.

For such an L, $\prod_{n\geq 1}(1+\lambda_n T)$ is an entire function. We define its spectral determinant, denoted $\det(1+TL|\mathcal{H})$, by

$$\det(1+TL|\mathcal{H}) := \prod_{n\geq 1}(1+\lambda_n T).$$

Scholie 7.9.6 0) *When $\mathcal{H}$ is $L_2([a,b],dx)$ for a compact interval $[a,b]$, and L is a positive self-adjoint integral operator given by a continuous kernel $L(x,y)$, then L is PTC, and its Fredholm determinant as integral operator coincides with its spectral determinant.*

1) *The zero operator on any $\mathcal{H}$ is PTC, with spectral determinant identically* 1.

2) *If L_i on $\mathcal{H}_i$ is PTC for $i=1,2$, then $L_1\oplus L_2$ on $\mathcal{H}_1\oplus\mathcal{H}_2$ is PTC, and we have the product formula*

$$\det(1+T(L_1\oplus L_2)|\mathcal{H}_1\oplus\mathcal{H}_2) = \det(1+TL_1|\mathcal{H}_1)\det(1+TL_2|\mathcal{H}_2).$$

3) *If L_i on $\mathcal{H}_i$ is a bounded operator for $i=1,2$, and if $L_1\oplus L_2$ on $\mathcal{H}_1\oplus\mathcal{H}_2$ is PTC, then so is each L_i on $\mathcal{H}_i$, and we have the product formula*

$$\det(1+T(L_1\oplus L_2)|\mathcal{H}_1\oplus\mathcal{H}_2) = \det(1+TL_1|\mathcal{H}_1)\det(1+TL_2|\mathcal{H}_2).$$

4) *If L on $\mathcal{H}$ is PTC, and if L_1 on $\mathcal{H}_1$ is isometrically isomorphic to L on $\mathcal{H}$, then L_1 on $\mathcal{H}_1$ is PTC, and their spectral determinants are equal.*

7.10. Application to $E(T,s)$ and $E_\pm(T,s)$

Lemma 7.10.1. *For each real $s>0$, the operators*

$$K_{2s} \text{ on } L_2([-s,s],dx),$$
$$K_{\pm,s} \text{ on } L_2([0,s],dx),$$
$$K_{2s} \text{ on } L_2([-s,s],dx)_{parity\ \pm},$$
$$K(2s) \text{ on } L_2(\mathbb{R},dx),$$
$$K(\pm,s) \text{ on } L_2(\mathbb{R},dx)$$

are all PTC. The Fredholm determinants

$$E(T,2s) := \det(1+TK_{2s}|L_2([-s,s],dx)),$$
$$E_\pm(T,s) := \det(1+TK_{\pm,s}|L_2([0,s],dx)),$$

are related to each other and to the spectral determinants by
1) $E_\pm(T,s) := \det(1+TK_{2s}|L_2([-s,s],dx)_{parity\ \pm})$,
2) $E(T,2s) = \det(1+TK(2s)|L_2(\mathbb{R},dx))$,
3) $E_\pm(T,s) := \det(1+TK(\pm,s)|L_2(\mathbb{R},dx))$,
4) $E(T,2s) = E_+(T,s)E_-(T,s)$.

PROOF. Once we know that all the operators in question are PTC, 1) follows via Scholie 7.9.6, 4), from the isometric isomorphism

$$K_{\pm,s}|L_2([0,s],dx) = K_{2s}|L_2([-s,s],dx)_{\text{parity }\pm},$$

and 2), 3) and 4) follow from the direct sum decompositions of 7.8.8 and 7.8.9, using parts 1) and 2) of Scholie 7.9.6.

For real $s>0$, each of the compact operators $K(2s)$ and $K(\pm,s)$ is self-adjoint and positive, by 7.8.10.1. From the direct sum decomposition

$$K(2s) = K_{2s}|L_2([-s,s],dx) \oplus 0$$

it follows that K_{2s} on $L_2([-s,s],dx)$ is positive (as well as compact and self-adjoint), so by 7.9.4 it is PTC. Applying Scholie 7.9.6, 2), we get that $K(2s)$ is PTC. From the orthogonal decomposition $K(2s) = K(+,s) \oplus K(-,s)$ we get, using Scholie 7.9.6, 3), that $K(\pm,s)$ is PTC. Then from

$$K(+,s) = K_{2s}|L_2([-s,s],dx)_{\text{even}} \oplus 0 \oplus 0,$$
$$K(-,s) = 0 \oplus K_{2s}|L_2([-s,s],dx)_{\text{odd}} \oplus 0,$$

we infer, via Scholie 7.9.6, 3), that each $K_{2s}|L_2([-s,s],dx)_{\text{parity }\pm}$ is PTC. Using the isometric isomorphism

$$K_{\pm,s}|L_2([0,s],dx) = K_{2s}|L_2([-s,s],dx)_{\text{parity }\pm},$$

we get, via Scholie 7.9.6, 4), that $K_{\pm,s}$ on $L_2([0,s],dx)$ is PTC. QED

7.11. Appendix: Large N limits of multi-eigenvalue location measures and of static and offset spacing measures on $U(N)$

7.11.1. We begin with a result valid for all the $G(N)$.

Proposition 7.11.2. *Let $N\geq 1$, $G(N)$ any of $U(N), SO(2N+1), USp(2N)$, $SO(2N), O_-(2N+2), O_-(2N+1)$. Given any integer $r\geq 1$ and an offset vector*

c in $\mathbb{Z}^r$, the multi-eigenvalue location measure $\nu(c, G(N))$ on $\mathbb{R}^r$ has the following tail estimate. For real $s \geq 0$,

$$\nu(c, G(N))(\{x \text{ in } \mathbb{R}^r \text{ with some } |x(i)| > s\}) \leq \left(\sum_i 2^{c(i)}\right) \exp(-\sigma s/4).$$

PROOF. The set $\{x \text{ in } \mathbb{R}^r \text{ with some } |x(i)| > s\}$ is the union of the r sets $\{x \text{ in } \mathbb{R}^r \text{ with } |x(i)| > s\}$, $i = 1, \ldots, r$. Since $\mathrm{pr}[i]_*\nu(c, G(N)) = \nu(c(i), G(N))$, we are reduced to the case $r = 1$, which is 6.9.8. QED

7.11.3. In the special case of $U(N)$, we have a stronger result.

Proposition 7.11.4. *Let $N \geq 1$. Given an integer $r \geq 1$ and a vector c in $\mathbb{Z}^r$ with*

$$c(1) < c(2) < \cdots < c(r),$$

the multi-eigenvalue location measure $\nu(c, U(N))$ on $\mathbb{R}^r$ has the following tail estimate. Denote

$$\begin{aligned} d(i) &:= \max(c(i), 1 - c(i)), \\ D &:= \max_i(d(i)). \end{aligned}$$

For real $s \geq 0$, we have

$$\nu(c, U(N))(\{x \text{ in } \mathbb{R}^r \text{ with some } |x(i)| > s\}) \leq rD^2(8/3)\,\mathrm{Sqrt}(\pi)e^{-s^2/16D^2}.$$

PROOF. Just as above, we reduce to treating the one variable case $\nu(c(i), G(N))$ for any $c(i)$ in $\mathbb{Z}$. To treat $c(i) \leq 0$, we use the fact that under complex conjugation $A \mapsto \overline{A}$ on $U(N)$ we have

$$\vartheta(1-k)(A) = -\vartheta(k)(\overline{A}), \quad \text{for every integer } k.$$

So for the one variable measures we have

$$\nu(c(i), U(N)) = [x \mapsto -x]^*\nu(1 - c(i), U(N)).$$

So we are reduced to treating the one-variable measure $\nu(b, U(N))$ for a single $b \geq 1$. To do this, use the relations 6.12.6

$$\sum_{j=1}^{b} \nu(j, U(N)) = \mathrm{Tail}_{\mu(U(N),\ \text{step } b)}(s)\, ds$$

and the estimates 6.13.6,

$$\mathrm{Tail}_{\mu(U(N),\ \text{step } b)}(s) \leq b(4/3)e^{-s^2/8b^2}.$$

Then for $s \geq 0$, we have

$$\begin{aligned}
\mathrm{Tail}_{\nu(b,U(N))}(s) &:= \int_{(s,\infty)} \nu(b,U(N)) \\
&\leq \sum_{j=1}^{b} \int_{(s,\infty)} \nu(j,U(N)) \\
&= \int_{(s,\infty)} \mathrm{Tail}_{\mu(U(N),\text{ step } b)}(x)\,dx \\
&\leq b(4/3)\int_{(s,\infty)} e^{-x^2/8b^2}\,dx \\
&= b(4/3)\int_{(s,\infty)} e^{-x^2/16b^2}e^{-x^2/16b^2}\,dx \\
&\leq b(4/3)e^{-s^2/16b^2}\int_{(0,\infty)} e^{-x^2/16b^2}\,dx
\end{aligned}$$

(now set $t := x/4b$)

$$\begin{aligned}
&= b(4/3)e^{-s^2/16b^2}4b\int_{(0,\infty)} e^{-t^2}\,dt \\
&= b(4/3)e^{-s^2/16b^2}2b\int_{\mathbb{R}} e^{-t^2}\,dt \\
&= b^2(8/3)\,\mathrm{Sqrt}(\pi)e^{-s^2/16b^2}. \quad \text{QED}
\end{aligned}$$

7.11.5. We now give some large N limit results whose proofs do not depend on the previous Fredholm theory.

Proposition 7.11.6. *Let $a \geq 0$ and $b \geq 0$ be integers, $r := a+b+2$. Consider the integer vector*

$$[-a, b+1] := (-a, 1-a, 2-a, \ldots, 0, 1, \ldots, b+1) \text{ in } \mathbb{Z}^{a+b+2},$$

and the associated multi-eigenvalue location measure

$$\nu([-a,b+1], U(N)) \text{ on } \mathbb{R}^r.$$

For any Borel measurable function f on $\mathbb{R}^r$ of polynomial growth, the limit

$$\lim_{N\to\infty}\int_{\mathbb{R}^r} f\,d\nu([-a,b+1],U(N))$$

exists. Moreover, if we denote by F the function on $\mathbb{R}^{r-1}$ defined by the Lebesgue integral

$$F(x) := \int_{[0,x(0)]} f(L(x) - t\mathbb{1})\,dt := |x(0)|\int_{[0,1]} f(L(x) - tx(0)\mathbb{1})\,dt,$$

then this limit is equal to $\int_{\mathbb{R}^{r-1}} F\,d\mu(\text{univ, sep. } 0_{r-1})$.

There exists a Borel probability measure $\nu([-a,b+1])$ on $\mathbb{R}^r$ which has exponential decay of type $(A,B,2)$ and for which we have

$$\lim_{N\to\infty}\int_{\mathbb{R}^r} f\,d\nu([-a,b+1],U(N)) = \int_{\mathbb{R}^r} f\,d\nu([-a,b+1])$$

for every Borel measurable f of polynomial growth. The measure $\nu([-a,b+1])$ is unique with these properties, and is the unique measure with its moments.

PROOF. We adapt the notations of 6.14.12. Given a Borel measurable f on $\mathbb{R}^r$ of polynomial growth, we have the function on $\mathbb{R}^{r-1}$

$$F(x) := \int_{[0,x(0)]} f(L(x) - t\mathbb{1})\,dt := |x(0)| \int_{[0,1]} f(L(x) - tx(0)\mathbb{1})\,dt,$$

which is itself a Borel measurable function of polynomial growth, and we have the identity

$$\int_{\mathbb{R}^r} f\,d\nu([-a,b+1],U(N)) = \int_{\mathbb{R}^{r-1}} F\,d\mu(U(N),\ \text{sep. } 0_{r-1}).$$

Let us admit temporarily the following assertion:

7.11.7. *For any Borel measurable function F of polynomial growth, we have the limit formula*

$$\lim_{N\to\infty} \int_{\mathbb{R}^{r-1}} F\,d\mu(U(N),\ \textit{sep. } 0_{r-1}) = \int_{\mathbb{R}^{r-1}} F\,d\mu(\textit{univ, sep. } 0_{r-1}).$$

Then we have

$$\lim_{N\to\infty} \int_{\mathbb{R}^r} f\,d\nu([-a,b+1],U(N)) = \int_{\mathbb{R}^{r-1}} F\,d\mu(\text{univ, sep. } 0_{r-1}),$$

which proves the first statement. The assignment

$$f \mapsto \int_{\mathbb{R}^{r-1}} F\,d\mu(\text{univ, sep. } 0_{r-1}),$$

restricted to f's which are characteristic functions χ_E of Borel sets in $\mathbb{R}^r$, defines a Borel measure, which we define to be $\nu([-a,b+1])$. All positive Borel functions f of polynomial growth are integrable against this measure, precisely because the same is true for the measure $\mu(\text{univ, sep. } 0_{r-1})$ in virtue of its exponential decay at ∞.

To see that $\nu([-a,b+1])$ has exponential decay of type $(A,B,2)$, we use the general shape of the transformation

$$F(x) := \int_{[0,x(0)]} f(L(x) - t\mathbb{1})\,dt,$$

and the fact that $\mu(\text{univ, sep. } 0_{r-1})$ has exponential decay of this same type.

Fix $s > 0$ real. If for some i in $[-a,b+1]$ the function f is supported in the set

$$E_{i,s} := \{x \text{ in } \mathbb{R}^r \text{ with } |x(i)| > s(a+b+2)\},$$

then the function F vanishes in

$$\{x \text{ in } \mathbb{R}^{r-1} \text{ with } |x(j)| < s \text{ for all } j \text{ in } [-a,b]\}.$$

If $|f| \leq 1$ everywhere, then we have

$$|F(x)| \leq |x(0)|.$$

Taking f to be the characteristic function of $E_{i,s}$, we get

$$\begin{aligned}\nu([-a,b+1])(E_{i,s}) &= \int_{\mathbb{R}^{r-1}} F\,d\mu(U(N),\ \text{sep. } 0_{r-1}) \\ &= \int_{\mathbb{R}^{r-1},\ \text{some } |x(j)|\geq s} F\,d\mu(U(N),\ \text{sep. } 0_{r-1}) \\ &\leq \sum_j \int_{\mathbb{R}^{r-1},|x(j)|\geq s} |x(0)|\,d\mu(U(N),\ \text{sep. } 0_{r-1}).\end{aligned}$$

To estimate these integrals, think of the region $|x(j)| > s$ as the disjoint union over $n \geq 0$ of the regions $\{|x(j)| > s, n \leq |x(0)| < n+1\}$, say $R(j,s,n)$. On $R(j,s,n)$ the integrand is bounded by $n+1$, so we have

$$\begin{aligned}&\int_{\mathbb{R}^{r-1},|x(j)|\geq s} |x(0)|\,d\mu(U(N),\ \text{sep. } 0_{r-1}) \\ &\quad\leq \sum_{n\geq 0}(n+1)\mu(U(N),\ \text{sep. } 0_{r-1})(R(j,s,n)). \\ &\quad= \sum_{n\geq 0, n<s+1} (n+1)\mu(U(N),\ \text{sep. } 0_{r-1})(R(j,s,n)) \\ &\quad\quad+ \sum_{n\geq s+1} (n+1)\mu(U(N),\ \text{sep. } 0_{r-1})(R(j,s,n)).\end{aligned}$$

Now use the fact that $\mu(U(N),\ \text{sep. } 0_{a+b+2})$ has exponential decay of explicit type $(4/3, 1/8, 2)$, by 6.13.2. The first sum is bounded by

$$(s+3)^2\mu(U(N),\ \text{sep. } 0_{r-1})(\{x \text{ with } |x(j)| > s\}) \leq (s+3)^2(4/3)e^{-s^{2/8}}.$$

The second sum is bounded by

$$\begin{aligned}&\sum_{n\geq s+1} (n+1)\mu(U(N),\ \text{sep. } 0_{r-1})(\{x \text{ with } |x(0)| > n\}) \\ &\quad\leq \sum_{n\geq s+1} (n+1)(4/3)e^{-n^{2/8}} \leq (4/3)\int_{x\geq s} (x+2)e^{-x^2}\,dx.\end{aligned}$$

For any $B > 1/8$, there exists an explicitable A such that both of these upper bounds are bounded by Ae^{-Bs^2}. This shows that our limit measure has the asserted decay. It follows by 7.5.7 that our limit measure is determined by its moments, and hence is unique. QED modulo proving 7.11.7.

7.11.8. It remains to prove 7.11.7.

Lemma 7.11.9. *For any integer $r \geq 2$, and any Borel measurable function F on $\mathbb{R}^{r-1}$ of polynomial growth, we have the limit formula*

$$\lim_{N\to\infty} \int_{\mathbb{R}^{r-1}} F\,d\mu(U(N),\ \textit{sep. } 0_{r-1}) = \int_{\mathbb{R}^{r-1}} F\,d\mu(\textit{univ},\ \textit{sep. } 0_{r-1}).$$

PROOF. Because all the measures $\mu(U(N),\ \text{sep. } 0_{r-1})$ and $\mu(\text{univ},\ \text{sep. } 0_{r-1})$ have the **same** tail estimate, given any F of polynomial growth, and given any $\varepsilon > 0$, we can pick a single constant s such that the integral of F over the region $\sum_i |x(i)| > s$ is at most ε for any of these measures. On the compact set $\sum_i |x(i)| \leq s$, our function is bounded. On this set our integrals converge, by 1.2.2. QED for 7.11.9, and with it 7.11.7 and 7.11.6.

7.11.10. The same argument gives

Corollary 7.11.11 to Lemma 7.11.9. *For any integer $r \geq 1$, for any separation vector a in $\mathbb{Z}^r$, and for any Borel measurable F on $\mathbb{R}^r$ of polynomial growth, we have the limit formula*

$$\lim_{N\to\infty} \int_{\mathbb{R}^r} F\, d\mu(U(N),\ sep.\ a) = \int_{\mathbb{R}^r} F\, d\mu(univ,\ sep.\ a).$$

Proposition 7.11.12. *The limit measure $\nu([-a, b+1])$ on $\mathbb{R}^{a+b+2}$ is absolutely continuous with respect to Lebesgue measure.*

PROOF. We must show that if $E \subset \mathbb{R}^{a+b+2}$ is a Borel set of Lebesgue measure zero, then E has $\nu([-a, b+1])$-measure zero. Passing to characteristic functions, it suffices to show that for any nonnegative Borel function f on $\mathbb{R}^{a+b+2}$ for which $\int_{\mathbb{R}^{a+b+2}} f(y)\, dy = 0$, we also have $\int_{\mathbb{R}^{a+b+2}} f\, d\nu([-a, b+1]) = 0$. But this last integral is equal to

$$\int_{\mathbb{R}^{a+b+1}} F\, d\mu(\text{univ, sep. } 0_{a+b+1}),$$

where $F(x)$ is the function on $\mathbb{R}^{a+b+1}$ defined by

$$F(x) = \int_{[0,x(0)]} f(L(x) - t\mathbb{1})\, dt := |x(0)| \int_{[0,1]} f(L(x) - tx(0)\mathbb{1})\, dt.$$

The measure $\mu(\text{univ, sep. } 0_{a+b+1})$ is absolutely continuous with respect to Lebesgue measure on $\mathbb{R}^{a+b+1}$ (by 7.0.13), so it suffices to show that F vanishes almost everywhere for Lebesgue measure. This is equivalent, F being nonnegative, to showing that

$$\int_{\mathbb{R}^{a+b+1}} F(x)\, dx = 0.$$

To see this, notice that we have

$$\begin{aligned}
\int_{\mathbb{R}^{a+b+1}} F(x)\, dx &= \int_{\mathbb{R}^{a+b+1}} \left(\int_{[0,x(0)]} f(L(x) - t\mathbb{1})\, dt \right) dx \\
&= \int_{\mathbb{R}^{a+b+1}, x(0)\geq 0} \left(\int_{[0,x(0)]} f(L(x) - t\mathbb{1})\, dt \right) dx \\
&\quad + \int_{\mathbb{R}^{a+b+1}, x(0)\leq 0} \left(\int_{[0,x(0)]} f(L(x) - t\mathbb{1})\, dt \right) dx.
\end{aligned}$$

In the second term, the inner integral is $\int_{[0,-x(0)]} f(L(x) + t\mathbb{1})\, dt$, so all in all we have

$$\int_{\mathbb{R}^{a+b+1}} F(x)\, dx = I_- + I_+,$$

where

$$I_\pm := \int_{\mathbb{R}^{a+b+1}, x(0)\geq 0} \left(\int_{[0,x(0)]} f(L(x) \pm t\mathbb{1})\, dt \right) dx.$$

We now rewrite $I_\pm$ as

$$\int_{\mathbb{R}^{a+b+1}\times\mathbb{R}, x(0)\geq t\geq 0} f(L(x) \pm t\mathbb{1})\, dt\, dx.$$

Because f is nonnegative, we have

$$0 \leq I_{\pm} \leq \int_{\mathbb{R}^{a+b+1}\times\mathbb{R}} f(L(x) \pm t\mathbb{1})\, dt\, dx.$$

The key observation now is that for either choice of sign $\pm$, the map

$$\mathbb{R}^{a+b+1} \times \mathbb{R} \to \mathbb{R}^{a+b+2},$$
$$(x, t) \mapsto L(x) \pm t\mathbb{1},$$

is a linear isomorphism. Thus the direct image of Haar measure $dt\, dx$ on $\mathbb{R}^{a+b+1}\times\mathbb{R}$ by this map is a Haar measure dy on $\mathbb{R}^{a+b+2}$, and we tautologically have

$$\int_{\mathbb{R}^{a+b+1}\times\mathbb{R}} f(L(x) \pm t\mathbb{1})\, dt\, dx = \int_{\mathbb{R}^{a+b+2}} f(y)\, dy.$$

But our hypothesis is precisely that this last integral vanishes. Therefore each of $I_{\pm} = 0$, and so $\int_{\mathbb{R}^{a+b+1}} F(x)\, dx = 0$. QED

Proposition 7.11.13. *Given an integer $r \geq 1$ and a vector c in $\mathbb{Z}^r$ with*

$$c(1) < c(2) < \cdots < c(r),$$

consider the multi-eigenvalue location measure $\nu(c, U(N))$ on $\mathbb{R}^r$. For any Borel measurable function f on $\mathbb{R}^r$ of polynomial growth, the limit

$$\lim_{N\to\infty} \int_{\mathbb{R}^r} f\, d\nu(c, U(N))$$

exists. There exists a Borel probability measure $\nu(c)$ on $\mathbb{R}^r$ which has exponential decay of type $(A, B, 2)$ and for which we have

$$\lim_{N\to\infty} \int_{\mathbb{R}^r} f\, d\nu(c, U(N)) = \int_{\mathbb{R}^r} f\, d\nu(c)$$

for every Borel measurable f of polynomial growth. The measure $\nu(c)$ is unique with these properties, and is the unique measure with its moments. Moreover, the measure $\nu(c)$ on $\mathbb{R}^r$ is absolutely continuous with respect to Lebesgue measure.

PROOF. This is immediate from the previous result. Indeed, if we pick integers a, b both ≥ 0 such that $-a \leq c(1), c(r) \leq b+1$, then the measures $\nu(c, U(N))$ on $\mathbb{R}^r$ are the direct images of the measures $\nu([-a, b+1], U(N))$ on $\mathbb{R}^{a+b+2}$ by the partial coordinate projection $\mathrm{pr}[c] : \mathbb{R}^{a+b+2} \to \mathbb{R}^r$. We define $\nu(c)$ to be $\mathrm{pr}[c]_*\nu([-a, b+1])$. Then by 7.5.12, all the measures $\nu(c, U(N))$ and $\nu(c)$ are of the same exponential type $(A, B, 2)$, because this was the case for $\nu([-a, b+1], U(N))$.

With this definition of $\nu(c)$, the convergence of integrals for f a Borel function of polynomial growth on $\mathbb{R}^r$ is immediately reduced to the previously treated case when c is $[-a, b+1]$, since for such an f, the composite $f \circ \mathrm{pr}[c]$ is such a function on $\mathbb{R}^{a+b+2}$. The uniqueness follows exactly as in 7.11.6 above.

That $\nu(c)$ is absolutely continuous with respect to Lebesgue measure on $\mathbb{R}^r$ follows from the formula

$$v(c) = \mathrm{pr}[c]_*\nu([-a, b+1]),$$

together with the absolute continuity, proven above, of $\nu([-a, b+1])$ with respect to Lebesgue measure dy on $\mathbb{R}^{a+b+2}$. So $\mathrm{pr}[c]_*\nu([-a, b+1])$ is absolutely continuous with respect to $\mathrm{pr}[c]_*(dy)$. But $\mathrm{pr}[c]_*(dy)$ is absolutely continuous with respect to Lebesgue measure on $\mathbb{R}^r$, because up to an $(\mathbb{R}_{>0})^\times$ factor, Lebesgue measure on

$$\mathbb{R}^{a+b+2} \cong \mathbb{R}^r \times \mathbb{R}^{a+b+2-r}$$

is the product of Lebesgue measure on the factors, so $\mathrm{pr}[c]^{-1}$(any Borel set of Lebesgue measure zero in $\mathbb{R}^r$) has Lebesgue measure zero in $\mathbb{R}^{a+b+2}$. QED

Proposition 7.11.14. *Given an integer $r \geq 2$ and a vector c in $\mathbb{Z}^r$ with*

$$c(1) < c(2) < \cdots < c(r),$$

consider the static spacing measure $\xi(c, U(N))$ on $\mathbb{R}^{r-1}$.

For any Borel measurable function f on $\mathbb{R}^{r-1}$ of polynomial growth, the limit

$$\lim_{N\to\infty} \int_{\mathbb{R}^{r-1}} f\, d\xi(c, U(N))$$

exists. There exists a Borel probability measure $\xi(c)$ on $\mathbb{R}^{r-1}$ which has exponential decay of type $(A, B, 2)$ and for which we have

$$\lim_{N\to\infty} \int_{\mathbb{R}^{r-1}} f\, d\xi(c, U(N)) = \int_{\mathbb{R}^r} f\, d\xi(c)$$

for every Borel measurable f of polynomial growth. The measure $\xi(c)$ is unique with these properties, and is the unique measure with its moments.

PROOF. We know that $\xi(c, U(N))$ is obtained from $\nu(c, U(N))$ by taking direct image by the map $\mathrm{SuccSub} : \mathbb{R}^r \to \mathbb{R}^{r-1}$, cf. 6.4.9. We define $\xi(c)$ to be $\mathrm{SuccSub}_* \nu(c)$. The proof is now identical to that of the result above. QED

7.11.15. With these limit results, we can now take the large N limits of various relations proven in 6.14 for finite N.

Proposition 7.11.16 (large N limit of 6.14.16). *The measure*

$$x\mu(\textit{univ, sep. } 0)$$

on $\mathbb{R}$ is equal to the static spacing measure $\xi((0,1))$ on $\mathbb{R}$.

Proposition 7.11.17 (large N limit of 6.14.18). *For any $r \geq 1$, the measure $x\mu(\textit{univ, step } r)$ on $\mathbb{R}$ is the sum of the static spacing measures $\xi((a-r, a))$, $a = 1, \ldots, r$.*

Proposition 7.11.18 (large N limit of 6.14.21). *Let $r \geq 1$ be an integer, and c in $\mathbb{Z}^r$ an offset vector,*

$$0 < c(1) < c(2) < \cdots < c(r).$$

Denote by $0 \oplus c$ in $\mathbb{Z}^{r+1}$ the vector $(0, c(1), \ldots, c(r))$, and define

$$c(0) := 0.$$

For each $i = 1, \ldots, r$, define

$$d(i) := c(i) - c(i-1).$$

Then for each $i = 1, \ldots, r$ the measure $x(i)\mu(\textit{univ, offset } c)$ on $\mathbb{R}^r$ is the sum of $d(i)$ static spacing measures

$$x(i)\mu(\textit{univ, offset } c) = \sum_{j=c(i-1)}^{c(i)-1} \xi(0 \oplus c - j\mathbb{1}_{r+1}).$$

Corollary 7.11.18.1. *Notations and hypotheses as in* 7.11.18, *suppose in addition that* $c(1) = 1$. *Then we have*

$$x(1)\mu(univ,\ offset\ c) = \xi(0 \oplus c).$$

7.11.19. We can also take the large N limit of 6.16.7.

Proposition 7.11.20 (large N limit of 6.16.7). *For any integer* $r \geq 1$, *and for any offset vector* c *in* $\mathbb{Z}^r$, *the offset spacing measure*

$$\mathrm{Off}\,\mu(univ,\ offsets\ c) := \mathrm{Off}_*(\mu(univ,\ offsets\ c))$$

on $\mathbb{R}^r$ *is related to the measures* $\nu(c - k\mathbb{1})$, $0 \leq k \leq c(1) - 1$, *as follows. For every Borel measurable function* g *on* $\mathbb{R}^r$ *of polynomial growth, we have the function*

$$G(x) := \int_{[0,x(1)]} g(x - t\mathbb{1})\,dt,$$

which is itself a Borel measurable function of polynomial growth, and we have the identity

$$\int_{\mathbb{R}^r} G\,d\,\mathrm{Off}\,\mu(univ,\ offsets\ c) = \sum_{0 \leq k \leq c(1)-1} \int_{\mathbb{R}^r} g\,d\nu(c - k\mathbb{1}).$$

PROOF. By 7.11.13, we have

$$\sum_{0 \leq k \leq c(1)-1} \int_{\mathbb{R}^r} g\,d\nu(c - k\mathbb{1}) = \lim_{N \to \infty} \sum_{0 \leq k \leq c(1)-1} \int_{\mathbb{R}^r} g\,d\nu(c - k\mathbb{1}, U(N)).$$

By 7.11.11 we have

$$\begin{aligned}
\int_{\mathbb{R}^r} G\,d\,\mathrm{Off}\,\mu(\text{univ, offsets } c) &= \int_{\mathbb{R}^r} (G \circ \mathrm{Off})d\mu(\text{univ, offsets } c) \\
&= \lim_{N \to \infty} \int_{\mathbb{R}^r} (G \circ \mathrm{Off})d\mu(U(N),\ \text{offsets } c) \\
&= \lim_{N \to \infty} \int_{\mathbb{R}^r} G\,d\,\mathrm{Off}\,\mu(U(N),\ \text{offsets } c).
\end{aligned}$$

Thus the asserted result is indeed the limit of its finite N analogue 6.16.7. QED

CHAPTER 8

Several Variables

8.0. Fredholm determinants in several variables and their measure-theoretic meaning (cf. [T-W])

8.0.1. Suppose we are given a topological space J, a Borel measure μ on J of finite total mass, and a bounded measurable function $F(x,y)$ on $J \times J$. [In practice J will be a closed interval in $\mathbb{R}$ and μ will be a constant multiple of usual Lebesgue measure dx.] We denote by F, or F_J, the integral operator

$$f \mapsto \left(\text{the function } x \mapsto \int_J F(x,y)f(y)\,d\mu(y)\right)$$

on $L_2(J, d\mu)$ with kernel $F(x,y)$. The operator $1 + F_J$ on $L_2(J, d\mu)$ has a well-defined Fredholm determinant, given by the explicit formula

$$\det(1 + F_J) = 1 + \sum_{n \geq 1} (1/n!) \int_{J^n} \det_{n \times n}(F(x(i), x(j))) \prod_i d\mu(i). \tag{8.0.1.1}$$

8.0.2. Fix an integer $r \geq 1$, and r disjoint (possibly empty) Borel measurable subsets $I_1, I_2, \ldots, I_r$ of J. For $j = 1, \ldots, r$, we denote by $x \mapsto \chi_j(x)$ the characteristic function of I_j, viewed as a function on J. Given r complex numbers $T_1, \ldots, T_r$, we consider the bounded measurable function $\sum_j T_j \chi_j(y) F(x,y)$ on $J \times J$, and the corresponding integral operator $\sum_j T_j \chi_j(y) F_J$ on $L_2(J, d\mu)$.

Lemma 8.0.3 (cf. [**T-W**]). *The Fredholm determinant*

$$\det\left(1 + \sum_j T_j \chi_j(y) F_J\right)$$

is an entire function of $(T_1, \ldots, T_r)$, *given (in usual multinomial notation) by the power series*

$$\det\left(1 + \sum_j T_j \chi_j(y) F_J\right)$$
$$= 1 + \sum_{n \geq 0, n \neq 0 \text{ in } \mathbb{Z}^r} (T^n/n!) \int_{\prod_k (I_k)^{n(k)}} \det_{\Sigma(n) \times \Sigma(n)}(F(x(i), x(j))) \prod_i d\mu(i)$$

PROOF. That this series is an entire function of T results from the Hadamard estimate [**W-W**, page 213, 11.1]

$$|\det_{\Sigma(n) \times \Sigma(n)}(F(x(i), x(j)))| \leq \Sigma(n)^{\Sigma(n)/2} (\operatorname{Sup}_{J \times J}(|F(x,y)|))^{\Sigma(n)}$$

and Stirling's formula.

To prove the identity, we first reduce to treating the case when all the I_j's are nonempty.

If all the I_j are empty, we are asserting that $1 = 1$, which is correct. In general, if say $I_1, I_2, \ldots, I_k$ are nonempty and I_j is empty for $j > k$, let $\mathcal{I}_{\leq k}$ denote the k-tuple $(I_1, I_2, \ldots, I_k)$. Then both

$$\det\left(1 + \sum_j T_j \chi_j(y) F_J\right)$$

and

$$1 + \sum_{n \geq 0, n \neq 0 \text{ in } \mathbb{Z}^r} (T^n/n!) \int_{\prod_k (I_k)^{n(k)}} \det_{\Sigma(n)\times\Sigma(n)}(F(x(i), x(j))) \prod_i d\mu(i)$$

are functions of $T_1, \ldots, T_k$ alone, and these functions are

$$\det\left(1 + \sum_{j \leq k} T_j \chi_j(y) F_J\right)$$

and

$$1 + \sum_{n \geq 0, n \neq 0 \text{ in } \mathbb{Z}^k} (T^n/n!) \int_{\prod_{j\leq k} (I_j)^{n(j)}} \det_{\Sigma(n)\times\Sigma(n)}(F(x(i), x(j))) \prod_i d\mu(i)$$

respectively. So it suffices to treat the case, possibly with lower r, when all the I_j are nonempty.

Thus we now suppose all the I_j are nonempty. We use the general formula for a Fredholm determinant recalled above, which gives

$$\det\left(1 + \sum_j T_j \chi_j(y) F_J\right)$$
$$= 1 + \sum_{l \geq 1} (1/l!) \int_{J^l} \det_{l\times l}\left(\sum_k T_k \chi_k(x(j)) F(x(i), x(j))\right) \prod_i d\mu(i).$$

So what we must show is that for each integer $l \geq 1$, we have the identity

$$(1/l!) \int_{J^l} \det_{l\times l}\left(\sum_k T_k \chi_k(x(j)) F(x(i), x(j))\right) \prod_i d\mu(i)$$
$$= \sum_{n \geq 0 \text{ in } \mathbb{Z}^r, \Sigma(n) = l} (T^n/n!) \int_{\prod_k (I_k)^{n(k)}} \det_{\Sigma(n)\times\Sigma(n)}(F(x(i), x(j))) \prod_i d\mu(i).$$

To see this, we compute the value of

$$\det_{l\times l}\left(\sum_k T_k \chi_k(x(j)) F(x(i), x(j))\right)$$

at a point x in J^l. We distinguish cases, according to which coordinates $x(j)$ of the point x lie in which of the sets I_k, and which lie in none of the I_k.

First of all, if some $x(j_0)$ lies in none of the I_k, then for every i we have

$$\sum_k T_k \chi_k(x(j_0)) F(x(i), x(j_0)) = 0,$$

and hence

$$\det_{l\times l}\left(\sum_k T_k\chi_k(x(j))F(x(i),x(j))\right)=0,$$

because the j_0'th column is identically zero.

It remains to treat those points x where every component $x(j)$ lies in some (necessarily unique, by disjointness) I_k. Given such a point, for each k we denote by J_k the set of those indices j for which $x(j)$ lies in I_k, and for each j we denote by $k(j)$ the unique index such that $x(j)$ lies in $I_{k(j)}$. The sets $\{J_k\}$ attached to x form a partition of the set $\{1,2,\dots,l\}$, and the cardinalities

$$n(k):=\mathrm{Card}(J_k)$$

form a partition of l.

At x, we have

$$\sum_k T_k\chi_k(x(j))F(x(i),x(j))=T_{k(j)}F(x(i),x(j)),$$

and hence, at x we have

$$\begin{aligned}\det_{l\times l}\left(\sum_k T_k\chi_k(x(j))F(x(i),x(j))\right)&=\det_{l\times l}(T_{k(j)}F(x(i),x(j)))\\&=\left(\prod_j T_{k(j)}\right)\det_{l\times l}(F(x(i),x(j)))\\&=\left(\prod_k (T_k)^{\mathrm{Card}(J_k)}\right)\det_{l\times l}(F(x(i),x(j)))\\&=T^n\det_{l\times l}(F(x(i),x(j))),\end{aligned}$$

for n in $\mathbb{Z}^r$ the vector of cardinalities $n(k)=\mathrm{Card}(J_k)$.

Thus we obtain

$$\begin{aligned}&(1/l!)\int_{J^l}\det_{l\times l}\left(\sum_k T_k\chi_k(x(j))F(x(i),x(j))\right)\prod_i d\mu(i)\\&=(1/l!)\sum_{\text{partitions }\{J_k\}_k}T^n\int_{\prod_k(I_k)^{n(k)}}\det_{l\times l}(F(x(i),x(j)))\prod_i d\mu(i),\end{aligned}$$

the sum over all partitions $\{J_k\}$ of the set $\{1,2,\dots,l\}$, and in which we have written $\int_{\prod_k(I_k)^{n(k)}}$ to denote integration over the subset of J^l consisting of those x for which $x(j)$ lies in I_k for j in J_k. However, the integrand $\det_{l\times l}(F(x(i),x(j)))$ is a symmetric function of its variables $x(1),\dots,x(l)$, so by symmetry we may group together all the $l!/n!$ partitions $\{J_k\}$ which give rise to the same cardinality vector n. Thus we obtain

$$\begin{aligned}&(1/l!)\int_{J^l}\det_{l\times l}\left(\sum_k T_k\chi_k(x(j))F(x(i),x(j))\right)\prod_i d\mu(i)\\&=\sum_{n\geq 0\text{ in }\mathbb{Z}^r,\Sigma(n)=l}(T^n/n!)\int_{\prod_k(I_k)^{n(k)}}\det_{l\times l}(F(x(i),x(j)))\prod_i d\mu(i).\end{aligned}$$ QED

Corollary 8.0.4. *For any Borel measurable subset $J_0 \subset J$ which contains all the I_j's, in particular for $J_0 := \bigsqcup_j I_j$, we have an equality of Fredholm determinants*

$$\det\left(1 + \sum_j T_j \chi_j(y) F_J\right) = \det(1 + \sum_j T_j \chi_j(y) F_{J_0}),$$

with the convention that the empty determinant is 1.

PROOF. This is obvious from the explicit formula. QED

Remark 8.0.5. More conceptually, if we take $J_0 := \bigsqcup_j I_j$ and denote by I_0 the complement in J of J_0, then we have an orthogonal direct sum decomposition

$$L_2(J, d\mu) = L_2(J_0, d\mu) \oplus L_2(I_0, d\mu),$$

in which $\sum_j T_j \chi_j(y) F_J$ kills $L_2(I_0, d\mu)$. Thus the "2×2 matrix" of F_J in this decomposition is lower triangular, of the form

$$\begin{pmatrix} F_{J_0} & 0 \\ * & 0 \end{pmatrix}.$$

So the result **should** follow from a reasonable spectral interpretation of these determinants.

8.1. Measure-theoretic application to the $G(N)$

8.1.1. We fix an integer $N \geq 2$, and take for $G(N)$ any of $U(N)$, $SO(2N+1), USp(2N), SO(2N), O_-(2N+2), O_-(2N+1)$. We have the associated kernel $L_N(x, y)$ attached to $G(N)$ in 5.2 and 6.4.

8.1.2. Fix an integer $r \geq 1$, and an ordered r-tuple $\mathcal{I}$ of disjoint (possibly empty) measurable subsets $I_1, \ldots, I_r$ in the space

$$[-\pi, \pi), \quad \text{for } U(N),$$
$$[0, \pi], \quad \text{for any other } G(N).$$

Given $n \geq 0$ in $\mathbb{Z}^r$, recall from 6.3.7, 6.3.15, and 6.4.6 that we denote by

$$\mathrm{Eigen}(n, \mathcal{I}, G(N)) \subset G(N)$$

the set of those elements A in $G(N)$ such that for each $j = 1, \ldots, r$, exactly $n(j)$ of its angles

$$0 \leq \varphi(1) \leq \varphi(2) \leq \cdots \leq \varphi(N) < 2\pi \quad \text{if } G(N) = U(N),$$
$$0 \leq \varphi(1) \leq \varphi(2) \leq \cdots \leq \varphi(N) \leq \pi \quad \text{for the other } G(N),$$

lie in the prescribed set I_j. If some I_k is empty, then $\mathrm{Eigen}(n, \mathcal{I}, G(N))$ is empty unless $n(k) = 0$.

8.1.3. It is convenient to be able to speak of $\mathrm{Eigen}(n, \mathcal{I}, G(N))$, defined by the above recipe, for any vector n in $\mathbb{Z}^r$. This set is empty unless $n \geq 0$. We have

$$G(N) = \coprod_n \mathrm{Eigen}(n, \mathcal{I}, G(N)),$$

the elements of $G(N)$ apportioned according to how many of their angles lie in the various I_k's. For example, in the extreme case that all the I_k are empty, $G(N) = \mathrm{Eigen}(0, \mathcal{I}, G(N))$, and for $n \neq 0$ we have $\mathrm{Eigen}(n, \mathcal{I}, G(N)) = \varnothing$. The set $\mathrm{Eigen}(n, \mathcal{I}, G(N))$ is empty if $\Sigma(n) > N$. We denote by

$$\mathrm{eigen}(n, \mathcal{I}, G(N)) := \text{ Haar measure of } \mathrm{Eigen}(n, \mathcal{I}, G(N)),$$

the Haar measure normalized to give $G(N)$ total mass one.

Lemma 8.1.4 ([**T-W**]). *For $N \geq 2$, and $G(N)$ any of $U(N), SO(2N+1)$, $USp(2N), SO(2N), O_-(2N+2), O_-(2N+1)$, we have the identity*

$$\det\left(1 + \sum_j T_j \chi_j(y) L_N | L_2\left(\coprod_j I_j, dx/\sigma\pi\right)\right)$$
$$= \det\left(1 + \sum_j T_j \chi_j(y) L_N | L_2([0,, \sigma\pi], dx/\sigma\pi)\right)$$
$$= \sum_{n \geq 0 \text{ in } \mathbb{Z}^r} (1+T)^n \,\mathrm{eigen}(n, \mathcal{I}, G(N)).$$

PROOF. That the first two Fredholm determinants coincide has already been remarked above in 8.0.4. That the last two agree is 6.3.7 for $U(N)$, 6.3.15 and 6.4.6 for the other $G(N)$. QED

8.2. Several variable Fredholm determinants for the $\sin(\pi x)/\pi x$ kernel and its $\pm$ variants

8.2.1. Fix an integer $r \geq 1$. Given two vectors s, t in $(\mathbb{R}_{\geq 0})^r$ which are "intertwined" in the sense that

$$0 \leq s(1) \leq t(1) \leq s(2) \leq t(2) \leq \cdots \leq s(r) \leq t(r),$$

we denote by $\mathcal{I}(s,t)$ the r-tuple of disjoint (and some possibly empty, if r is at least 2) intervals in $[0, t(r)]$ given by

$$I_1 := [s(1), t(1)],$$
$$I_j := (s(j), t(j)], \quad j = 2, \ldots, r.$$

8.2.2. Recall the kernels

$$K(x,y) := \sin(\pi(x-y))/\pi(x-y),$$
$$K_+(x,y) := K(x,y) + K(-x,y),$$
$$K_-(x,y) := K(x,y) - K(-x,y).$$

8.2.3. We define several variable Fredholm determinants

$$E(T,s,t) := \det\left(1+\sum_j T_j\chi_j(y)K(x,y)|L_2([0,t(r)],dx)\right)$$

and

$$E_\pm(T,s,t) := \det\left(1+\sum_j T_j\chi_j(y)K_\pm(x,y)|L_2([0,t(r)],dx)\right).$$

8.2.4. For fixed s,t, we know (by 8.0.3) that these are entire functions of T. For ease of later reference, we denote by $E_n(s,t)$ and by $E_{\pm,n}(s,t)$ their expansion coefficients around the point all $T_i = -1$:

$$E(T,s,t) = \sum_{n\geq 0 \text{ in } \mathbb{Z}^r} (1+T)^n E_n(s,t),$$

$$E_\pm(T,s,t) = \sum_{n\geq 0 \text{ in } \mathbb{Z}^r} (1+T)^n E_{\pm,n}(s,t).$$

Lemma 8.2.5. *For each $r \geq 1$, $E(T,s,t)$ and $E_\pm(T,s,t)$ are (the restrictions to $\mathbb{C}^r \times$ (the subset of $(\mathbb{R}_{\geq 0})^{2r}$ consisting of pairs of intertwined vectors) of) entire functions on $\mathbb{C}^{3r}$.*

PROOF. We will do the case of $E(T,s,t)$, the $E_\pm$ case being entirely similar. Expand $E(T,s,t)$ as a series in T, say

$$E(T,s,t) = 1 + \sum_{n\geq 0 \text{ in } \mathbb{Z}^r, n\neq 0} B_n(s,t)T^n.$$

The coefficient $B_n(s,t)$ is, by 8.0.3, the integral

$$(1/n!)\int_{\prod_k (I_k)^{n(k)}} \det_{\Sigma(n)\times\Sigma(n)}(K(x(i),x(j)))\prod_i dx(i),$$

where the variables are taken in the various I_k by some choice of an r-tuple of disjoint subsets J_i of $\{1,\ldots,\Sigma(n)\}$ such that $\mathrm{Card}(J_i) = n(i)$ for each $i = 1,\ldots,r$. To parameterize the intervals I_k, we make the following change of variable in this integral:

$$x(k) = s(l_k) + (t(l_k) - s(l_k))z(k) \quad \text{if } k \text{ lies in } J_{l_k}.$$

In terms of these variables, $B_n(s,t)$ is the integral

$$(1/n!)\prod_i (t(i)-s(i))^{n(i)} \int_{[0,1]^{\Sigma(n)}} V(n,s,t,z)\prod_i dz(i),$$

where $V(n,s,t,z)$ is the entire function of (s,t,z) defined by

$$\begin{aligned} &V(n,s,t,z) \\ &\quad := \det_{\Sigma(n)\times\Sigma(n)}(K(s(l_i)+(t(l_i)-s(l_i))z(i), s(l_j)+(t(l_j)-s(l_j))z(j))). \end{aligned}$$

This makes obvious that each individual coefficient $B_n(s,t)$ is an entire function of (s,t). It remains to show that the series

$$1 + \sum_{n\geq 0 \text{ in } \mathbb{Z}^r, n\neq 0} B_n(s,t)T^n$$

converges uniformly on compact subsets of $\mathbb{C}^{3r}$. This amounts to a decent estimate for the coefficients $B_n(s,t)$. As recalled in the proof of 7.0.9, we have the inequality

$$|\sin(\pi x)/\pi x| \leq \exp(\pi|x|) \quad \text{for any } x \text{ in } \mathbb{C}.$$

From this, we get that for z in $[0,1]^{\Sigma(n)}$, we have the estimate

$$\begin{aligned}
&|K(s(l_i)+(t(l_i)-s(l_i))z(i), s(l_j)+(t(l_j)-s(l_j))z(j))| \\
&\quad \leq \exp(\pi|(s(l_i)+(t(l_i)-s(l_i))z(i))-(t(l_j)-s(l_j))z(j))|) \\
&\quad \leq \exp\left(\pi\left(4\sum_i |s(i)| + 2\sum_i |t(i)|\right)\right),
\end{aligned}$$

from which by Hadamard's determinant inequality we get

$$|V(n,s,t,z)| \leq \Sigma(n)^{\Sigma(n)/2} \exp\left(\pi\left(4\sum_i |s(i)| + 2\sum_i |t(i)|\right)\right)^{\Sigma(n)}.$$

Thus we get

$$\begin{aligned}
&|B_n(s,t)T^n| \\
&\quad \leq (\Sigma(n)^{\Sigma(n)/2}/n!)\prod_i |t(i)-s(i)|^{n(i)} \exp\left(\pi\left(4\sum_i |s(i)| + 2\sum_i |t(i)|\right)\right)^{\Sigma(n)}, \\
&\quad \leq (\Sigma(n)^{\Sigma(n)/2}/n!)A^n,
\end{aligned}$$

For A the r-tuple $A(i) = |T_i|\,|t(i)-s(i)|\exp(\pi(4\sum_j |s(j)| + 2\sum_j |t(j)|))$. So it remains only to observe that, by Stirling's formula, the series

$$\sum_{n\geq 0 \text{ in } \mathbb{Z}^r} (\Sigma(n)^{\Sigma(n)/2}/n!)A^n$$

is an entire function of A in $\mathbb{C}^r$. QED

8.3. Large N scaling limits

8.3.1. Fix one of the series $U(N)$, $SO(2N+1)$, $USp(2N)$, $SO(2N)$, $O_-(2N+2), O_-(2N+1)$. Fix also an integer $r \geq 1$, and two vectors s,t in $(\mathbb{R}_{\geq 0})^r$ which are "intertwined" in the sense of 8.2.1. For $N \gg 0$, we have $\sigma\pi t(r)/(N+\lambda) < \sigma\pi$, so the intertwined vectors $\sigma\pi s/(N+\lambda), \sigma\pi t/(N+\lambda)$ give us an r-tuple of disjoint intervals in $[0,\sigma\pi]$,

$$\begin{aligned}
\mathcal{I}(s,t,G(N)) &:= \mathcal{I}(\sigma\pi s/(N+\lambda), \sigma\pi t/(N+\lambda)), \\
I_1 &= [\sigma\pi s(1)/(N+\lambda), \sigma\pi t(1)/(N+\lambda)], \\
I_j &= (\sigma\pi s(j)/(N+\lambda), \sigma\pi t(j)/(N+\lambda)] \quad \text{for } j = 2,\ldots,r.
\end{aligned}$$

8.3.2. We form, using the L_N kernel appropriate to $G(N)$,

$G(N)$	$L_N(x,y)$
$U(N)$	$\sum_{n=0}^{N-1} e^{in(x-y)}$ $= S_N(x-y)e^{i(N-1)(x-y)/2}$
other $G(N)$, i.e.,	$(\sigma/2)[S_{\rho N+\tau}(x-y)+\varepsilon S_{\rho N+\tau}(x+y)]$, i.e.,
$SO(2N+1)$	$(1/2)(S_{2N}(x-y)-S_{2N}(x+y))$
$USp(2N)$ or $O_-(2N+2)$	$(1/2)(S_{2N+1}(x-y)-S_{2N+1}(x+y))$
$SO(2N)$	$(1/2)(S_{2N-1}(x-y)+S_{2N-1}(x+y))$
$O_-(2N+1)$	$(1/2)(S_{2N}(x-y)+S_{2N}(x+y))$

the corresponding several variable Fredholm determinant

$$E(T,s,t,G(N)) := \det\left(1+\sum_j T_j\chi_j(y)L_N|L_2([0,\sigma\pi],dx/\sigma\pi)\right).$$

Each such determinant is a polynomial in T whose coefficients, we will show below, are entire functions of (s,t). According to 8.1.4., its expansion around the point all $T_i=-1$ is

$$E(T,s,t,G(N)) = \sum_{n\geq 0 \text{ in } \mathbb{Z}^r, \Sigma(n)\leq N} (1+T)^n \operatorname{eigen}(n,\mathcal{I}(s,t,G(N)),G(N)).$$

According to 8.0.3, its expansion around the origin is

$$E(T,s,t,G(N)) = 1 + \sum_{n\geq 0 \text{ in } \mathbb{Z}^r, 1\leq\Sigma(n)\leq N} B_n(s,t,G(N))T^n,$$

where $B_n(s,t,G(N))$ is given by the integral

$$(1/n!)\int_{\prod_k (I_k)^{n(k)}} \det_{\Sigma(n)\times\Sigma(n)}(L_N(x(i),x(j)))\prod_i (dx(i)/\sigma\pi).$$

Lemma 8.3.3. *For $G(N)$ one of $U(N), SO(2N+1), USp(2N), SO(2N)$, $O_-(2N+2), O_-(2N+1)$ the coefficients $B_n(s,t,G(N))$ are (the restriction to $\{$points (s,t) in $(\mathbb{R}_{\geq 0})^{2r}$ which are intertwined$\}$ of) entire functions, which for (s,t) in $\mathbb{C}^{2r}$ are bounded by*

$$|B_n(s,t,G(N))|$$
$$\leq |1/n!|\left(\prod_i |t(i)-s(i)|^{n(i)}\right)(\Sigma(n)^{\Sigma(n)/2})\left(3\exp\left(3\pi\left(4\sum_i|s(i)|+2\sum_i|t(i)|\right)\right)\right)^{\Sigma(n)}.$$

PROOF. The coefficient $B_n(s,t,G(N))$ is the integral

$$(1/n!)\int_{\prod_k (I_k)^{n(k)}} \det_{\Sigma(n)\times\Sigma(n)}(L_N(x(i),x(j)))\prod_i (dx(i)/\sigma\pi),$$

where

$$I_1 = [\sigma\pi s(1)/(N+\lambda), \sigma\pi t(1)/(N+\lambda)],$$
$$I_j = (\sigma\pi s(1)/(N+\lambda), \sigma\pi t(1)/(N+\lambda)] \quad \text{for } j=2,\ldots,r,$$

and where the variables are taken in the various I_k by some choice of an r-tuple of disjoint subsets J_i of $\{1,\ldots,\Sigma(n)\}$ such that $\mathrm{Card}(J_i)=n(i)$ for each $i=1,\ldots,r$. Let us temporarily define

$$A=\sigma\pi/(N+\lambda),\qquad B=A/\sigma\pi=1/(N+\lambda).$$

To parameterize the intervals I_k, make the following change of variable in this integral:

$$x(k)=A(s(l_k)+(t(l_k)-s(l_k))z(k))\quad\text{if } k \text{ lies in } J_{l_k},$$

i.e.,

$$x(k)/\sigma\pi=B(s(l_k)+(t(l_k)-s(l_k))z(k))\quad\text{if } k \text{ lies in } J_{l_k}.$$

In terms of these variables, $B_n(s,t,G(N))$ is the integral

$$(1/n!)\prod_i(t(i)-s(i))^{n(i)}\int_{[0,1]^{\Sigma(n)}}V(n,s,t,z,G(N))\prod_i dz(i),$$

where $V(n,s,t,z,G(N))$ is the entire function of (s,t,z) defined by

$$V(n,s,t,z,G(N))$$
$$:=\det_{\Sigma(n)\times\Sigma(n)}(BL_N(A(s(l_i)+(t(l_i)-s(l_i))z(i)),A(s(l_j)+(t(l_j)-s(l_j))z(j)))).$$

[Strictly speaking, this formula for $B_n(s,t,G(N))$ is a priori correct only if $s(i)<t(i)$ for each i. However, in the case $s(i)=t(i)$ for some i, it is still correct, since both

$$(1/n!)\int_{\prod_k(I_k)^{n(k)}}\det_{\Sigma(n)\times\Sigma(n)}(L_N(x(i),x(j)))\prod_i(dx(i)/\sigma\pi)$$

and

$$(1/n!)\prod_i(t(i)-s(i))^{n(i)}\int_{[0,1]^{\Sigma(n)}}V(n,s,t,z,G(N))\prod_i dz(i)$$

vanish if $n(i)>0$, while if $n(i)=0$ the interval I_i is irrelevant to both.]

By the Hadamard determinant inequality, it remains to show that for z in $[0,1]^{\Sigma(n)}$, we have

$$|BL_N(A(s(l_i)+(t(l_i)-s(l_i))z(i)),A(s(l_j)+(t(l_j)-s(l_j))z(j)))|$$
$$\le 3\exp\left(3\pi\left(4\sum_i|s(i)|+2\sum_i|t(i)|\right)\right).$$

To see this, recall from 7.4.5 that for x in $\mathbb{C}$, we have

$$|S_N(x)|\le N\exp(N|x|/2).$$

In view of the formulas defining L_N in terms of $S_{\rho N+\tau}$, we have

$$|L_N(x,y)|\le(2N+1)\exp((\rho N+1)(|x|+|y|))\quad\text{for } (x,y) \text{ in } \mathbb{C}^2,$$

and hence

$$\begin{aligned}|BL_N(Ax,Ay)|&=|(1/(N+\lambda))L_N(\sigma\pi x/(N+\lambda),\sigma\pi y/(N+\lambda))|\\&\le((2N+1)/(N+\lambda))\exp(((\rho N+1)/(N+\lambda))(\sigma\pi)(|x|+|y|))\\&\le 3\exp(3\pi(|x|+|y|)).\end{aligned}$$

This gives

$$\begin{aligned}&|BL_N(A(s(l_i)+(t(l_i)-s(l_i))z(i)),A(s(l_j)+(t(l_j)-s(l_j))z(j)))|\\&\quad\le 3\exp\left(3\pi\left(4\sum_i|s(i)|+2\sum_i|t(i)|\right)\right),\end{aligned}$$

as required. QED

8.3.4. In order to be able to state more easily the estimate given in the next result, it is convenient to denote by $\mathbb{D}_0(x)$ the entire function of one complex variable

$$\mathbb{D}_0(x) := 1+\sum_{n\ge1}x^n n^{n/2}/n!,$$

and to names its "tails"

$$\mathbb{D}_k(x) := \sum_{n\ge k}x^n n^{n/2}/n!$$

for each integer $k\ge1$.

Proposition 8.3.5. *Fix one of the series $G(N) = U(N)$, $SO(2N+1)$, $USp(2N)$, $SO(2N)$, $O_-(2N+2)$, or $O(2N+1)$. As $N\to\infty$, the sequence $E(T,s,t,G(N))$ of entire functions of (T,s,t) converges, uniformly on compact subsets of $\mathbb{C}^{3r}$, to*

$$\begin{aligned}&E(T,s,t),\ \text{if } G(N)=U(N),\\&E_+(T,s,t),\ \text{if } G(N)=SO(2N)\text{ or }O_-(2N+1),\\&E_-(T,s,t),\ \text{if } G(N)=USp(2N)\text{ or }SO(2N+1)\text{ or }O_-(2N+2).\end{aligned}$$

Moreover, if we define

$$\|x\| := \sum_i|x(i)|$$

for x in $\mathbb{C}^r$, then each of the entire functions $E(T,s,t,G(N))$, as well as each of the limits $E(T,s,t)$ and $E_\pm(T,s,t)$, has the following property: for each integer $l\ge0$, its power series $\sum_{n\ge0\text{ in }\mathbb{Z}^r}C_n(s,t)T^n$ around $T=0$ satisfies

$$\begin{aligned}&\sum_{n\ge0\text{ in }\mathbb{Z}^r,\Sigma(n)\ge l}|C_n(s,t)T^n|\\&\qquad\le \mathbb{D}_l(r\|T\|(\|t\|+\|s\|)(3\exp(3\pi(4\|s\|+2\|t\|))))\end{aligned}$$

for all (T,s,t) in $\mathbb{C}^{3r}$.

PROOF. The asserted majorization is immediate from 8.3.3, which majorizes the n'th term $|C_n(s,t)T^n|$ by

$$\begin{aligned}&|T^n/n!|\prod_i|t(i)-s(i)|^{n(i)}(\Sigma(n)^{\Sigma(n)/2})\left(3\exp\left(3\pi\left(4\sum_i|s(i)|+2\sum_i|t(i)|\right)\right)\right)^{\Sigma(n)}\\&\quad\le(\|T\|(\|t\|+\|s\|)(3\exp(3\pi(4\|s\|+2\|t\|))))^{\Sigma(n)}(\Sigma(n)^{\Sigma(n)/2})/n!.\end{aligned}$$

Grouping together all terms with a given value k of $\Sigma(n)$, and taking into account the identity (equate coefficients in $\exp(x)^r=\exp(rx)$)

$$r^k/k! = \sum_{n\ge0\text{ in }\mathbb{Z}^r,\Sigma(n)=k}1/n!,$$

gives the asserted estimate. The proof of convergence is entirely analogous to the proof of 7.4.6, and is left to the reader. QED

8.3.6. We denote by $E_n(s,t,G(N))$ the entire functions of (s,t) defined as the expansion coefficients of $E(T,s,t,G(N))$ around the point all $T_i = -1$:

$$E(T,s,t,G(N)) = \sum_{n\geq 0 \text{ in } \mathbb{Z}^r, \Sigma(n)\leq N} (1+T)^n E_n(s,t,G(N)).$$

Thus (by 8.1.4) for s and t in $(\mathbb{R}_{\geq 0})^r$ and intertwined, we have

$$E_n(s,t,G(N)) = \text{eigen}(n, \mathcal{I}(s,t,G(N)), G(N)).$$

Applying the differential operator argument used in the proof of 7.5.1, we obtain the following corollary.

Corollary 8.3.7. *Fix $n \geq 0$ in $\mathbb{Z}^r$. As $N \to \infty$, the sequence of entire functions $E_n(s,t,G(N))$ converges, uniformly on compact subsets of $\mathbb{C}^{2r}$, to*

$$\begin{array}{ll} E_n(s,t) & \textit{if } G(N) = U(N), \\ E_{+,n}(s,t) & \textit{if } G(N) = SO(2N) \textit{ or } O_-(2N+1), \\ E_{-,n}(s,t) & \textit{if } G(N) = USp(2N) \textit{ or } SO(2n+1) \textit{ or } O_-(2N+2). \end{array}$$

In particular, for s and t in $(\mathbb{R}_{\geq 0})^r$ and intertwined, we have the limit formulas

$$E_n(s,t) = \lim_{N\to\infty} \text{eigen}(n, \mathcal{I}(s,t,U(N)), U(N)),$$
$$E_{+,n}(s,t) = \lim_{N\to\infty} \text{eigen}(n, \mathcal{I}(s,t,G(N)), G(N))$$
$$\textit{for } G(N) = SO(2N) \textit{ or } O_-(2N+1),$$

and

$$E_{-,n}(s,t) = \lim_{N\to\infty} \text{eigen}(n, \mathcal{I}(s,t,G(N)), G(N))$$
$$\textit{for } G(N) = USp(2N) \textit{ or } SO(2N+1) \textit{ or } O_-(2N+2).$$

8.3.8. It will be convenient to have a more uniform version of this convergence.

Lemma 8.3.9. *Hypotheses and notations as in Proposition 8.3.5 above, each one of the entire functions $E(T,s,t,G(N))$, as well as each of the entire functions $E(T,s,t)$ and $E_\pm(T,s,t)$, has the following property: for every integer $l \geq 0$, its power series $\sum_{n\geq 0 \text{ in } \mathbb{Z}^r}(1+T)^n A_n(s,t)$ around the point all $T_i = -1$ satisfies*

$$\sum_{n\geq 0 \text{ in } \mathbb{Z}^r, \Sigma(n)\geq l} (1+|T|)^n |A_n(s,t)|$$
$$\leq \mathbb{D}_l(r(2r+\|T\|)(\|t\|+\|s\|)(3\exp(3\pi(4\|s\|+2\|t\|)))),$$

where by $(1+|T|)^n$ we mean $\prod_i (1+|T_i|)^{n(i)}$.

PROOF. Fix s and t, and denote by $f(T) := f(T,s,t)$ one of the entire functions $E(T,s,t,G(N))$, or $E(T,s,t)$ or $E_\pm(T,s,t)$. We consider the two expansions

$$f(T) = \sum_{n\geq 0 \text{ in } \mathbb{Z}^r} C_n T^n = \sum_{n\geq 0 \text{ in } \mathbb{Z}^r} A_n (1+T)^n.$$

We first solve for A_n's in terms of C_n's. We have

$$\begin{aligned}\sum_{n\geq 0 \text{ in } \mathbb{Z}^r} A_n T^n &= \sum_{n\geq 0 \text{ in } \mathbb{Z}^r} C_n (T-1)^n \\ &= \sum_{n\geq 0 \text{ in } \mathbb{Z}^r} C_n \sum_{0\leq m\leq n} (-1)^{n-m} \operatorname{Binom}(n,m) T^m \\ &= \sum_{m\geq 0 \text{ in } \mathbb{Z}^r} T^m \sum_{n\geq m} (-1)^{n-m} \operatorname{Binom}(n,m) C_n,\end{aligned}$$

and hence we obtain

$$A_m = \sum_{n\geq m} (-1)^{n-m} \operatorname{Binom}(n,m) C_n,$$

and hence

$$|A_m| \leq \sum_{n\geq m} \operatorname{Binom}(n,m) |C_n|,$$

so

$$\begin{aligned}&\sum_{m\geq 0 \text{ in } \mathbb{Z}^r, \Sigma(m)\geq l} (1+|T|)^m |A_m| \\ &\leq \sum_{m\geq 0 \text{ in } \mathbb{Z}^r, \Sigma(m)\geq l} (1+|T|)^m \sum_{n\geq m} \operatorname{Binom}(n,m)|C_n| \\ &= \sum_{n\geq 0 \text{ in } \mathbb{Z}^r, \Sigma(n)\geq l} |C_n| \sum_{0\leq m\leq n, \Sigma(m)\geq l} \operatorname{Binom}(n,m)(1+|T|)^m \\ &\leq \sum_{n\geq 0 \text{ in } \mathbb{Z}^r, \Sigma(n)\geq l} |C_n| \sum_{0\leq m\leq n} \operatorname{Binom}(n,m)(1+|T|)^m \\ &= \sum_{n\geq 0 \text{ in } \mathbb{Z}^r, \Sigma(n)\geq l} |C_n|(2+|T|)^n.\end{aligned}$$

According to the previous result, we have

$$\sum_{n\geq 0 \text{ in } \mathbb{Z}^r, \Sigma(n)\geq l} |C_n|\,|T|^n \leq \mathbb{D}_l(r\|T\|(\|t\|+\|s\|)(3\exp(3\pi(4\|s\|+2\|t\|)))).$$

Replacing T by $2+|T|$:= the point with coordinates $2+|T_i|$, and noting that $2+|T|$ has $\|(2+|T|)\| = 2r + \|T\|$, we get the asserted estimate. QED

Proposition 8.3.10. *Fix a nonempty subset W of $\{n \geq 0$ in $\mathbb{Z}^r\}$, and $G(N)$ one of $U(N), SO(2N+1), USp(2N), SO(2N), O_-(2N+2), O_-(2N+1)$.*

1) *The series*

$$E_W(s,t,G(N)) := \sum_{n \text{ in } W} E_n(s,t,G(N))$$

as well as the series

$$E_W(s,t) := \sum_{n \text{ in } W} E_n(s,t)$$

and

$$E_{\pm,W}(s,t) := \sum_{n \text{ in } W} E_{\pm,n}(s,t),$$

converge uniformly on compact subsets of $\mathbb{C}^{2r}$, to entire functions of (s,t).

2) As $N \to \infty$, the sequence of entire functions $E_W(s,t,G(N))$ converges, uniformly on compact subsets of $\mathbb{C}^{2r}$, to

$$\begin{array}{ll} E_W(s,t) & \textit{if } G(N) = U(N), \\ E_{+,W}(s,t) & \textit{if } G(N) = SO(2N) \textit{ or } O_-(2N+1), \\ E_{-,W}(s,t) & \textit{if } G(N) = USp(2N) \textit{ or } SO(2N+1) \textit{ or } O_-(2N+2). \end{array}$$

PROOF. For 1), the uniform convergence on compacta results immediately from the growth estimates of the previous result, with $T = 0$. To prove 2), we use these estimates to reduce, just as in the proof of 7.4.6, to treating the case when W consists of a single n, in which case it results from 8.3.7. QED

8.4. Large N limits of multi-eigenvalue location measures attached to $G(N)$

8.4.1. Exactly as in 6.14.1 and 6.20.1, for $G(N)$ any of $U(N), SO(2N+1), USp(2N), SO(2N), O_-(2N+2), O_-(2N+1)$, and A in $G(N)$, we have its sequence of angles

$$\begin{array}{ll} 0 \le \varphi(1) \le \varphi(2) \le \cdots \le \varphi(N) < 2\pi & \text{if } G(N) = U(N), \\ 0 \le \varphi(1) \le \varphi(2) \le \cdots \le \varphi(N) \le \pi & \text{for the other } G(N) \end{array}$$

and its sequence of **normalized angles**

$$0 \le \vartheta(1) \le \vartheta(2) \le \cdots \le \vartheta(N) \le N + \lambda,$$

defined by

$$\vartheta(n) := (N+\lambda)\varphi(n)/\sigma\pi.$$

Concretely,

$$\begin{array}{ll} \vartheta(n) := N\varphi(n)/2\pi & \text{for } U(N) \\ \vartheta(n) := (N+1/2)\varphi(n)/\pi = (2N+1)\varphi(n)/2\pi & \text{for } SO(2N+1) \text{ or } O_-(2N+1) \\ \vartheta(n) := N\varphi(n)/\pi = 2N\varphi(n)/2\pi & \text{for } USp(2N) \text{ or } SO(2N) \\ \vartheta(n) := (N+1)\varphi(n)/\pi = (2N+2)\varphi(n)/2\pi & \text{for } O_-(2N+2). \end{array}$$

8.4.2. Given an integer $r \ge 1$, an offset vector c in $\mathbb{Z}^r$,

$$0 < c(1) < c(2) < \cdots < c(r),$$

and an integer $N \ge c(r)$, recall that we denote by $\nu(c, G(N))$ the probability measure on $\mathbb{R}^r$ which is the direct image of total mass one Haar measure on $G(N)$ by the map $G(N) \to \mathbb{R}^r$ defined by the normalized angles

$$A \mapsto (\vartheta(c(1)), \ldots, \vartheta(c(r))).$$

Thus

$$\nu(c, G(N)) := (\vartheta(c(1)), \ldots, \vartheta(c(r)))_* \operatorname{Haar}_{G(N)}.$$

The measure $\nu(c, G(N))$ is supported in $(\mathbb{R}_{\ge 0})^r(\text{order})$.

8.4.3. Given a point s in $(\mathbb{R}_{\geq 0})^r$(order), we denote by $\mathcal{J}(s)$ the r-tuple of disjoint intervals

$$I_1 = [0, s(1)],$$
$$I_j = (s(j-1), s(j)] \quad \text{for } j = 2, \ldots, r.$$

For n in $\mathbb{Z}^r$, we define the set $\text{Eigen}(n, s, G(N)) \subset G(N)$ to be the set of elements A in $G(N)$ such that for each $i = 1, \ldots, r$, A has exactly $n(i)$ normalized angles in the interval I_i. We define $\text{eigen}(n, s, G(N))$ to be the normalized Haar measure of the set $\text{Eigen}(n, s, G(N))$. The set $\text{Eigen}(n, s, G(N))$ is empty unless $n \geq 0$.

8.4.4. If $N > s(r)$, the set $\text{Eigen}(n, s, G(N))$ is related to the sets of 8.1.2 as follows. We denote

$$\mathcal{J}(s, G(N)) := \mathcal{J}(\sigma\pi s/(N + \lambda)),$$

an r-tuple of disjoint intervals in $[0, \sigma\pi]$. Then

$$\text{Eigen}(n, s, G(N)) := \text{Eigen}(n, \mathcal{J}(s, G(N)), G(N)),$$
$$\text{eigen}(n, s, G(N)) := \text{eigen}(n, \mathcal{J}(s, G(N)), G(N))$$

for every n in $\mathbb{Z}^r$.

Lemma 8.4.5 (CDF's and tails of $\nu(c, G(N))$). *Suppose given an integer $r \geq 1$, an offset vector c in $\mathbb{Z}^r$, a point s in $(\mathbb{R}_{\geq 0})^r$(order), an integer $N \geq c(r)$, and $G(N)$ one of $U(N), SO(2N+1), USp(2N), SO(2N), O_-(2N+2), O_-(2N+1)$.*

1) *We have the identity*

$$\text{CDF}_{\nu(c,G(N))}(s) = \sum_{n \geq 0 \text{ in } \mathbb{Z}^r, \text{Off}(n) \geq c} \text{eigen}(n, s, G(N)).$$

2) *We have the identity*

$$\text{Tail}_{\nu(c,G(N))}(s) = \sum_{n \geq 0 \text{ in } \mathbb{Z}^r, \text{Off}(n) < c} \text{eigen}(n, s, G(N)).$$

PROOF. The key point is that, because s lies in $(\mathbb{R}_{\geq 0})^r$(order), for each $i = 1, \ldots, r$, the closed interval $[0, s(i)]$ is the disjoint union of the intervals I_k, $1 \leq k \leq i$, in $\mathcal{J}(s)$. Therefore we can describe the set $\text{Eigen}(n, s, G(N))$ as the set of those A in $G(N)$ such that for each $i = 1, \ldots, r$, A has exactly $n(1) + n(2) + \cdots + n(i)$ $(= \text{Off}(n)(i))$ normalized angles in $[0, s(i)]$.

Using this, we prove 1) and 2). By definition, $\text{CDF}_{\nu(c,G(N))}(s)$ is the Haar measure of the set

$$\begin{aligned}
&\{A \text{ in } G(N) | \vartheta(c(i))(A) \leq s(i) \text{ for } i = 1, \ldots, r\} \\
&\quad = \{A \text{ in } G(N) | A \text{ has } \geq c(i) \text{ normalized angles in } [0, s(i)], \text{ for } i = 1, \ldots, r\} \\
&\quad = \coprod_{n \text{ in } \mathbb{Z}^r, \text{Off}(n) \geq c} \text{Eigen}(n, s, G(N)).
\end{aligned}$$

Taking the Haar measures of both sides gives 1).

Similarly, $\text{Tail}_{\nu(c,G(N))}(s)$ is the Haar measure of the set

$$\begin{aligned}
&\{A \text{ in } G(N) | \vartheta(c(i))(A) > s(i) \text{ for } i = 1, \ldots, r\} \\
&\quad = \{A \text{ in } G(N) | A \text{ has } < c(i) \text{ normalized angles in } [0, s(i)], \text{ for } i = 1, \ldots, r\} \\
&\quad = \coprod_{n \text{ in } \mathbb{Z}^r, \text{Off}(n) < c} \text{Eigen}(n, s, G(N)).
\end{aligned}$$

Taking the Haar measures of both sides gives 2). QED

Remark 8.4.6. In the above result, the CDF is given as an infinite sum of eigen's, but the tail function is a **finite** sum of eigen's.

8.4.7. The previous result computed the CDF and tail of $\nu(c, G(N))$ at points s in $(\mathbb{R}_{\geq 0})^r$(order). Because the measure $\nu(c, G(N))$ is supported in $(\mathbb{R}_{\geq 0})^r$(order), and has a continuous CDF (by 6.20.5), both its CDF and its tail are determined by their values there, as the next lemma shows.

Lemma 8.4.8. *Let ν be any Borel measure of finite total mass on $\mathbb{R}^r$ which is supported in $(\mathbb{R}_{\geq 0})^r$(order).*

1) If s in $\mathbb{R}^r$ has some $s(i) < 0$, then $\mathrm{CDF}_\nu(s) = 0$.

2) For s in $(\mathbb{R}_{\geq 0})^r$, denote by $\mathrm{SuccMin}(s)$ *the point*

$$(\mathrm{Min}_{j \geq 1} s(j), \mathrm{Min}_{j \geq 2} s(j), \ldots, \mathrm{Min}_{j \geq r} s(j))$$

in $(\mathbb{R}_{\geq 0})^r$(order). Then

$$\mathrm{CDF}_\nu(s) = \mathrm{CDF}_\nu(\mathrm{SuccMin}(s)).$$

3) If s in $\mathbb{R}^r$, denote by $\mathrm{Max}(0, s)$ *the componentwise maximum. If* Tail_ν *is continuous, then*

$$\mathrm{Tail}_\nu(s) = \mathrm{Tail}_\nu(\mathrm{Max}(0, s)).$$

4) For s in $(\mathbb{R}_{\geq 0})^r$, denote by $\mathrm{SuccMax}(s)$ *the point*

$$(\mathrm{Max}_{j \leq 1} s(j), \mathrm{Max}_{j \leq 2} s(j), \ldots, \mathrm{Max}_{j \leq r} s(j))$$

in $(\mathbb{R}_{\geq 0})^r$(order). Then

$$\mathrm{Tail}_\nu(s) = \mathrm{Tail}_\nu(\mathrm{SuccMax}(s)).$$

PROOF. 1) For s not ≥ 0, the rectangle $R(s)$ has empty intersection with $(\mathbb{R}_{\geq 0})^r$, so a fortiori it has empty intersection with $(\mathbb{R}_{\geq 0})^r$(order). 2) For $s \geq 0$, the rectangles $R(s)$ and $R(\mathrm{SuccMin}(s))$ have the same intersection with $(\mathbb{R}_{\geq 0})^r$(order). 3) If Tail_ν is continuous, $\mathrm{Tail}_\nu(s)$ is also (by 6.17.4) the measure of the "closed tail rectangle" $\overline{T}(s) := \{x \text{ in } \mathbb{R}^r \text{ with } x(i) \geq s(i) \text{ for } i = 1, \ldots, r\}$. For any s, the closed tail rectangles $\overline{T}(s)$ and $\overline{T}(\mathrm{Max}(0, s))$ have the same intersection with $(\mathbb{R}_{\geq 0})^r$, so a fortiori with $(\mathbb{R}_{\geq 0})^r$(order). 4) For $s \geq 0$, the rectangles $T(s)$ and $T(\mathrm{SuccMax}(s))$ have the same intersection with $(\mathbb{R}_{\geq 0})^r$(order). QED

8.4.9. Fix an integer $r \geq 1$. Given s and t intertwined in $(\mathbb{R}_{\geq 0})^r$,

$$0 \leq s(1) \leq t(1) \leq s(2) \leq t(2) \leq \cdots \leq s(r) \leq t(r),$$

we defined in 8.2.3 the several variable Fredholm determinants

$$E(T, s, t) \text{ and } E_{\pm}(T, s, t).$$

8.4.10. We now specialize this general choice of r successive intervals to a choice of r adjacent intervals, which begin at the origin. Thus we give ourselves a point s in $(\mathbb{R}_{\geq 0})^r$(order),

$$0 \leq s(1) \leq s(2) \leq \cdots \leq s(r).$$

We wish to consider the r-tuple of intervals

$$I_1 = [0, s(1)],$$
$$I_i = (s(i-1), s(i)] \quad \text{for } i = 2, \ldots, r.$$

To put this in the $E(T,s,t)$ context, we denote

$$s_{\text{trunc}} := (0, s(1), s(2), \ldots, s(r-1)).$$

Then s_{trunc} and s are intertwined, and we define

$$E(T,s) := E(T, s_{\text{trunc}}, s)$$
$$E_{\pm}(T,s) := E_{\pm}(T, s_{\text{trunc}}, s).$$

8.4.11. We denote by $E_n(s)$ and by $E_{\pm,n}(s)$ their expansion coefficients around the point all $T_i = -1$:

$$E(T,s) = \sum_{n\geq 0 \text{ in } \mathbb{Z}^r} (1+T)^n E_n(s),$$
$$E_{\pm}(T,s) = \sum_{n\geq 0 \text{ in } \mathbb{Z}^r} (1+T)^n E_{\pm,n}(s).$$

In the case $r = 1$, this notation is compatible with that of 7.3.6 and 7.5.1.

8.4.12. Similarly, for each $G(N)$, we denote by

$$E(T,s,G(N)) := E(T, s_{\text{trunc}}, s, G(N)),$$

and by $E_n(s, G(N))$ its expansion coefficients around the point all $T_i = -1$:

$$E(T,s,G(N)) = \sum_{n\geq 0 \text{ in } \mathbb{Z}^r} (1+T)^n E_n(s, G(N)).$$

Thus for s in $(\mathbb{R}_{\geq 0})^r(\text{order})$, we have

$$E_n(s, G(N)) = \text{eigen}(n, s, G(N))$$

for every $N > s(r)$.

8.4.13. Since the map $s \mapsto (s_{\text{trunc}}, s)$ is holomorphic, we get from 8.3.5 the following result.

Lemma 8.4.14. *The functions $E(T,s,G(N)), E(T,s)$ and $E_{\pm}(T,s)$ are the restrictions to $\mathbb{C}^r \times (\mathbb{R}_{\geq 0})^r(\text{order})$ of entire functions on $\mathbb{C}^{2r}$. As $N \to \infty$, the sequence $E(T,s,G(N))$ of entire functions of (T,s) converges, uniformly on compact subsets of $\mathbb{C}^{2r}$, to*

$$\begin{aligned} &E(T,s), && \text{if } G(N) = U(N),\\ &E_{+}(T,s), && \text{if } G(N) = SO(2N) \text{ or } O_{-}(2N+1),\\ &E_{-}(T,s), && \text{if } G(N) = USp(2N) \text{ or } SO(2N+1) \text{ or } O_{-}(2N+2). \end{aligned}$$

From 8.3.10, we get

Proposition 8.4.15. *Fix a nonempty subset W of $\{n \geq 0 \text{ in } \mathbb{Z}^r\}$, and $G(N)$ one of $U(N), SO(2N+1), USp(2N), SO(2N), O_{-}(2N+2), O_{-}(2N+1)$.*

1) *The series*

$$E_W(s, G(N)) := \sum_{n \text{ in } W} E_n(s, G(N))$$

as well as the series

$$E_W(s) := \sum_{n \text{ in } W} E_n(s)$$

and

$$E_{\pm,W}(s) := \sum_{n \text{ in } W} E_{\pm,n}(s),$$

converge, uniformly on compact subsets of $\mathbb{C}^r$, *to entire functions of* s.

2) *As* $N \to \infty$, *the sequence of entire functions* $E_W(s, G(N))$ *converges, uniformly on compact subsets of* $\mathbb{C}^r$, *to*

$$\begin{aligned} &E_W(s) && \text{if } G(N) = U(N), \\ &E_{+,W}(s) && \text{if } G(N) = SO(2N) \text{ or } O_-(2N+1), \\ &E_{-,W}(s) && \text{if } G(N) = USp(2N) \text{ or } SO(2N+1) \text{ or } O_-(2N+2). \end{aligned}$$

8.4.16. We now turn to the question of taking large N limits of the measures $\nu(c, G(N))$, for c an offset vector.

Proposition 8.4.17. *Fix an integer* $r \geq 1$, *and an offset vector* c *in* $\mathbb{Z}^r$. *There exist Borel probability measures* $\nu(c)$ *and* $\nu(\pm, c)$ *on* $\mathbb{R}^r$, *supported in* $(\mathbb{R}_{\geq 0})^r(\text{order})$ *and having continuous CDF's, such that*

$$\begin{aligned} \lim_{N\to\infty} \nu(c, G(N)) &= \nu(c), && \text{if } G(N) = U(N), \\ &= \nu(+, c), && \text{if } G(N) = SO(2N) \text{ or } O_-(2N+1), \\ &= \nu(-, c), && \text{if } G(N) = USp(2N), SO(2N+1), O_-(2N+2), \end{aligned}$$

in the sense of uniform convergence of both CDF's and of tail functions on $\mathbb{R}^r$. *As functions on* $(\mathbb{R}_{\geq 0})^r(\text{order})$, *the CDF's and tail functions of these measures are (the restrictions to* $(\mathbb{R}_{\geq 0})^r(\text{order})$ *of) entire functions on* $\mathbb{C}^r$, *given by the explicit formulas*

$$\begin{aligned} \mathrm{CDF}_{\nu(c,G(N))}(s) &= \sum_{n \geq 0 \text{ in } \mathbb{Z}^r, \mathrm{Off}(n) \geq c} E_n(s, G(N)), \\ \mathrm{CDF}_{\nu(c)}(s) &= \sum_{n \geq 0 \text{ in } \mathbb{Z}^r, \mathrm{Off}(n) \geq c} E_n(s), \\ \mathrm{CDF}_{\nu(\pm,c)}(s) &= \sum_{n \geq 0 \text{ in } \mathbb{Z}^r, \mathrm{Off}(n) \geq c} E_{\pm,n}(s), \\ \mathrm{Tail}_{\nu(c,G(N))}(s) &= \sum_{n \geq 0 \text{ in } \mathbb{Z}^r, \mathrm{Off}(n) < c} E_n(s, G(N)), \\ \mathrm{Tail}_{\nu(c)}(s) &= \sum_{n \geq 0 \text{ in } \mathbb{Z}^r, \mathrm{Off}(n) < c} E_n(s), \\ \mathrm{Tail}_{\nu(\pm,c)}(s) &= \sum_{n \geq 0 \text{ in } \mathbb{Z}^r, \mathrm{Off}(n) < c} E_{\pm,n}(s). \end{aligned}$$

The measures $\nu(c)$ *and* $\nu(\pm, c)$ *have exponential decay* (A, B, λ) *with* $A = 2^{c(r)}$, $B = \sigma/4$, *and* $\lambda = 1$. *For any continuous function* f *of polynomial growth on* $\mathbb{R}^r$, *we have the limit formulas*

$$\int f\,d\nu(c) = \lim_{N\to\infty} \int f\,d\nu(c, U(N)),$$

$$\int f\,d\nu(+,c) = \lim_{N\to\infty} \int f\,d\nu(c, G(N)) \quad \textit{for } G(N) = SO(2N) \textit{ or } O_-(2N+1),$$

$$\int f\,d\nu(-,c) = \lim_{N\to\infty} \int f\,d\nu(c, G(N))$$
$$\textit{for } G(N) = USp(2N), SO(2N+1), \textit{ or } O_-(2N+2).$$

The measures $\nu(c)$ and $\nu(\pm, c)$ are determined by their moments.

PROOF. Consider the measure $\nu(c, G(N))$ on $\mathbb{R}^r$. Its one variable projections are given by 6.20.3,

$$\mathrm{pr}[i]_*\nu(c, G(N)) = \nu(c(i), G(N)),$$

and these satisfy the tail estimate 6.9.8,

$$\mathrm{Tail}_{\nu(c(i),G(N))}(s) \le 2^{c(i)}e^{-\sigma s/4} \le 2^{c(i)}e^{-s/4}.$$

So $\nu(c, G(N))$ has exponential decay $(A, B, \lambda) = (2^{c(r)}, \sigma/4, 1)$.

We first claim that as $N \to \infty$, the CDF's of the $\nu(c, G(N))$ converge uniformly on $\mathbb{R}^r$. In view of their uniform exponential decay, it suffices to show uniform convergence on compact subsets of $\mathbb{R}^r$. In view of 8.4.8, it suffices for this to show uniform convergence on compact subsets of $(\mathbb{R}_{\ge 0})^r$. For s in $(\mathbb{R}_{\ge 0})^r$, and $N > s(r)$, we have (by 8.4.5)

$$\begin{aligned} \mathrm{CDF}_{\nu(c,G(N))}(s) &= \sum_{n\ge 0 \text{ in } \mathbb{Z}^r, \mathrm{Off}(n)\ge c} \mathrm{eigen}(n, s, G(N)) \\ &:= \sum_{n\ge 0 \text{ in } \mathbb{Z}^r, \mathrm{Off}(n)\ge c} E_n(s, G(N)) \\ &= E_W(s, G(N)) \end{aligned}$$

for W the set $\{n \ge 0 \text{ in } \mathbb{Z}^r, \mathrm{Off}(n) \ge c\}$, so the required convergence is 8.4.15. A similar argument, with W now taken as the finite set $\{n \ge 0 \text{ in } \mathbb{Z}^r, \mathrm{Off}(n) < c\}$, shows that as $N \to \infty$, the tails of the $\nu(c, G(N))$ converge uniformly on $\mathbb{R}^r$.

The result now follows from 7.5.11 and 6.20.4 (combined with 2.11.18). QED

Remark 8.4.18. As proven in 7.11.12, the measure $\nu(c)$ on $\mathbb{R}^r$ is absolutely continuous with respect to Lebesgue measure. That this also holds for the measures $\nu(\pm, c)$ will be proven in AD.4.4.1. It amounts to the statement that these measures give measure zero to the hyperplane $x(1) = 0$ and to the hyperplanes $x(i) - x(i-1) = 0$ for $i = 2, \ldots, r$, but the proof is along different lines.

Corollary 8.4.19. *Suppose given an integer $r \ge 1$, and an offset vector c in $\mathbb{Z}^r$. For any nonempty subset*

$$S = \{1 \le s_1 < \cdots < s_k \le r\}$$

of the index set $\{1, \ldots, r\}$, we denote by

$$\mathrm{pr}[S] : \mathbb{R}^r \to \mathbb{R}^{\mathrm{Card}(S)}$$

the projection onto the named coordinates. We have equalities

$$\mathrm{pr}[S]_*\nu(c) = \nu(\mathrm{pr}[S](c))$$
$$\mathrm{pr}[S]_*\nu(\pm, c) = \nu(\pm, \mathrm{pr}[S](c))$$

of measures on $\mathbb{R}^{\mathrm{Card}(S)}$.

PROOF. Use the previous result together with 6.20.3, which gives the finite N analogue, to see that both of the measures in question agree on bounded continuous functions. QED

8.5. Relation of the limit measure Off μ(univ, offsets c) with the limit measures $\nu(c)$

Proposition 8.5.1. *Fix an integer* $r \geq 1$, *and denote*

$$\mathbb{1} := \mathbb{1}_r := (1, 1, \ldots, 1) \text{ in } \mathbb{R}^r.$$

For any Borel measurable function g *on* $\mathbb{R}^r$ *of polynomial growth, denote by* G *the Borel measurable function of polynomial growth defined by the Lebesgue integral*

$$G(x) := \int_{[0,x(1)]} g(x - t\mathbb{1})\, dt := |x(1)| \int_{[0,1]} g(x - tx(1)\mathbb{1})\, dt.$$

Fix an offset vector c *in* $\mathbb{Z}^r$:

$$1 \leq c(1) < c(2) < \cdots < c(r).$$

For each integer k *with* $0 \leq k \leq c(1) - 1$, $c - k\mathbb{1}$ *is again an offset vector, and we have the identity*

$$\int_{\mathbb{R}^r} G\, d\,\mathrm{Off}\,\mu(\textit{univ, offsets } c) = \sum_{0 \leq k \leq c(1)-1} \int_{\mathbb{R}^r} g\, d\nu(c - k\mathbb{1}).$$

PROOF. By 7.11.11 and 8.4.17 applied to the left and right sides respectively, this is the large N limit of 6.16.7. QED

8.5.2. Once we have this result, we get

Proposition 8.5.3. *Fix an integer* $r \geq 1$ *and an offset vector* c *in* $\mathbb{Z}^r$,

$$1 \leq c(1) < c(2) < \cdots < c(r).$$

For any s *in* $\mathbb{R}^r$ *with* $s(1) \geq 0$, *we have the relation*

$$\sum_{0 \leq k \leq c(1)-1} \mathrm{Tail}_{\nu(c-k\mathbb{1})}(s) = \int_{[0,\infty]} \mathrm{Tail}_{\mathrm{Off}\,\mu(\textit{univ, offsets } c)}(s + t\mathbb{1})\, dt.$$

PROOF. Repeat the proof of 6.18.1, using the previous result in place of 6.16.6. QED

Corollary 8.5.4. *Hypotheses and notations as in* 8.5.3 *above, for any* s *in* $\mathbb{R}^r$ *with* $s(1) > 0$, *we have*

$$\mathrm{Tail}_{\mathrm{Off}\,\mu(\textit{univ, offsets } c)}(s) = -\left(\sum_i \partial/\partial s(i)\right) \sum_{0 \leq k \leq c(1)-1} \mathrm{Tail}_{\nu(c-k\mathbb{1})}(s).$$

PROOF. Repeat the proof of 6.18.2. QED

Proposition 8.5.5. *Hypotheses and notations as in 8.5.3 above, we have the following relations among moments: for any polynomial function $H(x)$ on $\mathbb{R}^r$ which is divisible by $x(1)$ as a polynomial, denote by $h(x)$ the polynomial*

$$h := \left(\sum_i \partial/\partial x(i)\right) H.$$

Then we have the relation

$$\sum_{0\leq k\leq c(1)-1} \int_{\mathbb{R}^r} h\, d\nu(c - k\mathbb{1}) = \int_{\mathbb{R}^r} H\, d\,\mathrm{Off}\,\mu(\mathit{univ},\ \mathit{offsets}\ c).$$

PROOF. Repeat the proof of 6.19.1. QED

8.5.6. The following lemma is standard.

Density Lemma 8.5.7. *Let $r \geq 1$ be an integer, $U \subset \mathbb{R}^r$ an open set, and μ a Borel probability measure on $\mathbb{R}^r$. Suppose that*

1) *μ is supported in the closure $\overline{U}$ of U.*
2) *The boundary $\partial(U) := \overline{U} - U$ has Lebesgue measure zero.*
3) *The measure μ on $\mathbb{R}^r$ is absolutely continuous with respect to Lebesgue measure.*
4) *The restriction to U of the function Tail_μ is $\mathcal{C}^\infty$.*

Then μ is the extension by zero of its restriction to U, and on U the measure μ is given in coordinates $x(1), \ldots, x(r)$ on the ambient $\mathbb{R}^r$ by

$$(-1)^r \left(\prod_{i=1}^r \partial/\partial x(i)\right) (\mathrm{Tail}_\mu) \prod_{i=1}^r dx(i).$$

PROOF. Already hypotheses 1), 2) and 3) imply that μ is the extension by zero of its restriction to U. It remains to see that if in addition Tail_μ is $\mathcal{C}^\infty$ on U, then μ on U is given by the asserted formula. To see this, we use the fact that for s in U and for $t \geq 0$ in $\mathbb{R}^r$, the measure of the rectangle $(s, s+t]$ is equal to

$$\sum_{\text{subsets } S \text{ of } \{1,2,\ldots,r\}} (-1)^{\mathrm{Card}(S)}\, \mathrm{Tail}_\mu \left(s + \sum_{i \text{ in } S} t(i)e(i)\right).$$

For t sufficiently small, all of the points $s + \sum_{i \text{ in } S} t(i)e(i)$ lie in U.

Because Tail_μ is $\mathcal{C}^\infty$ in U, we see by iterating the fundamental theorem of calculus r times that this alternating sum is equal to the integral over the rectangle $(s, s+t]$ of the differential form

$$(-1)^r \left(\prod_{i=1}^r \partial/\partial x(i)\right) (\mathrm{Tail}_\mu) \prod_{i=1}^r dx(i).$$

Therefore both $\mu|U$ and $(-1)^r(\prod_{i=1}^r \partial/\partial x(i))(\mathrm{Tail}_\mu) \prod_{i=1}^r dx(i)|U$ agree as Borel measures on U, because they give the same measure to all rectangles in U. QED

Proposition 8.5.8. *Let $r \geq 1$ be an integer, c in $\mathbb{Z}^r$ an offset vector. Denote by $U \subset \mathbb{R}^r$ the open set $0 < x(1) < x(2) < \cdots < x(r)$, whose closure $\overline{U}$ is*

$(\mathbb{R}_{\geq 0})^r$(*order*). *Let us define functions* $F(c,x)$ *and* $P(c,x)$ *of* x *in* U *by*

$$F(c,x) := (-1)^r \left(\prod_{i=1}^{r} \partial/\partial x(i)\right) (\mathrm{Tail}_{\nu(c)}),$$

$$P(c,x) := (-1)^r \left(\prod_{i=1}^{r} \partial/\partial x(i)\right) (\mathrm{Tail}_{\mathrm{Off}\,\mu(univ,\ offsets\ c)}).$$

Then

1) $\nu(c)$ *is the extension by zero of the measure* $F(c,x)\prod_{i=1}^{r} dx(i)$ *on* U.

2) $\mathrm{Off}\,\mu$(*univ, offsets* c) *is the extension by zero of the measure* $P(c,x)\prod_{i=1}^{r} dx(i)$ *on* U.

3) *Both* $F(c,x)$ *and* $P(c,x)$ *are the restrictions to* U *of entire functions on* $\mathbb{C}^r$, *which are related by*

$$-\left(\sum_i \partial/\partial s(i)\right) P(c,x) = \sum_{0\leq k\leq c(1)-1} F(c-k\mathbb{1},x).$$

PROOF. Apply the previous lemma to $\nu(c)$ and to $\mathrm{Off}\,\mu$(univ, offsets c) to get statements 1) and 2). Statement 3) then follows by applying the differential operator $(-1)^r(\prod_{i=1}^{r} \partial/\partial x(i))$ to the identity

$$\mathrm{Tail}_{\mathrm{Off}\,\mu(\mathrm{univ,\ offsets}\ c)}(x) = -\left(\sum_i \partial/\partial x(i)\right) \sum_{0\leq k\leq c(1)-1} \mathrm{Tail}_{\nu(c-k\mathbb{1})}(x)$$

proven in 8.5.4. QED

Corollary 8.5.9. *Let* $r \geq 1$ *be an integer,* c *in* $\mathbb{Z}^r$ *an offset vector. The measure* μ(*univ, offsets* c) *is the extension by zero of its restriction to the open set* $(\mathbb{R}_{>0})^r$, *where it is given by* $P(c,\mathrm{Off}(x))\prod_{i=1}^{r} dx(i)$.

PROOF. Apply Off^* to the previous description of

$$\mathrm{Off}\,\mu(\mathrm{univ,\ offsets}\ c) := \mathrm{Off}_*\,\mu(\mathrm{univ,\ offsets}\ c),$$

and use the fact that $\mathrm{Off}^*\,\mathrm{Off}_*$ is the identity. Since $\mathrm{Off}^{-1}(U)$ is $(\mathbb{R}_{>0})^r$, we find that μ(univ, offsets c) is the extension by zero of its restriction to the open set $(\mathbb{R}_{\geq 0})^r$, on which it is given by

$$\mathrm{Off}^*\left(P(c,x)\prod_{i=1}^{r} dx(i)\right) = P(c,\mathrm{Off}(x))\,\mathrm{Off}^*\left(\prod_{i=1}^{r} dx(i)\right)$$
$$= P(c,\mathrm{Off}(x))\prod_{i=1}^{r} dx(i),$$

the last equality because Off is unipotent. QED

Remark 8.5.10. If we combine these last results with 8.4.17, according to which for x in U we have the finite sum formula

$$\mathrm{Tail}_{\nu(c)}(x) = \sum_{n\geq 0 \text{ in } \mathbb{Z}^r,\,\mathrm{Off}(n)<c} E_n(x),$$

we get explicit formulas for both $F(c,x)$ and $P(c,x)$ as finite sums of higher derivatives of the functions $E_n(x)$. For F, we get

$$(-1)^r F(c,x) = \sum_{n\geq 0 \text{ in } \mathbb{Z}^r, \mathrm{Off}(n)<c} \left(\prod_{i=1}^{r} \partial/\partial x(i)\right)(E_n(x)).$$

To state the result for P, let us introduce, for $n \geq 0$ in $\mathbb{Z}^r$ and for c an offset vector in $\mathbb{Z}^r$, the nonnegative integer

$$\begin{aligned} N(n,c) := &\text{ the number of integers } k \text{ with } 0 \leq k \leq c(1)-1 \\ &\text{ such that } \mathrm{Off}(n) < c - k\mathbb{1}. \end{aligned}$$

Then we get

$$\begin{aligned} &(-1)^{r+1} P(c,x) \\ &= \left(\prod_{i=1}^{r} \partial/\partial x(i)\right)\left(\sum_i \partial/\partial x(i)\right) \sum_{n\geq 0 \text{ in } \mathbb{Z}^r, \mathrm{Off}(n)<c} N(n,c) E_n(x). \end{aligned}$$

In the case $r = 1$, we recover the formulas tabulated in 7.7, where we labeled by the separation, whereas here we label by the offset, i.e., μ_a there is $\mu(\text{univ, offset } a+1)$ here.

CHAPTER 9

Equidistribution

9.0. Preliminaries

9.0.1. In this chapter, we will review the basic equidistribution results proven by Deligne in [**De-Weil II**, 3.5].

9.0.2. Recall [**SGA 1**] that for any connected scheme X, and any geometric point ξ of X (i.e., ξ is a point of X with values in some algebraically closed field), we have the profinite fundamental group $\pi_1(X,\xi)$, which classifies finite etale coverings of X. For variable (X,ξ) as above, formation of π_1 is a covariant functor. As in classical topology, for any two geometric points ξ_1 and ξ_2 of X, there is the notion of a "path" from ξ_1 to ξ_2, which induces an isomorphism from $\pi_1(X,\xi_1)$ to $\pi_1(X,\xi_2)$. This isomorphism is independent of the chosen path up to inner automorphism of either source or target. In the special case when X is the spec of a field k, a geometric point ξ is an algebraically closed overfield L of k, and $\pi_1(X,\xi)$ is just the galois group $\mathrm{Gal}(k^{\mathrm{sep}}/k)$, k^{sep} denoting the separable closure of k in L. In the particular case of a finite field, or more generally of a connected scheme X whose π_1 is abelian, the group $\pi_1(X,\xi)$ is canonically independent of base point, and may be denoted $\pi_1(X)$ with no ambiguity.

9.0.3. When the connected scheme X is normal, it is irreducible, say with generic point $\eta :=$ the spec of its function field K. In terms of an algebraic closure $\overline{K}$ of K, viewed as a geometric generic point $\overline{\eta}$ of X, the group $\pi_1(X,\overline{\eta})$ is the quotient of $\mathrm{Gal}(K^{\mathrm{sep}}/K)$ which classifies those finite separable L/K with the property that the normalization of X in L is finite etale over X.

9.0.4. When k is a finite field, say of cardinality q, then $\mathrm{Gal}(k^{\mathrm{sep}}/k)$ is canonically the group $\widehat{\mathbb{Z}}$, with generator the "arithmetic Frobenius" automorphism $x \mapsto x^q$ of k^{sep}. The **inverse** of this generator is called the "geometric Frobenius", denoted F_k.

9.0.5. Suppose now that X is a connected scheme, with geometric point ξ, that k is a finite field, and that x in $X(k)$ is a k-valued point of X, which we view as a morphism $x : \mathrm{Spec}(k) \to X$. So if we pick a separable closure k^{sep} of k, and denote by $\overline{x}$ the $\overline{k}$-valued point lying over x, we get a canonical homomorphism

$$\pi_1(\mathrm{Spec}(k)) \to \pi_1(X,\overline{x}).$$

If we compose this with any isomorphism $\pi_1(X,\overline{x}) \cong \pi_1(X,\xi)$ given by a path, we get a homomorphism

$$\pi_1(\mathrm{Spec}(k)) \to \pi_1(X,\xi)$$

which is well-defined up to inner automorphism of the target. The image of F_k in $\pi_1(X,\xi)$ is thus well-defined up to conjugacy: its conjugacy class in $\pi_1(X,\xi)$ is called the **Frobenius conjugacy class**, denoted $F_{k,x}$, attached to the finite field k and to the k-valued point x in $X(k)$.

9.0.6. Let us now recall the basic set-up for Deligne's equidistribution theorem. We are given a finite field k, a smooth, geometrically connected scheme X/k of dimension $d \geq 0$, a prime number l invertible in k, and a lisse $\overline{\mathbb{Q}}_l$-sheaf $\mathcal{F}$ on X of rank $r \geq 1$, which for our purposes we may define to be an r-dimensional, continuous $\overline{\mathbb{Q}}_l$-representation ρ of $\pi_1(X, \overline{\eta})$ (intrinsically in $GL(\mathcal{F}_{\overline{\eta}})$). [This definition of a lisse $\overline{\mathbb{Q}}_l$-sheaf, as a continuous $\overline{\mathbb{Q}}_l$-representation ρ of $\pi_1(X, \overline{\eta})$, makes sense on any normal connected scheme X, and later in this chapter (in 9.3.1) such more general X's will come into play.]

Remark 9.0.7. We should mention here the fact that any compact subgroup of $GL(r, \overline{\mathbb{Q}}_l)$, in particular the image $\rho(\pi_1(X, \overline{\eta}))$, lies in $GL(r, E_\lambda)$ for some finite extension E_λ of $\mathbb{Q}_l$, so this definition agrees with the one given in [**De-Weil II**, 1.1.1], where it was required that ρ land in some $GL(r, E_\lambda)$. This fact, which does not seem to appear in the literature, has been repeatedly rediscovered. We first learned it from Sinnott in 1989, and then again from Pop in 1995. Here is a Haar measure version of Pop's argument, which was based on Baire category. Since $\mathbb{Q}_l$ has only finitely many extensions of any given degree n inside a fixed $\overline{\mathbb{Q}}_l$, we can find a countable sequence of finite extensions of $\mathbb{Q}_l$, say $E_1 := \mathbb{Q}_l \subset E_2 \subset E_3 \subset \cdots$, whose union is $\overline{\mathbb{Q}}_l$ (e.g., take for E_n the compositum inside $\overline{\mathbb{Q}}_l$ of all finite extensions of $\mathbb{Q}_l$ of degree $\leq n$). Then for each n, $G_n := GL(r, E_n)$ is a closed subgroup of $G = GL(r, \overline{\mathbb{Q}}_l)$, we have $G_n \subset G_{n+1}$, and G is the union of the G_n. Now apply the following lemma.

Lemma 9.0.8 (Pop). *Let G be a topological group, which is the increasing union of a countable sequence of closed subgroups G_n. Then any compact subgroup K of G lies in some G_n.*

PROOF. For each n, $K_n := K \cap G_n$ is a closed subgroup of K, we have $K_n \subset K_{n+1}$, and K is the union of the K_n. We claim that K is equal to some K_n. To see this, consider the Haar measure μ on K of total mass one. Since K is the increasing union of the measurable (because closed) subsets K_n, we have $1 = \mu(K) = \lim_{n\to\infty} \mu(K_n)$. Pick n large enough that $\mu(K_n) > 1/2$. Then $K = K_n$, for if not there exists a left coset γK_n of K_n in K which is disjoint from K_n, and then K contains $K_n \bigsqcup \gamma K_n$, and so

$$1 = \mu(K) \geq \mu(K_n) + \mu(\gamma K_n) = 2\mu(K_n)$$

(the last equality by the translation invariance of μ), contradiction. QED

9.0.9. We denote $\pi_1^{\mathrm{geom}}(X, \overline{\eta}) := \pi_1(X \otimes_k \overline{k}, \overline{\eta})$, or just π_1^{geom} if X/k is understood, and call it the geometric fundamental group. Writing π_1 for $\pi_1(X, \overline{\eta})$, we have the fundamental exact sequence

$$1 \to \pi_1^{\mathrm{geom}} \to \pi_1 \xrightarrow{\deg} \mathrm{Gal}(\overline{k}/k)(\cong \widehat{\mathbb{Z}} \text{ via } F_k) \to 1.$$

9.0.10. We fix an embedding ι of fields $\overline{\mathbb{Q}}_l \to \mathbb{C}$. For any real number w, we say that $\mathcal{F}$ is ι-pure of weight w if, for every finite overfield E of k, and for every point x in $X(E)$, all the eigenvalues of $\rho(F_{E,x})$ have, under the complex embedding ι, complex absolute value $\mathrm{Card}(E)^{w/2}$. We say that $\mathcal{F}$ is pure of weight w if it is ι-pure of this same weight w for every choice of ι.

9.0.11. For any integer n, we denote by $\overline{\mathbb{Q}}_l(n)$ the lisse, rank one $\overline{\mathbb{Q}}_l$-sheaf on $\mathrm{Spec}(k)$ which, as character of $\mathrm{Gal}(\overline{k}/k)$, takes the value $\mathrm{Card}(k)^{-n}$ on F_k. Thus $\overline{\mathbb{Q}}_l(n)$ is ι-pure of weight $-2n$, for any ι. More generally, for any unit α in the ring of integers of $\overline{\mathbb{Q}}_l$, we denote by "$\alpha^{\deg}$" the lisse, rank one $\overline{\mathbb{Q}}_l$-sheaf on $\mathrm{Spec}(k)$ which, as character of $\mathrm{Gal}(\overline{k}/k)$, takes the value α on F_k. Thus $\overline{\mathbb{Q}}_l(-n)$ is "$\alpha^{\deg}$" for $\alpha = \mathrm{Card}(k)^n$. For each embedding $\iota : \overline{\mathbb{Q}}_l \to \mathbb{C}$, "$\alpha^{\deg}$" is ι-pure of weight $2\log(|\iota(\alpha)|)/\log(\mathrm{Card}(k))$, which might (e.g., if α is transcendental) vary wildly with ι. If we choose an N'th root of $\mathrm{Card}(k)$ for some N, we may speak of the lisse rank one sheaves $\overline{\mathbb{Q}}_l(M/N)$ for any rational number M/N in $(1/N)\mathbb{Z}$. For any X/k as above, we denote by $\overline{\mathbb{Q}}_l(-n)$ and "$\alpha^{\deg}$" the lisse, rank one $\overline{\mathbb{Q}}_l$-sheaf on X obtained by pullback. As characters, their values on Frobenius conjugacy classes $F_{E,x}$ are given by $\mathrm{Card}(E)^n$ and by $\alpha^{\deg(E/k)}$ respectively. Given any lisse $\mathcal{F}$ as above, we denote by $\mathcal{F}(-n)$ and $\mathcal{F} \otimes \alpha^{\deg}$ the tensor products of $\mathcal{F}$ with $\overline{\mathbb{Q}}_l(-n)$ and "$\alpha^{\deg}$" respectively.

9.0.12. Given $\mathcal{F}$ on X as above, we denote by G_{geom} the $\overline{\mathbb{Q}}_l$-algebraic group defined as the Zariski closure of $\rho(\pi_1(X \otimes_k \overline{k}))$ in $GL(r)$. It is a fundamental result of Deligne [**De-Weil II**, 3.4.1 (iii)] that if $\mathcal{F}$ is ι-pure of some weight w, then $\mathcal{F}$ is completely reducible as a representation of $\pi_1(X \otimes_k \overline{k})$, and hence that G_{geom} is reductive. It then follows from a result of Grothendieck [**De-Weil II**, 1.3.9 and the first four lines of its proof] that G_{geom} is in fact semisimple (in the sense that its identity component is semisimple, or equivalently that its Lie algebra is semisimple).

9.0.13. Given $\mathcal{F}$ on X as above, there are compact(ly supported) cohomology groups $H^i_c(X \otimes_k \overline{k}, \mathcal{F})$, which are finite-dimensional $\overline{\mathbb{Q}}_l$-spaces on which $\mathrm{Gal}(\overline{k}/k)$ acts continuously, and which vanish unless i lies in $[0, 2d]$. For any integer n, or any l-adic unit α, we have

$$H^i_c(X \otimes_k \overline{k}, \mathcal{F}(n)) \cong H^i_C(X \otimes_k \overline{k}, \mathcal{F})(n),$$
$$H^i_c(X \otimes_k \overline{k}, \mathcal{F} \otimes \alpha^{\deg}) \cong H^i_c(X \otimes_k \overline{k}, \mathcal{F}) \otimes \alpha^{\deg},$$

a kind of trivial projection formula. Moreover, $H^{2d}_c(X \otimes_k \overline{k}, \mathcal{F}(d))$ is canonically the coinvariants $\mathcal{F}_{\pi_1^{\mathrm{geom}}}$ under π_1^{geom} in $\mathcal{F}$, i.e.,

$$H^{2d}_c(X \otimes_k \overline{k}, \mathcal{F}) \cong (\mathcal{F}_{\pi_1^{\mathrm{geom}}})(-d).$$

Moreover, if X is in addition affine, then $H^i_c(X \otimes_k \overline{k}, \mathcal{F})$ vanishes unless i lies in $[d, 2d]$; this is the Poincaré dual of the Lefschetz affine theorem, that a d-dimensional affine variety over an algebraically closed field has cohomological dimension $\leq d$.

9.0.14. The diophantine interest of these compact cohomology groups is given by the Lefschetz trace formula, in which for γ a conjugacy class in π_1 we write $\mathrm{Trace}(\gamma|\mathcal{F}) := \mathrm{Trace}(\rho(\gamma))$: for any finite overfield E of k, we have

$$\begin{aligned}\sum_{x \text{ in } X(E)} \mathrm{Trace}(F_{E,x}|\mathcal{F}) &:= \sum_{x \text{ in } X(E)} \mathrm{Trace}(\rho(F_{E,x})) \\ &= \sum_i (-1)^i \,\mathrm{Trace}((F_k)^{\deg(E/k)}|H^i_c(X \otimes_k \overline{k}, \mathcal{F})).\end{aligned} \tag{9.0.14.1}$$

9.0.15. Suppose now in addition that $\mathcal{F}$ is ι-pure of weight w. Then the fundamental result of Deligne in [**De-Weil II**, 3.3.1 and 3.3.10] is that for each i, the compact cohomology group $H^i_c(X \otimes_k \overline{k}, \mathcal{F})$ is mixed of weight $\leq w+i$, in the sense that every eigenvalue of F_k on this cohomology group has, via ι, complex absolute value $\leq \mathrm{Card}(k)^{(w+i)/2}$. Taking for $\mathcal{F}$ the constant sheaf $\overline{\mathbb{Q}}_l$, which is ι-pure of weight zero for any choice of ι, we see that $H^i_c(X \otimes_k \overline{k}, \overline{\mathbb{Q}}_l)$ is mixed of weight $\leq i$, while the description of H^{2d}_c in terms of π_1^{geom}-coinvariants shows that $H^{2d}_c(X \otimes_k \overline{k}, \overline{\mathbb{Q}}_l) = \overline{\mathbb{Q}}_l(-d)$, the one-dimensional space on which F_k acts as $\mathrm{Card}(k)^d$. The Lefschetz trace formula then gives that for every finite extension E of k,

$$\mathrm{Card}(X(E)) = \sum_i (-1)^i \,\mathrm{Trace}((F_k)^{\deg(E/k)} | H^i_c(X \otimes_k \overline{k}, \overline{\mathbb{Q}}_l)). \tag{9.0.15.1}$$

Putting the H^{2d}_c term to the other side, and denoting

$$h^i_c(X \otimes_k \overline{k}, \overline{\mathbb{Q}}_l) := \dim_{\overline{\mathbb{Q}}_l} H^i_c(X \otimes_k \overline{k}, \overline{\mathbb{Q}}_l),$$

we get the Lang-Weil estimate [**Lang-Weil**]

$$\begin{aligned} |\,\mathrm{Card}(X(E)) - \mathrm{Card}(E)^d| &\leq \sum_{i<2d} h^i_c(X \otimes_k \overline{k}, \overline{\mathbb{Q}}_l)\,\mathrm{Card}(E)^{i/2} \\ &\leq \left(\sum_{i<2d} h^i_c(X \otimes_k \overline{k}, \overline{\mathbb{Q}}_l)\right) \mathrm{Card}(E)^{(2d-1)/2}. \end{aligned} \tag{9.0.15.2}$$

9.0.16. Consider now the case when $\mathcal{F}$ is ι-pure of weight zero, and in which π_1^{geom} acts irreducibly and nontrivially, or more generally without nonzero coinvariants, in the representation ρ corresponding to $\mathcal{F}$. In view of the description (cf. 9.0.13) of H^{2d}_c in terms of π_1^{geom}-coinvariants, this means precisely that $H^{2d}_c(X \otimes_k \overline{k}, \mathcal{F}) = 0$. The Lefschetz trace formula then gives, for every finite extension E of k,

$$\begin{aligned} &\sum_{x \text{ in } X(E)} \mathrm{Trace}(F_{E,x} | \mathcal{F}) \\ &\qquad = \sum_{i<2d} (-1)^i \,\mathrm{Trace}((F_k)^{\deg(E/k)} | H^i_c(X \otimes_k \overline{k}, \mathcal{F})), \end{aligned} \tag{9.0.16.1}$$

with H^i_c ι-mixed of weight $\leq i$. This gives the following inequality of complex (via ι) absolute values, for every finite extension E of k:

$$\begin{aligned} \left| \sum_{x \text{ in } X(E)} \mathrm{Trace}(\rho(F_{E,x})) \right| &\leq \sum_{i<2d} h^i_c(X \otimes_k \overline{k}, \mathcal{F})\,\mathrm{Card}(E)^{i/2} \\ &\leq \left(\sum_{i<2d} h^i_c(X \otimes_k \overline{k}, \mathcal{F})\right) \mathrm{Card}(E)^{(2d-1)/2}. \end{aligned} \tag{9.0.16.2}$$

9.1. Interlude: zeta functions in families: how lisse pure $\mathcal{F}$'s arise in nature

9.1.1. Before embarking on an exposition of Deligne's equidistribution theorem for lisse, ι-pure sheaves $\mathcal{F}$, we first explain the most fundamental way such $\mathcal{F}$ arise in nature.

9.1.2. To begin, let us recall the most basic facts about zeta functions of general varieties over finite fields, and their expression by l-adic cohomology. We begin with a finite field k, and a scheme X/k which is separated and of finite type. Inside (any choice of) $\overline{k}$, k has a unique extension k_n of each degree $n \geq 1$. We denote

$$N_n := N_n(X/k) := \mathrm{Card}(X(k_n)).$$

The sequence of integers $N_n(X/k)$ is the most basic diophantine invariant of X/k. They are most meaningfully packaged in the zeta function of X/k, defined as the formal series in one variable T

$$\mathrm{Zeta}(X/k, T) := \exp\left(\sum_{n\geq 1} N_n T^n/n\right) \quad \text{in } 1 + T\mathbb{Q}[[T]].$$

This series also has an Euler product expression, over the closed points of X. For X/k as above, a closed point x of X is an orbit of $\mathrm{Gal}(\overline{k}/k)$ acting on the set $X(\overline{k})$ of $\overline{k}$-valued points of X. The degree of a closed point x, denoted $\deg(x)$, is the cardinality of the corresponding orbit. Thus the closed points of degree dividing a given integer $n \geq 1$ are precisely the orbits of $\mathrm{Gal}(\overline{k}/k)$ on the set $X(k_n)$. So if we denote by

$$B_n := B_n(X/k) := \text{ the number of closed points of degree } n,$$

we have the identities

$$N_n = \sum_{r|n} rB_r.$$

These identities are then equivalent to the Euler product expression

$$\mathrm{Zeta}(X/k, T) = \prod_{\text{closed points } x \text{ of } X} (1 - T^{\deg(x)})^{-1},$$

which in turn shows that $\mathrm{Zeta}(X/k, T)$ lies in $1 + T\mathbb{Z}[[T]]$. By a fundamental result of Dwork [**Dw**], $\mathrm{Zeta}(X/k, T)$ is a rational function of T, i.e., lies in $\mathbb{Q}(T)$, and, in lowest terms, is of the form $A(T)/B(T)$ with A and B in $1 + T\mathbb{Z}[T]$. In particular, the reciprocal zeroes and poles of the zeta function are algebraic integers.

9.1.3. We now turn to Grothendieck's l-adic cohomological approach to the zeta function. For any prime number l invertible in k, Grothendieck et al. defined [**SGA 1, 4, 4$\frac{1}{2}$, 5**], for such an X/k, compact cohomology groups $H^i_c(X \otimes_k \overline{k}, \overline{\mathbb{Q}}_l)$, which are finite-dimensional $\overline{\mathbb{Q}}_l$-spaces on which $\mathrm{Gal}(\overline{k}/k)$ acts continuously, and which vanish unless i lies in $[0, 2\dim(X)]$.

9.1.4. Before going further, notice that each $H^i_c(X \otimes_k \overline{k}, \overline{\mathbb{Q}}_l)$, thought of as a finite-dimensional $\overline{\mathbb{Q}}_l$-space on which $\mathrm{Gal}(\overline{k}/k)$ acts continuously, may be thought of, tautogously, as a lisse $\overline{\mathbb{Q}}_l$-sheaf on $\mathrm{Spec}(k)$. This point of view will be of vital conceptual importance below.

9.1.5. The Lefschetz trace formula [**Gro-FL, SGA $4\frac{1}{2}$**, Rapport] for X/k, recalled in 9.0.15.1 above, asserts that for each $n \geq 1$,

$$N_n = \sum_i (-1)^i \operatorname{Trace}((F_k)^n | H^i_c(X \otimes_k \overline{k}, \overline{\mathbb{Q}}_l)) \quad \text{in } \overline{\mathbb{Q}}_l. \tag{9.1.5.1}$$

This formula for given X/k and all n is equivalent to the formula

$$\operatorname{Zeta}(X/k, T) = \prod_{i=0}^{2\dim(X)} \det(1 - TF_k | H^i_c(X \otimes_k \overline{k}, \overline{\mathbb{Q}}_l))^{(-1)^{i+1}}, \tag{9.1.5.2}$$

an identity in $1 + T\overline{\mathbb{Q}}_l[[T]]$ which we call the l-adic cohomological expression of the zeta function. It is known [**SGA 7**, Exposé XXI, Appendice, 5.2.2] that the individual factors have coefficients in the ring of all algebraic integers, i.e., that the eigenvalue α of F_k on $H^i_c(X \otimes_k \overline{k}, \overline{\mathbb{Q}}_l)$ are algebraic integers. Deligne's main result in Weil II, [**De-Weil II**, 3.3.4], applied to X/k, gives that $H^i_c(X \otimes_k \overline{k}, \overline{\mathbb{Q}}_l)$ as lisse $\overline{\mathbb{Q}}_l$-sheaf on $\operatorname{Spec}(k)$ is mixed of integer weights w in $[0, i]$, in the sense that it is a successive extension of lisse $\overline{\mathbb{Q}}_l$-sheaves on $\operatorname{Spec}(k)$ each of which is pure of some integer weight in $[0, i]$. Concretely, this means that for any eigenvalue α of F_k on $H^i_c(X \otimes_k \overline{k}, \overline{\mathbb{Q}}_l)$, there is an integer w in $[0, i]$, called the weight of α, such that $|\iota(\alpha)| = \operatorname{Card}(k)^{w/2}$ for any complex embedding ι of the field of all algebraic numbers. We should remind the reader that, for a general X/k, it is (conjectured but) not known that for each i, the individual l-adic factor

$$\det(1 - TF_k | H^i_c(X \otimes_k \overline{k}, \overline{\mathbb{Q}}_l))$$

lies in $\mathbb{Z}[T]$ and is independent of the auxiliary choice of l invertible in k.

9.1.6. We now assume that in addition X/k is proper, smooth, and geometrically connected, of dimension $d \geq 0$. In this case, we have Poincaré duality: the cup-product pairing

$$H^i_c(X \otimes_k k, \overline{\mathbb{Q}}_l) \times H^{2d-i}_c(X \otimes_k \overline{k}, \overline{\mathbb{Q}}_l) \to H^{2d}_c(X \otimes_k \overline{k}, \overline{\mathbb{Q}}_l) \cong \overline{\mathbb{Q}}_l(-d)$$

is a perfect pairing which is $\operatorname{Gal}(\overline{k}/k)$-equivariant. Since $\overline{\mathbb{Q}}_l(-d)$ is pure of weight $2d$, and each H^i_c is mixed of weight $\leq i$, it follows from this duality that each $H^i_c(X \otimes_k \overline{k}, \overline{\mathbb{Q}}_l)$ is pure of weight i, cf. [**De-Weil II**, 3.3.9].

9.1.7. Thus for X/k proper, smooth and geometrically connected of dimension $d \geq 0$, each $H^i_c(X \otimes_k \overline{k}, \overline{\mathbb{Q}}_l)$, thought of as a finite-dimensional $\overline{\mathbb{Q}}_l$-space on which $\operatorname{Gal}(\overline{k}/k)$ acts continuously, is a lisse $\overline{\mathbb{Q}}_l$-sheaf on $\operatorname{Spec}(k)$ which is pure of weight i.

9.1.8. From this purity, we see that there can be no cancellation in the cohomological expression of the zeta function, that all the reciprocal zeroes (respectively poles) of the zeta function are pure of some odd (resp. even) weight, and that for each i the individual factor

$$\det(1 - TF_k | H^i_c(X \otimes_k \overline{k}, \overline{\mathbb{Q}}_l))$$

is independent of l and lies in $\mathbb{Z}[T]$, because it can be recovered intrinsically from the zeta function by looking at the zeroes (for i odd) or poles (for i even) of the zeta function which lie on the complex circle of radius $\operatorname{Card}(k)^{-i/2}$.

9.1.9. If we fix a choice α_k of $q^{1/2}$ in $\overline{\mathbb{Q}}_l$, we may form the lisse rank one $\overline{\mathbb{Q}}_l$-sheaf

$$\overline{\mathbb{Q}}_l(-1/2) := (\alpha_k)^{\deg} \tag{9.1.9.1}$$

on $\mathrm{Spec}(k)$, and its tensor powers $\overline{\mathbb{Q}}_l(-i/2) := (\overline{\mathbb{Q}}_l(-1/2))^{\otimes i}$, i in $\mathbb{Z}$. We then form the Tate-twisted cohomology groups $H^i_c(X \otimes_k \overline{k}, \overline{\mathbb{Q}}_l)(i/2)$, which are pure of weight zero. On these twisted groups, cup product induces $\mathrm{Gal}(\overline{k}/k)$-equivariant perfect pairings

$$\begin{aligned} &H^i_c(X \otimes_k \overline{k}, \overline{\mathbb{Q}}_l)(i/2) \times H^{2d-i}_c(X \otimes_k \overline{k}, \overline{\mathbb{Q}}_l)((2d-i)/2) \\ &\quad \to H^{2d}_c(X \otimes_k \overline{k}, \overline{\mathbb{Q}}_l)(d) \cong \overline{\mathbb{Q}}_l. \end{aligned} \tag{9.1.9.2}$$

In particular, the twisted middle dimensional cohomology group

$$H^d_c(X \otimes_k \overline{k}, \overline{\mathbb{Q}}_l)(d/2)$$

is autodual. Because cup product obeys the usual sign rules, this autoduality is symplectic if d is odd, orthogonal if d is even.

9.1.10. Now let us consider not a single X/k which is proper, smooth, and geometrically connected, but rather a family of such, by which we mean a proper smooth morphism $\pi : X \to S$ with geometrically connected fibres. We fix a prime number l, and we assume that S is a connected normal $\mathbb{Z}[1/l]$-scheme which is separated and of finite type over $\mathbb{Z}[1/l]$. For each field k, and each k-valued point s in $S(k)$, the fibre X_s/k of X/S at the point s is proper, smooth and geometrically connected. For each i, the compact cohomology group $H^i_c(X_s \otimes_k \overline{k}, \overline{\mathbb{Q}}_l)$ with its continuous action of $\mathrm{Gal}(\overline{k}/k)$ is a lisse $\overline{\mathbb{Q}}_l$-sheaf on $\mathrm{Spec}(k)$, which, if k is finite, is pure of weight i. How are these sheaves related for various fields k, and for various k-valued points s of S?

9.1.11. The answer is given by the specialization theorem for the cohomology of the fibres of a proper smooth morphism [**SGA 4**, XVI, 2.2], which in turn depends on the proper base change theorem [**SGA 4**, XII, 5.1 (iii) and 5.2 (iii)] and the smooth base change theorem [**SGA 4**, XVI, 1.1 and 1.2]. For each i there is a lisse $\overline{\mathbb{Q}}_l$-sheaf, denoted $R^i\pi_!\overline{\mathbb{Q}}_l$, of rank denoted b_i, on S, i.e., a continuous finite dimensional $\overline{\mathbb{Q}}_l$-representation $\rho_{i,l}$ of $\pi_1(S)$ to $GL(b_i, \overline{\mathbb{Q}}_l)$, with the following interpolation property:

9.1.11.1. For every field k, and every point s in $S(k)$, viewed as a morphism $s : \mathrm{Spec}(k) \to S$, the pullback $s^*(R^i\pi_!\overline{\mathbb{Q}}_l)$, i.e., the composite representation

$$\mathrm{Gal}(\overline{k}/k) = \pi_1(\mathrm{Spec}(k)) \xrightarrow{s_*} \pi_1(S) \xrightarrow{\rho_{i,l}} GL(b_i, \overline{\mathbb{Q}}_l)$$

of $\mathrm{Gal}(\overline{k}/k)$, is isomorphic to the sheaf on $\mathrm{Spec}(k)$ given by $H^i_c(X_s \otimes_k \overline{k}, \overline{\mathbb{Q}}_l)$ with its continuous action of $\mathrm{Gal}(\overline{k}/k)$.

9.1.12. Let us make several remarks here. First of all, if we take for k the function field K_S of S, and for s in $S(K_S)$ the generic point η of S, then $\pi_1(S)$ is a quotient of $\mathrm{Gal}(\overline{K}_S/K_S)$. So a representation of $\pi_1(S)$ is determined by its pullback to $\mathrm{Gal}(\overline{K}_S/K_S)$. Therefore there is at most one lisse $\overline{\mathbb{Q}}_l$-sheaf on S which has the stated interpolation property (9.1.11.1), even for the single test case (K_S, η). So we could rephrase the theorem in to parts, as saying that

1) $H^i_c(X_\eta \otimes_{K_S} \overline{K}_S, \overline{\mathbb{Q}}_l)$ as representation of $\mathrm{Gal}(\overline{K}_S/K_S)$ is "unramified on S" in the sense that it is trivial on the kernel of the surjection $\mathrm{Gal}(\overline{K}_S/K_S) \twoheadrightarrow \pi_1(S)$, and hence factors through this surjection, and thus defines a lisse $\overline{\mathbb{Q}}_l$-sheaf on S,

2) the lisse sheaf on S defined in part 1) has the stated interpolation property (9.1.11.1).

9.1.13. The second remark is this. Take for k a finite field, s in $S(k)$. On the one hand, we defined a Frobenius conjugacy class $F_{k,s}$ in $\pi_1(S)$, so we may speak of the characteristic polynomial

$$\det(1 - TF_{k,s}|R^i\pi_!\overline{\mathbb{Q}}_l) := \det(1 - T\rho_{i,l}(F_{k,s})).$$

On the other hand, we defined $s^*(R^i\pi_!\overline{\mathbb{Q}}_l)$ as a lisse sheaf on $\mathrm{Spec}(k)$, so we may speak of

$$\det(1 - TF_k|s^*(R^i\pi_!\overline{\mathbb{Q}}_l)).$$

It is tautologous that we have the equality

$$\det(1 - TF_{k,s}|R^i\pi_!\overline{\mathbb{Q}}_l) = \det(1 - TF_k|s^*(R^i\pi_!\overline{\mathbb{Q}}_l)).$$

It results from the specialization theorem that we have the equality

$$\det(1 - TF_k|s^*(R^i\pi_!\overline{\mathbb{Q}}_l)) = \det(1 - TF_k|H^i_c(X_s \otimes_k \overline{k}, \overline{\mathbb{Q}}_l)).$$

Putting these last two equalities together, we find the fundamental compatibility

$$\det(1 - TF_{k,s}|R^i\pi_!\overline{\mathbb{Q}}_l) = \det(1 - TF_k|H^i_c(X_s \otimes_k \overline{k}, \overline{\mathbb{Q}}_l)). \tag{9.1.13.1}$$

In the extreme cases $i = 0$ and $i = 2d$, for each fibre X_s/k, we have

$$H^0_c(X_s \otimes_k \overline{k}, \overline{\mathbb{Q}}_l) = \overline{\mathbb{Q}}_l,$$
$$H^{2d}_c(X_s \otimes_k \overline{k}, \overline{\mathbb{Q}}_l)(d) = \overline{\mathbb{Q}}_l,$$

as representations of $\mathrm{Gal}(\overline{k}/k)$. We may infer that the lisse rank one $\overline{\mathbb{Q}}_l$-sheaves $R^0\pi_!\overline{\mathbb{Q}}_l$ and $R^{2d}\pi_!\overline{\mathbb{Q}}_l(d)$ are both isomorphic to $\overline{\mathbb{Q}}_l$, i.e., trivial as characters of $\pi_1(S)$. To check this, it suffices by Chebotarev to show that both are trivial on all Frobenius conjugacy classes $F_{k,s}$, and this triviality is given by 9.1.13.1 above.

9.1.14. The cup product pairings

$$(R^i\pi_!\overline{\mathbb{Q}}_l)(i/2) \times (R^{2d-i}\pi_!\overline{\mathbb{Q}}_l)((2d-i)/2) \to (R^{2d}\pi_!\overline{\mathbb{Q}}_l)(d) \cong \overline{\mathbb{Q}}_l$$

are perfect pairings of lisse sheaves, pure of weight zero, which induce on fibres the cup product pairings of 9.1.9.2. In particular, $(R^d\pi_!\overline{\mathbb{Q}}_l)(d/2)$ is autodual, symplectically for d odd and orthogonally for d even.

Scholie 9.1.15 *Let X/S be proper and smooth with geometrically connected fibres all of dimension $d \geq 0$, with S normal, connected, and of finite type over $\mathbb{Z}[1/l]$. The lisse $\overline{\mathbb{Q}}_l$-sheaves $R^i\pi_!\overline{\mathbb{Q}}_l$ on S vanish for i outside $[0, 2d]$, $R^i\pi_!\overline{\mathbb{Q}}_l$ is of pure weight i, $R^0\pi_!\overline{\mathbb{Q}}_l \cong \overline{\mathbb{Q}}_l$, $R^{2d}\pi_!\overline{\mathbb{Q}}_l \cong \overline{\mathbb{Q}}_l(-d)$, and $R^d\pi_!\overline{\mathbb{Q}}_l(d/2)$ is autodual, symplectically for d odd and orthogonally for d even. For each finite field k, and each point s in $S(k)$, we have*

$$\det(1 - TF_{k,s}|R^i\pi_!\overline{\mathbb{Q}}_l) = \det(1 - TF_k|H^i_c(X_s \otimes_k \overline{k}, \overline{\mathbb{Q}}_l)).$$

Taking the alternating product over i, we get

$$\mathrm{Zeta}(X_s/k, T) = \prod_{i=0}^{2\dim(X)} \det(1 - TF_{k,s}|R^i\pi_!\overline{\mathbb{Q}}_l)^{(-1)^{i+1}}.$$

9.1.16. Just to make this entirely down to earth, let us consider the case when X/S is a family of proper smooth geometrically connected curves C/S of genus g, i.e., the case $d = 1$ of the above discussion. The above formula for the zeta functions of the fibres boils down to this: for any finite field k, and any s in $S(k)$, we have

$$\mathrm{Zeta}(C_s/k, T) = \det(1 - TF_{k,s}|R^1\pi_!\overline{\mathbb{Q}}_l)/(1-T)(1-\mathrm{Card}(k)T),$$
$$\det(1 - TF_{k,s}|R^1\pi_!\overline{\mathbb{Q}}_l) = \det(1 - TF_k|H^1_c(C_s \otimes_k \overline{k}, \overline{\mathbb{Q}}_l)),$$

and the twisted sheaf $R^1\pi_!\overline{\mathbb{Q}}_l(1/2)$ is symplectically self-dual of rank $2g$, and pure of weight zero. This sheaf $R^1\pi_!\overline{\mathbb{Q}}_l(1/2)$ is perhaps the archetypical example of a lisse $\overline{\mathbb{Q}}_l$-sheaf which is pure of weight zero.

9.2. A version of Deligne's equidistribution theorem

9.2.1. We can now state and prove a useful version of Deligne's equidistribution theorem. We return to our basic setup: k is a finite field, X/k is smooth and geometrically connected of dimension $d \geq 0$, l is invertible in k, ι is a complex embedding of $\overline{\mathbb{Q}}_l$, and $\mathcal{F}$ is a lisse $\overline{\mathbb{Q}}_l$-sheaf on X, which we suppose to be ι-pure of weight zero. As above, we denote by G_{geom} the Zariski closure of $\rho(\pi_1^{\mathrm{geom}})$, which is a semisimple algebraic group over $\overline{\mathbb{Q}}_l$ (by [**De-Weil II**, 1.3.9 and the first four lines of its proof] and the ι-purity of $\mathcal{F}$). We now make the following assumption:

$$\text{under } \rho, \text{we have } \rho(\pi_1) \subset G_{\mathrm{geom}}(\overline{\mathbb{Q}}_l). \tag{9.2.1.1}$$

9.2.2. We will use this assumption in the following way. For any finite-dimensional $\overline{\mathbb{Q}}_l$-representation

$$\Lambda : G_{\mathrm{geom}} \to GL(m)$$

of the algebraic group G_{geom}, the composite representation $\Lambda \circ \rho$ of π_1,

$$\pi_1 \to G_{\mathrm{geom}}(\overline{\mathbb{Q}}_l) \to GL(m, \overline{\mathbb{Q}}_l)$$

"is" a lisse $\overline{\mathbb{Q}}_l$-sheaf, denoted $\Lambda(\mathcal{F})$, on X, with the property that for every element γ in π_1, we have

$$\mathrm{Trace}(\gamma|\Lambda(\mathcal{F})) = \mathrm{Trace}(\Lambda(\rho(\gamma))).$$

9.2.3. Via ι, we may speak of the group $G_{\mathrm{geom}}(\mathbb{C})$, which we may view as a complex Lie group. Because G_{geom} is semisimple, there exists in $G_{\mathrm{geom}}(\mathbb{C})$ a maximal compact subgroup K, and any two such are conjugate.

9.2.4. Our next task is to define Frobenius conjugacy classes in the compact group K. For any finite extension E of k, for any x in $X(E)$, and for any choice of a Frobenius element $F_{E,x}$ in π_1, the image $\iota\rho(F_{E,x})$ in $G_{\mathrm{geom}}(\mathbb{C})$ has all its eigenvalues on the unit circle ($\mathcal{F}$ being ι-pure of weight zero). We do not know that $\rho(F_{E,x})$ is semisimple, i.e., diagonalizable, but if we pass to its semisimple part $\rho(F_{E,x})^{ss}$ is the sense of Jordan decomposition, we obtain an element $\iota(\rho(F_{E,x})^{ss})$ of $G_{\mathrm{geom}}(\mathbb{C})$ which is semisimple with unitary eigenvalues, and which hence lies in a compact subgroup of $G_{\mathrm{geom}}(\mathbb{C})$ (e.g., it lies in the topological closure of the cyclic group it generates, which is such a compact subgroup). Therefore $\iota(\rho(F_{E,x})^{ss})$ is $G_{\mathrm{geom}}(\mathbb{C})$-conjugate to an element of K. Using the fact that G_{geom} and $G_{\mathrm{geom}}(\mathbb{C})$ have the "same" finite-dimensional representation theory, the unitarian trick (that $G_{\mathrm{geom}}(\mathbb{C})$ and K have the "same" finite-dimensional representation theory) and the Peter-Weyl theorem (that K-conjugacy classes are separated by the traces of finite-dimensional representations of K), one shows easily [**De-Weil II**, proof of 2.2.2] that the element of K obtained this way is well-defined up to K-conjugacy. In other words, the $G_{\text{geom}}(\mathbb{C})$-conjugacy class of $\iota(\rho(F_{E,x})^{ss})$ meets K in a single K-conjugacy class. We denote this K-conjugacy class $\vartheta(E,x)$, and think of it as the generalized "angle" of $\iota\rho(F_{E,x})$.

9.2.5. We can now state a version of Deligne's equidistribution theorem. This version is similar to the one given in [**Ka-GKM**, 3.6], where we assumed in addition that X/k was an affine curve. There we used the Euler-Poincaré formula to prove 3) below, giving an expression for the constant here denoted $C(X\otimes_k \overline{k}, \mathcal{F})$ in terms of the breaks of $\mathcal{F}$ at each of the points at infinity. In that case, denoting by g the genus of the complete nonsingular model $\overline{X}$ of X, by N the number of $\overline{k}$-points in $\overline{X}-X$, say x_i for $i=1,\ldots,N$, and by r_i the largest break of $\mathcal{F}$ at x_i, we found that

$$C(X\otimes_k \overline{k}, \mathcal{F}) = 2g-2+N+\sum_i r_i$$

would work in 3). The idea of bounding the constant in the general case, as we do here in 4) below, we learned from Ofer Gabber.

Theorem 9.2.6 (compare [**De-Weil II**, 3.5.3], [**Ka-GKM**, 3.6]). *Let k be a finite field, X/k a smooth, geometrically connected scheme of dimension $d\geq 0$, l a prime number invertible in k, ι a field embedding of $\overline{\mathbb{Q}}_l$ into $\mathbb{C}$, and $\mathcal{F}$ a lisse $\overline{\mathbb{Q}}_l$-sheaf on X of rank $r\geq 1$ corresponding to an r-dimensional continuous $\overline{\mathbb{Q}}_l$-representation ρ of π_1. Denote by G_{geom} the Zariski closure of $\rho(\pi_1^{\text{geom}})$. Assume that $\mathcal{F}$ is ι-pure of weight zero, and that $\rho(\pi_1)$ lies in $G_{\text{geom}}(\overline{\mathbb{Q}}_l)$. Denote by K a maximal compact subgroup of $G_{\text{geom}}(\mathbb{C})$. Then we have the following results.*

1) *The Frobenius conjugacy classes $\vartheta(E,x)$ defined in 9.2.4 above are equidistributed in the space of conjugacy classes of K, in the sense that for any continuous $\mathbb{C}$-valued central function f on K, we have the limit formula*

$$\int_K f\, d\,\text{Haar} = \lim_{\text{Card}(E)\to\infty} (1/\text{Card}(X(E))) \sum_{x \text{ in } X(E)} f(\vartheta(E,x)),$$

the limit taken over finite extensions E of k large enough that $X(E)$ is nonempty.

2) *For any finite-dimensional irreducible nontrivial $\mathbb{C}$-representation Λ of K, deduced via ι and restriction to K from a finite-dimensional irreducible nontrivial $\overline{\mathbb{Q}}_l$-representation Λ of G_{geom}, we have the inequality*

$$\left|\sum_{x \text{ in } X(E)} \text{Trace}(\Lambda(\vartheta(E,x)))\right| \leq \sum_i h_c^i(X\otimes_k \overline{k}, \Lambda(\mathcal{F}))\,\text{Card}(E)^{(2d-1)/2}$$

for all finite extensions E of k.

3) *There exists an integer $C(X\otimes_k \overline{k}, \mathcal{F})$ with the property that for any finite-dimensional $\overline{\mathbb{Q}}_l$-representation Λ of G_{geom}, we have the inequality*

$$\sum_i h_c^i(X\otimes_k \overline{k}, \Lambda(\mathcal{F})) \leq \dim(\Lambda) C(X\otimes_k \overline{k}, \mathcal{F}).$$

4) *Pick a finite extension E_λ of $\mathbb{Q}_l$ such that ρ lands in $GL(r, E_\lambda)$. Denote by $\mathcal{O}_\lambda$ the ring of λ-adic integers in E_λ, and by $\mathbb{F}_\lambda$ the residue field of $\mathcal{O}_\lambda$. Pick an*

$\mathcal{O}_\lambda$-form $\mathcal{F}_{\mathcal{O}_\lambda}$ of $\mathcal{F}$, and consider the lisse $\mathbb{F}_\lambda$-sheaf $\mathcal{F}_{\mathrm{mod}\,\lambda} := \mathcal{F}_{\mathcal{O}_\lambda}/\lambda\mathcal{F}_{\mathcal{O}_\lambda}$. Choose any finite etale galois covering (not necessarily connected) Y of $X \otimes_k \overline{k}$, say

$$\varphi : Y \to X \otimes_k \overline{k},$$

such that $\varphi^*(\mathcal{F}_{\mathrm{mod}\,\lambda})$ on Y is trivial. Then $C(X \otimes_k \overline{k}, \mathcal{F}) := \sum_i h^i_c(Y, \mathbb{F}_\lambda)$, the sum of the mod λ *Betti numbers of Y, "works" in 3) above.*

5) *For $A(X \otimes_k \overline{k}) := \sum_{i<2d} h^i_c(X \otimes_k \overline{k}, \overline{\mathbb{Q}}_l)$ and $C(X \otimes_k \overline{k}, \mathcal{F})$ as in 4) above, for every irreducible nontrivial representation Λ of K, and every finite extension E of k with* $\mathrm{Card}(E) \geq 4A(X \otimes_k \overline{k})^2$, *we have the estimate*

$$\left| (1/\mathrm{Card}(X(E))) \sum_{x \text{ in } X(E)} \mathrm{Trace}(g(\vartheta(E,x))) \right|$$
$$\leq 2\dim(\Lambda) C(X \otimes_k \overline{k}, \mathcal{F}) / \mathrm{Card}(E)^{1/2}.$$

PROOF. Statement 2) follows from [**De-Weil II**] and the Lefschetz trace formula for the sheaf $\Lambda(\mathcal{F})$ on X/k, namely that for every finite extension E of k we have

$$\sum_{x \text{ in } X(E)} \mathrm{Trace}(\Lambda(\rho(F_{E,x})))$$
$$= \sum_i (-1)^i \mathrm{Trace}((F_k)^{\deg(E/k)} | H^i_c(X \otimes_k \overline{k}, \Lambda(\mathcal{F}))).$$

In this equality, the left hand side we may rewrite as

$$\sum_{x \text{ in } X(E)} \mathrm{Trace}(\Lambda(\rho(F_{E,x}))^{ss}) = \sum_{x \text{ in } X(E)} \mathrm{Trace}(\Lambda(\rho(F_{E,x})^{ss})),$$

whose image under ι is precisely $\sum_{x \text{ in } X(E)} \mathrm{Trace}(\Lambda(\vartheta(E,x)))$. What about the right hand side? For Λ irreducible and nontrivial, we have

$$H^{2d}_c(X \otimes_k \overline{k}, \Lambda(\mathcal{F})) = 0,$$

and the sheaf $\Lambda(\mathcal{F})$ is ι-pure of weight zero (G_{geom} being semisimple, $\Lambda(\mathcal{F})$ is a subquotient of some tensor power $\mathcal{F}^{\otimes n}$ of $\mathcal{F}$). Thus by the main result of [**De-Weil II**], each group $H^i_c(X \otimes_k \overline{k}, \Lambda(\mathcal{F}))$ is ι-mixed of weight $\leq i$. So we get

$$\left| \sum_{x \text{ in } X(E)} \mathrm{Trace}(\Lambda(\vartheta(E,x))) \right| \leq \sum_i h^i_c(X \otimes_k \overline{k}, \Lambda(\mathcal{F}))\,\mathrm{Card}(E)^{i/2}$$
$$\leq \sum_{i \leq 2d-1} h^i_c(X \otimes_k \overline{k}, \Lambda(\mathcal{F}))\,\mathrm{Card}(E)^{i/2}$$
$$\leq \sum_i h^i_c(X \otimes_k \overline{k}, \Lambda(\mathcal{F}))\,\mathrm{Card}(E)^{(2d-1)/2}.$$

Now that we have proven 2), we turn to proving 1). We first analyze when we can be sure $X(E)$ will be nonempty. By the Lang-Weil estimate (9.0.15.2), if we denote by $A(X \otimes_k \overline{k}) := \sum_{i<2d} h^i_c(X \otimes_k \overline{k}, \overline{\mathbb{Q}}_l)$ the sum of all but the highest l-adic Betti numbers of X, we have the inequality

$$|\,\mathrm{Card}(X(E)) - \mathrm{Card}(E)^d| \leq A(X \otimes_k \overline{k})\,\mathrm{Card}(E)^{(2d-1)/2}$$

for every finite extension E of k. So $X(E)$ is nonempty provided that

$$\mathrm{Card}(E)^{1/2} > A(X \otimes_k \overline{k}),$$

and if $\operatorname{Card}(E)^{1/2} > 2A(X \otimes_k \overline{k})$ we have

$$\operatorname{Card}(X(E)) \geq \operatorname{Card}(E)^d/2.$$

So by 2) we have

$$\left|(1/\operatorname{Card}(X(E))) \sum_{x \text{ in } X(E)} \operatorname{Trace}(\Lambda(\vartheta(E,x)))\right|$$
$$\leq \sum_i h_c^i(X \otimes_k \overline{k}, \Lambda(\mathcal{F}))[\operatorname{Card}(E)^d/\operatorname{Card}(X(E))]/\operatorname{Sqrt}(\operatorname{Card}(E))$$
$$\leq \sum_i h_c^i(X \otimes_k \overline{k}, \Lambda(\mathcal{F}))2/\operatorname{Sqrt}(\operatorname{Card}(E)).$$

Once we have this estimate, then 1) follows by the Peter-Weyl theorem, which assures us that every continuous $\mathbb{C}$-valued central function f on K is a uniform limit of finite sums of trace functions of irreducible, finite-dimensional representations of K. To prove 1), it suffices to check individually the case that f is either the constant function 1, in which case there is no need to pass to a limit, or the case when f is the trace function of an irreducible nontrivial representation Λ, in which case the integral $\int_K \operatorname{Trace}(\Lambda(k))\, dk = 0$, and the finite E-sum is $O(1/\operatorname{Sqrt}(\operatorname{Card}(E)))$ by this last estimate.

To prove 3) and 4), we argue as follows. We must show that for any finite-dimensional $\overline{\mathbb{Q}}_l$-representation Λ of G_{geom}, we have the inequality

$$\sum_i h_c^i(X \otimes_k \overline{k}, \Lambda(\mathcal{F})) \leq \dim(\Lambda) \sum_i h_c^i(Y, \mathbb{F}_\lambda).$$

Fix one such Λ. It is defined over some finite extension H of our already chosen E_λ. If we work over $\mathcal{O}_H$, using the $\mathcal{O}_H$-form $\mathcal{F}_{\mathcal{O}_\lambda} \otimes_{\mathcal{O}_\lambda} \mathcal{O}_H$, its reduction $(\mathcal{F}_{\mathcal{O}_\lambda} \otimes_{\mathcal{O}_\lambda} \mathcal{O}_H) \otimes_{\mathcal{O}_H} \mathbb{H}_\lambda$ is just the extension of scalars $(\mathcal{F}_{\operatorname{mod} \lambda}) \otimes_{\mathbb{F}_\lambda} \mathbb{H}_\lambda$, and hence itself becomes trivial as $\mathbb{H}_\lambda$-sheaf on the same Y. Since

$$H_c^i(Y, \mathbb{H}_\lambda) = H_c^i(Y, \mathbb{F}_\lambda) \otimes_{\mathbb{F}_\lambda} \mathbb{H}_\lambda,$$

we have $\sum_i h_c^i(Y, \mathbb{F}_\lambda) = \sum_i h_c^i(Y, \mathbb{H}_\lambda)$. So it suffices to treat universally the case in which the representation Λ is also defined over E_λ.

Since $\varphi : Y \to X \otimes_k \overline{k}$ is finite etale galois, say with group Γ, for any $\overline{\mathbb{Q}}_l$-sheaf $\mathcal{G}$ on $X \otimes_k \overline{k}$ we have

$$H_c^i(X \otimes_k \overline{k}, \mathcal{G}) \cong H_c^i(Y, \varphi^*\mathcal{G})^\Gamma,$$

so we have a trivial inequality $h_c^i(X \otimes_k \overline{k}, \mathcal{G}) \leq h_c^i(Y, \varphi^*\mathcal{G})$. The sheaf $\Lambda(\mathcal{F})$ is a subquotient of some tensor power $\mathcal{F}^{\otimes n}$ of $\mathcal{F}$, so $\varphi^*\Lambda(\mathcal{F})$ is a subquotient of $(\varphi^*\mathcal{F})^{\otimes n}$. Pick an $\mathcal{O}_\lambda$-form $\Lambda(\mathcal{F})_{\mathcal{O}_\lambda}$ of $\Lambda(\mathcal{F})$. We claim first that the lisse $\mathbb{F}_\lambda$-sheaf $\Lambda(\mathcal{F})_{\operatorname{mod} \lambda}$ defined as the reduction $\operatorname{mod} \lambda$ of $\Lambda(\mathcal{F})_{\mathcal{O}_\lambda}$ has $\varphi^*(\Lambda(\mathcal{F})_{\operatorname{mod} \lambda})$ a successive extension of the constant sheaf $\mathbb{F}_\lambda$ by itself on Y. To see this, let U be any connected component of Y, and γ any element of $\pi_1(U$, some base point). By hypothesis, the characteristic polynomial of γ on $\varphi^*\mathcal{F}$, which is a priori a monic polynomial of degree $r = \operatorname{rank}(\mathcal{F})$ with $\mathcal{O}_\lambda$-coefficients, is certainly congruent $\operatorname{mod} \lambda$ to $(T-1)^r$, because γ acts trivially $\operatorname{mod} \lambda$. Therefore the characteristic polynomial of γ on $\varphi^*\mathcal{F}^{\otimes n}$ is monic with $\mathcal{O}_\lambda$-coefficients and $\operatorname{mod} \lambda$ is $(T-1)^{r^n}$. Consider the characteristic polynomial of γ on $\varphi^*\Lambda(\mathcal{F})$. It is monic with $\mathcal{O}_\lambda$-coefficients, and it divides the characteristic polynomial of γ on $\varphi^*\mathcal{F}^{\otimes n}$. Therefore the characteristic polynomial of γ on $\varphi^*\Lambda(\mathcal{F}), \bmod \lambda$, is a divisor of $(T-1)^{r^n}$, and hence must itself be congruent $\bmod \lambda$ to $(T-1)^{\dim(\Lambda)}$. Since a representation is determined up to semisimplification by its characteristic polynomials, it follows that on each connected component U of Y, $\varphi^*\Lambda(\mathcal{F})_{\bmod \lambda}$ is a successive extension of constant sheaves $\mathbb{F}_\lambda$. Writing the long exact cohomology sequence attached to a short exact sequence of $\mathbb{F}_\lambda$-sheaves on Y, say

$$0 \to A \to B \to C \to 0,$$

we see that

$$\sum_i h_c^i(Y, B) \le \sum_i h_c^i(Y, A) + \sum_i h_c^i(Y, C).$$

This yields

$$\sum_i h_c^i(Y, \varphi^*\Lambda(\mathcal{F})_{\bmod \lambda}) \le \dim(\Lambda) \sum_i h_c^i(Y, \mathbb{F}_\lambda).$$

By universal coefficients, i.e., by the long exact cohomology sequence attached to the short exact sequence of sheaves on Y

$$0 \to \varphi^*\Lambda(\mathcal{F})_{\mathcal{O}_\lambda} \xrightarrow{\lambda} \varphi^*\Lambda(\mathcal{F})_{\mathcal{O}_\lambda} \to \varphi^*\Lambda(\mathcal{F})_{\bmod \lambda} \to 0,$$

we have for each i the inequality

$$h_c^i(Y, \varphi^*\Lambda(\mathcal{F})) \le h_c^i(Y, \varphi^*\Lambda(\mathcal{F})_{\bmod \lambda}).$$

Combining this with the already noted inequality

$$\sum_i h_c^i(X \otimes_k \bar{k}, \Lambda(\mathcal{F})) \le \sum_i h_c^i(Y, \varphi^*\Lambda(\mathcal{F})),$$

we find the asserted inequality

$$\sum_i h_c^i(X \otimes_k \bar{k}, \Lambda(\mathcal{F})) \le \dim(\Lambda) \sum_i h_c^i(Y, \mathbb{F}_\lambda).$$

Statement 5) is immediate from 2), 3), 4) and [**Lang-Weil**]. QED

9.3. A uniform version of Theorem 9.2.6

9.3.1. In the version of Deligne's equidistribution theorem given above, we were concerned with questions of equidistribution for a suitable $\mathcal{F}$ on a single X/k. We now turn to the question of uniformity, when we allow X/k to vary in a family. Thus we fix a prime number l, and a connected normal $\mathbb{Z}[1/l]$-scheme S which is separated and of finite type over $\mathbb{Z}[1/l]$. In fact, our principal application will be to the case when S is $\mathrm{Spec}(\mathbb{Z}[1/l])$ itself. Over S we give ourselves a smooth X/S all of whose geometric fibres are connected of common dimension $d \ge 0$. On X, which is normal and connected, we pick a geometric point ξ, and give ourselves a lisse $\overline{\mathbb{Q}}_l$-sheaf $\mathcal{F}$ of rank $r \ge 1$, corresponding to an r-dimensional continuous $\overline{\mathbb{Q}}_l$-representation ρ of the fundamental group $\pi_1(X, \xi)$. For each finite field k of characteristic different from l, and each point s in $S(k)$, the restriction $\mathcal{F}_s$ of $\mathcal{F}$ to the fibre X_s of X/S at s is a lisse $\overline{\mathbb{Q}}_l$-sheaf of rank r on a smooth, geometrically connected scheme X_s/k of dimension d over a finite field k in which l is invertible.

9.3.2. We first give two standard uniformity lemmas.

Lemma 9.3.3. *There exists an integer $A(X/S)$ such that for all finite fields k, and all k-valued points s in $S(k)$, we have the inequality*

$$A(X_s \otimes_k \overline{k}) := \sum_{i<2d} h_c^i(X_s \otimes_k \overline{k}, \overline{\mathbb{Q}}_l) \leq A(X/S).$$

PROOF. This is immediate from the constructibility of the higher direct images with compact support $R^i f_! \overline{\mathbb{Q}}_l$ for $f : X \to S$ the structural morphism, together with proper base change. QED

Lemma 9.3.4. *Pick a finite extension E_λ of $\mathbb{Q}_l$ such that $\rho(\pi_1(X,\xi))$ lies in $GL(r,E_\lambda)$, an $\mathcal{O}_\lambda$-form $\mathcal{F}_{\mathcal{O}_\lambda}$ of $\mathcal{F}$, and a finite etale galois $\varphi : Y \to X$ such that $\varphi^*(\mathcal{F}_{\mathcal{O}_\lambda}/\lambda\mathcal{F}_{\mathcal{O}_\lambda})$ is trivial as lisse $\mathbb{F}_\lambda$-sheaf on Y. [Thus for all finite fields k, and all k-valued points s in $S(k)$, the sheaf $\varphi_s^*((\mathcal{F}_{\mathcal{O}_\lambda})_s/\lambda(\mathcal{F}_{\mathcal{O}_\lambda})_s)$ is trivial as lisse $\mathbb{F}_\lambda$-sheaf on Y_s, so a fortiori is trivial on $Y_s \otimes_k \overline{k}$.] There exists an integer $C(X/S,\mathcal{F})$ such that for all finite fields k, and all k-valued points s in $S(k)$, we have the inequality*

$$C(X_s \otimes_k \overline{k}, \mathcal{F}_s) := \sum_i h_c^i(Y_s \otimes_k \overline{k}, \mathbb{F}_\lambda) \leq C(X/S,\mathcal{F}).$$

PROOF. This is immediate from the constructibility of the higher direct images with compact support $R^i g_! \mathbb{F}_\lambda$ for $g : Y \to S$ the structural map. QED

9.3.5. We now fix a semisimple $\overline{\mathbb{Q}}_l$-algebraic subgroup G of $GL(r)$, which we assume satisfies the following condition (9.3.5.1):

9.3.5.1. The given r-dimensional representation of G is irreducible ("G is an irreducible subgroup of $GL(r)$"), and the normalizer of G in $GL(r)$ is $\mathbb{G}_m \cdot G$.

9.3.6. The standard examples of such G in $GL(r)$ are the special linear group $SL(r)$, for $r \geq 3$ the full orthogonal group $O(r)$, for r odd the special orthogonal group $SO(r)$, and for r even the symplectic group $Sp(r)$.

9.3.7. Fix a field embedding ι of $\overline{\mathbb{Q}}_l$ into $\mathbb{C}$. We assume that the sheaf $\mathcal{F}$ on X satisfies the following conditions (9.3.7.1) and (9.3.7.2):

9.3.7.1. For every finite field k and every k-valued point s in $S(k)$, there exists a real number w_s such that $\mathcal{F}_s$ on X_s/k is ι-pure of weight w_s.

9.3.7.2. For every finite field k and every k-valued point s in $S(k)$, the geometric monodromy group $G_{\text{geom},s}$ attached to $\mathcal{F}_s$ on X_s/k is conjugate in $GL(r)$ to G.

9.4. Interlude: Pathologies around (9.3.7.1)

In all known examples which "occur in nature" and in which (9.3.7.1) holds, the weight w_s is independent of s. Here is an artificial example, worked out in a discussion with Bill McCallum, where the weight w_s does in fact vary with s. In this example, the rank r will be one, X/S will be S/S (thus each $\pi_1^{\text{geom}}(X_s)$ will be trivial), and G will be the identity subgroup $\{1\}$ of $GL(1)$. To begin, we pick a prime number l which does **not** split completely in the Gaussian field $\mathbb{Q}(i)$. Thus l is either 2 or a prime $l \equiv 3 \bmod 4$, and $\mathbb{Q}_l(i)$ is a field which is quadratic over $\mathbb{Q}_l$, with ring of integers $\mathbb{Z}_l[i]$. The scheme S will be the spec of the ring $R := \mathbb{Z}[1/2l, i]$. Over this ring, we have the elliptic curve E/R given by the (affine) equation $y^2 = x^3 - x$. Each maximal ideal of R is principal and has a unique generator π which lies in $\mathbb{Z}[i]$, is prime to l, and which satisfies $\pi \equiv 1 \bmod (1+i)^3$ in $\mathbb{Z}[i]$. As was already known implicitly to Gauss (cf. [**Weil-NS**]), the curve E/R has complex multiplication by $\mathbb{Z}[i]$, and the Frobenius endomorphism of $E \otimes_R (R/\pi R)$ is (the reduction mod π of) complex multiplication by π. Now consider the l-adic Tate module $T_l(E)$: it is a free of rank one $\mathbb{Z}_l[i]$-module on which $\pi_1(S)$ acts linearly and continuously by a continuous $(\mathbb{Z}_l[i])^\times$-valued character, say $\chi_{l,\,E/R}$, of $\pi_1(S)$. For each prime π in $\mathbb{Z}[i]$ which is prime to $2l$ and with $\pi \equiv 1 \bmod (1+i)^3$, the arithmetic Frobenius at π, call it φ_π, acts on $T_l(E)$ as π. Thus $\chi_{l,\,E/R}(\varphi_\pi) = \pi$ for every maximal ideal (π) of S. On the **geometric** Frobenius F_π at π, we have $\chi_{l,\,E/R}(F_\pi) = 1/\pi$.

The character $\chi_{2,\,E/R} : \pi_1(S) \to (\mathbb{Z}_2[i])^\times$ has image the subgroup

$$1 + (1+i)^3\mathbb{Z}_2[i].$$

To see this, notice that because the image is closed, it suffices to show that the image contains the dense subset consisting of the elements in $\mathbb{Z}[i]$ which are $1 \bmod (1+i)^3$. But by unique factorization in $\mathbb{Z}[i]$, these are precisely the products of gaussian primes π with the same property. We define

$$L_{2,E/R} := \log_2 \circ \chi_{2,\,E/R}.$$

Thus $L_{2,E/R} : \pi_1(S) \to (1+i)^3\mathbb{Z}_2[i]$ is a continuous surjective homomorphism onto the free rank two $\mathbb{Z}_2$-module $M_2 := (1+i)^3\mathbb{Z}_2[i]$.

For $l \equiv 3 \bmod 4$, the character $\chi_{l,\,E/R} : \pi_1(S) \to (\mathbb{Z}_l[i])^\times$ is surjective. To see this, notice that, by unique factorization, a dense subgroup of $(\mathbb{Z}_l[i])^\times$ is generated by

$$\{i, 1+i, \text{ all Gaussian primes } \pi \equiv 1 \bmod (1+i)^3 \text{ and prime to } l\}.$$

Since the image of $\chi_{l,\,E/R}$ contains all of the latter elements, it suffices to show that i and $1+i$ are each congruent mod any power l^n of l to an element of $\mathbb{Z}[i]$ which is $1 \bmod (1+i)^3$ and prime to l. To do i, use $i + l^{2n}(1-i)$. To do $1+i$, use $1 + i + l^{2n+1}i$. We define

$$L_{l,E/R} := \log_l \circ (\chi_{l,\,E/R})^{l^2-1}.$$

Thus $L_{l,E/R} : \pi_1(S) \to l\mathbb{Z}_l[i]$ is a continuous surjective homomorphism onto the free rank two $\mathbb{Z}_l$-module $M_l := l\mathbb{Z}_l[i]$.

Since $L_{l,E/R} \bmod l$ is onto M_l/lM_l, and Frobenius elements attached to closed points fill any finite abelian quotient of $\pi_1(S)$, there exist gaussian primes π and ρ such that $L_{l,E/R}(\pi)$ and $L_{l,E/R}(\rho) \bmod l$ form an $\mathbb{F}_l$-basis of M_l/lM_l. Hence $L_{l,E/R}(\pi)$ and $L_{l,E/R}(\rho)$ form a $\mathbb{Z}_l$-basis of M_l. Now consider the continuous $1+l\mathbb{Z}_l$-valued character $\tau_l : M_l \to 1 + l\mathbb{Z}_l$ defined by

$$\tau_l(aL_{l,E/R}(\pi) + bL_{l,E/R}(\rho)) = (1+l)^{-b} \quad \text{for } a, b \text{ in } \mathbb{Z}_l.$$

Finally, consider the composite homomorphism

$$\tau_l \circ L_{l,E/R} : \pi_1(S) \to 1 + l\mathbb{Z}_l \subset \mathbb{Z}_l^\times \subset \overline{\mathbb{Q}}_l^\times.$$

If we view this homomorphism as a lisse, rank one $\overline{\mathbb{Q}}_l$-sheaf $\mathcal{L}$ on S, then for **any** choice of ι, $\mathcal{L}_s$ is ι-pure of weight zero at the closed point s corresponding to (π), and $\mathcal{L}_s$ is ι-pure of nonzero weight $2\log(1+l)/\log(\rho\overline{\rho})$ ("log" here the logarithm of Napier) at the closed point s corresponding to (ρ).

Once we have this $\mathcal{L}$ on S, we can take an l-adic $\mathcal{F}$ on an X/S which satisfies (9.3.5.1), (9.3.7.1), and (9.3.7.2) with a single w, and twist by $\mathcal{L}$, i.e., pass to $\mathcal{F} \otimes f^*\mathcal{L}$ for $f : X \to S$ the structural map. The resulting situation $\mathcal{F} \otimes f^*\mathcal{L}$ on X/S still satisfies (9.3.5.1), (9.3.7.1), and (9.3.7.2), but now with a w_s which depends nontrivially on s. For a concrete example of such an $\mathcal{F}$ on an X/S, take X/S to be the spec of the ring $R[T, 1/(T(1-T))]$. Consider over X the Legendre family of elliptic curves $\mathcal{E}/X$, in which $\mathcal{E} - \{0_{\mathcal{E}}\}$ is defined by the equation $y^2 = x(x-1)(x-T)$, with structural map $g : \mathcal{E} \to X$. Take for $\mathcal{F}$ the sheaf $R^1 g_! \overline{\mathbb{Q}}_l$. This $\mathcal{F}$ is lisse on X of rank two [**SGA 4**, XVI, 2.2] ι-pure of weight one for every choice of ι [**Hasse**], and on each geometric fibre of X/S the G_{geom} is $SL(2)$ [**Ig**].

There is another, even simpler, example of such pathology, worked out in a discussion with Ofer Gabber. Take a prime number l, take $S := \mathrm{Spec}(\mathbb{Z}[1/l])$, and consider the l-cyclotomic character $\chi_l : \pi_1(S)^{ab} \cong \mathbb{Z}_l^\times$. Define $\rho := (\chi_l)^{l(1-l)}$. Then ρ maps $\pi_1(S)^{ab}$ onto the subgroup $1 + 2l^2\mathbb{Z}_l$ of $\mathbb{Z}_l^\times$. We have $\rho(F_p) = p^{l(l-1)}$ for each prime $p \neq l$. Because ρ takes values in $1 + 2l^2\mathbb{Z}_l$, which is a $\mathbb{Z}_l$-module, we can speak of the powers ρ^a for any a in $\mathbb{Z}_l$. Let us enumerate in increasing order the primes p_i other than l, $p_1 < p_2 < \cdots$. We claim that for all but at most countably many choices of a in $\mathbb{Z}_l$, the values $\{\rho^a(F_{p_i})\}_{i \geq 1}$ in $1 + 2l^2\mathbb{Z}_l$ are algebraically independent over $\mathbb{Q}$. The following argument is due to Ofer Gabber. It suffices to show that for each integer $n \geq 1$, there are at most countably many a in $\mathbb{Z}_l$ for which the n values $\{\rho^a(F_{p_i})\}_{1 \leq i \leq n}$ are algebraically dependent over $\mathbb{Q}$. To prove this, we argue as follows. To say that the n values $\{\rho^a(F_{p_i})\}_{1 \leq i \leq n}$ are algebraically dependent over $\mathbb{Q}$ is to say there exists a nonzero polynomial P in $\mathbb{Z}[X_1, \ldots, X_n]$ which vanishes at the point $\{\rho^a(F_{p_i})\}_{1 \leq i \leq n}$. For fixed i, the map $a \mapsto \rho^a(p_i)$ is l-adically analytic in the sense that there exists a power series

$$f_i(T) \text{ in } \mathbb{Z}_l[[lT]]$$

such that for a in $\mathbb{Z}_l$, we have $f_i(a) = \rho^a(F_{p_i})$. (Indeed, the l-adic log of $\rho(F_{p_i})$ lies in $2l^2\mathbb{Z}_l$, so $(1/l)\log(\rho(F_{p_i}))$ lies in $2l\mathbb{Z}_l$, call it b_i, and we have

$$\rho^a(p_i) = \exp(a \cdot \log((p_i)^{l(l-1)})) = \exp(lab_i) \quad \text{with } b_i \text{ in } 2l\mathbb{Z}_l.$$

For any b_i in $2l\mathbb{Z}_l$, the series $\exp(Xb_i)$ lies in $\mathbb{Z}_l[[X]]$.)

Therefore $P(\{\rho^a(F_{p_i})\}_{1 \leq i \leq n})$ is the value at $T = a$ of a power series $f_P(T)$ in $\mathbb{Z}_l[[lT]]$. If such a power series is not identically zero, then (by Weierstrass preparation in the ring $\mathbb{Z}_l[[T]]$) $f_P(T)$ has at most finitely many zeroes in $\{t$ in $\mathbb{C}_l$ with $\|lt\|_l < 1\}$. In particular, $f_P(T)$ is either identically zero or it has at most finitely many zeroes in $\mathbb{Z}_l$. Since there are only countably many P's, it remains only to show that there are no nonzero P's for which $f_P(T)$ vanishes identically. So suppose we have such a nonzero P in $\mathbb{Z}[X_1, \ldots, X_n]$, say $P = \sum_w A_w X^w$ with A_w in $\mathbb{Z}$. Then for every a in $\mathbb{Z}_l$, we are to have

$$\sum_w A_w \prod_i (p_i)^{w_i l(l-1)a} = 0.$$

In particular, for each integer $k \geq 0$ we are to have

$$\sum_w A_w \left(\prod_i (p_i)^{w_i l(l-1)} \right)^k = 0.$$

Now by unique factorization in $\mathbb{Z}$, the integers

$$n(w) := \prod_i (p_i)^{w_i l(l-1)}$$

are all distinct, i.e., $n(w) \neq n(v)$ if $v \neq w$ in $\mathbb{Z}^n$. So if we have at most N distinct values of w, say $w_0, \ldots, w_{N-1}$, for which A_w is possibly nonzero, then taking $k = 0, 1, 2, \ldots, N-1$ we see that the $N \times N$ matrix whose k'th row is $\{n(w_i)^k\}_i$ annihilates the column vector $\{A_{w_i}\}_i$. But this matrix is invertible, because its transpose is the Vandermonde matrix attached to the N distinct integers $n(w_i)$ for $i = 0, \ldots, N-1$. Therefore all A_{w_i} vanish, and hence P vanishes identically.

Pick an arbitrary sequence r_i of strictly positive real numbers. There exists a sequence x_i of algebraically independent (over $\mathbb{Q}$) elements in $\mathbb{C}$ with $|x_i| = r_i$ (proof: for each n, the algebraic closure of $\mathbb{Q}(x_i$ with $i < n)$ in $\mathbb{C}$ is countable, but the circle of radius r_n is uncountable). Now pick one of the uncountably many a in $\mathbb{Z}_l$ for which the values $\rho^a(F_{p_i})$ are algebraically independent over $\mathbb{Q}$. For any sequence x_i of algebraically independent (over $\mathbb{Q}$) elements in $\mathbb{C}$, there exists an embedding ι of $\overline{\mathbb{Q}}_l$ into $\mathbb{C}$ which maps $\rho^a(F_{p_i})$ to x_i for every i. For such an ι, the $\mathcal{L}$ on S corresponding to ρ^a will have ι-weight $2\log(r_i)/\log(p_i)$ at p_i. Thus by varying ι we can achieve any sequence w_{p_i} of real numbers as the sequence of the ι-weights of $\mathcal{L}$ at the p_i.

9.5. Interpretation of (9.3.7.2)

Here is another way to think about the condition (9.3.7.2). According to a result of Pink [**Ka-ESDE**, 8.18.2], there is a dense open set U in S such that for any geometric point u of U, the geometric monodromy group $G_{\mathrm{geom},u}$ attached to the restriction of $\mathcal{F}$ to the geometric fibre X_u is equal to $G_{\mathrm{geom},\overline{\eta}}$, the group attached to the restriction of $\mathcal{F}$ to the geometric generic fibre $X_{\overline{\eta}}$ of X/S. Moreover, for any geometric point $\overline{s}$ of S, $G_{\mathrm{geom},\overline{s}}$ is conjugate in $GL(r)$ to a **subgroup** of $G_{\mathrm{geom},\overline{\eta}}$.

If (9.3.7.2) holds, we claim that $G_{\mathrm{geom},\overline{\eta}}$ is itself (a $GL(r)$ conjugate of) G. Indeed, since U is a nonempty scheme of finite type over $\mathbb{Z}[1/l]$, it contains k-valued points for some finite field k in which l is invertible, e.g., any closed point of U underlies such a point, with k the residue field. Taking for u a geometric point in U lying over such a k-valued point, we see from (9.3.7.2) and Pink's result that $G_{\mathrm{geom},\overline{\eta}}$ is (a $GL(r)$ conjugate of) G. We see further that for every closed point s of S, $G_{\mathrm{geom},\overline{s}}$ is conjugate to $G_{\mathrm{geom},\overline{\eta}}$. Conversely, suppose that $G_{\mathrm{geom},\overline{\eta}}$ is conjugate to G, and that for every closed point s of S, the group $G_{\mathrm{geom},\overline{s}}$, a priori conjugate to a subgroup of $G_{\mathrm{geom},\overline{\eta}}$, is in fact conjugate to $G_{\mathrm{geom},\overline{\eta}}$ itself. Then (9.3.7.2) holds.

9.6. Return to a uniform version of Theorem 9.2.6

9.6.1. Suppose now that both (9.3.5.1) and (9.3.7.2) hold. Pick a geometric point ξ on $X_{\overline{\eta}}$, and consider the fundamental groups of X, of the generic fibre X_η of X/S, and of the geometric generic fibre $X_{\overline{\eta}}$ of X/S, namely

$$\pi_1(X, \xi), \quad \pi_1(X_\eta, \xi), \quad \pi_1(X_{\overline{\eta}}, \xi).$$

Both of the first two groups are quotients of the absolute galois group of the function field of X, so the inclusion of X_η into X induces a surjective homomorphism

$$\pi_1(X_\eta, \xi) \twoheadrightarrow \pi_1(X, \xi).$$

On the other hand, the short exact sequence

$$1 \to \pi_1(X_{\overline{\eta}}, \xi) \to \pi_1(X_\eta, \xi) \to \pi_1(\eta, \overline{\eta}) \to 1$$

shows that $\pi_1(X_{\overline{\eta}}, \xi)$ is a normal subgroup of $\pi_1(X_\eta, \xi)$. Therefore $\rho(\pi_1(X_\eta, \xi))$ normalizes $\rho(\pi_1(X_{\overline{\eta}}, \xi))$ and hence also normalizes its Zariski closure $G_{\mathrm{geom},\overline{\eta}} = G$. So by (9.3.5.1) we have

$$\rho(\pi_1(X_\eta, \xi)) \subset (\mathbb{G}_m \cdot G)(\overline{\mathbb{Q}}_l).$$

Because the map $\pi_1(X_\eta, \xi) \twoheadrightarrow \pi_1(X, \xi)$ is surjective, we must have

$$\rho(\pi_1(X, \xi)) = \rho(\pi_1(X_\eta, \xi)) \subset (\mathbb{G}_m \cdot G)(\overline{\mathbb{Q}}_l).$$

Therefore for any finite-dimensional $\overline{\mathbb{Q}}_l$-representation Λ of the algebraic group $\mathbb{G}_m \cdot G$, we may form the lisse $\overline{\mathbb{Q}}_l$-sheaf $\Lambda(\mathcal{F})$ on X.

9.6.2. We now assume that (9.3.5.1), (9.3.7.1) and (9.3.7.2) all hold. Because G is semisimple, its center $Z(G)$ is finite. Since G is an irreducible subgroup of $GL(r)$, $Z(G)$ consists entirely of scalars, i.e., $Z(G) = G \cap \mathbb{G}_m$, intersection inside $GL(r)$. So if we denote by N the order of $Z(G)$, we have

$$Z(G) = G \cap \mathbb{G}_m = \mu_N.$$

We may now define a homomorphism of $\overline{\mathbb{Q}}_l$-algebraic groups

$$\begin{aligned} \mathrm{multip}_G : \mathbb{G}_m \cdot G &\to \mathbb{G}_m, \\ \beta g &\mapsto \beta^N. \end{aligned}$$

[We use the notation "multip_G" because when G is either the orthogonal or symplectic group, then $N = 2$, $\mathbb{G}_m \cdot G$ is the group of orthogonal or symplectic similitudes, and "multip_G" is precisely the "multiplicator" character of the similitude group. Of course, if G is $SL(r)$, then $N = r$, $\mathbb{G}_m \cdot G$ is $GL(r)$, and in this case "multip_G" is the determinant.] The "multip_G" character sits in a short exact sequence

$$1 \to G \to \mathbb{G}_m \cdot G \to \mathbb{G}_m \to 1.$$

Lemma 9.6.3. *In the situation $\mathcal{F}$ on X/S above, with* (9.3.5.1), (9.3.7.1) *and* (9.3.7.2) *assumed to hold, fix a finite field k, and a k-valued point s of S. There exists an l-adic unit α_s in $\overline{\mathbb{Q}}_l$ such that on X_s/k, the representation $\rho \otimes \alpha_s^{\deg}$ of $\pi_1(X_s)$ corresponding to the twisted sheaf $\mathcal{F}_s \otimes \alpha_s^{\deg}$ maps the entire group $\pi_1(X_s)$, and not just $\pi_1(X_{\overline{s}})$, to G, i.e., we have $(\rho \otimes \alpha_s^{\deg})(\pi_1(X_s)) \subset G(\overline{\mathbb{Q}}_l)$. The choice of α_s is unique up to multiplication by an N'th root of unity. For any such α_s, $\mathcal{F}_s \otimes \alpha_s^{\deg}$ is ι-pure of weight zero.*

PROOF. Pick any element $\widetilde{F}_k$ in $\pi_1(X_s)$ which maps onto F_k in $\mathrm{Gal}(\overline{k}/k)$. We are given that ρ maps $\pi_1(X_{\overline{s}})$ to $G(\overline{\mathbb{Q}}_l)$. Since $(\widetilde{F}_k)^{\mathbb{Z}} \cdot \pi_1(X_{\overline{s}})$ is dense in $\pi_1(X_s)$, the requirement on α_s is precisely that $(\rho \otimes \alpha_s^{\deg})(\widetilde{F}_k)$ lie in $G(\overline{\mathbb{Q}}_l)$, i.e., that $\alpha_s \rho(\widetilde{F}_k)$ lie in $G(\overline{\mathbb{Q}}_l)$. Since $\alpha_s \rho(\widetilde{F}_k)$ is in any case an element of $\mathbb{G}_m \cdot G$, this last condition holds if and only if $\mathrm{multip}_G(\alpha_s \rho(\widetilde{F}_k)) = 1$, i.e., if and only if $(\alpha_s)^N \mathrm{multip}_G(\rho(\widetilde{F}_k)) = 1$.

Because the composite $\mathrm{multip}_G \cdot \rho$ is a continuous $\overline{\mathbb{Q}}_l^{\times}$-valued character of the compact group $\pi_1(X)$, it has values in the subgroup of l-adic units. In particular, $\mathrm{multip}_G(\rho(\widetilde{F}_k))$ is an l-adic unit in $\overline{\mathbb{Q}}_l$. Thus α_s is any N'th root of the l-adic unit $1/\mathrm{multip}_G(\rho(\widetilde{F}_k))$.

It remains to see that $\mathcal{F}_s \otimes \alpha_s^{\deg}$ is ι-pure of weight zero. Since $\mathcal{F}_s$ is assumed ι-pure of some weight w_s, $\mathcal{F}_s \otimes \alpha_s^{\deg}$ is ι-pure of weight

$$w_{1,s} = w_s + 2\log(|\iota(\alpha_s)|)/\log(\mathrm{Card}(k)).$$

Therefore we can recover $rw_{1,s}$, which we claim to be zero, as the ι-weight of $\det(\mathcal{F}_s \otimes \alpha_s^{\deg})$. Since $(\rho \otimes \alpha_s^{\deg})(\pi_1(X_s)) \subset G(\overline{\mathbb{Q}}_l)$ and G, being semisimple, has finite image under the determinant, we see that $\det(\rho \otimes \alpha_s^{\deg})$ is of finite order as character of $\pi_1(X_s)$. Therefore $\det(\rho \otimes \alpha_s^{\deg})$ is ι-pure of weight zero for any ι, and hence $rw_{1,s} = 0$, as required. QED

Remark 9.6.4. Here is a more conceptual approach to the question addressed in Lemma 9.6.3. We claim that the composite character of $\pi_1(X)$

$$\mathrm{multip}_G \circ \rho : \pi_1(X) \to \overline{\mathbb{Q}}_l$$

factors through $\pi_1(S)$. To see this, denote by η the generic point of S, by $\overline{\eta}$ a geometric point of S lying over η, and by ξ a geometric point of $X_{\overline{\eta}}$. As proven in [**Ka-Lang**, Lemma 2], we have a right-exact sequence

$$\pi_1(X_{\overline{\eta}}, \xi) \to \pi_1(X, \xi) \to \pi_1(S, \overline{\eta}) \to 1.$$

So it suffices to show that $\mathrm{multip}_G \circ \rho$ is trivial on $\pi_1(X_{\overline{\eta}}, \xi)$. But we have already seen above that $\mathrm{multip}_G \circ \rho$ is trivial on every geometric fibre of X/S. Therefore $\mathrm{multip}_G \circ \rho$ factors through some character χ of $\pi_1(S)$.

9.6.5. For k a finite field and for s a k-valued point of S, the character $\alpha_s^{\deg}$ of $\pi_1(s)$ found in the lemma above is precisely an N'th root of $\chi^{-1}|\pi_1(s)$. So if χ^{-1} has an N'th root, say τ, as a character of $\pi_1(S)$, then $\rho \otimes \tau$ maps the entire group $\pi_1(X_s)$, and not just $\pi_1(X_{\overline{s}})$, to G, for every finite-field-valued point s of S.

9.6.6. Unfortunately, a $\overline{\mathbb{Q}}_l^\times$-valued character χ of $\pi_1(S)$, for S a normal connected scheme of finite type over $\mathbb{Z}[1/l]$, does not have an N'th root in general. For example, consider the case when S is the spec of $\mathbb{Z}[1/l]$, and χ is the l-cyclotomic character

$$\chi_l : \pi_1(S)^{ab} \cong \mathbb{Z}_l^\times.$$

Finding an N'th root of χ_l as character of $\pi_1(S)$ amounts to finding a continuous homomorphism τ from $\mathbb{Z}_l^\times$ to $\overline{\mathbb{Q}}_l^\times$ which maps each element of $\mathbb{Z}_l^\times$ to an N'th root of itself. If l is odd, the group $\mathbb{Z}_l^\times$ is the product of the cyclic group μ_{l-1} of $l-1$'st roots of unity with the pro-l cyclic group generated by $1+l$. There is no problem finding an N'th root of $1+l$ which lies in $1+\lambda\mathcal{O}_\lambda$ for a suitable finite extension E_λ of $\mathbb{Q}_l$. But the restriction of τ to μ_{l-1} must map μ_{l-1} to itself (just because τ is a group homomorphism), and this is possible if and only if the N'th power map is surjective and hence bijective on μ_{l-1}, i.e., it is possible if and only if N is prime to $l-1$. A similar analysis in the case $l=2$, where $\mathbb{Z}_2^\times$ is $\{\pm 1\} \times (1+4\mathbb{Z}_2)$ shows that an N'th root of χ_2 exists if and only if N is odd. In particular, for any prime l, χ_l does not have a square root as character of $\pi_1(\mathrm{Spec}(\mathbb{Z}[1/l]))$.

9.6.7. An elementary but useful fact is that one can always take an N'th root of χ up to a character of finite order. Let S be any connected scheme, $\mathcal{O}_\lambda$ the ring of integers in a finite extension E_λ of $\mathbb{Q}_l$, and $\chi : \pi_1(S) \to \mathcal{O}_\lambda^\times$ a continuous character. Write $\mathcal{O}_\lambda^\times$ as the product ("Teichmuller decomposition") of (roots of unity in $\mathcal{O}_\lambda$ of order prime to l) with the group $1+\lambda\mathcal{O}_\lambda$ of principal units, say $x = \mathrm{Teich}(x)\langle x\rangle$.

Correspondingly, we get a product decomposition of χ as $\mathrm{Teich}(\chi)$, a character of finite order prime to l, times $\langle\chi\rangle$, a character with values in $1+\lambda\mathcal{O}_\lambda$. Now $1+\lambda\mathcal{O}_\lambda$ is a $\mathbb{Z}_l$-module, and it is finitely generated, since it contains with finite index $1+l\lambda\mathcal{O}_\lambda$, which via the logarithm is isomorphic to $l\lambda\mathcal{O}_\lambda$, which is $\mathbb{Z}_l$-free of rank $r=\deg(E_l/\mathbb{Q}_l)$. The torsion subgroup of $1+\lambda\mathcal{O}_\lambda$ is the group of those l-power roots of unity which lie in E_λ, so is a finite cyclic group of order l^a for some integer $a\geq 0$. So as $\mathbb{Z}_l$-module, we have

$$1+\lambda\mathcal{O}_\lambda \cong (\mathbb{Z}/l^a\mathbb{Z}) \oplus (\mathbb{Z}_l)^r,$$

corresponding to some choice of a generator ζ of μ_{l^a} and a choice of r elements $u_1,\ldots,u_r$ of $1+\lambda\mathcal{O}_\lambda$ which project onto a $\mathbb{Z}_l$-basis of $(1+\lambda\mathcal{O}_\lambda)/\text{torsion}$. Via such a choice, we get a further product decomposition of $\langle\chi\rangle$ as $\langle\chi\rangle_{\text{finite}}\prod_{i=1}^r u_i^{\sigma_i}$, where σ_i is a character of $\pi_1(S)$ with values in $\mathbb{Z}_l$, and $\langle\chi\rangle_{\text{finite}}$ is a character of order dividing l^a. Choose for each i an N'th root v_i of u_i with v_i itself a principal unit in $\overline{\mathbb{Q}}_l$. Then $\tau := \prod_{i=1}^r v_i^{\sigma_i}$ is a $\overline{\mathbb{Q}}_l^\times$-valued character of $\pi_1(S)$, and $\chi/\tau^N = \mathrm{Teich}(\chi)\langle\chi\rangle_{\text{finite}}$ is of finite order, as required.

9.6.8. Choose a maximal compact subgroup K in $G(\mathbb{C})$. For each choice of a finite field k with $\mathrm{Card}(k) > A(X/S)^2$ in which l is invertible, of a k-valued point s in S, of an l-adic unit α_s such that $\alpha_s^{\deg}\otimes\mathcal{F}_s$ has its arithmetic monodromy inside G, and of a k-valued point x in $X_s(k)$ (a set which is nonempty by the hypothesis that $\mathrm{Card}(k) > A(X/S)^2$), the earlier discussion of $\alpha_s^{\deg}\otimes\mathcal{F}_s$ on X_s/k gives a Frobenius conjugacy class in K, which we denote $\vartheta(k,s,\alpha_s,x)$.

9.6.9. We denote by $\mu(k,s,\alpha_s)$ the probability measure on the space $K^\#$ of conjugacy classes in K defined by averaging over the Frobenius conjugacy classes $\vartheta(k,s,\alpha_s,x)$ as x runs over $X_s(k)$:

$$\mu(k,s,\alpha_s) := (1/\mathrm{Card}(X_s(k))) \sum_{x \text{ in } X_s(k)} \delta_{\vartheta(k,s,\alpha_s,x)}.$$

Theorem 9.6.10. *Suppose we are given $(l,X/S,\mathcal{F},\iota,G)$ as above, such that the conditions* (9.3.5.1), (9.3.7.1) *and* (9.3.7.2) *are all satisfied. For any sequence of choices of data (k_i,s_i,α_{s_i}) as above in which* $\mathrm{Card}(k_i)$ *increases to infinity, the measures $\mu(k_i,s_i,\alpha_{s_i})$ on $K^\#$ converge weak $*$ to the measure $\mu^\#$ on $K^\#$ which is the direct image of normalized (total mass one) Haar measure on K, i.e., for any continuous $\mathbb{C}$-valued central function f on K, we have*

$$\int_K f\,d\mathrm{Haar} = \lim_{i\to\infty}(1/\mathrm{Card}(X_{s_i}(k_i))) \sum_{x \text{ in } X_{s_i}(k_i)} f(\vartheta(k_i,s_i,\alpha_{s_i},x)).$$

More precisely, if Λ is any irreducible nontrivial representation of K, and (k,s,α_s) if any datum as above with $\mathrm{Card}(k)\geq 4A(X/S)^2$, *and $C(X/S,\mathcal{F})$ is the constant introduced in* 9.3.4, *then we have the estimate*

$$\left|\int_{K^\#}\mathrm{Trace}(\Lambda)\,d\mu(k,s,\alpha_s)\right| \leq 2C(X/S,\mathcal{F})\dim(\Lambda)/\mathrm{Card}(k)^{1/2}.$$

PROOF. Once we choose a finite etale galois Y/X which trivializes the reduction $\mathrm{mod}\,\lambda$ of an integral form $\mathcal{F}_{\mathcal{O}_\lambda}$ of $\mathcal{F}$, the geometric fibres $Y_{\bar{s}}/X_{\bar{s}}$ trivialize the reductions $\mathrm{mod}\,\lambda$ of the integral forms $(\alpha_s^{\deg}\otimes\mathcal{F}_{\mathcal{O}_\lambda})_s$ of $\alpha_s^{\deg}\otimes\mathcal{F}_s$ on $X_{\bar{s}}$, precisely because when we pull back $\alpha_s^{\deg}$ from X_s to $X_{\bar{s}}$, it becomes trivial. It now suffices to apply part 5) of Theorem 9.2.6 fibre by fibre. QED

Remark 9.6.11. The merit of the "more precise" estimate

$$\left|\int_{K\#} \mathrm{Trace}(\Lambda)\, d\mu(k,s,\alpha_s)\right| \leq 2C(X/S,\mathcal{F})\dim(\Lambda)/\operatorname{Card}(k)^{1/2}$$

is that the error term is $O(\dim(\Lambda)/\operatorname{Card}(k)^{1/2})$. A weaker assertion, but one still adequate (compare the proof of 9.2.6) to give the equidistribution as $\operatorname{Card}(k) \to \infty$, is that for each irreducible nontrivial representation Λ of K there exists a constant $D(X/S,\mathcal{F},\Lambda)$ such that for all (k,s,α_s) as above we have

$$\left|\int_{K\#} \mathrm{Trace}(\Lambda)\, d\mu(k,s,\alpha_s)\right| \leq 2D(X/S,\mathcal{F},\Lambda)/\operatorname{Card}(k)^{1/2}.$$

To obtain this weaker statement directly, observe that any irreducible representation Λ of G extends to a representation $\widetilde{\Lambda}$ of $\mathbb{G}_m \cdot G$. [Indeed, because Λ is irreducible it must map the center $Z(G) = \mathbb{G}_m \cap G = \mu_N$ to the scalars, hence $\Lambda|\mu_N$ must be of the form $\beta \mapsto \beta^n \mathbb{1}$ for a unique integer n in $[0, N-1]$. So we have $\Lambda(\beta g) = \beta^n \Lambda(g)$ for any β in μ_N. Therefore we may define $\widetilde{\Lambda}$ on $\mathbb{G}_m \cdot G$ by $\widetilde{\Lambda}(\beta g) := \beta^n \Lambda(g)$.] We then form the lisse $\overline{\mathbb{Q}}_l$-sheaf $\widetilde{\Lambda}(\mathcal{F})$ on X. On each geometric fibre $X_{\overline{s}}$ of X/S, we have $\widetilde{\Lambda}(\mathcal{F})_{\overline{s}} \cong \Lambda(\mathcal{F}_{\overline{s}}) \cong \Lambda((\alpha_s^{\deg} \otimes \mathcal{F}_s)_{\overline{s}})$. So for Λ irreducible nontrivial we may take for $D(X/S,\mathcal{F},\Lambda)$ the sup over geometric points $\overline{s}$ of S of the quantity $\sum_{0 \leq i < 2d} h_c^i(X_{\overline{s}}, \widetilde{\Lambda}(\mathcal{F})_{\overline{s}})$. This sup is finite by proper base change and the constructibility of the sheaves $R^i f_!(\widetilde{\Lambda}(\mathcal{F}))$ on S, for $f : X \to S$ the structural morphism.

9.7. Another version of Deligne's equidistribution theorem

9.7.1. We suppose given a prime number l, an embedding ι of $\overline{\mathbb{Q}}_l$ into $\mathbb{C}$, a connected normal $\mathbb{Z}[1/l]$-scheme S of finite type over $\mathbb{Z}[1/l]$ with generic point η and geometric generic point $\overline{\eta}$, a smooth X/S with all fibres geometrically connected of some common dimension $d \geq 0$, a geometric point ξ of $X_{\overline{\eta}}$, and a lisse $\overline{\mathbb{Q}}_l$-sheaf $\mathcal{F}$ on X, of rank $r \geq 1$, corresponding to an r-dimensional continuous representation

$$\rho : \pi_1(X,\xi) \to GL(\mathcal{F}_\xi) \cong GL(r, \overline{\mathbb{Q}}_l).$$

9.7.2. We suppose given two semisimple $\overline{\mathbb{Q}}_l$-algebraic subgroups of $GL(r)$,

$$G \subset G_{\mathrm{arith}} \subset GL(r),$$

and we suppose that G is a normal subgroup of G_{arith} of finite index. We denote by Γ the finite group $G_{\mathrm{arith}}(\overline{\mathbb{Q}}_l)/G(\overline{\mathbb{Q}}_l)$. We further suppose

(1) $\rho(\pi_1(X,\xi)) \subset G_{\mathrm{arith}}(\overline{\mathbb{Q}}_l)$, and $\rho(\pi_1(X,\xi))$ is Zariski dense in G_{arith}.

(2) $\rho(\pi_1(X_{\overline{\eta}},\xi)) \subset G(\overline{\mathbb{Q}}_l)$.

(3) For every finite field k, and every k-valued point s of S, the group G_{geom} for $\mathcal{F}_s := \mathcal{F}|X_s$ is (conjugate in $GL(r)$ to) G.

(4) $\mathcal{F}$ is ι-pure of weight zero.

9.7.3. By (1), the composite homomorphism

$$\pi_1(X,\xi) \to G_{\mathrm{arith}}(\overline{\mathbb{Q}}_l) \to \Gamma$$

is surjective. By the right exact homotopy sequence

$$\pi_1(X_{\overline{\eta}},\xi) \to \pi_1(X,\xi) \to \pi_1(S,\overline{\eta}) \to 1,$$

we see as above that the surjective map $\pi_1(X,\xi) \twoheadrightarrow \Gamma$ factors through $\pi_1(S,\overline{\eta})$, say as $A : \pi_1(S,\overline{\eta}) \twoheadrightarrow \Gamma$. For a finite field k and a k-valued point s of S, we denote by

$\gamma(k,s)$ the conjugacy class in Γ of $A(F_{k,s})$. By Chebotarev, each conjugacy class in Γ is a $\gamma(k,s)$ for suitable data (k,s). Moreover, among all (k,s) with given $\gamma(k,s)$, $\mathrm{Card}(k)$ is unbounded. [Proof: Write $N := 1 + \mathrm{Card}(\Gamma)$, so $\gamma^N = \gamma$ for any γ in Γ. For a finite field k, denote by k_N/k the extension of k of degree N. For s in $S(k) \subset S(k_N)$, we have $(F_{k,s})^N = F_{k_N,s}$ and hence $\gamma(k,s) = \gamma(k_N,s)$.]

9.7.4. Pick a maximal compact subgroup K of $G(\mathbb{C})$. Since K is a compact subgroup of $G_{\mathrm{arith}}(\mathbb{C})$, we may pick a maximal compact subgroup K_{arith} of $G_{\mathrm{arith}}(\mathbb{C})$ which contains K. Since $K \subset K_{\mathrm{arith}} \cap G$, and $K_{\mathrm{arith}} \cap G$ is a compact subgroup of G, we have $K = K_{\mathrm{arith}} \cap G$ by the maximality of K. Because K_{arith} is Zariski dense in G, it maps onto the finite quotient Γ, and hence $K_{\mathrm{arith}}/K \cong \Gamma$.

9.7.5. For any finite field k, any k-valued point s of S, and any k-valued point x of X_s, $\iota(\rho(F_{k,x})^{ss})$ is a conjugacy class in $G_{\mathrm{arith}}(\mathbb{C})$ which is semisimple with unitary eigenvalues, and hence defines a conjugacy class, which we denote $\vartheta(k,s,x)$, in K_{arith}. The image of this conjugacy class in Γ we temporarily denote $\gamma(k,s,x)$. The class $\gamma(k,s,x)$ in Γ is also the image in Γ viewed as G_{arith}/G of the conjugacy class of $\rho(F_{k,x})$. As noted above, the image in Γ of any class in $\pi_1(X)$ depends only on its image in $\pi_1(S)$. So the class in Γ of $\rho(F_{k,x})$ depends only on the image $F_{k,s}$ of $F_{k,x}$ in $\pi_1(S)$. Hence this class $\gamma(k,s,x)$ is equal to the class denoted $\gamma(k,s)$ in 9.7.4 above.

9.7.6. The surjective homomorphism $K_{\mathrm{arith}} \to \Gamma$ induces a surjective map of the spaces of conjugacy classes $(K_{\mathrm{arith}})^\# \to \Gamma^\#$. For each conjugacy class γ in $\Gamma^\#$, we denote by $\mathrm{Card}(\gamma)$ its cardinality, viewing γ as a subset of Γ. We denote by $(K_{\mathrm{arith},\gamma})^\# \subset (K_{\mathrm{arith}})^\#$ the inverse image of γ, and we denote by $K_{\mathrm{arith},\gamma} \subset K_{\mathrm{arith}}$ the set of those elements in K_{arith} whose image in Γ lies in the class γ. Thus $(K_{\mathrm{arith}})^\#$ is the disjoint union of the open and closed sets $(K_{\mathrm{arith},\gamma})^\#$, and K_{arith} is the disjoint union of the open and closed sets $K_{\mathrm{arith},\gamma}$, each of which is open and closed in K_{arith}.

9.7.7. We denote by μ the Haar measure on K_{arith} of total mass $\mathrm{Card}(\Gamma)$; this is the Haar measure on K_{arith} which gives K total mass one. We denote by $\mu^\#$ the direct image of μ on $(K_{\mathrm{arith}})^\#$, and for each γ in $\Gamma^\#$, we denote by $\mu(\gamma)^\#$ the restriction of $\mu^\#$ to $(K_{\mathrm{arith},\gamma})^\#$. Thus $\mu(\gamma)^\#$ gives $(K_{\mathrm{arith},\gamma})^\#$ total mass $\mathrm{Card}(\gamma)$. We may also view $\mu(\gamma)^\#$ as the direct image on $(K_{\mathrm{arith},\gamma})^\#$ of the restriction to $K_{\mathrm{arith},\gamma}$ of the Haar measure μ on K_{arith}, by the natural map $K_{\mathrm{arith},\gamma} \to (K_{\mathrm{arith},\gamma})^\#$.

Lemma 9.7.8. *Let $\Lambda : K_{\mathrm{arith}} \to GL(V)$ be an irreducible finite-dimensional $\mathbb{C}$-representation of K_{arith}. Denote by $V^K \subset V$ the space of K-invariants.*

1) *Either $V^K = 0$ (in which case the restriction of Λ to K is a sum of irreducible nontrivial representations of K) or $V^K = V$ (in which case Λ is a representation of the quotient Γ).*

2) *If $V^K = 0$, then for any element g in K_{arith}, we have*

$$\int_{gK} \mathrm{Trace}(\Lambda)\, d\mu = 0,$$

and for any γ in $\Gamma^\#$, we have

$$\int_{(K_{\mathrm{arith},\gamma})^\#} \mathrm{Trace}(\Lambda)\, d\mu(\gamma)^\# = 0.$$

3) *If $V^K \neq 0$, i.e. if Λ is a representation of Γ, then for any γ in $\Gamma^\#$, we have*

$$\int_{(K_{\mathrm{arith},\gamma})^\#} \mathrm{Trace}(\Lambda)\, d\mu(\gamma)^\# = \mathrm{Card}(\gamma)\,\mathrm{Trace}(\Lambda(\gamma)).$$

PROOF. For 1), observe that since K is normal in K_{arith}, V^K is a K_{arith}-stable subspace of V, so is either 0 or V by irreducibility. For 2), denoting by dA the normalized Haar measure on K, we have

$$\int_{gK} \mathrm{Trace}(\Lambda)\, d\mu = \int_K \mathrm{Trace}(\Lambda(gA))\, dA = \int_K \mathrm{Trace}(\Lambda(Ag))\, dA$$
$$= \mathrm{Trace}\left(\text{the endomorphism } v \mapsto \int_K \Lambda(A)(\Lambda(g)(v))\, dA\right).$$

But for each v in V, the integral $\int_K \Lambda(A)(\Lambda(g)(v))\, dA$ lands in V^K, hence vanishes. The second assertion of 2) is the sum of instances of the first, applied to a set of lifts g in K_{arith} of the elements of γ in Γ. Assertion 3) is a tautology. QED

9.7.9. For each finite field k with $\mathrm{Card}(k) > A(X/S)^2$ and each k-valued point s in S, we denote by $\mu(k,s)$ the measure of total mass $\mathrm{Card}(\gamma(k,s))$ on $(K_{\mathrm{arith},\gamma(k,s)})^\#$ defined by averaging over the Frobenius conjugacy classes $\vartheta(k,s,x)$, each of which lies in $(K_{\mathrm{arith},\gamma(k,s)})^\#$, as x runs over $X_s(k)$, and then multiplying by $\mathrm{Card}(\gamma(k,s))$:

$$\mu(k,s) := (\mathrm{Card}(\gamma(k,s))/\mathrm{Card}(X_s(k))) \sum_{x \text{ in } X_s(k)} \delta_{\vartheta(k,s,x)}.$$

Theorem 9.7.10. *Suppose we are given $(l, X/S, \mathcal{F}, \iota, G, G_{\mathrm{arith}})$ as above, such that the conditions (1), (2), (3) and (4) of 9.7.2 are all satisfied. Fix a conjugacy class γ in $\Gamma^\#$. For any sequence of choices of data (k_i, s_i) as above with each $\mathrm{Card}(k_i) > A(X/S)^2$, with each $\gamma(k_i, s_i) = \gamma$ and in which $\mathrm{Card}(k_i)$ increases to infinity, the measures $\mu(k_i, s_i)$ on $(K_{\mathrm{arith},\gamma})^\#$ converge weak $*$ to the measure $\mu(\gamma)^\#$ on $(K_{\mathrm{arith},\gamma})^\#$, i.e., for any continuous $\mathbb{C}$-valued central function f on K_{arith}, we have*

$$\int_{(K_{\mathrm{arith},\gamma})^\#} f\, d\mu(\gamma)^\# := \int_{K_{\mathrm{arith},\gamma}} f\, d\mu$$
$$= \lim_{i\to\infty} \int_{(K_{\mathrm{arith},\gamma})^\#} f\, d\mu(k_i, s_i)$$
$$:= \lim_{i\to\infty} (\mathrm{Card}(\gamma)/\mathrm{Card}(X_{s_i}(k_i))) \sum_{x \text{ in } X_{s_i}(k_i)} f(\vartheta(k_i, s_i, x)).$$

More precisely, if (Λ, V) is any irreducible representation of K_{arith}, if (k,s) is any datum as above with $\mathrm{Card}(k) \geq 4A(X/S)^2$ and $\gamma(k,s) = \gamma$, and if $C(X/S, \mathcal{F})$ is the constant introduced in 9.3.4 above, then we have the estimate

$$\left| \int_{(K_{\mathrm{arith},\gamma})^\#} \mathrm{Trace}(\Lambda)\, d\mu(k,s) - \int_{(K_{\mathrm{arith},\gamma})^\#} \mathrm{Trace}(\Lambda)\, d\mu(\gamma)^\# \right|$$
$$\leq 2C(X/S, \mathcal{F}) \dim(V/V^K)/\mathrm{Card}(k)^{1/2}.$$

PROOF. By Peter-Weyl, every continuous $\mathbb{C}$-valued central function f on K_{arith} is the uniform limit of finite $\mathbb{C}$-linear combinations of traces of irreducible representations of K_{arith}, so it suffices to prove the "more precisely" estimate. In the case when $V^K = V$, i.e., when Λ is a representation of Γ, we are asserting the identity

$$\int_{(K_{\text{arith},\gamma})^\#} \operatorname{Trace}(\Lambda)\, d\mu(k,s) = \int_{(K_{\text{arith},\gamma})^\#} \operatorname{Trace}(\Lambda)\, d\mu(\gamma)^\#.$$

This holds because both sides are equal to $\operatorname{Card}(\gamma)\operatorname{Trace}(\Lambda(\gamma))$. In the case when $V^K = 0$, we are asserting that

$$\left|\int_{(K_{\text{arith},\gamma})^\#} \operatorname{Trace}(\Lambda)\, d\mu(k,s)\right| \leq 2C(X/S,\mathcal{F})\dim(V)/\operatorname{Card}(k)^{1/2}.$$

This results from the uniformity Lemma 9.3.4 applied to $\mathcal{F}$ on X/S, and from the Lefschetz trace formula and Deligne's main result [**De-Weil II**, 3.3.1 and 3.3.10], applied to the lisse sheaf $\Lambda(\mathcal{F})$ on X, which on each geometric fibre of X/S is ι-pure of weight zero and has its $H^{2d}_c(X_{\bar{s}}, \Lambda(\mathcal{F})_s) = 0$. QED

Remark 9.7.11. When G_{arith} is a **finite** group, i.e., when K_{arith} is finite, this theorem is simply an effective version, which is uniform in a family, of the Chebotarev density theorem [**Lang-LSer**, Thm. 1] for finite etale galois coverings of smooth varieties over finite fields, cf. [**Chav**] for a recent application of that theorem. To see this, fix a conjugacy class γ of Γ, and fix a subset W of $K_{\text{arith},\gamma}$ which is stable by K_{arith}-conjugation. Taking for f the characteristic function of W, we find that for all (k,s) with $\gamma(k,s) = \gamma$ and with $\operatorname{Card}(k) \gg 0$, the percentage of k-valued points x in $X_s(k)$ whose Frobenius conjugacy class under ρ lands in W is approximately $\operatorname{Card}(W)/\operatorname{Card}(K_{\text{arith},\gamma})$. Let us make explicit the constants which emerge.

Lemma 9.7.12. *Let G be a finite group, and $W \subset G$ a subset stable by conjugation. Denote by $x \mapsto \operatorname{char}_W(x)$ the characteristic function of W. Then in the expression of the central function char_W as a $\mathbb{C}$-linear sum of traces of irreducible representations Λ of G, say*

$$\operatorname{char}_W(x) = \sum_{\text{irred }\Lambda} c(W,\Lambda)\operatorname{Trace}(\Lambda(x)),$$

we have the inequalities

$$|c(W,\Lambda)| \leq \dim(\Lambda).$$

PROOF. Use the orthonormality of the functions $\operatorname{Trace}(\Lambda)$ on G for the total mass one Haar measure $d\,\text{Haar}$ on G to compute

$$c(W,\Lambda) = \int_G \operatorname{char}_W \operatorname{Trace}(\overline{\Lambda})\, d\,\text{Haar}.$$

The inequality $|c(W,\Lambda)| \leq \dim(\Lambda)$ is now obvious, since the integrand is itself pointwise bounded in absolute value by $\dim(\Lambda)$. QED

Theorem 9.7.13. *Hypotheses and notations as in Theorem* 9.7.10 *above, suppose in addition that K_{arith} is finite. Let γ be an element of $\Gamma^\#$, and W a subset of $K_{\text{arith},\gamma}$ which is stable by K_{arith}-conjugation. Given (k,s) with* $\operatorname{Card}(k) \geq 4A(X/S)^2$ *and with $\gamma(k,s) = \gamma$, we have*

$$\begin{aligned}
&|\operatorname{Card}(W)/\operatorname{Card}(K_{\text{arith},\gamma}) - \operatorname{Card}(\{x \text{ in } X_s(k) \mid \vartheta(k,s,x) \text{ in } W\})/\operatorname{Card}(X_s(k))| \\
&\quad \leq 2C(X/S,\mathcal{F})\operatorname{Card}(K_{\text{arith}})/\operatorname{Card}(\gamma)\operatorname{Card}(k)^{1/2} \\
&\quad \leq 2C(X/S,\mathcal{F})\operatorname{Card}(K_{\text{arith}})/\operatorname{Card}(k)^{1/2}.
\end{aligned}$$

PROOF. We apply Theorem 9.7.10, taking for f the characteristic function of W. Since W lies in $K_{\mathrm{arith},\gamma}$, it is tautologous that

$$\int_{(K_{\mathrm{arith},\gamma})^{\#}} f\, d\mu(\gamma)^{\#} = \mathrm{Card}(W)/\,\mathrm{Card}(K),$$

and that

$$\int_{(K_{\mathrm{arith},\gamma})^{\#}} f\, d\mu(k,s)$$
$$= \mathrm{Card}(\gamma)\,\mathrm{Card}(\{x \text{ in } X_s(k) \mid \vartheta(k,s,x) \text{ in } W\})/\,\mathrm{Card}(X_s(k)).$$

Write f as $\sum_{\text{irred rep's } \Lambda \text{ of } K_{\mathrm{arith}}} c(W,\Lambda)\,\mathrm{Trace}(\Lambda)$. By Lemma 9.7.12, we have $|c(W,\Lambda)| \le \dim(\Lambda)$. The individual integrals are given by

$$\int_{(K_{\mathrm{arith},\gamma})^{\#}} \mathrm{Trace}(\Lambda)\, d\mu(\gamma)^{\#} = \begin{cases} \mathrm{Card}(\gamma)\,\mathrm{Trace}(\Lambda(\gamma)), & \Lambda \text{ trivial on } K, \\ 0, & \Lambda \text{ nontrivial on } K. \end{cases}$$

For the measure $\mu(k,s)$, we have the equality

$$\int_{(K_{\mathrm{arith},\gamma})^{\#}} \mathrm{Trace}(\Lambda)\, d\mu(k,s) = \mathrm{Card}(\gamma)\,\mathrm{Trace}(\Lambda(\gamma))$$

if Λ is trivial on K, and we have the estimate

$$\left| \int_{(K_{\mathrm{arith},\gamma})^{\#}} \mathrm{Trace}(\Lambda)\, d\mu(k,s) \right| \le 2C(X/S,\mathcal{F})\dim(\Lambda)/\,\mathrm{Card}(k)^{1/2}$$

if Λ is nontrivial on K. Using the estimate

$$|c(W,\Lambda)| \le \dim(\Lambda),$$

and the identity

$$\sum_{\text{irred } \Lambda} \dim(\Lambda)^2 = \mathrm{Card}(K_{\mathrm{arith}}),$$

we get

$$|\,\mathrm{Card}(W)/\,\mathrm{Card}(K) - \mathrm{Card}(\gamma)\,\mathrm{Card}(\{x \text{ in } X_s(k) \mid \vartheta(k,s,x) \text{ in } W\})/\,\mathrm{Card}(X_s(k))|$$
$$= \left| \int_{(K_{\mathrm{arith},\gamma})^{\#}} f\, d\mu(\gamma)^{\#} - \int_{(K_{\mathrm{arith},\gamma})^{\#}} f\, d\mu(k,s) \right|$$
$$\le 2C(X/S,\mathcal{F})\,\mathrm{Card}(K_{\mathrm{arith}})/\,\mathrm{Card}(k)^{1/2}.$$

Dividing through by $\mathrm{Card}(\gamma)$, we find the asserted inequality. QED

CHAPTER 10

Monodromy of Families of Curves

10.0. Explicit families of curves with big G_{geom}

10.0.1. Our first task is to give an example, for every genus $g \geq 1$, and for every finite field k, of a one parameter family of curves of genus g in characteristic p whose geometric monodromy is as big as possible. In all the examples, the parameter space will be a nonempty open set U of $\mathrm{Spec}(k[T])$, the affine T-line $\mathbb{A}^1$ over k, and the family, a proper smooth map $\pi : \mathcal{C} \to U$ all of whose geometric fibres are connected, proper smooth curves of genus g, will have the property that for every prime number $l \neq p$, the group G_{geom} for the lisse $\overline{\mathbb{Q}}_l$-sheaf $R^1\pi_!\overline{\mathbb{Q}}_l$ on U is $Sp(2g)$.

10.1. Examples in odd characteristic

10.1.1. We first give, for each genus $g \geq 1$, a construction valid in any characteristic $p \neq 2$. Fix a finite field k of characteristic $p \neq 2$, and a polynomial $f(X)$ in $k[X]$ of degree $2g$ which has all distinct roots in $\overline{k}$. [For example, we may take for f an irreducible monic polynomial of degree $2g$, i.e., the monic irreducible over k for any field generator of the extension of degree $2g$.] Take as family of curves, with parameter T, the family of hyperelliptic curves with affine equation (there is a single point at infinity)

$$Y^2 = f(X)(X - T).$$

[To get an affine neighborhood of the point at infinity, one passes to the coordinates $U := 1/X$ and $V := Y/X^{g+1}$. In terms of the polynomial $f_{\text{rev}}(U) := f(X)/X^{2g}$ in $k[U]$ of degree $2g$ (or $2g - 1$ if $f(0) = 0$) with all distinct roots in $\overline{k}$, the equation becomes

$$V^2 = Uf_{\text{rev}}(U)(1 - TU),$$

with the origin $(U = 0, V = 0)$ as the point at infinity.]

10.1.2. The curve $Y^2 = f(X)(X - t)$ is nonsingular of genus g so long as t avoids being a zero of f, so we may and will take U to be $\mathrm{Spec}(k[T][1/f(T)])$.

10.1.3. However, it will be important for the proof that G_{geom} is big to think about this curve for all values of T. Thus we denote

$$\begin{aligned} \mathcal{C}^{\text{aff}} &:= \text{ the hypersurface } Y^2 = f(X)(X - T) \text{ in } \mathbb{A}^2 \times \mathbb{A}^1, \\ \mathrm{pr} : \mathcal{C}^{\text{aff}} &\to \mathbb{A}^1 \text{ the map } (X, Y, T) \mapsto T. \end{aligned}$$

This will be our main object to study. We also denote

$$\begin{aligned} \mathcal{C}_{\text{aff}} &:= \text{ the hypersurface } V^2 = Uf_{\text{rev}}(U)(1 - TU) \text{ in } \mathbb{A}^2 \times \mathbb{A}^1, \\ \pi_{\text{aff}} : \mathcal{C}_{\text{aff}} &\to \mathbb{A}^1 \text{ the map } (U, V, T) \mapsto T. \end{aligned}$$

The open set of $\mathcal{C}^{\text{aff}}$ where X is invertible is isomorphic, as scheme over $\mathbb{A}^1$, to the open set of $\mathcal{C}_{\text{aff}}$ where U is invertible, by $(X, Y, T) \mapsto (1/X, Y/X^{g+1}, T)$. We define $\pi : \mathcal{C} \to \mathbb{A}^1$ by glueing together $\mathcal{C}^{\text{aff}}$ and $\mathcal{C}_{\text{aff}}$ along this common open set. Over U, π is a proper smooth map whose geometric fibres are connected curves of genus g.

10.1.4. We now turn to a detailed study of the sheaf $\mathcal{F} := R^1 \operatorname{pr}_! \overline{\mathbb{Q}}_l$ on $\mathbb{A}^1$. In it, we will make use of the following two general lemmas.

Lemma 10.1.5. *Let k be an algebraically closed field in which l is invertible. For any integer $n \geq 1$, any affine smooth connected S/k of dimension $n+1$, e.g., $S = \mathbb{A}^{n+1}$, any hypersurface V in S, (i.e., V is defined by the vanishing of a nonzero global function on S) and any $i \leq n-1$, we have $H^i_c(V, \mathbb{F}_l) = 0$ and $H^i_c(V, \overline{\mathbb{Q}}_l) = 0$.*

PROOF. With either $\mathbb{F}_l$ or $\overline{\mathbb{Q}}_l$ as coefficients, we have the excision long exact sequence

$$\cdots \to H^i_c(S - V) \to H^i_c(S) \to H^i_c(V) \to H^{i+1}_c(S - V) \to \cdots .$$

Because both S and $S - V$ are affine, smooth and connected of dimension $n+1$, we have $H^i_c(S - V) = H^i_c(S) = 0$ for $i \leq n$, by the Poincaré dual of the Lefschetz affine theorem [**SGA 4**, XIV, Thm. 3.1 and Cor. 3.2]. So the result is immediate from the excision sequence. QED

Lemma 10.1.6. *Let k be an algebraically closed field in which l is invertible. Let C/k be an affine, smooth, connected curve. Suppose we are given an integer $n \geq 0$ and a k-morphism $f : V \to C$, such that*

a) *V is a hypersurface in an affine smooth connected k-scheme of dimension $n+2$,*

b) *for each algebraically closed overfield L of k, and each L-valued point t of C, there exists an affine smooth connected L-scheme of dimension $n+1$ in which $f^{-1}(t)$ is a hypersurface.*

Then

1) *the sheaves $R^i f_! \overline{\mathbb{Q}}_l$ on C vanish for $i < n$,*

2) *the sheaf $R^n f_! \overline{\mathbb{Q}}_l$ on C has no nonzero punctual sections, i.e., denoting by $j : U \to C$ the inclusion of a nonvoid open set on which $R^n f_! \overline{\mathbb{Q}}_l$ is lisse, the natural map*

$$R^n f_! \overline{\mathbb{Q}}_l \to j_* j^* R^n f_! \overline{\mathbb{Q}}_l$$

is injective, or equivalently, $H^0_c(C, R^n f_! \overline{\mathbb{Q}}_l) = 0$, cf. [**Ka-SE**, 4.5.2].

PROOF. The first assertion is immediate from proper base change and the previous lemma applied to each geometric fibre. For the second, we consider the Leray spectral sequence for $Rf_!$,

$$E_2^{a,b} = H^a_c(C, R^b f_! \overline{\mathbb{Q}}_l) \Rightarrow H^{a+b}_c(V, \overline{\mathbb{Q}}_l).$$

The only possibly nonzero E_2 terms have a in $[0, 2]$ (cohomological dimension of curves) and $b \geq n$ (by part 1). The only such term with $a + b = n$ is $E_2^{0,n}$. The only possibly nonzero differentials are $d_2^{0,n+m+1} : E_2^{0,n+m+1} \to E_2^{2,n+m}$ for $m \geq 0$. Thus we have

$$H^0_c(C, R^n f_! \overline{\mathbb{Q}}_l) = E_2^{0,n} = E_\infty^{0,n} = H^n_c(V, \overline{\mathbb{Q}}_l) = 0,$$

the last vanishing by the previous lemma, applied to V. QED

Remark 10.1.7. A useful special case of 10.1.6 is a nonconstant function $f : V \to \mathbb{A}^1$ with V/k affine smooth and connected of dimension $n+1$. [V is a hypersurface in $V \times \mathbb{A}^1$, and each geometric fibre of f is a hypersurface in V, since $f - t$ never vanishes identically.]

10.1.8. Applying this lemma 10.1.6 to $\mathcal{C}^{\mathrm{aff}}$, which is a hypersurface in $\mathbb{A}^3$, and the morphism $\mathrm{pr} : \mathcal{C}^{\mathrm{aff}} \to \mathbb{A}^1$, whose fibres are hypersurfaces in $\mathbb{A}^2$, we find that $R^0 \mathrm{pr}_! \overline{\mathbb{Q}}_l = 0$, and that $R^1 \mathrm{pr}_! \overline{\mathbb{Q}}_l$ has no nonzero punctual sections.

Lemma 10.1.9. *The restriction to U of the sheaf $\mathcal{F} := R^1 \mathrm{pr}_! \overline{\mathbb{Q}}_l$ on $\mathbb{A}^1$ is lisse of rank $2g$ and pure of weight one. At geometric points of $\mathbb{A}^1 - U$, the stalk of $\mathcal{F}$ has dimension $2g-1$. The restriction to U of $R^2 \mathrm{pr}_! \overline{\mathbb{Q}}_l$ is the sheaf $\overline{\mathbb{Q}}_l(-1)$, and the cup-product pairing*

$$\mathcal{F}|U \times \mathcal{F}|U \to R^2 \mathrm{pr}_! \overline{\mathbb{Q}}_l|U = \overline{\mathbb{Q}}_l(-1)$$

is a perfect symplectic autoduality of $\mathcal{F}$ with values in $\overline{\mathbb{Q}}_l(-1)$. In particular, G_{geom} for $\mathcal{F}|U$ is a subgroup of $Sp(2g)$.

PROOF. Since $\mathcal{F}$ has no nonzero punctual sections, it is lisse precisely at those points of $\mathbb{A}^1$ where its stalk has maximum dimension. By proper base change, the stalk $\mathcal{F}_t$ at a geometric point t in $L = \mathbb{A}^1(L)$, L some algebraically closed extension of k, is the compact cohomology group

$$\mathcal{F}_t = H^1_c(\mathrm{pr}^{-1}(t), \overline{\mathbb{Q}}_l) = H^1_c((Y^2 = f(X)(X-t)), \overline{\mathbb{Q}}_l).$$

If t lies in U, i.e., if $f(t) \neq 0$, this curve is the complement of a single rational point ∞ in a proper smooth geometrically connected curve $\mathcal{C}_t$ of genus g. Since there is a single missing point, the inclusion induces an isomorphism

$$H^1_c(Y^2 = f(X)(X-t), \overline{\mathbb{Q}}_l) \cong H^1_c(\mathcal{C}_t, \overline{\mathbb{Q}}_l).$$

Thus the stalks of $\mathcal{F}$ have dimension $2g$ at points of U. This shows that $\mathcal{F}|U$ is lisse of rank $2g$, and also that $\mathcal{F}|U$ is pure of weight one, by the Riemann hypothesis for curves over finite fields [**Weil-CA**, §IV, No. 22, page 70].

If t does not lie in U, then $f(t) = 0$, so t lies in $\overline{k}$, f is divisible by $X - t$, say $f(X) = g(X)(X-t)$, and g is a polynomial of degree $2g-1$ with $g(t) \neq 0$, and g has $2g-1$ distinct roots in $\overline{k}$. Then $\mathrm{pr}^{-1}(t)$ is the curve of equation

$$Y^2 = (X-t)^2 g(X).$$

Let us temporarily denote by C the curve of equation $Y^2 = g(X)$. Then we have

$$C - \{\text{the two points } (X = t, Y = \pm g(t)^{1/2})\} \cong \mathrm{pr}^{-1}(t) - \{(X = t, Y = 0)\}.$$

But C is the complement of a single point ∞ in a proper smooth connected curve of genus $g-1$, so

$$h^1_c(C - \{\text{the two points } (X = t, Y = \pm g(t)^{1/2})\}, \overline{\mathbb{Q}}_l) = 2g - 2 + 2 = 2g.$$

Thus we have

$$h^1_c(\mathrm{pr}^{-1}(t) - \{\text{one point } (X = t, Y = 0)\}, \overline{\mathbb{Q}}_l) = 2g.$$

In the excision sequence, say with $\overline{\mathbb{Q}}_l$-coefficients, for the inclusion of $\mathrm{pr}^{-1}(t) - \{\text{one point}\}$ into $\mathrm{pr}^{-1}(t)$, we have $H^0_c(\mathrm{pr}^{-1}(t)) = 0$ by 10.1.5, as $\mathrm{pr}^{-1}(t)$ is a hypersurface in $\mathbb{A}^2$, and $H^1_c(\{\text{point}\}) = 0$ trivially, so we get a short exact sequence

$$0 \to H^0_c(\{\text{point}\}) \to H^1_c(\mathrm{pr}^{-1}(t) - \{\text{point}\}) \to H^1_c(\mathrm{pr}^{-1}(t)) \to 0.$$

This gives the asserted value $2g-1$ for $h^1_c(\mathrm{pr}^{-1}(t))$.

Over U, the morphism is smooth of relative dimension one, with geometrically connected fibres, so the sheaf $R^2\,\mathrm{pr}_!\,\overline{\mathbb{Q}}_l|U$ is the geometrically constant sheaf $\overline{\mathbb{Q}}_l(-1)$, cf. 9.1.13. The cup-product pairing

$$\mathcal{F}|U \times \mathcal{F}|U \to R^2\,\mathrm{pr}_!\,\overline{\mathbb{Q}}_l|U = \overline{\mathbb{Q}}_l(-1)$$

is an alternating pairing of lisse sheaves. To show that it is nondegenerate, it suffices to check nondegeneracy on a single (or on every) geometric fibre of pr over U. But for t in U, we have already noted that $\mathrm{pr}^{-1}(t)$ is of the form $\mathcal{C}_t - \{\infty\}$. The inclusion induces an isomorphism

$$H^1_c(\mathrm{pr}^{-1}(t), \overline{\mathbb{Q}}_l) \cong H^1_c(\mathcal{C}_t, \overline{\mathbb{Q}}_l),$$

and, dually, restriction induces an isomorphism of cohomology without support,

$$H^1(\mathcal{C}_t, \overline{\mathbb{Q}}_l) \cong H^1(\mathrm{pr}^{-1}(t), \overline{\mathbb{Q}}_l).$$

Thus the canonical "forget supports" map is an isomorphism

$$H^1_c(\mathrm{pr}^{-1}(t), \overline{\mathbb{Q}}_l) \cong H^1(\mathrm{pr}^{-1}(t), \overline{\mathbb{Q}}_l),$$

and so our cup product becomes the cup product pairing

$$H^1_c(\mathrm{pr}^{-1}(t), \overline{\mathbb{Q}}_l) \times H^1(\mathrm{pr}^{-1}(t), \overline{\mathbb{Q}}_l) \to \overline{\mathbb{Q}}_l,$$

which is nondegenerate by Poincaré duality. QED

10.1.10. We could also have deduced directly all the assertions about $\mathcal{F}|U$ by relating $\mathcal{F}$ directly to the sheaf $R^1\pi_!\overline{\mathbb{Q}}_l$ for $\pi : \mathcal{C} \to \mathbb{A}^1$, which over U is a proper smooth relative curve of genus g.

Lemma 10.1.11. *The inclusion of $\mathcal{C}^{\mathrm{aff}}$ into $\mathcal{C}$ as schemes over $\mathbb{A}^1$ induces isomorphisms $R^i\,\mathrm{pr}_!\,\overline{\mathbb{Q}}_l \cong R^i\pi_!\overline{\mathbb{Q}}_l$ for $i=1$ and $i=2$.*

PROOF. The complement of $\mathcal{C}^{\mathrm{aff}}$ in $\mathcal{C}$ is the ∞ section, $\mathrm{id} : \mathbb{A}^1 \to \mathbb{A}^1$ as scheme over $\mathbb{A}^1$. The lemma is immediate from the excision sequence

$$\cdots \to R^i\,\mathrm{pr}_!\,\overline{\mathbb{Q}}_l \to R^i\pi_!\overline{\mathbb{Q}}_l \to R^i\,\mathrm{id}_!\,\overline{\mathbb{Q}}_l \to \cdots,$$

in which the restriction map $R^0\pi_!\overline{\mathbb{Q}}_l \to R^0\,\mathrm{id}_!\,\overline{\mathbb{Q}}_l = \overline{\mathbb{Q}}_l$ is surjective (check fibre by fibre: each fibre of π is nonvoid), and in which $R^i\,\mathrm{id}_!\,\overline{\mathbb{Q}}_l = 0$ for $i > 0$. QED

Lemma 10.1.12. *The lisse sheaf $\mathcal{F}$ on U is everywhere tame, i.e., it is tamely ramified at all the points at ∞ of U.*

PROOF. The Euler-Poincaré formula [**Ray**] for a lisse $\mathcal{F}$ on U asserts

$$\chi(U \otimes_k \overline{k}, \mathcal{F}) = \mathrm{rank}(\mathcal{F})\chi(U \otimes_k \overline{k}, \overline{\mathbb{Q}}_l) - \sum_{\text{points } x \text{ at } \infty} \mathrm{Swan}_x(\mathcal{F}).$$

Now $\mathcal{F}$ is tame at x if and only if $\mathrm{Swan}_x(\mathcal{F}) = 0$. As each term $\mathrm{Swan}_x(\mathcal{F})$ is a nonnegative integer, $\mathcal{F}$ is everywhere tame if and only if

$$\chi(U \otimes_k \overline{k}, \mathcal{F}) = \mathrm{rank}(\mathcal{F})\chi(U \otimes_k \overline{k}, \overline{\mathbb{Q}}_l),$$

i.e., if and only if

$$\chi_c(U \otimes_k \overline{k}, R^1\,\mathrm{pr}_!\,\overline{\mathbb{Q}}_l) = 2g(1-2g).$$

By the Leray spectral sequence, we have

$$\chi_c(\mathrm{pr}^{-1}(U) \otimes_k \overline{k}, \overline{\mathbb{Q}}_l) = \chi_c(U \otimes_k \overline{k}, R^2\,\mathrm{pr}_!\,\overline{\mathbb{Q}}_l) - \chi_c(U \otimes_k \overline{k}, R^1\,\mathrm{pr}_!\,\overline{\mathbb{Q}}_l).$$

As noted in the proof of 10.1.9, the sheaf $R^2 \operatorname{pr}_! \overline{\mathbb{Q}}_l | U$ is the geometrically constant sheaf $\overline{\mathbb{Q}}_l(-1)$, so we have

$$\begin{aligned}\chi_c(\operatorname{pr}^{-1}(U) \otimes_k \overline{k}, \overline{\mathbb{Q}}_l) &= \chi_c(U \otimes_k \overline{k}, \overline{\mathbb{Q}}_l) - \chi_c(U \otimes_k \overline{k}, R^1 \operatorname{pr}_! \overline{\mathbb{Q}}_l) \\ &= 1 - 2g - \chi_c(U \otimes_k \overline{k}, R^1 \operatorname{pr}_! \overline{\mathbb{Q}}_l).\end{aligned}$$

So $\mathcal{F}$ is everywhere tame if and only if we have

$$\chi_c(\operatorname{pr}^{-1}(U) \otimes_k \overline{k}, \overline{\mathbb{Q}}_l) = (1-2g)^2.$$

But we have a disjoint union decomposition of the total space $\mathcal{C}^{\mathrm{aff}} \otimes_k \overline{k}$ into $\operatorname{pr}^{-1}(U) \otimes_k \overline{k}$, and the $2g$ singular fibres $\operatorname{pr}^{-1}(t)$ as t runs over the zeroes of f. Each singular fibre $\operatorname{pr}^{-1}(t)$ has $h_c^2 = 1$, because h_c^2 does not see points, and $\operatorname{pr}^{-1}(t) - \{(X = t, Y = 0)\}$ is of the form (complete smooth connected curve of genus $g-1$)$-\{3 \text{ points}\}$. So each singular fibre $\operatorname{pr}^{-1}(t)$ has $\chi_c = 1-(2g-1) = 2-2g$. Thus we have

$$\chi_c(\mathcal{C}^{\mathrm{aff}} \otimes_k \overline{k}, \overline{\mathbb{Q}}_l) = \chi_c(\operatorname{pr}^{-1}(U) \otimes_k \overline{k}, \overline{\mathbb{Q}}_l) + 2g(2-2g).$$

So $\mathcal{F}$ is everywhere tame if and only if we have

$$\chi_c(\mathcal{C}^{\mathrm{aff}} \otimes_k \overline{k}, \overline{\mathbb{Q}}_l) = (1-2g)^2 + 2g(2-2g) = 1.$$

So we are reduced to proving that the hypersurface $Y^2 = f(X)(X-T)$ in $\mathbb{A}^3 \otimes_k \overline{k}$ has $\chi_c = 1$. By the change of variable $Z := X - T$, this becomes the hypersurface $Y^2 = Zf(X)$ in $\mathbb{A}^3 \otimes_k \overline{k}$ with coordinates X, Y, Z. On the open set where $f(X)$ is invertible, Y is free, and we can solve for Z as $Y^2/f(X)$. Thus this open set is the product $U \times \mathbb{A}^1$, with coordinates X, Y. On the (reduced) closed set where $f(X) = 0$, we have $Y = 0$, and Z is free, so this set is an $(f = 0 \text{ in } \mathbb{A}^1) \times \mathbb{A}^1$, with coordinates X, Z. Thus we find that $Y^2 = Zf(X)$ has

$$\chi_c = \chi_c(U \times \mathbb{A}^1) + \chi_c((f=0 \text{ in } \mathbb{A}^1) \times \mathbb{A}^1) = (1-2g) \times 1 + (2g) \times 1 = 1. \quad \text{QED}$$

Lemma 10.1.13. *At each geometric point t of $\mathbb{A}^1 - U$, the local monodromy of $\mathcal{F}$ is either trivial or is a unipotent pseudoreflection.*

PROOF. Given a $\overline{k}$-valued zero t of f, pick a geometric point $\overline{\eta}$ in its formal punctured neighborhood $\operatorname{Spec}(\overline{k}((T-t)))$. The inertia group (local monodromy group at t) $I(t) := \operatorname{Gal}(\overline{k}((T-t))^{\mathrm{sep}}/\overline{k}((T-t))) = \pi_1(\operatorname{Spec}(\overline{k}((T-t))), \overline{\eta})$ acts on $\mathcal{F}_{\overline{\eta}}$. Denoting by $j : U \to \mathbb{A}^1$ the inclusion, it is tautological that the stalk of $j_*j^*\mathcal{F}$ at t is the space $(\mathcal{F}_{\overline{\eta}})^{I(t)}$ of invariants under $I(t)$. Since $\mathcal{F}$ has no nonzero punctual sections, we have $\mathcal{F} \subset j_*j^*\mathcal{F}$, so we get

$$\mathcal{F}_t \subset (\mathcal{F}_{\overline{\eta}})^{I(t)} \subset \mathcal{F}_{\overline{\eta}}.$$

Since $\mathcal{F}_t$ has codimension one in $\mathcal{F}_{\overline{\eta}}$, either $I(t)$ acts trivially or $(\mathcal{F}_{\overline{\eta}})^{I(t)}$ is a codimension one subspace.

In the latter case, $I(t)$ acts through pseudoreflections with this fixed space. Since $I(t)$ acts through $Sp(2g)$ and $Sp(2g) \subset SL(2g)$, these pseudoreflections must be unipotent. Therefore the monodromy is tame (being pro-l), and the action is through the maximal pro-l quotient $\mathbb{Z}_l(1)$ of $I(t)$, any generator γ_t of which acts as a unipotent pseudoreflection. QED

Lemma 10.1.14. *The group G_{geom} for $\mathcal{F}|U$ is either trivial or is a connected subgroup of $Sp(2g)$ generated by unipotent pseudoreflections.*

PROOF. Because $\mathcal{F}$ is everywhere tame, it is a representation of the tame quotient $\pi_1^{\text{tame}}(\mathbb{A}^1 \otimes_k \overline{k} - \{\text{zeroes of } f\})$. This group is topologically generated by all conjugates of all local monodromy groups $I(t)$ at all zeroes of f, since the corresponding quotient classifies finite etale coverings of $\mathbb{A}^1 \otimes_k \overline{k}$ which are tame at ∞, and any such covering is trivial. Although G_{geom} is defined as the Zariski closure of $\rho(\pi_1^{\text{geom}})$, it also the Zariski closure of any dense subgroup of $\rho(\pi_1^{\text{geom}})$. Hence G_{geom} is the Zariski closure in $Sp(2g)$ of a group generated by some (possibly empty) list of pseudoreflections. For each pseudoreflection g on the list, consider the one parameter subgroup $X_g := \exp(t\log(g))$ of G_{geom}. Then G_{geom} is equal to the group generated by finitely many of these one parameter subgroups, and, being generated by connected subgroups, is itself connected. QED

Lemma 10.1.15. *The lisse sheaf $\mathcal{F}|U$ is irreducible as a $\overline{\mathbb{Q}}_l$-representation of $\pi_1^{\text{geom}}(U)$.*

PROOF. The question is geometric, so we may make a preliminary extension of finite fields to reduce to the case when f has all its roots in k. We will apply the diophantine criterion for irreducibility, cf. [**Ka-RLS**, 7.0.3]. Since $\mathcal{F}|U$ is pure of weight one, it is semisimple as representation of π_1^{geom}. So by Schur's lemma, we must show that $\underline{\text{End}}(\mathcal{F})$ has precisely a one-dimensional space of π_1^{geom}-invariants, or equivalently ($\underline{\text{End}}(\mathcal{F})$ being semisimple) a one-dimensional space of π_1^{geom}-coinvariants, i.e., we must show $h_c^2(U\otimes_k \overline{k}, \underline{\text{End}}(\mathcal{F})) = 1$. Since $\mathcal{F}$ is self-dual toward $\overline{\mathbb{Q}}_l(-1)$, $\underline{\text{End}}(\mathcal{F}) = \mathcal{F}\otimes\mathcal{F}^\vee \cong \mathcal{F}\otimes\mathcal{F}(1)$, and so we must show

$$h_c^2(U \otimes_k \overline{k}, \mathcal{F}\otimes\mathcal{F}) = 1.$$

As $\mathcal{F}\otimes\mathcal{F}$ on U is pure of weight 2, the group $H_c^2(U\otimes_k \overline{k}, \mathcal{F}\otimes\mathcal{F})$ is pure of weight 4. The symplectic pairing $\mathcal{F}\otimes\mathcal{F} \to \overline{\mathbb{Q}}_l(-1)$ induces a surjective map of $\text{Gal}(\overline{k}/k)$-modules

$$H_c^2(U\otimes_k \overline{k}, \mathcal{F}\otimes\mathcal{F}) \to H_c^2(U\otimes_k \overline{k}, \overline{\mathbb{Q}}_l(-1)) \cong \overline{\mathbb{Q}}_l(-2),$$

which we must prove to be an isomorphism. Since $H_c^1(U\otimes_k \overline{k}, \mathcal{F}\otimes\mathcal{F})$ is mixed of weight ≤ 3, and $H_c^0(U\otimes_k \overline{k}, \mathcal{F}\otimes\mathcal{F}) = 0$, this amounts, via the Lefschetz trace formula, to proving that for any finite extension E of k, we have an estimate

$$\text{Card}(E)^2 - \sum_{t \text{ in } E, f(t)\neq 0} (\text{Trace}(F_{E,t}|\mathcal{F}))^2 = O(\text{Card}(E)^{3/2}).$$

As $\mathcal{F}$ is mixed of weight ≤ 1 on all of $\mathbb{A}^1$, if we add the terms $(\text{Trace}(F_{E,t}|\mathcal{F}))^2$ for the $2g$ possible bad t values in E, we change our sum by $O(\text{Card}(E))$, so it is equivalent to prove that

$$\text{Card}(E)^2 - \sum_{t \text{ in } E} (\text{Trace}(F_{E,t}|\mathcal{F}))^2 = O(\text{Card}(E)^{3/2}).$$

In fact, we will see that this is $O(\text{Card}(E))$, cf. [**Ka-RLS**, 7.1.1] for the meaning of this improvement.

For each t in E, the curve $\text{pr}^{-1}(t)$ is $Y^2 = f(X)(X-t)$, which over $\overline{k}$ has $H_c^2 = \overline{\mathbb{Q}}_l(-1)$, so by the Lefschetz trace formula on $\text{pr}^{-1}(t)$ and proper base change, we have

$$\text{Card}(E) - \text{Trace}(F_{E,t}|\mathcal{F}) = \text{Card}(\text{pr}^{-1}(t)(E)).$$

But we can count the E-points on $Y^2 = f(X)(X-t)$ by summing over X in E and seeing for each how many square roots $f(X)(X-t)$ has in E. So if we denote by χ the quadratic character of $E^\times$, extended to be 0 at 0, we have

$$\mathrm{Card}(\mathrm{pr}^{-1}(t)(E)) = \sum_{x \text{ in } E} (1 + \chi((x-t)f(x))),$$

and hence we get

$$-\mathrm{Trace}(F_{E,t}|\mathcal{F}) = \sum_{x \text{ in } E} \chi((x-t)f(x)).$$

Therefore we have

$$\begin{aligned}\sum_{t \text{ in } E} (\mathrm{Trace}(F_{E,t}|\mathcal{F}))^2 &= \sum_{t \text{ in } T} \left(\sum_{x \text{ in } E} \chi((x-t)f(x)) \right)^2 \\ &= \sum_{t \text{ in } E} \sum_{x,z \text{ in } E} \chi((x-t)f(x)(z-t)f(z)) \\ &= \sum_{x,z \text{ in } E} \chi(f(x)f(z)) \sum_{t \text{ in } E} \chi((x-t)(z-t)).\end{aligned}$$

The innermost sum is, for given x and z in E, related to the number of E-points on the quadric curve $Q_{x,z}$ of equation $Y^2 = (T-x)(T-z)$ by the same character sum method used above:

$$\mathrm{Card}(Q_{x,z}(E)) = \mathrm{Card}(E) + \sum_{t \text{ in } E} \chi((x-t)(z-t)).$$

If $x = z$, then $Q_{x,x}$ is $Y^2 = (T-x)^2$, two lines crossing at $(x,0)$, and has $2\,\mathrm{Card}(E) - 1$ E-points. If $x \neq z$, then $Q_{x,z}$ is the finite part of a smooth quadric curve with 2 rational points at ∞, and so has $\mathrm{Card}(E) - 1$ E-points. Thus we find

$$\begin{aligned}\sum_{t \text{ in } E} \chi((x-t)(z-t)) &= \mathrm{Card}(E) - 1, \quad \text{if } x = z, \\ &= -1, \quad \text{if } x \neq z.\end{aligned}$$

Thus we get

$$\begin{aligned}&\sum_{x,z \text{ in } E} \chi(f(x)f(z)) \sum_{t \text{ in } E} \chi((x-t)(z-t)) \\ &\quad = \sum_{x=z \text{ in } E} \chi(f(x)f(z))\,\mathrm{Card}(E) - \sum_{x,z \text{ in } E} \chi(f(x)f(z)) \\ &\quad = \mathrm{Card}(E) \sum_{x \text{ in } E} \chi(f(x))^2 - \sum_{x,z \text{ in } E} \chi(f(x)f(z)) \\ &\quad = \mathrm{Card}(E)(\mathrm{Card}(E) - 2g) - \left(\sum_{x \text{ in } E} \chi(f(x)) \right)^2.\end{aligned}$$

By the Riemann hypothesis for the curve $Y^2 = f(X)$ over E, the last term $(\sum_{x \text{ in } E} \chi(f(x)))^2$ is itself $O(\mathrm{Card}(E))$. QED

Theorem 10.1.16. *The group G_{geom} for $\mathcal{F}|U$ is $Sp(2g)$, i.e., the one-parameter family of curves $Y^2 = f(X)(X-T)$ with parameter T has biggest possible geometric monodromy group, namely $Sp(2g)$.*

PROOF. Combining the previous two lemmas, we find that G_{geom} is a connected irreducible subgroup of $Sp(2g)$ which contains a unipotent pseudoreflection. Thus $\text{Lie}(G_{\text{geom}})$ is an irreducible Lie-subalgebra of the symplectic Lie algebra $\mathcal{SP}(2g)$ which is normalized by a unipotent pseudoreflection. By the Kazhdan-Margulis theorem [**Ka-ESDE**, 1.5], $\text{Lie}(G_{\text{geom}})$ is all of $\mathcal{SP}(2g)$, and hence G_{geom}, a priori a subgroup of $Sp(2g)$, is all of $Sp(2g)$. QED

Remark 10.1.17. A faster but more "high tech" approach to $\mathcal{F}$ is to view it as the restriction to U of the middle convolution $\mathcal{L}_{\chi^*\,\text{mid}}\mathcal{L}_{\chi(f)}$ on $\mathbb{A}^1$ in the sense of [**Ka-RLS**, 2.6]. It follows [**Ka-RLS**, 3.3.6] from this description and its relation to Fourier transform that $\mathcal{F}$ is geometrically irreducible, that at each finite singularity the local monodromy is a unipotent pseudoreflection, and that at ∞ the local monodromy is (-1) times a unipotent pseudoreflection. In the next section, we will give examples in characteristic two, where we do not know any way to avoid the Fourier transform theory.

Variant 10.1.18 (universal families of hyperelliptic curves).

10.1.18.1. For each integer $d \geq 1$, let us denote by $\mathcal{H}_d$ the space of monic polynomials of degree d with all distinct roots. Concretely, $\mathcal{H}_d$ is the open set in $\mathbb{A}^d$ with coordinates $a_0, a_1, \ldots, a_{d-1}$ (thought of as the coefficients of the monic polynomial $f(X) := X^d + \sum_{i<d} a_i X^i$) where the discriminant of that polynomial $f(X)$ is invertible. Thus for any field k, $\mathcal{H}_d(k)$ is precisely the set of square-free monic polynomials f of degree d in one variable over k.

10.1.18.2. For each genus $g \geq 1$, we have a family of genus g hyperelliptic curves over $\mathcal{H}_{2g+1}[1/2]$, the fibre over a point f being the curve of equation $Y^2 = f(X)$. [We need to invert the prime 2 for this curve to be nonsingular.] In view of 10.1.17, which tells us that various one-parameter subfamilies of this family already have geometric monodromy as large as possible, namely $Sp(2g)$, we find

Theorem 10.1.18.3. *In any odd characteristic p, and for any genus $g \geq 1$, the family of genus g hyperelliptic curves $Y^2 = f(X)$ over $\mathcal{H}_{2g+1} \otimes \mathbb{F}_p$ has largest possible geometric monodromy group $Sp(2g)$.*

10.1.18.4. For the sake of completeness, let us also spell out what happens if we consider hyperelliptic curves of genus g of equation $Y^2 = f(X)$ with f monic and square-free of degree $2g+2$. Let us denote by $\widetilde{\mathcal{H}}_d$ the space of not-necessarily monic degree d polynomials $\sum_{i \leq d} a_i X^i$ with a_d invertible and with discriminant invertible. If we "factor out" a_d, i.e., write f as $a_d f_{\text{monic}}$, we see that $\widetilde{\mathcal{H}}_d = \mathbb{G}_m \times \mathcal{H}_d$.

Theorem 10.1.18.5. 1) *In any odd characteristic p, and for any genus $g \geq 1$, the family of genus g hyperelliptic curves $Y^2 = f(X)$ over $\widetilde{\mathcal{H}}_{2g+2} \otimes \mathbb{F}_p$ has largest possible geometric monodromy group $Sp(2g)$.*

2) *In any odd characteristic p, and for any genus $g \geq 1$, the family of genus g hyperelliptic curves $Y^2 = f(X)$ over $\mathcal{H}_{2g+2} \otimes \mathbb{F}_p$ has largest possible geometric monodromy group $Sp(2g)$.*

PROOF. We first show that 1) implies 2). Indeed, if 1) holds, then because $Sp(2g)$ is connected, if we pull back our family to the finite etale double convering of $\widetilde{\mathcal{H}}_{2g+2} \otimes \mathbb{F}_p$ defined by taking a square root, say α_d, of a_d, $d := 2g+2$, the geometric monodromy group remains $Sp(2g)$. But over this double covering, the universal curve $Y^2 = (\alpha_d)^2 f_{\text{monic}}(X)$ is isomorphic (replace Y by Y/α_d) to $Y^2 = f_{\text{monic}}(X)$.

In other words, the universal family here becomes isomorphic to the pullback from $\mathcal{H}_d \otimes \mathbb{F}_p$ via the map $f \mapsto f_{\text{monic}}$ of $\widetilde{\mathcal{H}}_d[\alpha_d] \otimes \mathbb{F}_p$ to $\mathcal{H}_d \otimes \mathbb{F}_p$. Since the geometric monodromy group can only decrease after pullback, it must be at least $Sp(2g)$ for the family over $\mathcal{H}_d \otimes \mathbb{F}_p$. But G_{geom} is a subgroup of $Sp(2g)$ in any case, so it must be $Sp(2g)$ over $\mathcal{H}_d \otimes \mathbb{F}_p$.

It remains to prove 1). For this, we argue as follows. Denote by $U \subset \mathcal{H}_{2g+1}$ the open set consisting of polynomials f which in addition have constant term $f(0)$ invertible. Then for every prime p, $U \otimes \mathbb{F}_p$ is a nonempty,and hence dense, open set of $\mathcal{H}_{2g+1} \otimes \mathbb{F}_p$. Therefore the restriction to $U \otimes \mathbb{F}_p$ of the universal family $Y^2 = f(X)$ over $\mathcal{H}_{2g+1} \otimes \mathbb{F}_p$ has the same G_{geom}, namely $Sp(2g)$, thanks to 10.1.16.

We then rewrite the universal curve over U, namely $Y^2 = f(X)$, as

$$(Y/X^{g+1})^2 = (1/X)(1/X^{2g+1})f(X) = (1/X)f_{\text{reversed}}(1/X).$$

If we make the change of variable $\widetilde{Y} = Y/X^{g+1}, \widetilde{X} = 1/X$, we get

$$(\widetilde{Y})^2 = \widetilde{X} f_{\text{reversed}}(\widetilde{X}).$$

Thus we have embedded U in $\widetilde{\mathcal{H}}_{2g+2}$ as the closed set consisting of polynomials whose constant term vanishes. Since the restriction from $\widetilde{\mathcal{H}}_{2g+2} \otimes \mathbb{F}_p$ to this embedded $U \otimes \mathbb{F}_p$ already has biggest possible G_{geom}, namely $Sp(2g)$, the G_{geom} on all of $\widetilde{\mathcal{H}}_{2g+2} \otimes \mathbb{F}_p$ can be no smaller, and, being a subgroup of $Sp(2g)$, must be $Sp(2g)$. QED

10.2. Examples in characteristic two

10.2.1. For each genus $g \geq 1$, we work over the parameter space

$$\mathbb{G}_{m,\mathbb{F}_2} := \text{Spec}(\mathbb{F}_2[T, 1/T]),$$

and we consider the one-parameter family of Artin-Schreier curves

$$Y^2 - Y = X^{2g-1} + T/X.$$

This is (the complement of two disjoint sections $(0, \infty)$ and (∞, ∞) in) a proper smooth relative curve $\pi : \mathcal{C} \to \mathbb{G}_{m,\mathbb{F}_2}$ of genus g over $\mathbb{G}_{m,\mathbb{F}_2}$. We denote by $\mathcal{F}$ the sheaf $R^1\pi_!\overline{\mathbb{Q}_l}$.

Theorem 10.2.2 ([**Sut**]). *For each genus $g \geq 1$, the G_{geom} for $\mathcal{F}$ on $\mathbb{G}_{m,\mathbb{F}_2}$ is $Sp(2g)$, i.e., the family of genus g curves over $\mathbb{G}_{m,\mathbb{F}_2}$ given by the affine equation $Y^2 - Y = X^{2g-1} + T/X$ has biggest possible geometric monodromy group, namely $Sp(2g)$.*

PROOF. We will sketch the proof, and refer the reader to [**Sut**] for full details. For each $t \neq 0$, the curve in question is a finite etale connected $\mathbb{Z}/2\mathbb{Z}$-covering of the $\mathbb{G}_m$ of X, which is fully ramified over both 0 and ∞. Denote by ψ the unique nontrivial additive character of $\mathbb{F}_2$ $(\psi(0) = 1, \psi(1) = -1)$. For each finite extension E of $\mathbb{F}_2$, denote $\psi_E := \psi \circ \text{Trace}_{E/\mathbb{F}_2}$. Then for each point t in $E^\times$, we have

$$\text{Card}(\mathcal{C}_t(E)) = 2 + \sum_{x \text{ in } E^\times} (1 + \psi_E(x^{2g-1} + t/x)),$$

and hence

$$\text{Trace}(F_{E,t}|\mathcal{F}) = - \sum_{x \text{ in } E^\times} \psi_E(x^{2g-1} + t/x).$$

Since $\mathrm{Card}(E^\times)$ is odd, and each value of ψ_E is ± 1, this trace is an odd integer. In particular, it is never zero.

From the trace expression, we see that $\mathcal{F}$ has the same trace function as $\mathcal{G} :=$ the Fourier transform $FT_\psi(\mathcal{L}_{\psi(1/x^{2g-1})})|\mathbb{G}_{m,\mathbb{F}_2}$. By Fourier theory [**Lau-TF**], $\mathcal{G}$ is lisse on $\mathbb{G}_{m,\mathbb{F}_2}$ and irreducible as π_1^{geom}-representation, hence a fortiori irreducible as π_1-representation. Because $\mathcal{F}$ and $\mathcal{G}$ are lisse sheaves on $\mathbb{G}_{m,\mathbb{F}_2}$ with the same trace function, their π_1-semisimplifications are isomorphic. Therefore $\mathcal{F}$ is itself π_1-irreducible, and hence $\mathcal{F}$ is isomorphic to $\mathcal{G}$ as a lisse sheaf on $\mathbb{G}_{m,\mathbb{F}_2}$.

By the known local monodromy structure of a Fourier transform due to Laumon, the local monodromy of $\mathcal{F}$ at 0 is a unipotent pseudoreflection [**Ka-ESDE**, 7.6.3.1]. So again by the Kazhdan-Margulis theorem, it suffices to show that $\mathcal{F}$ is Lie-irreducible.

To show that $\mathcal{F}$ is Lie-irreducible, we rule out the other possibilities. If $\mathcal{F}$ is not Lie-irreducible, then by [**Ka-MG**, Prop. 1] it is either geometrically induced, or is geometrically a tensor product $\mathcal{G} \otimes \mathcal{H}$ of lisse sheaves, with $\mathcal{G}$ Lie-irreducible and with $\mathcal{H}$ of rank ≥ 2 and having finite G_{geom}. If $\mathcal{F}$ were geometrically induced, then it would be arithmetically induced, at least after possibly passing to a larger finite field, and this would force its trace to vanish on a set of positive density. Since its trace function is nowhere vanishing, $\mathcal{F}$ cannot be geometrically induced. A unipotent pseudoreflection cannot be a tensor product $A \otimes B$ with both A and B of size > 1, so if $\mathcal{F}$ is $\mathcal{G} \otimes \mathcal{H}$, either $\mathcal{G}$ or $\mathcal{H}$ must be of rank one. As $\mathcal{H}$ has rank ≥ 2, we must have $\mathcal{G}$ of rank one. But any $\mathcal{G}$ of rank one has finite G_{geom}, and so $\mathcal{F} = \mathcal{G} \otimes \mathcal{H}$ has finite G_{geom}, which is impossible, because a unipotent pseudoreflection is not of finite order. QED

Remark 10.2.3. Instead of the family $Y^2 - Y = X^{2g-1} + T/X$, we could consider the family $Y^2 - Y = f(X) + T/X$ for any monic polynomial f of degree $2g-1$. Sutor's result remains true, and the proof is identical, cf. [**Sut**].

10.3. Other examples in odd characteristic

10.3.1. Fix an odd prime p and an integer $d \geq 3$ prime to p. Choose a monic polynomial $f(X)$ in $\mathbb{F}_p$ of degree d whose second derivative f'' does not vanish identically, e.g., $X^d + X^2$. Let A and B be indeterminates, and denote by Δ in $\mathbb{F}_p[A, B]$ the discriminant of the polynomial $f(X) + AX + B$. One can show that Δ is nonzero, cf. [**Ka-ACT**, 3.5]. Over the open set $\mathrm{Spec}(\mathbb{F}_p[A, B][1/\Delta])$ of $\mathbb{A}^2$ where Δ is invertible, consider the two parameter family of hyperelliptic curves of affine equation

$$Y^2 = f(X) + AX + B.$$

If d is odd, say $d = 2g + 1$, this is (the complement of a single section at ∞ of) a proper smooth family of connected curves of genus g. If d is even, say $d = 2g + 2$, this is (the complement of two disjoint sections at ∞ of) a proper smooth family of connected curves of genus g. In both cases, it is proven in [**Ka-ACT**, 5.4 (1) and 5.17 (1) respectively] that for any $l \neq p$, (the sheaf $R^1\pi_!\overline{\mathbb{Q}}_l$ attached to) this family has $G_{\mathrm{geom}} = Sp(2g)$.

10.3.2. We can give one-parameter examples with big monodromy by specializing A. Recall [**Ka-ACT**, 5.5.2] that a polynomial $g(X)$ over a field k whose degree d is invertible in k is called "weakly supermorse" if its derivative $g'(X)$ has $d - 1$ distinct zeroes in $\overline{k}$, and if g separates the zeroes of g'. It is proven in

[**Ka-ACT**, 5.15] that, given a polynomial f over $\mathbb{F}_p$ as above (degree prime to p, f'' nonzero), there is a nonzero polynomial $D(A)$ in $\mathbb{F}_p[A]$ such that for any extension field k of $\mathbb{F}_p$, and for any a in k with $D(a) \neq 0$, the polynomial $f(X) + aX$ over k is weakly supermorse. It is further proven [**Ka-ACT**, 5.7 for d odd, 5.18 for d even] that if $g(X)$ in $k[X]$ is weakly supermorse of degree d prime to p, the one parameter family of hyperelliptic, genus $g := [(d-1)/2]$ curves

$$Y^2 = g(X) - T$$

over $\mathrm{Spec}(k[T][1/\prod_{\text{zeroes } \alpha \text{ of } g'}(T - g(\alpha))])$ has, for any $l \neq p$, its sheaf $R^1\pi_!\overline{\mathbb{Q}}_l$ everywhere tame, with $G_{\text{geom}} = Sp(2g)$.

10.3.3. Given an odd prime p and an integer $g \geq 1$, at least one of $2g+1$ or $2g+2$ is prime to p. Take one such as d, say take $d = 2g+1$ unless p divides $2g+1$, in which case take $d = 2g+2$. We get a two parameter family of genus g curves over $\mathrm{Spec}(\mathbb{F}_p[A, B][1/\Delta])$ such that, for any $l \neq p$, we have $G_{\text{geom}} = Sp(2g)$. If we specialize A to a nonzero a of the polynomial $D(A)$ in an extension field k of $\mathbb{F}_p$, we get a one-parameter family of genus g curves over a nonvoid open set of $\mathbb{A}^1$ over k such that, for any $l \neq p$, we have $G_{\text{geom}} = Sp(2g)$.

10.3.4. Here is a very simple example of such a one parameter family. Given an integer $d \geq 3$, consider the one parameter family

$$Y^2 = X^d - dX - T.$$

In any characteristic p which does not divide $d(d-1)$, the polynomial $X^d - dX$ is weakly supermorse. So for any such p, this is a one parameter family of genus $g := [(d-1)/2]$ curves over the nonvoid open set $\mathrm{Spec}(\mathbb{F}_p[T, 1/(T^{d-1} - (1-d)^{d-1})])$ of $\mathbb{A}^1$ over $\mathbb{F}_p$ such that, for any $l \neq p$, we have $G_{\text{geom}} = Sp(2g)$.

Remark 10.3.5. Suppose we are given any proper smooth family of geometrically connected curves of genus $g \geq 1, \pi : \mathcal{C} \to X$, where X is smooth, with geometrically connected fibres, over a normal connected S of finite type over $\mathbb{Z}[1/l]$. We put $\mathcal{F} := R^1\pi_!\overline{\mathbb{Q}}_l$. For each finite field k, and each s in $S(k)$, either choice of $\mathrm{Card}(k)^{1/2}$ in $\overline{\mathbb{Q}}_l$ allows us to define $\mathcal{F}_s(1/2)$ on X_s. This sheaf on X_s is, via cup product, symplectically self-dual toward the constant sheaf $\overline{\mathbb{Q}}_l$, so we automatically have that for it, $\rho(\pi_1(X_s))$ lies in $Sp(2g)$. Thus if we know that G_{geom} for every $\mathcal{F}|X_s$ is $Sp(2g)$, we are automatically in a position to apply to $\mathcal{F}$ Theorem 9.6.1 with α_s taken to be any choice of $1/\mathrm{Card}(k)^{1/2}$.

10.4. Effective constants in our examples

10.4.1. In our characteristic $p \neq 2$ examples of type $Y^2 = f(X)(X-T)$, the sheaf $\mathcal{F}$ on U is tame, and $U \otimes_k \overline{k}$ is $\mathbb{P}^1 - \{2g+1 \text{ points}\}$. So we may take the constant $C(U \otimes_k \overline{k}, \mathcal{F})$ in part 3) of 9.2.6 to be $2g-1$.

10.4.2. In our characteristic two example, the sheaf $\mathcal{F}$ on $\mathbb{G}_{m,\mathbb{F}_2}$ is tame at 0, and at ∞ all its breaks are $(2g-1)/2g$, as follows from the Fourier transform theory, cf. [**Sut**] or [**Ka-ESDE**, 7.5.4]. So we may take the constant $C(\mathbb{G}_{m,\overline{\mathbb{F}}_2}, \mathcal{F})$ to be $(2g-1)/2g$.

10.4.3. In our characteristic $p \neq 2$ examples of type $Y^2 = g(X) - T$ with g weakly supermorse of degree d prime to p, the sheaf $\mathcal{F}$ on U is tame, and $U \otimes_k \overline{k}$ is $\mathbb{A}^1 - \{d-1 \text{ points}\}$. So we may take the constant $C(U \otimes_k \overline{k}, \mathcal{F})$ to be $d-2$. For the two parameter examples of type $Y^2 = f(X) + AX + B$, we do not know an explicit upper bound for the constant $C(\mathbb{A}^2[1/\Delta] \otimes_{\mathbb{F}_p} \overline{\mathbb{F}}_p, \mathcal{F})$.

10.5. Universal families of curves of genus $g \geq 2$

10.5.1. In this section, we fix an integer $g \geq 2$. Given an arbitrary scheme S, by a "curve of genus g over S" we mean a proper smooth morphism $\pi : \mathcal{C} \to S$ whose geometric fibres are connected curves of genus g. It is well-known that (because some curves have nontrivial automorphisms, e.g., hyperelliptic curves) there is no "universal family" of curves of genus g, i.e., the functor $\mathcal{M}_g$ from (Schemes) to (Sets) defined by

$$\mathcal{M}_g(S) := \{S\text{-isomorphism classes of curves of genus } g \text{ over } S\}$$

cannot be representable. For if $\mathcal{M}_g$ existed as a scheme, then for any extension L/K of fields, the natural map from $\mathcal{M}_g(K)$ to $\mathcal{M}_g(L)$ would be injective. This would imply that if two curves over K become isomorphic over L, they must already be isomorphic over K. This is nonsense, already for $\mathbb{R} \subset \mathbb{C}$, or $\mathbb{Q} \subset \mathbb{Q}(i)$. Consider the two genus g curves given by the affine equations

$$Y^2 = X^{2g+2} + 1 \quad \text{and} \quad -Y^2 = X^{2g+2} + 1$$

over $\mathbb{Q}$. Over $\mathbb{C}$, indeed over $\mathbb{Q}(i)$, these two curves become isomorphic: just replace Y by iY. But they are not $\mathbb{R}$-isomorphic (and hence not $\mathbb{Q}$-isomorphic), because the first curve has a plethora of $\mathbb{R}$-valued points (two for each choice of X in $\mathbb{R} \cup \{\infty\}$) while the second curve has none.

10.5.2. To get a similar example for $\mathbb{F}_p \subset \mathbb{F}_{p^2}$ with p an odd prime, take $f(X)$ in $\mathbb{F}_p[X]$ an irreducible polynomial of degree $2g+1$, and take α in $\mathbb{F}_p^\times$ a nonsquare. Consider the genus g curves over $\mathbb{F}_p$ given by the affine equations

$$Y^2 = f(X) \quad \text{and} \quad Y^2 = \alpha f(X).$$

These become isomorphic over $\mathbb{F}_{p^2}$, but they are not $\mathbb{F}_p$-isomorphic, because they have different numbers of $\mathbb{F}_p$-points. To see this, compare the quadratic character sum expressions for their numbers of $\mathbb{F}_p$-points, namely

$$p + 1 + \sum_{x \text{ in } \mathbb{F}_p} \chi_2(f(x)) \quad \text{and} \quad p + 1 - \sum_{x \text{ in } \mathbb{F}_p} \chi_2(f(x)).$$

We must see that $\sum_{x \text{ in } \mathbb{F}_p} \chi_2(f(x))$ is nonzero. Since f is irreducible over $\mathbb{F}_p$, $f(x)$ is nonzero for x in $\mathbb{F}_p$, and each term $\chi_2(f(x))$ is ± 1. As the number p of terms is odd, $\sum_{x \text{ in } \mathbb{F}_p} \chi_2(f(x))$ is odd, hence nonzero.

10.5.3. To get a similar example for $\mathbb{F}_2 \subset \mathbb{F}_4$, take for $f(X)$ any irreducible polynomial of degree $2g+1$ in $\mathbb{F}_2[X]$ (or indeed any polynomial of degree $2g+1$ in $\mathbb{F}_2[X]$ with no zeroes in $\mathbb{F}_2$), and consider the genus g curves over $\mathbb{F}_2$ given by the affine equations

$$Y^2 - Y = f(X) \quad \text{and} \quad Y^2 - Y + 1 = f(X).$$

These curves become isomorphic over $\mathbb{F}_4$ (replace Y by $Y + \beta$, with $\beta^2 - \beta = 1$). They are not $\mathbb{F}_2$-isomorphic because the first has only one $\mathbb{F}_2$-point (the point at

∞), while the second has five $\mathbb{F}_2$-points (the point at ∞ and the four points (x, y) in $\mathbb{A}^2(\mathbb{F}_2)$).

10.5.4. It is also well known that this sort of problem, arising from automorphisms, is the only thing that "keeps" $\mathcal{M}_g$ from being representable. Roughly speaking, for any reasonable ["relatively representable" in the sense of [**Ka-Maz**, 4.2]] notion of "auxiliary structure", say $\mathcal{P}$, on curves of genus g, and which is rigidifying (in the sense that a pair (C, p) consisting of a curve of genus g together with a $\mathcal{P}$-structure has no nontrivial automorphisms), the functor $\mathcal{M}_{g,\mathcal{P}}$ from (Schemes) to (Sets) defined by

$$S \mapsto \{S\text{-isomorphism classes of pairs}$$
$$(C/S \text{ a curve of genus } g, p \text{ a } \mathcal{P}\text{-structure on } C/S)\}$$

is representable. Let us explain this general principle.

10.5.5. Fix an integer $n \geq 1$. Given a scheme S on which n is invertible, and a genus g curve C/S, denote by J/S its Jacobian. Recall that a (raw) level n structure on J/S is a list of $2g$ points $e_1, \ldots, e_{2g}$ in $J(S)[n]$, which, on each geometric fibre of J/S, forms a $\mathbb{Z}/n\mathbb{Z}$-basis of the group of points of order n on that fibre. Equivalently, a "raw level n structure" on J/S is an isomorphism $\alpha : (\mathbb{Z}/n\mathbb{Z})^{2g}_S \cong J[n]$ of group-schemes over S. We define a level n structure on C/S to be a raw level n structure on its Jacobian.

10.5.6. One knows that for $n \geq 3$, a curve $C/S/\mathbb{Z}[1/n]$ of genus $g \geq 2$ together with a level n structure has no nontrivial automorphisms. Let us recall the proof. Because $n \geq 3$, n has a divisor which is either an odd prime l or is 4. So it suffices to treat the case when n is either an odd prime l, or is l^2 with $l = 2$. One reduces successively to the case when S is the spectrum of a ring R which is noetherian, then complete noetherian local, then artin local, then a field k, the last reduction using the fact that, because $g \geq 2$, C/k has no nonzero global vector fields. One further reduces to the case when k is algebraically closed. Again because $g \geq 2$, any automorphism γ of C/k is of finite order. If γ is nontrivial, it has at most finitely many fixed points, so in $C \times C$ its graph Γ_γ and the diagonal Δ intersect properly, with intersection multiplicity $\Delta \cdot \Gamma_\gamma$ equal to the sum, over the fixed points of γ, of the strictly positive multiplicity of each fixed point. In particular, we have

$$\Delta \cdot \Gamma_\gamma \geq 0.$$

But we can use the Lefschetz trace formula to calculate this intersection multiplicity:

$$\Delta \cdot \Gamma_\gamma = \sum_i (-1)^i \operatorname{Trace}(\gamma | H^i(C, \mathbb{Z}_l)).$$

Both H^0 and H^2 are $\mathbb{Z}_l$-free of rank one, and γ acts trivially. The assumption that γ fixes the points of order l (or l^2 if $l = 2$) on the Jacobian means precisely that γ acts trivially on H^1/lH^1 (or on H^1/l^2H^1 if $l = 2$). Since γ is of finite order on H^1, which is $\mathbb{Z}_l$-free of rank $2g$, Serre's lemma [**Serre-Rig**] shows that γ acts as the identity on H^1. Thus we have

$$\Delta \cdot \Gamma_\gamma = \sum_i (-1)^i \operatorname{Trace}(\gamma | H^i(C, \mathbb{Z}_l)) = 2 - 2g < 0,$$

contradiction. Therefore γ is trivial, as required.

10.5.7. A fundamental fact, cf. [**De-Mum**], is that the functor ${}_n\mathcal{M}_g$ on (Schemes/$\mathbb{Z}[1/n]$) defined by

$$S \mapsto \{S\text{-isomorphism classes of pairs}$$
$$(C/S \text{ a curve of genus } g, \alpha_n \text{ a level } n \text{ structure on } C/S)\}$$

is representable, by a scheme quasi-projective over $\mathbb{Z}[1/n]$.

10.5.8. Once we have this basic representability result, the representability of $\mathcal{M}_{g,\mathcal{P}}$ for any relatively representable, rigidifying $\mathcal{P}$ follows easily. One considers, over $\mathbb{Z}[1/n]$, the moduli problem ${}_n\mathcal{M}_{g,\mathcal{P}}$ where one imposes both a level n structure and a $\mathcal{P}$-structure on the curve. It is representable over $\mathbb{Z}[1/n]$, because ${}_n\mathcal{M}_g$ is representable, and $\mathcal{P}$ is relatively representable. The finite group $GL(2g, \mathbb{Z}/n\mathbb{Z})$ acts on ${}_n\mathcal{M}_{g,\mathcal{P}}$ through its action $(g, \alpha_n) \mapsto \alpha_n \cdot g^{-1}$ on the level n structure. The quotient ${}_n\mathcal{M}_{g,\mathcal{P}}/GL(2g, \mathbb{Z}/n\mathbb{Z})$ can be shown (use descent theory to descend the universal curve and its $\mathcal{P}$-structure, cf. [**Ka-Maz**, 4.7]) to represent $\mathcal{M}_{g,\mathcal{P}} \otimes_{\mathbb{Z}} \mathbb{Z}[1/n]$. If we repeat this construction with two relatively prime values n_1 and n_2 of $n \geq 3$, and glue together over $\mathbb{Z}[1/n_1 n_2]$, we get the required $\mathcal{M}_{g,\mathcal{P}}$.

10.5.9. The problem with working with an auxiliary level n structure when thinking about curves of genus g over, say, a finite field $k = \mathbb{F}_q$ in which n is invertible, is this. It is almost never the case that a curve of genus g over $\mathbb{F}_q$ admits a level n structure over the same $\mathbb{F}_q$, even if n is a small prime l. First of all, the Weil pairing would force $\mathbb{F}_q$ to contain the l'th roots of unity, i.e., would force $q \equiv 1 \bmod l$. Even if $q \equiv 1 \bmod l$, there is still a serious obstruction: the numerator of the zeta function of $C/\mathbb{F}_q$ is forced to be congruent to $(1-T)^{2g} \bmod l$, a highly unlikely event. Let us illustrate by example just how unlikely.

10.5.10. Suppose l is an odd prime. Fix $g \geq 1$, fix a finite field of odd characteristic $\mathbb{F}_q$ with $q \equiv 1 \bmod l$, and fix a polynomial $f(X)$ in $\mathbb{F}_q[X]$ of degree $2g$ with all distinct roots in $\overline{\mathbb{F}}_q$. Consider the family of genus g curves of equation $Y^2 = f(X)(X-A)$ over $U := \operatorname{Spec}(\mathbb{F}_q[A][1/f(A)])$. According to [**Yu**], the "mod l representation" ρ_l of $\pi_1(U)$ (defined by $R^1\pi_1(\mathbb{F}_l)$) maps π_1^{geom} onto $Sp(2g, \mathbb{F}_l)$. It also maps the entire π_1 to $Sp(2g, \mathbb{F}_l)$, because $q \equiv 1 \bmod l$. By 9.7.10, applied to $K_{\mathrm{arith}} :=$ the image of $Sp(2g, \mathbb{F}_l)$ in any faithful $\overline{\mathbb{Q}}_l$-representation Λ, and to the composite $\overline{\mathbb{Q}}_l$-representation $\Lambda \circ \rho_l$ of π_1, we know that for large finite extensions E of $\mathbb{F}_q$, the fraction of points a in $U(E)$ for which the curve of equation $Y^2 = f(X)(X-a)$ over E has its numerator of zeta equal $\bmod l$ to $(1-T)^{2g}$ is approximately the ratio

$$\begin{aligned}
&\operatorname{Card}\{g \text{ in } Sp(2g, \mathbb{F}_l) \,|\, \det(1-Tg) \equiv (1-T)^{2g} \bmod l\} / \operatorname{Card}(Sp(2g, \mathbb{F}_l)) \\
&\quad = \operatorname{Card}(\{\text{unipotent elements in } Sp(2g, \mathbb{F}_l)\} / \operatorname{Card}(Sp(2g, \mathbb{F}_l)) \\
&\quad = l^g / \prod_{i=0}^{g-1} ((l^{2g-2i} - 1)(l^{2g-2i} - l^{2g-2i-1})/(l-1)) \\
&\quad = l^g / l^{g^2} \prod_{i=0}^{g-1} (l^{2g-2i} - 1) \\
&\quad \leq l^g / l^{g^2} (l^2 - 1)^{g(g+1)/2} = 1/l^{g^2 - g}(l^2 - 1)^{g(g+1)/2} \\
&\quad \leq 1/(l^2)^{g(g-1)/2}(l^2 - 1)^{g(g+1)/2} \leq 1/(l^2 - 1)^{g^2}.
\end{aligned}$$

10.5.11. Here is an even simpler example, again valid in any odd characteristic p, which shows the rarity of having a rational level two structure. Fix a genus $g \geq 1$, and introduce $2g+1$ indeterminates A_i, $i = 0, \ldots, 2g$. Denote by Δ the discriminant of the universal monic polynomial of degree $2g+1$,

$$f_{\text{univ}}(X) := X^{2g+1} + \sum_{i=0}^{2g} A_i X^i.$$

Consider the family of genus g curves

$$Y^2 = f_{\text{univ}}(X)$$

over the open set $U_{2g+1} := \mathbb{A}^{2g+1}[1/\Delta] := \mathrm{Spec}(\mathbb{Z}[1/2][\text{the } A_i][1/\Delta])$ of the space of coefficients where f_{univ} has all distinct roots and where 2 is invertible. For any field k of odd characteristic, a k-valued point s of U_{2g+1} is precisely a monic polynomial $f_{k,s}(X)$ of degree $2g+1$ over k with $2g+1$ distinct roots in $\overline{k}$. The curve $Y^2 = f_{k,s}(X)$ over k admits a level two structure over k if and only if the polynomial $f_{k,s}(X)$ splits completely over k.

10.5.12. If k is a finite field $\mathbb{F}_q$, there are exactly $q^{2g+1} - q^{2g}$ points in $U_{2g+1}(\mathbb{F}_q)$, and exactly $\mathrm{Binom}(q, 2g+1)$ of them split completely over $\mathbb{F}_q$. So the percentage of points s in $U_{2g+1}(\mathbb{F}_q)$ for which the corresponding curve $Y^2 = f_{k,s}(X)$ admits a level two structure over k is the ratio

$$\begin{aligned}\mathrm{Binom}(q, 2g+1)/(q^{2g+1} - q^{2g}) &= (1/(2g+1)!) \prod_{i=2}^{2g} (1 - i/q) \\ &\leq 1/(2g+1)!.\end{aligned}$$

10.6. The moduli space $\mathcal{M}_{g,3K}$ for $g \geq 2$

10.6.1. In order to avoid the irrationality problems encountered with level n structures, we will work systematically with the notion of a "$3K$ structure". Given a curve C/S of genus $g \geq 2$, say $\pi : C \to S$, one has on C the relative canonical bundle $K_{C/S}$, namely the line bundle, sometimes denoted $\omega_{C/S}$ or $\Omega^1_{C/S}$, of relative one-forms. One knows that in genus $g \geq 2$, $3K$ is very ample, cf. [**De-Mum**, Thm. 1.2 and Corollary], for the harder case of stable curves. More precisely, $\pi_*((K_{C/S})^{\otimes 3})$ is a locally free $\mathcal{O}_S$-module of rank $5g-5$ whose formation commutes with arbitrary change of base $T \to S$, and sections of $\pi_*((K_{C/S})^{\otimes 3})$ define a closed S-immersion of C/S into the relative projective space $\mathbb{P}_S(\pi_*((K_{C/S})^{\otimes 3}))$ over S.

10.6.2. Let us also recall that for C/S of genus $g \geq 2$, the functor from (Schemes/S) to (Groups) defined by

$$T/S \mapsto \text{the group } \mathrm{Aut}(C \times_S T/T)$$

is represented by a group-scheme $\underline{\mathrm{Aut}}_{C/S}$ which is finite and unramified over S, cf. [**De-Mum**, Thm. 1.11] for the harder case of stable curves. In particular, for k a field and C/k a curve of genus $g \geq 2$, $\underline{\mathrm{Aut}}_{C/k}$ is a finite etale group-scheme over k, and $\mathrm{Aut}(C/k) := \underline{\mathrm{Aut}}_{C/k}(k)$ is a finite group.

10.6.3. Given C/S of genus $g \geq 2$, we define a $3K$ structure on C/S to be an $\mathcal{O}_S$-basis of $\pi_*((K_{C/S})^{\otimes 3})$, or equivalently an isomorphism of $\mathcal{O}_S$-modules $(\mathcal{O}_S)^{5g-5} \cong \pi_*((K_{C/S})^{\otimes 3})$. Since $\pi_*((K_{C/S})^{\otimes 3})$ is locally free on S, any C/S admits a $3K$ structure Zariski locally on S (rather than etale locally, as was the case with level structures). In particular, a level $3K$ structure on a genus $g \geq 2$ curve over a field, C/k, is simply a choice of a k-basis of the $5g-5$ dimensional k-vector space $H^0(C, (\Omega^1_{C/k})^{\otimes 3})$. Thus not only does C/k admit a $3K$ structure over k, but the set of all its $3K$ structures over k is principal homogeneous under the group $GL(5g-5, k)$.

10.6.4. We claim that a $3K$ structure rigidifies a curve of genus $g \geq 2$. Because $K_{C/S} := \Omega^1_{C/S}$ and hence $(K_{C/S})^{\otimes 3}$ are canonically attached to C/S, any automorphism φ of C/S induces an automorphism of $\pi_*((K_{C/S})^{\otimes 3})$, which in turn induces an automorphism of $\mathbb{P}(\pi_*((K_{C/S})^{\otimes 3}))$ under which the tri-canonically embedded C/S is stable, and undergoes the automorphism φ with which we began. So if φ induces the identity on $\pi_*((K_{C/S})^{\otimes 3})$, φ must be the identity on C/S.

10.6.5. Therefore for each genus $g \geq 2$ the functor $\mathcal{M}_{g,3K}$ is representable, thanks to the preceding discussion 10.5.8. But we should point out here that it is a serious anachronism to use the representability of ${}_n\mathcal{M}_g$ to deduce that of $\mathcal{M}_{g,3K}$, since the representability of the latter is historically prior to that of ${}_n\mathcal{M}_g$. Indeed, $\mathcal{M}_{g,3K}$ is a minor variant of the moduli problem that is the starting point in Mumford's construction [**Mum-GIT**, Chapter 5, §2, Prop. 5.1] of $\mathcal{M}_g$ as a coarse moduli space. Mumford **begins** with the **representable** (by what he calls H_3 in [**Mum-GIT**], but H^0_g in [**De-Mum**]) moduli problem of tri-canonically embedded curves of genus g. Mumford's H^0_g is the (Zariski sheafification of the) moduli problem of genus g curves C/S together with a $3K$ structure on C/S given only up to multiplication by a unit in $\Gamma(S, \mathcal{O}_S)^\times$. Since H^0_g is representable [**Mum-GIT**, Prop. 5.1], our moduli problem $\mathcal{M}_{g,3K}$, being a $\mathbb{G}_m$-bundle over H^0_g, is itself representable. To construct $\mathcal{M}_g$ (denoted $\mathcal{M}^0_g$ in [**De-Mum**]) as coarse moduli space, Mumford has to pass to the quotient of H^0_g by the group $PGL(5g-5)$, or equivalently pass to the quotient of $\mathcal{M}_{g,3K}$ by the group $GL(5g-5)$. It is exactly this hard step which we are not taking, and indeed do not want to take.

10.6.6. Given any field k, we have the following tautological interpretation of the set $\mathcal{M}_{g,3K}(k)$ of its k-valued points:

$$\mathcal{M}_{g,3K}(k) = \{k\text{-isomorphism classes of pairs} \quad (C/k \text{ of genus } g, \alpha \text{ a } k\text{-basis of } H^0(C, (\Omega^1_{C/k})^{\otimes 3}))\}.$$

10.6.7. We have a morphism of functors, "forget the $3K$ structure",

$$\mathcal{M}_{g,3K} \to \mathcal{M}_g,$$

so for any field k a map

$$\mathcal{M}_{g,3K}(k) \to \mathcal{M}_g(k),$$
$$k\text{-isomorphism class of } (C/k, \alpha) \mapsto k\text{-isomorphism class of } C/k.$$

Lemma 10.6.8. *Let k be a finite field, $g \geq 2$ an integer, and C/k a (proper smooth and geometrically connected) genus g curve over k. There are exactly*

$$\mathrm{Card}(GL(5g-5, k))/\mathrm{Card}(\mathrm{Aut}(C/k))$$

points in $\mathcal{M}_{g,3K}(k)$ *whose underlying curve is* k*-isomorphic to* C/k*, i.e., the fibre of the map*

$$\mathcal{M}_{g,3K}(k) \to \mathcal{M}_g(k),$$

over the point given by C/k *has* $\mathrm{Card}(GL(5g-5,k))/\mathrm{Card}(\mathrm{Aut}(C/k))$ *points.*

PROOF. The group $GL(5g-5,k)$ acts transitively on the fibre, and the stabilizer of any point $(C/k,\alpha)$ in the fibre is the group $\mathrm{Aut}(C/k)$, viewed inside $GL(5g-5,k)$ by its action on α. QED

Corollary 10.6.9. *Let* k *be a finite field, and* $g \geq 2$ *an integer. We have the "mass formula"*

$$\sum_{C/k \text{ in } \mathcal{M}_g(k)} 1/\mathrm{Card}(\mathrm{Aut}(C/k)) = \mathrm{Card}(\mathcal{M}_{g,3K}(k))/\mathrm{Card}(GL(5g-5,k)).$$

Theorem 10.6.10 (Deligne-Mumford). *For* $g \geq 2$*,* $\mathcal{M}_{g,3K}$ *is smooth over* $\mathbb{Z}$ *of relative dimension* $3g-3+(5g-5)^2$*, with geometrically connected fibres.*

PROOF. It is proven in [**Mum-GIT**, Prop. 5.3] that H_g^0 is smooth over $\mathbb{Z}$. That it is of relative dimension $3g-3+(5g-5)^2-1$ is immediate from the deformation theory of curves. It is proven in [**De-Mum**, §3] that the geometric fibres of $H_g^0/\mathbb{Z}$ are connected. So the asserted theorem results from the fact that $\mathcal{M}_{g,3K}$ is a $\mathbb{G}_m$-bundle over H_g^0. QED

Theorem 10.6.11. *Fix a genus* $g \geq 2$*, and denote by* $\pi : \mathcal{C} \to \mathcal{M}_{g,3K}$ *the universal curve with* $3K$ *structure. For any prime number* l*, consider the lisse sheaf* $\mathcal{F}_l := R^1\pi_!\overline{\mathbb{Q}}_l|\mathcal{M}_{g,3K} \otimes_{\mathbb{Z}} \mathbb{Z}[1/l]$*. On any geometric fibre of* $\mathcal{M}_{g,3K} \otimes_{\mathbb{Z}} \mathbb{Z}[1/l]/\mathbb{Z}[1/l]$*, its geometric monodromy group is* $Sp(2g)$*.*

PROOF. By Pink's specialization theorem [**Ka-ESDE**, 8.18.2], and the a priori inclusion of G_{geom} in $Sp(2g)$ noted above in 10.3.5, the characteristic zero case follows from the characteristic p case. In the characteristic p case, it suffices to treat the $\overline{\mathbb{F}}_p$ fibre, since for any connected scheme X over an algebraically closed field k, and any algebraically closed overfield E of k, and any geometric point x of $X \otimes_k E$, the map $\pi_1(X \otimes_k E, x) \to \pi_1(X,x)$ is surjective.

As noted above, it suffices to show that $\mathcal{F}_l|\mathcal{M}_{g,3K} \otimes_{\mathbb{Z}} \overline{\mathbb{F}}_p$ has $G_{\mathrm{geom}} \supset Sp(2g)$. For this, it trivially suffices to exhibit a connected $U/\overline{\mathbb{F}}_p$ and a map

$$f : U \to \mathcal{M}_{g,3K} \otimes_{\mathbb{Z}} \overline{\mathbb{F}}_p,$$

such that $f^*\mathcal{F}_l$ on U has $G_{\mathrm{geom}} = Sp(2g)$. We have given examples of families $\varphi : C \to U$ of genus g curves whose parameter space U is a nonvoid open set in $\mathbb{A}^1$ over $\overline{\mathbb{F}}_p$ such that for all $l \neq p$, G_{geom} for $R^1\varphi_!\overline{\mathbb{Q}}_l$ is $Sp(2g)$. Since any vector bundle on an open set of $\mathbb{A}^1$ over a field is trivial, C/U admits a $3K$ structure. Choosing one, we get a classifying map, say $f : U \to \mathcal{M}_{g,3K} \otimes_{\mathbb{Z}} \overline{\mathbb{F}}_p$, such that C/U is the pullback of the universal family. By proper base change, $R^1\varphi_!\overline{\mathbb{Q}}_l$ is the pullback $f^*\mathcal{F}_l$. QED

Lemma 10.6.12. *Fix a genus* $g \geq 2$*. For any algebraically closed field* k*, and for any curve* C/k *of genus* g*,* $\mathrm{Card}(\mathrm{Aut}(C/k)) \leq 4^{4g^2}$*.*

PROOF. If k is not of characteristic 2, the points of order 4 on the Jacobian rigidify C/k, so we get $\mathrm{Aut}(C/k) \subset GL(2g, \mathbb{Z}/4\mathbb{Z})$. If k has characteristic two, the points of order 3 give $\mathrm{Aut}(C/k) \subset GL(2g, \mathbb{F}_3)$. QED

Lemma 10.6.13. *Let S be a noetherian scheme, C/S a curve of genus $g \geq 2$. There is an open set $U_{\leq 1}$ of S which is characterized by the following property: a point s in S lies in $U_{\leq 1}$ if and only if for the corresponding curve $C_s/\kappa(s)$ we have $\underline{\mathrm{Aut}}_{C_s/\kappa(s)} = \{e\}$. More generally, for any integer $i \geq 1$, there is an open set $U_{\leq i}$ of S which is characterized by the following property: a point s in S lies in $U_{\leq i}$ if and only if the curve $C_s/\kappa(s)$ has $\underline{\mathrm{Aut}}_{C_s/\kappa(s)}$ of* rank $\leq i$, *i.e., if and only if* $\mathrm{Card}(\mathrm{Aut}(C_s \otimes \overline{\kappa}(s)/\overline{\kappa}(s))) \leq i$.

PROOF. Consider the group-scheme $G := \underline{\mathrm{Aut}}_{C/S}$. It is finite over S, so it is the relative Spec of a coherent sheaf, say $\mathcal{G}$, of $\mathcal{O}_S$-algebras. For each point s of S, $\underline{\mathrm{Aut}}_{C_s/\kappa(s)}$ is the Spec of the finite etale $\kappa(s)$-algebra $\mathcal{G} \otimes_{\mathcal{O}_S} \kappa(s)$. So to say that $\underline{\mathrm{Aut}}_{C_s/\kappa(s)} = \{e\}$ is precisely to say that $\dim(\mathcal{G} \otimes_{\mathcal{O}_S} \kappa(s)) = 1$. But for any point s, $\dim(\mathcal{G} \otimes_{\mathcal{O}_S} \kappa(s)) \geq 1$, just because the coordinate ring of a group is nonzero. So our $U_{\leq 1}$ is the set of points s at which $\dim(\mathcal{G} \otimes_{\mathcal{O}_S} \kappa(s)) \leq 1$. Similarly, the set $U_{\leq i}$ is the set of points s at which $\dim(\mathcal{G} \otimes_{\mathcal{O}_S} \kappa(s)) \leq i$. Let us recall why such a set is open. The general fact is that for any integer $i \geq 0$, and for any coherent sheaf $\mathcal{G}$ on any noetherian scheme X, the set $U := \{x \text{ in } X \text{ at which } \dim(\mathcal{G} \otimes_{\mathcal{O}_X} \kappa(x)) \leq i\}$ is Zariski open in X. This question is Zariski local on X, so it suffices to treat the case in which there exists a presentation of $\mathcal{G}$ as the cokernel of a map of free $\mathcal{O}_X$-modules, say $A : (\mathcal{O}_X)^p \to (\mathcal{O}_X)^q$. Then U is the set of points at which the rank of A is $\geq q - i$, and its complement is the closed set of X defined by the vanishing of all the $(q-i) \times (q-i)$ minors of A. QED

Theorem 10.6.14. *Let $g \geq 3$. Over any algebraically closed field k, there exists a genus g curve C/k with no nontrivial automorphisms. Equivalently, the open set $U_{\leq 1}$ in $\mathcal{M}_{g,3K}$ of curves with no nontrivial automorphisms meets every geometric fibre of $\mathcal{M}_{g,3K}/\mathbb{Z}$.*

PROOF. We first reduce to treating the case $k = \overline{\mathbb{F}}_p$ for each prime p. This will obviously take care of any k of positive characteristic. It will also show that $U_{\leq 1}$ is nonempty, and hence that $U_{\leq 1}$ contains the generic point of $\mathcal{M}_{g,3K}$, which lies in $\mathcal{M}_{g,3K} \otimes \mathbb{Q}$. Therefore $U_{\leq 1}$ meets $\mathcal{M}_{g,3K} \otimes \overline{\mathbb{Q}}$ as well, and this takes care of any k of characteristic zero.

So we may suppose that k is $\overline{\mathbb{F}}_p$. Since $\mathcal{M}_{g,3K} \otimes k$ is smooth and connected, it has a generic point η, and a corresponding generic curve $C_\eta/\kappa(\eta)$. Let us denote by r the rank of the finite etale $\kappa(\eta)$-group-scheme $\underline{\mathrm{Aut}}_{C_\eta/\kappa(\eta)}$, i.e., r is the cardinality of $\mathrm{Aut}(C_{\overline{\eta}}/\kappa(\overline{\eta}))$. Now consider the dense open set $U_{\leq r} \otimes k$ of $\mathcal{M}_{g,3K} \otimes k$. Over it, the coherent sheaf $\mathcal{G}$ (whose relative Spec is $\underline{\mathrm{Aut}}$) has constant fibre dimension r. As $U_{\leq r} \otimes k$ is reduced, being smooth over a field, $\mathcal{G}$ is locally free of rank r on $U_{\leq r} \otimes k$. Therefore $\underline{\mathrm{Aut}}|U_{\leq r} \otimes k$ is flat, and hence finite etale, over $U_{\leq r} \otimes k$. So over a finite etale connected covering $V/U_{\leq r} \otimes k$, $\underline{\mathrm{Aut}}$ becomes a constant group-scheme G_V for some finite group G of order r.

The pullback C/V of the universal curve still has $G_{\mathrm{geom}} = Sp(2g)$, because restriction from $\mathcal{M}_{g,3K} \otimes k$ to the dense open set $U_{\leq r} \otimes k$ does not change G_{geom}, and pullback to a finite etale cover does not change $(G_{\mathrm{geom}})^0$. The finite group G acts by functoriality as automorphisms of $\mathcal{F}_l|V$, and hence G commutes with the action of $\pi_1(V, v)$, and hence with the action of $Sp(2g)$, on the stalk $\mathcal{F}_{l,v}$ at any chosen geometric point v in V. Pick one such v, and denote by C_v the corresponding curve. Since the standard representation of $Sp(2g)$ is irreducible, G acts as scalars on $H^1(C_v, \overline{\mathbb{Q}}_l)$, say γ in G acts as the scalar $\chi(\gamma)$ in $\mu_r(\overline{\mathbb{Q}}_l)$. As the action is faithful, it identifies G with the cyclic group $\mu_r(\overline{\mathbb{Q}}_l)$. For any automorphism γ of any curve C over an algebraically closed field in which l is invertible, one knows, by the Lefschetz trace formula for powers of γ, that the polynomial $\det(T-\gamma|H^1(C,\overline{\mathbb{Q}}_l))$ lies in $\mathbb{Z}[T]$. Therefore $(T-\chi(\gamma))^{2g}$ lies in $\mathbb{Z}[T]$. Looking at the coefficient of T^{2g-1}, we see that $2g\chi(\gamma)$ lies in $\mathbb{Z}$. Thus $\chi(\gamma)$ is a root of unity in $\mathbb{Q}$, and hence $\chi(\gamma) = \pm 1$. Taking γ to be a generator of G, we see that r is either 1 or 2. If $r = 1$, we are done.

If $r = 2$, the nontrivial element γ in G acts on $H^1(C_v, \overline{\mathbb{Q}}_l)$ as -1. This means precisely that C_v is hyperelliptic. [The quotient C_v/G has genus zero (its H^1 is the G-invariants in $H^1(C_v, \overline{\mathbb{Q}}_l)$), and hence C_v is hyperelliptic, in the sense of being a generically etale double covering of $\mathbb{P}^1$. Conversely, given a generically etale double covering $C/\mathbb{P}^1$ over an algebraically closed field of characteristic $\neq l$, the attached involution σ of C must act as -1 on $H^1(C, \overline{\mathbb{Q}}_l)$, since the quotient $C/\langle\sigma\rangle$ is a $\mathbb{P}^1$.] Since the curve C_v depends only on the point u in $U_{\leq 2} \otimes k$ lying under v in V, we find that **every** curve in the open dense set $U_{\leq 2} \otimes k$ of $\mathcal{M}_{g,3K} \otimes k$ is hyperelliptic. But the hyperelliptic locus is closed under specialization (use the above characterization of hyperelliptics as those admitting an involution which is -1 on H^1), so **every** curve in $\mathcal{M}_{g,3K} \otimes k$ is hyperelliptic. This is well known to be impossible, because we have $2g-1 < 3g-3$ provided $g \geq 3$.

Let us recall the argument in odd characteristic. Every hyperelliptic curve of genus g over an algebraically closed field of odd characteristic is isomorphic to $Y^2 = X(X-1)f(X)$, with f monic of degree $2g-1$, and with $f(0)f(1)\Delta(f) \neq 0$. The space of such f's is a dense open set in the $\mathbb{A}^{2g-1}$ of coefficients. On such a curve, the holomorphic one-form dX/Y has divisor $(2g-2)\infty$ as section of the canonical bundle K. So $(dX/Y)^{\otimes 3}$ as section of $3K$ has divisor $(6g-6)\infty$. Using the Riemann-Roch notation $L(D) := H^0(I(D)^{-1})$, we have

$$H^0(3K) = L((6g-6)\infty)(dX/Y)^{\otimes 3},$$

a basis of which is $\{X^i \text{ for } 0 \leq i \leq 3g-3,\ YX^i \text{ for } 0 \leq i \leq 2g-4\}(dX/Y)^{\otimes 3}$. Using this basis as a base point α, to be moved by $GL(5g-5)$, we get a classifying map between k-schemes,

$$(\text{open in } \mathbb{A}^{2g-1}) \times GL(5g-5) \to \mathcal{M}_{g,3K} \otimes k,$$
$$(f, g) \mapsto (Y^2 = X(X-1)f(X),\ \text{basis } \alpha \circ g^{-1})$$

which is surjective on geometric points. Comparing dimensions of source and target, which are $2g-1+(5g-5)^2$ and $3g-3+(5g-5)^2$ respectively, we see that this is impossible for $g \geq 3$.

In characteristic two, the argument is similar, except that now every hyperelliptic curve of genus $g \geq 2$ is isomorphic to a member of one of a finite number of explicit families of hyperelliptic curves. These are the families

$$Y^2 - Y = P_\infty(X) + P_0(1/X) + P_1(1/(X-1)) + \sum_{i=2}^{g-1} P_i(1/(X-a_i)),$$

with each P a (possibly zero) polynomial whose only possibly nonvanishing terms are of odd degree, and with degrees satisfying

$$\sum_{\text{nonzero } P} (1 + \deg(P)) = 2g + 2$$

and

$$\deg(P_\infty) \geq \deg(P_0) \geq \deg(P_1) \geq \cdots \geq \deg(P_{g-1}).$$

One verifies easily that each allowed choice of degrees leads to a family which depends on $\leq 2g-1$ parameters. For instance, the most generic family is

$$Y^2 - Y = b_\infty X + b_0/X + b_1/(X-1) + \sum_{i=2}^{g-1} b_i/(X-a_i),$$

which depends on $g+1$ b_i's and on $g-2$ a_i's. The most special family is

$$Y^2 - Y = P_\infty(X),$$

with P_∞ an odd polynomial of degree $2g+1$, which has only $g+1$ coefficients. QED

Remark 10.6.15. A more constructive proof of this theorem would be to write down, for any genus $g \geq 3$ and in any characteristic p, an **explicit** curve which has no nontrivial automorphisms. Recently, this was done by Poonen [**Poon-Curves**], who exhibits curves of every genus $g \geq 3$ over every prime field $\mathbb{F}_p$ with no nontrivial automorphisms over $\overline{\mathbb{F}}_p$. His curves are (the complete nonsingular models of) cubic branched covers of $\mathbb{P}^1$. [The idea of looking at such trigonal curves was also suggested independently by Mochizuki.]

10.6.16. For some but not all genera g, one can use the theory of Lefschetz pencils to write down, in every characteristic p, an **explicit family** of genus g curves whose general member has no nontrivial automorphisms.

10.6.17. Let us first illustrate the simplest case, that of smooth plane curves. For g a triangular number $0, 1, 3, 6, 10, \ldots, (d-1)(d-2)/2$ with $d \geq 3$, a smooth hypersurface C of degree d in $\mathbb{P}^2$ has genus g, and its canonical bundle K_C is the restriction to C of $\mathcal{O}(d-3)$. For $d \geq 4$, K_C is very ample, already being so on $\mathbb{P}^2$, and hence C is not hyperelliptic. Moreover, for any $d \geq 1$, $\mathcal{O}(d)$ defines a Lefschetz embedding of $\mathbb{P}^2$ [**SGA 7**, XVII, 2.5.1 for the case $d \geq 2$, the case $d = 1$ being trivial], so there exist Lefschetz pencils of degree d curves in $\mathbb{P}^2$.

10.6.18. It is a general fact that for a Lefschetz pencil of curves of genus g on a simply connected projective smooth surface S, the group G_{geom} is $Sp(2g)$. [To see this, notice first that S, being simply connected, trivially satisfies the hard Lefschetz theorem LV of [**SGA 7**, XVIII, 5.2.2]. Then apply [**SGA 7**, XVIII, 6.5.2.1, 6.6 and 6.7] to conclude that G_{geom} is the Zariski closure of an irreducible subgroup of $Sp(2g)$ generated by unipotent pseudoreflections, hence by Kazhdan-Margulis G_{geom} must be $Sp(2g)$]. So by the monodromy argument above, for any Lefschetz pencil of nonhyperelliptic curves on a simply connected surface S, all but at most finitely many of the curves in the pencil have no nontrivial automorphisms. To avoid having to write an explicit Lefschetz pencil on S, we consider the family of all smooth hyperplane sections of S. In this family, which is as explicit as S is, all members in a dense open set have no nontrivial automorphisms. Taking $S := \mathbb{P}^2$ in its d-fold embedding, we find that "most" smooth curves of degree $d \geq 4$ in $\mathbb{P}^2$ have no nontrivial automorphisms.

10.6.19. Given a genus $g \geq 4$, is there an explicit projective, smooth, simply connected surface S whose general hyperplane section is a nonhyperelliptic curve of genus g? We have just seen that for g a triangular number $(d-1)(d-2)/2$, the answer is yes. As Shin Mochizuki pointed out to us, the answer is also yes if $g \geq 4$ is **not prime**. For we may write g as $(a-1)(b-1)$ with a and b both at least 3, and take S to be $\mathbb{P}^1 \times \mathbb{P}^1$ in its $\mathcal{O}(a,b)$ embedding into $\mathbb{P}^{(a+1)(b+1)-1}$. A general hyperplane section C of S, i.e., a general hypersurface of bidegree (a,b) in $\mathbb{P}^1 \times \mathbb{P}^1$, has genus $(a-1)(b-1)$, and its K_C is the restriction to C of $\mathcal{O}(a-2,b-2)$. As both $a,b \geq 3$, K_C is very ample and hence C is not hyperelliptic. [If a or b were 2, C would be hyperelliptic, being of degree 2 over one of the $\mathbb{P}^1$ factors.] For any a,b each ≥ 1, $\mathcal{O}(a,b)$ defines a Lefschetz embedding. [To check this, it suffices by [**SGA 7**, XVII, 3.7] to show that over any algebraically closed field k, given any k-valued point x_0 of $\mathbb{P}^1 \times \mathbb{P}^1$, there is a hypersurface of bidegree (a,b) in $\mathbb{P}^1 \times \mathbb{P}^1$ which has an ordinary double point at x_0. Moving x_0 by an automorphism of $\mathbb{P}^1 \times \mathbb{P}^1$, we may assume x_0 to be $(0,1) \times (0,1)$. In projective coordinates (X,A) and (Y,B) on the two factors, the equation $XA^{a-1}YB^{b-1} = 0$ does the job.] Thus there exist Lefschetz pencils of smooth curves of bidegree (a,b) in $\mathbb{P}^1 \times \mathbb{P}^1$. As explained above, all but at most finitely many curves in such a pencil have no nontrivial automorphisms. In particular, for any genus $g = (a-1)(b-1)$ with a and b both at least 3, a general smooth curve of bidegree (a,b) in $\mathbb{P}^1 \times \mathbb{P}^1$ is a curve of genus g with no nontrivial automorphisms.

Remark 10.6.20. In genus two, every curve is hyperelliptic, and hence $U_{\leq 1}$ is empty. The first part of the argument shows that the generic value of r is either 1 or 2. Thus for $g = 2$, $U_{\leq 1}$ is empty, and $U_{\leq 2}$ meets every geometric fibre of $\mathcal{M}_{2,3K}/\mathbb{Z}$.

Corollary 10.6.21. *Fix a genus $g \geq 3$, and denote by $\pi : \mathcal{C} \to \mathcal{M}_{g,3K}$ the universal curve with $3K$ structure. Denote by*

$$\mathcal{M}_{g,3K,\mathrm{aut}\leq 1} \subset \mathcal{M}_{g,3K}$$

the open set parameterizing curves which geometrically have no nontrivial automorphisms, i.e., $\mathcal{M}_{g,3K,\mathrm{aut}\leq 1}$ is the open set $U_{\leq 1}$ of 10.6.13 for the universal curve. Then we have

1) *$\mathcal{M}_{g,3K,\mathrm{aut}\leq 1}$ meets every geometric fibre of $\mathcal{M}_{g,3K}/\mathbb{Z}$ in a dense open set which is smooth and connected.*

2) *For any prime number l, consider the lisse sheaf*

$$\mathcal{F}_l := R^1\pi_!\overline{\mathbb{Q}}_l|\mathcal{M}_{g,3K,\mathrm{aut}\leq 1} \otimes_{\mathbb{Z}} \mathbb{Z}[1/l].$$

On any geometric fibre of $\mathcal{M}_{g,3K,\mathrm{aut}\leq 1} \otimes_{\mathbb{Z}} \mathbb{Z}[1/l]/\mathbb{Z}[1/l]$, its geometric monodromy group is $Sp(2g)$.

PROOF. Since $\mathcal{M}_{g,3K,\mathrm{aut}\leq 1}$ is open in $\mathcal{M}_{g,3K}$, and the geometric fibres of $\mathcal{M}_{g,3K}/\mathbb{Z}$ are smooth and connected, hence irreducible, the intersection of $\mathcal{M}_{g,3K,\mathrm{aut}\leq 1}$ with a geometric fibre is either empty or is open dense and irreducible, so itself smooth and connected. The intersection is not empty, by 10.6.14. Assertion 2) results from 1) and from 10.6.1, because G_{geom} for $\mathcal{F}_l$ is the same on a geometric fibre of $\mathcal{M}_{g,3K} \otimes_{\mathbb{Z}} \mathbb{Z}[1/l]/\mathbb{Z}[1/l]$ as it is a nonvoid open set of that geometric fibre. QED

10.6.22. For each $g \geq 3$, we denote by

$$\mathcal{M}_{g,3K,\mathrm{aut}\geq 2} := \mathcal{M}_{g,3K} - \mathcal{M}_{g,3K,\mathrm{aut}\leq 1} \subset \mathcal{M}_{g,3K}$$

the (reduced) Zariski closed subset of $\mathcal{M}_{g,3K}$ parameterizing curves which geometrically have nontrivial automorphisms. Thanks to 10.6.14, the geometric fibres of $\mathcal{M}_{g,3K,\mathrm{aut}\geq 2}/\mathbb{Z}$ are of strictly positive codimension in those of $\mathcal{M}_{g,3K}/\mathbb{Z}$. Applying the Lang-Weil method, we get the following lemma.

Lemma 10.6.23. *For each integer $g \geq 3$, denote by*

$$d(g) := 3g - 3 + (5g-5)^2,$$

the relative dimension of $\mathcal{M}_{g,3K}/\mathbb{Z}$. There exists constants $A(g)$ and $B(g)$ such that for any finite field $\mathbb{F}_q$, we have the inequalities

$$|\operatorname{Card}(\mathcal{M}_{g,3K,\mathrm{aut}\leq 1}(\mathbb{F}_q)) - q^{d(g)}| \leq A(g)q^{d(g)-1/2},$$
$$\operatorname{Card}(\mathcal{M}_{g,3K,\mathrm{aut}\geq 2}(\mathbb{F}_q)) \leq B(g)q^{d(g)-1},$$
$$|\operatorname{Card}(\mathcal{M}_{g,3K}(\mathbb{F}_q)) - q^{d(g)}| \leq (A(g) + B(g)q^{-1/2})q^{d(g)-1/2}.$$

PROOF. Since $\mathcal{M}_{g,3K,\mathrm{aut}\leq 1}/\mathbb{Z}$ has smooth, geometrically connected fibres of dimension $d(g)$, we get the first inequality if we take for $A(g)$ the maximum of the constants

$$A(g,l) := \operatorname*{Max}_{p \neq l} \left\{ \sum_{i<2d(g)} h_c^i(\mathcal{M}_{g,3K,\mathrm{aut}\leq 1} \otimes_{\mathbb{Z}} \overline{\mathbb{F}}_p, \overline{\mathbb{Q}}_l) \right\}$$

for two different values of l, say 2 and 3. Each constant $A(g,l)$ is finite by the constructibility of the $R^i\pi[1/l]_!\overline{\mathbb{Q}}_l$ for $\pi[1/l]$ the structural map

$$\mathcal{M}_{g,3K,\mathrm{aut}\leq 1} \otimes_{\mathbb{Z}} \mathbb{Z}[1/l] \to \mathbb{Z}[1/l].$$

Similarly, since $\mathcal{M}_{g,3K,\mathrm{aut}\geq 2}/\mathbb{Z}$ has geometric fibres of dimension $< d(g)$, we get the second inequality if we take for $B(g)$ the maximum of the constants

$$B(g,l) := \operatorname*{Max}_{p \neq l} \left\{ \sum_{i \leq 2d(g)-2} h_c^i(\mathcal{M}_{g,3K,\mathrm{aut}\geq 2} \otimes_{\mathbb{Z}} \overline{\mathbb{F}}_p, \overline{\mathbb{Q}}_l) \right\}$$

for two different values of l. The third inequality results by adding the first two. QED

Remark 10.6.24. Let $g \geq 3$. For $q > A(g)^2$, $\mathcal{M}_{g,3K,\mathrm{aut}\leq 1}(\mathbb{F}_q)$ is nonempty, i.e., over every finite field $\mathbb{F}_q$ of sufficiently large cardinality there exists a genus g curve which geometrically has no nontrivial automorphisms. As noted above, Poonen [**Poon-Curves**] has shown that there is such a curve over every prime field $\mathbb{F}_p$.

Lemma 10.6.25. *Let $g \geq 3$. Denote by $D(g)$ any upper bound for the order of the automorphism group of a genus g curve (e.g., we have seen in* 10.6.12 *that 4^{4g^2} is such an upper bound). For any finite field k, denote by $\mathcal{M}_g(k)$ the set of k-isomorphism classes of genus g curves C/k, by $\mathcal{M}_{g,\mathrm{aut}\leq 1}(k)$ the subset of $\mathcal{M}_g(k)$ consisting of those C/k which geometrically admit no nontrivial automorphism, and by $\mathcal{M}_{g,\mathrm{aut}\geq 2}(k)$ the complement $\mathcal{M}_g(k) - \mathcal{M}_{g,\mathrm{aut}\leq 1}(k)$.*

1) *We have the mass formulas*

$$\sum_{C/k \text{ in } \mathcal{M}_{g,\mathrm{aut}\leq 1}(k)} 1/\operatorname{Card}(\operatorname{Aut}(C/k)) \left(= \sum_{C/k \text{ in } \mathcal{M}_{g,\mathrm{aut}\leq 1}(k)} 1 \right)$$
$$= \operatorname{Card}(\mathcal{M}_{g,3K,\mathrm{aut}\leq 1}(k)) / \operatorname{Card}(GL(5g-5,k))$$

and

$$\sum_{C/k \text{ in } \mathcal{M}_{g,\mathrm{aut}\geq 2}(k)} 1/\operatorname{Card}(\operatorname{Aut}(C/k))$$
$$= \operatorname{Card}(\mathcal{M}_{g,3K,\mathrm{aut}\geq 2}(k))/\operatorname{Card}(GL(5g-5,k)).$$

2) *We have the equality*

$$\operatorname{Card}(\mathcal{M}_{g,\mathrm{aut}\leq 1}(k)) = \operatorname{Card}(\mathcal{M}_{g,3K,\mathrm{aut}\leq 1}(k))/\operatorname{Card}(GL(5g-5,k)),$$

and the inequalities

$$\operatorname{Card}(\mathcal{M}_{g,\mathrm{aut}\geq 2}(k)) \leq D(g)\operatorname{Card}(\mathcal{M}_{g,3K,\mathrm{aut}\geq 2}(k))/\operatorname{Card}(GL(5g-5,k)),$$
$$\operatorname{Card}(\mathcal{M}_{g,\mathrm{aut}\geq 2}(k)) \geq \operatorname{Card}(\mathcal{M}_{g,3K,\mathrm{aut}\geq 2}(k))/\operatorname{Card}(GL(5g-5,k)).$$

PROOF. The mass formulas of 1) are immediate from 10.6.8. They trivially imply 2), since for C/k in $\mathcal{M}_{g,\mathrm{aut}\leq 1}(k)$, we have $1/\operatorname{Card}(\operatorname{Aut}(C/k)) = 1$, and for C/k in $\mathcal{M}_{g,\mathrm{aut}\geq 2}(k)$ we have the inequality $1 \geq 1/\operatorname{Card}(\operatorname{Aut}(C/k)) \geq 1/D(g)$. QED

Corollary 10.6.26. *Notations as in the above three lemmas, fix $g \geq 3$. Denote by $E(g)$ the constant $E(g) := 2D(g)B(g)$. For any finite field $\mathbb{F}_q$ with $q \geq 4A(g)^2$, we have*

$$\operatorname{Card}(\mathcal{M}_{g,\mathrm{aut}\geq 2}(\mathbb{F}_q))/\operatorname{Card}(\mathcal{M}_{g,\mathrm{aut}\leq 1}(\mathbb{F}_q)) \leq E(g)/q.$$

PROOF. By the two previous lemmas, we have

$$\begin{aligned}
&\operatorname{Card}(\mathcal{M}_{g,\mathrm{aut}\geq 2}(\mathbb{F}_q))/\operatorname{Card}(\mathcal{M}_{g,\mathrm{aut}\leq 1}(\mathbb{F}_q)) \\
&\quad \leq D(g)\operatorname{Card}(\mathcal{M}_{g,3K,\mathrm{aut}\geq 2}(\mathbb{F}_q))/\operatorname{Card}(\mathcal{M}_{g,3K,\mathrm{aut}\leq 1}(\mathbb{F}_q)) \\
&\quad \leq D(g)B(g)q^{d(g)-1}/(q^{d(g)} - A(g)q^{d(g)-1/2}) \\
&\quad = D(g)B(g)/q(1 - A(g)q^{-1/2}) \\
&\quad \leq D(g)B(g)/q(1-1/2) = 2D(g)B(g)/q = E(g)/q. \quad \text{QED}
\end{aligned}$$

10.7. Naive and intrinsic measures on $USp(2g)^{\#}$ attached to universal families of curves

10.7.1. Fix an integer $g \geq 1$, a finite field $k = \mathbb{F}_q$, and a choice α_k of a real square root of $\operatorname{Card}(k) = q$. [Since we are choosing here a real square root, we could, of course, choose the positive one, and the reader is welcome to make that choice. But it is well to keep in mind that either choice works just as well. Moreover, if we first make a choice in $\overline{\mathbb{Q}}_l$ of a square root of $\operatorname{Card}(k)$, and then transport it to $\mathbb{R}$ via a field embedding $\iota : \overline{\mathbb{Q}}_l \to \mathbb{C}$, we certainly have no idea of which real square root we end up with.]

10.7.2. Given a genus g curve C/k, its zeta function has the form

$$P(T)/(1-T)(1-qT),$$

with $P(T)$ of degree $2g$. By the Riemann Hypothesis for curves over finite fields, we know that there exists a conjugacy class $\vartheta(k, \alpha_k, C/k)$ in $USp(2g)^{\#}$ with the property that

$$P(T) = \det(1 - \alpha_k T \vartheta(k, \alpha_k, C/k)).$$

Because elements of $USp(2g2)^{\#}$ are uniquely determined by their characteristic polynomials, the conjugacy class $\vartheta(k, \alpha_k, C/k)$ in $USp(2g)^{\#}$ is uniquely determined by this property. We call it the unitarized Frobenius conjugacy class attached to C/k.

10.7.3. We define the "naive" probability measure $\mu(\text{naive}, g, k, \alpha_k)$ on $USp(2g)^{\#}$ by averaging over the unitarized Frobenius conjugacy classes of all the k-isomorphism classes of genus g curves C/k, each counted with multiplicity one:

$$\mu(\text{naive}, g, k, \alpha_k) := (1/\operatorname{Card}(\mathcal{M}_g(k))) \sum_{C/k \text{ in } \mathcal{M}_g(k)} \delta_{\vartheta(k,\alpha_k,C/k)}.$$

10.7.4. We define the "intrinsic" probability measure $\mu(\text{intrin}, g, k, \alpha_k)$ on $USp(2g)^{\#}$ by averaging over the unitarized Frobenius conjugacy classes of all the k-isomorphism classes of genus g curves C/k, but now counting C/k with multiplicity $1/\operatorname{Card}(\operatorname{Aut}(C/k))$:

$$\begin{aligned} &\mu(\text{intrin}, g, k, \alpha_k) \\ &\quad := (1/\operatorname{Intrin}\operatorname{Card}(\mathcal{M}_g(k))) \sum_{C/k \text{ in } \mathcal{M}_g(k)} (1/\operatorname{Card}(\operatorname{Aut}(C/k)))\delta_{\vartheta(k,\alpha_k,C/k)}, \end{aligned}$$

where we have put

$$\begin{aligned} \operatorname{Intrin}\operatorname{Card}(\mathcal{M}_g(k)) &:= \sum_{C/k \text{ in } \mathcal{M}_g(k)} 1/\operatorname{Card}(\operatorname{Aut}(C/k)) \\ &= \operatorname{Card}(\mathcal{M}_{g,3K}(k))/\operatorname{Card}(GL(5g-5, k)), \end{aligned}$$

the last equality by 10.6.9.

Lemma 10.7.5. (1) *The rational number* $\operatorname{Intrin}\operatorname{Card}(\mathcal{M}_g(k))$ *is an integer, namely*

$$\operatorname{Intrin}\operatorname{Card}(\mathcal{M}_g(k)) = \operatorname{Card}(\operatorname{Image}(\mathcal{M}_g(k) \to \mathcal{M}_g(\overline{k}))).$$

(2) *More precisely, for each point ξ in this image, we have the identity*

$$1 = \sum_{C/k \text{ in } \mathcal{M}_g(k) \text{ having image } \xi \text{ in } \mathcal{M}_g(\overline{k})} 1/\operatorname{Card}(\operatorname{Aut}(C/k)).$$

(3) *The image of $\mathcal{M}_g(k)$ in $\mathcal{M}_g(\overline{k})$ consists precisely of the points in $\mathcal{M}_g(\overline{k})$ which are fixed by the action of* $\operatorname{Gal}(\overline{k}/k)$.

Proof. Consider the natural map $\mathcal{M}_g(k) \to \mathcal{M}_g(\overline{k})$, and collect terms $1/\operatorname{Card}(\operatorname{Aut}(C/k))$ according to which fibre C/k lies in. Then we see that (2) implies (1). To prove (2), we argue as follows. Pick one C/k, and denote by $\overline{C}/\overline{k}$ the curve gotten from C/k by extension of scalars. Denote by F the Frobenius endomorphism $\operatorname{Frob}_{C/k}$ of $\overline{C}/\overline{k}$. As explained in [**Serre-GACC**, VI, §1, Prop. 1 and §2, Prop. 2], the k-isomorphism class of C/k is determined by the $\overline{k}$-isomorphism class of the pair $(\overline{C}/\overline{k}, F)$. But any A in $\operatorname{Aut}(\overline{C}/\overline{k})$ defines an isomorphism from $(\overline{C}/\overline{k}, F)$ to $(\overline{C}/\overline{k}, AFA^{-1}) = (\overline{C}/\overline{k}, AA^{-(q)}F)$, q denoting $\operatorname{Card}(k)$. Any other form C_1/k gives rise to a Frobenius F_1 of $\overline{C}/\overline{k}$ which must be of the form BF, for some B in $\operatorname{Aut}(\overline{C}/\overline{k})$, and the k-isomorphism class of C_1/k is the class of B mod the twisted inner action of $\operatorname{Aut}(\overline{C}/\overline{k})$ on itself which has A map B to $ABA^{-(q)}$.

Thus the k-isomorphism classes of the k-forms C_1/k of $\overline{C}/\overline{k}$ are in bijective correspondence with the twisted conjugacy classes in $\operatorname{Aut}(\overline{C}/\overline{k})$, and for C_1/k

with Frobenius FB, $\mathrm{Aut}(C_1/k)$ is the twisted centralizer of B, i.e., $\mathrm{Aut}(C_1/k)$ is $\{A \text{ in } \mathrm{Aut}(\overline{C}/\overline{k}) | ABA^{-(q)} = B\}$. So our asserted identity is a familiar one from finite groups: we have an action of the finite group $G = \mathrm{Aut}(\overline{C}/\overline{k})$ on the finite set $X = G$ (by twisted conjugation), and we are writing

$$\begin{aligned} \mathrm{Card}(X)/\mathrm{Card}(G) &= \sum_{\text{orbits of } G \text{ in } X} \mathrm{Card}(\text{orbit})/\mathrm{Card}(G) \\ &= \sum_{\text{rep's } x \text{ in } X \text{ of } X/G} 1/\mathrm{Card}(\text{Fixer in } G \text{ of } x). \end{aligned}$$

To prove (3), we argue as follows. It is obvious that the image of $\mathcal{M}_g(k)$ lies in the $\mathrm{Gal}(\overline{k}/k)$-fixed points of $\mathcal{M}_g(\overline{k})$. Suppose that a point $X/\overline{k}$ is fixed by $\mathrm{Gal}(\overline{k}/k)$, i.e., suppose there exists a $\overline{k}$-isomorphism $\varphi : X \to X^{(q)}$. All the data $(X/\overline{k}, \varphi, \mathrm{Aut}(X/\overline{k}))$ is defined over some finite extension k_ν of k. Let us denote by d the order of $\mathrm{Aut}(X/\overline{k})$. Then $\varphi : X \to X^{(q)}$ defines a descent from $k_{\nu d}$ to k of (X viewed over k_ν) $\otimes_{k_\nu} k_{\nu d}$. To see this, just notice that the composite

$$X \xrightarrow{\varphi} X^{(q)} \xrightarrow{\varphi^{(q)}} \cdots \xrightarrow{\varphi^{(q^{\nu-1})}} X^{(q^\nu)} = X$$

is an automorphism of X, call it α, and hence is defined over k_ν. The composite

$$X \xrightarrow{\varphi} X^{(q)} \xrightarrow{\varphi^{(q)}} \cdots \xrightarrow{\varphi^{(q^{d\nu-1})}} X^{(q^{\nu d})} = X$$

is $\alpha^{(q^{(d-1)\nu})}\alpha^{(q^{(d-2)\nu})} \cdots \alpha^{(q^\nu)}\alpha = \alpha^d = \mathrm{id}$, as required. QED

10.7.6. We now return to the measures

$$\mu(\text{naive}, g, k, \alpha_k) \quad \text{and} \quad \mu(\text{intrin}, g, k, \alpha_k).$$

For $g \geq 3$, it is natural to compare both of these measures to a third one

$$\mu(\text{aut} \leq 1, g, k, \alpha_k),$$

where we average only over the curves which geometrically have no nontrivial automorphisms. For k such that $\mathcal{M}_{g,\text{aut}\leq 1}(k)$ is nonempty, e.g., if $\mathrm{Card}(k) > A(g)^2$, we define

$$\mu(\text{aut} \leq 1, g, k, \alpha_k) := (1/\mathrm{Card}(\mathcal{M}_{g,\text{aut}\leq 1}(k))) \sum_{C/k \text{ in } \mathcal{M}_{g,\text{aut}\leq 1}(k)} \delta_{\vartheta(k,\alpha_k,C/k)}.$$

10.7.7. It is a simple matter to express these last two measures

$$\mu(\text{intrin}, g, k, \alpha_k) \quad \text{and} \quad \mu(\text{aut} \leq 1, g, k, \alpha_k)$$

in terms of $\mathcal{M}_{g,3K}$ and $\mathcal{M}_{g,3K,\text{aut}\leq 1}$ respectively.

Lemma 10.7.8. 1) *Suppose that* ≥ 2. *The measure* $\mu(\textit{intrin}, g, k, \alpha_k)$ *is given by the formula*

$$\mu(\textit{intrin}, g, k, \alpha_k) = (1/\mathrm{Card}(\mathcal{M}_{g,3K}(k))) \sum_{(C/k,\alpha) \textit{ in } \mathcal{M}_{g,3K}(k)} \delta_{\vartheta(k,\alpha_k,C/k)}.$$

2) *Suppose that* $g \geq 3$, *and that* $\mathcal{M}_{g,\text{aut}\leq 1}(k)$ *is nonempty. The measure* $\mu(\text{aut} \leq 1, g, k, \alpha_k)$ *is given by the formula*

$$\begin{aligned} &\mu(\text{aut} \leq 1, g, k, \alpha_k) \\ &\quad = (1/\mathrm{Card}(\mathcal{M}_{g,3K,\text{aut}\leq 1}(k))) \sum_{(C/k,\alpha) \textit{ in } \mathcal{M}_{g,3K,\text{aut}\leq 1}(k)} \delta_{\vartheta(k,\alpha_k,C/k)}. \end{aligned}$$

PROOF. This is immediate from 10.6.8 and 10.6.25. QED

10.7.9. Suppose now that $g \geq 3$. Using the previous lemma, we can make explicit the relation of these measures to those which occur in 9.6.10. Fix a prime number l, and an embedding ι of $\overline{\mathbb{Q}}_l$ into $\mathbb{C}$. We take S to be $\mathrm{Spec}(\mathbb{Z}[1/l])$, we take X/S to be either $\mathcal{M}_{g,3K}[1/l]/\mathbb{Z}[1/l]$ or $\mathcal{M}_{g,3K,\mathrm{aut}\leq 1}[1/l]/\mathbb{Z}[1/l]$ respectively, we denote by $\pi : \mathcal{C} \to X$ the corresponding universal curve, we take $\mathcal{F}$ on X to be $R^1\pi_!\overline{\mathbb{Q}}_l$, and we take G to be $Sp(2g)$. The constant $A(X/S)$ may be taken to be $A(g)$ in the case of $\mathcal{M}_{g,3K,\mathrm{aut}\leq 1}[1/l]/\mathbb{Z}[1/l]$, and may be taken to be $A(g) + B(g)$ in the case of $\mathcal{M}_{g,3K}[1/l]/\mathbb{Z}[1/l]$, cf. 10.6.23.

10.7.10. For a finite field k in which l is invertible, there is a unique point s in $S(k)$. An l-adic unit α_s such that $(\alpha_s)^{\deg} \otimes \mathcal{F}_s$ has its arithmetic monodromy inside G is precisely a choice of $1/\mathrm{Sqrt}(\mathrm{Card}(k))$ inside $\overline{\mathbb{Q}}_l$. Under the chosen embedding ι of $\overline{\mathbb{Q}}_l$ into $\mathbb{C}$, $1/\alpha_s$ goes to a choice of α_k. [The other choice $-\alpha_s$ of α_s gives the other choice $-\alpha_k$ of α_k.] In view of the basic compatibility 9.1.13.1, it is tautologous that for $\mathrm{Card}(k) > A(X/S)^2$, the measure $\mu(k, s, \alpha_s)$ occurring in 9.6.10 is the measure on $USp(2g)^{\#}$ given by

$$\mu(\mathrm{intrin}, g, k, \iota(1/\alpha_s)), \quad \text{for } X/S = \mathcal{M}_{g,3K}[1/l]/\mathbb{Z}[1/l],$$
$$\mu(\mathrm{aut} \leq 1, g, k, \iota(1/\alpha_s)), \quad \text{for } X/S = \mathcal{M}_{g,3K,\mathrm{aut}\leq 1}[1/l]/\mathbb{Z}[1/l].$$

10.7.11. We define constants $C(g)$ and $C_{\mathrm{aut}\leq 1}(g)$ as follows. Pick two different values of l, say 2 and 3. The constant $C(g)$ is the maximum of the constants $C(\mathcal{M}_{g,3K}[1/l]/\mathbb{Z}[1/l], R^1\pi_!\overline{\mathbb{Q}}_l)$ for $l = 2$ and $l = 3$. The constant $C_{\mathrm{aut}\leq 1}(g)$ is the maximum of the constants $C(\mathcal{M}_{g,3K,\mathrm{aut}\leq 1}[1/l]/\mathbb{Z}[1/l], R^1\pi_!\overline{\mathbb{Q}}_l)$ for $l = 2$ and $l = 3$.

Theorem 10.7.12. *Fix a genus $g \geq 3$. For any finite field k with*

$$\mathrm{Card}(k) > (A(g) + B(g))^2 \quad (\textit{resp.}\ \mathrm{Card}(k) > A(g)^2)$$

and any choice α_k of $\mathrm{Sqrt}(\mathrm{Card}(k))$ in $\mathbb{R}$, consider the measure $\mu(\mathit{intrin}, g, k, \alpha_k)$ (resp. the measure $\mu(\mathrm{aut} \leq 1, g, k, \alpha_k)$) on $USp(2g)^{\#}$. In any sequence of data (k_i, α_{k_i}) with $\mathrm{Card}(k_i)$ increasing to infinity, the sequence of measures

$$\mu(\mathit{intrin}, g, k_i, \alpha_{k_i}) \quad (\textit{resp.}\ \mu(\mathrm{aut} \leq 1, g, k_i, \alpha_{k_i}))$$

converges weak $*$ to the measure $\mu^{\#}$ on $USp(2g)^{\#}$ which is the direct image from $USp(2g)$ of normalized Haar measure, i.e., for any continuous $\mathbb{C}$-valued central function f on $USp(2g)$, we have

$$\begin{aligned}\int_{USp(2g)} f\, d\,\mathrm{Haar} &= \lim_{i\to\infty} \int_{USp(2g)^{\#}} f\, d\mu(\mathit{intrin}, g, k_i, \alpha_{k_i}) \\ &= \lim_{i\to\infty} \int_{USp(2g)^{\#}} f\, d\mu(\mathrm{aut} \leq 1, g, k_i, \alpha_{k_i}).\end{aligned}$$

More precisely, if Λ is any irreducible nontrivial representation of $USp(2g)$, and (k, α_k) is as above with $\mathrm{Card}(k) \geq 4(A(g) + B(g))^2$ (resp. $\mathrm{Card}(k) \geq 4A(g)^2$), we have the estimates

$$\left| \int_{USp(2g)^{\#}} \mathrm{Trace}(\Lambda)\, d\mu(\mathit{intrin}, g, k, \alpha_k) \right| \leq 2C(g) \dim(\Lambda)/\mathrm{Card}(k)^{1/2},$$

respectively

$$\left|\int_{USp(2g)^{\#}} \mathrm{Trace}(\Lambda)\, d\mu(\mathrm{aut} \le 1, g, k, \alpha_k)\right| \le 2C_{\mathrm{aut}\le 1}(g)\dim(\Lambda)/\operatorname{Card}(k)^{1/2}.$$

PROOF. The "more precisely" estimates trivially imply the weak $*$ convergence, and they result in turn from the "more precisely" estimates of 9.6.10, applied to $\mathcal{M}_{g,3K}[1/l]/\mathbb{Z}[1/l]$ (respectively to $\mathcal{M}_{g,3K,\mathrm{aut}\le 1}[1/l]/\mathbb{Z}[1/l]$) with $l = 2$ for k of odd characteristic, and with $l = 3$ for k of characteristic two. QED

10.7.13. We now compare both the naive measure $\mu(\mathrm{naive}, g, k, \alpha_k)$ and the intrinsic measure $\mu(\mathrm{intrin}, g, k, \alpha_k)$ to the measure $\mu(\mathrm{aut} \le 1, g, k, \alpha_k)$, using Lemma 10.6.23.

Lemma 10.7.14. *Let $g \ge 3$. For any $(k = \mathbb{F}_q, \alpha_k)$ with $q \ge 4A(g)^2$, and any $\mathbb{C}$-valued function f on $USp(2g)^{\#}$, we have the inequalities*

$$\left|\int f\, d\mu(\mathit{naive}, g, k, \alpha_k) - \int f\, d\mu(\mathrm{aut} \le 1, g, k, \alpha_k)\right| \le 2E(g)\|f\|_{\sup}/q,$$

$$\left|\int f\, d\mu(\mathit{intrin}, g, k, \alpha_k) - \int f\, d\mu(\mathrm{aut} \le 1, g, k, \alpha_k)\right| \le 2E(g)\|f\|_{\sup}/q.$$

PROOF. Fix a choice of $(k = \mathbb{F}_q, \alpha_k)$ with $q \ge 4A(g)^2$. Denote by Z the finite set $\mathcal{M}_g(X)$, and by X and Y its nonempty subsets $\mathcal{M}_{g,\mathrm{aut}\le 1}(k)$ and $\mathcal{M}_{g,\mathrm{aut}\ge 2}(k)$ respectively. For $z = C/k$ in $Z = \mathcal{M}_g(k)$, denote by

$$\vartheta(z) := \text{ the conjugacy class } \vartheta(k, \alpha_k, C/k) \text{ in } USp(2g)^{\#},$$
$$a(z) := \text{ the real number } 1/\operatorname{Card}(\operatorname{Aut}(C/k)) \text{ in } (0, 1].$$

According to 10.6.23, we have

$$\#Y/\#X \le E(g)/q.$$

We readily compute

$$\begin{aligned}
&\left|\int f\, d\mu(\mathrm{naive}, g, k, \alpha_k) - \int f\, d\mu(\mathrm{aut} \le 1, g, k, \alpha_k)\right| \\
&\quad = \left|(1/\#Z)\sum_z f(\vartheta(z)) - (1/\#X)\sum_x f(\vartheta(x))\right| \\
&\quad = \left|(1/\#Z)\sum_x f(\vartheta(x)) + (1/\#Z)\sum_y f(\vartheta(y)) - (1/\#X)\sum_x f(\vartheta(x))\right| \\
&\quad \le \left|(1/\#Z)\sum_x f(\vartheta(x)) - (1/\#X)\sum_x f(\vartheta(x))\right| + \left|(1/\#Z)\sum_y f(\vartheta(y))\right| \\
&\quad = (\#Y/\#Z)\left|(1/\#X)\sum_x f(\vartheta(x))\right| + (\#Y/\#Z)\left|(1/\#Y)\sum_y f(\vartheta(y))\right| \\
&\quad \le 2(\#Y/\#Z)\|f\|_{\sup} \le 2(\#Y/\#X)\|f\|_{\sup} \le 2E(g)\|f\|_{\sup}/q.
\end{aligned}$$

Similarly, we have

$$\left|\int f\, d\mu(\mathrm{intrin}, g, k, \alpha_k) - \int f\, d\mu(\mathrm{aut} \le 1, g, k, \alpha_k)\right|$$

$$= \left| (1/(\#X + \sum_y a(y)))(\sum_x f(\vartheta(x)) + \sum_y a(y) f(\vartheta(y))) - (1/\#X) \sum_x f(\vartheta(x)) \right|$$

$$\leq \left| \left(1 \Big/ \left(\#X + \sum_y a(y)\right)\right) \left(\sum_x f(\vartheta(x))\right) - (1/\#X) \sum_x f(\vartheta(x)) \right|$$

$$+ \left| \left(1 \Big/ \left(\#X + \sum_y a(y)\right)\right) \sum_y a(y) f(\vartheta(y)) \right|$$

$$= \left(\left(\sum_y a(y)\right) \Big/ \left(\#X + \sum_y a(y)\right)\right) \left| (1/\#X) \sum_x f(\vartheta(x)) \right|$$

$$+ \left(\left(\sum_y a(y)\right) \Big/ \left(\#X + \sum_y a(y)\right)\right) \left| \left(1/\sum_y a(y)\right) \sum_y a(y) f(\vartheta(y)) \right|$$

$$\leq 2 \left(\left(\sum_y a(y)\right) \Big/ \left(\#X + \sum_y a(y)\right)\right) \|f\|_{\mathrm{sup}}$$

$$\leq 2(\#Y/\#X)\|f\|_{\mathrm{sup}} \leq 2E(g)\|f\|_{\mathrm{sup}}/q. \quad \text{QED}$$

Theorem 10.7.15. *Fix a genus $g \geq 3$. For any finite field k with* $\mathrm{Card}(k) > A(g)^2$, *and any choice α_k of* $\mathrm{Sqrt}(\mathrm{Card}(k))$ *in* $\mathbb{R}$, *consider the measure*

$$\mu(naive, g, k, \alpha_k)$$

on $USp(2g)^{\#}$. In any sequence of data (k_i, α_{k_i}) with $\mathrm{Card}(k_i)$ *increasing to infinity, the sequence of measures $\mu(naive, g, k_i, \alpha_{k_i})$ converges weak $*$ to the measure $\mu^{\#}$ on $USp(2g)^{\#}$ which is the direct image from $USp(2g)$ of normalized Haar measure, i.e., for any continuous $\mathbb{C}$-valued central function f on $USp(2g)$, we have*

$$\int_{USp(2g)} f \, d\,\mathrm{Haar} = \lim_{i \to \infty} \int_{USp(2g)^{\#}} f \, d\mu(naive, g, k_i, \alpha_{k_i}).$$

More precisely, if Λ is any irreducible nontrivial representation of $USp(2g)$, and (k, α_k) is as above with $\mathrm{Card}(k) \geq \mathrm{Max}(4A(g)^2, E(g)^2)$, *we have the estimate*

$$\left| \int_{USp(2g)^{\#}} \mathrm{Trace}(\Lambda) \, d\mu(naive, g, k, \alpha_k) \right| \leq (2C_{\mathrm{aut} \leq 1}(g) + 2) \dim(\Lambda)/\mathrm{Card}(k)^{1/2}.$$

PROOF. The weak convergence results from the asserted estimate. The estimate results in turn from the corresponding estimate for $\mu(\mathrm{aut} \leq 1, g, k, \alpha_k)$ proven above, its comparison to $\mu(\mathrm{naive}, g, k, \alpha_k)$ and the fact that for $f := \mathrm{Trace}(\Lambda)$, $\|f\|_{\mathrm{sup}} = \dim(\Lambda)$. QED

10.8. Measures on $USp(2g)^{\#}$ attached to universal families of hyperelliptic curves

10.8.1. Fix an integer $g \geq 1$, a finite field $k = \mathbb{F}_q$, and a choice α_k of a real square root of $\mathrm{Card}(k) = q$. Fix an integer $d \geq 3$. For each square-free monic polynomial over k of degree d, say f in $\mathcal{H}_d(k)$, we have the hyperelliptic curve of genus $g := [(d-1)/2]$ over k of equation $Y^2 = f(X)$. Exactly as in 10.7.2, this curve, call it C_f/k, has a unitarized Frobenius conjugacy class $\vartheta(k, \alpha_k, C_f/k)$ in $USp(2g)^\#$. We define the hyperelliptic probability measure $\mu(\text{hyp}, d, g, k, \alpha_k)$ on $USp(2g)^\#$ by averaging over the unitarized Frobenius conjugacy classes of all the curves C_f/k, f in $\mathcal{H}_d(k)$:

$$\mu(\text{hyp}, d, g, k, \alpha_k) := (1/\operatorname{Card}(\mathcal{H}_d(k))) \sum_{C/k \text{ in } \mathcal{H}_d(k)} \delta_{\vartheta(k,\alpha_k,C_f/k)}.$$

Applying 9.6.10 to the universal family over $\mathcal{H}_d$, and using 10.1.18.2 for odd d and 10.1.18.5 part 2) for even d, we find

Theorem 10.8.2. *Fix an integer $d \geq 3$. Define $g := [(d-1)/2]$. For any finite field k of odd characteristic p, and any choice α_k of* Sqrt(Card(k)) *in $\mathbb{R}$, consider the measure $\mu(hyp, d, g, k, \alpha_k)$ on $USp(2g)^\#$. In any sequence of data (k_i, α_{k_i}) with* Card(k_i) *odd and increasing to infinity, the sequence of measures $\mu(hyp, d, g, k_i, \alpha_{k_i})$ converges weak $*$ to the measure $\mu^\#$ on $USp(2g)^\#$ which is the direct image from $USp(2g)$ of normalized Haar measure, i.e., for any continuous $\mathbb{C}$-valued central function f on $USp(2g)$, we have*

$$\begin{aligned}\int_{USp(2g)} f\, d\,\text{Haar} &= \lim_{i\to\infty} \int_{USp(2g)^\#} f\, d\mu(intrin, g, k_i, \alpha_{k_i}) \\ &= \lim_{i\to\infty} \int_{USp(2g)^\#} f\, d\mu(\text{aut} \leq 1, g, k_i, \alpha_{k_i}).\end{aligned}$$

More precisely, there exist constants $A(d)$ and $C(d)$ with the following property: if Λ is any irreducible nontrivial representation of $USp(2g)$, and (k, α_k) is as above with Card(k) $\geq A(d)$, *we have the estimate*

$$\left| \int_{USp(2g)^\#} \operatorname{Trace}(\Lambda)\, d\mu(hyp, d, g, k, \alpha_k) \right| \leq 2C(d) \dim(\Lambda)/\operatorname{Card}(k)^{1/2}.$$

CHAPTER 11

Monodromy of Some Other Families

11.0. Universal families of principally polarized abelian varieties

11.0.1. Fix an integer $g \geq 1$. Over an arbitrary scheme S, we consider an abelian scheme $\pi : A \to S$, or A/S for short, of relative dimension g, its dual abelian scheme A^t/S, and a normalized Poincaré line bundle $\mathcal{P}$ on $A \times_S A^t$, i.e., a Poincaré bundle given with a trivialization $\iota : \mathcal{O}_{A^t} \cong \mathcal{P}|\{0_S\} \times A^t$ of $\mathcal{P}|\{0_S\} \times A^t$. Recall that a principal polarization (respectively a polarization of degree d^2, for an integer $d \geq 1$, the case $d = 1$ being precisely a principal polarization) of A/S is an isomorphism (respectively an isogeny of degree d) $\varphi : A \to A^t$ of abelian schemes over S such that the line bundle $\mathcal{L}(\varphi) := (\mathrm{id} \times \varphi)^*\mathcal{P}$ on A/S is relatively ample. One knows that $\mathcal{L}(\varphi)$ induces the map 2φ, that $\mathcal{L}(\varphi)^{\otimes 3}$ is relatively very ample on A/S, that $\pi_*(\mathcal{L}(\varphi)^{\otimes 3})$ is a locally free $\mathcal{O}_S$-module of rank 6^g (resp. $(6d)^g$) and hence that $\mathcal{L}(\varphi)^{\otimes 3}$ embeds A/S into the relative $\mathbb{P}^{6^g-1}$ (resp. $\mathbb{P}^{(6d)^g-1}$) given by $\mathbb{P}_S(\pi_*(\mathcal{L}(\varphi)^{\otimes 3}))$ [**Mum-GIT**, Ch. 6, §2, 6.13]. An isomorphism $\rho : (A/S, \varphi_A) \cong (B/S, \varphi_B)$ of polarized abelian schemes over S is an S-isomorphism $\rho : A \to B$ of abelian schemes such that $\rho^*(\mathcal{L}(\varphi)) \cong \mathcal{L}(\varphi)$ as line bundles trivialized along the zero section of A (such an isomorphism, respecting the given trivializations, is unique if it exists). In particular, an automorphism ρ of a polarized $(A/S, \varphi)$ induces an automorphism $\tilde{\rho}$ of $\pi_*(\mathcal{L}(\varphi)^{\otimes 3})$, and ρ is determined by $\tilde{\rho}$. Therefore we may rigidify a polarized $(A/S, \varphi)$ by imposing the additional structure of an $\mathcal{O}_S$-basis α of $\pi_*(\mathcal{L}(\varphi)^{\otimes 3})$: we call α a $3\mathcal{L}$ structure. The merit of a $3\mathcal{L}$ structure, as opposed to a usual level structure via points of finite order, is that when S is a field k, any polarized $(A/k, \varphi)$ admits a $3\mathcal{L}$ structure α over the same field k.

11.0.2. We will be particularly interested in the moduli of principally polarized abelian schemes. We denote by $\mathrm{Ab}_{g,\mathrm{prin},3\mathcal{L}}$ the functor on (Schemes)

$$\mathrm{Ab}_{g,\mathrm{prin},3\mathcal{L}}(S) := \{S\text{-isomorphism classes of triples } (A/S, \varphi, \alpha)\}$$

with A/S an abelian scheme of relative dimension g, φ a principal polarization of A/S, and α a $3\mathcal{L}$ structure on A/S.

Theorem 11.0.3 ([**Mum-GIT, C-F**]). *The functor* $\mathrm{Ab}_{g,\mathrm{prin},3\mathcal{L}}$ *is representable by a scheme which is smooth over* $\mathbb{Z}$ *with geometrically connected fibres of dimension* $g(g+1)/2 + 6^{2g}$.

PROOF. To prove the representability, it suffices to do so over $\mathbb{Z}[1/n]$ for two relatively prime values of n. For $n \geq 3$, consider the "raw" level n moduli problem $\mathrm{Ab}_{g,\mathrm{prin},\mathrm{raw}\ n}$, principally polarized $A/S/\mathbb{Z}[1/n]$ together with a raw (ignoring the e_n-pairing) level n structure, namely $2g$ sections of $A[n](S)$ which, on each geometric fibre, form a $\mathbb{Z}/n\mathbb{Z}$-basis. It is representable by a scheme smooth and quasiprojective over $\mathbb{Z}[1/n]$, everywhere of relative dimension $g(g+1)/2$, cf. [**Mum-GIT**, Thm. 7.9 and the remark following it] and [**C-F**, 6.8]. Over this, the moduli problem $\mathrm{Ab}_{b,\mathrm{prin},3\mathcal{L},\mathrm{raw}\ n}$, where one imposes both a raw level n structure and a basis α of $\pi_*(\mathcal{L}(\varphi)^{\otimes 3})$, is represented by the total space of a $GL(6^g)$ torsor. One constructs $\mathrm{Ab}_{b,\mathrm{prin},3\mathcal{L}} \otimes_{\mathbb{Z}} \mathbb{Z}_{1/n}$ by dividing $\mathrm{Ab}_{b,\mathrm{prin},3\mathcal{L},\mathrm{raw}\ n}$ by the free (because the $3\mathcal{L}$ structure rigidifies) action of $GL(2g, \mathbb{Z}/n\mathbb{Z})$. This construction shows that $\mathrm{Ab}_{b,\mathrm{prin},3\mathcal{L}} \otimes_{\mathbb{Z}} \mathbb{Z}[1/n]$ is smooth over $\mathbb{Z}[1/n]$, everywhere of the asserted dimension, namely $g(g+1)/2 + 6^{2g}$. Taking $n = 3$ and $n = 4$, and patching over $\mathbb{Z}[1/12]$, we get $\mathrm{Ab}_{b,\mathrm{prin},3\mathcal{L}}$ over $\mathbb{Z}$, smooth and everywhere of the asserted dimension.

To show that $\mathrm{Ab}_{b,\mathrm{prin},3\mathcal{L}}$ over $\mathbb{Z}$ has geometrically connected fibres, it suffices to do so after we extend scalars to $\mathbb{Z}[1/n, \zeta_n]$ for two relatively prime values of $n \geq 3$. There we consider the more usual notion of a symplectic level n structure on a principally polarized A/S, namely the giving of $2g$ sections $e_1, f_1, \ldots, e_g, f_g$ of $A[n](S)$ which, under the e_n-pairing, satisfy

$$(e_i, e_j) = 1 = (f_i, f_j) \quad \text{for all } i, j,$$
$$(e_i, f_j) = 1 \quad \text{if } i \neq j, \qquad (e_i, f_i) = \zeta_n \quad \text{for all } i.$$

The corresponding moduli problem $\mathrm{Ab}_{g,\mathrm{prin},n}/\mathbb{Z}[1/n, \zeta_n]$ is known to be smooth and quasiprojective with geometrically connected fibres of dimension $g(g+1)/2$ [**C-F**, 6.8]. Therefore the combined moduli problem $\mathrm{Ab}_{g,\mathrm{prin},3\mathcal{L},n}/\mathbb{Z}[1/n, \zeta_n]$, represented by the total space of a $GL(6^g)$ torsor over $\mathrm{Ab}_{g,\mathrm{prin},n}$, itself is smooth over $\mathbb{Z}[1/n, \zeta_n]$ with geometrically connected fibres of dimension $g(g+1)/2 + 6^{2g}$. The space $\mathrm{Ab}_{g,\mathrm{prin},3\mathcal{L}} \otimes_{\mathbb{Z}} \mathbb{Z}[1/n, \zeta_n]$ is the quotient of $\mathrm{Ab}_{g,\mathrm{prin},3\mathcal{L},n}/\mathbb{Z}[1/n, \zeta_n]$ by the free action of the group $Sp(2g, \mathbb{Z}/n\mathbb{Z})$, so a fortiori its geometric fibres are geometrically connected as well. QED

Theorem 11.0.4. *Consider the universal family*

$$\pi_{\mathrm{univ}} : \mathcal{A} \to \mathrm{Ab}_{g,\mathrm{prin},3\mathcal{L}}$$

of principally polarized abelian varieties with a $3\mathcal{L}$ structure. For every prime l, consider the lisse sheaf

$$\mathcal{F}_l := R^1(\pi_{\mathrm{univ}})_! \overline{\mathbb{Q}}_l | \mathrm{Ab}_{g,\mathrm{prin},3\mathcal{L}}[1/l].$$

On every geometric fibre of $\mathrm{Ab}_{g,\mathrm{prin},3\mathcal{L}}[1/l]/\mathbb{Z}[1/l]$, the geometric monodromy group of $\mathcal{F}_l$ is $Sp(2g)$.

PROOF. The e_{l^n}-pairings from the principal polarization define an alternating autoduality

$$\mathcal{F}_l \times \mathcal{F}_l \to \overline{\mathbb{Q}}_l(-1),$$

so we have an a priori inclusion of G_{geom} in $Sp(2g)$. So it suffices to exhibit, for each $g \geq 1$ and each prime p, a family A/U of principally polarized abelian varieties over a nonvoid open set U in $\mathbb{A}^1$ over $\mathbb{F}_p$ whose geometric monodromy is $Sp(2g)$ for every $l \neq p$, cf. the proof of 10.6.1. For this we may use the Jacobians of our one-parameter families of curves ($y^2 = f_{2g}(x)(x-t)$ in odd characteristic, or of our family $y^2 - y = x^{2g-1} + t/x$ in characteristic 2), whose G_{geom} we have proven to be $Sp(2g)$. QED

11.1. Other "rational over the base field" ways of rigidifying curves and abelian varieties

11.1.1. Given an abelian scheme A/S of relative dimension g, one knows that $A[n]$ is a finite flat group scheme over S of rank n^{2g}. Let us denote by $\mathrm{Aff}(A[n])$ its coordinate ring, the $\mathcal{O}_S$-locally free of rank n^{2g} sheaf of $\mathcal{O}_S$-algebras whose $\underline{\mathrm{Spec}}_S$ is $A[n]$. Suppose now that $(A/S, \varphi)$ is a (not necessarily principally) polarized abelian scheme. The following lemma is well-known.

Lemma 11.1.2 ("Serre's lemma", cf. [**Serre-Rig**], [**Mum-AV**, IV.21, Thm. 5]). *For any $n \geq 3$, $A[n]$ rigidifies $(A/S, \varphi)$, in the sense that any automorphism σ of $(A/S, \varphi)$ which induces the identity on $A[n]$ is the identity.*

PROOF. By a standard spreading out of the data $(A/S, \varphi, \sigma)$ we reduce to the case when S is of finite type over $\mathbb{Z}$. Then by rigidity [**Mum-GIT**, Ch. 6, §1, 6.1 applied to $X = Y = A$, $f = \sigma - 1$], we reduce to the case when S is the spectrum of a finite field $\mathbb{F}_q$. Then $\pi_*(\mathcal{L}(\varphi)^{\otimes 3}) = H^0(A, \mathcal{L}(\varphi)^{\otimes 3})$ is a finite-dimensional $\mathbb{F}_q$-vector space. Since it rigidifies $(A/\mathbb{F}_q, \varphi)$, we have an inclusion of groups

$$\mathrm{Aut}(A/\mathbb{F}_q, \varphi) \subset \mathrm{Aut}_{\mathbb{F}_q}(H^0(A, \mathcal{L}(\varphi)^{\otimes 3})).$$

Therefore the entire group $\mathrm{Aut}(A/\mathbb{F}_q, \varphi)$ is finite, and hence our polarized automorphism σ is of finite order, say $\sigma^k = 1$ in $\mathrm{End}(A)$. The hypothesis is that $\sigma - 1$ kills $A[n]$, and therefore $\sigma - 1 = nb$ for some b in $\mathrm{End}(A)$. But any endomorphism b has algebraic integer eigenvalues (:= roots of its "characteristic polynomial", the unique monic $\mathbb{Z}$-polynomial $P_b(T)$ whose values at integers m are given by $P_b(m) = \deg(m - b)$, cf. [**Mum-AV**, IV.19, Thm. 4]). Therefore the eigenvalues of $\sigma = 1 + nb$ are both roots of unity and lie in $1 + n\mathcal{O}_{\overline{\mathbb{Q}}}$, hence are 1. But an endomorphism satisfies its characteristic polynomial [**Mum-AV**, IV.19, Thm. 4], and hence $(\sigma - 1)^{2g} = 0$ in $\mathrm{End}(A)$. Thus σ satisfies both $\sigma^k = 1$ and $(\sigma - 1)^{2g} = 0$ in $\mathrm{End}(A)$. But in $\mathbb{Q}[T]$ the g.c.d. of the polynomials $T^k - 1$ and $(T - 1)^{2g}$ is $T - 1$, so $\sigma = 1$ in $\mathrm{End}(A) \otimes \mathbb{Q}$. But $\mathrm{End}(A)$ is flat over $\mathbb{Z}$, so $\sigma = 1$. QED

11.1.3. Let us call an $\mathcal{O}_S$-basis of $\mathrm{Aff}(A[n])$ an $\mathrm{aff}[n]$-structure. Repeating mutatis mutandis the arguments given above in the case of $3\mathcal{L}$-structures, we find

Theorem 11.1.4. *For any integers $g \geq 1$ and $n \geq 3$, the moduli problem on (Schemes) given by $S \mapsto$ {S-isomorphism classes of principally polarized abelian schemes A/S of relative dimension g together with an $\mathrm{aff}[n]$-structure} is representable by a scheme $\mathrm{Ab}_{g,\mathrm{prin},\mathrm{aff}[n]}$ which is smooth over $\mathbb{Z}$ with geometrically connected fibres of dimension $g(g+1)/2 + n^{4g}$. For each pair $l \neq p$ of distinct prime numbers, the sheaf $\mathcal{F}_l := R^1(\pi_{\mathrm{univ}})_! \overline{\mathbb{Q}}_l$ on $\mathrm{Ab}_{g,\mathrm{prin},\mathrm{aff}[n]} \otimes \overline{\mathbb{F}}_p$ has $G_{\mathrm{geom}} = Sp(2g)$.*

11.1.5. Another way to rigidify rationally is by means of de Rham cohomology.

Lemma 11.1.6. *Over any $\mathbb{Z}[1/2]$-scheme S, a polarized abelian scheme $(A/S, \varphi)$ is rigidified by its $H^1_{DR}(A/S)$, in the sense that a polarized automorphism σ which induces the identity on $H^1_{DR}(A/S)$ is the identity.*

PROOF. To see this, we argue as follows. By spreading out and rigidity, it suffices to treat the case when S is the spectrum of a finite field k of characteristic $p > 2$. In this case, we know by Oda's thesis [**Oda**] that $H^1_{DR}(A/k)$ is the Dieudonné module $\mathbb{D}(A[p])$ of $A[p]$, a result which today we may think as the reduction mod p of the equality of $H^1_{\mathrm{cris}}(A/W(k))$ and $\mathbb{D}(A[p^\infty])$, cf. [**Messing**]. So if σ fixes $H^1_{DR}(A/k)$, it induces the identity on $A[p]$, so by 11.1.2 we have $\sigma = 1$ on A. QED

11.1.7. So again we get

Theorem 11.1.8. *For any integer* $g \geq 1$, *the moduli problem on*

$$(Schemes/\mathbb{Z}[1/2])$$

given by $S \mapsto \{S$*-isomorphism classes of principally polarized abelian schemes* A/S *of relative dimension* g *together with an* $\mathcal{O}_S$*-basis of* $H^1_{DR}(A/S)\}$ *is representable by a scheme* $\mathrm{Ab}_{g,\mathrm{prin},DR}$ *which is smooth over* $\mathbb{Z}[1/2]$ *with geometrically connected fibres of dimension* $g(g+1)/2 + 4g^2$. *For each prime* $p \neq 2$ *and each prime* $l \neq p$, *the sheaf* $\mathcal{F}_l := R^1(\pi_{\mathrm{univ}})_!\overline{\mathbb{Q}}_l$ *on* $\mathrm{Ab}_{g,\mathrm{prin},DR} \otimes \overline{\mathbb{F}}_p$ *has* $G_{\mathrm{geom}} = Sp(2g)$.

11.1.9. An even simpler way to rigidify rationally is via the invariant one-forms $\omega_{A/S} := \pi_*\Omega^1_{A/S}$.

Lemma 11.1.10. *Over any* $\mathbb{Z}[1/6]$*-scheme* S, *a principally polarized abelian scheme* $(A/S, \varphi)$ *is rigidified by its* $\omega_{A/S}$, *in the sense that a polarized automorphism* σ *which induces the identity on* $\omega_{A/S}$ *is the identity.*

PROOF. Again we reduce to the case when S is the spectrum of a finite field k of characteristic $p > 3$. The idea is to reduce to the previous case by making use of the Hodge filtration

$$0 \to \omega_{A/k} \to H^1_{DR}(A/k) \to H^1(A, \mathcal{O}_A) \to 0.$$

To do this, we use the fact that the principal polarization provides an alternating autoduality $\langle\ ,\ \rangle$ on $H^1_{DR}(A/k)$ for which σ is an isometry, under which $\omega_{A/k}$ is isotropic, and which makes $H^1(A, \mathcal{O}_A)$ the dual of $\omega_{A/k}$. Therefore if σ is 1 on $\omega_{A/k}$, by duality we also have that σ is 1 on $H^1(A, \mathcal{O}_A)$. Looking at the Hodge filtration, we see that $(\sigma - 1)^2$ kills $H^1_{DR}(A/k)$. Viewing $H^1_{DR}(A/k)$ as $\mathbb{D}(A[p])$, we get that $(\sigma - 1)^2 = pb$ for some b in $\mathrm{End}(A)$. Looking at eigenvalues, we see that those of σ are roots of unity which are $1 \bmod \mathrm{Sqrt}(p)\mathcal{O}_{\overline{\mathbb{Q}}}$. As p is at least 5, any such root of unity is 1, and we conclude as above that $\sigma = 1$. QED

11.1.11. The same arguments as above now give

Theorem 11.1.12. *For any integer* $g \geq 1$, *the moduli problem on*

$$(Schemes/\mathbb{Z}[1/6])$$

given by $S \mapsto \{S$*-isomorphism classes of principally polarized abelian schemes* A/S *of relative dimension* g *together with an* $\mathcal{O}_S$*-basis of* $\omega_{A/S}\}$ *is representable by a scheme* $\mathrm{Ab}_{g,\mathrm{prin},\omega}$ *which is smooth over* $\mathbb{Z}[1/6]$ *with geometrically connected fibres of dimension* $g(g+1)/2 + g^2$. *For each prime* $p \neq 2, 3$ *and each prime* $l \neq p$, *the sheaf* $\mathcal{F}_l := R^1(\pi_{\mathrm{univ}})_!\overline{\mathbb{Q}}_l$ *on* $\mathrm{Ab}_{g,\mathrm{prin},\omega} \otimes \overline{\mathbb{F}}_p$ *has* $G_{\mathrm{geom}} = Sp(2g)$.

11.1.13. In the case $g = 1$, we recover Weierstrass normal form. For $g = 1$, the principal polarization is unique, given by $I^{-1}(0)$. So the moduli problem $\mathrm{Ab}_{1,\mathrm{prin},\omega}/\mathbb{Z}[1/6]$ is that of elliptic curves together with a nowhere vanishing differential. It is represented by the spectrum of the ring $\mathbb{Z}[1/6, g_2, g_3][1/\Delta]$, Δ the discriminant $(g_2)^3 - 27(g_3)^2$, and over it the universal (E, ω) is

$$(y^2 = 4x^3 - g_2 x - g_3, dx/2y),$$

cf. **[De-CEF]**. It would be interesting to see explicitly what the moduli space $\mathrm{Ab}_{g,\mathrm{prin},\omega}/\mathbb{Z}[1/6]$ looks like for higher g.

11.1.14. Since a curve of genus $g \geq 2$ is rigidified by its Jacobian, we can also make variants $\mathcal{M}_{g,DR}/\mathbb{Z}[1/2]$ and $\mathcal{M}_{g,\omega}/\mathbb{Z}[1/6]$ respectively of $\mathcal{M}_{g,3K}/\mathbb{Z}$, where instead of imposing a $3K$ structure we impose an $\mathcal{O}_S$-basis of $H^1_{DR}(C/S)$, or an $\mathcal{O}_S$-basis of $\pi_*\Omega^1_{C/S}$ respectively. Again the same arguments as in 10.6.11 above give

Theorem 11.1.15. *For any integer $g \geq 2$, the moduli problem on*

$$(Schemes/\mathbb{Z}[1/2])$$

given by $S \mapsto \{S$-isomorphism classes of curves C/S of genus g together with an $\mathcal{O}_S$-basis of $H^1_{DR}(C/S)\}$ is representable by a scheme $\mathcal{M}_{g,DR}$ which is smooth over $\mathbb{Z}[1/2]$ with geometrically connected fibres of dimension $3g-3+4g^2$. For each prime $p \neq 2$ and each prime $l \neq p$, the sheaf $\mathcal{F}_l := R^1(\pi_{\text{univ}})_!\overline{\mathbb{Q}}_l$ on $\mathcal{M}_{g,DR} \otimes \overline{\mathbb{F}}_p$ has $G_{\text{geom}} = Sp(2g)$.

Theorem 11.1.16. *For any integer $g \geq 2$, the moduli problem on*

$$(Schemes/\mathbb{Z}[1/6])$$

given by $S \mapsto \{S$-isomorphism classes of curves C/S of genus g together with an $\mathcal{O}_S$-basis of $\pi_*\Omega^1_{C/S}\}$ is representable by a scheme $\mathcal{M}_{g,\omega}$ which is smooth over $\mathbb{Z}[1/6]$ with geometrically connected fibres of dimension $3g-3+g^2$. For each prime $p \neq 2,3$ and each prime $l \neq p$, the sheaf $\mathcal{F}_l := R^1(\pi_{\text{univ}})_!\overline{\mathbb{Q}}_l$ on $\mathcal{M}_{g,\omega} \otimes \overline{\mathbb{F}}_p$ has $G_{\text{geom}} = Sp(2g)$.

11.2. Automorphisms of polarized abelian varieties

11.2.1. Suppose we are given an abelian scheme A/S. The functor in groups $\underline{\text{Aut}}_S(A/S)$ on (Schemes/S) defined by $T \mapsto \text{Aut}(A \times_S T/T)$ is representable by a group scheme locally of finite type over S, thanks to the existence of the Hilbert scheme, cf. [**FGA**, Exposé 221] and [**Mum-GIT**, Ch. 0, §5]. This group scheme satisfies the valuative criterion for unramifiedness (thanks to rigidity [**Mum-GIT**, Ch. 6, §1, 6.1]), as well as that for properness (because over a discrete valuation ring, an abelian scheme is the Néron model of its general fibre), but $\underline{\text{Aut}}_S(A/S)$ is very far from being of finite type over S in general. For example, start with any S, and with any A/S. Consider the n-fold product $A^n/S := A \times_S A \times_S \cdots \times_S A$ for any $n \geq 2$. We have an obvious action of the infinite discrete group $GL(n,\mathbb{Z})$ on A^n/S.

11.2.2. On the other hand, if we take a polarized abelian scheme $(A/S, \varphi)$, the functor of polarized automorphisms is represented by a closed sub-group scheme $\underline{\text{Aut}}_S(A/S,\varphi)$ of $\underline{\text{Aut}}_S(A/S)$. The following lemma is well-known.

Lemma 11.2.3. *The group scheme $\underline{\text{Aut}}_S(A/S,\varphi)$ is finite and unramified over S, with geometric fibres of rank $\leq 4^{4g^2}$.*

PROOF. We first reduce to the case when S is noetherian, or even of finite type over $\mathbb{Z}$. The question is Zariski local on S, so we may assume further that the polarization φ is everywhere of some degree d^2 on S, and that we have chosen an $\mathcal{O}_S$-basis of $\pi_*(\mathcal{L}(\varphi)^{\otimes 3})$. Then $\underline{\text{Aut}}_S(A/S,\varphi)$ is a closed subgroup of

$$\underline{\text{Aut}}_S(\pi_*(\mathcal{L}(\varphi)^{\otimes 3})) \approx GL((6d)^g)_S,$$

so certainly is of finite type over S. Since it is valuatively proper over S, it is in fact proper over S. So it suffices that it have finite geometric fibres over S. By rigidity, the geometric fibres are valuatively unramified. As they are of finite type, they are unramified, and thus (being over a field) etale. So it suffices that for any point s of S with values in an algebraically closed field k, the group

$$\underline{\mathrm{Aut}}_S(A/S, \varphi)(k) = \mathrm{Aut}(A_s/k, \varphi_s)$$

be finite. But for any $n \geq 3$ invertible in k, $A_s[n](k)$ rigidifies $(A_s/k, \varphi_s)$, so $\#\mathrm{Aut}(A_s/k, \varphi_s)$ divides $\#GL(2g, \mathbb{Z}/n\mathbb{Z})$. Taking $n = 3$ or 4, we get the asserted bound. QED

11.2.4. For curves of genus $g \geq 3$, most have no nontrivial automorphisms. Of course, every polarized abelian scheme (of relative dimension > 0) admits ± 1 as automorphisms. In the next two lemmas, we show that in general ± 1 are the only automorphisms. Exactly as in 10.6.13, we have

Lemma 11.2.5. *Let S be a noetherian scheme, $(A/S, \varphi)$ a polarized abelian scheme over S. There is an open set $U_{\leq 2}$ of S which is characterized by the following property: a point s in S lies in $U_{\leq 2}$ if and only if for the corresponding $(A_s/\kappa(s), \varphi_s)$ we have $\underline{\mathrm{Aut}}_{\kappa(s)}(A_s/\kappa(s), \varphi_s) = \{\pm 1\}$. More generally, for any integer $i \geq 2$, there is an open set $U_{\leq i}$ of S which is characterized by the following property: a point s in S lies in $U_{\leq i}$ if and only if for the corresponding $(A_s/\kappa(s), \varphi_s)$ the finite etale $\kappa(s)$-group scheme $\underline{\mathrm{Aut}}_{\kappa(s)}(A_s/\kappa(s), \varphi_s)$ has rank $\leq i$.*

Lemma 11.2.6. *Let $g \geq 1$. Over every algebraically closed field k there exists a g-dimensional principally polarized abelian variety $(A/k, \varphi)$, in fact a Jacobian, whose automorphism group* $\mathrm{Aut}(A/k, \varphi)$ *is* $\{\pm 1\}$.

PROOF. Pick a prime l_0 invertible in k. Take one of our families C/U of curves of genus g over a nonvoid open set of $\mathbb{A}^1$ over k whose G_{geom} is $Sp(2g)$ for every l invertible in k. [If k is of characteristic zero, take the family of curves $y^2 = (x^{2g} - 1)(x - t)$. It has $G_{\mathrm{geom}} = Sp(2g)$ in every characteristic p not dividing $2g$, and hence, thanks to Pink [**Ka-ESDE**, 8.18], it also has big G_{geom} in characteristic zero.] At the cost of passing to a nonvoid open set U_0 of U, and then passing to a finite etale covering V of U_0, we get a family C/V with G_{geom} still $Sp(2g)$ and in which the finite group scheme $\underline{\mathrm{Aut}}_V(\mathrm{Jac}(C/V)$, canonical polarization) is finite etale and constant on V. We claim that this constant group is the group $\{\pm 1\}$. Once this is proven, then for any k-valued point v in V, the Jacobian of C_v has $\{\pm 1\}$ as its polarized automorphism group.

Because $Sp(2g)$ is absolutely irreducible in its standard representation, any polarized automorphism σ of $\mathrm{Jac}(C/V)$ must act on $H^1(C_v/\overline{\mathbb{Q}}_{l_0})$ as a scalar, call it ζ. Because σ has finite order, this scalar is a root of unity. But the trace of σ on $H^1(C_v, \overline{\mathbb{Q}}_{l_0})$ lies in $\mathbb{Z}$ [**Mum-AV**, IV.19, Thm. 4], so $2g\zeta$ lies in $\mathbb{Z}$, so ζ is a root of unity in $\mathbb{Q}$, so ζ is ± 1, and hence σ is ± 1. QED

11.3. Naive and intrinsic measures on $USp(2g)^{\#}$ attached to universal families of principally polarized abelian varieties

11.3.1. Fix an integer $g \geq 1$, a finite field $k = \mathbb{F}_q$, and a choice α_k of a real square root of $\mathrm{Card}(k) = q$. Given a g-dimensional abelian variety A/k, with Frobenius endomorphism denoted $F_{A/k}$, or simply F, its (reversed) characteristic polynomial of Frobenius is a $\mathbb{Z}$-polynomial $P(T)$ of degree $2g$, with constant term 1, uniquely determined by the property that for every integer m we have $P(m) = \deg(1 - mF_{A/k})$. For every l invertible in k, we have

$$P(T) = \det(1 - TF|H^1(A \otimes_k \overline{k}, \overline{\mathbb{Q}}_l)).$$

Its relation to the zeta function of A/k is slightly clumsy to express in naive terms. If we factor $P(T) = \prod(1 - T\beta_i)$, then for each $n \geq 1$ we have

$$\begin{aligned}\#A(\mathbb{F}_{q^n}) &= \deg(1 - F^n) = \prod(1 - (\beta_i)^n) \\ &= \sum_i (-1)^i \operatorname{Trace}(\Lambda^i(F^n)|\Lambda^i(H^1)).\end{aligned}$$

More conceptually, the zeta function $\operatorname{Zeta}(A/k, T)$ is the alternating product

$$\prod_i \det(1 - TF|H^i(A \otimes_k \overline{k}, \overline{\mathbb{Q}}_l))^{(-1)^{i+1}},$$

and $H^i = \Lambda^i(H^1)$.

11.3.2. By the Riemann Hypothesis for abelian varieties over finite fields, we know that there exists a conjugacy class $\vartheta(k, \alpha_k, A/k)$ in $USp(2g)^\#$ with the property that

$$P(T) = \det(1 - \alpha_k T\vartheta(k, \alpha_k, A/k)).$$

Because elements of $USp(2g)^\#$ are uniquely determined by their characteristic polynomials, the conjugacy class $\vartheta(k, \alpha_k, A/k)$ in $USp(2g)^\#$ is uniquely determined by this property. We call it the unitarized Frobenius conjugacy class attached to A/k.

11.3.3. We denote by $\mathrm{Ab}_{g,\mathrm{prin}}$ the functor $S \mapsto \{S\text{-isomorphism classes of principally polarized } (A/S, \varphi) \text{ of relative dimension } g\}$. This functor is not representable, because it tries to classify objects with automorphisms. Nonetheless, for k a finite field, the set $\mathrm{Ab}_{g,\mathrm{prin}}(k)$ is finite, and we have the following mass formula, entirely analogous to that of 10.6.9:

$$\#\,\mathrm{Ab}_{g,\mathrm{prin},3\mathcal{L}}(k)/\#GL(6^g, k) = \sum_{(A/k,\varphi) \text{ in } \mathrm{Ab}_{g,\mathrm{prin}}(k)} 1/\#\,\mathrm{Aut}(A/k, \varphi).$$

If k has odd characteristic, we have the de Rham variant

$$\#\,\mathrm{Ab}_{g,\mathrm{prin},DR}(k)/\#GL(2g, k) = \sum_{(A/k,\varphi) \text{ in } \mathrm{Ab}_{g,\mathrm{prin}}(k)} 1/\#\,\mathrm{Aut}(A/k, \varphi).$$

If k has characteristic ≥ 5, we have the ω variant

$$\#\,\mathrm{Ab}_{g,\mathrm{prin},\omega}(k)/\#GL(g, k) = \sum_{(A/k,\varphi) \text{ in } \mathrm{Ab}_{g,\mathrm{prin}}(k)} 1/\#\,\mathrm{Aut}(A/k, \varphi).$$

11.3.4. Exactly as we did in the case of curves, we define the "naive" probability measure $\mu(\text{naive}, g, \text{prin}, k, \alpha_k)$ on $USp(2g)^\#$ by averaging over the unitarized Frobenius conjugacy classes of all the k-isomorphism classes of principally polarized g-dimensional A/k, each counted with multiplicity one:

$$\begin{aligned}&\mu(\text{naive}, g, \text{prin}, k, \alpha_k) \\ &\quad := (1/\#(\mathrm{Ab}_{g,\mathrm{prin}}(k))) \sum_{(A/k,\varphi) \text{ in } \mathrm{Ab}_{g,\mathrm{prin}}(k)} \delta_{\vartheta(k,\alpha_k,A/k)}.\end{aligned}$$

11.3.5. We define the "intrinsic" probability measure $\mu(\text{intrin}, g, \text{prin}, k, \alpha_k)$ on $USp(2g)^{\#}$ by averaging over the unitarized Frobenius conjugacy classes, but now counting $(A/k, \varphi)$ with multiplicity $1/\#\operatorname{Aut}(A/k, \varphi)$:

$$\mu(\text{intrin}, g, \text{prin}, k, \alpha_k)$$
$$:= (1/\operatorname{Intrin}\operatorname{Card}(\mathrm{Ab}_{g,\text{prin}}(k))) \sum_{(A/k,\varphi) \text{ in } \mathrm{Ab}_{g,\text{prin}}(k)} (1/\#\operatorname{Aut}(A/k,\varphi))\delta_{\vartheta(k,\alpha_k,A/k)},$$

where we have put

$$\begin{aligned}\operatorname{Intrin}\operatorname{Card}(\mathrm{Ab}_{g,\text{prin}}(k)) &:= \sum_{(A/k,\varphi) \text{ in } \mathrm{Ab}_{g,\text{prin}}(k)} 1/\#\operatorname{Aut}(A/k,\varphi) \\ &= \#\mathrm{Ab}_{g,\text{prin},3\mathcal{L}}(k)/\#GL(6^g, k).\end{aligned}$$

11.3.6. Exactly as in the case of curves we have the following two lemmas, cf. 10.7.5 and 10.7.8.

Lemma 11.3.7. (1) *The rational number* $\operatorname{Intrin}\operatorname{Card}(\mathrm{Ab}_{g,\text{prin}}(k))$ *is an integer, namely*

$$\operatorname{Intrin}\operatorname{Card}(\mathrm{Ab}_{g,\text{prin}}(k)) = \operatorname{Card}(\operatorname{Image}(\mathrm{Ab}_{g,\text{prin}}(k) \to \mathrm{Ab}_{g,\text{prin}}(\overline{k}))).$$

(2) *More precisely, for each point* ξ *in this image, we have the identity*

$$1 = \sum_{(A/k,\varphi) \text{ in } \mathrm{Ab}_{g,\text{prin}}(k) \mapsto \xi \text{ in } \mathrm{Ab}_{g,\text{prin}}(\overline{k})} 1/\operatorname{Card}(\operatorname{Aut}(A/k,\varphi)).$$

(3) *The image of* $\mathrm{Ab}_{g,\text{prin}}(k)$ *in* $\mathrm{Ab}_{g,\text{prin}}(\overline{k})$ *consists precisely of the points in* $\mathrm{Ab}_{g,\text{prin}}(\overline{k})$ *fixed by* $\operatorname{Gal}(\overline{k}/k)$.

Lemma 11.3.8. *The measure* $\mu(\textit{intrin}, g, \text{prin}, k, \alpha_k)$ *is given by the formulas*

$$\begin{aligned}&\mu(\textit{intrin}, g, \text{prin}, k, \alpha_k)\\ &= (1/\operatorname{Card}(\mathrm{Ab}_{g,\text{prin},3\mathcal{L}}(k))) \sum_{(A/k,\varphi,\alpha) \textit{ in } \mathrm{Ab}_{g,\text{prin},3\mathcal{L}}(k)} \delta_{\vartheta(k,\alpha_k,A/k)}\\ &= (1/\operatorname{Card}(\mathrm{Ab}_{g,\text{prin},\text{aff}[n]}(k)))\\ &\quad\times \sum_{(A/k,\varphi,\alpha) \textit{ in } \mathrm{Ab}_{g,\text{prin},\text{aff}[n]}(k)} \delta_{\vartheta(k,\alpha_k,A/k)}, \quad \textit{for } n \geq 3,\\ &= (1/\operatorname{Card}(\mathrm{Ab}_{g,\text{prin},DR}(k)))\\ &\quad\times \sum_{(A/k,\varphi,\alpha) \textit{ in } \mathrm{Ab}_{g,\text{prin},DR}(k)} \delta_{\vartheta(k,\alpha_k,A/k)}, \quad \textit{if } \operatorname{char}(k) > 2,\\ &= (1/\operatorname{Card}(\mathrm{Ab}_{g,\text{prin},\omega}(k)))\\ &\quad\times \sum_{(A/k,\varphi,\alpha) \textit{ in } \mathrm{Ab}_{g,\text{prin},\omega}(k)} \delta_{\vartheta(k,\alpha_k,A/k)}, \quad \textit{if } \operatorname{char}(k) > 3.\end{aligned}$$

11.3.9. Exactly as explained above so laboriously in the case of curves, for each integer $g \geq 1$ there exist constants $A(g)$ and $C(g)$ so that we have the following theorem.

Theorem 11.3.10. *Fix an integer* $g \geq 1$. *For each finite field* k, *and each choice* α_k *of* $\operatorname{Sqrt}(\operatorname{Card}(k))$ *in* $\mathbb{R}$, *consider the measures on* $USp(2g)^{\#}$ *given by*

$$\mu(\textit{intrin}, g, \text{prin}, k, \alpha_k) \quad \textit{and} \quad \mu(\textit{naive}, g, \text{prin}, k, \alpha_k).$$

In any sequence of data (k_i, α_{k_i}) *with* $\mathrm{Card}(k_i)$ *increasing to infinity, the two sequences of measures*

$$\mu(intrin, g, \mathrm{prin}, k_i, \alpha_{k_i}) \quad and \quad \mu(naive, g, \mathrm{prin}, k_i, \alpha_{k_i})$$

both converge weak $*$ *to the measure* $\mu^{\#}$ *on* $USp(2g)^{\#}$ *which is the direct image from* $USp(2g)$ *of normalized Haar measure, i.e., for any continuous* $\mathbb{C}$*-valued central function* f *on* $USp(2g)$*, we have*

$$\begin{aligned}\int_{USp(2g)} f\, d\,\mathrm{Haar} &= \lim_{i\to\infty}\int_{USp(2g)^{\#}} \mathrm{Trace}(\Lambda)\, d\mu(intrin, g, \mathrm{prin}, k_i, \alpha_{k_i})\\ &= \lim_{i\to\infty}\int_{USp(2g)^{\#}} f\, d\mu(naive, g, \mathrm{prin}, k_i, \alpha_{k_i}).\end{aligned}$$

More precisely, if Λ *is any irreducible nontrivial representation of* $USp(2g)$*, and* (k, α_k) *is as above with* $\mathrm{Card}(k) \geq 4A(g)^2$*, we have the estimates*

$$\left|\int_{USp(2g)^{\#}} \mathrm{Trace}(\Lambda)\, d\mu(intrin, g, \mathrm{prin}, k, \alpha_k)\right| \leq 2C(g)\dim(\Lambda)/\operatorname{Card}(k)^{1/2},$$

and

$$\left|\int_{USp(2g)^{\#}} \mathrm{Trace}(\Lambda)\, d\mu(naive, g, \mathrm{prin}, k, \alpha_k)\right| \leq 2C(g)\dim(\Lambda)/\operatorname{Card}(k)^{1/2}.$$

11.4. Monodromy of universal families of hypersurfaces

11.4.1. In this section, we will consider smooth hypersurfaces X of degree $d \geq 2$ in $\mathbb{P}^{n+1}$, $n \geq 1$. Recall that over a finite field $k = \mathbb{F}_q$, the zeta function of such an X/k has the form:

$$\text{if } n \text{ is odd:} \qquad P(T)/\prod_{i=0}^{n}(1 - q^i T),$$

$$\text{if } n \text{ is even:} \qquad 1/P(T)\prod_{i=0}^{n}(1 - q^i T),$$

where $P(T)$ is a $\mathbb{Z}$-polynomial with constant term one, of degree

$$\mathrm{prim}(n, d) := (d-1)((d-1)^{n+1} - (-1)^{n+1})/d.$$

Pick a choice α_k of $\mathrm{Sqrt}(\mathrm{Card}(k)^n)$ in $\mathbb{R}$. For n odd (resp. even), we know by [**De-Weil I**] that there is a conjugacy class $\vartheta(k, \alpha_k, X/k)$ in $USp(\mathrm{prim}(n,d))^{\#}$ (resp. in $O(\mathrm{prim}(n,d))^{\#}$) such that

$$P(T) = \det(1 - \alpha_k T \vartheta(k, \alpha_k, X/k)).$$

In both of the groups USp and O, conjugacy classes are determined by their characteristic polynomials, so $\vartheta(k, \alpha_k, X/k)$ is uniquely determined by $P(T)$ and by the choice of α_k. We call it the unitarized Frobenius conjugacy class attached to X/k.

11.4.2. The cohomological reason for these facts is this. Fix a prime l invertible in k, which now could be any field. For n odd, the only nonzero odd-dimensional cohomology group $H^i(X \otimes_k \overline{k}, \overline{\mathbb{Q}}_l)$ is H^n, it has dimension $\mathrm{prim}(n,d)$, and the cup-product is a nondegenerate alternating form

$$H^n(X \otimes_k \overline{k}, \overline{\mathbb{Q}}_l) \times H^n(X \otimes_k \overline{k}, \overline{\mathbb{Q}}_l) \to H^{2n}(X \otimes_{k}, \overline{k}, \overline{\mathbb{Q}}_l) \cong \overline{\mathbb{Q}}_l(-n).$$

The only nonzero even-dimensional cohomology groups $H^i(X \otimes_k \overline{k}, \overline{\mathbb{Q}}_l)$ are the groups H^{2k} for $k = 0, \ldots, n$, with each $H^{2k} \cong \overline{\mathbb{Q}}_l(-k)$.

11.4.3. For n even, the situation is just slightly more complicated. There is no nonzero odd-dimensional cohomology, and the nonzero even-dimensional cohomology other than H^n, which we will discuss in a moment, is $H^{2k} \cong \overline{\mathbb{Q}}_l(-k)$ for $k = 0, \ldots, n$, $k \neq n/2$. For H^n, the restriction map

$$H^n((\mathbb{P}^{n+1})_{\overline{k}}, \overline{\mathbb{Q}}_l) \to H^n(X \otimes_k \overline{k}, \overline{\mathbb{Q}}_l)$$

is injective. Its target, by means of cup-product, is equipped with a nondegenerate symmetric pairing

$$H^n(X \otimes_k \overline{k}, \overline{\mathbb{Q}}_l) \times H^n(X \otimes_k \overline{k}, \overline{\mathbb{Q}}_l) \to H^{2n}(X \otimes_k \overline{k}, \overline{\mathbb{Q}}_l) \cong \overline{\mathbb{Q}}_l(-n).$$

Its source $H^n((\mathbb{P}^{n+1})_{\overline{k}}, \overline{\mathbb{Q}}_l)$ is canonically $\overline{\mathbb{Q}}_l(-n/2)$, and any nonzero element η in its image in $H^n(X \otimes_k \overline{k}, \overline{\mathbb{Q}}_l)$ has $\eta \cdot \eta \neq 0$. We define the subspace

$$\mathrm{Prim}^n(X \otimes_k \overline{k}, \overline{\mathbb{Q}}_l) \subset H^n(X \otimes_k \overline{k}, \overline{\mathbb{Q}}_l)$$

to be the orthogonal of any such η. Thus we have a direct sum decomposition

$$H^n(X \otimes_k \overline{k}, \overline{\mathbb{Q}}_l) = \mathrm{Prim}^n(X \otimes_k \overline{k}, \overline{\mathbb{Q}}_l) \oplus H^n((\mathbb{P}^{n+1})_{\overline{k}}, \overline{\mathbb{Q}}_l),$$

and the restriction to $\mathrm{Prim}^n(X \otimes_k \overline{k}, \overline{\mathbb{Q}}_l)$ of the cup-product pairing is a nondegenerate symmetric form

$$\mathrm{Prim}^n \times \mathrm{Prim}^n \to H^{2n}(X \otimes_k \overline{k}, \overline{\mathbb{Q}}_l) \cong \overline{\mathbb{Q}}_l(-n).$$

The dimension of $\mathrm{Prim}^n(X \otimes_k \overline{k}, \overline{\mathbb{Q}}_l)$ is $\mathrm{prim}(n,d)$.

11.4.4. Now let us consider the universal family $\pi : \mathcal{X}_{n,d} \to \mathcal{H}_{n,d}$ of smooth, degree d hypersurfaces in $\mathbb{P}^{n+1}$. To write this family explicitly, we begin with the universal form of degree d in $n+2$ homogeneous variables $X_1, \ldots, X_{n+2}$, say $F(X) := \sum A_w X^w$ with indeterminate coefficients A_w, one for each of the $\mathrm{Binom}(n+1+d, d)$ monomials X^w of degree d in the $n+2$ variables X_i. As was well known in the last century, there exists a universal homogeneous form Δ in the A_w, with $\mathbb{Z}$-coefficients, the "discriminant" of the form $\sum A_w X^w$, whose nonvanishing at a field-valued point is equivalent to the smoothness of the corresponding hypersurface.

11.4.5. Let us recall the interpretation of Δ as the equation of the "dual variety". Think of a hypersurface of degree d in $\mathbb{P}^{n+1}$ as being a hyperplane section of the image, call it V, of the d-fold Segre embedding of $\mathbb{P}^{n+1}$ into $\mathbb{P}^N$, $N := \mathrm{Binom}(n+1+d, d) - 1$, by means of monomials of degree d, and consider the dual projective space $\mathbb{P}^\vee$ to $\mathbb{P}^N$. Inside $V \times \mathbb{P}^\vee$ we have the incidence variety Z consisting of those pairs (v in V, H a hyperplane in $\mathbb{P}^N$) such that H is tangent to V at v. Viewed over V, Z is a $\mathbb{P}^{N-1-\dim V}$ bundle, so as a scheme Z is irreducible, and it is proper and smooth over $\mathbb{Z}$ with geometrically connected fibres of dimension $N-1$. The dual variety $V^\vee$ is defined to be the image in $\mathbb{P}^\vee$ of Z by the second projection. Thus for any field k, a k-valued point of $\mathbb{P}^\vee$, i.e. a degree d

hypersurface in $(\mathbb{P}^{n+1}) \otimes_{\mathbb{Z}} k$, lies in $V^\vee$ if and only if it is singular. Because Z is irreducible and proper over $\mathbb{P}^\vee$, $V^\vee$ is a closed irreducible subset of $\mathbb{P}^\vee$. We endow $V^\vee$ with its reduced structure. So endowed, $V^\vee$ is a reduced and irreducible closed subscheme of $\mathbb{P}^\vee$, which surjects onto $\mathrm{Spec}(\mathbb{Z})$ (because Z did), and hence, being reduced and irreducible, is flat over $\mathbb{Z}$. So $V^\vee$ as a subscheme of $\mathbb{P}^\vee$ must be the schematic closure in $\mathbb{P}^\vee$ of the subscheme $(V^\vee) \otimes_{\mathbb{Z}} \mathbb{Q}$ of $(\mathbb{P}^\vee) \otimes_{\mathbb{Z}} \mathbb{Q}$, which is known [**SGA 7**, XVII, 3.3, 3.5, and 3.7.1] to be an irreducible hypersurface. So concretely, to obtain $V^\vee$, one writes the equation over $\mathbb{Q}$ for this hypersurface, and then one scales the equation by a $\mathbb{Q}^\times$-factor so that it has integer coefficients which generate the unit ideal (1) in $\mathbb{Z}$. This scaled equation, well-defined up to ± 1, is the equation for $V^\vee$, and is the desired Δ.

11.4.6. The point of view of the dual variety makes obvious the existence, in any characteristic, of smooth hypersurfaces in $\mathbb{P}^{n+1}$ of any degree $d \geq 2$. Alternatively, one can show the existence by writing down the following explicit equations, the last of which we learned from Ofer Gabber:

$$d = 2, n+2 = 2k \text{ even:} \qquad \sum_{i=1}^{k} X_i X_{i+k},$$

$$d = 2, n+2 = 2k+1 \text{ odd:} \qquad (X_{2k+1})^2 + \sum_{i=1}^{k} X_i X_{i+k},$$

$$d \geq 3 \text{ prime to } p: \qquad \sum_{i=1}^{n+2} (X_i)^d,$$

$$d \geq 3 \text{ divisible by } p: \qquad (X_1)^d + \sum_{i=1}^{n+1} X_i (X_{i+1})^{d-1}.$$

11.4.7. The universal family has parameter space $\mathcal{H}_{n,d} :=$ the open set in the projective space of the A_w where Δ is invertible, and over it the universal family $\pi : \mathcal{X}_{n,d} \to \mathcal{H}_{n,d}$ is the hypersurface of equation $\sum A_w X^w$ in $\mathbb{P}^{n+2} \times_{\mathbb{Z}} \mathcal{H}_{n,d}$, mapping to $\mathcal{H}_{n,d}$ by the second projection. Because there are smooth hypersurfaces of every degree and dimension in every characteristic, the discriminant Δ is not identically zero in any characteristic. So each geometric fibre of $\mathcal{H}_{n,d}/\mathbb{Z}$ is a nonvoid open set in a projective space of dimension $\mathrm{Binom}(n+1+d, d) - 1$, so in particular is smooth and connected of dimension $\mathrm{Binom}(n+1+d, d) - 1$.

11.4.8. For each prime l, we define a lisse $\overline{\mathbb{Q}}_l$-sheaf Prim_l^n on $\mathcal{H}_{n,d}[1/l]$ as follows. For n odd, we take

$$\mathrm{Prim}_l^n := R^n \pi_! \overline{\mathbb{Q}}_l | \mathcal{H}_{n,d}[1/l].$$

For n even, the sheaf $R^n \pi_! \overline{\mathbb{Q}}_l | \mathcal{H}_{n,d}[1/l]$ receives injectively the geometrically constant sheaf $\overline{\mathbb{Q}}_l(-n/2)_{\mathcal{H}_{n,d}[1/l]}$ which is the H^n of the ambient $\mathbb{P}^{n+2}$, and we define $\mathrm{Prim}_l^n \subset R^n \pi_! \overline{\mathbb{Q}}_l | \mathcal{H}_{n,d}[1/l]$ to be its orthogonal. So for n even, we have

$$R^n \pi_! \overline{\mathbb{Q}}_l | \mathcal{H}_{n,d}[1/l] = \overline{\mathbb{Q}}_l(-n/2)_{\mathcal{H}_{n,d}[1/l]} \oplus \mathrm{Prim}_l^n.$$

In both cases, cup product provides an autoduality

$$\mathrm{Prim}_l^n \times \mathrm{Prim}_l^n \to \overline{\mathbb{Q}}_l(-n),$$

which is alternating for n odd, and symmetric for n even.

Theorem 11.4.9 ([**De-Weil II**, 4.4.1]). *Suppose that $d \geq 3$, that $n \geq 1$, and that $(n,d) \neq (2,3)$. For any prime l, the lisse sheaf Prim_l^n on $\mathcal{H}_{n,d}[1/l]$ has, on every geometric fibre of $\mathcal{H}_{n,d}[1/l]/\mathbb{Z}[1/l]$, the group G_{geom} given by*

$$Sp(\mathrm{prim}(n,d)), \quad \text{if } n \text{ odd},$$
$$O(\mathrm{prim}(n,d)), \quad \text{if } n \text{ even}.$$

PROOF. In view of the cup-product autoduality, G_{geom} is a priori a subgroup of the named group. Let us examine the situation in a given characteristic $p \neq l$. By a standard Bertini argument, e.g. using [**Ka-ACT**, 3.11.1], if we restrict the lisse sheaf Prim_l^n on $\mathcal{H}_{n,d} \otimes \overline{\mathbb{F}}_p$ to a sufficiently general line, we do not change its monodromy, so in particular we do not change its G_{geom}. But there exist Lefschetz pencils of degree d hypersurfaces in $\mathbb{P}^{n+1}$ for every $d \geq 2$ (by [**SGA 7**, XVII, 2.5.1, 4.1 and 4.2]), and any sufficiently general line in $\mathcal{H}_{n,d} \otimes \overline{\mathbb{F}}_p$ is a Lefschetz pencil, so it suffices to show that a sufficiently general Lefschetz pencil has the asserted G_{geom}. For n odd, that G_{geom} is Sp is proven in [**De-Weil II**, 4.4.1]. For n even, it is proven there that either G_{geom} is the full orthogonal group O, or that it is a finite irreducible subgroup of O. It is this last case we must rule out, and it is in order to rule it out that we have to exclude the case $n = 2$, $d = 3$ of cubic surfaces, where we do in fact get a finite group.

So we consider the situation n even $\geq 2, d \geq 3, n(d-2) \geq 4$. If a general Lefschetz pencil has finite G_{geom}, call it G, then the sheaf Prim_l^n on $\mathcal{H}_{n,d} \otimes \mathbb{F}_p$ has this same finite G as its G_{geom}. But G is irreducible, and it is normalized by all Frobenii $\mathrm{Frob}_{k,x}$ at all finite-field valued points x of $\mathcal{H}_{n,d} \otimes \mathbb{F}_p$. As G is finite, the automorphism $\mathrm{Ad}(\mathrm{Frob}_{k,x})$ of G is of finite order, so some power of $\mathrm{Frob}_{k,x}$ commutes with G, hence is scalar. But $(\mathrm{Card}(k))^{-n/2}\,\mathrm{Frob}_{k,x}$ is orthogonal, and a power of it is scalar. As the only orthogonal scalars are ± 1, we see that all eigenvalues of $\mathrm{Frob}_{k,x}$ are of the form $(\mathrm{Card}(k))^{n/2}$(a root of unity). This in turn forces the p-adic Newton polygon to be a straight line, for every smooth hypersurface of degree d in $\mathbb{P}^{n+2}$ in characteristic p. By a result of Deligne and Illusie [**Ill-Ord**, Thm. 0.1], we know that for a sufficiently general such hypersurface, its p-adic Newton polygon is equal to its Hodge polygon, and hence can be a straight line only if, over $\mathbb{C}$, a smooth hypersurface X of degree d in $\mathbb{P}^{n+1}$ has its Prim^n entirely of Hodge type $(n/2, n/2)$. There is a simple recipe for calculating the dimensions $\mathrm{prim}^{a,n-a}$ of the Hodge pieces $\mathrm{Prim}^{a,n-a}$ of Prim^n, namely $\mathrm{prim}^{a,n-a}$ is the number of monomials X^w in $n+2$ variables which satisfy

$$1 \leq w_i \leq d-1 \quad \text{for all } i,$$
$$\sum w_i = (a+1)d.$$

[In this description, $w_i \mapsto d - w_i$ implements Hodge symmetry.]

It follows from this description that the least a for which $\mathrm{prim}^{a,n-a} \neq 0$, the "Hodge co-level" of Prim^n, is given by the recipe [**SGA 7**, XI, 2.8], cf. also [**Ka-TA**, 2.7],

$$a + 1 = \text{ the least integer } \geq (n+2)/d.$$

Thus for Prim^n not to be entirely of type $(n/2, n/2)$, we need the Hodge co-level $\leq n/2 - 1$, i.e., we need $n/2 \geq (n+2)/d$, or equivalently $n(d-2) \geq 4$, and this is precisely what we have assumed. QED

11.5. Projective automorphisms of hypersurfaces

11.5.1. By a projective automorphism of a degree d hypersurface X/S in $\mathbb{P}^{n+1}$, we mean one induced by an automorphism of the ambient $\mathbb{P}^{n+1}$, i.e., by an S-valued point of $PGL(n+2)$ which stabilizes X/S. We denote by

$$\operatorname{Proj}\operatorname{Aut}(X/S) \subset PGL(n+2)(S)$$

the group of projective automorphisms of X/S. Over the base $\mathcal{H}_{n,d}$, we may form the group scheme $\underline{\operatorname{Proj}\operatorname{Aut}}_{\mathcal{X}_{n,d}/\mathcal{H}_{n,d}}$, which represents the functor on (Schemes/$\mathcal{H}_{n,d}$) which attaches to an S-valued point $\pi : S \to \mathcal{H}_{n,d}$ the group $\operatorname{Proj}\operatorname{Aut}(X_\pi/S)$ for X_π/S the pullback by π of the universal family. Clearly $\underline{\operatorname{Proj}\operatorname{Aut}}_{\mathcal{X}_{n,d}/\mathcal{H}_{n,d}}$ is a closed sub-group scheme of $PGL(n+2)_{\mathcal{H}_{n,d}}$. Our aim is to show that $\underline{\operatorname{Proj}\operatorname{Aut}}_{\mathcal{X}_{n,d}/\mathcal{H}_{n,d}}$ is finite and unramified over $\mathcal{H}_{n,d}$, for $n \geq 0$ and $d \geq 3$ with the single exception $(n,d) = (1,3)$, in which case we have the finiteness, but must invert 3 to have the unramifiedness. The unramifiedness will come from the following result.

Theorem 11.5.2 (Bott, Deligne, Kodaira-Spencer, Matsumura-Monsky). *Suppose $n \geq 0$ and $d \geq 3$, and that $(n,d) \neq (1,3)$. Then for any field k, and any smooth hypersurface X/k of degree d in $\mathbb{P}^{n+1}$, we have $H^0(X, T_{X/k}) = 0$.*

PROOF. For $n = 0$ there is nothing to prove, since $T_X = 0$. Suppose next that $n = 1$. Then X is a smooth plane curve of degree $d \geq 4$, of genus $g = (d-1)(d-2)/2 \geq 3$, and its tangent bundle $T_{X/k}$ is a line bundle of degree $2 - 2g < 0$, hence has vanishing H^0.

For $n \geq 2$, there are two different proofs, both of which are well-known to the experts, but for neither of which do we know an explicit reference.

11.6. First proof of 11.5.2

11.6.1. In characteristic zero, this was proven by Kodaira-Spencer [**K-S**, Lemma 14.2], by a beautiful inductive argument which, as Ofer Gabber remarked, is valid over any field, except that it requires as input a form of Kodaira vanishing on $\mathbb{P}^{n+1}$ due to Bott [**Bott**]. This same vanishing was later proven in arbitrary characteristic by Deligne [**De-CCI**, Thm. 1.1]. Let us recall briefly the Kodaira-Spencer argument, and the statement of the required input.

11.6.2. We work over a field k. Given a smooth S/k everywhere of relative dimension $\nu + 1 \geq 1$, and a divisor $D \subset S$ with D/k smooth, $\underline{\operatorname{Der}}_D(S/k)$ is the subsheaf of $T_{S/k} := \underline{\operatorname{Der}}(S/k)$ consisting of those derivations which map the ideal sheaf $I(D)$ of D to itself. In local coordinates $x_1, \ldots, x_{\nu+1}$ on S/k in which D is $x_1 = 0$, $\underline{\operatorname{Der}}_D(S/k)$ is $\mathcal{O}_S$-free on $x_1\partial/\partial x_1$ and on the $\partial/\partial x_j$, $2 \leq j \leq \nu+1$. The $\mathcal{O}_S$-dual of $\underline{\operatorname{Der}}_D(S/k)$ is denoted $\Omega^1_{S/k}(\log D)$. Dual to the inclusion of $\underline{\operatorname{Der}}_D$ in $T_{S/k}$ is the inclusion

$$\Omega^1_{S/k} \subset \Omega^1_{S/k}(\log D).$$

In local coordinates, $\Omega^1_{S/k}(\log D)$ is $\mathcal{O}_S$-free on $d\log(x_1) := dx_1/x_1$ and on the dx_j, $2 \leq j \leq \nu+1$. For $i \geq 0$, we define the locally free $\mathcal{O}_S$-modules

$$\Omega^i_{S/k}(\log D) := \Lambda^i(\Omega^1_{S/k}(\log D)).$$

In local coordinates, we have

$$\Omega^i_{S/k}(\log D) = \Omega^i_{S/k} + (dx_1/x_1)\Omega^{i-1}_{S/k},$$
$$I(D) \otimes_{\mathcal{O}_S} \Omega^i_{S/k}(\log D) = x_1\Omega^i_{S/k} + (dx_1)\Omega^{i-1}_{S/k}.$$

We have natural inclusions

$$I(D) \otimes_{\mathcal{O}_S} \Omega^i_{S/k}(\log D) \subset \Omega^i_{S/k} \subset \Omega^i_{S/k}(\log D).$$

For every i, we have both the Poincaré residue short exact sequence

residue$_i$ $$0 \to \Omega^i_{S/k} \to \Omega^i_{S/k}(\log D) \to \Omega^{i-1}_{D/k} \to 0,$$

and the restriction short exact sequence

restrict$_i$ $$0 \to I(D) \otimes_{\mathcal{O}_S} \Omega^i_{S/k}(\log D) \to \Omega^i_{S/k} \to \Omega^i_{D/k} \to 0,$$

(with the usual convention that for $i < 0$, any flavor of Ω^i is 0). As D/k is smooth, everywhere of dimension ν, we have

$$T_{D/k} \cong \Omega^{\nu-1}_{D/k} \otimes (K_{D/k})^{\otimes -1},$$

because both sides are $\underline{\mathrm{Hom}}_{\mathcal{O}_D}(\Omega^1_{D/k}, \mathcal{O}_D)$.

11.6.3. We apply this to $D =$ our smooth, degree d hypersurface X in $S = \mathbb{P} := \mathbb{P}^{n+1}$, in which case $I(D)$ is $\mathcal{O}_{\mathbb{P}}(-d)$, $K_{X/k}$ is $\mathcal{O}_X(d-n-2)$, and

$$T_{X/k} \cong \Omega^{n-1}_{X/k}(n+2-d).$$

So what we wish to prove is that

$$H^0(X, \Omega^{n-1}_{X/k}(n+2-d)) = 0.$$

This is the case $k = 0$ of the following statement $C(k)$,

$C(k)$: $$H^k(X, \Omega^{n-1-k}_{X/k}(n+2-(k+1)d)) = 0.$$

The Kodaira-Spencer idea is to prove $C(k)$ for all k by descending induction, and to prove simultaneously another statement $A(k)$,

$A(k)$: $$H^k(\mathbb{P}^{n+1}, \Omega^{n-k}_{\mathbb{P}/k}(\log X)(n+2-(k+1)d)) = 0,$$

by deducing them both from two other vanishing statements $B(k)$ and $D(k)$ on $\mathbb{P} := \mathbb{P}^{n+1}/k$,

$B(k)$: $$H^k(\mathbb{P}^{n+1}, \Omega^{n-1-k}_{\mathbb{P}/k}(n+2-(k+1)d)) = 0,$$

$D(k)$: $$H^k(\mathbb{P}^{n+1}, \Omega^{n-k}_{\mathbb{P}/k}(n+2-(k+1)d)) = 0.$$

11.6.4. Indeed, the restriction short exact sequence, twisted termwise by

$$\mathcal{O}_{\mathbb{P}}(n+2-(k+1)d),$$

namely

$$0 \to \Omega^{n-k-1}_{\mathbb{P}/k}(\log X)(n+2-(k+2)d) \to \Omega^{n-k-1}_{\mathbb{P}/k}(n+2-(k+1)d)$$
$$\to \Omega^{n-k-1}_{X/k}(n+2-(k+1)d) \to 0,$$

gives a long exact cohomology sequence which shows that

$$B(k) \text{ and } A(k+1) \text{ together imply } C(k).$$

The residue short exact sequence residue$_{n-k}$, twisted termwise by

$$\mathcal{O}_{\mathbb{P}}(n+2-(k+1)d),$$

namely

$$\begin{aligned} 0 &\to \Omega^{n-k}_{\mathbb{P}/k}(n+2-(k+1)d) \to \Omega^{n-k}_{\mathbb{P}/k}(\log X)(n+2-(k+1)d) \\ &\to \Omega^{n-k-1}_{X/k}(n+2-(k+1)d) \to 0, \end{aligned}$$

gives a long exact cohomology sequence which shows that

$$C(k) \text{ and } D(k) \text{ together imply } A(k).$$

So **if** $B(k)$ and $D(k)$ hold for all k, then we are left with the implications

$$A(k+1) \Rightarrow C(k) \Rightarrow A(k).$$

Since $A(k)$ holds trivially for large k, e.g. for $k > n$, we conclude that both $A(k)$ and $C(k)$ hold for all k.

11.6.5. That both $B(k)$ and $D(k)$ hold for all k, provided that $n \geq 2$ and $d \geq 3$ (or that $n = 1$ and $d \geq 4$), is easily checked to be a special case of the vanishing theorem of Bott [**Bott**] over $\mathbb{C}$ and Deligne [**De-CCI**, Thm. 1.1] over any field: given three integers (a, b, c), the group $H^a(\mathbb{P}^{n+1}, \Omega^b_{\mathbb{P}/k}(c))$ vanishes except in the following cases:

$$\begin{aligned} &a = 0, \text{ and either } c > b \geq 0 \text{ or } c = b = 0, \\ &1 \leq a \leq n, b = a \text{ and } c = 0, \\ &a = n+1, b \geq 0 \text{ and either } b - c > n+1 \text{ or both } b = n+1 \text{ and } c = 0. \end{aligned}$$

This concludes the first proof.

11.7. Second proof of 11.5.2

11.7.1. We continue to suppose that $n \geq 2$, and that $F(X_1, \ldots, X_{n+2})$ is the form of degree d which defines X. For each $i = 1, \ldots, n-2$ we write

$$F_i := \partial F / \partial X_i.$$

On $\mathbb{P} := \mathbb{P}^{n+1}$ over k, we have the basic short exact sequence

$$0 \to \Omega^1_{\mathbb{P}} \to \mathcal{O}_{\mathbb{P}}(-1)^{n+2} \to \mathcal{O}_{\mathbb{P}} \to 0,$$

with the dual short exact sequence

$$\text{(i)} \qquad 0 \to \mathcal{O}_{\mathbb{P}} \to \mathcal{O}_{\mathbb{P}}(1)^{n+2} \to T_{\mathbb{P}} \to 0.$$

Concretely, this sequence says that global vector fields on $\mathbb{P}$ are of the form

$$\sum \mathrm{Lin}_i \, \partial / \partial X_i,$$

with Lin_i a linear form in the X_j's, with the single relation that $\sum X_i \partial / \partial X_i = 0$ as vector field on $\mathbb{P}$.

11.7.2. This sequence, being locally split, restricts to any closed subscheme X to give an exact sequence

$$0 \to \mathcal{O}_X \to \mathcal{O}_X(1)^{n+2} \to T_{\mathbb{P}}|X \to 0.$$

For smooth X, denoting by $I(X)$ the ideal sheaf of X, we have the normal bundle short exact sequence on X

$$0 \to T_X \to T_{\mathbb{P}}|X \to \underline{\mathrm{Hom}}_{\mathcal{O}_X}(I(X)/I(X)^2, \mathcal{O}_X) \to 0.$$

For X our smooth hypersurface of degree d, the map "multiplication by F" gives $\mathcal{O}_{\mathbb{P}}(-d) \cong I(X)$, i.e., we have a short exact sequence

(ii) $$0 \to \mathcal{O}_{\mathbb{P}}(-d) \to \mathcal{O}_{\mathbb{P}} \to \mathcal{O}_X \to 0.$$

The normal bundle short exact sequence on X becomes

$$0 \to T_X \to T_{\mathbb{P}}|X \to \mathcal{O}_X(d) \to 0.$$

On global sections, we have an exact sequence

(iii) $$0 \to H^0(X, T_X) \to H^0(X, T_{\mathbb{P}}|X) \to H^0(X, \mathcal{O}_X(d)).$$

To compute $H^0(X, T_{\mathbb{P}}|X)$, we use the short exact sequences on $\mathbb{P}$

(iv) $$0 \to T_{\mathbb{P}}(-d) \to T_{\mathbb{P}} \to T_{\mathbb{P}}|X \to 0$$

and the $(-d)$ twist of (i)

(v) $$0 \to \mathcal{O}_{\mathbb{P}}(-d) \to \mathcal{O}_{\mathbb{P}}(1-d)^{n+2} \to T_{\mathbb{P}}(-d) \to 0.$$

11.7.3. Since $\mathbb{P}$ is $\mathbb{P}^{n+1}$ with $n \geq 2$, $H^i(\mathbb{P}, \mathcal{O}_{\mathbb{P}}(j))$ vanishes for $i = 1$ and $i = 2$, for every integer j, and it vanishes for $i = 0$ if $j < 0$. So we infer from (v) that $H^i(\mathbb{P}, T_{\mathbb{P}}(-d))$ vanishes for $i = 0$ or 1, and then from (iv) that

$$H^0(\mathbb{P}, T_{\mathbb{P}}) \cong H^0(X, T_{\mathbb{P}}|X).$$

From (ii)$(-d)$, we get

$$H^0(\mathbb{P}, \mathcal{O}_{\mathbb{P}}(d))/kF \cong H^0(X, \mathcal{O}_X(d)).$$

So all in all, (iii) becomes

$$0 \to H^0(X, T_X) \to H^0(\mathbb{P}, T_{\mathbb{P}}) \to H^0(\mathbb{P}, \mathcal{O}_{\mathbb{P}}(d))/kF,$$

the map given by

$$\sum_i \mathrm{Lin}_i\, \partial/\partial X_i \mapsto \sum_i \mathrm{Lin}_i\, F_i \bmod kF.$$

11.7.4. So for $n \geq 2$ and $d \geq 3$, the vanishing of $H^0(X, T_X)$ amounts exactly to the statement that the only solutions $(\{\mathrm{Lin}_i\}_i, \lambda)$ in $H^0(\mathbb{P}, \mathcal{O}_{\mathbb{P}}(1)^{n+2}) \oplus k$ of the equation

$$\sum_i \mathrm{Lin}_i\, F_i = \lambda F$$

are k-proportional to the Euler identity solution $\sum_i X_i F_i = dF$. We will see that this statement is in fact valid for any $n \geq 0, d \geq 3$, with the single exception of $n = 1, d = 3$, where every smooth plane cubic in characteristic 3 is a counterexample.

11.7.5. It is this last statement which is proven, for $n \geq 2$ and $d \geq 3$, by Matsumura and Monsky in [**Mat-Mon**, pages 348–352], although they do not relate it there to the vanishing of $H^0(X, T_X)$, and indeed in a remark they say that they do not know if this vanishing holds in positive characteristic! For the convenience of the reader, we will recall their argument.

Lemma 11.7.6 ([**Mat-Mon**, pages 348–352]). *Let $n \geq 0, d \geq 3$. Over any field k, let F be a form of degree d in $n+2$ variables X_i, which defines a nonsingular hypersurface in $\mathbb{P}^{n+1}$. If $n = 1$ and $d = 3$, suppose also that* $\mathrm{char}(k) \neq 3$. *Then the only solutions $(\{\mathrm{Lin}_i\}_i, \lambda)$ in $H^0(\mathbb{P}^{n+1}, \mathcal{O}_{\mathbb{P}}(1)^{n+2}) \oplus k$ of the equation*

$$\sum_i \mathrm{Lin}_i\, F_i = \lambda F$$

are k-proportional to the Euler identity solution $\sum_i X_i F_i = dF$. If $n = 1$ and $d = 3$ and $\mathrm{char}(k) = 3$, *then for every smooth place cubic F, there are solutions which are not proportional to the Euler solution.*

PROOF. The general fact which we need is Macaulay's unmixedness theorem, that if $G_1, \ldots, G_r$ are $1 \leq r \leq n+2$ homogeneous forms of degrees $d_i \geq 1$ in $n+2$ variables over a field k whose zero locus in $\mathbb{P}^{n+1}$ has dimension $\leq n+1-r$, then $(G_1, \ldots, G_r)$ is a regular sequence in the polynomial ring $k[X_1, \ldots, X_{n+2}]$: G_1 is nonzero, and for $2 \leq i \leq r$, G_i is not a zero divisor in $k[X\text{'s}]/(G_1, \ldots, G_{i-1})$. [For a local algebra proof, apply [**A-K**, III, 4.3 and 4.12] successively to infer that the G_i form a regular sequence in the power series ring $k[[X\text{'s}]]$, and then infer the polynomial result by looking degree by degree.] The corollary of Macaulay's unmixedness theorem that we need is the following. If in addition all the G_i are homogeneous of the same degree $d - 1 \geq 2$, then for any $0 \leq k \leq d-2$, there are no nontrivial relations $\sum H_i G_i = 0$ with H_i forms of degree k (since in any such relation, each H_i must lie in the ideal generated by the G_j with $j \neq i$, and every nonzero homogeneous element of this ideal has degree $\geq d-1$).

Suppose first that the degree d is invertible in k. Then correcting any solution of $\sum_i \mathrm{Lin}_i\, F_i = \lambda F$ by (λ/d)(the Euler solution), we must prove that

$$\sum_i \mathrm{Lin}_i\, F_i = 0$$

has no nonzero solutions in linear forms Lin_i. But in this case, the Euler identity shows that F is in the ideal of the F_i, so the smoothness of F shows that $F_1, \ldots, F_{n+2}$ have no common zero in $\mathbb{P}^{n+1}$. As the F_i have degree $d-1 \geq 2$, there are no nontrivial linear relations $\sum_i \mathrm{Lin}_i\, F_i = 0$, by Macaulay.

Suppose now that $d = 0$ in k. The Euler relation becomes

$$\sum_i X_i F_i = 0.$$

The existence of this relation shows that the F_i must in fact have some common zeroes in $\mathbb{P}^{n+1}$. Therefore in any relation

$$\sum_i \mathrm{Lin}_i\, F_i = \lambda F,$$

we must have $\lambda = 0$, for otherwise F would vanish at any common zero of the F_i, contradicting its nonsingularity. We must show that any solution $\{\mathrm{Lin}_i\}_i$ of $\sum_i \mathrm{Lin}_i\, F_i = 0$ is k-proportional to $\{X_i\}_i$. For this k-linear question, we may extend scalars from k to any overfield, and so we may assume that over k there are smooth hyperplane sections of our hypersurface. By a linear change of coordinates, we may assume that $X_{n+2} = 0$ is one such. Then $(F_1, \ldots, F_{n+1}, F)$ have no common zeroes in $\mathbb{P}^{n+1}$. For by assumption there are none with $X_{n+2} = 0$, but at any with X_{n+2} nonzero the Euler relation shows that F_{n+2} also vanishes, contradicting the nonsingularity of F. Therefore $(F_1, \ldots, F_{n+1}, F)$, and hence also $(F_1, \ldots, F_{n+1})$, form a regular sequence in $k[X_1, \ldots, X_{n+2}]$.

Now suppose we have a nontrivial relation $\sum_i \mathrm{Lin}_i F_i = 0$. We first remark that we must have Lin_{n+2} nonzero by Macaulay, because $(F_1, \ldots, F_{n+1})$ form a regular sequence. If Lin_{n+2} is proportional to X_{n+2}, then subtracting a multiple of the Euler identity, we get a new relation in which $\mathrm{Lin}_{n+2} = 0$, so by the first remark the new relation is trivial, which means our original relation was proportional to the Euler relation.

Suppose Lin_{n+2} is not proportional to X_{n+2}. Look at the relation with degree 2 coefficients among $F_1, \ldots, F_{n+2}$ given by

$$0 = \mathrm{Lin}_{n+2} \sum X_i F_i - X_{n+2} \sum \mathrm{Lin}_i F_i = \sum_{i \leq n+1} (\mathrm{Lin}_{n+2} X_i - X_{n+2} \mathrm{Lin}_i) F_i.$$

If $d \geq 4$, then by Macaulay there are no nontrivial relations among $(F_1, \ldots, F_{n+1})$ with degree two coefficients, which means precisely that we have

$$\mathrm{Lin}_{n+2} X_i = X_{n+2} \mathrm{Lin}_i \quad \text{in } k[X\text{'s}]$$

for $i = 1, \ldots, n+1$. Taking $i = 1$, we see that X_{n+2} divides $\mathrm{Lin}_{n+2} X_1$, and hence that Lin_{n+2} is proportional to X_{n+2}, contradiction.

If $d = 3$, we argue as follows. By Macaulay the only degree two relations $\sum_{i \leq n+1} H_i F_i = 0$ among $(F_1, \ldots, F_{n+1})$ have each H_i in the ideal generated by the F_j with $j \neq i$, $j \leq n+1$. By degree, each H_i is a k-linear combination of these F_j. Thus we have

$$H_i := \mathrm{Lin}_{n+2} X_i - X_{n+2} \mathrm{Lin}_i = \sum_{j \neq i, j \leq n+1} a_{i,j} F_j, \quad \text{coef's } a_{i,j} \text{ in } k,$$

for $i = 1, \ldots, n+1$. We next show that the H_i are linearly independent over k. To prove this, observe that as Lin_{n+2} is not proportional to X_{n+2}, Lin_{n+2} contains some other variable X_l, for some $l \leq n+1$. Then $\mathrm{Lin}_{n+2} X_i$ contains the monomial $X_l X_i$, but $X_{n+2} \mathrm{Lin}_i$ cannot contain it, because $X_{n+2} \mathrm{Lin}_i$ contains only monomials divisible by X_{n+2}. For $j \neq i$, $j \leq n+1$, neither $\mathrm{Lin}_{n+2} X_j$ nor $X_{n+2} \mathrm{Lin}_j$ can contain $X_l X_i$. Thus among the H_j, H_i is the unique one which contains $X_l X_i$, and hence the H_i, $1 \leq i \leq n+1$, are linearly independent over k. Since the H_i are in the k-span of the F_i, they are a basis of that k-span. Hence each F_i, $1 \leq i \leq n+1$, is in the k-span of, and hence in the ideal generated by, the H_i. But each H_i lies in the ideal $(X_{n+2}, \mathrm{Lin}_{n+2})$, so we find that all the F_i, $1 \leq i \leq n+1$, vanish on the codimension 2 linear space $\mathrm{Lin}_{n+2} = X_{n+2} = 0$ in $\mathbb{P}^{n+1}$. If $n \geq 2$, this is a variety of strictly positive dimension. But $(F_1, \ldots, F_{n+1}, F)$ have no common zeroes in $\mathbb{P}^{n+1}$, and hence $(F_1, \ldots, F_{n+1})$ can have only a finite set of common zeroes in $\mathbb{P}^{n+1}$, contradiction.

If $d = 3$ and $n = 1$ (and $d = 0$ in k, the case we are in the middle of treating), we argue as follows. We may replace k by its algebraic closure. Thus k is algebraically closed of characteristic 3. Then every nonsingular plane cubic over k is projectively isomorphic either to a member of the family

$$\mu(X^3 - Y^3 + Z^3) + XYZ = 0, \quad \text{any } \mu \text{ in } k^\times,$$

or to the single curve

$$Y^2Z - X^3 + XZ^2 = 0,$$

the two cases according to whether the elliptic curve we get after choosing an origin is ordinary or supersingular.

In the case of the μ family, $F_1 = YZ, F_2 = XZ, F_3 = XY$, and so we have the non-Euler relation

$$YF_2 - ZF_3 = 0.$$

In the case of the supersingular curve $Y^2Z - X^3 + XZ^2 = 0$ (which the attentive reader will notice is essentially the $n = 1, d = 3$ case of Gabber's example 11.4.6 of a nonsingular hypersurface of degree d in $\mathbb{P}^{n+2}$ over an $\mathbb{F}_p$ with $p|d$), we have $F_1 = Z^2, F_2 = 2YZ, F_3 = Y^2 + 2XZ$. But here we have the relation

$$YF_1 + ZF_2 = 0.$$

In other words, for any smooth plane cubic in characteristic three, there is a nonzero global vector field on $\mathbb{P}^2$ ($Y\partial/\partial Y - Z\partial/\partial Z$ and $Y\partial/\partial X + Z\partial/\partial Y$ respectively in our examples), i.e., a nontrivial automorphism of $\mathbb{P}^2$ over $k[\varepsilon]/\varepsilon^2$ which is the identity $\bmod \varepsilon$, and which induces a nontrivial automorphism of our curve over $k[\varepsilon]/\varepsilon^2$ which is the identity $\bmod \varepsilon$.

Let us explore this characteristic 3 phenomenon a bit further. In the ordinary case, this pathology arises from the action of the group scheme $\boldsymbol{\mu}_3$ on $\mathbb{P}^2$ in which ζ maps $(X, Y, Z) \mapsto (X, \zeta Y, \zeta^2 Z)$. This action preserves the equation

$$\mu(X^3 - Y^3 + Z^3) + XYZ = 0.$$

For μ nonzero in k, if we take the point $(1, 1, 0)$ as origin, we get an elliptic curve E_μ. In this curve, the points $(1, \zeta, 0)$ with ζ in $\boldsymbol{\mu}_3$ form a sub-group scheme. The above action of $\boldsymbol{\mu}_3$ on $\mathbb{P}^2$ induces the translation action of $\boldsymbol{\mu}_3$ on E_μ. [To see the truth of the last two statements simultaneously, pick any k-algebra A, and any element ζ in $\boldsymbol{\mu}_3(A)$. The action of ζ on $\mathbb{P}^2$ carries $(1, 1, 0)$ to the point $P(\zeta) := (1, \zeta, 0)$ and induces an automorphism, call it $t(\zeta)$, of E_μ, which reduces to the identity over A^{red} (because ζ becomes 1 in A^{red}). If we follow $t(\zeta)$ with translation by $-P(\zeta)$, we get an automorphism of E_μ **as elliptic curve** (because it fixes the origin) which is the identity over A^{red}, so by ridigity is the identity over A. Thus $t(\zeta)$ on E_μ is translation by $P(\zeta)$. Therefore $t(\zeta_1\zeta_2) = t(\zeta_1) \circ t(\zeta_2)$ is both translation by $P(\zeta_1\zeta_2)$ and translation by $P(\zeta_1) + P(\zeta_2)$, and hence $P(\zeta_1\zeta_2) = P(\zeta_1) + P(\zeta_2)$.] Taking A to be $k[\varepsilon]/\varepsilon^2$, and $\zeta = 1 + \varepsilon$, we see that the vector field $Y\partial/\partial Y - Z\partial/\partial Z$ acts on E_μ as translation by the point $(1, 1 + \varepsilon, 0)$.

In the supersingular case, the curve $Y^2Z - X^3 + XZ^2 = 0$ over $\mathbb{F}_3$, we take as origin the point $(0, 1, 0)$, and call this elliptic curve E. We have an action of $\boldsymbol{\alpha}_3 := \operatorname{Spec}(\mathbb{F}_3[\Lambda]/\Lambda^3)$, with usual addition as the group law, on $\mathbb{P}^2$, λ in $\boldsymbol{\alpha}_3(A)$ acting as

$$(X, Y, Z) \mapsto (X + \lambda Y + (\lambda^2/2)Z, Y + \lambda Z, Z).$$

This action fixes $Y^2Z - X^3 + XZ^2$. We also have a subgroup scheme $\boldsymbol{\alpha}_3$ in E, namely the points $(\lambda, 1, 0)$ with λ in $\boldsymbol{\alpha}_3$. Just as in the ordinary case, we see that the above action of $\boldsymbol{\alpha}_3$ on $\mathbb{P}^2$ induces the translation action of $\boldsymbol{\alpha}_3$ on E. Taking A

to be $k[\varepsilon]/\varepsilon^2$, and $\lambda = \varepsilon$, we see that the vector field $Y\partial/\partial X + Z\partial/\partial Y$ acts on E as translation by the point $(\varepsilon, 1, 0)$.

There is a unified way of presenting these two cases à la Mumford [**Mum-AV**, §13], which Ofer Gabber pointed out to us. Over any field k, start with an elliptic curve E/k, and a line bundle $\mathcal{L}$ on E which has some strictly positive degree $d \geq 3$. Global sections of such an $\mathcal{L}$ embed E in $\mathrm{Proj}(H^0(E, \mathcal{L})) = \mathbb{P}^{d-1}$. On the other hand, for any line bundle $\mathcal{L}$ we have the homomorphism $\varphi_{\mathcal{L}}$ from E to $\mathrm{Pic}^0_{E/k}$ defined on S-valued points, S any k-scheme, by

$$P \mapsto \text{ the class of } \mathcal{L}^{-1} \otimes (\text{trans. by } P)^* \mathcal{L} \quad \text{in } \mathrm{Pic}^0(E \times_k S)/\mathrm{Pic}(S).$$

For $\mathcal{L} = I(0)$, the ideal sheaf of the origin, this map is the usual identification of E with $\mathrm{Pic}^0_{E/k}$. For $\mathcal{L}$ of degree 0, $\varphi_{\mathcal{L}}$ is zero. Since $\varphi_{\mathcal{L}_1 \otimes \mathcal{L}_2} = \varphi_{\mathcal{L}_1} + \varphi_{\mathcal{L}_2}$, we see that for $\mathcal{L}$ of any nonzero degree d, $\varphi_{\mathcal{L}}$ is the endomorphism $[-d]$ of E. So for $\mathcal{L}$ of degree $d \geq 3$ on E, any k-scheme S, and any P in $E(S)[d]$, we have $\varphi_{\mathcal{L}}(P) = 0$, i.e., translation by P on $E \times_k S$ preserves the class of $\mathcal{L}$ in $\mathrm{Pic}(E \times_k S)/\mathrm{Pic}(S)$. Take $S = \mathrm{Spec}(A)$ affine, with $\mathrm{Pic}(A)$ trivial to fix ideas. Once we choose an isomorphism φ_P from $(\text{trans. by } P)^*\mathcal{L}$ to $\mathcal{L}$, we get an A-linear automorphism $\varphi_P \circ (\text{trans. by } P)^*$ of $H^0(E \otimes_k A, \mathcal{L} \otimes_k A) \cong H^0(E, \mathcal{L}) \otimes_k A$. Since the choice of φ_P is indeterminate up to an $A^\times$-factor, when we vary P in $E(A)[d]$, we get a projective representation of the group $E(A)[d]$ on $H^0(E, \mathcal{L}) \otimes_k A$. The induced action of $E(A)[d]$ on $\mathrm{Proj}_A(H^0(E, \mathcal{L}) \otimes_k A = \mathbb{P}^{d-1}$ over A then maps E to itself (and induces on E the action of $E(A)[d]$ by translation).

When we take d divisible by a prime p, and take our field to be of characteristic p, then $\mathrm{Ker}(F) \subset E[p] \subset E[d]$, so we get a nontrivial action of $\mathrm{Ker}(F)$ on $\mathbb{P}^{d-1}$ which maps E to itself. But $\mathrm{Lie}(\mathrm{Ker}(F)) \cong \mathrm{Lie}(E)$, so every global vector field on E is induced by the tangent action of $\mathrm{Ker}(F)$ on $\mathbb{P}^{d-1}$. The case we encountered above, namely smooth plane cubics in characteristic 3, is precisely the case $d = 3$ of this general discussion.

After this long digression, we return to the last remaining case, namely $n = 0$, $d = 3, \mathrm{char}(k) = 3$, of the lemma we are in the course of proving. We may pass to $\overline{k}$. There our cubic has three distinct zeroes in $\mathbb{P}^1$, which we may assume by a projective transformation to be $0, 1, \infty$, i.e., F is projectively equivalent to $XY(X - Y)$. So $F_1 = 2XY - Y^2 = -Y(X+Y)$, and $F_2 = X^2 - 2XY = X(X+Y)$, and any relation $\mathrm{Lin}_1 F_1 + \mathrm{Lin}_2 F_2 = 0$ gives $\mathrm{Lin}_1(-Y) + \mathrm{Lin}_2 X = 0$, so $(\mathrm{Lin}_1, \mathrm{Lin}_2)$ is proportional to (X, Y), as required.

This concludes the proof of Lemma 11.7.6, and with it the second proof of Theorem 11.5.2. QED

11.8. A properness result

11.8.1. The following result is proven in [**Mum-GIT**, Ch. 0, §4, Prop. 0.8 and Ch. 4, §2, Prop. 4.2], as a consequence of the general theory. We thank Ofer Gabber for explaining to us what the general theory comes down to in this concrete case.

Proposition 11.8.2 ([**Mum-GIT**]). *Suppose that $n \geq 0$ and $d \geq 3$. Then the action of $PGL(n+2)$ on $\mathcal{H}_{n,d}$ is proper in the sense of* [**Mum-GIT**], *i.e., the*

morphism

$$PGL(n+2)\times_{\mathbb{Z}}\mathcal{H}_{n,d}\to\mathcal{H}_{n,d}\times_{\mathbb{Z}}\mathcal{H}_{n,d},$$
$$(g,h)\mapsto(h,g(h)),$$

is a proper morphism.

PROOF. We use the valuative criterion. Thus R is a discrete valuation ring, with residue field k and fraction field K, and π is a uniformizing parameter for R. We must show that if F and G are equations of smooth hypersurfaces over R, of degree d and dimension n, and if an element $\overline{g}$ in $PGL(n+2,K)$ transforms F into a $K^{\times}$-multiple of G, then $\overline{g}$ lies in $PGL(n+2,R)$. Equivalently, we must show that any lift g in $GL(n+2,K)$ of $\overline{g}$ lies in $K^{\times}GL(n+2,R)$.

By the theory of elementary divisors, every g in $GL(n+2,K)$ can be written as a product $\alpha t\beta$, with α and β in $GL(n+2,R)$, and with t a diagonal matrix $\operatorname{diag}(\pi^{e(1)},\dots,\pi^{e(n+2)})$. Such an element lies in $K^{\times}GL(n+2,R)$ if and only if $e(1)=e(2)=\cdots=e(n+2)$.

Since $g=\alpha t\beta$ carries F to a $K^{\times}$ multiple of G, t carries $\beta(F)$ to a $K^{\times}$ multiple of $\alpha^{-1}(G)$. Both $\beta(F)$ and $\alpha^{-1}(G)$ are smooth over R, so, renaming them F and G, we are reduced to showing that if $t:=\operatorname{diag}(\pi^{e(1)},\dots,\pi^{e(n+2)})$ carries F to a $K^{\times}$-multiple of G, then all $e(i)$'s are equal. For this, we may pass from R to any larger discrete valuation ring $R[\pi^{1/N}]$, any integer $N\geq 1$. Taking N to be $n+2$, and scaling t by $\pi^{-\sum e(i)/N}$, we may further assume that $\sum e(i)=0$, i.e., that $t:=\operatorname{diag}(\pi^{e(1)},\dots,\pi^{e(n+2)})$ lies in $SL(n+2,K)$. We must then prove that all $e(i)$ vanish.

We first show that t transforms F into an $R^{\times}$-multiple of G. The key point for doing this is that the discriminant Δ is invariant under $SL(n+2)$. [Proof: Over $\mathbb{C}$, Δ is an irreducible equation for the $SL(n+2,\mathbb{C})$-stable set of singular hypersurfaces, so Δ must transform under $SL(n+2,\mathbb{C})$ by a character. But $SL(n+2,\mathbb{C})$ is its own commutator subgroup (in fact, for any connected semisimple G, every element in $G(\mathbb{C})$ is a commutator, cf. [**Shoda**] for $G=SL(n)$, and [**P-W**] and [**Ree**] for the general case), so Δ is invariant under $SL(n+2,\mathbb{C})$. Therefore Δ is invariant under $SL(n+2,A)$ for any subring A of $\mathbb{C}$, e.g., for any integral domain A of generic characteristic zero which is finitely generated as a $\mathbb{Z}$-algebra, and so for A the coordinate ring of $SL(n+2)$ over $\mathbb{Z}$.]

Let us write $t(F)=\lambda G$, with λ in $K^{\times}$. To show that λ lies in $R^{\times}$, we note that Δ is homogeneous of some strictly positive degree M, so we have

$$\Delta(t(F))=\Delta(\lambda G)=\lambda^{M}\Delta(G),$$

but we also have $\Delta(t(F))=\Delta(F)$ because t lies in SL. Since $\Delta(F)$ and $\Delta(G)$ lie in $R^{\times}$, $\lambda^{M}=\Delta(F)/\Delta(G)$ lies in $R^{\times}$. As $M\neq 0$, taking ord's shows that λ is in $R^{\times}$.

So replacing G by λG, we may further assume that $t(F)=G$. Now pick any monomial X^{w} which occurs in F with unit coefficient A_{w} in $R^{\times}$. Then $A_{w}\pi^{\sum e(i)w(i)}$ is the coefficient of X^{w} in $t(F)=G$. Since G has coefficients in R, we infer that $\sum e(i)w(i)\geq 0$ for every monomial X^{w} which occurs in F with unit coefficient. Now consider the reduction $\bmod\,\pi$ of F, call it $\overline{F}$. This is the equation of a smooth hypersurface over the field k, and we have shown that for it, we have $\sum e(i)w(i)\geq 0$ for every monomial which occurs in it. Now introduce an indeterminate z and pass to the Laurent polynomial ring $k[z, 1/z]$. Consider the action of the element

$$T := \operatorname{diag}(z^{e(1)}, \dots, z^{e(n+2)})$$

on $\overline{F}$. It transforms $\overline{F}$ into

$$T(\overline{F}) = \sum \overline{A}_w z^{\sum e(i)w(i)} X^w,$$

which has coefficients in $k[z]$ precisely because $\sum e(i)w(i) \geq 0$ for every monomial X^w in $\overline{F}$. But T lies in $SL(n+2, k[z, 1/z])$, so $\Delta(T(\overline{F})) = \Delta(\overline{F})$ lies in $k^\times \subset k[z]^\times$. Therefore if we consider $T(\overline{F}) \bmod z$, we get a smooth hypersurface over k, whose equation, call it H, involves only those monomials X^w with $\sum e(i)w(i) = 0$. Thus we have a smooth hypersurface equation H over k which is fixed by the action of $\mathbb{G}_{m,k} = \operatorname{Spec}(k[z, 1/z])$ on the ambient $\mathbb{P}^{n+1}$ defined by z in $\mathbb{G}_m$ mapping X_i to $z^{e(i)}X_i$. We wish to infer that all the $e(i) = 0$. A direct and elementary proof that all the $e(i)$ are equal, and hence zero, is in [**Mat-Mon**, page 350]. For the convenience of the reader, we give below a (variant) proof.

We consider the tangent action. Thus we pass to the ring of dual numbers $k[\varepsilon]/\varepsilon^2$, and consider the action of $1 + \varepsilon$ in $\mathbb{G}_m(k[\varepsilon]/\varepsilon^2)$. It maps X_i to

$$(1 + \varepsilon)^{e(i)} X_i = X_i + \varepsilon e(i) X_i,$$

and fixes H:

$$H((X_i)_i) = H((X_i + \varepsilon e(i) X_i)_i).$$

Expanding and equating coefficients of ε, and putting $H_i := \partial H/\partial X_i$, we get

$$\sum e(i) X_i H_i = 0.$$

If k has characteristic zero, the nonsingularity of H together with the Euler relation shows that the H_i have no common zero, so by Macaulay $\sum_i \operatorname{Lin}_i H_i = 0$ has no nontrivial solutions for H of degree $d \geq 3$, and hence all $e(i) = 0$ in k, and hence in $\mathbb{Z}$, as required.

If k has characteristic $p > 0$, lift H to an equation $\widetilde{H}$ over the Witt vectors $W := W(k^{\text{perf}})$ by lifting each coefficient in H arbitrarily, subject only to the rule that if a coefficient is zero in k, we lift it to zero in W. Then $\widetilde{H}$ is smooth over W (since $\Delta(\widetilde{H})$ lies in W, and reduces $\bmod p$ to $\Delta(H)$, which lies in $k^\times$), and so $\widetilde{H}$ is smooth over the fraction field K of W, which is a field of characteristic zero. As $\widetilde{H}$ contains exactly the same monomials X^w that H did, the above discussion shows that $\sum e(i) X_i \widetilde{H}_i = 0$, and thus all $e(i) = 0$. QED

Corollary 11.8.3. *Suppose $n \geq 0$ and $d \geq 3$. Then the "proper action morphism" of the previous proposition*

$$PGL(n+2) \times_{\mathbb{Z}} \mathcal{H}_{n,d} \to \mathcal{H}_{n,d} \times_{\mathbb{Z}} \mathcal{H}_{n,d},$$
$$(g, h) \mapsto (h, g(h)),$$

is a finite morphism.

PROOF. The scheme $\mathcal{H}_{n,d}$ over $\mathbb{Z}$ is affine, being the open set, in the projective space of coefficients, where the discriminant Δ is invertible. So both source and target are affine, so the map itself is affine. As the map is also proper, it is finite. QED

Corollary 11.8.4. *Suppose $n \geq 0$ and $d \geq 3$. Then the group scheme*

$$\underline{\mathrm{Proj\,Aut}}_{\mathcal{X}_{n,d}/\mathcal{H}_{n/d}}$$

is finite over $\mathcal{H}_{n,d}$. If $(n,d) \neq (1,3)$, this group scheme is unramified over $\mathcal{H}_{n,d}$. If $n = 1$ and $d = 3$, it is unramified over the open set $\mathcal{H}_{1,3}[1/3]$, and it is ramified at every point in characteristic 3.

PROOF. The group scheme $\underline{\mathrm{Proj\,Aut}}_{\mathcal{X}_{n,d}/\mathcal{H}_{n,d}}$ is tautologically the pullback to the diagonal of the "proper action map"

$$PGL(n+2) \times_{\mathbb{Z}} \mathcal{H}_{n,d} \to \mathcal{H}_{n,d} \times_{\mathbb{Z}} \mathcal{H}_{n,d}, \qquad (g,h) \mapsto (h, g(h)),$$

which we have just seen is finite. Therefore $\underline{\mathrm{Proj\,Aut}}_{\mathcal{X}_{n,d}/\mathcal{H}_{n,d}}$ is finite. Its unramifiedness over a field valued point of $\mathcal{H}_{n,d}$ which "is" a smooth degree d hypersurface X/k in $\mathbb{P}^{n+1}$, say of equation $F = 0$, is the statement that any automorphism of $\mathbb{P}^{n+1}$ over $k[\varepsilon]/\varepsilon^2$ which is the identity $\mathrm{mod}\,\varepsilon$ and which maps $F = 0$ to itself must be the identity, or equivalently that no nonzero global vector field D on $\mathbb{P}^{n+1}$ maps F to a multiple of itself. We have seen above in 11.7.6 that this holds if $(n,d) \neq (1,3)$. For $n = 1$ and $d = 3$, we have seen that it holds outside characteristic 3, but that it becomes false at every elliptic curve in characteristic 3. QED

Exactly as in 10.6.13, we have

Lemma 11.8.5. *Fix $n \geq 2$ and $d \geq 3$, or $n = 1$ and $d \geq 4$. Let S be a noetherian scheme, and X/S in $\mathbb{P}^{n+1}/S$ a smooth hypersurface of degree d. For every integer $i \geq 1$, there is an open set $U_{\leq i}$ of S which is characterized by the following property: a point s in S lies in $U_{\leq i}$ if and only if the finite etale $\kappa(s)$-group scheme $\underline{\mathrm{Proj\,Aut}}_{\kappa(s)}(X_s/\kappa(s))$ has rank $\leq i$.*

Theorem 11.8.6. [**Mat-Mon**, Thm. 5, page 355] *Fix $n \geq 2$ and $d \geq 3$, or $n = 1$ and $d \geq 4$. The open set $U_{\leq 1}$ of $\mathcal{H}_{n,d}$ meets every geometric fibre of $\mathcal{H}_{n,d}/\mathbb{Z}$.*

PROOF. If $n = 1$ and $d \geq 4$, this has already been noted in 10.6.17. If $n \geq 2$ and $d \geq 3$, Matsumura-Monsky proved that for every prime p, the geometric generic point of $\mathcal{H}_{n,d} \otimes \mathbb{F}_p$ lies in $U_{\leq 1}$. Since $U_{\leq 1}$ is smooth, and hence flat, over $\mathrm{Spec}(\mathbb{Z})$, its image in $\mathrm{Spec}(\mathbb{Z})$ is open, and contains all primes, so $U_{\leq 1}$ maps onto $\mathrm{Spec}(\mathbb{Z})$, as required. QED

Corollary 11.8.7. *Given $n \geq 2$ and $d \geq 3$, or $n = 1$ and $d \geq 4$, the open set $U_{\leq 1}$ of $\mathcal{H}_{n,d}$ is smooth over $\mathbb{Z}$ with geometrically connected fibres of dimension* $\mathrm{Binom}(n+1+d, d) - 1$. *There exists a constant $G(n,d)$ such that for every finite field k with* $\mathrm{Card}(k) \geq G(n,d)$, *the set $U_{\leq 1}(k)$ is nonempty.*

PROOF. The first statement is obvious from the previous theorem and the corresponding fact about $\mathcal{H}_{n,d}/\mathbb{Z}$. The second then follows from Lang-Weil in its uniform form (9.0.15.1 and 9.3.3), applied to $U_{\leq 1}/\mathbb{Z}$. QED

Remark 11.8.8. By the above corollary, $U_{\leq 1}(\mathbb{F}_p)$ is nonempty for all primes $p \gg 0$. Poonen [**Poon-Hy**] has recently shown that $U_{\leq 1}(\mathbb{F}_p)$ is nonempty for all primes p.

11.9. Naive and intrinsic measures on $USp(\mathrm{prim}(n,d))^{\#}$ (if n is odd) or on $O(\mathrm{prim}(n,d))^{\#}$ (if n is even) attached to universal families of smooth hypersurfaces of degree d in $\mathbb{P}^{n+1}$

11.9.1. We fix $n \geq 2$ and $d \geq 3$, or $n = 1$ and $d \geq 4$. For a finite field k, denote by α_k a choice of $\mathrm{Sqrt}(\mathrm{Card}(k)^n)$ in $\mathbb{R}$. Given X/k a smooth hypersurface of degree d in $\mathbb{P}^{n+1}$, we have defined in 11.4.1 its unitarized Frobenius conjugacy class $\vartheta(k, \alpha_k, X/k)$ in $USp(\mathrm{prim}(n,d))^{\#}$ for n odd, and in $O(\mathrm{prim}(n,d))^{\#}$ for n even.

11.9.2. We denote by $\mathrm{Iso}\,\mathcal{H}_{n,d}$ the functor

$$S \mapsto \{PGL(n+2)(S)\text{-orbits in } \mathcal{H}_{n,d}(S)\}$$

of projective isomorphism classes of smooth hypersurfaces of dimension n and degree d. This functor is not representable, because it tries to classify objects with automorphisms. Nonetheless, for k a finite field, the set $\mathrm{Iso}\,\mathcal{H}_{n,d}(k)$ is finite, and we have the following mass formula, entirely analogous to that of 10.6.9:

$$\#\mathcal{H}_{n,d}(k)/\#PGL(n+2)(k) = \sum_{X/k \text{ in } \mathrm{Iso}\,\mathcal{H}_{n,d}(k)} 1/\#\,\mathrm{Proj}\,\mathrm{Aut}(X/k).$$

11.9.3. Exactly as we did in the case of curves, we define the "naive" probability measure

$$\mu(\mathrm{naive}, n, d, k, \alpha_k)$$

on $USp(\mathrm{prim}(n,d))^{\#}$ for n odd, and on $O(\mathrm{prim}(n,d))^{\#}$ for n even, by averaging over the unitarized Frobenius conjugacy classes of all the k-isomorphism classes X/k in $\mathrm{Iso}\,\mathcal{H}_{n,d}$:

$$\mu(\mathrm{naive}, n, d, k, \alpha_k) := (1/\#(\mathrm{Iso}\,\mathcal{H}_{n,d}(k))) \sum_{X/k \text{ in } \mathrm{Iso}\,\mathcal{H}_{n,d}(k)} \delta_{\vartheta(k,\alpha_k,X/k)}.$$

11.9.4. We define the "intrinsic" probability measure $\mu(\mathrm{intrin}, n, d, k, \alpha_k)$ on $USp(\mathrm{prim}(n,d))^{\#}$ for n odd, and on $O(\mathrm{prim}(n,d))^{\#}$ for n even, by averaging over the unitarized Frobenius conjugacy classes, but now counting X/k with multiplicity $1/\#\,\mathrm{Proj}\,\mathrm{Aut}(X/k)$:

$$\begin{aligned}\mu(\mathrm{intrin}, n, d, k, \alpha_k) &:= (1/\,\mathrm{Intrin}\,\mathrm{Card}(\mathrm{Iso}\,\mathcal{H}_{n,d}(k))) \\ &\quad\times \sum_{X/k \text{ in } \mathrm{Iso}\,\mathcal{H}_{n,d}(k)} (1/\#\,\mathrm{Proj}\,\mathrm{Aut}(X/k))\delta_{\vartheta(k,\alpha_k,X/k)},\end{aligned}$$

where we have put

$$\begin{aligned}\mathrm{Intrin}\,\mathrm{Card}(\mathrm{Iso}\,\mathcal{H}_{n,d}(k)) &:= \sum_{X/k \text{ in } \mathrm{Iso}\,\mathcal{H}_{n,d}(k)} 1/\#\,\mathrm{Proj}\,\mathrm{Aut}(X/k) \\ &= \#\mathcal{H}_{n,d}(k)/\#PGL(n+2)(k).\end{aligned}$$

11.9.5. Exactly as in the cases of curves and of abelian varieties, we have the following two lemmas.

Lemma 11.9.6. (1) *The rational number* $\operatorname{Intrin}\operatorname{Card}(\operatorname{Iso}\mathcal{H}_{n,d}(k))$ *is an integer, namely*

$$\operatorname{Intrin}\operatorname{Card}(\operatorname{Iso}\mathcal{H}_{n,d}(k)) = \operatorname{Card}(\operatorname{Image}(\operatorname{Iso}\mathcal{H}_{n,d}(k) \to \operatorname{Iso}\mathcal{H}_{n,d}(\overline{k}))).$$

(2) *More precisely, for each point* ξ *in this image, we have the identity*

$$1 = \sum_{X/k \text{ in } \operatorname{Iso}\mathcal{H}_{n,d}(k) \mapsto \xi \text{ in } \operatorname{Iso}\mathcal{H}_{n,d}(\overline{k})} 1/\operatorname{Card}(\operatorname{Proj}\operatorname{Aut}(X/k)).$$

(3) *The image of* $\operatorname{Iso}\mathcal{H}_{n,d}(k)$ *in* $\operatorname{Iso}\mathcal{H}_{n,d}(\overline{k})$ *consists precisely of the points in* $\operatorname{Iso}\mathcal{H}_{n,d}(\overline{k})$ *fixed by* $\operatorname{Gal}(\overline{k}/k)$.

Lemma 11.9.7. *The measure* $\mu(\mathit{intrin}, n, d, k, \alpha_k)$ *is given by the formula*

$$\mu(\mathit{intrin}, n, d, k, \alpha_k) = (1/\operatorname{Card}(\mathcal{H}_{n,d}(k))) \sum_{X/k \text{ in } \mathcal{H}_{n,d}(k)} \delta_{\vartheta(k,\alpha_k,X/k)}.$$

11.9.8. Exactly as in the cases of curves and of abelian varieties, for each $n \geq 1$ and $d \geq 3$, except for the two cases $(n,d) = (1,3)$ or $(2,3)$, there exist constants $A(n,d)$ and $C(n,d)$ so that we have the following theorem.

Theorem 11.9.9. *Fix* $n \geq 1$ *and* $d \geq 3$. *If* $d = 3$, *assume* $n \geq 3$. *Denote by* K *the compact group*

$$K := USp(\operatorname{prim}(n,d)), \quad \text{if } n \text{ is odd},$$
$$K := O(\operatorname{prim}(n,d)), \quad \text{if } n \text{ is even}.$$

Suppose that n *is odd (resp. even). For each finite field* k, *and each choice* α_k *of* $\operatorname{Sqrt}(\operatorname{Card}(k)^n)$ *in* $\mathbb{R}$, *consider the measures on* $K^\#$ *given by* $\mu(\mathit{intrin}, n, d, k, \alpha_k)$ *and* $\mu(\mathit{naive}, n, d, k, \alpha_k)$. *In any sequence of data* (k_i, α_{k_i}) *with* $\operatorname{Card}(k_i)$ *increasing to infinity, the sequence of measures* $\mu(\mathit{intrin}, n, d, k_i, \alpha_{k_i})$ *and the sequence of measures* $\mu(\mathit{naive}, n, d, k_i, \alpha_{k_i})$ *both converge weak* $*$ *to the measure* $\mu^\#$ *on* $K^\#$ *which is the direct image from* K *of normalized Haar measure, i.e., for any continuous* $\mathbb{C}$*-valued central function* f *on* K, *we have*

$$\int_K f\, d\operatorname{Haar} = \lim_{i\to\infty} \int_{K^\#} f\, d\mu(\mathit{intrin}, n, d, k_i, \alpha_{k_i})$$
$$= \lim_{i\to\infty} \int_{K^\#} f\, d\mu(\mathit{naive}, n, d, k_i, \alpha_{k_i}).$$

More precisely, if Λ *is any irreducible nontrivial representation of* K, *and* (k, α_k) *is as above with* $\operatorname{Card}(k) \geq 4A(n,d)^2$, *we have the estimates*

$$\left| \int_{K^\#} \operatorname{Trace}(\Lambda)\, d\mu(\mathit{intrin}, n, d, k, \alpha_k) \right| \leq 2C(n,d) \dim(\Lambda)/\operatorname{Card}(k)^{1/2},$$

and

$$\left| \int_{K^\#} \operatorname{Trace}(\Lambda)\, d\mu(\mathit{naive}, n, d, k, \alpha_k) \right| \leq 2C(n,d) \dim(\Lambda)/\operatorname{Card}(k)^{1/2}.$$

11.10. Monodromy of families of Kloosterman sums

11.10.1. In this section, we summarize the results of [**Ka-GKM**, 13.5] concerning the monodromy groups attached to Kloosterman sums in several variables.

11.10.2. Fix an integer $n \geq 2$, a finite field k, a nontrivial $\mathbb{C}$-valued additive character $\psi : (k, +) \to \mathbb{C}^\times$, and an element a in $k^\times$. The Kloosterman sum $\mathrm{Kl}_n(k, \psi, a)$ is the complex number defined by

$$\mathrm{Kl}(k, \psi, a) := \sum_{x_1 x_2 \cdots x_n = a, \text{ all } x_i \text{ in } k} \psi\left(\sum_i x_i\right).$$

For each integer $d \geq 1$, denote by k_d the unique extension of degree d of k (inside some chosen $\overline{k}$). Then $\psi \circ \mathrm{Trace}_{k_d/k}$ is a nontrivial additive character of k_d, and we may speak of the Kloosterman sum $\mathrm{Kl}_n(k_d, \psi \circ \mathrm{Trace}_{k_d/k}, a)$. Varying d, we put all these sums together in a single Kloosterman L-function, defined as the formal series in $\mathbb{C}[[T]]$ given by

$$L_{n,k,\psi,a}(T) := \exp\left((-1)^n \sum_{d \geq 1} \mathrm{Kl}_n(k_d, \psi \circ \mathrm{Trace}_{k_d/k}, a) T^d/d\right).$$

11.10.3. This L-function is a polynomial in T, of degree n. To say more about it, we denote by K the compact group

$$K := \begin{cases} USp(n) & \text{if } n \text{ is even,} \\ SU(n) & \text{if } n \text{ is odd and } \mathrm{char}(k) \text{ is odd,} \\ SO(n) & \text{if } n \text{ is odd and } \mathrm{char}(k) = 2. \end{cases}$$

[Notice that in all cases, conjugacy classes in K are determined by their characteristic polynomials.] Denote by α_k a choice in $\mathbb{C}$ of $\mathrm{Sqrt}(\mathrm{Card}(k)^{n-1})$, with the proviso that if n is odd, then we take $\alpha_k = \mathrm{Card}(k)^{(n-1)/2}$. There exists a unique conjugacy class

$$\vartheta(n, k, \psi, a, \alpha_k) \quad \text{in } K^\#$$

with the property that

$$L_{n,k,\psi,a}(T) = \det(1 - \alpha_k T \vartheta(n, k, \psi, a, \alpha_k)).$$

We call $\vartheta(n, k, \psi, a, \alpha_k)$ the unitarized Frobenius conjugacy attached to the L-function in question.

11.10.4. Using the unitarized Frobenius conjugacy classes, we define a probability measure $\mu(n, k, \psi, \alpha_k)$ on $K^\#$ by averaging over a in $k^\times$:

$$\mu(n, k, \psi, \alpha_k) := (1/\mathrm{Card}(k^\times)) \sum_{a \text{ in } k^\times} \delta_{\vartheta(n,k,\psi,a,\alpha_k)}.$$

Because we average over all a in $k^\times$, the measure $\mu(n, k, \psi, \alpha_k)$ is independent of the particular choice of nontrivial ψ, so we write

$$\mu(n, k, \alpha_k) := \mu(n, k, \psi, \alpha_k).$$

If n is odd, then $\alpha_k = \mathrm{Card}(k)^{(n-1)/2}$ is already determined by the finite field k, so we may denote $\mu(n, k, \alpha_k)$ simply as $\mu(n, k)$ for n odd.

Theorem 11.10.5. [**Ka-GKM**, 11.1, 11.4, 13.5.3] *Fix an integer $n \geq 2$. Fix any sequence of data (k_i, α_{k_i}) in which the cardinalities of the finite fields k_i tend to infinity, and in which for n odd all the k_i are of odd [resp. even] characteristic. If $n = 7$, suppose in addition that all the k_i are of odd characteristic. Consider the* *sequence of measures $\mu(n, k_i, \alpha_{k_i})$ [or simply $\mu(n, k_i)$ if n is odd] on $K^{\#}$ for K the compact group*

$$\begin{aligned} &USp(n) \quad \textit{if } n \textit{ is even}, \\ &SU(n) \quad \textit{if } n \textit{ is odd and } \operatorname{char}(k_i) \textit{ is odd}, \\ &SO(n) \quad \textit{if } n \textit{ is odd and } \operatorname{char}(k_i) = 2. \end{aligned}$$

This sequence of measures converges weak $*$ to the measure $\mu^{\#}$ on $K^{\#}$ which is the direct image from K of normalized Haar measure, i.e., for any continuous $\mathbb{C}$-valued central function f on K, we have

$$\begin{aligned} \int_K f \, d\,\mathrm{Haar} &= \lim_{i \to \infty} \int_{K^{\#}} \mathrm{Trace}(\Lambda) \, d\mu(\textit{intrin}, n, d, k_i, \alpha_{k_i}) \\ &= \lim_{i \to \infty} \int_{K^{\#}} f \, d\mu(n, k_i, \alpha_{k_i}). \end{aligned}$$

More precisely, if Λ is any irreducible nontrivial representation of K, and (k, α_k) is as above, we have the estimate

$$\left| \int_{K^{\#}} \mathrm{Trace}(\Lambda) \, d\mu(n, k, \alpha_k) \right| \leq (\dim(\Lambda)/n)(\mathrm{Card}(k)^{1/2} / \mathrm{Card}(k^{\times})).$$

CHAPTER 12

GUE Discrepancies in Various Families

12.0. A basic consequence of equidistribution: axiomatics

12.0.1. In this section, we consider the following axiomatic situation. We are given a compact group K, its normalized (total mass one) Haar measure μ, and the direct image $\mu^{\#}$ of μ on the space $K^{\#}$ of conjugacy classes in K. For each integer $n \geq 1$, we are given a finite nonvoid set X_n, a probability measure μ_n on X_n, and a mapping

$$\vartheta_n : X_n \to K^{\#}.$$

For each x in X_n, we denote by $\mu_n(x)$ the measure of the set $\{x\}$: thus

$$\sum_{x \text{ in } X_n} \mu_n(x) = 1.$$

We assume that the sequence of measures

$$(\vartheta_n)_* \mu_n := \sum_{x \text{ in } X_n} \mu_n(x) \delta_{\vartheta_n(x)}$$

on $K^{\#}$ converges weak $*$ to $\mu^{\#}$, i.e., for every continuous $\mathbb{C}$-valued central function f on K, we have

$$\int_{K^{\#}} f\, d\mu^{\#} = \lim_{n \to \infty} \int_{K^{\#}} f\, d(\vartheta_n)_* \mu_n,$$

or, more concretely,

$$\int_K f\, d\mu = \lim_{n \to \infty} \sum_{x \text{ in } X_n} \mu_n(x) f(\vartheta_n(x)).$$

Lemma 12.0.2. *In the axiomatic situation* 12.0.1 *above, suppose that* $f : K \to \mathbb{R}_{\geq 0}$ *is a continuous* $\mathbb{R}$*-valued central function on* K *which is nonnegative. Let* $\varepsilon > 0$ *be real, and suppose* $\int_K f\, d\mu \leq \varepsilon$. *Then there exists an integer* $M = M(K, \{X_n, \mu_n, \vartheta_n\}_n, f, \varepsilon)$ *such that for all* $n \geq M$, *we have the inequality*

$$\sum_{x \textit{ in } X_n} \mu_n(x) f(\vartheta_n(x)) \leq 2\varepsilon.$$

Proof. For $n \gg 0$, we have $|\int_{K^{\#}} f\, d\mu^{\#} - \int_{K^{\#}} f\, d(\vartheta_n)_* \mu_n| \leq \varepsilon$. QED

Corollary 12.0.3. *Hypotheses and notations as in* 12.0.2 *above, for any* $n \geq M$ *and for any two positive real constants* A *and* B *with* $AB = \varepsilon$, *we have the inequality*

$$\mu_n(\{x \textit{ in } X_n \textit{ such that } f(\vartheta_n(x)) \geq A\}) \leq 2B.$$

PROOF. Fix A and B, fix an $n > M$, and denote temporarily by Z_n the set of x in X_n such that $f(\vartheta_n(x)) \geq A$. Because f is nonnegative, we have

$$2AB = 2\varepsilon \geq \sum_{x \text{ in } X_n} \mu_n(x) f(\vartheta_n(x)) \geq \sum_{x \text{ in } Z_n} \mu_n(x) f(\vartheta_n(x))$$
$$\geq A \sum_{x \text{ in } Z_n} \mu_n(x) = A\mu_n(Z_n). \quad \text{QED}$$

12.1. Application to GUE discrepancies

12.1.1. Recall that we have proven

Theorem 1.7.6. *Let $r \geq 1$ be an integer, b in $\mathbb{Z}^r$ a step vector with corresponding separation vector a and offset vector c. Denote*

$$\mu := \mu(\textit{univ, offsets } c).$$

Suppose given an integer k with $1 \leq k \leq r$, and a surjective linear map

$$\pi : \mathbb{R}^r \to \mathbb{R}^k.$$

1) *The measure $\pi_*\mu$ on $\mathbb{R}^k$ is absolutely continuous with respect to Lebesgue measure, and (consequently) has a continuous CDF.*

2) *Given any real $\varepsilon > 0$, there exists an explicit constant $N(\varepsilon, r, c, \pi)$ with the following property: For $G(N)$ any of the compact classical groups in their standard representations,*

$$U(N), SU(N), SO(2N+1), O(2N+1), USp(2N), SO(2N), O(2N),$$

and for

$$\mu(A, N) := \mu(A, G(N), \textit{ offsets } c), \quad \textit{for each } A \textit{ in } G(N),$$

we have the inequality

$$\int_{G(N)} \text{discrep}(\pi_*\mu(A, N), \pi_*\mu) dA \leq N^{\varepsilon - 1/(2r+4)},$$

provided that $N \geq N(\varepsilon, r, c, \pi)$.

In slightly greater generality, we had

Theorem 1.7.7. *Let $G(N) \subset H(N)$ be compact groups in one of the following four cases:*
a) *$G(N) = SU(N) \subset H(N) \subset$ normalizer of $G(N)$ in $U(N)$,*
b) *$G(N) = SO(2N+1) \subset H(N) \subset$ normalizer of $G(N)$ in $U(2N+1)$,*
c) *$G(N) = USp(2N) \subset H(N) \subset$ normalizer of $G(N)$ in $U(2N)$,*
d) *$G(N) = SO(2N) \subset H(N) \subset$ normalizer of $G(N)$ in $U(2N)$.*
For ε, r, c, π as in Theorem 3, with explicit constant $N(\varepsilon, r, c, \pi)$, we have the inequality

$$\int_{H(N)} \text{discrep}(\pi_*\mu(A, H(N), \textit{ offsets } c), \pi_*\mu(\textit{univ, offsets } c))\, dA \leq N^{\varepsilon - 1/(2r+4)}$$

provided that $N \geq N(\varepsilon, r, c, \pi)$.

12.1.2. We now take the K of the axiomatic situation of the previous section 12.0.1 to be $H(N)$, and we take f to be the continuous central function on $H(N)$ given by

$$A \mapsto \text{discrep}(\pi_*\mu(A, H(N), \text{ offsets } c), \pi_*\mu(\text{univ, offsets } c)).$$

This f has values in the closed interval $[0,1]$, and according to Theorem 1.7.7 we have $\int_{H(N)} f\, d\text{Haar} \le N^{\varepsilon - 1/(2r+4)}$, provided that $N \ge N(\varepsilon, r, c, \pi)$. So we get

Corollary 12.1.3. *Hypotheses and notations as in Theorems* 1.7.6 *and* 1.7.7, *let* $\varepsilon > 0$, *and suppose that* $N \ge N(\varepsilon, r, c, \pi)$. *Fix any of the groups* $H(N)$, *and denote by* μ *is normalized Haar measure. Suppose that for each integer* $n \ge 1$, *we are given a finite nonvoid set* $X_{n,N}$, *a probability measure* $\mu_{n,N}$ *on* $X_{n,N}$, *and a mapping*

$$\vartheta_{n,N} : X_{n,N} \to H(N)^{\#},$$

such that the sequence of measures

$$(\vartheta_{n,N})_*\mu_{n,N} := \sum_{x \text{ in } X_{n,N}} \mu_{n,N}(x)\delta_{\vartheta_{n,N}(x)}$$

on $H(N)^{\#}$ *converges weak* $*$ *to* $\mu^{\#} :=$ *the direct image of normalized Haar measure from* $H(N)$.

(1) *There exists an integer* $M_N = M(H(N), \{X_{n,N}, \mu_{n,N}, \vartheta_{n,N}\}_n, \varepsilon, r, c, \pi)$ *such that for all* $n \ge M_N$, *we have*

$$\int_{X_{n,N}} \text{discrep}(\pi_*\mu(\vartheta_{n,N}(x), H(N), \textit{ offsets } c), \pi_*\mu(\textit{univ, offsets } c))\, d\mu_{n,N}$$
$$\le 2N^{\varepsilon - 1/(2r+4)}.$$

(2) *For any real numbers* α *and* β *with*

$$\alpha + \beta = 1/(2r+4) - \varepsilon,$$

and for any $n \ge M_N$, *the subset of* $X_{n,N}$ *where*

$$\text{discrep}(\pi_*\mu(\vartheta_{n,N}(x), H(N), \textit{ offsets } c), \pi_*\mu(\textit{univ, offsets } c)) \ge N^{-\alpha}$$

has $\mu_{n,N}$*-measure* $\le 2N^{-\beta}$.

12.2. GUE discrepancies in universal families of curves

12.2.1. For each prime power q, fix a choice $q^{1/2}$ of $\text{Sqrt}(q)$ in $\mathbb{R}$. As explained in 10.7.2, for each curve $C/\mathbb{F}_q$ of genus $g \ge 1$, there is a unique conjugacy class $\vartheta(C/\mathbb{F}_q)$ in $USp(2g)^{\#}$ such that the zeta function of $C/\mathbb{F}_q$ is given by

$$\text{Zeta}(C/\mathbb{F}_q, T) = \det(1 - q^{1/2}T\vartheta(C/\mathbb{F}_q))/((1-T)(1-qT)).$$

We have the finite nonvoid set $\mathcal{M}_g(\mathbb{F}_q)$ of isomorphism classes of genus g curves $C/\mathbb{F}_q$, the map $\vartheta_{g,q} : \mathcal{M}_g(\mathbb{F}_q) \to USp(2g)^{\#}$ defined by

$$\vartheta_{g,q}(C/\mathbb{F}_q) = \vartheta(C/\mathbb{F}_q),$$

and the choice of two probability measures $\mu_{g,q,\text{ naive}}$ or $\mu_{g,q,\text{ intrin}}$ on $\mathcal{M}_g(\mathbb{F}_q)$, defined by

$$\mu_{g,q,\text{ naive}}(C/\mathbb{F}_q) := 1/\text{Card}(\mathcal{M}_g(\mathbb{F}_q))$$

and by

$$\mu_{g,q,\ \mathrm{intrin}}(C/\mathbb{F}_q) := 1/\operatorname{Card}(\operatorname{Aut}(C/\mathbb{F}_q))\operatorname{Intrin}\operatorname{Card}(\mathcal{M}_g(\mathbb{F}_q)),$$

where

$$\operatorname{Intrin}\operatorname{Card}(\mathcal{M}_g(\mathbb{F}_q)) := \sum_{C/\mathbb{F}_q \text{ in } \mathcal{M}_g(\mathbb{F}_q)} 1/\operatorname{Card}(\operatorname{Aut}(C/\mathbb{F}_q)).$$

12.2.2. We know from 10.7.12 that for fixed $g \geq 3$, both of the sequences of measures on $USp(2g)^{\#}$ given by $(\vartheta_{g,q})_*\mu_{g,q,\ \mathrm{intrin}}$ and by $(\vartheta_{g,q})_*\mu_{g,q,\ \mathrm{naive}}$, indexed by prime powers q, converge weak $*$ to the measure $\mu^{\#}$ on $USp(2g)^{\#}$. So applying the above corollary, with $N := g, H(N) := USp(2g)$, and $X_{n,N}$ the set $\mathcal{M}_N(\mathbb{F}_q)$ with q the n'th prime power, we find

Theorem 12.2.3. *Let $r \geq 1$ be an integer, b in $\mathbb{Z}^r$ a step vector with corresponding separation vector a and offset vector c. Denote*

$$\mu := \mu(\mathit{univ},\ \mathit{offsets}\ c).$$

Suppose given an integer k with $1 \leq k \leq r$, and a surjective linear map

$$\pi : \mathbb{R}^r \to \mathbb{R}^k.$$

Given any real $\varepsilon > 0$, there exists an explicit constant $N(\varepsilon, r, c, \pi)$ with the following properties:

(1) *For each $g \geq N(\varepsilon, r, c, \pi)$, there exists a constant $M(g, \varepsilon, r, c, \pi)$ such that for $q \geq M(g, \varepsilon, r, c, \pi)$, putting*

$$\mu(A, g) := \mu(A, USp(2g),\ \mathit{offsets}\ c), \quad \text{for each } A \text{ in } USp(2g),$$

we have the inequalities

$$\int_{\mathcal{M}_g(\mathbb{F}_q)} \operatorname{discrep}(\pi_*\mu(\vartheta(C/\mathbb{F}_q), g), \pi_*\mu)\, d\mu_{g,q,\ \mathit{naive}} \leq 2g^{\varepsilon - 1/(2r+4)},$$

$$\int_{\mathcal{M}_g(\mathbb{F}_q)} \operatorname{discrep}(\pi_*\mu(\vartheta(C/\mathbb{F}_q), g), \pi_*\mu)\, d\mu_{g,q,\ \mathit{intrin}} \leq 2g^{\varepsilon - 1/(2r+4)}.$$

(2) *For any real numbers α and β with*

$$\alpha + \beta = 1/(2r+4) - \varepsilon,$$

for any $g \geq N(\varepsilon, r, c, \pi)$, and for any $q \geq M(g, \varepsilon, r, c, \pi)$, the subset of $\mathcal{M}_g(\mathbb{F}_q)$ where

$$\operatorname{discrep}(\pi_*\mu(\vartheta(C/\mathbb{F}_q), USp(2g),\ \mathit{offsets}\ c), \pi_*\mu(\mathit{univ},\ \mathit{offsets}\ c)) \geq g^{-\alpha}$$

has $\mu_{g,q,\ naive}$-measure $\leq 2g^{-\beta}$ and $\mu_{g,q,intrin}$-measure $\leq 2g^{-\beta}$.

Remark 12.2.4. This theorem should be thought of as a slightly effective version of the following statement: if we pick a real number $\delta > 0$ and ask what is the probability (in either the $\mu_{g,q,\ \mathrm{naive}}$ or $\mu_{g,q,\ \mathrm{intrin}}$ senses) that a curve of genus g over $\mathbb{F}_q$ has the property that the CDF of the discrete measure formed out of any prechosen spacing statistic of the unitarized zeroes of its zeta function is uniformly within δ of the CDF for the corresponding limit measure, that probability will be $\geq 1 - \delta$ provided both that $g \gg 0$ (with the notion of $\gg$ depending both on δ and on which spacing statistic) and that $q \gg 0$, but here the notion of $\gg$ depends not only on δ and on which statistic but also on g. More seriously, although the constant $N(\varepsilon, r, c, \pi)$ is effective, the constant $M(g, \varepsilon, r, c, \pi)$ is not, at present, effective. This same problem will recur in the other instances (abelian varieties, hypersurfaces, Kloosterman sums) of this same theorem that we make explicit in the following pages.

12.3. GUE discrepancies in universal families of abelian varieties

12.3.1. For each prime power q, fix a choice $q^{1/2}$ of $\mathrm{Sqrt}(q)$ in $\mathbb{R}$. As explained in 11.3.1–2, for each abelian variety $A/\mathbb{F}_q$ of dimension $g \geq 1$, the $\mathbb{Z}$-valued function on $\mathbb{Z}$ defined by

$$m \mapsto \deg(1 - mF_{A/\mathbb{F}_q})$$

is a polynomial function $P_{A/\mathbb{F}_q}(m)$ of m, and there is a unique conjugacy class $\vartheta(A/\mathbb{F}_q)$ in $USp(2g)^{\#}$ such that

$$P_{A/\mathbb{F}_q}(T) = \det(1 - q^{1/2}T\vartheta(A/\mathbb{F}_q)).$$

The zeta function of $A/\mathbb{F}_q$ is given by the exterior powers of $\vartheta(A/\mathbb{F}_q)$:

$$\mathrm{Zeta}(A/\mathbb{F}_q, T) = \prod_{i=0}^{2g} \det(1 - q^{i/2}T\Lambda^i\vartheta(A/\mathbb{F}_q))^{(-1)^{i+1}}.$$

12.3.2. We have the finite nonvoid set $\mathrm{Ab}_{g,\mathrm{prin}}(\mathbb{F}_q)$ of isomorphism classes of principally polarized g-dimensional abelian varieties $(A/\mathbb{F}_q, \varphi)$ over $\mathbb{F}_q$, the map $\vartheta_{g,q} : \mathrm{Ab}_{g,\mathrm{prin}}(\mathbb{F}_q) \to USp(2g)^{\#}$ defined by

$$\vartheta_{g,q}(A/\mathbb{F}_q, \varphi) = \vartheta(A/\mathbb{F}_q),$$

and the choice of two probability measures $\mu_{g,q,\ \mathrm{naive}}$ or $\mu_{g,q,\ \mathrm{intrin}}$ on $\mathrm{Ab}_{g,\mathrm{prin}}(\mathbb{F}_q)$, defined by

$$\mu_{g,q,\ \mathrm{naive}}(A/\mathbb{F}_q, \varphi) := 1/\,\mathrm{Card}(\mathrm{Ab}_{g,\mathrm{prin}}(\mathbb{F}_q))$$

and by

$$\mu_{g,q,\ \mathrm{intrin}}(A/\mathbb{F}_q, \varphi) := 1/\,\mathrm{Card}(\mathrm{Aut}(A/\mathbb{F}_q, \varphi))\,\mathrm{Intrin\,Card}(\mathrm{Ab}_{g,\mathrm{prin}}(\mathbb{F}_q)),$$

where

$$\mathrm{Intrin\,Card}(\mathrm{Ab}_{g,\mathrm{prin}}(\mathbb{F}_q)) := \sum_{(A/\mathbb{F}_q,\varphi)\ \mathrm{in}\ \mathrm{Ab}_{g,\mathrm{prin}}(\mathbb{F}_q)} 1/\,\mathrm{Card}(\mathrm{Aut}(A/\mathbb{F}_q, \varphi)).$$

12.3.3. We know from 11.3.10 that for fixed $g \geq 1$, both of the sequences of measures on $USp(2g)^{\#}$ given by $(\vartheta_{g,q})_*\mu_{g,q,\ \mathrm{intrin}}$ and by $(\vartheta_{g,q})_*\mu_{g,q,\ \mathrm{naive}}$, indexed by prime powers q, converge weak $*$ to the measure $\mu^{\#}$ on $USp(2g)^{\#}$. We apply the above Corollary 12.1.3, with $N := g, H(N) := USp(2g)$, and $X_{n,N}$ the set $\mathrm{Ab}_{N,\mathrm{prin}}(\mathbb{F}_q)$ with q the n'th prime power.

Theorem 12.3.4. *Let $r \geq 1$ be an integer, b in $\mathbb{Z}^r$ a step vector with corresponding separation vector a and offset vector c. Denote*

$$\mu := \mu(univ,\ offsets\ c).$$

Suppose given an integer k with $1 \leq k \leq r$, and a surjective linear map

$$\pi : \mathbb{R}^r \to \mathbb{R}^k.$$

Given any real $\varepsilon > 0$, there exists an explicit constant $N(\varepsilon, r, c, \pi)$ with the following properties:

(1) *For each* $g \geq N(\varepsilon, r, c, \pi)$, *there exists a constant* $M(g, \varepsilon, r, c, \pi)$ *such that for* $q \geq M(g, \varepsilon, r, c, \pi)$, *putting*

$$\mu(A, g) := \mu(A, USp(2g),\ \textit{offsets } c), \quad \textit{for each } A \textit{ in } USp(2g),$$

we have the inequalities

$$\int_{\mathrm{Ab}_{g,\mathrm{prin}}(\mathbb{F}_q)} \mathrm{discrep}(\pi_*\mu(\vartheta(A/\mathbb{F}_q), g), \pi_*\mu)\, d\mu_{g,q,\ naive} \leq 2g^{\varepsilon - 1/(2r+4)},$$

$$\int_{\mathrm{Ab}_{g,\mathrm{prin}}(\mathbb{F}_q)} \mathrm{discrep}(\pi_*\mu(\vartheta(A/\mathbb{F}_q), g), \pi_*\mu)\, d\mu_{g,q,\ intrin} \leq 2g^{\varepsilon - 1/(2r+4)}.$$

(2) *For any real numbers* α *and* β *with*

$$\alpha + \beta = 1/(2r+4) - \varepsilon,$$

for any $g \geq N(\varepsilon, r, c, \pi)$, *and for any* $q \geq M(g, \varepsilon, r, c, \pi)$, *the subset of* $\mathrm{Ab}_{g,\mathrm{prin}}(\mathbb{F}_q)$ *where*

$$\mathrm{discrep}(\pi_*\mu(\vartheta(A/\mathbb{F}_q), USp(2g),\ \textit{offsets } c), \pi_*\mu(\textit{univ},\ \textit{offsets } c)) \geq g^{-\alpha}$$

has $\mu_{g,q,\ naive}$*-measure* $\leq 2g^{-\beta}$ *and* $\mu_{g,q,\ intrin}$*-measure* $\leq 2g^{-\beta}$.

12.4. GUE discrepancies in universal families of hypersurfaces

12.4.1. Fix a pair of integers (n, d) with $n \geq 1, d \geq 3$, but not $(1,3)$ or $(2,3)$. For each prime power q, fix a choice $q^{n/2}$ of $\mathrm{Sqrt}(q^n)$. Recall (11.4.1) that, given $X/\mathbb{F}_q$ a smooth hypersurface of degree d in $\mathbb{P}^{n+1}$, its zeta function $\mathrm{Zeta}(X/\mathbb{F}_q, T)$ has the form

$$\text{if } n \text{ is odd:} \qquad P(T)/\prod_{i=0}^{n}(1 - q^i T),$$

$$\text{if } n \text{ is even:} \qquad 1/P(T)\prod_{i=0}^{n}(1 - q^i T),$$

where $P(T)$ is a $\mathbb{Z}$-polynomial with constant term one, of degree

$$\mathrm{prim}(n, d) := (d-1)((d-1)^{n+1} - (-1)^{n+1})/d.$$

For n odd (resp. even), there is a unique conjugacy class

$$\vartheta(X/\mathbb{F}_q) \text{ in } USp(\mathrm{prim}(n,d))^{\#} \quad (\text{resp. in } O(\mathrm{prim}(n,d))^{\#})$$

such that

$$P(T) = \det(1 - q^{n/2} T \vartheta(X/\mathbb{F}_q)).$$

12.4.2. Let us denote

$$K_{n,d} := \begin{cases} USp(\mathrm{prim}(n,d)) & \text{if } n \text{ is odd,} \\ O(\mathrm{prim}(n,d)) & \text{if } n \text{ is even.} \end{cases}$$

We have the finite nonvoid set $\mathrm{Iso}\,\mathcal{H}_{n,d}(\mathbb{F}_q)$ of projective isomorphism classes of all such hypersurfaces over $\mathbb{F}_q$, the map

$$\vartheta_{n,d,q} : \mathrm{Iso}\,\mathcal{H}_{n,d}(\mathbb{F}_q) \to (K_{n,d})^{\#}$$

defined by

$$\vartheta_{n,d,q}(X/\mathbb{F}_q) := \vartheta(X/\mathbb{F}_q),$$

and the choice of two probability measures

$$\mu_{n,d,q,\text{ naive}} \quad \text{or} \quad \mu_{n,d,q,\text{ intrin}}$$

on $\operatorname{Iso}\mathcal{H}_{n,d}(\mathbb{F}_q)$, defined by

$$\mu_{n,d,q,\text{ naive}}(X/\mathbb{F}_q) := 1/\operatorname{Card}(\operatorname{Iso}\mathcal{H}_{n,d}(\mathbb{F}_q))$$

and by

$$\mu_{g,q,\text{ intrin}}(X/\mathbb{F}_q) := 1/\operatorname{Card}(\operatorname{Proj}\operatorname{Aut}(X/\mathbb{F}_q))\operatorname{Intrin}\operatorname{Card}(\operatorname{Iso}\mathcal{H}_{n,d}(\mathbb{F}_q)),$$

where

$$\operatorname{Intrin}\operatorname{Card}(\operatorname{Iso}\mathcal{H}_{n,d}(\mathbb{F}_q)) := \sum_{X/\mathbb{F}_q \text{ in } \operatorname{Iso}\mathcal{H}_{n,d}(\mathbb{F}_q)} 1/\operatorname{Card}(\operatorname{Proj}\operatorname{Aut}(X/\mathbb{F}_q)).$$

12.4.3. We know from 11.9.9 that for fixed (n,d) as above, both of the sequences of measures on $(K_{n,d})^{\#}$ given by

$$(\vartheta_{n,d,q})_*\mu_{n,d,q,\text{ intrin}}$$

and by

$$(\vartheta_{n,d,q})_*\mu_{n,d,q,\text{ naive}},$$

indexed by prime powers q, converge weak $*$ to the measure $\mu^{\#}$ on $(K_{n,d})^{\#}$.

Just as in the cases of curves and of abelian varieties, we find

Theorem 12.4.4. *Let $r \geq 1$ be an integer, b in $\mathbb{Z}^r$ a step vector with corresponding separation vector a and offset vector c. Denote*

$$\mu := \mu(\textit{univ, offsets } c).$$

Suppose given an integer k with $1 \leq k \leq r$, and a surjective linear map

$$\pi : \mathbb{R}^r \to \mathbb{R}^k.$$

Given any real $\varepsilon > 0$, there exists an explicit constant $N(\varepsilon, r, c, \pi)$ with the following properties:

(1) *For each (n,d) with $n \geq 1, d \geq 3$ other than $(1,3)$ or $(2,3)$ and with $\operatorname{prim}(n,d) \geq N(\varepsilon, r, c, \pi)$, there exists a constant*

$$M(n,d,\varepsilon,r,c,\pi)$$

such that for $q \geq M(n,d,\varepsilon,r,c,\pi)$, putting

$$K_{n,d} := \begin{cases} USp(\operatorname{prim}(n,d)) & \textit{if } n \textit{ is odd,} \\ O(\operatorname{prim}(n,d)) & \textit{if } n \textit{ is even,} \end{cases}$$

and putting

$$\mu(A,n,d) := \mu(A, K_{n,d}, \textit{ offsets } c), \quad \textit{for each } A \textit{ in } K_{n,d},$$

we have the inequalities

$$\int_{\operatorname{Iso}\mathcal{H}_{n,d}(\mathbb{F}_q)} \operatorname{discrep}(\pi_*\mu(\vartheta(X/\mathbb{F}_q),n,d), \pi_*\mu)\, d\mu_{n,d,q,\textit{ naive}} \leq 2[\operatorname{prim}(n,d)/2]^{\varepsilon-1/(2r+4)},$$

$$\int_{\operatorname{Iso}\mathcal{H}_{n,d}(\mathbb{F}_q)} \operatorname{discrep}(\pi_*\mu(\vartheta(X/\mathbb{F}_q),n,d), \pi_*\mu)\, d\mu_{n,d,q,\textit{ intrin}} \leq 2[\operatorname{prim}(n,d)/2]^{\varepsilon-1/(2r+4)}.$$

(2) *For any real numbers* α *and* β *with*

$$\alpha + \beta = 1/(2r+4) - \varepsilon,$$

for any (n, d) *as above with*

$$\mathrm{prim}(n, d) \geq N(\varepsilon, r, c, \pi),$$

and for any $q \geq M(n, d, r, c, \pi)$, *the subset of* $\mathrm{Iso}\,\mathcal{H}_{n,d}(\mathbb{F}_q)$ *where*

$$\mathrm{discrep}(\pi_*\mu(\vartheta(X/\mathbb{F}_q), K_{n,d},\ \textit{offsets}\ c), \pi_*\mu(\textit{univ},\ \textit{offsets}\, c)) \geq [\mathrm{prim}(n, d)/2]^{-\alpha}$$

has

$$\mu_{g,q,\ naive}\text{-}\textit{measure}\ \leq 2[\mathrm{prim}(n, d)/2]^{-\beta}$$

and

$$\mu_{g,q,\ intrin}\text{-}\textit{measure}\ \leq 2[\mathrm{prim}(n, d)/2]^{-\beta}.$$

12.5. GUE discrepancies in families of Kloosterman sums

12.5.1. Fix an integer $n \geq 2$. For each prime power q, pick a nontrivial $\mathbb{C}$-valued additive character ψ of $\mathbb{F}_q$. If n is even, then for each prime power q, make a choice $q^{(n-1)/2}$ in $\mathbb{R}$ of $\mathrm{Sqrt}(q^{n-1})$. For each a in $\mathbb{F}_q^\times$, we have defined (11.10.3) the Kloosterman sum, the Kloosterman L-function, and the unitarized Frobenius conjugacy class

$$\vartheta(n, \mathbb{F}_q, \psi, a) \text{ in } K_{n,q}^{\#},$$

for $K_{n,q}$ the compact group

$$K_{n,q} := \begin{cases} USp(n) & \text{if } n \text{ is even,} \\ SU(n) & \text{if } n \text{ is odd and } q \text{ is odd,} \\ SO(n) & \text{if } n \text{ is odd and } q \text{ is even.} \end{cases}$$

Denote by μ_{q-1} the normalized Haar measure on $\mathbb{F}_q^\times$ which gives each point mass $1/(q-1)$, and denote by

$$\vartheta_{n,q} : \mathbb{F}_q^\times \to K_{n,q}^{\#}$$

the map $a \mapsto \vartheta(n, \mathbb{F}_q, \psi, a)$. We know (11.10.5) that

(1) if n is even, the sequence of measures $(\vartheta_{n,q})_*\mu_{q-1}$ on $USp(n)^{\#}$ indexed by prime powers q converges weak $*$ to the measure $\mu^{\#}$,

(2) if n is odd, the sequence of measures $(\vartheta_{n,q})_*\mu_{q-1}$ on $SU(n)^{\#}$ indexed by *odd* prime powers q converges weak $*$ to the measure $\mu^{\#}$ on $SU(n)^{\#}$,

(3) if n is odd, the sequence of measures $(\vartheta_{n,q})_*\mu_{q-1}$ on $SO(N)^{\#}$ indexed by *even* prime powers q converges weak $*$ to the measure $\mu^{\#}$ on $SO(n)^{\#}$.

12.5.2. So just as in the cases of curves, abelian varieties, and hypersurfaces already discussed, we have

Theorem 12.5.3. *Let* $r \geq 1$ *be an integer,* b *in* $\mathbb{Z}^r$ *a step vector with corresponding separation vector* a *and offset vector* c. *Denote*

$$\mu := \mu(\textit{univ},\ \textit{offsets}\ c).$$

Suppose given an integer k *with* $1 \leq k \leq r$, *and a surjective linear map*

$$\pi : \mathbb{R}^r \to \mathbb{R}^k.$$

Given any real $\varepsilon > 0$, *there exists an explicit constant* $N(\varepsilon, r, c, \pi)$, *with the following properties*:

(1) *For each* $n \geq 2$ *with* $n \geq N(\varepsilon, r, c, \pi)$, *there exists a constant* $M(n, \varepsilon, r, c, \pi)$ *such that for* $q \geq M(n, \varepsilon, r, c, \pi)$, *putting*

$$K_{n,q} := \begin{cases} USp(n) & \text{if } n \text{ is even,} \\ SU(n) & \text{if } n \text{ is odd and } q \text{ is odd,} \\ SO(n) & \text{if } n \text{ is odd and } q \text{ is even,} \end{cases}$$

and putting

$$\mu(A, n, q) := \mu(A, K_{n,q}, \ \textit{offsets } c), \quad \textit{for each } A \textit{ in } K_{n,q},$$

we have the inequality

$$\int_{(\mathbb{F}_q)^\times} \mathrm{discrep}(\pi_*\mu(\vartheta_{n,q}(a), n, q), \pi_*\mu)\, d\mu_{q-1} \leq 2[n/2]^{\varepsilon - 1/(2r+4)}.$$

(2) *For any real numbers* α *and* β *with*

$$\alpha + \beta = 1/(2r+4) - \varepsilon,$$

for any $n \geq 2$ *with* $n \geq N(\varepsilon, r, c, \pi)$, *and for any* $q \geq M(n, d, \varepsilon, r, c, \pi)$, *the subset of* $\mathbb{F}_q^\times$ *where*

$$\mathrm{discrep}(\pi_*\mu(\vartheta_{n,q}(a), K_{n,q}, \ \textit{offsets } c), \pi_*\mu(\textit{univ}, \ \textit{offsets } c)) \geq [n/2]^{-\alpha}$$

has μ_{q-1}*-measure* $\leq 2[n/2]^{-\beta}$.

CHAPTER 13

Distribution of Low-lying Frobenius Eigenvalues in Various Families

13.0. An elementary consequence of equidistribution

13.0.1. In this section, we consider the following axiomatic situation, which is a slight generalization of that considered in 12.0.1. The extra generality is needed (only) to take care of orthogonal groups. We are given a compact group K, a finite abelian group Γ, and a continuous, surjective homomorphism $\rho : K \to \Gamma$. [The situation we have in mind is $K = O(N), \Gamma = \{\pm 1\}$, and ρ is the determinant. If we take Γ to be the trivial group, we recover the situation 12.0.1.] For each element γ in Γ, we denote by $K_\gamma \subset K$ the open and closed set $\rho^{-1}(\gamma)$. Because Γ is abelian and ρ is a group homomorphism, the sets K_γ are each stable by K-conjugation, and ρ induces a map $\rho^\# : K^\# \to \Gamma$. We denote by $K^\#_\gamma \subset K^\#$ the open and closed set $(\rho^\#)^{-1}(\gamma)$ in $K^\#$: concretely, $K^\#_\gamma$ is the image of K_γ in $K^\#$. We denote by μ the normalized (total mass one) Haar measure μ on K, and by $\mu^\#$ its direct image on $K^\#$. For each element γ in Γ, we denote by $\mu^\#_\gamma$ the probability measure on $K^\#_\gamma$ defined as

$$\mu^\#_\gamma := \mathrm{Card}(\Gamma) \times (\text{the restriction of } \mu^\# \text{ to } K^\#_\gamma).$$

We suppose that for each integer $n \geq 1$, we are given a finite nonvoid set X_n, a probability measure μ_n on X_n, and a mapping

$$\vartheta_n : X_n \to K^\#.$$

For each x in X_n, we denote by $\mu_n(x)$ the measure of the set $\{x\}$: thus

$$\sum_{x \text{ in } X_n} \mu_n(x) = 1.$$

We assume that the sequence of measures

$$(\vartheta_n)_*\mu_n := \sum_{x \text{ in } X_n} \mu_n(x)\delta_{\vartheta_n(x)}$$

on $K^\#$ converges weak $*$ to $\mu^\#$, i.e., for every continuous $\mathbb{C}$-valued central function f on K, we have

$$\int_{K^\#} f\, d\mu^\# = \lim_{n\to\infty} \int_{K^\#} f d(\vartheta_n)_*\mu_n,$$

or, concretely,

$$\int_K f\, d\mu = \lim_{n\to\infty} \sum_{x \text{ in } X_n} \mu_n(x) f(\vartheta_n(x)).$$

For each element γ in Γ, we denote by $X_{n,\gamma} \subset X_n$ the subset $(\vartheta_n)^{-1}(K_\gamma^\#)$. If $\mu_n(X_{n,\gamma}) := \sum_{x \text{ in } X_{n,\gamma}} \mu_n(x)$ is nonzero (which implies in particular that $X_{n,\gamma}$ is nonempty), we denote by $\mu_{n,\gamma}$ the probability measure on $X_{n,\gamma}$ defined as

$$\begin{aligned}\mu_{n,\gamma} &:= (1/\mu_n(X_{n,\gamma})) \times (\text{the restriction of } \mu_n \text{ to } X_{n,\gamma}) \\ &= (1/\mu_n(X_{n,\gamma})) \times \sum_{x \text{ in } X_{n,\gamma}} \mu_n(x)\delta_x.\end{aligned}$$

We denote by

$$\vartheta_{n,\gamma} : X_{n,\gamma} \to K_\gamma^\#$$

the restriction of ϑ_n to $X_{n,\gamma}$.

Lemma 13.0.2. *In the axiomatic situation* 13.0.1 *above, we have*:

1) *For n sufficiently large, $\mu_n(X_{n,\gamma})$ is nonzero for every element γ in Γ.*

2) *For each element γ in Γ, the sequence of probability measures on $K_\gamma^\#$, indexed by sufficiently large n,*

$$(\vartheta_{n,\gamma})_*\mu_{n,\gamma} := (1/\mu_n(X_{n,\gamma})) \times \sum_{x \text{ in } X_{n,\gamma}} \mu_n(x)\delta_{\vartheta_n(x)},$$

converges weak $*$ to the measure $\mu_\gamma^\#$ on $K_\gamma^\#$.

PROOF. For each element γ in Γ, denote by χ_γ the characteristic function of the set K_γ. Then χ_γ is a continuous central function on K. If we apply to $f := \chi_\gamma$ the limit formula

$$\int_K f\, d\mu = \lim_{n\to\infty} \sum_{x \text{ in } X_n} \mu_n(x) f(\vartheta_n(x))$$

we find

$$1/\operatorname{Card}(\Gamma) = \lim_{n\to\infty} \sum_{x \text{ in } X_{n,\gamma}} \mu_n(x), \tag{13.0.2.1}$$

which proves 1). To prove 2), we argue as follows. A continuous $\mathbb{C}$-valued function g on $K_\gamma^\#$ may be thought of as a continuous function g on $K^\#$ which is supported in $K_\gamma^\#$. For this g, the limit formula

$$\int_{K^\#} g\, d\mu^\# = \lim_{n\to\infty} \int_{K^\#} g d(\vartheta_n)_*\mu_n$$

says precisely that for $n \gg 0$ we have

$$(1/\operatorname{Card}(\Gamma)) \times \int_{K_\gamma^\#} g\, d\mu_\gamma^\# = \lim_{n\to\infty} \mu_n(X_{n,\gamma}) \int_{K_\gamma^\#} g d(\vartheta_{n,\gamma})_*\mu_{n,\gamma}.$$

Crossmultiply by $\operatorname{Card}(\Gamma)$ and use (13.0.2.1) to turn this into

$$\int_{K_\gamma^\#} g\, d\mu_\gamma^\# = \lim_{n\to\infty} \int_{K_\gamma^\#} g d(\vartheta_{n,\gamma})_*\mu_{n,\gamma},$$

which proves 2). QED

Lemma 13.0.3. *In the axiomatic situation* 13.0.1, *suppose we are given an element γ in Γ, a locally compact topological space S, and a continuous map $F : K_\gamma^\# \to S$. Denote by $\nu := F_*\mu_\gamma^\#$ the probability measure on S obtained by* *taking the direct image via F of the (total mass one) Haar measure $\mu_\gamma^\#$ on $K_\gamma^\#$. Consider the sequence of probability measures on S, indexed by $n \gg 0$, given by*

$$F_*(\vartheta_{n,\gamma})_*\mu_{n,\gamma} := (1/\mu_n(X_{n,\gamma})) \sum_{x \text{ in } X_{n,\gamma}} \mu_n(x)\delta_{F(\vartheta_n(x))}.$$

Then this sequence of measures on S converges weak $*$ to ν, i.e., for every continuous $\mathbb{C}$-valued function f on S, we have

$$\int_S f\,d\nu = \lim_{n\to\infty} (1/\mu_n(X_{n,\gamma})) \sum_{x \text{ in } X_{n,\gamma}} \mu_n(x)f(F(\vartheta_n(x))).$$

PROOF. Since the map $F : K_\gamma^\# \to S$ is continuous, the function $g := f \circ F$ on $K_\gamma^\#$ is continuous. By the previous lemma, for any continuous central function g on $K_\gamma^\#$, we have the limit formula

$$\int_{K_\gamma^\#} g\,d\mu_\gamma^\# = \lim_{n\to\infty} \int_{K_\gamma^\#} g\,d(\vartheta_{n,\gamma})_*\mu_{n,\gamma}.$$

For $g := f \circ F$, this says precisely that

$$\int_S f\,d\nu = \lim_{n\to\infty} (1/\mu_n(X_{n,\gamma})) \sum_{x \text{ in } X_{n,\gamma}} \mu_n(x)f(F(\vartheta_n(x))). \quad \text{QED}$$

13.1. Review of the measures $\nu(c, G(N))$

13.1.1. Recall from 8.4 that for $G(N)$ any of $U(N), SO(2N+1), USp(2N), SO(2N), O_-(2N+2), O_-(2N+1)$, and A in $G(N)$, we have its sequence of angles

$$0 \le \varphi(1) \le \varphi(2) \le \cdots \le \varphi(N) < 2\pi \quad \text{if } G(N) = U(N),$$
$$0 \le \varphi(1) \le \varphi(2) \le \cdots \le \varphi(N) \le \pi \quad \text{for the other } G(N),$$

and its sequence of **normalized angles**

$$0 \le \vartheta(1) \le \vartheta(2) \le \cdots \le \vartheta(N) \le N + \lambda,$$

defined by

$$\vartheta(n) := (N+\lambda)\varphi(n)/\sigma\pi.$$

Concretely,

$$\begin{aligned}
&\vartheta(n) := N\varphi(n)/2\pi && \text{for } U(N),\\
&\vartheta(n) := (N+1/2)\varphi(n)/\pi = (2N+1)\varphi(n)/2\pi && \text{for } SO(2N+1) \text{ or } O_-(2N+1),\\
&\vartheta(n) := N\varphi(n)/\pi = 2N\varphi(n)/2\pi && \text{for } USp(2N) \text{ or } SO(2N),\\
&\vartheta(n) := (N+1)\varphi(n)/\pi = (2N+2)\varphi(n)/2\pi && \text{for } O_-(2N+2).
\end{aligned}$$

13.1.2. As already remarked in 6.9, for $G(N)$ any of $SO(2N+1), USp(2N), SO(2N), O_-(2N+2), O_-(2N+1)$ [but **not** for $U(N)$], formation of each normalized angle $\vartheta(n)$ is a continuous function $A \mapsto \vartheta(n)(A)$ on $G(N)$, which is central when $G(N)$ is a group, and which in the O_- cases is invariant by conjugation by elements of the ambient O group.

13.1.3. Given an integer $r \geq 1$, an offset vector c in $\mathbb{Z}^r$,

$$0 < c(1) < c(2) < \cdots < c(r),$$

and an integer $N \geq c(r)$, recall that we denote by $\nu(c, G(N))$ the probability measure on $\mathbb{R}^r$ which is the direct image of total mass one Haar measure on $G(N)$ by the map $F_c : G(N) \to \mathbb{R}^r$ defined by the normalized angles

$$A \mapsto F_c(A) := (\vartheta(c(1))(A), \ldots, \vartheta(c(r))(A)).$$

Thus

$$\nu(c, G(N)) := (F_c)_* \operatorname{Haar}_{G(N)} := (\vartheta(c(1)), \ldots, \vartheta(c(r)))_* \operatorname{Haar}_{G(N)}.$$

13.1.4. For $G(N)$ any of $SO(2N+1), USp(2N), SO(2N), O_-(2N+2)$, $O_-(2N+1)$ [but *not* for $U(N)$], the map $F_c : G(N) \to \mathbb{R}^r$ is continuous. So for these $G(N)$, the measures $\nu(c, G(N))$ are precisely measures ν of the type discussed in 13.0.3.

13.2. Equidistribution of low-lying eigenvalues in families of curves according to the measure $\nu(c, USp(2g))$

13.2.1. Fix a genus $g \geq 3$. Recall that by looking at "all" genus g curves we defined, for each finite field k and each choice α_k of a square root of $\operatorname{Card}(k)$, two probability measures on $USp(2g)^\#$, by looking at the normalized (by α_k) Frobenius conjugacy classes of all curves over the given field k:

$$\mu(\text{naive}, g, k, \alpha_k) := (1/\operatorname{Card}(\mathcal{M}_g(k))) \sum_{C/k \text{ in } \mathcal{M}_g(k)} \delta_{\vartheta(k, \alpha_k, C/k)},$$

and

$$\mu(\text{intrin}, g, k, \alpha_k)$$
$$:= (1/\operatorname{Intrin}\operatorname{Card}(\mathcal{M}_g(k))) \sum_{C/k \text{ in } \mathcal{M}_g(k)} (1/\operatorname{Card}(\operatorname{Aut}(C/k)))\delta_{\vartheta(k, \alpha_k, C/k)},$$

cf. 10.7.3 and 10.7.4.

13.2.2. Now fix an integer $r \geq 1$, and an offset vector c in $\mathbb{Z}^r$ with $c(r) \leq g$. We have the continuous map $F_c : USp(2g)^\# \to \mathbb{R}^r$ of 13.1.3. Given a genus g curve over a finite field, say C/k, we denote by

$$F_c(\vartheta(k, \alpha_k, C/k)) := \vartheta(k, \alpha_k, C/k)(c(1), \ldots, c(r))$$

the r-tuple of its normalized angles named by the offset vector c. When we combine the deep results 10.7.12 and 10.7.15 with the trivial result 13.0.3, we find

Theorem 13.2.3. *Fix a genus $g \geq 3$, an integer $r \geq 1$, and an offset vector c in $\mathbb{Z}^r$ with $c(r) \leq g$. In any sequence of data (k_i, α_{k_i}) with* $\operatorname{Card}(k_i)$ *increasing to infinity, the two sequences of measures*

$$(F_c)_*\mu(\textit{naive}, g, k_i, \alpha_{k_i})$$
$$:= (1/\operatorname{Card}(\mathcal{M}_g(k_i))) \sum_{C/k \textit{ in } \mathcal{M}_g(k_i)} \delta_{\vartheta(k_i, \alpha_{k_i}, C/k_i)(c(1), \ldots, c(r))}$$

and

$$\begin{aligned}(F_c)_*\mu(intrin, g, k_i, \alpha_k)\\ &:= (1/\operatorname{Intrin}\operatorname{Card}(\mathcal{M}_g(k_i)))\\ &\quad\times \sum_{C/k_i \text{ in } \mathcal{M}_g(k_i)} (1/\operatorname{Card}(\operatorname{Aut}(C/k_i)))\delta_{\vartheta(k,\alpha_{k_i},C/k_i)(c(1),\ldots,c(r))},\end{aligned}$$

both converge weak $*$ to the measure $\nu(c, USp(2g))$ on $\mathbb{R}^r$.

13.2.4. We now turn to universal families of hyperelliptic curves. For each integer $d \geq 3$, we have the space $\mathcal{H}_d$ of monic, degree d polynomials f with invertible discriminant, and, over $\mathcal{H}_d[1/2]$, the family of hyperelliptic, genus $g := [(d-1)/2]$, curves $Y^2 = f(X)$, cf. 10.1.18. In 10.8.1 we defined, for every finite field k of odd characteristic, and every choice of a square root of $\operatorname{Card}(k)$, a probability measure $\mu(\text{hyp}, d, g, k, \alpha_k)$ on $USp(2g)^\#$, defined as

$$\mu(\text{hyp}, d, g, k, \alpha_k) := (1/\operatorname{Card}(\mathcal{H}_d(k))) \sum_{C/k \text{ in } \mathcal{H}_d(k)} \delta_{\vartheta(k,\alpha_k,C_f/k)}.$$

Theorem 13.2.5. *Fix a degree $d \geq 3$, and define $g := [(d-1)/2]$. Fix an integer $r \geq 1$, and an offset vector c in $\mathbb{Z}^r$ with $c(r) \leq g$. In any sequence of data (k_i, α_{k_i}) with $\operatorname{Card}(k_i)$ odd and increasing to infinity, the sequence of measures*

$$\begin{aligned}(F_c)_*\mu(hyp, d, g, k_i, \alpha_{k_i})\\ &:= (1/\operatorname{Card}(\mathcal{H}_d(k_i))) \sum_{f \text{ in } \mathcal{H}_d(k)} \delta_{\vartheta(k_i,\alpha_{k_i},C_f/k_i)(c(1),\ldots,c(r))}\end{aligned}$$

converges weak $*$ to the measure $\nu(c, USp(2g))$ on $\mathbb{R}^r$.

13.3. Equidistribution of low-lying eigenvalues in families of abelian varieties according to the measure $\nu(c, USp(2g))$

13.3.1. Fix an integer $g \geq 1$. Recall from 11.3.4 and 11.3.5 that by looking at "all" principally polarized g-dimensional abelian varieties we defined, for each finite field k and each choice α_k of a square root of $\operatorname{Card}(k)$, two probability measures on $USp(2g)^\#$, by looking at the normalized (by α_k) Frobenius conjugacy classes of all principally polarized abelian varieties over the given field k:

$$\mu(\text{naive}, g, \text{prin}, k, \alpha_k) := (1/\#(\text{Ab}_{g,\text{prin}}(k))) \sum_{(A/k,\varphi) \text{ in } \text{Ab}_{g,\text{prin}}(k)} \delta_{\vartheta(k,\alpha_k,A/k)}$$

and

$$\begin{aligned}\mu(\text{intrin}, g, \text{prin}, k, \alpha_k)\\ &:= (1/\operatorname{Intrin}\operatorname{Card}(\text{Ab}_{g,\text{prin}}(k))) \sum_{(A/k,\varphi) \text{ in } \text{Ab}_{g,\text{prin}}(k)} (1/\#\operatorname{Aut}(A/k,\varphi))\delta_{\vartheta(k,\alpha_k,A/k)}.\end{aligned}$$

Exactly as in the case above of curves, we combine 11.3.10 with 13.0.3 to obtain

Theorem 13.3.2. *Fix an integer $g \geq 3$, an integer $r \geq 1$, and an offset vector c in $\mathbb{Z}^r$ with $c(r) \leq g$. In any sequence of data (k_i, α_{k_i}) with $\operatorname{Card}(k_i)$ increasing to*

infinity, the two sequences of measures

$$(F_c)_*\mu(naive, g, \mathrm{prin}, k_i, \alpha_{k_i}) := (1/\#(\mathrm{Ab}_{g,\mathrm{prin}}(k_i))) \sum_{(A/k_i,\varphi)\ in\ \mathrm{Ab}_{g,\mathrm{prin}}(k_i)} \delta_{\vartheta(k_i,\alpha_{k_i},A/k_i)(c(1),\dots,c(r))}$$

and

$$(F_c)_*\mu(intrin, g, \mathrm{prin}, k_i, \alpha_{k_i}) := (1/\,\mathrm{Intrin\,Card}(\mathrm{Ab}_{g,\mathrm{prin}}(k_i))) \times \sum_{(A/k_i,\varphi)\ in\ \mathrm{Ab}_{g,\mathrm{prin}}(k_i)} (1/\#\,\mathrm{Aut}(A/k_i,\varphi))\delta_{\vartheta(k_i,\alpha_{k_i},A/k_i)(c(1),\dots,c(r))}$$

both converge weak $*$ to the measure $\nu(c, USp(2g))$ on $\mathbb{R}^r$.

13.4. Equidistribution of low-lying eigenvalues in families of odd-dimensional hypersurfaces according to the measure $\nu(c, USp(\mathrm{prim}(n,d)))$

13.4.1. Fix an odd integer n, and a degree $d \geq 3$. If $n = 1$, suppose that $d \geq 4$. Recall from 11.9.3 and 11.9.4 that by looking at "all" projective smooth hypersurfaces in $\mathbb{P}^{n+1}$ of degree d, we defined, for each finite field k and each choice α_k of a square root of $\mathrm{Card}(k)$, two probability measures on $USp(\mathrm{prin}(n,d))^\#$, by looking at the normalized (by α_k) Frobenius conjugacy classes of all such hypersurfaces over the given field k:

$$\mu(\mathrm{naive}, n, d, k, \alpha_k) := (1/\#(\mathrm{Iso}\,\mathcal{H}_{n,d}(k))) \sum_{X/k \text{ in } \mathrm{Iso}\,\mathcal{H}_{n,d}(k)} \delta_{\vartheta(k,\alpha_k,X/k)},$$

and

$$\mu(\mathrm{intrin}, n, d, k, \alpha_k) := (1/\,\mathrm{Intrin\,Card}(\mathrm{Iso}\,\mathcal{H}_{n,d}(k))) \sum_{X/k \text{ in } \mathrm{Iso}\,\mathcal{H}_{n,d}(k)} (1/\#\,\mathrm{Proj\,Aut}(X/k))\delta_{\vartheta(k,\alpha_k,X/k)}.$$

Exactly as in the cases above of curves and abelian varieties, we combine 11.9.9 with 13.0.3 to obtain

Theorem 13.4.2. *Fix an odd integer n, and a degree $d \geq 3$. If $n = 1$, suppose that $d \geq 4$. Fix an integer $r \geq 1$, and an offset vector c in $\mathbb{Z}^r$ with $c(r) \leq \mathrm{prim}(n,d)/2$. In any sequence of data (k_i, α_{k_i}) with $\mathrm{Card}(k_i)$ increasing to infinity, the two sequences of measures*

$$(F_c)_*\mu(naive, n, d, k_i, \alpha_{k_i}) := (1/\#(\mathrm{Iso}\,\mathcal{H}_{n,d}(k_i))) \sum_{X/k_i\ in\ \mathrm{Iso}\,\mathcal{H}_{n,d}(k_i)} \delta_{\vartheta(k_i,\alpha_{k_i},X/k_i)(c(1),\dots,c(r))}$$

and

$$(F_c)_*\mu(intrin, n, d, k_i, \alpha_{k_i}) := (1/\,\mathrm{Intrin\,Card}(\mathrm{Iso}\,\mathcal{H}_{n,d}(k))) \times \sum_{X/k\ in\ \mathrm{Iso}\,\mathcal{H}_{n,d}(k_i)} (1/\#\,\mathrm{Proj\,Aut}(X/k_i))\delta_{\vartheta(k_i,\alpha_{k_i},X/k_i)(c(1),\dots,c(r))}$$

both converge weak $*$ to the measure $\nu(c, USp(\mathrm{prim}(n,d)))$ on $\mathbb{R}^r$.

13.5. Equidistribution of low-lying eigenvalues of Kloosterman sums in evenly many variables according to the measure $\nu(c, USp(2n))$

13.5.1. Fix an even integer $2n$. Recall from 11.10.4 that by looking at all the Kloosterman sums in $2n$ variables, we defined, for each finite field k and each choice α_k of a square root of Card(k), a probability measure on $USp(2n)^{\#}$, by looking at the normalized (by α_k) Frobenius conjugacy classes of all such Kloosterman sums over the given field k:

$$\mu(2n, k, \alpha_k) := \mu(2n, k, \psi, \alpha_k) := (1/\operatorname{Card}(k^\times)) \sum_{a \text{ in } k^\times} \delta_{\vartheta(2n,k,\psi,a,\alpha_k)}.$$

Exactly as in the above cases of curves, abelian varieties, and odd-dimensional hypersurfaces, we combine 11.10.5 with 13.0.3 to obtain

Theorem 13.5.2. *Fix an even integer $2n$, an integer $r \geq 1$, and an offset vector c in $\mathbb{Z}^r$ with $c(r) \leq n$. In any sequence of data (k_i, α_{k_i}) with* Card(k_i) *increasing to infinity, the sequence of measures*

$$(F_c)_*\mu(2n, k_i, \alpha_{k_i}) := (1/\operatorname{Card}(k_i^\times)) \sum_{a \text{ in } k_i^\times} \delta_{\vartheta(2n,k_i,\psi,a,\alpha_{k_i})(c(1),\ldots,c(r))}$$

converges weak $*$ to the measure $\nu(c, USp(2n))$ on $\mathbb{R}^r$.

13.6. Equidistribution of low-lying eigenvalues of characteristic two Kloosterman sums in oddly many variables according to the measure $\nu(c, SO(2n+1))$

13.6.1. Fix an odd integer $2n+1 \geq 3, 2n+1 \neq 7$. Recall from 11.10.4 that by looking at all the characteristic two Kloosterman sums in $2n+1$ variables, we defined, for each finite field k of characteristic two and each choice α_k of a square root of Card(k), a probability measure on $SO(2n+1)^{\#}$, by looking at the normalized (by $\alpha_k := \operatorname{Card}(k)^{(n-1)/2}$) Frobenius conjugacy classes of all such Kloosterman sums over the given field k:

$$\begin{aligned}\mu(2n+1, k, \alpha_k) &:= \mu(2n+1, k, \psi, \alpha_k)\\ &:= (1/\operatorname{Card}(k_i^\times)) \sum_{a \text{ in } k_i^\times} \delta_{\vartheta(2n+1,k,\psi,a,\alpha_k)}.\end{aligned}$$

Exactly as in the above cases of curves, abelian varieties, and odd-dimensional hypersurfaces, we combine 11.10.5 with 13.0.3 to obtain

Theorem 13.6.2. *Fix an odd integer $2n+1 \geq 3, 2n+1 \neq 7$, an integer $r \geq 1$, and an offset vector c in $\mathbb{Z}^r$ with $c(r) \leq n$. In any sequence of finite fields k_i with k_i of characteristic two and with* Card(k_i) *increasing to infinity, the sequence of measures*

$$(F_c)_*\mu(2n+1, k, \alpha_k) := (1/\operatorname{Card}(k^\times)) \sum_{a \text{ in } k^\times} \delta_{\vartheta(2n+1,k,\psi,a,\alpha_k)(c(1),\ldots,c(r))}$$

converges weak $*$ to the measure $\nu(c, SO(2n+1))$ on $\mathbb{R}^r$.

13.7. Equidistribution of low-lying eigenvalues in families of even-dimensional hypersurfaces according to the measures $\nu(c, SO(\mathrm{prim}(n,d)))$ and $\nu(c, O_-(\mathrm{prim}(n,d)))$

13.7.1. Fix an even integer $n \geq 2$, and a degree $d \geq 3$. If $n = 2$, assume $d \geq 4$. Recall from 11.9.3 and 11.9.4 that by looking at "all" projective smooth hypersurfaces in $\mathbb{P}^{n+1}$ of degree d, we defined, for each finite field k and the choice $\alpha_k = \mathrm{Card}(k)^{n/2}$, two probability measures on $O(\mathrm{prim}(n,d))^\#$, by looking at the normalized (by $\alpha_k = \mathrm{Card}(k)^{n/2}$) Frobenius conjugacy classes of all such hypersurfaces over the given field k:

$$\mu(\mathrm{naive}, n, d, k, \alpha_k) := (1/\#(\mathrm{Iso}\,\mathcal{H}_{n,d}(k))) \sum_{X/k \text{ in } \mathrm{Iso}\,\mathcal{H}_{n,d}(k)} \delta_{\vartheta(k,\alpha_k,X/k)}$$

and

$$\mu(\mathrm{intrin}, n, d, k, \alpha_k)$$
$$:= (1/\,\mathrm{Intrin\,Card}(\mathrm{Iso}\,\mathcal{H}_{n,d}(k))) \sum_{X/k \text{ in } \mathrm{Iso}\,\mathcal{H}_{n,d}(k)} (1/\#\,\mathrm{Proj\,Aut}(X/k))\delta_{\vartheta(k,\alpha_k,X/k)}.$$

We now combine 11.9.9 with 13.0.2, applied to $K = O(\mathrm{prim}(n,d))$, to $\Gamma = \{\pm 1\}$, and to the determinant homomorphism. For each finite field k, we denote by $\mathrm{Iso}\,\mathcal{H}_{n,d,+}(k)$ and $\mathrm{Iso}\,\mathcal{H}_{n,d,-}(k)$ respectively the two subsets of $\mathrm{Iso}\,\mathcal{H}_{n,d}(k)$ where the determinant of the normalized Frobenius takes the values $+1$ and -1 respectively. [These are the subsets $X_{n,\gamma}$ in this instance of 13.0.2.] For all k of sufficiently large cardinality, 13.0.2 tells us that both of these subsets $\mathrm{Iso}\,\mathcal{H}_{n,d,\pm}(k)$ are nonempty. Denote by $O_\pm(\mathrm{prim}(n,d))^\#$ the two subsets of $O(\mathrm{prim}(n,d))^\#$ where the determinant takes the values ± 1. For k sufficiently large we may form on each of them the two probability measures

$$\mu(\pm,\ \mathrm{naive}, n, d, k, \alpha_k) := (1/\#(\mathrm{Iso}\,\mathcal{H}_{n,d,\pm}(k))) \sum_{X/k \text{ in } \mathrm{Iso}\,\mathcal{H}_{n,d,\pm}(k)} \delta_{\vartheta(k,\alpha_k,X/k)}$$

and

$$\mu(\pm,\ \mathrm{intrin}, n, d, k, \alpha_k) := (1/\,\mathrm{Intrin\,Card}(\mathrm{Iso}\,\mathcal{H}_{n,d,\pm}(k)))$$
$$\times \sum_{X/k \text{ in } \mathrm{Iso}\,\mathcal{H}_{n,d,\pm}(k)} (1/\#\,\mathrm{Proj\,Aut}(X/k))\delta_{\vartheta(k,\alpha_k,X/k)},$$

where we define

$$\mathrm{Intrin\,Card}(\mathrm{Iso}\,\mathcal{H}_{n,d,\pm}(k)) := \sum_{X/k \text{ in } \mathrm{Iso}\,\mathcal{H}_{n,d,\pm}(k)} (1/\#\,\mathrm{Proj\,Aut}(X/k)).$$

According to 13.0.2, for any sequence of finite fields k_i whose cardinalities increase to infinity, the corresponding sequences of measures $\{\mu(\pm,\ \mathrm{naive}, n, d, k_i, \alpha_{k_i})\}$ and $\{\mu(\pm,\ \mathrm{intrin}, n, d, k_i, \alpha_{k_i})\}$ both converge weak $*$ to the normalized Haar measure on $O_\pm(\mathrm{prim}(n,d))^\#$. If we now further apply 13.0.3, we obtain

Theorem 13.7.2. *Hypotheses and notations as in* 13.7.1 *above, fix an integer* $r \geq 1$, *and an offset vector* c *in* $\mathbb{Z}^r$ *with* $c(r) < \mathrm{prim}(n,d)/2$. *Fix a choice of sign* $\pm$. *In any sequence of finite fields* k_i *with* $\mathrm{Card}(k_i)$ *increasing to infinity, the two* *sequences of measures*

$$(F_c)_*\mu(\pm,\ naive, n, d, k, \alpha_k)$$
$$:= (1/\#(\mathrm{Iso}\,\mathcal{H}_{n,d,\pm}(k))) \sum_{X/k \text{ in } \mathrm{Iso}\,\mathcal{H}_{n,d,\pm}(k)} \delta_{\vartheta(k,\alpha_k,X/k)(c(1),\ldots,c(r))}$$

and

$$(F_c)_*\mu(\pm,\ intrin, n, d, k, \alpha_k)$$
$$:= (1/\operatorname{Intrin\,Card}(\mathrm{Iso}\,\mathcal{H}_{n,d,\pm}(k)))$$
$$\times \sum_{X/k \text{ in } \mathrm{Iso}\,\mathcal{H}_{n,d,\pm}(k)} (1/\#\operatorname{Proj\,Aut}(X/k))\delta_{\vartheta(k,\alpha_k,C/k)(c(1),\ldots,c(r))}$$

both converge weak $*$ to the measure

$$\nu(c, SO(\mathrm{prim}(n,d))) \quad \textit{on } \mathbb{R}^r,\ \textit{if the chosen sign is } +,$$
$$\nu(c, O_-(\mathrm{prim}(n,d))) \quad \textit{on } \mathbb{R}^r,\ \textit{if the chosen sign is } -.$$

13.8. Passage to the large N limit

13.8.1. We now combine the results of the last sections 13.2–7 with 8.4.17. In applying 8.4.17 with the orthogonal group $SO(N)$ and with $O_-(N)$, the parity of N is of vital importance. In the monodromy of even-dimensional hypersurfaces of dimension n and degree d, it is the orthogonal group $O(\mathrm{prim}(n,d))$ which occurs. Thus it will be important to know the parity of $\mathrm{prim}(n,d)$.

Lemma 13.8.2. *Given integers $n \geq 1$ and $d \geq 1$, the parity of*

$$\mathrm{prim}(n,d) := (d-1)((d-1)^{n+1} - (-1)^{n+1})/d$$

is

even, if n is odd or if d is odd

odd, if both n and d are even.

PROOF. First observe that

$$p_n(X) := ((X-1)^{n+1} - (-1)^{n+1})/X$$

lies in $\mathbb{Z}[X]$, and that its constant term is $(-1)^n(n+1)$. Now $\mathrm{prim}(n,d)$ is the value at $X = d$ of $(X-1)p_n(X)$. Thus $\mathrm{prim}(n,d)$ is divisible by $d-1$, hence is even if d is odd. Suppose now that d is even. Then $\mathrm{prim}(n,d)$ has the same parity as $p_n(d)$, and $p_n(d) \bmod d$ is equal to $(-1)^n(n+1)$. Thus for d even, $\mathrm{prim}(n,d)$ has the same parity as $n+1$. QED

13.8.3. It is now a simple matter to combine 13.2–7, which told us how to compute the measures $\nu(c, G(N))$ as limits over larger and larger finite fields, with 8.4.17, which tells us how to recover the measures $\nu(\pm, c)$ as limits of the measures $\nu(c, G(N))$ in suitable sequences of $G(N)$'s. Using 13.8.2 to keep track of parities among the orthogonal groups, we find the following Theorems 13.8.4 and 13.8.5.

Theorem 13.8.4. *Fix an integer $r \geq 1$, an offset vector c in $\mathbb{Z}^r$, and a bounded, continuous $\mathbb{C}$-valued function f on $\mathbb{R}^r$. Then $\int_{\mathbb{R}^r} f\,d\nu(-,c)$ can be computed by means of each of the following double limits*:

1) (*via curves*) *Pick any sequence of genera g_i, all $> c(r)/2$ and increasing to infinity, and for each g_i choose a sequence $k_{i,j}$ of finite fields whose cardinalities*

increase to infinity. For each $k_{i,j}$*, choose* $\alpha_{k_{i,j}}$ *a square root of* $\mathrm{Card}(k_{i,j})$*. Then we have the double limit formulas*

$$\int_{\mathbb{R}^r} f\,d\nu(-,c) = \lim_i \lim_j \int_{\mathbb{R}^r} fd(F_c)_*\mu(naive, g_i, k_{i,j}, \alpha_{k_{i,j}})$$
$$= \lim_i \lim_j \int_{\mathbb{R}^r} fd(F_c)_*\mu(intrin, g_i, k_{i,j}, \alpha_{k_{i,j}}).$$

2) (*via hyperelliptic curves*) *Pick any sequence of degrees* d_i*, all* $> c(r)+2$ *and increasing to infinity, put* $g_i := [(d_i - 1)/2]$*, and for each* d_i *choose a sequence* $k_{i,j}$ *of finite fields whose odd cardinalities increase to infinity. For each* $k_{i,j}$*, choose* $\alpha_{k_{i,j}}$ *a square root of* $\mathrm{Card}(k_{i,j})$*. Then we have the double limit formula*

$$\int_{\mathbb{R}^r} f\,d\nu(-,c) = \lim_i \lim_j \int_{\mathbb{R}^r} fd(F_c)_*\mu(hyp, d_i, g_i, k_{i,j}, \alpha_{k_{i,j}})$$

3) (*via abelian varieties*) *Pick any sequence of dimensions* g_i*, all* $> c(r)/2$ *and increasing to infinity, and for each* g_i *choose a sequence* $k_{i,j}$ *of finite fields whose cardinalities increase to infinity. For each* $k_{i,j}$*, choose* $\alpha_{k_{i,j}}$ *a square root of* $\mathrm{Card}(k_{i,j})$*. Then we have the double limit formulas*

$$\int_{\mathbb{R}^r} f\,d\nu(-,c) = \lim_i \lim_j \int_{\mathbb{R}^r} fd(F_c)_*\mu(naive, g_i, \mathrm{prin}, k_{i,j}, \alpha_{k_{i,j}})$$
$$= \lim_i \lim_j \int_{\mathbb{R}^r} fd(F_c)_*\mu(intrin, g_i, \mathrm{prin}, k_{i,j}, \alpha_{k_{i,j}}).$$

4) (*via odd-dimensional hypersurfaces*) *Pick a sequence of pairs* (n_i, d_i) *with* n_i *an odd integer* ≥ 3*,* d_i *an integer* ≥ 3*, such that each* $\mathrm{prim}(n_i, d_i) > c(r)/2$ *and such that the* $\mathrm{prim}(n_i, d_i)$ *increase to infinity. For each* (n_i, d_i) *choose a sequence* $k_{i,j}$ *of finite fields whose cardinalities increase to infinity. For each* $k_{i,j}$*, choose* $\alpha_{k_{i,j}}$ *a square root of* $\mathrm{Card}(k_{i,j})^{n_i}$*. Then we have the double limit formulas*

$$\int_{\mathbb{R}^r} f\,d\nu(-,c) = \lim_i \lim_j \int_{\mathbb{R}^r} fd(F_c)_*\mu(naive, n_i, d_i, k_{i,j}, \alpha_{k_{i,j}})$$
$$= \lim_i \lim_j \int_{\mathbb{R}^r} fd(F_c)_*\mu(intrin, n_i, d_i, k_{i,j}, \alpha_{k_{i,j}}).$$

5) (*via Kloosterman sums in evenly many variables*) *Pick a sequence of even integers* n_i*, each* $> c(r)/2$ *and increasing to infinity. For each* n_i *choose a sequence* $k_{i,j}$ *of finite fields whose cardinalities increase to infinity. For each* $k_{i,j}$*, choose* $\alpha_{k_{i,j}}$ *a square root of* $\mathrm{Card}(k_{i,j})^{n_i-1}$*. Then we have the double limit formula*

$$\int_{\mathbb{R}^r} f\,d\nu(-,c) = \lim_i \lim_j \int_{\mathbb{R}^r} fd(F_c)_*\mu(n_i, k_{i,j}, \alpha_{k_{i,j}}).$$

6) (*via characteristic two Kloosterman sums in an odd number of variables*) *Pick a sequence of odd integers* n_i*, each* $> 7 + c(r)/2$ *and increasing to infinity. For each* n_i *choose a sequence* $k_{i,j}$ *of finite fields of even characteristic whose cardinalities increase to infinity. For each* $k_{i,j}$*, choose* $\alpha_{k_{i,j}}$ *to be* $\mathrm{Card}(k_{i,j})^{(n_i-1)/2}$*. Then we have the double limit formula*

$$\int_{\mathbb{R}^r} f\,d\nu(-,c) = \lim_i \lim_j \int_{\mathbb{R}^r} fd(F_c)_*\mu(n_i, k_{i,j}, \alpha_{k_{i,j}}).$$

7) (*via certain even-dimensional hypersurfaces*) *Pick a sequence of pairs* (n_i, d_i) *with* n_i *an even integer* ≥ 4*,* d_i *an even integer* ≥ 4*. [These parity choices mean*

precisely that prim(n_i, d_i) *is* ***odd***.] *Suppose that each* prim$(n_i, d_i) > c(r)/2$ *and that the* prim(n_i, d_i) *increase to infinity. For each* (n_i, d_i) *choose a sequence* $k_{i,j}$ *of finite fields whose cardinalities increase to infinity. For each* $k_{i,j}$, *choose* $\alpha_{k_{i,j}}$ *a square root of* Card$(k_{i,j})^{n_i}$. *Then we have the double limit formulas*

$$\int_{\mathbb{R}^r} f\, d\nu(-,c) = \lim_i \lim_j \int_{\mathbb{R}^r} fd(F_c)_*\mu(+,\ \textit{naive}, n_i, d_i, k_{i,j}, \alpha_{k_{i,j}})$$
$$= \lim_i \lim_j \int_{\mathbb{R}^r} fd(F_c)_*\mu(+,\ \textit{intrin}, n_i, d_i, k_{i,j}, \alpha_{k_{i,j}}).$$

8) (*via certain other even-dimensional hypersurfaces*) *Pick a sequence of pairs* (n_i, d_i) *with* n_i *an even integer* ≥ 4, d_i *an odd integer* ≥ 3. [*This choice of parities insures that* prim(n_i, d_i) *is* ***even***.] *Suppose that each* prim$(n_i, d_i) > c(r)/2$ *and that the* prim(n_i, d_i) *increase to infinity. For each* (n_i, d_i) *choose a sequence* $k_{i,j}$ *of finite fields whose cardinalities increase to infinity. For each* $k_{i,j}$, *choose* $\alpha_{k_{i,j}}$ *a square root of* Card$(k_{i,j})^{n_i}$. *Then we have the double limit formulas*

$$\int_{\mathbb{R}^r} f\, d\nu(-,c) = \lim_i \lim_j \int_{\mathbb{R}^r} fd(F_c)_*\mu(-,\ \textit{naive}, n_i, d_i, k_{i,j}, \alpha_{k_{i,j}})$$
$$= \lim_i \lim_j \int_{\mathbb{R}^r} fd(F_c)_*\mu(-,\ \textit{intrin}, n_i, d_i, k_{i,j}, \alpha_{k_{i,j}}).$$

Theorem 13.8.5. *Fix an integer* $r \geq 1$, *an offset vector* c *in* $\mathbb{Z}^r$, *and a bounded, continuous* $\mathbb{C}$*-valued function* f *on* $\mathbb{R}^r$. *Then* $\int_{\mathbb{R}^r} f\, d\nu(+,c)$ *can be computed by means of each of the following double limits*:

1) (*via certain even-dimensional hypersurfaces*) *Pick a sequence of pairs* (n_i, d_i) *with* n_i *an even integer* ≥ 4, d_i *an even integer* ≥ 4. [*These parity choices mean precisely that* prim(n_i, d_i) *is* ***odd***.] *Suppose that each* prim$(n_i, d_i) > c(r)/2$ *and that the* prim(n_i, d_i) *increase to infinity. For each* (n_i, d_i) *choose a sequence* $k_{i,j}$ *of finite fields whose cardinalities increase to infinity. For each* $k_{i,j}$, *choose* $\alpha_{k_{i,j}}$ *a square root of* Card$(k_{i,j})^{n_i}$. *Then we have the double limit formulas*

$$\int_{\mathbb{R}^r} f\, d\nu(+,c) = \lim_i \lim_j \int_{\mathbb{R}^r} fd(F_c)_*\mu(-,\ \textit{naive}, n_i, d_i, k_{i,j}, \alpha_{k_{i,j}})$$
$$= \lim_i \lim_j \int_{\mathbb{R}^r} fd(F_c)_*\mu(-,\ \textit{intrin}, n_i, d_i, k_{i,j}, \alpha_{k_{i,j}}).$$

2) (*via certain other even-dimensional hypersurfaces*) *Pick a sequence of pairs* (n_i, d_i) *with* n_i *an even integer* ≥ 4, d_i *an odd integer* ≥ 3. [*These parity choices insure that* prim(n_i, d_i) *is* ***even***.] *Suppose that each* prim$(n_i, d_i) > c(r)/2$ *and that the* prim(n_i, d_i) *increase to infinity. For each* (n_i, d_i) *choose a sequence* $k_{i,j}$ *of finite fields whose cardinalities increase to infinity. For each* $k_{i,j}$, *choose* $\alpha_{k_{i,j}}$ *a square root of* Card$(k_{i,j})^{n_i}$. *Then we have the double limit formulas*

$$\int_{\mathbb{R}^r} f\, d\nu(+,c) = \lim_i \lim_j \int_{\mathbb{R}^r} fd(F_c)_*\mu(+,\ \textit{naive}, n_i, d_i, k_{i,j}, \alpha_{k_{i,j}})$$
$$= \lim_i \lim_j \int_{\mathbb{R}^r} fd(F_c)_*\mu(+,\ \textit{intrin}, n_i, d_i, k_{i,j}, \alpha_{k_{i,j}}).$$

Appendix: Densities

AD.0. Overview

AD.0.1. In this appendix, we define, for each integer $n \geq 1$, the n-level scaling density of eigenvalues near one for $G(N)$ any of $U(N), SO(2N+1), USp(2N)$, $SO(2N), O_-(2N+2), O_-(2N+1)$, and we determine their large N limits. We then give the relation of the eigenvalue location measures $\nu(c, G(N))$, c any offset vector, to the scaling densities for $G(N)$, and use this relation to give a second proof of the existence of the large N limits of the eigenvalue location measures for these $G(N)$.

We then consider the scaling densities for $SU(N)$, and more generally for any of the groups

$$U_k(N) := \{A \text{ in } U(N) \text{ with } \det(A)^k = 1\} = \mu_{Nk} SU(N),$$

which for variable integers $k \geq 1$ are all the proper closed subgroups of $U(N)$ which contain $SU(N)$. Harold Widom has shown that for these groups, the large N limits of the scaling densities exist and are equal to the large N limits in the $U(N)$ case. We thank him for allowing us to present his result. The relation of the eigenvalue location measures to the scaling densities allows us to prove that for any offset vector c, and any sequence of integers $k_N \geq 1$, the large N limit of the eigenvalue location measures $\nu(c, U_{k_N}(N))$ exists and is equal to the measure $\nu(c)$ obtained in 7.11.13 and 8.4.17 as the large N limit of the measures $\nu(c, U(N))$. Using this result in the special case "all $k_N = 1$" of $SU(N)$, we then explain how to compute $\nu(c)$ via low-lying eigenvalues of Kloosterman sums in oddly many variables in odd characteristic.

In a final section, we define a variant of the 1-level scaling density for $G(N)$ which is defined in a more symmetric way and which is more amenable to analysis in certain problems, cf. [**Ka-Sar**].

AD.1. Basic definitions: $W_n(f, A, G(N))$ and $W_n(f, G(N))$

AD.1.1. Exactly as in 8.4.1, for $G(N)$ any of $U(N), U_k(N) :=$ the kernel of $\det^k$ in $U(N), SO(2N+1), USp(2N), SO(2N), O_-(2N+2), O_-(2N+1)$, and A in $G(N)$, we have its sequences of angles

$$0 \leq \varphi(1) \leq \varphi(2) \leq \cdots \leq \varphi(N) < 2\pi \quad \text{if } G(N) = U(N) \text{ or } U_k(N),$$
$$0 \leq \varphi(1) \leq \varphi(2) \leq \cdots \leq \varphi(N) \leq \pi \quad \text{for the other } G(N),$$

and its sequence of **normalized angles**

$$0 \leq \vartheta(1) \leq \vartheta(2) \leq \cdots \leq \vartheta(N) \leq N + \lambda,$$

defined by

$$\vartheta(n) := (N+\lambda)\varphi(n)/\sigma\pi.$$

Concretely,

$$\begin{aligned}
\vartheta(n) &:= N\varphi(n)/2\pi && \text{for } U(N) \text{ or } U_k(N),\\
\vartheta(n) &:= (N+1/2)\varphi(n)/\pi = (2N+1)\varphi(n)/2\pi && \text{for } SO(2N+1) \text{ or } O_-(2N+1),\\
\vartheta(n) &:= N\varphi(n)/\pi = 2N\varphi(n)/2\pi && \text{for } USp(2N) \text{ or } SO(2N),\\
\vartheta(n) &:= (N+1)\varphi(n)/\pi = (2N+2)\varphi(n)/2\pi && \text{for } O_-(2N+2).
\end{aligned}$$

AD.1.2. Fix an integer $r \geq 1$, an integer $N \geq r$, and one of the $G(N)$ above. Given a $\mathbb{C}$-valued, bounded, Borel measurable function f on $\mathbb{R}^r$, we define a $\mathbb{C}$-valued function on $G(N)$, $A \mapsto W_n(f, A, G(N))$ as follows. Denote by

$$0 \leq \varphi(1)(A) \leq \varphi(2)(A) \leq \cdots \leq \varphi(N)(A) \leq \sigma\pi$$

the sequence of angles of A in $G(N)$, and define

$$\begin{aligned}
&W_r(f, A, G(N))\\
&\quad := \sum_{1 \leq i(1) < i(2) < \cdots < i(r) \leq N} f((N+\lambda)\varphi(i(1))(A)/\sigma\pi, \ldots, (N+\lambda)\varphi(i(r))(A)/\sigma\pi).
\end{aligned}$$

Since formation of each angle $\varphi(i)$ is a bounded, Borel measurable function on $G(N)$, the function $A \mapsto W_r(f, A, G(N))$ is a bounded, Borel measurable function on $G(N)$. We may then define

$$W_r(f, G(N)) := \int_{G(N)} W_r(f, A, G(N))\, dA,$$

where dA denotes the normalized (total mass one) Haar measure on $G(N)$.

For $r > N$, we define $W_r(f, A, G(N)) = 0$ and $W_r(f, G(N)) = 0$.

AD.2. Large N limits: the easy case

AD.2.1. Recall the three kernels

$$K(x, y) := \sin(\pi(x-y))/\pi(x-y)$$

and

$$\begin{aligned}
K_\pm(x, y) &:= K(x, y) \pm K(-x, y)\\
&= \sin(\pi(x-y))/\pi(x-y) \pm \sin(\pi(x+y))/\pi(x+y).
\end{aligned}$$

For each integer $r > 1$, we define functions W_r and $W_{r,\pm}$ on $\mathbb{R}^r$ by

$$\begin{aligned}
W_r(x(1), \ldots, x(r)) &:= \det_{r\times r}(K(x(i), x(j))),\\
W_{r,\pm}(x(1), \ldots, x(r)) &:= \det_{r\times r}(K_\pm(x(i), x(j))).
\end{aligned}$$

Theorem AD.2.2. *For any integer $r \geq 1$, denote by dx usual Lebesgue measure on $\mathbb{R}^r$. Let f be a $\mathbb{C}$-valued, bounded, Borel measurable function f on $\mathbb{R}^r$ which is symmetric and whose restriction to $(\mathbb{R}_{\geq 0})^r$ is of compact support. For $G(N)$ any of $U(N), SO(2N+1), USp(2N), SO(2N), O_-(2N+2), O_-(2N+1)$, we have the* *large N limit formulas*

$$\begin{aligned}
&\lim_{N\to\infty} W_r(f, G(N)) \\
&= (1/r!)\int_{(\mathbb{R}_{\geq 0})^r} f(x)W_r(x)\,dx, \quad \textit{if } G(N) \textit{ is } U(N), \\
&= (1/r!)\int_{(\mathbb{R}_{\geq 0})^r} f(x)W_{r,+}(x)\,dx, \quad \textit{if } G(N) \textit{ is } SO(2N) \textit{ or } O_-(2N+1), \\
&= (1/r!)\int_{(\mathbb{R}_{\geq 0})^r} f(x)W_{r,-}(x)\,dx, \quad \textit{if } G(N) \textit{ is } USp(2N), O_-(2N+2) \\
&\qquad\qquad\qquad\qquad\qquad\qquad\qquad\qquad \textit{or } SO(2N+1).
\end{aligned}$$

PROOF. Recall (8.3.1) the L_N kernels attached to the various $G(N)$, expressed in terms of the function $S_N(x) := \sin(Nx/2)/\sin(x/2)$:

$G(N)$	$L_n(x,y)$
$U(N)$	$\sum_{n=0}^{N-1} e^{in(x-y)} = S_N(x-y)e^{i(N-1)(x-y)/2}$
other $G(N)$,	$(\sigma/2)[S_{\rho N+\tau}(x-y) + \varepsilon S_{\rho N+\tau}(x+y)]$, i.e.,
$SO(2N+1)$	$(1/2)(S_{2N}(x-y) - S_{2N}(x+y))$
$USp(2N)$ or $O_-(2N+2)$	$(1/2)(S_{2N+1}(x-y) - S_{2N+1}(x+y))$
$SO(2N)$	$(1/2)(S_{2N-1}(x-y) + S_{2N-1}(x+y))$
$O_-(2N+1)$	$(1/2)(S_{2N}(x-y) + S_{2N}(x+y))$.

Fix an integer $r \geq 1$, a function f on $\mathbb{R}^r$ as in the statement of the theorem, an integer $N \geq r$, and a choice of $G(N)$. Denote by $f_{G(N)}$ the function on $[0, \sigma\pi]^r$ defined by

$$f_{G(N)}(\varphi(1), \ldots, \varphi(r)) := f((N+\lambda)\varphi(1)/\sigma\pi, \ldots, (N+\lambda)\varphi(r)/\sigma\pi).$$

In the notation of 5.1.3, part 5), the function $f_{G(N)}[r, N]$ on $[0, \sigma\pi]^N$ is the function

$$\begin{aligned}
&(\varphi(1), \ldots, \varphi(N)) \\
&\mapsto \sum_{1 \leq i(1) < i(2) < \cdots < i(r) \leq N} f((N+\lambda)\varphi(i(1))/\sigma\pi, \ldots, (N+\lambda)\varphi(i(r))/\sigma\pi).
\end{aligned}$$

Because f is symmetric in its n variables, the function $f_{G(N)}[r, N]$ on $[0, \sigma\pi]^N$ is symmetric, so can be thought of as a central function on $G(N)$. Given an element A in $G(N)$, denote by $\varphi(A) := (\varphi(1)(A), \ldots, \varphi(N)(A))$ in $[0, \sigma\pi]^N$ its vector of angles. Then we have the tautologous identity

$$f_{G(N)}[r, N](\varphi(A)) = W_r(f, A, G(N)).$$

Thus we have

$$W_r(f, G(N)) := \int_{G(N)} W_r(f, A, G(N))\, dA = \int_{G(N)} f_{G(N)}[r, N](\varphi(A))\, dA,$$

which in the notations of 5.0 we may rewrite as

$$= \int_{[0,\sigma\pi]^N} f_{G(N)}[r, N]\, d\mu(G(N)).$$

From 5.1.3, part 5) and the L_N form (5.2 and 6.4) of the Weyl integration formula, we have the identity

$$\text{(AD.2.2.1)} \quad \begin{aligned} &\int_{[0,\sigma\pi]^N} f_{G(N)}[r, N]\, d\mu(G(N)) \\ &\quad = (1/r!) \int_{[0,\sigma\pi]^r} f_{G(N)}(x) \det_{r\times r}(L_N(x(i), x(j))) \prod_{i=1}^{r} (dx(i)/\sigma\pi). \end{aligned}$$

If we now make the change of variable $x(i) \mapsto \sigma\pi x(i)/(N+\lambda)$, this last integral becomes $(1/r!)$ times

$$\int_{[0,N+\lambda]^r} f(x) \det_{r\times r}(L_N(\sigma\pi x(i)/(N+\lambda), \sigma\pi x(j)/(N+\lambda))) \prod_i (dx(i)/(N+\lambda)).$$

Because $f|(\mathbb{R}_{\geq 0})^r$ is of compact support, for N large the support of $f|(\mathbb{R}_{\geq 0})^r$ is contained in the region where all $x(i) \leq N$, so for large N the integral (AD.2.2.1) is equal to $(1/r!)$ times

$$\int_{(\mathbb{R}_{\geq 0})^r} f(x) \det_{r\times r}(L_N(\sigma\pi x(i)/(N+\lambda), \sigma\pi x(j)/(N+\lambda))) \prod_i (dx(i)/(N+\lambda)),$$

i.e., (AD.2.2.1) is equal to

$$(1/r!) \int_{(\mathbb{R}_{\geq 0})^r} f(x) \det_{r\times r}((1/(N+\lambda)) L_N(\sigma\pi x(i)/(N+\lambda), \sigma\pi x(j)/(N+\lambda)))\, dx.$$

Thus for N large, the above integral is $W_r(f, G(N))$. To see that the large N limit of $W_r(f, G(N))$ is as asserted, we exploit the fact that $f|(\mathbb{R}_{\geq 0})^r$ is bounded and of compact support: it suffices to show that the continuous function

$$\det_{r\times r}((1/(N+\lambda)) L_N(\sigma\pi x(i)/(N+\lambda), \sigma\pi x(j)/(N+\lambda)))$$

converges pointwise,as $N \to \infty$, to $W_r(x)$ in the $U(N)$ case, and to the asserted choice of $W_{r,\pm}(x)$ in the remaining cases. This convergence results immediately from the explicit S_N formulas, recalled above, for the L_N kernel, together with the fact that for fixed x, $\sin(x)/N\sin(x/N)$ converges to $\sin(x)/x$ as $N \to +\infty$. QED

AD.2.3. Here is a minor variant, where we drop the requirement that the function f be symmetric, but change the domain of integration from $(\mathbb{R}_{\geq 0})^r$ to $(\mathbb{R}_{\geq 0})^r(\text{order})$, the closed set defined by the inequalities

$$0 \leq x(1) \leq x(2) \leq \cdots \leq x(r).$$

Theorem AD.2.4 (variant of AD.2.2). *For any integer $r \geq 1$, denote by dx the usual Lebesgue measure on $\mathbb{R}^r$. For any $\mathbb{C}$-valued, bounded, Borel measurable function f on $\mathbb{R}^r$ whose restriction to $(\mathbb{R}_{\geq 0})^r(\text{order})$ is of compact support, and for*

$G(N)$ any of $U(N), SO(2N+1), USp(2N), SO(2N), O_-(2N+2), O_-(2N+1)$, we have the large N limit formulas

$$\lim_{N\to\infty} W_r(f, G(N))$$
$$= \int_{(\mathbb{R}_{\geq 0})^r(order)} f(x)W_r(x)\,dx, \text{ if } G(N) \text{ is } U(N),$$
$$= \int_{(\mathbb{R}_{\geq 0})^r(order)} f(x)W_{r,+}(x)\,dx, \text{ if } G(N) \text{ is } SO(2N) \text{ or } O_-(2N+1),$$
$$= \int_{(\mathbb{R}_{\geq 0})^r(order)} f(x)W_{r,-}(x)\,dx, \text{ if } G(N) \text{ is } USp(2N), O_-(2N+2)$$
$$\text{or } SO(2N+1).$$

PROOF. Suppose first that f is symmetric. Because any $r\times r$ determinant of the shape $\det_{r\times r}(F(x(i),x(j)))$ is a symmetric function, each of $W_r(x)$ and $W_{r,\pm}(x)$ is symmetric, so in AD.2.2 we may rewrite the limits as

$$= \int_{(\mathbb{R}_{\geq 0})^r(\text{order})} f(x)W_r(x)\,dx, \quad \text{if } G(N) \text{ is } U(N),$$
$$= \int_{(\mathbb{R}_{\geq 0})^r(\text{order})} f(x)W_{r,+}(x)\,dx, \quad \text{if } G(N) \text{ is } SO(2N) \text{ or } O_-(2N+1),$$
$$= \int_{(\mathbb{R}_{\geq 0})^r(\text{order})} f(x)W_{r,-}(x)\,dx, \quad \text{if } G(N) \text{ is } USp(2N), O_-(2N+2)$$
$$\text{or } SO(2N+1).$$

Given any function f on $\mathbb{R}^r$, denote by $\tilde{f}$ the unique symmetric function on $\mathbb{R}^r$ which agrees with f on the set $\mathbb{R}^r(\text{order})$. Because $W_r(f,G(N))$ depends only on the restriction of f to $(\mathbb{R}_{\geq 0})^r(\text{order})$, we have $W_r(f,G(N)) = W_r(\tilde{f},G(N))$. Now apply to $\tilde{f}$ the above rewriting of AD.2.2, and remember that $\tilde{f}$ agrees with f on $(\mathbb{R}_{\geq 0})^r(\text{order})$. QED

We also record, for later use, the following lemma.

Lemma AD.2.5. *For any integer $r \geq 1$, denote by dx usual Lebesgue measure on $\mathbb{R}^r$. For any $\mathbb{C}$-valued, bounded, Borel measurable function f on $\mathbb{R}^r$, and for $G(N)$ any of $U(N), SO(2N+1), USp(2N), SO(2N), O_-(2N+2), O_-(2N+1)$, we have the formula*

$$W_r(f, G(N)) = \int_{[0,N+\lambda]^r(order)} f(x)W_{r,G(N)}(x)\,dx,$$

where $W_{r,G(N)}(x)$ denotes the $r\times r$ determinant

$$W_{r,G(N)}(x) := \det_{r\times r}((1/(N+\lambda))L_N(\sigma\pi x(i)/(N+\lambda), \sigma\pi x(j)/(N+\lambda)))$$

made from the L_N kernel for $G(N)$.

PROOF. In the proof of AD.2.2, we saw that for f as above and symmetric, we have

$$W_r(f, G(N)) = (1/r!)\int_{[0,N+\lambda]^r} f(x)W_{r,G(N)}(x)\,dx.$$

By the symmetry of both f and $W_{r,G(N)}(x)$, we thus have

$$W_r(f, G(N)) = \int_{[0,N+\lambda]^r(\text{order})} f(x) W_{r,G(N)}(x)\, dx.$$

Given any function f on $\mathbb{R}^r$, denote by $\tilde{f}$ the unique symmetric function on $\mathbb{R}^r$ which agrees with f on the set $\mathbb{R}^r(\text{order})$. Because $W_r(f, G(N))$ depends only on the restriction of f to $(\mathbb{R}_{\geq 0})^r(\text{order})$, we have $W_r(f, G(N)) = W_r(\tilde{f}, G(N))$. Now apply to $\tilde{f}$ the above, and remember that $\tilde{f}$ agrees with f on $(\mathbb{R}_{\geq 0})^r(\text{order})$. QED

AD.3. Relations between eigenvalue location measures and densities: generalities

AD.3.1. Fix an integer $r \geq 1$. Fix an offset vector c in $\mathbb{Z}^r$,

$$c = (c(1), \ldots, c(r)), \qquad 1 \leq c(1) < c(2) < \cdots < c(r),$$

and denote by a in $\mathbb{Z}^r$ the corresponding separation vector, cf. 1.0.5. Thus if we denote by $\mathbb{1}$ in $\mathbb{Z}^r$ the vector $(1, 1, \ldots, 1)$, we have

$$c = \text{Off}(\mathbb{1} + a).$$

AD.3.1.1. For each $N \geq c(r)$, and $G(N)$ any of $U(N), U_k(N), SO(2N+1)$, $USp(2N), SO(2N), O_-(2N+2), O_-(2N+1)$, we have a map

$$F_c : G(N) \to \mathbb{R}^r,$$
$$A \mapsto (\vartheta(c(1))(A), \ldots, \vartheta(c(r))(A))$$

defined by attaching to A in $G(N)$ the vector of those of its normalized angles named by the offset vector c. Just as in 8.4.1, we define the multi-eigenvalue location measure $\nu(c, G(N))$ on $\mathbb{R}^r$ to be the direct image by the map F_c of normalized Haar measure on $G(N)$.

AD.3.2. Given an offset vector c in $\mathbb{Z}^r$, we denote by

$$[\text{offset } c] : \mathbb{R}^{c(r)} \to \mathbb{R}^r$$

the projection onto the coordinates named by c: thus for x in $\mathbb{R}^{c(r)}$,

$$[\text{offset } c](x) := (x(c(1)), \ldots, x(c(r))).$$

Given a function f on $\mathbb{R}^r$, we denote by $f \circ [\text{offset } c]$ the composite function on $\mathbb{R}^{c(r)}$, defined by

$$f \circ [\text{offset } c](x) := f([\text{offset } c](x)).$$

[Notice that if the restriction of f to $(\mathbb{R}_{\geq 0})^r(\text{order})$ is of compact support, supported in $x(r) \leq \alpha$, then the restriction of $f \circ [\text{offset } c]$ to $(\mathbb{R}_{\geq 0})^{c(r)}(\text{order})$ is of compact support, supported in $x(c(r)) \leq \alpha$.]

AD.3.2.1. In terms of the corresponding separation vector a in $\mathbb{Z}^r$, we have $c(r) = r + \Sigma(a)$, and we also write the map [offset c] as

$$[\mathrm{sep}\, a] : \mathbb{R}^{r+\Sigma(a)} \to \mathbb{R}^r.$$

Thus for f a function on $\mathbb{R}^r$, we have

$$f \circ [\mathrm{sep}\, a](x) := f([\mathrm{sep}\, a](x)) := f([\text{offset } \mathrm{Off}(\mathbb{1} + a)](x)).$$

For any integer $N \geq c(r)$, any choice of $G(N)$ as in AD.3.1.1 above, and any element A in $G(N)$, we define

$$W(\mathrm{sep}\, a, f, A, G(N)) := W_{r+\Sigma(a)}(f \circ [\mathrm{sep}\, a], A, G(N)).$$

If f is $\mathbb{C}$-valued, bounded, and Borel measurable, we define

$$W(\mathrm{sep}\, a, f, G(N)) := \int_{G(N)} W(\mathrm{sep}\, a, f, A, G(N))\, dA.$$

Lemma AD.3.3. *Fix an integer $r \geq 1$ and an offset vector c in $\mathbb{Z}^r$ with corresponding separation vector a. Let f be a $\mathbb{C}$-valued function on $\mathbb{R}^r$. For any $N \geq r + \Sigma(a)$, any $G(N)$ as in* AD.3.1.1 *above, and any A in $G(N)$, we have*

$$\begin{aligned} W(\mathrm{sep}\, a, f, A, G(N)) &= \sum_{b \geq 0 \text{ in } \mathbb{Z}^r \text{ with } r+\Sigma(b) \leq N} \mathrm{Binom}(b, a) f(\vartheta(\mathrm{Off}(\mathbb{1} + b))(A)) \\ &= \sum_{b \geq a \text{ in } \mathbb{Z}^r \text{ with } r+\Sigma(b) \leq N} \mathrm{Binom}(b, a) f(\vartheta(\mathrm{Off}(\mathbb{1} + b))(A)). \end{aligned}$$

PROOF. The quantity $W(\mathrm{sep}\, a, f, A, G(N))$ is a sum of values of f at certain ordered r-tuples of normalized angles of A. A given offset vector $y := \mathrm{Off}(\mathbb{1} + b)$ of locations occurs as many times as there is a sequence of $c(r)$ indices

$$1 \leq i(1) < i(2) < \cdots < i(c(r)) \leq N$$

for which the subsequence of r indices

$$i(c(1)) < i(c(2)) < \cdots < i(c(r))$$

is the sequence

$$y(1) < y(2) < \cdots < y(r).$$

We claim this number is $\mathrm{Binom}(b, a)$. To see this, notice that in the sought for sequence of length $c(r)$, the indices named by y are given: It is in choosing the indices which go in between that we have some choice. We must pick $a(1)$ indices out of the $b(1)$ possible indices less than $y(1)$, we must pick $a(2)$ indices out of the $b(2)$ possible indices strictly between $y(1)$ and $y(2), \ldots,$ we must pick $a(r)$ indices out of the $b(r)$ possible indices strictly between $y(r-1)$ and $y(r)$. There are $\mathrm{Binom}(b, a)$ such choices. Since $\mathrm{Binom}(b, a)$ vanishes unless $b \geq a$, we may sum only over $b \geq a$. QED

Corollary AD.3.3.1. *Fix an integer $r \geq 1$ and an offset vector c in $\mathbb{Z}^r$ with corresponding separation vector a. Let f be a bounded, Borel measurable $\mathbb{C}$-valued*

function on $\mathbb{R}^r$. *For any* $N \geq r + \Sigma(a)$, *and any* $G(N)$ *as in* AD.3.1.1 *above, we have*

$$W(\operatorname{sep} a, f, G(N)) = \sum_{b \geq 0 \text{ in } \mathbb{Z}^r \text{ with } r+\Sigma(b) \leq N} \operatorname{Binom}(b, a) \int f \, d\nu(\operatorname{Off}(\mathbb{1} + b), G(N))$$

$$= \sum_{b \geq a \text{ in } \mathbb{Z}^r \text{ with } r+\Sigma(b) \leq N} \operatorname{Binom}(b, a) \int f \, d\nu(\operatorname{Off}(\mathbb{1} + b), G(N)).$$

PROOF. Integrate the previous result over $G(N)$. QED

AD.3.3.2. If we adopt the convention that both $W(\operatorname{sep} a, f, G(N))$ and $\nu(c, G(N)) = 0$ whenever $c(r) > N$, we may rewrite the above corollary as

Corollary AD.3.3.3. *Fix an integer* $r \geq 1$ *and an offset vector* c *in* $\mathbb{Z}^r$ *with corresponding separation vector* a. *Let* f *be a bounded, Borel measurable* $\mathbb{C}$*-valued function on* $\mathbb{R}^r$. *For any* $G(N)$ *as in* AD.3.1.1 *above, we have*

$$W(\operatorname{sep} a, f, G(N)) = \sum_{b \geq 0 \text{ in } \mathbb{Z}^r} \operatorname{Binom}(b, a) \int f \, d\nu(\operatorname{Off}(\mathbb{1} + b), G(N)).$$

Corollary AD.3.3.4. *Fix an integer* $r \geq 1$. *Let* f *be a bounded, Borel measurable* $\mathbb{C}$*-valued function on* $\mathbb{R}^r$. *For any* $G(N)$ *as above, the* r*-variable generating series (which are in fact polynomials)*

$$\sum_{a \geq 0 \text{ in } \mathbb{Z}^r} W(\operatorname{sep} a, f, G(N)) T^a$$

and

$$\sum_{b \geq 0 \text{ in } \mathbb{Z}^r} \int f \, d\nu(\operatorname{Off}(\mathbb{1} + b), G(N)) T^b$$

are related by

$$\sum_{a \geq 0 \text{ in } \mathbb{Z}^r} W(\operatorname{sep} a, f, G(N)) T^a = \sum_{b \geq 0 \text{ in } \mathbb{Z}^r} (1 + T)^b \int f \, d\nu(\operatorname{Off}(\mathbb{1} + b), G(N)),$$

and

$$\sum_{b \geq 0 \text{ in } \mathbb{Z}^r} T^b \int f \, d\nu(\operatorname{Off}(\mathbb{1} + b), G(N)) = \sum_{a \geq 0 \text{ in } \mathbb{Z}^r} W(\operatorname{sep} a, f, G(N)) (T - 1)^a.$$

Equating coefficients of T^a in the second identity, we get

Corollary AD.3.3.5. *Fix an integer* $r \geq 1$. *Let* f *be a bounded, Borel measurable* $\mathbb{C}$*-valued function on* $\mathbb{R}^r$. *For any* $G(N)$ *as in* AD.3.1.1 *above, we have*

$$\int f \, d\nu(\operatorname{Off}(\mathbb{1} + a), G(N)) = \sum_{b \geq a \text{ in } \mathbb{Z}^r} (-1)^{b-a} \operatorname{Binom}(b, a) W(\operatorname{sep} b, f, G(N)).$$

Remark AD.3.4. The reader will no doubt have noticed that the relation between the $W(\operatorname{sep} a, f, G(N))$ and the $\int f \, d\nu(\operatorname{Off}(\mathbb{1} + b), G(N))$ is precisely that between the $\operatorname{Clump}(a)$ and the $\operatorname{Sep}(b)$ of 2.3.8 and 2.4.9. Proceeding exactly as in 2.3.11 and 2.4.12, whose proofs depend only on these relations and on the fact that for real $f \geq 0$ both $W(\operatorname{sep} a, f, G(N))$ and $\int f \, d\nu(\operatorname{Off}(\mathbb{1} + b), G(N))$ are real and ≥ 0, we find

Proposition AD.3.5. *Let f be a nonnegative, $\mathbb{R}$-valued, bounded Borel measurable function on $\mathbb{R}^r$. For each $a \geq 0$ in $\mathbb{Z}^r$, and each integer $m \geq \Sigma(a)$, and each $G(N)$ as in* AD.3.1.1 *above, we have the following inequalities*:

If $m - \Sigma(a)$ is odd,

$$\sum_{b \geq a \text{ in } \mathbb{Z}^r, \Sigma(b) \leq m} (-1)^{b-a} \operatorname{Binom}(b, a) W(\operatorname{sep} b, f, G(N)) \leq \int f \, d\nu(\operatorname{Off}(\mathbb{1} + a), G(N)).$$

If $m - \Sigma(a)$ is even,

$$\int f \, d\nu(\operatorname{Off}(\mathbb{1} + a), G(N)) \leq \sum_{b \geq a \text{ in } \mathbb{Z}^r, \Sigma(b) \leq m} (-1)^{b-a} \operatorname{Binom}(b, a) W(\operatorname{sep} b, f, G(N)).$$

AD.4. Second construction of the large N limits of the eigenvalue location measures $\nu(c, G(N))$ for $G(N)$ one of $U(N)$, $SO(2N+1), USp(2N), SO(2N), O_-(2N+2), O_-(2N+1)$

AD.4.1. In the previous section AD.3, the results took place at finite level N, and were valid for $U_k(N)$ as well. But in this section, we must exclude the $U_k(N)$ case. We will return to the $U_k(N)$ case in the next section AD.5.

Lemma AD.4.2. *Let f be a $\mathbb{C}$-valued, bounded Borel measurable function on $\mathbb{R}^r$. Suppose that the restriction of f to $(\mathbb{R}_{\geq 0})^r(\text{order})$ is of compact support, supported in the region $x(r) \leq \alpha$. For any $a \geq 0$ in $\mathbb{Z}^r$, and for $G(N)$ any of $U(N), SO(2N+1), USp(2N), SO(2N), O_-(2N+2), O_-(2N+1)$, we have the estimate*

$$|W(\operatorname{sep} a, f, G(N))| \leq \|f\|_{\sup} (2\alpha)^{r+\Sigma(a)} / (r + \Sigma(a))!.$$

PROOF. Since $W(\operatorname{sep} a, f, G(N))$ is determined by the restriction of f to $(\mathbb{R}_{\geq 0})^r(\text{order})$, we may first replace f by the unique function $\tilde{f}$ which vanishes outside $(\mathbb{R}_{\geq 0})^r$, agrees with f on $(\mathbb{R}_{\geq 0})^r(\text{order})$, and is symmetric.

Thus we suppose that f is symmetric, supported in $(\mathbb{R}_{\geq 0})^r$ with support contained in the region $\operatorname{Sup}_i x(i) \leq \alpha$. Let us denote

$$R := r + \Sigma(a),$$
$$F := f \circ [\operatorname{sep} a].$$

We then have

$$W(\operatorname{sep} a, f, G(N)) := W_R(F, G(N)).$$

The restriction to $(\mathbb{R}_{\geq 0})^R(\text{order})$ of F has compact support contained in the domain $x(R) \leq \alpha$. Since the value of $W_R(F, G(N))$ depends only on the restriction of F to $(\mathbb{R}_{\geq 0})^R(\text{order})$, we may replace F by the unique symmetric function $\widetilde{F}$ which is supported in $(\mathbb{R}_{\geq 0})^R$ and which agrees with F on $(\mathbb{R}_{\geq 0})^R(\text{order})$. Notice that $\widetilde{F}$ is supported in the region $\operatorname{Sup}_i x(i) \leq \alpha$, i.e. in the region $[0, \alpha]^R$. Thus we have

$$W(\operatorname{sep} a, f, G(N)) := W_R(F, G(N)) = W_R(\widetilde{F}, G(N)).$$

As explained in the proof of AD.2.2, $W_R(\widetilde{F}, G(N))$ is equal to $(1/R!)$ times

$$\int_{[0,N+\lambda]^R} \widetilde{F}(x)\det_{R\times R}(L_N(\sigma\pi x(i)/(N+\lambda), \sigma\pi x(j)/(N+\lambda)))\prod_i (dx(i)/(N+\lambda)).$$

According to 5.3.3, we have the estimate

$$|\det_{R\times R}(L_N(x(i), x(j)))| \leq (2N)^R.$$

[This estimate holds also for the L_N kernel for $O_-(2N+1)$ because that kernel is the L_N kernel for $SO(2N+1)$ evaluated at $(\pi - x, \pi - y)$, cf. 6.4.8]. Thus we obtain

$$\begin{aligned}|W_R(\widetilde{F}, G(N))| &\leq \|f\|_{\sup}(1/R!)(2N/(N+\lambda))^R(\text{measure of } [0,\alpha]^R)\\ &\leq \|f\|_{\sup}(2\alpha/R!),\end{aligned}$$

which, tracing back, is precisely the required estimate

$$|W(\operatorname{sep} a, f, G(N))| \leq \|f\|_{\sup}(2\alpha)^{r+\Sigma(a)}/(r+\Sigma(a))!. \quad \text{QED}$$

Corollary AD.4.2.1. *Let f be a $\mathbb{C}$-valued, bounded Borel measurable function on $\mathbb{R}^r$. Suppose that the restriction of f to $(\mathbb{R}_{\geq 0})^r(\text{order})$ is of compact support, supported in the region $x(r) \leq \alpha$. For $G(N)$ any of $U(N), SO(2N+1)$, $USp(2N), SO(2N), O_-(2N+2), O_-(2N+1)$, the generating series polynomial*

$$\sum_{a\geq 0 \text{ in } \mathbb{Z}^r} W(\operatorname{sep} a, f, G(N))T^a$$

is dominated, coefficient by coefficient, by the entire function

$$\|f\|_{\sup}\exp\left(2\alpha\sum_{i=1}^r T_i\right).$$

Proof. Immediate from the previous estimate. QED

Taking the large N limit via AD.2.4, we get

Corollary AD.4.2.2. *Let f be a $\mathbb{C}$-valued, bounded Borel measurable function on $\mathbb{R}^r$. Suppose that the restriction of f to $(\mathbb{R}_{\geq 0})^r(\text{order})$ is of compact support, supported in the region $x(r) \leq \alpha$. For $G(N)$ any of $U(N), SO(2N+1)$, $USp(2N), SO(2N), O_-(2N+2), O_-(2N+1)$, the generating series*

$$W_G(f, T) := \sum_{a\geq 0 \text{ in } \mathbb{Z}^r}\left(\lim_{N\to\infty} W(\operatorname{sep} a, f, G(N))\right)T^a$$

is entire, and is dominated, coefficient by coefficient, by the entire function

$$\|f\|_{\sup}\exp\left(2\alpha\sum_{i=1}^r T_i\right).$$

Theorem AD.4.3. *Let f be a $\mathbb{C}$-valued, bounded Borel measurable function on $\mathbb{R}^r$. Suppose that the restriction of f to $(\mathbb{R}_{\geq 0})^r(\text{order})$ is of compact support. Denote by $G(N)$ any of $U(N), SO(2N+1), USp(2N), SO(2N), O_-(2N+2), O_-(2N+1)$.*

1) *For any $a \geq 0$ in $\mathbb{Z}^r$, the large N limit*

$$\lim_{N\to\infty}\int_{\mathbb{R}^r} f\, d\nu(\operatorname{Off}(\mathbb{1}+a), G(N))$$

exists and is equal to the sum of the convergent series

$$\sum_{b \geq a \text{ in } \mathbb{Z}^r} (-1)^{b-a} \operatorname{Binom}(b, a) \Big(\lim_{N \to \infty} W(\operatorname{sep} b, f, G(N)) \Big).$$

2) *For any offset vector c in $\mathbb{Z}^r$, consider the large N limit measures $\nu(c)$ and $\nu(\pm, c)$ of* 8.4.17. *We have*

$$\begin{aligned} &\lim_{N \to \infty} \int_{\mathbb{R}^r} f \, d\nu(c, G(N)) \\ &\quad = \int_{\mathbb{R}^r} f \, d\nu(c), \qquad \text{if } G(N) = U(N), \\ &\quad = \int_{\mathbb{R}^r} f \, d\nu(+, c), \quad \text{if } G(N) = SO(2N) \text{ or } O_-(2N+1), \\ &\quad = \int_{\mathbb{R}^r} f \, d\nu(-, c), \quad \text{if } G(N) = USp(2N), SO(2N+1), O_-(2N+2). \end{aligned}$$

3) *The generating series*

$$\nu_G(f, T) := \sum_{a \geq 0 \text{ in } \mathbb{Z}^r} \Big(\lim_{N \to \infty} \int_{\mathbb{R}^r} f \, d\nu(\operatorname{Off}(\mathbb{1} + a), G(N)) \Big) T^a$$

is entire, and it is related to the entire function

$$W_G(f, T) := \sum_{a \geq 0 \text{ in } \mathbb{Z}^r} \Big(\lim_{N \to \infty} W(\operatorname{sep} a, f, G(N)) \Big) T^a$$

by the inversion formulas

$$W_G(f, T) = \nu_G(f, 1 + T), \qquad \nu_G(f, T) = W_G(f, T - 1).$$

PROOF. The proof of 1) is similar to that of 2.9.1. We reduce immediately to the case when f is real valued and everywhere ≥ 0. For fixed N and each integer $m \geq \Sigma(a)$ with $m - \Sigma(a)$ odd, we have the inequalities

$$\begin{aligned} &\sum_{b \geq a \text{ in } \mathbb{Z}^r, \Sigma(b) \leq m} (-1)^{b-a} \operatorname{Binom}(b, a) W(\operatorname{sep} b, f, G(N)) \\ &\qquad \leq \int f \, d\nu(\operatorname{Off}(\mathbb{1} + a), G(N)) \\ &\qquad \leq \sum_{b \geq a \text{ in } \mathbb{Z}^r, \Sigma(b) \leq m+1} (-1)^{b-a} \operatorname{Binom}(b, a) W(\operatorname{sep} b, f, G(N)). \end{aligned}$$

Taking the lim sup and lim inf over N, and using AD.2.4, we find

$$\begin{aligned} &\sum_{b \geq a \text{ in } \mathbb{Z}^r, \Sigma(b) \leq m} (-1)^{b-a} \operatorname{Binom}(b, a) \Big(\lim_{N \to \infty} W(\operatorname{sep} b, f, G(N)) \Big) \\ &\qquad \leq \liminf_{N \to \infty} \int f \, d\nu(\operatorname{Off}(\mathbb{1} + a), G(N)) \\ &\qquad \leq \limsup_{N \to \infty} \int f \, d\nu(\operatorname{Off}(\mathbb{1} + a), G(N)) \\ &\qquad \leq \sum_{b \geq a \text{ in } \mathbb{Z}^r, \Sigma(b) \leq m+1} (-1)^{b-a} \operatorname{Binom}(b, a) \Big(\lim_{N \to \infty} W(\operatorname{sep} b, f, G(N)) \Big). \end{aligned}$$

Taking the limit over m such that $m - \Sigma(a)$ is odd, and using the fact that the series

$$\sum_{b \geq a \text{ in } \mathbb{Z}^r} (-1)^{b-a} \operatorname{Binom}(b,a) \Big(\lim_{N \to \infty} W(\operatorname{sep} b, f, G(N)) \Big)$$

is absolutely convergent, we get

$$\begin{aligned}
&\sum_{b \geq a \text{ in } \mathbb{Z}^r} (-1)^{b-a} \operatorname{Binom}(b,a) \Big(\lim_{N \to \infty} W(\operatorname{sep} b, f, G(N)) \Big) \\
&\qquad \leq \liminf_{N \to \infty} \int f \, d\nu(\operatorname{Off}(\mathbb{1} + a), G(N)) \\
&\qquad \leq \limsup_{N \to \infty} \int f \, d\nu(\operatorname{Off}(\mathbb{1} + a), G(N)) \\
&\qquad \leq \sum_{b \geq a \text{ in } \mathbb{Z}^r} (-1)^{b-a} \operatorname{Binom}(b,a) \Big(\lim_{N \to \infty} W(\operatorname{sep} b, f, G(N)) \Big),
\end{aligned}$$

and this proves 1).

To prove 2), we argue as follows. By 1), there is a Borel measure

$$\text{“} \lim_{N \to \infty} \nu(c, G(N)) \text{”}$$

on $\mathbb{R}^r$ of total mass ≤ 1 with the property that for all bounded, Borel measurable $\mathbb{C}$-valued functions f of compact support on $\mathbb{R}^r$, we have

$$\lim_{N \to \infty} \int_{\mathbb{R}^r} f \, d\nu(c, G(N)) = \int_{\mathbb{R}^r} f d\text{“} \lim_{N \to \infty} \nu(c, G(N)) \text{”}.$$

Taking for f the characteristic functions of finite rectangles, we see from 8.4.17 that “$\lim_{N \to \infty} \nu(c, G(N))$” has the same CDF as

$$\begin{aligned}
&\nu(c), && \text{if } G(N) = U(N), \\
&\nu(+,c), && \text{if } G(N) = SO(2N) \text{ or } O_-(2N+1), \\
&\nu(-,c), && \text{if } G(N) = USp(2N), SO(2N+1), O_-(2N+2).
\end{aligned}$$

Since measures are determined by their CDF's, we find that

$$\text{“} \lim_{N \to \infty} \nu(c, G(N)) \text{”} = \begin{cases} \nu(c), & \text{if } G(N) = U(N), \\ \nu(+,c), & \text{if } G(N) = SO(2N) \text{ or } O_-(2N+1), \\ \nu(-,c), & \text{if } G(N) = USp(2N), SO(2N+1), O_-(2N+2). \end{cases}$$

This proves 2).

To prove 3), we use the fact that $W_G(f,T)$ is entire, and consider the entire function $V_G(f,T) := W_G(f, T-1)$. According to 1), the formal series $\nu_G(f,T)$ agrees term by term with $V_G(f,T)$. Therefore $\nu_G(f,T)$ is itself entire, and we have the asserted relations. QED

AD.4.4. We can use this last result AD.4.3 to settle a question which we mentioned in 8.4.18, at which point we had only treated the $U(N)$ case. The method is similar to that used in 7.0.3.

Corollary AD.4.4.1. *Denote by $G(N)$ any of $U(N), SO(2N+1), USp(2N), SO(2N), O_-(2N+2), O_-(2N+1)$. Fix an integer $r \geq 1$. For any offset vector c in $\mathbb{Z}^r$, each of the measures $\nu(c, G(N))$ and the limit measure $\lim_{N \to \infty} \nu(c, G(N))$ on $\mathbb{R}^r$ is absolutely continuous with respect to Lebesgue measure.*

PROOF. These measures are supported in $(\mathbb{R}_{\geq 0})^r(\text{order})$, so we must show that any set E in $(\mathbb{R}_{\geq 0})^r(\text{order})$ of Lebesgue measure zero lies in a Borel set which has measure zero both for $\nu(c, G(N))$ and for $\lim_{N\to\infty} \nu(c, G(N))$. For this, it suffices to treat the case when E is bounded (write E as the increasing union of its intersections with the sets $x(r) \leq M$ as M grows). Once E lies in $x(r) \leq M$, use the fact that for every $\varepsilon > 0$, E is contained in a Borel set E_ε inside $x(r) \leq M$ of Lebesgue measure $\leq \varepsilon$. So E is contained in the bounded Borel set $\bigcap_n E_{1/n}$, which has Lebesgue measure zero. Thus it suffices to show that bounded Borel sets E in $(\mathbb{R}_{\geq 0})^r(\text{order})$ of Lebesgue measure zero have measure zero both for $\nu(c, G(N))$ and for $\lim_{N\to\infty} \nu(c, G(N))$. Let f on $\mathbb{R}^r$ be the characteristic function of such a set E.

It suffices to show now that $\int f\, d\nu(c, G(N)) = 0$ for every offset vector c and every finite N: the limit case follows because we have

$$\int fd\Big(\lim_{N\to\infty} \nu(c, G(N))\Big) = \lim_{N\to\infty} \int f\, d\nu(c, G(N)).$$

In view of the expression of $\int f\, d\nu(c, G(N))$ as a finite sum of terms

$$W(\operatorname{sep} a, f, G(N)),$$

it suffices to show that $W(\operatorname{sep} a, f, G(N)) = 0$ for each $a \geq 0$ in $\mathbb{Z}^r$.

For each $a \geq 0$ in $\mathbb{Z}^r$, the function $f \circ [\operatorname{sep} a]$ on $\mathbb{R}^{r+\Sigma(a)}$ is the characteristic function of $[\operatorname{sep} a]^{-1}(E)$. Because $[\operatorname{sep} a]$ is a projection onto an $\mathbb{R}^r$ inside $\mathbb{R}^{r+\Sigma(a)}$, after a permutation of coordinates the set $[\operatorname{sep} a]^{-1}(E)$ is the product $E \times \mathbb{R}^{\Sigma(a)}$ inside $\mathbb{R}^r \times \mathbb{R}^{\Sigma(a)}$, and hence is a set of Lebesgue measure zero in $\mathbb{R}^{r+\Sigma(a)}$. According to AD.2.5, we have

$$W(\operatorname{sep} a, f, G(N)) := W_{r+\Sigma(a)}(f \circ [\operatorname{sep} a], G(N))$$

$$(\text{by AD.2.5}) = \int_{[0,N+\lambda]^{r+\Sigma(a)}(\text{order})} f([\operatorname{sep} a](x)) W_{r+\Sigma(a),G(N)}(x)\, dx.$$

This integral vanishes because its integrand vanishes outside the set $[\operatorname{sep} a]^{-1}(E)$, which has Lebesgue measure zero. QED

AD.5. Large N limits for the groups $U_k(N)$: Widom's result

Theorem AD.5.1 (Widom). *Fix an integer $r \geq 1$, and denote by dx usual Lebesgue measure on $\mathbb{R}^r$. For any $\mathbb{C}$-valued, bounded Borel measurable function f on $\mathbb{R}^r$ which is of compact support, and for any sequence $\{k_N\}_{N\geq r}$ of strictly positive integers, we have the large N limit formulas*

$$\lim_{N\to\infty} W_r(f, U_{k_N}(N)) = \lim_{N\to\infty} W_r(f, U(N)) = \int_{(\mathbb{R}_{\geq 0})^r(order)} f(x) W_r(x)\, dx.$$

More precisely, we have

Theorem AD.5.2. *For any integer $r \geq 1$, for any integer $k \geq 1$, for any integer $N > r$, and for f any bounded, Borel measurable $\mathbb{C}$-valued function on $\mathbb{R}^r$ with $\|f\|_{\sup} \leq 1$ and with support contained in the set $\operatorname{Sup}_i |x(i)| \leq \alpha$, we have the equality*

$$W_r(f, U_k(N)) = W_r(f, U(N)) \quad \text{if } k > r,$$

and, more generally the estimate

$$|W_r(f, U_k(N)) - W_r(f, U(N))| \leq 2[r/k](\alpha^r/r!)r^r e^{r+r^2/2} N^{-k/2},$$

where $[r/k]$ *denotes the largest integer* $\leq r/k$.

PROOF. Fix r, k, and N.

AD.5.2.1. Because both $W_r(f, U_k(N))$ and $W_r(f, U(N))$ depend only on the restriction of f to the set $[0, N)^r$(order), we may replace f by (the extension by zero to all of $\mathbb{R}^r$ of) its restriction to $[0, N)^r$(order), say f_1. We may then replace f_1 by the unique symmetric function $\tilde{f}_1$ on $\mathbb{R}^r$ which agrees with f_1 on $\mathbb{R}^r$(order). Thus it suffices to treat the case where the function f on $\mathbb{R}^r$ is symmetric and supported in $[0, N)^r$.

AD.5.2.2. Because f is supported in $[0, N)^r$, the function

$$f_N(x) := f(Nx/2\pi)$$

is supported in $[0, 2\pi)^r$, and may be seen as a function on the r-torus $(S^1)^r$. More intrinsically, the series

$$f_{N,\mathrm{per}}(x) := \sum_{n \text{ in } \mathbb{Z}^r} f_N(x + 2\pi n) \tag{AD.5.2.3}$$

has, for each x in $\mathbb{R}^r$, precisely one possibly nonzero term, so converges pointwise to a $2\pi\mathbb{Z}^r$-periodic function and thus is a function on $\mathbb{R}^r/2\pi\mathbb{Z}^r \cong (S^1)^r$. For x in $[0, 2\pi)^r$, we have the relation

$$f_{N,\mathrm{per}}(x) = f_N(x). \tag{AD.5.2.4}$$

The function $f_{N,\mathrm{per}}$ on $(S^1)^r$ is thus bounded, Borel measurable, and symmetric.

We now interrupt the proof of AD.5.2 for two "interludes", AD.6 and AD.7. The proof will be concluded in AD.8–9.

AD.6. Interlude: The quantities $V_r(\varphi, U_k(N))$ and $V_r(\varphi, U(N))$

AD.6.1. Fix integers $r \geq 1$ and $N \geq r$. Given a bounded, Borel measurable function F on $(S^1)^r$, viewed as a $2\pi\mathbb{Z}^r$-periodic function on $\mathbb{R}^r$, and an element A in $U(N)$ with angles φ_j and eigenvalues $\alpha_j := \exp(i\varphi_j)$, $j = 1, \ldots, N$, we define

$$V_r(F, A, U(N)) := (1/r!) \sum_{i_1, i_2, \ldots, i_r \text{ all distinct in } [1,N]} F(\varphi_{i_1}, \ldots, \varphi_{i_r}). \tag{AD.6.1.1}$$

Given $k \geq 1$ and A in $U_k(N)$, we define

$$V_r(F, A, U_k(N)) := V_r(F, A, U(N)). \tag{AD.6.1.2}$$

We then define

$$V_r(F, U_k(N)) := \int_{U_k(N)} V_r(F, A, U_k(N))\, dA, \tag{AD.6.1.3}$$

$$V_r(F, U(N)) := \int_{U(N)} V_r(F, A, U(N))\, dA, \tag{AD.6.1.4}$$

in both integrals using the total mass one Haar measure.

AD.6.2. Before going on, let us point out the relation of these quantities to the W_r quantities defined previously, for functions f on $\mathbb{R}^r$. If we begin with a function f on $\mathbb{R}^r$ which is **symmetric** and supported in $[0, N)^r$, and take for F the function $f_{N,\text{per}}$, we have the tautological relations

$$W_r(f, A, U(N)) = V_r(f_{N,\text{per}}, A, U(N)), \tag{AD.6.2.1}$$

$$W_r(f, U(N)) = V_r(f_{N,\text{per}}, U(N)), \tag{AD.6.2.2}$$

and

$$W_r(f, A, U_k(N)) = V_r(f_{N,\text{per}}, A, U_k(N)), \tag{AD.6.2.3}$$

$$W_r(f, U_k(N)) = V_r(f_{N,\text{per}}, U_k(N)). \tag{AD.6.2.4}$$

The advantage of the V_r is that they are intrinsic to $(S^1)^r$, whereas the W_r depend on the particular choice of $[0, 2\pi)$ as fundamental domain in which to order the N angles φ_j of an element A in $U(N)$.

AD.6.3. Our next task is to develop integral formulas for $V_r(F, U_k(N))$ and for $V_r(F, U(N))$. To clarify what is involved here, we first give another interpretation of $V_r(F, U(N))$ and $V_r(F, U_k(N))$.

AD.6.3.1. An element A in $U(N)$ with angles φ_j, $j = 1, \ldots, N$, and eigenvalues $\alpha_j := \exp(i\varphi_j)$, gives rise to the Borel measure $V_r(A, U(N))$ on $(S^1)^r$ of total mass $\text{Binom}(N, r)$ defined as

$$V_r(A, U(N)) := (1/r!) \sum_{i_1, i_2, \ldots, i_r \text{ all distinct in } [1,N]} \delta(\varphi_{i_1}, \ldots, \varphi_{i_r}). \tag{AD.6.3.2}$$

In terms of this measure on $(S^1)^r$, we have, for any bounded, Borel measurable function F on $(S^1)^r$,

$$V_r(F, A, U(N)) = \int_{(S^1)^r} F \, dV_r(A, U(N)). \tag{AD.6.3.3}$$

If A lies in $U_k(N)$, we define the Borel measure $V_r(A, U_k(N))$ on $(S^1)^r$ to be

$$V_r(A, U_k(N)) := V_r(A, U(N)). \tag{AD.6.3.4}$$

We then define Borel measures $V_r(U_k(N))$ and $V_r(U(N))$ on $(S^1)^r$ of total mass $\text{Binom}(N, r)$ as the expected values of these measures:

$$V_r(U_k(N)) := \int_{U_k(N)} V_r(A, U_k(N)) \, dA, \tag{AD.6.3.5}$$

$$V_r(U(N)) := \int_{U(N)} V_r(A, U(N)) \, dA, \tag{AD.6.3.6}$$

in both integrals using the total mass one Haar measure. Exactly as in 1.1, we have, for any bounded, Borel measurable function F on $(S^1)^r$, the integration formulas

$$V_r(F, U_k(N)) = \int_{(S^1)^r} F \, dV_r(U_k(N)), \tag{AD.6.3.7}$$

$$V_r(F, U(N)) = \int_{(S^1)^r} F \, dV_r(U(N)). \tag{AD.6.3.8}$$

AD.6.4. Thus we see that giving integral formulas for $V_r(F, U_k(N))$ and for $V_r(F, U(N))$ amounts to giving formulas for the Borel measures $V_r(U_k(N))$ and $V_r(U(N))$ on $(S^1)^r$.

AD.6.4.1. We first explain the idea which underlies these formulas. Let us begin with an element A in $U(N)$, whose angles are φ_j, $j = 1, \ldots, N$, and whose eigenvalues are $\alpha_j := \exp(i\varphi_j)$. For any Borel measurable F on $(S^1)^r$, viewed as a $2\pi\mathbb{Z}^r$-periodic function on $\mathbb{R}^r$, we have

$$(\text{AD.6.4.1.1}) \quad V_r(F, A, U(N)) := (1/r!) \sum_{i_1, i_2, \ldots, i_r \text{ all distinct in } [1,N]} F(\varphi_{i_1}, \ldots, \varphi_{i_r}).$$

AD.6.4.2. Choose an "approximate identity" on S^1, i.e., a sequence of non-negative $\mathcal{C}^\infty$ functions $\Phi_n(\vartheta)$ on S^1, each of total integral one over S^1 against normalized Haar measure $d\vartheta$, with the property that for any open neighborhood U of 0 in S^1 and for any real $\varepsilon > 0$, we have $\int_U \Phi_n(\vartheta)\, d\vartheta \geq 1 - \varepsilon$ for $n \gg 0$. One knows that for any continuous function H on S^1, viewed as a periodic continuous function on $\mathbb{R}$, and any point x in $\mathbb{R}$, we have

$$(\text{AD.6.4.2.1}) \qquad H(x) = \lim_{n \to \infty} \int_{S^1} H(\vartheta)\Phi_n(x - \vartheta)\, d\vartheta,$$

and the convergence is uniform, i.e., the convolutions $H * \Phi_n$ converge uniformly to H.

AD.6.4.2.2. For example, we may take for Φ_n the Fejér kernel $(S_n(\vartheta))^2$, with $S_n(\vartheta) := \sin(n\vartheta/2)/\sin(\vartheta/2)$. Or we may take for Φ_n the Poisson kernel $\sum_{m \text{ in } \mathbb{Z}} (1 - 1/n)^{|m|} e^{im\vartheta}$ with parameter $1 - 1/n$.

AD.6.4.3. Once we pick an approximate identity Φ_n on S^1, the product functions $\Phi_{n,r}(\vartheta) := \Phi_n(\vartheta_1)\Phi_n(\vartheta_2) \cdots \Phi_n(\vartheta_r)$ on $(S^1)^r$ form an approximate identity on $(S^1)^r$: for any continuous F on $(S^1)^r$, and any point $\varphi = (\varphi_1, \ldots, \varphi_r)$ in $(S^1)^r$, we have

$$(\text{AD.6.4.3.1}) \qquad F(\varphi) = \lim_{n \to \infty} \int_{(S^1)^r} F(\vartheta)\Phi_{n,r}(\varphi - \vartheta)\, d\vartheta,$$

where we write $d\vartheta$ for the normalized Haar measure $\prod_i (d\vartheta_i/2\pi)$. Again the convergence is uniform as φ varies in $(S^1)^r$.

AD.6.4.4. Given A in $U(N)$ with angles φ_j, $j = 1, \ldots, N$, the coefficient of $T_1 T_2 \cdots T_r$ in

$$\int_{(S^1)^r} F(\vartheta) \prod_{i=1}^{N} \left(1 + \sum_{j=1}^{r} T_j \Phi_n(\varphi_i - \vartheta_j)\right) d\vartheta$$

is

$$\sum_{i_1, i_2, \ldots, i_r \text{ all distinct in } [1,N]} \prod_{j=1}^{r} \Phi_n(\varphi_{i_j} - \vartheta_j).$$

Thus for F a continuous function on $(S^1)^r$, we have the identity

$$(\text{AD.6.4.5}) \qquad \begin{aligned} V_r(F, A, U(N)) = \lim_{n \to \infty} \Bigg(&\text{the coefficient of } T_1 T_2 \cdots T_r \text{ in} \\ &(1/r!) \int_{(S^1)^r} F(\vartheta) \prod_{i=1}^{N} \left(1 + \sum_{j=1}^{r} T_j \Phi_n(\varphi_i - \vartheta_j)\right) d\vartheta \Bigg), \end{aligned}$$

and the convergence is uniform as A varies in $U(N)$.

AD.6.5. Recall from 6.3.2 that for f any bounded, Borel measurable function on S^1, and A in $U(N)$, we may form the operator $f(A)$ in $M_N(\mathbb{C})$. Applying this construction to the function $x \mapsto \Phi_n(x - \vartheta_j)$, we may speak of the operator $\Phi_n(A - \vartheta_j)$. Because Φ_n is $\mathcal{C}^\infty$, it has a rapidly convergent Fourier series, say

$$\Phi_n(x) = \sum_{m \text{ in } \mathbb{Z}} \widehat{\Phi_n}(m) e^{imx},$$

and $\Phi_n(A - \vartheta)$ is the operator

$$\Phi_n(A - \vartheta) := \sum_{m \text{ in } \mathbb{Z}} \widehat{\Phi_n}(m) A^m e^{-im\vartheta}.$$

AD.6.6. In terms of these operators $\Phi_n(A - \vartheta_j)$, we have the identity

$$\prod_{i=1}^{N} \left(1 + \sum_{j=1}^{r} T_j \Phi_n(\varphi_i - \vartheta_j)\right) = \det\left(1 + \sum_{j=1}^{r} T_j \Phi_n(A - \vartheta_j)\right). \tag{AD.6.6.1}$$

Thus for F continuous on $(S^1)^r$, we have the identity

$$\begin{aligned} V_r(F, A, U(N)) = \lim_{n\to\infty} \Bigg(&\text{the coefficient of } T_1 T_2 \cdots T_r \text{ in} \\ &(1/r!) \int_{(S^1)^r} F(\vartheta) \det\left(1 + \sum_{j=1}^{r} T_j \Phi_n(A - \vartheta_j)\right) d\vartheta \Bigg), \end{aligned} \tag{AD.6.6.2}$$

and convergence is uniform as A varies in $U(N)$. Because the function $A \mapsto V_r(F, A, U(N))$ is bounded (by $\mathrm{Binom}(N, r)\|F\|_{\mathrm{sup}}$) as A varies in $U(N)$, and the convergence is uniform in A, the n'th approximant will be bounded by

$$2\,\mathrm{Binom}(N, r)\|F\|_{\mathrm{sup}}$$

for $n \gg 0$. So we may apply dominated convergence to integrate over $U(N)$ or over $U_k(N)$. For F continuous on $(S^1)^r$, we find the formulas

$$\begin{aligned} V_r(F, U(N)) = \lim_{n\to\infty} \Bigg(&\text{the coefficient of } T_1 T_2 \cdots T_r \text{ in} \\ &(1/r!) \int_{U(N)} \int_{(S^1)^r} F(\vartheta) \det\left(1 + \sum_{j=1}^{r} T_j \Phi_n(A - \vartheta_j)) \, d\vartheta\right) dA \Bigg), \end{aligned} \tag{AD.6.6.3}$$

and, for each $k \geq 1$,

(AD.6.6.4)

$$\begin{aligned} V_r(F, U_k(N)) = \lim_{n\to\infty} \Bigg(&\text{the coefficient of } T_1 T_2 \cdots T_r \text{ in} \\ &(1/r!) \int_{U_k(N)} \int_{(S^1)^r} F(\vartheta) \det\left(1 + \sum_{j=1}^{r} T_j \Phi_n(A - \vartheta_j)) \, d\vartheta\right) dA \Bigg). \end{aligned}$$

Using Fubini to interchange the order of integration, we get

(AD.6.6.5)

$$V_r(F, U(N)) = \lim_{n\to\infty} \Bigg(\text{the coefficient of } T_1T_2\cdots T_r \text{ in}$$

$$(1/r!)\int_{(S^1)^r} F(\vartheta)\left[\int_{U(N)} \det\left(1+\sum_{j=1}^{r} T_j\Phi_n(A-\vartheta_j)\right) dA\right] d\vartheta\Bigg)$$

and

(AD.6.6.6)

$$V_r(F, U_k(N)) = \lim_{n\to\infty} \Bigg(\text{the coefficient of } T_1T_2\cdots T_r \text{ in}$$

$$(1/r!)\int_{(S^1)^r} F(\vartheta)\left[\int_{U_k(N)} \det\left(1+\sum_{j=1}^{r} T_j\Phi_n(A-\vartheta_j)\right) dA\right] d\vartheta\Bigg).$$

AD.6.7. We now study the coefficient of $T_1T_2\cdots T_r$ in the integrals

(AD.6.7.1)
$$I_{r,k,N,n}(\vartheta, T) := \int_{U_k(N)} \det\left(1+\sum_{j=1}^{r} T_j\Phi_n(A-\vartheta_j)\right) dA$$

and

(AD.6.7.2)
$$I_{r,N,n}(\vartheta, T) := \int_{U(N)} \det\left(1+\sum_{j=1}^{r} T_j\Phi_n(A-\vartheta_j)\right) dA.$$

For each fixed (ϑ, T) in $(S^1)^r \times \mathbb{C}^r$, the integrand

$$\det\left(1+\sum_{j=1}^{r} T_j\Phi_n(A-\vartheta j)\right)$$

is a $\mathcal{C}^\infty$ central function on $U(N)$.

AD.7. Interlude: Integration formulas on $U(N)$ and on $U_k(N)$

Lemma AD.7.1. *Let $N \geq 1$, f a $\mathcal{C}^\infty$ $\mathbb{C}$-valued central function on $U(N)$. For any integer $k \geq 1$, the integral of f over the subgroup $U_k(N)$ of $U(N)$ is given by the absolutely convergent series*

$$\int_{U_k(N)} f(A)\, dA = \sum_{l \text{ in } \mathbb{Z}} \int_{U(N)} f(A)\det(A)^{lk}\, dA.$$

PROOF. Let K be a compact Lie group, and f a $\mathcal{C}^\infty$ $\mathbb{C}$-valued central function on K. In terms of the irreducible unitary representations ρ of K, compute the representation-theoretic Fourier coefficients of f,

$$a(\rho) := \int_K f(A)\,\mathrm{Trace}(\overline{\rho}(A))\, dA.$$

It is well known, cf. [**Sug**, Chapter II, Theorem 8.1], that the series

$$\sum_{\text{irred. } \rho} a(\rho)\, \mathrm{Trace}(\rho(A))$$

converges absolutely and uniformly to $f(A)$. In particular, taking A to be the identity element, the series

$$\sum_{\text{irred. } \rho} |a(\rho)| \dim(\rho)$$

converges, and hence so does the series $\sum_{\text{irred. } \rho} |a(\rho)|$.

If we take now K to be $U(N)$, then

$$\int_{U(N)} f(A) \det(A)^{lk}\, dA = a(\det^{-lk}),$$

and so the series

$$\sum_{l \text{ in } \mathbb{Z}} \int_{U(N)} f(A) \det(A)^{lk}\, dA = \sum_{l \text{ in } \mathbb{Z}} a(\det^{-lk})$$

is absolutely convergent. To show that its sum is $\int_{U_k(N)} f(A)\, dA$, we argue as follows. Because $\sum_{\text{irred. } \rho} a(\rho)\, \mathrm{Trace}(\rho(A))$ converges absolutely and uniformly to f on $U(N)$, it does so a fortiori on $U_k(N)$, so we may integrate term by term:

$$\int_{U_k(N)} f(A)\, dA = \sum_{\text{irred. } \rho \text{ of } U(N)} a(\rho) \int_{U_k(N)} \mathrm{Trace}(\rho(A))\, dA.$$

For ρ any finite-dimensional representation of $U(N)$, we have

$$\int_{U_k(N)} \mathrm{Trace}(\rho(A))\, dA = \dim(\text{space of } U_k(N)\text{-invariants in } \rho).$$

Because $U_k(N)$ is a normal subgroup of $U(N)$, the space of $U_k(N)$-invariants in ρ is a $U(N)$-subrepresentation of ρ. So for ρ irreducible, either the space of $U_k(N)$-invariants in ρ is reduced to zero, or it is all of ρ, in which case ρ is trivial on $U_k(N)$. Thus we find

$$\int_{U_k(N)} f(A)\, dA = \sum_{\text{irred. } \rho \text{ of } U(N) \text{ trivial on } U_k(N)} a(\rho).$$

Because the quotient $U(N)/U_k(N)$ is abelian, isomorphic to S^1 by the map $A \mapsto \det(A)^k$, the irreducible representations ρ of $U(N)$ which are trivial on $U_k(N)$ are precisely the powers $\det^{lk}$ of $\det^k$. QED

Lemma AD.7.2 ("Heine's formula" [**Szego**, pages 23 and 26]). *Let f be a bounded, Borel measurable $\mathbb{C}$-valued function on S^1, with Fourier series*

$$\sum_n a(n) e^{in\vartheta}.$$

Then we have the integration formula

$$\int_{U(N)} \det(f(A))\, dA = \det_{N \times N}(a(i-j)).$$

PROOF. Recall from 6.3.5 that

$$\int_{U(N)} \det(1+Tf(A))\,dA = \det(1+TK_N(x,y)f(y)|L^2(S^1,\mu)),$$

where μ is normalized Haar measure on S^1, $K_N(x,y)$ is the kernel $\sum_{n=0}^{N-1} e^{in(x-y)}$, and $K_N(x,y)f(y)$ is the integral operator on $L^2(S^1,\mu)$ given by

$$g \mapsto \text{ the function } x \mapsto \int_{S^1} K_N(x,y)f(y)g(y)\,d\mu(y).$$

Let us denote by $P(N) \subset L^2(S^1,\mu)$ the span of the functions e^{inx} for $n=0,\ldots,N-1$, and by $\pi(N)$ the orthogonal projection of $L^2(S^1,\mu)$ onto $P(N)$. The integral operator $K_N(x,y)$ is precisely $\pi(N)$, so the integral operator $K_N(x,y)f(y)$ is the composite operator

$$g \mapsto \pi(N)(fg).$$

Since this composite operator has image contained in $P(N)$, we have (cf. 6.0.4)

$$\begin{aligned}\det(1+TK_N(x,y)f(y)|L^2(S^1,\mu)) &= \det(1+TK_N(x,y)f(y)|P(N))\\ &= \det(1+T(g\mapsto \pi(N)(fg))|P(N)).\end{aligned}$$

In the obvious basis $e_j := e^{ijx}$ of $P(N)$, $j=0,\ldots,N-1$, we have

$$\pi(N)(fe_j) = \pi(N)\left(\sum_n a(n-j)e_n\right) = \sum_{k=0}^{N-1} a(k-j)e_k,$$

and so the matrix of $g \mapsto \pi(N)(fg)$ on $P(N)$ is the $N\times N$ matrix whose (j,k) entry is $a(j-k)$, for j,k running from 0 to $N-1$, or, what is the same, for j,k running from 1 to N.

Thus we find

$$\int_{U(N)} \det(1+Tf(A))\,dA = \det_{N\times N}(1+T(a(i-j))).$$

Introduce a second indeterminate X, and multiply both sides by X^N: we find

$$\int_{U(N)} \det(X+XTf(A))\,dA = \det_{N\times N}(X+XT(a(i-j))),$$

an identity in $\mathbb{C}[X,T]$, hence in $\mathbb{C}[X,X^{-1},T]$. Replacing T by T/X, we find an identity

$$\int_{U(N)} \det(X+Tf(A))\,dA = \det_{N\times N}(X+T(a(i-j)))$$

in $\mathbb{C}[X,X^{-1},T]$, where both sides lie in $\mathbb{C}[X,T]$. So this identity holds in $\mathbb{C}[X,T]$, and putting $X=0, T=1$, we find the asserted identity. QED

AD.8. Return to the proof of Widom's theorem

AD.8.1. We now return to the integrals of AD.6.7,

$$I_{r,k,N,n}(\vartheta,T) := \int_{U_k(N)} \det\left(1+\sum_{j=1}^{r} T_j\Phi_n(A-\vartheta_j)\right) dA$$

and

$$I_{r,N,n}(\vartheta, T) := \int_{U(N)} \det\left(1 + \sum_{j=1}^{r} T_j \Phi_n(A - \vartheta_j)\right) dA.$$

Because Φ_n is $\mathcal{C}^\infty$ on S^1, the integrand $\det(1 + \sum_{j=1}^{r} T_j \Phi_n(A - \vartheta_j))$ is, for fixed (ϑ, T), a $\mathcal{C}^\infty$ central function on $U(N)$. So by Lemma AD.7.1, we have

$$\begin{aligned} &\int_{U_k(N)} \det\left(1 + \sum_{j=1}^{r} T_j \Phi_n(A - \vartheta_j)\right) dA \\ &= \sum_{l \text{ in } \mathbb{Z}} \int_{U(N)} \det\left(1 + \sum_{j=1}^{r} T_j \Phi_n(A - \vartheta_j)\right) \det(A)^{lk}\, dA \\ &= \sum_{l \text{ in } \mathbb{Z}} \int_{U(N)} \det\left(A^{lk}\left(1 + \sum_{j=1}^{r} T_j \Phi_n(A - \vartheta_j)\right)\right) dA \\ &= \sum_{l \text{ in } \mathbb{Z}} \int_{U(N)} \det\left(e^{ilkx}\left(1 + \sum_{j=1}^{r} T_j \Phi_n(x - \vartheta_j)\right)\Bigg|_{x=A}\right) dA. \end{aligned} \tag{AD.8.1.1}$$

If we now apply Lemma AD.7.2, we find

$$\begin{aligned} &\int_{U(N)} \det\left(e^{ilkx}\left(1 + \sum_{j=1}^{r} T_j \Phi_n(x - \vartheta_j)\right)\Bigg|_{x=A}\right) dA \\ &= \det{}_{N\times N}\Bigg((a,b) \mapsto \text{ the } a-b \text{ Fourier coef. of} \\ &\qquad \text{the function } e^{ilkx} + \sum_{j=1}^{r} T_j \Phi_n(x - \vartheta_j) e^{ilkx}\Bigg). \end{aligned} \tag{AD.8.1.2}$$

Lemma AD.8.2. *Suppose $N > r$. View the integral*

$$\begin{aligned} &\int_{U(N)} \det\left(1 + \sum_{j=1}^{r} T_j \Phi_n(A - \vartheta_j)\right) \det(A)^{lk}\, dA \\ &= \int_{U(N)} \det\left(e^{ilkx}\left(1 + \sum_{j=1}^{r} T_j \Phi_n(x - \vartheta_j)\right)\Bigg|_{x=A}\right) dA \end{aligned}$$

as a polynomial in the variables T_j, with coefficient functions on $(S^1)^r$. Every nonzero monomial in the T_j's has total degree $\geq \mathrm{Min}(|lk|, N)$. In particular, if $|lk| > r$, then the coefficient of $T_1 T_2 \cdots T_r$ vanishes.

Proof. This integral is, as noted above, the determinant of the $N \times N$ matrix whose (a, b) entry is the $a - b$ Fourier coefficient of the function

$$e^{ilkx} + \sum_{j=1}^{r} T_j \Phi_n(x - \vartheta_j) e^{ilkx}.$$

Thus the (a,b) entry is

$$\delta_{a-b,lk} + \sum_{j=1}^{r} T_j \widehat{\Phi_n}(a-b-lk)e^{-i(a-b-lk)\vartheta_j}.$$

If the term $\delta_{a-b,lk}$ were absent, we would be looking at an $N \times N$ matrix of linear forms in the T_j's, and the determinant would be homogeneous of total degree N in the T_j's. The idea is that for most (a,b), the term $\delta_{a-b,lk}$ vanishes. Since a and b run from 1 to N, their difference $a-b$ cannot exceed $N-1$ in absolute value. Thus if $|lk| \geq N$, the determinant is in fact homogeneous of total degree N. Suppose now that $|lk| \leq N-1$. Then there are precisely $N-|lk|$ pairs (a,b) in $[1,N]^2$ with $a-b=lk$: if $lk \geq 0$, these are the pairs $(lk+j,j)$ for $j=1$ to $N-lk$, and if $lk<0$ they are the pairs $(j,j-lk)$ with $j=1+lk$ to N. Thus there are at most $N-|lk|$ entries which are not linear forms in the T_j's. So in expanding out the determinant as the alternating sum of $N!$ products of N distinct entries, each of the $N!$ terms is divisible by a product of at least $|lk|$ linear forms, and hence each of the $N!$ terms contains only monomials of degree $\geq |lk|$. QED

Corollary AD.8.3. *Suppose $N > r$. For any $k \geq 1$, we have a congruence modulo the ideal $(T_1,\ldots,T_r)^{r+1}$ in $\mathbb{C}[T_1,\ldots,T_r]$ generated by all monomials of degree $> r$,*

$$\int_{U_k(N)} \det\left(1+\sum_{j=1}^{r} T_j\Phi_n(A-\vartheta_j)\right) dA$$
$$\equiv \sum_{l \text{ in } \mathbb{Z}, |lk|\leq r} \int_{U(N)} \det\left(1+\sum_{j=1}^{r} T_j\Phi_n(A-\vartheta_j)\right) \det(A)^{lk}\, dA.$$

In particular, both sides have the same coefficient of $T_1T_2\cdots T_r$.

AD.8.4. For each integer l, let us denote consider the integral

$$J_{r,l,N,n}(\vartheta,T) := \int_{U(N)} \det\left(1+\sum_{j=1}^{r} T_j\Phi_n(A-\vartheta_j)\right) \det(A)^{l}\, dA \tag{AD.8.4.1}$$

= the determinant of the $N \times N$ matrix whose (a,b) entry is

$$\delta_{a-b,l} + \sum_{j=1}^{r} T_j \widehat{\Phi_n}(a-b-l)e^{-i(a-b-l)\vartheta_j}.$$

Define

(AD.8.4.2) $\quad J_{r,l,N,n}(\vartheta) :=$ the coefficient of $T_1T_2\cdots T_r$ in $J_{r,l,N,n}(\vartheta,T)$.

We will also have occasion to look at the integral

$$J_{r,l,N}(\vartheta,T) := \det_{N\times N}\left((a,b) \mapsto \delta_{a-b,l} + \sum_{j=1}^{r} T_j e^{-i(a-b-l)\vartheta_j}\right), \tag{AD.8.4.3}$$

and at the function

(AD.8.4.4) $\quad J_{r,l,N}(\vartheta) :=$ the coefficient of $T_1T_2\cdots T_r$ in $J_{r,l,N}(\vartheta,T)$.

Lemma AD.8.5. *Fix $r, l, N > r$, and an approximate identity $\{\Phi_n\}$ on S^1. The integrals $J_{r,l,N,n}(\vartheta, T)$ and $J_{r,l,N}(\vartheta, T)$ are polynomials in the T_j whose coefficients are trigonometric polynomials in the ϑ_j's. For (ϑ, T) in $(S^1)^r \times$ (a compact set in $\mathbb{C}^r$), we have uniform convergence of $J_{r,l,N,n}(\vartheta, T)$ to $J_{r,l,N}(\vartheta, T)$ as $n \to \infty$. In particular, we have uniform convergence of $J_{r,l,N,n}(\vartheta)$ to $J_{r,l,N}(\vartheta)$ as $n \to \infty$.*

PROOF. Indeed, already the individual (a, b) entries in the $N \times N$ matrices whose determinants we are taking have this uniform convergence, since they involve only finitely many Fourier coefficients of the Φ_n, and any given Fourier coefficient of Φ_n tends to 1 as $n \to \infty$. This gives the asserted convergence of the determinants. To infer the uniform convergence of the coefficient of $T_1 T_2 \cdots T_r$, use the Cauchy formulas

$$J_{r,l,N,n}(\vartheta) = (1/2\pi i)^r \int_{(S^1)^r} (J_{r,l,N,n}(\vartheta, T)/T_1 T_2 \cdots T_r) \prod_{j=1}^{r} (dT_j/T_j)$$

and

$$J_{r,l,N}(\vartheta) = (1/2\pi i)^r \int_{(S^1)^r} (J_{r,l,N}(\vartheta, T)/T_1 T_2 \cdots T_r) \prod_{j=1}^{r} (dT_j/T_j). \quad \text{QED}$$

AD.8.6. Let us now recapitulate what we have so far. For $N > r$, and for $F(\vartheta)$ a continuous $\mathbb{C}$-valued function on $(S^1)^r$, we have, from AD.6.6.3 and AD.6.6.4, the formulas

(AD.8.6.1)

$$V_r(F, U(N)) = \lim_{n \to \infty} \Bigg(\text{the coefficient of } T_1 T_2 \cdots T_r \text{ in}$$
$$(1/r!) \int_{(S^1)^r} F(\vartheta) \left[\int_{U(N)} \det \left(1 + \sum_{j=1}^{r} T_j \Phi_n(A - \vartheta_j) \right) dA \right] d\vartheta \Bigg),$$

and

(AD.8.6.2)

$$V_r(F, U_k(N)) = \lim_{n \to \infty} \Bigg(\text{the coefficient of } T_1 T_2 \cdots T_r \text{ in}$$
$$(1/r!) \int_{(S^1)^r} F(\vartheta) \left[\int_{U_k(N)} \det \left(1 + \sum_{j=1}^{r} T_j \Phi_n(A - \vartheta_j) \right) dA \right] d\vartheta \Bigg).$$

The first formula we rewrite, using the above results, as

$$\text{(AD.8.6.3)} \qquad V_r(F, U(N)) = \lim_{n \to \infty} (1/r!) \int_{(S^1)^r} F(\vartheta) J_{r,0,N,n}(\vartheta) \, d\vartheta,$$

the second as

$$\text{(AD.8.6.4)} \qquad V_r(F, U_k(N)) = \lim_{n \to \infty} \sum_{l \text{ in } \mathbb{Z}, |lk| \le r} (1/r!) \int_{(S^1)^r} F(\vartheta) J_{r,lk,N,n}(\vartheta) \, d\vartheta.$$

Using the uniform convergence of $J_{r,lk,N,n}(\vartheta)$ to $J_{r,lk,N}(\vartheta)$, we may apply dominated convergence to again rewrite these as

$$V_r(F, U(N)) = (1/r!) \int_{(S^1)^r} F(\vartheta) J_{r,0,N}(\vartheta)\, d\vartheta \tag{AD.8.6.5}$$

and

$$V_r(F, U_k(N)) = \sum_{l \text{ in } \mathbb{Z}, |lk| \le r} (1/r!) \int_{(S^1)^r} F(\vartheta) J_{r,lk,N}(\vartheta)\, d\vartheta. \tag{AD.8.6.6}$$

Proposition AD.8.7. *For any integers $r \ge 1, k \ge 1$, and $N > r$, and for any bounded, Borel measurable $\mathbb{C}$-valued function F on $(S^1)^r$, we have the integration formulas*

$$V_r(F, U(N)) = (1/r!) \int_{(S^1)^r} F(\vartheta) J_{r,0,N}(\vartheta)\, d\vartheta$$

and

$$V_r(F, U_k(N)) = \sum_{l \text{ in } \mathbb{Z}, |lk| \le r} (1/r!) \int_{(S^1)^r} F(\vartheta) J_{r,lk,N}(\vartheta)\, d\vartheta.$$

PROOF. We have defined, in AD.6.3, Borel measures $V_r(U(N))$ and $V_r(U_k(N))$ on $(S^1)^r$ of total mass $\mathrm{Binom}(N, r)$ such that for any bounded, Borel measurable $\mathbb{C}$-valued function F on $(S^1)^r$, we have the integration formulas

$$V_r(F, U(N)) = \int_{(S^1)^r} F\, dV_r(U(N)),$$

$$V_r(F, U_k(N)) = \int_{(S^1)^r} F\, dV_r(U_k(N)).$$

We have shown that for F continuous, we have

$$V_r(F, U(N)) = (1/r!) \int_{(S^1)^r} F(\vartheta) J_{r,0,N}(\vartheta)\, d\vartheta$$

and

$$V_r(F, U_k(N)) = \sum_{l \text{ in } \mathbb{Z}, |lk| \le r} (1/r!) \int_{(S^1)^r} F(\vartheta) J_{r,lk,N}(\vartheta)\, d\vartheta.$$

So the general situation is this. We have a (positive) Borel measure ν of finite total mass on $(S^1)^r$. We have a continuous $\mathbb{C}$-valued function $J(\vartheta)$ on $(S^1)^r$, and we know that

$$\int_{(S^1)^r} F\, d\nu = \int_{(S^1)^r} F(\vartheta) J(\vartheta)\, d\vartheta$$

for every continuous F. This uniquely determines the continuous function J: if K were another, then $\int_{(S^1)^r} F(\vartheta)(J(\vartheta) - K(\vartheta))\, d\vartheta = 0$ for every continuous F, and taking F to be $J - K$ we see that $J = K$. Because ν is a positive measure, the function J is real and nonnegative, and so $J(\vartheta)\, d\vartheta$ is itself a positive Borel measure of finite mass on $(S^1)^r$ which agrees with ν on all continuous functions. But $(S^1)^r$ is a compact Hausdorff space in which every open set is σ-compact. So by [**Rudin**, 2.14 and 2.18], $\nu = J(\vartheta)\, d\vartheta$ as a Borel measure. Hence we have

$$\int_{(S^1)^r} F\, d\nu = \int_{(S^1)^r} F(\vartheta) J(\vartheta)\, d\vartheta$$

for every bounded, Borel measurable $\mathbb{C}$-valued function F on $(S^1)^r$. QED

For ease of later reference, we here make explicit what the previous result gives in the case $r = 1$.

Corollary AD.8.7.1. *For any integers $k \geq 1$ and $N > 1$, and for any bounded, Borel measurable $\mathbb{C}$-valued function F on S^1, denoting by $d\vartheta$ the normalized Haar measure on S^1, we have the integration formulas*

$$V_1(F, U(N)) = N \int_{S^1} F(\vartheta)\, d\vartheta = (N/2\pi) \int_{[0,2\pi]} F(x)\, dx,$$
$$V_1(F, U_k(N)) = V_1(F, U(N)) \quad \textit{for } k \geq 2,$$

and

$$V_1(F, SU(N)) = V_1(F, U(N)) + (-1)^{N-1} 2 \int_{S^1} F(\vartheta) \cos(N\vartheta)\, d\vartheta$$
$$= (N/2\pi) \int_{[0,2\pi]} F(x)(1 - (-1)^N (2/N) \cos(Nx))\, dx.$$

PROOF. The $r = 1$ case of AD.8.7 gives us the formulas

$$V_1(F, U(N)) = \int_{S^1} F(\vartheta) J_{1,0,N}(\vartheta)\, d\vartheta,$$
$$V_1(F, U_k(N)) = V_1(F, U(N)) \quad \text{for } k \geq 2,$$

and

$$V_1(F, SU(N)) = V_1(F, U(N)) + \int_{S^1} F(\vartheta)(J_{1,1,N}(\vartheta) + J_{1,-1,N}(\vartheta))\, d\vartheta.$$

By definition, $J_{1,l,N}(\vartheta)$ is the coefficient of T in

$$\det_{N \times N}((a,b) \mapsto \delta_{a-b,l} + Te^{-i(a-b-l)\vartheta}),$$

and we readily calculate

$$J_{1,0,N}(\vartheta) = N,$$
$$J_{1,\pm 1,N}(\vartheta) = (-1)^{N-1} e^{\pm iN\vartheta}. \quad \text{QED}$$

We now return to our main task.

Proposition AD.8.8. *For any integers $r \geq 1$, l arbitrary, and $N > r$, we have the following estimate for the* sup *norm of the function $N^{-r} J_{r,l,N}(\vartheta)$:*

$$\|N^{-r} J_{r,l,N}(\vartheta)\|_{\sup} = 0 \quad \textit{if } |l| > r,$$
$$\|N^{-r} J_{r,l,N}(\vartheta)\|_{\sup} \leq (r^l / N^{l/2}) \exp(r + r^2/2) \quad \textit{if } |l| \leq r.$$

PROOF. According to Lemma AD.8.2, if $|l| > r$ the function $J_{r,l,N}(\vartheta)$ vanishes identically.

If $|l| \leq r$, we argue as follows. The function $J_{r,l,N}(\vartheta)$ is the coefficient of $T_1 T_2 \cdots T_r$ in $J_{r,l,N}(\vartheta, T)$. Therefore $N^{-r} J_{r,l,N}(\vartheta)$ is the coefficient of $T_1 T_2 \cdots T_r$ in $J_{r,l,N}(\vartheta, T/N)$. Since we can recover $N^{-r} J_{r,l,N}(\vartheta)$ from $J_{r,l,N}(\vartheta, T/N)$ by the Cauchy formula as the integral against Haar measure over the $(S^1)^r$ of T's of the function $J_{r,l,N}(\vartheta, T/N)/(T_1 T_2 \cdots T_r)$, it suffices to prove the estimate

$$|J_{r,l,N}(\vartheta, T/N)| \leq (r^l / N^{l/2}) \exp(r + r^2/2)$$

if $1 \leq |l| \leq r$ and $|T_j| = 1$ for $j = 1, \ldots, r$.

To see this, we make use of the interpretation of $J_{r,l,N}(\vartheta, T/N)$ as a determinant, and apply to it the Hadamard estimate. QED

Lemma AD.8.9 (Hadamard). *Let $n \geq 1$ be an integer. In an n-dimensional complex Hilbert space, suppose we are given two sets of $n \geq 1$ vectors, say*

$$\{v_1, \ldots, v_n\} \quad \textit{and} \quad \{w_1, \ldots, w_n\}.$$

Then we have the estimate

$$|\det_{n\times n}(\langle v_i, w_j\rangle)| \leq \left(\prod_i \|v_i\|\right)\left(\prod_i \|w_i\|\right).$$

PROOF. Pick an orthonormal basis $\{e_1, \ldots, e_n\}$. Then

$$\langle v_i, w_l\rangle = \sum_j \langle v_i, e_j\rangle\langle e_j, w_l\rangle,$$

and so we have an identity of $n \times n$ matrices

$$(\langle v_i, w_l\rangle) = (\langle v_i, e_j\rangle)(\langle e_j, w_l\rangle).$$

Thus

$$|\det_{n\times n}(\langle v_i, w_j\rangle)| = |\det_{n\times n}(\langle v_i, e_j\rangle)| \times |\det_{n\times n}(\langle e_i, w_j\rangle)|.$$

If we take $\{w_1, \ldots, w_n\}$ to coincide with $\{v_1, \ldots, v_n\}$, this gives

$$\begin{aligned}|\det_{n\times n}(\langle v_i, v_j\rangle)| &= |\det_{n\times n}(\langle v_i, e_j\rangle)| \times |\det_{n\times n}(\langle e_i, v_j\rangle)| \\ &= |\det_{n\times n}(\langle v_i, e_j\rangle)|^2.\end{aligned}$$

The Hadamard inequality 5.3.4,

$$|\det_{n\times n}(\langle v_i, v_j\rangle)| \leq \prod_i \|v_i\|^2,$$

now gives

$$|\det_{n\times n}(\langle v_i, e_j\rangle)| \leq \prod_i \|v_i\|.$$

Similarly, we infer that

$$|\det_{n\times n}(\langle e_i, w_j\rangle)| = |\det_{n\times n}(\langle w_i, e_j\rangle)| \leq \prod_i \|w_i\|.$$

Combining these last two inequalities with the matrix factorization

$$(\langle v_i, w_l\rangle) = (\langle v_i, e_j\rangle)(\langle e_j, w_l\rangle)$$

noted above gives the asserted estimate. QED

Corollary AD.8.9.1 (Hadamard). *Let $n \geq 1$ be an integer, $(a(i,j))$ an $n \times n$ matrix over $\mathbb{C}$. Then we have the estimate*

$$|\det_{n\times n}(a(i,j))| \leq \prod_{i=1}^{n}\left(\sum_{j=1}^{n}|a(i,j)|^2\right)^{1/2}.$$

PROOF. This is the previous inequality in the Hilbert space $\mathbb{C}^n$ with the standard inner produce $\sum x_i\bar{y}_j$, with $\{v_j, \ldots, v_n\}$ taken to be the columns of the matrix $(a(i,j))$ and $\{w_1, \ldots, w_n\}$ taken to be the standard basis. QED

AD.8.10. We now conclude the proof of Proposition AD.8.8. We must show

$$|J_{r,l,N}(\vartheta, T/N)| \leq (r^l/N^{l/2}) \exp(r + r^2/2)$$

if $|l| \leq r$ and $|T_j| = 1$ for $j = 1, \ldots, r$.

To prove this, we notice that the $N \times N$ matrix, say A, of which $J_{r,l,N}(\vartheta, T/N)$ is the determinant has its entries given by

$$A(a, b) := \delta_{a-b,l} + \sum_{j=1}^{r} (T_j/N) e^{-i(a-b-l)\vartheta_j}.$$

If $|T_j| = 1$ for $j = 1, \ldots, r$, we have the estimate

$$|A(a, b)| \leq \delta_{a-b,l} + r/N.$$

There are exactly $N - |l|$ pairs (a, b) in $[1, N]^2$ with $a - b = |l|$, and these pairs have $N - |l|$ distinct values of a. Thus the matrix A has $|l|$ columns in which every entry is bounded in absolute value by r/N, and has $N - |l|$ columns in which one entry is bounded in absolute value by $1 + r/N$ and the other $N - 1$ entries are each bounded in absolute value by r/N. Each of the $|l|$ columns has length at most $r/\operatorname{Sqrt}(N)$, and each of the $N - |l|$ columns has length at most $\operatorname{Sqrt}((1 + r/N)^2 + (N - 1) \times (r/N)^2) = \operatorname{Sqrt}(1 + 2r/N + r^2/N)$. So by the previous Corollary AD.8.9.1, we have

$$\begin{aligned} &|J_{r,l,N}(\vartheta, T/N)| \\ &\quad \leq (r^l/N^{l/2})(1 + (2r + r^2)/N)^{(N-|l|)/2} \\ &\quad \leq (r^l/N^{l/2})(1 + (2r + r^2)/N)^{N/2} \\ &\quad \leq (r^l/N^{l/2}) \exp(r + r^2/2). \end{aligned}$$

QED for Proposition AD.8.8.

AD.9. End of the proof of Theorem AD.5.2

Let us recall its statement.

Theorem AD.5.2. *For any integer $r \geq 1$, for any integer $k \geq 1$, for any integer $N > r$, and for f any bounded, Borel measurable $\mathbb{C}$-valued function on $\mathbb{R}^r$ with $\|f\|_{\sup} \leq 1$ and with support contained in the set $\operatorname{Sup}_i |x(i)| \leq \alpha$, we have the equality*

$$W_r(f, U_k(N)) = W_r(f, U(N)) \quad \textit{if } k > r,$$

and, more generally the estimate

$$|W_r(f, U_k(N)) - W_r(f, U(N))| \leq 2[r/k](\alpha^r/r!) r^r e^{r + r^2/2} N^{-k/2},$$

where $[r/k]$ denotes the largest integer $\leq r/k$.

PROOF. Fix r, k, and N. We have already seen that it suffices to treat the case where the function f on $\mathbb{R}^r$ is symmetric and supported in $[0, N)^r$. Recall that we defined the periodic function

$$f_{N,\mathrm{per}}(x) := \sum_{n \text{ in } \mathbb{Z}^r} f_N(x + 2\pi n),$$

in terms of which we have

$$W_r(f, U(N)) = V_r(f_{N,\mathrm{per}}, U(N)),$$

and

$$W_r(f, U_k(N)) = V_r(f_{N,\mathrm{per}}, U_k(N)).$$

Thus we have

$$W_r(f, U_k(N)) - W_r(f, U(N)) = V_r(f_{N,\mathrm{per}}, U_k(N)) - V_r(f_{N,\mathrm{per}}, U(N)).$$

According to AD.8.7, we have, for any bounded, Borel measurable function F on $(S^1)^r$,

$$V_r(F, U(N)) = (1/r!) \int_{(S^1)^r} F(\vartheta) J_{r,0,N}(\vartheta)\, d\vartheta$$

and

$$V_r(F, U_k(N)) = \sum_{l \text{ in } \mathbb{Z}, |lk| \le r} (1/r!) \int_{(S^1)^r} F(\vartheta) J_{r,lk,N}(\vartheta)\, d\vartheta.$$

Thus we have

$$\begin{aligned}
&W_r(f, U_k(N)) - W_r(f, U(N)) \\
&\quad = V_r(f_{N,\mathrm{per}}, U_k(N)) - V_r(f_{N,\mathrm{per}}, U(N)) \\
&\quad = \sum_{l \text{ in } \mathbb{Z}, 1 \le |lk| \le r} (1/r!) \int_{(S^1)^r} f_{N,\mathrm{per}}(\vartheta) J_{r,lk,N}(\vartheta)\, d\vartheta \\
&\quad = \sum_{l \text{ in } \mathbb{Z}, 1 \le |lk| \le r} (1/r!) \int_{[0,2\pi)^r} f_{N,\mathrm{per}}(x) J_{r,lk,N}(x) \prod (dx_i/2\pi).
\end{aligned}$$

Because f is supported in $[0, N)^r$, for x in $[0, 2\pi)^r$, we have the relation AD.5.2.4

$$f_{N,\mathrm{per}}(x) = f_N(x) := f(Nx/2\pi).$$

Thus we have

$$\begin{aligned}
&W_r(f, U_k(N)) - W_r(f, U(N)) \\
&\quad = \sum_{l \text{ in } \mathbb{Z}, 1 \le |lk| \le r} (1/r!) \int_{[0,2\pi)^r} f_N(x) J_{r,lk,N}(x) \prod (dx_i/2\pi) \\
&\quad = \sum_{l \text{ in } \mathbb{Z}, 1 \le |lk| \le r} (1/r!) \int_{[0,2\pi)^r} f(Nx/2\pi) J_{r,lk,N}(x) \prod (dx_i/2\pi) \\
&\quad = \sum_{l \text{ in } \mathbb{Z}, 1 \le |lk| \le r} (1/r!) \int_{[0,N)^r} f(x) J_{r,lk,N}(2\pi x/N) \prod (dx_i/N) \\
&\quad = \sum_{l \text{ in } \mathbb{Z}, 1 \le |lk| \le r} (1/r!) \int_{[0,N)^r} f(x) (N^{-r} J_{r,lk,N}(2\pi x/N))\, dx,
\end{aligned}$$

where we write dx for standard Lebesgue measure on $\mathbb{R}^r$.

Recall that by AD.8.8 we have

$$\|N^{-r} J_{r,l,N}(\vartheta)\|_{\sup} \le (r^l/N^{l/2}) \exp(r + r^2/2) \quad \text{if } |l| \le r.$$

Therefore for each l in $\mathbb{Z}$ with $1 \leq |lk| \leq r$, we have

$$\left|(1/r!) \int_{[0,N)^r} f(x)(N^{-r} J_{r,lk,N}(2\pi x/N))\, dx\right| \\ \leq (1/r!)(r^{lk}/N^{lk/2}) \exp(r + r^2/2) \int_{[0,N)^r} |f(x)|\, dx.$$

For f satisfying $\|f\|_{\sup} \leq 1$ and with support contained in the set $\mathrm{Sup}_i |x(i)| \leq \alpha$, we have the trivial estimate

$$\int_{[0,N)^r} |f(x)|\, dx \leq \alpha^r.$$

So we find

$$\begin{aligned} &|W_r(f, U_k(N)) - W_r(f, U(N))| \\ &\leq \sum_{l \text{ in } \mathbb{Z}, 1 \leq |lk| \leq r} |(1/r!) \int_{[0,N)^r} f(x)(N^{-r} J_{r,lk,N}(2\pi x/N))\, dx| \\ &\leq \sum_{l \text{ in } \mathbb{Z}, 1 \leq |lk| \leq r} (\alpha^r/r!)(r^{lk}/N^{lk/2}) \exp(r + r^2/2) \\ &\leq 2[r/k](\alpha^r/r!)(r^r/N^{k/2}) \exp(r + r^2/2), \end{aligned}$$

as required. QED

AD.10. Large N limits of the eigenvalue location measures on the $U_k(N)$

AD.10.1. Fix an integer $r \geq 1$. Fix an offset vector c in $\mathbb{Z}^r$,

$$c = (c(1), \ldots, c(r)), \qquad 1 \leq c(1) < c(2) < \cdots < c(r),$$

and denote by a in $\mathbb{Z}^r$ the corresponding separation vector, cf. 1.0.5. Thus if we denote by $\mathbb{1}$ in $\mathbb{Z}^r$ the vector $(1, 1, \ldots, 1)$, we have

$$c = \mathrm{Off}(\mathbb{1} + a).$$

For $N \geq c(r)$, and $G(N)$ either $U(N)$ or $U_k(N)$ for some $k \geq 1$, we have defined in AD.3.1.1 the multi-eigenvalue location measure $\nu(c, G(N))$ on $\mathbb{R}^r$. We have already constructed the large N limit $\nu(c)$ of the measures $\nu(c, U(N))$, first in 7.11.13 and then again in AD.4.3.

Theorem AD.10.2. *Notations as in* AD.10.1 *above, let f be a $\mathbb{C}$-valued, bounded Borel measurable function on $\mathbb{R}^r$. Suppose that the restriction of f to $(\mathbb{R}_{\geq 0})^r(\mathrm{order})$ is of compact support. For each integer $N \geq 1$, pick an integer $k_N \geq 1$. For any $a \geq 0$ in $\mathbb{Z}^r$, we have the large N limit formula*

$$\lim_{N \to \infty} \int_{\mathbb{R}^r} f\, d\nu(\mathrm{Off}(\mathbb{1} + a), U_{k_N}(N)) = \int_{\mathbb{R}^r} f\, d\nu(\mathrm{Off}(\mathbb{1} + a)).$$

PROOF. We reduce immediately to the case when f is real valued and everywhere ≥ 0. For fixed N and each integer $m \geq \Sigma(a)$ with $m - \Sigma(a)$ odd, we have, by AD.3.5, the inequalities

$$\sum_{b \geq a \text{ in } \mathbb{Z}^r, \Sigma(b) \leq m} (-1)^{b-a} \operatorname{Binom}(b,a) W(\operatorname{sep} b, f, U_{k_N}(N))$$
$$\leq \int f \, d\nu(\operatorname{Off}(\mathbb{1} + a), U_{k_N}(N))$$
$$\leq \sum_{b \geq a \text{ in } \mathbb{Z}^r, \Sigma(b) \leq m+1} (-1)^{b-a} \operatorname{Binom}(b,a) W(\operatorname{sep} b, f, U_{k_N}(N)).$$

Taking the lim sup and lim inf over N, and using Widom's result AD.5.1, we find

$$\sum_{b \geq a \text{ in } \mathbb{Z}^r, \Sigma(b) \leq m} (-1)^{b-a} \operatorname{Binom}(b,a) \Big(\lim_{N \to \infty} W(\operatorname{sep} b, f, U(N)) \Big)$$
$$\leq \liminf_{N \to \infty} \int f \, d\nu(\operatorname{Off}(\mathbb{1} + a), U_{k_N}(N))$$
$$\leq \limsup_{N \to \infty} \int f \, d\nu(\operatorname{Off}(\mathbb{1} + a), U_{k_N}(N))$$
$$\leq \sum_{b \geq a \text{ in } \mathbb{Z}^r, \Sigma(b) \leq m+1} (-1)^{b-a} \operatorname{Binom}(b,a) \Big(\lim_{N \to \infty} W(\operatorname{sep} b, f, U(N)) \Big).$$

Now take the limit over integers $m \geq \Sigma(a)$ with $m - \Sigma(a)$ odd. Using the fact [AD.4.3 parts 1) and 2), or 7.11.13] that the series

$$\sum_{b \geq a \text{ in } \mathbb{Z}^r} (-1)^{b-a} \operatorname{Binom}(b,a) \Big(\lim_{N \to \infty} W(\operatorname{sep} b, f, U(N)) \Big)$$

is absolutely convergent, with sum $\int_{\mathbb{R}^r} f \, d\nu(\operatorname{Off}(\mathbb{1} + a))$, we get

$$\int_{\mathbb{R}^r} f \, d\nu(\operatorname{Off}(\mathbb{1} + a))$$
$$= \sum_{b \geq a \text{ in } \mathbb{Z}^r} (-1)^{b-a} \operatorname{Binom}(b,a) \Big(\lim_{N \to \infty} W(\operatorname{sep} b, f, U(N)) \Big)$$
$$\leq \liminf_{N \to \infty} \int f \, d\nu(\operatorname{Off}(\mathbb{1} + a), G(N))$$
$$\leq \limsup_{N \to \infty} \int f \, d\nu(\operatorname{Off}(\mathbb{1} + a), G(N))$$
$$\leq \sum_{b \geq a \text{ in } \mathbb{Z}^r} (-1)^{b-a} \operatorname{Binom}(b,a) \Big(\lim_{N \to \infty} W(\operatorname{sep} b, f, U(N)) \Big)$$
$$= \int_{\mathbb{R}^r} f \, d\nu(\operatorname{Off}(\mathbb{1} + a)). \quad \text{QED}$$

Remark AD.10.3. Fix an integer $r \geq 1$ and an offset vector c in $\mathbb{Z}^r$. We have shown above theat for any sequence of integers k_N, the sequence of measures $\nu(c, U_{k_N}(N))$, $N \geq c(r)$, has as large N limit the measure $\nu(c)$. According to AD.4.4.1, the limit measure $\nu(c)$, as well as its approximants $\nu(c, U(N))$, are absolutely continuous on $\mathbb{R}^r$ with respect to Lebesgue measure. However, it is not the case that all of the measures $\nu(c, U_k(N))$ are absolutely continuous with respect to Lebesgue measure. Indeed, if we take $N = r$ and c the offset vector $(1, 2, 3, \ldots, N)$

in $\mathbb{Z}^N$, then $\nu(c, U_k(N))$ is not absolutely continuous with respect to Lebesgue measure. To see this, recall that $\nu(c, U_k(N))$ is the direct image of Haar measure on $U_k(N)$ by the map

$$U_k(N) \to [0, N)^N(\text{order}) \subset \mathbb{R}^N$$

which sends an element A to its N normalized angles, taken in increasing order. This map induces a bijection from the space of conjugacy classes in $U_k(N)$ to the subset of $[0, N)^N(\text{order})$ which is the disjoint union of its intersections with the Nk hyperplanes of equation $\sum x_i = Nb/k$ for $b = 0, 1, \ldots, Nk - 1$. This image set has Lebesgue measure zero in $\mathbb{R}^r$, and yet its characteristic function has integral one against $\nu(c, U_k(N))$.

It seems plausible that this example is the only one. If $N > r$, then for any offset vector c in $\mathbb{Z}^r$ with $c(r) \leq N$, the measure $\nu(c, U_k(N))$ should be absolutely continuous with respect to Lebesgue measure. The intuition is this. If $N > r$, then $\nu(c, U_k(N))$ depends only on the first r normalized angles. Because $r < N$, these first r normalized angles do not yet "know" that **all** the normalized angles are constrained to sum to $(N/k) \times$ (an integer), and so these angles "think" they came from an element of $U(N)$, and behave accordingly. Caveat emptor.

AD.11. Computation of the measures $\nu(c)$ via low-lying eigenvalues of Kloosterman sums in oddly many variables in odd characteristic

AD.11.1. Recall from 11.10.4 that for each odd integer $n \geq 3$, and for each finite field k of odd characteristic, we defined a probability measure $\mu(n, k)$ on $SU(n)^\#$ by averaging over the unitarized Frobenius conjugacy classes attached to all the n variable Kloosterman sums over the field k. According to 11.1.0.5, for fixed odd $n \geq 3$, in any sequence of finite fields k_i of odd characteristic whose cardinalities tend to infinity, the sequence of measures $\mu(n, k_i)$ on $SU(n)^\#$ converges weak $*$ to the measure $\mu^\#$ on $SU(n)^\#$ which is the direct image from $SU(n)$ of normalized Haar measure.

AD.11.2. Just as in 13.1.3, given an integer $r \geq 1$ and an offset vector c in $\mathbb{Z}^r$ with $c(r) \leq n$, we have the map $F_c : SU(n)^\# \to \mathbb{R}^r$ defined by the normalized angles named by c. The map is not continuous, but it is continuous on the open set U of $SU(n)^\#$ consisting of those elements A with $\det(A - 1) \neq 0$, cf. 1.8.5. Thus we cannot a priori conclude that the sequence of measures $(F_c)_*\mu(n, k_i)$ on $\mathbb{R}^r$ tends weak $*$ to $(F_c)_*\mu^\# := \nu(c, SU(N))$. The problem is that if we start with a continuous function f on $\mathbb{R}^r$, the function $f \circ F_c$ on $SU(n)^\#$ need not be continuous: all we can say is that it is bounded (because f is bounded in $[0, N]^r$, where F_c takes values), Borel measurable, and continuous on U. We also know that the closed complement

$$Z := \{A \text{ in } SU(n)^\# \text{ with } \det(A - 1) = 0\}$$

of U has measure zero for $\mu^\#$.

Proposition AD.11.3. *Let $n \geq 3$ be an odd integer, k_i a sequence of finite fields of odd characteristic whose cardinalities tend to infinity. Let $r \geq 1$, c an offset vector in $\mathbb{R}^r$ with $c(r) \leq n$. The sequence of measures $(F_c)_*\mu(n, k_i)$ on $\mathbb{R}^r$ tends weak $*$ to $\nu(c, SU(N))$.*

PROOF. Let f be a continuous function on $\mathbb{R}^r$. We must show that the function $f \circ F_c$ on $SU(n)^\#$ has $\int (f \circ F_c)\, du^\# = \lim_i \int (f \circ F_c)\, d\mu(n, k_i)$. This results from the following standard lemma, which we include for ease of reference, applied to the space $X = SU(n)^\#$ (see it as a metric space by the "characteristic polynomial" embedding into the space of monic polynomials of degree n), the measure $\mu^\#$, the sequence of measures $\{\mu(n, k_i)\}$, the closed set

$$Z := \{A \text{ in } SU(n)^\# \text{ with } \det(A - 1) = 0\}$$

and the function $f \circ F_c$.

Lemma AD.11.4. *Let X be a compact Hausdorff metric space, μ a Borel probability measure on X, and $\{\mu_i\}_i$ a sequence of Borel probability measures on X which tend weak $*$ to μ in the sense that for any continuous $\mathbb{C}$-valued function f on X, $\int f\, d\mu = \lim_i \int f\, d\mu_i$. Suppose that $Z \subset X$ is a closed set with $\mu(Z) = 0$. For any bounded, Borel measurable $\mathbb{C}$-valued function f on X which is continuous at every point of $X - Z$, we have $\int f\, d\mu = \lim_i \int f\, d\mu_i$.*

PROOF. We reduce first to the case when f has real values, then to the case when f is nonnegative, then, remembering that f is bounded, to the case when f has values in the closed interval $[0, 1]$. For every real $\varepsilon > 0$, let us denote by

$$Z(\leq \varepsilon) := \{x \text{ in } X | \operatorname{dist}(x, Z) \leq \varepsilon\}$$

the closed ε-neighborhood of Z, and by

$$Z(< 2\varepsilon) := \{x \text{ in } X | \operatorname{dist}(x, Z) < 2\varepsilon\}$$

the open ε-neighborhood of Z in X. By Urysohn, there exists a continuous function ψ_ε on X with values in $[0, 1]$ which is 1 on $Z(\leq \varepsilon)$ and which vanishes outside $Z(< 2\varepsilon)$. The function $f(1 - \psi_\varepsilon)$ is continuous on all of X and has values in $[0, 1]$. The function $f\psi_\varepsilon$ has values in $[0, 1]$ and is supported in $Z(< 2\varepsilon)$. Thus

$$f = f(1 - \psi_\varepsilon) + f\psi_\varepsilon.$$

We have $f \geq f(1 - \psi_\varepsilon)$ pointwise, so

$$\begin{aligned}
\liminf_i \int f\, d\mu_i &\geq \liminf_i \int f(1 - \psi_\varepsilon)\, d\mu_i = \lim_i \int f(1 - \psi_\varepsilon)\, d\mu_i \\
&= \int f(1 - \psi_\varepsilon)\, d\mu = \int f\, d\mu - \int f\psi_\varepsilon\, d\mu \\
&\geq \int f\, d\mu - \int \psi_\varepsilon\, d\mu \\
&\geq \int f\, d\mu - \mu(Z(< 2\varepsilon)).
\end{aligned}$$

Meanwhile, $f = f(1-\psi_\varepsilon) + f\psi_\varepsilon \le f(1-\psi_\varepsilon) + \psi_\varepsilon$ pointwise, so

$$\begin{aligned}\limsup_i \int f\,d\mu_i &\le \limsup_i \int (f(1-\psi_\varepsilon)+\psi_\varepsilon)\,d\mu_i\\ &= \lim_i \int (f(1-\psi_\varepsilon)+\psi_\varepsilon)\,d\mu_i\\ &= \int (f(1-\psi_\varepsilon)+\psi_\varepsilon)\,d\mu\\ &= \int f\,d\mu + \int (1-f)\psi_\varepsilon\,d\mu\\ &\le \int f\,d\mu + \int \psi_\varepsilon\,d\mu\\ &\le \int f\,d\mu + \mu(Z(<2\varepsilon)).\end{aligned}$$

Thus we obtain

$$\int f\,d\mu - \mu(Z(<\varepsilon)) \le \liminf_i \int f\,d\mu_i \le \limsup_i \int f\,d\mu_i \le \int f\,d\mu + \mu(Z(<2\varepsilon)).$$

Take now a sequence ε_l of ε's decreasing to zero. The sets $Z(<2\varepsilon_l)$ have shrinking intersection Z, with $\mu(Z)=0$, so $\mu(Z(<2\varepsilon_l)) \to 0$. QED

This concludes the proof of Proposition AD.11.3. By using it, together with Theorem AD.10.2, we get the following theorem, which does for $\nu(c)$ what Theorems 13.8.4 and 13.8.5 did for $\nu(-,c)$ and for $\nu(+,c)$ respectively.

Theorem AD.11.5. *Let $r \ge 1$ be an integer, c in $\mathbb{Z}^r$ an offset vector, and f a continuous function on $\mathbb{R}^r$ of compact support. We may compute the integral $\int_{\mathbb{R}^r} f\,d\nu(c)$ as follows. Pick a sequence of odd integers $n_i \ge 3$, each $\ge c(r)$ and increasing to infinity. For each n_i choose a sequence $k_{i,j}$ of finite fields whose cardinalities are odd and increase to infinity. Then we have the double limit formula*

$$\int_{\mathbb{R}^r} f\,d\nu(c) = \lim_i \lim_j \int_{\mathbb{R}^r} f d(F_c)_*\mu(n_i, k_{i,j}).$$

AD.12. A variant of the one-level scaling density

AD.12.1. Fix an integer $N \ge 1$. Given an element A in $U(N)$, with angles

$$0 \le \varphi(1)(A) \le \varphi(2)(A) \le \cdots \le \varphi(N)(A) < 2\pi,$$

we define a real number $\varphi(n)(A)$ for all n in $\mathbb{Z}$ by writing n as $j + lN$ with j in $[1, N]$ and setting

(AD.12.1.1) $$\varphi(j+lN)(A) := \varphi(j)(A) + 2\pi l.$$

We refer to the $\varphi(n)(A)$, n in $\mathbb{Z}$, as being "all" the real angles of A.

AD.12.2. Given a bounded, Borel measurable $\mathbb{C}$-valued function $f(x)$ of compact support on $\mathbb{R}$, we define

(AD.12.2.1) $$D(f, A, U(N)) := \sum_{n \text{ in } \mathbb{Z}} f(N\varphi(n)(A)/2\pi).$$

AD.12.3. Given a closed subgroup K of $U(N)$, endowed with its normalized Haar measure, and an element A in K, we define

(AD.12.3.1) $$D(f, A, K) := D(f, A, U(N)),$$

(AD.12.3.2) $$D(f, K) := \int_K D(f, A, K)\, dA.$$

We will apply these definitions to the situations

$$U_k(N) \subset U(N),$$
$$USp(2N) \subset U(2N),$$
$$SO(2N+1) \subset U(2N+1),$$
$$SO(2N+2) \subset U(2N+2).$$

AD.12.4. More generally, given a left translate γK of such a K by an element γ in $U(N)$ and an element A in γK, we endow γK with the left translate of Haar measure on K, and define

(AD.12.4.1) $$D(f, A, \gamma K) := D(f, A, U(N)),$$

(AD.12.4.2) $$D(f, \gamma K) := \int_{\gamma K} D(f, A, \gamma K)\, dA := \int_K D(f, \gamma A, \gamma K)\, dA.$$

We will apply these definitions to the situations

$$O_-(2N+1) \subset U(2N+1),$$
$$O_-(2N+2) \subset U(2N+2).$$

AD.12.4.3. Our first task is to relate these quantities (AD.12.4.1,2) to the quantities $W_1(f, A, G(N))$ and $W_1(f, G(N))$.

Lemma AD.12.5. *Let $N \geq 1$ be an integer, $f(x)$ a $\mathbb{C}$-valued, bounded, Borel measurable function on $\mathbb{R}$ with support in $(-N, N)$. Denote by $f_+(x)$ and $f_-(x)$ the functions on $\mathbb{R}$ defined by*

$$f_+(x) := \begin{cases} f(x) & \textit{if } x \geq 0, \\ 0 & \textit{if } x < 0, \end{cases}$$

and

$$f_-(x) := \begin{cases} f(-x) & \textit{if } x \geq 0, \\ 0 & \textit{if } x < 0. \end{cases}$$

1) *For any A in $U(N)$, denote by* $\operatorname{mult}(1, A)$ *the multiplicity of* 1 *as eigenvalue of A. We have*

$$D(f, A, U(N)) = W_1(f_+, A, U(N)) + W_1(f_-, \overline{A}, U(N)) - \operatorname{mult}(1, A) f(0).$$

1a) *If, for some integer $k \geq 1$, A lies in $U_k(N)$, we have*

$$D(f, A, U_k(N)) = W_1(f_+, A, U_k(N)) + W_1(f_-, \overline{A}, U_k(N)) - \operatorname{mult}(1, A) f(0).$$

2) *For any A in $USp(2N)$, we have*

$$D(f, A, USp(2N)) = W_1(f_+, A, USp(2N)) + W_1(f_-, A, U_k(N)).$$

3) *For any A in $SO(2N)$, we have*

$$D(f, A, SO(2N)) = W_1(f_+, A, SO(2N)) + W_1(f_-, A, SO(2N)).$$

4) *For any A in $O_-(2N+1)$, we have*

$$D(f, A, O_-(2N+1)) = W_1(f_+, A, O_-(2N+1)) + W_1(f_-, A, O_-(2N+)).$$

5) *For any A in $SO(2N+1)$, we have*

$$D(f, A, SO(2N+1)) = W_1(f_+, A, SO(2N+1)) + W_1(f_-, A, SO(2N+1)) + f(0).$$

6) *For any A in $O_-(2N+2)$, we have*

$$D(f, A, O_-(2N+2)) = W_1(f_+, A, O_-(2N+2)) + W_1(f_-, A, O_-(2N+2)) + f(0).$$

PROOF. In the $U(N)$ case, the fact that f is supported in the open interval $(-N, N)$ means that in $D(f, A, U(N))$ only the angles $\varphi(n)(A)$ which lie in the open interval $(-2\pi, 2\pi)$ occur in the sum. These are the angles $\varphi(j)(A)$ with $j = 1, \ldots, N$, together with the angles $-\varphi(j)(\overline{A})$ with $j = 1, \ldots, N$, except that we have counted doubly all instances of the angle 0. Similarly for the $U_k(N)$ case.

If $G(N)$ is $USp(2N)$ or $SO(2N)$, the fact that f is supported in the open interval $(-N, N)$ means that in $D(f, A, G(N))$ only the angles $\varphi(n)(A)$ which lie in $(-\pi, \pi)$ can contribute. These angles are among the $2N$ angles $\pm\varphi(j)(A)$ for $j = 1, \ldots, N$.

In the $SO(2N+1)$ case, every element A has an eigenvalue 1. The fact that f is supported in the open interval $(-N, N)$, and hence in the open interval $(-N-1/2, N+1/2)$ means that in $D(f, A, SO(2N+1))$ only the angles $\varphi(n)(A)$ which lie in the open interval $(-\pi, \pi)$ can contribute. These are among the $2N$ angles $\pm\varphi(j)(A)$ for $j = 1, \ldots, N$, together with the angle 0.

In the $O_-(2N+1)$ case, every element A has eigenvalue -1. The fact that f is supported in the open interval $(-N, N)$, hence in $(-N-1/2, N+1/2)$, means that in $D(f, A, O_-(2N+1))$ only the angles $\varphi(n)(A)$ which lie in the open interval $(-\pi, \pi)$ can contribute. These are among the $2N$ angles $\pm\varphi(j)(A)$ for $j = 1, \ldots, N$: the eigenvalue -1 of A cannot contribute.

In the $O_-(2N+2)$ case, every element has eigenvalue ± 1. The fact that f is supported in the open interval $(-N, N)$, hence in the open interval $(-N-1, N+1)$, means that in $D(f, A, O_-(2N+2))$ only the angles $\varphi(n)(A)$ which lie in the open interval $(-\pi, \pi)$ can contribute. These are among the $2N$ angles $\pm\varphi(j)(A)$ for $j = 1, \ldots, N$ together with the angle 0: the eigenvalue -1 of A cannot contribute. QED

Integrating the previous result over $G(N)$, and remembering that in the case of $U(N)$ and $U_k(N)$ the set of elements A which admit 1 as an eigenvalue has measure zero, we find

Corollary AD.12.5.1. *Let $N \geq 1$ be an integer, $f(x)$ a $\mathbb{C}$-valued, bounded, Borel measurable function on $\mathbb{R}$ with support in the open interval $(-N, N)$. Denote by $f_+(x)$ and $f_-(x)$ the functions on $\mathbb{R}$ defined by*

$$f_+(x) := \begin{cases} f(x) & \text{if } x \geq 0, \\ 0 & \text{if } x < 0, \end{cases}$$

and

$$f_-(x) := \begin{cases} f(-x) & \text{if } x \geq 0, \\ 0 & \text{if } x < 0. \end{cases}$$

1) $D(f, U(N)) = W_1(f_+, U(N)) + W_1(f_-, U(N))$.

1a) $D(f, U_k(N)) = W_1(f_+, U_k(N)) + W_1(f_-, U_k(N))$.
2) $D(f, USp(2N)) = W_1(f_+, USp(2N)) + W_1(f_-, U_k(N))$.
3) $D(f, SO(2N)) = W_1(f_+, SO(2N)) + W_1(f_-, SO(2N))$.
4) $D(f, O_-(2N+1)) = W_1(f_+, O_-(2N+1)) + W_1(f_-, O_-(2N+1))$.
5) $D(f, SO(2N+1)) = W_1(f_+, SO(2N+1)) + W_1(f_-, SO(2N+1)) + f(0)$.
6) $D(f, O_-(2N+2)) = W_1(f_+, O_-(2N+2)) + W_1(f_-, O_-(2N+2)) + f(0)$.

Corollary AD.12.5.2. *Let $N \geq 1$ be an integer, $f(x)$ a $\mathbb{C}$-valued bounded, Borel measurable function on $\mathbb{R}$ with support in the open interval $(-N, N)$. In terms of $S_N(x) := \sin(Nx/2)/\sin(x/2)$ we have the following explicit formulas:*
1) $D(f, U(N)) = \int_{\mathbb{R}} f(x)\,dx$.
1a) $D(f, U_k(N)) = \int_{\mathbb{R}} f(x)\,dx \quad$ *if* $k \geq 2$.
1b) $D(f, SU(N)) = \int_{\mathbb{R}} f(x)(1 + (2/N)\cos(x/2\pi))\,dx$.
2) $D(f, USp(2N)) = \int_{\mathbb{R}} f(x)(1 + 1/2N - S_{2N+1}(4\pi x/2N)/2N))\,dx$.
3) $D(f, SO(2N)) = \int_{\mathbb{R}} f(x)(1 - 1/2N + S_{2N-1}(4\pi x/2N)/2N)\,dx$.
4) $D(f, O_-(2N+1))$
$= \int_{\mathbb{R}} f(x)(1 - 1/(2N+1) + S_{2N}(4\pi x/(2N+1))/(2N+1))\,dx$.
5) $D(f, SO(2N+1))$
$= f(0) + \int_{\mathbb{R}} f(x)(1 - 1/(2N+1) - S_{2N}(4\pi x/(2N+1))/(2N+1))\,dx$.
6) $D(f, O_-(2N+2))$
$= f(0) + \int_{\mathbb{R}} f(x)(1 - 1/(2N+2) - S_{2N+1}(4\pi x/(2N+2))/(2N+2))\,dx$.
7) $D(f, O(2N+2)) = (1/2)f(0) + \int_{\mathbb{R}} f(x)(1 - 1/(2N+2))\,dx$.
8) $D(f, O(2N+1)) = (1/2)f(0) + \int_{\mathbb{R}} f(x)(1 - 1/(2N+1))\,dx$.

PROOF. For the $U(N)$ case, apply AD.5.2.4 and AD.6.2.2 to $f_\pm$, both of which are supported in $[0, N)$, and then apply AD.8.7.1 to get

$$\begin{aligned} D(f, U(N)) &= W_1(f_+, U(N)) + W_1(f_-, U(N)) \\ &= V_1((f_+)_{N,\mathrm{per}}, U(N)) + V_1((f_-)_{N,\mathrm{per}}, U(N)) \\ &= (N/2\pi)\int_{[0,2\pi]} f_+(Nx/2\pi)\,dx + (N/2\pi)\int_{[0,2\pi]} f_-(Nx/2\pi)\,dx \\ &= \int_{[0,N]} f_+(x)\,dx + \int_{[0,N]} f_-(x)\,dx = \int_{[-N,N]} f(x)\,dx = \int_{\mathbb{R}} f(x)\,dx. \end{aligned}$$

The $U_k(N)$ case with $k \geq 2$ results from 1a) of the previous result AD.12.5.1. The $U_1(N) = SU(N)$ case results from AD.5.2.4 and AD.6.2.4 applied to $f_\pm$, together with AD.8.7.1.

For the cases 2) through 6), we use AD.12.5.1 together with AD.2.5 in the case $r = 1$, applied to both of $f_\pm$, according to which

$$W_1(f, G(N)) = \int_{[0,N+\lambda]} f(x) W_{1,G(N)}(x)\,dx,$$

where

$$W_{1,G(N)}(x) := (1/(N+\lambda))L_N(\sigma\pi x/(N+\lambda), \sigma\pi x/(N+\lambda)),$$

$L_N(x, y)$ denoting the L_N kernel for $G(N)$. The calculations, entirely routine, are left to the reader. Case 7) results from averaging the results of cases 3) and 6), and case 8) results from averaging the results of cases 4) and 5). QED

Taking the large N limit in the previous result, we find the following theorem.

Theorem AD.12.6. *Let $f(x)$ be a $\mathbb{C}$-valued, bounded, Borel measurable function on $\mathbb{R}$ with compact support. We have the following large N limit formulas:*

1) $\lim_{N\to\infty} D(f, U(N)) = \int_{\mathbb{R}} f(x)\, dx$.

1a) $\lim_{N\to\infty} D(f, U_k(N)) = \int_{\mathbb{R}} f(x)\, dx$ *for all* $k \geq 1$.

2) $\lim_{N\to\infty} D(f, USp(2N)) = \int_{\mathbb{R}} f(x)(1 - \sin(2\pi x)/2\pi x)\, dx$.

3) $\lim_{N\to\infty} D(f, SO(2N)) = \int_{\mathbb{R}} f(x)(1 + \sin(2\pi x)/2\pi x)\, dx$.

4) $\lim_{N\to\infty} D(f, O_-(2N+1)) = \int_{\mathbb{R}} f(x)(1 + \sin(2\pi x)/2\pi x)\, dx$.

5) $\lim_{N\to\infty} D(f, SO(2N+1)) = f(0) + \int_{\mathbb{R}} f(x)(1 - \sin(2\pi x)/2\pi x)\, dx$.

6) $\lim_{N\to\infty} D(f, O_-(2N+2)) = f(0) + \int_{\mathbb{R}} f(x)(1 - \sin(2\pi x)/2\pi x)\, dx$.

7) $\lim_{N\to\infty} D(f, O(2N)) = (1/2)f(0) + \int_{\mathbb{R}} f(x)\, dx$.

8) $\lim_{N\to\infty} D(f, O(2N+1)) = (1/2)f(0) + \int_{\mathbb{R}} f(x)\, dx$.

Equivalently, we have

$$\lim_{N\to\infty} D(f, G(N)) = \int_{\mathbb{R}} f(x) D_G(x)\, dx,$$

where

$$\begin{aligned}
D_U(x) &= D_{U_k}(x) = 1, \\
D_{USp}(x) &= 1 - \sin(2\pi x)/2\pi x \\
D_{SO(even)}(x) &= D_{O_-(odd)}(x) = 1 + \sin(2\pi x)/2\pi x \\
D_{SO(odd)}(x) &= D_{O_-(even)}(x) = \delta_0(x) + 1 - \sin(2\pi x)/2\pi x, \\
D_O(x) &= (1/2)\delta_o(x) + 1.
\end{aligned}$$

Appendix: Graphs

AG.0. How the graphs were drawn, and what they show

AG.0.1. The CDF's of the one-variable measures $\nu(b)$ and $\nu(\pm, b)$, $b \geq 1$, on $\mathbb{R}_{\geq 0}$ can (by 7.5.5) all be computed from the Fredholm determinants $E(T, s)$ and $E_{\pm}(T, s)$, and hence (by 7.9.4) from the eigenvalues $\lambda_j(s)$ and the parities of the eigenfunctions $\varphi_{j,s}(x)$, $j = 0, 1, 2, \ldots$, of the integral operator K_s. The eigenfunctions are known to be prolate spheroidal functions, cf. [**Slep-Pol**] and [**Fuchs**], a fact already exploited by Gaudin [**Gaudin**]. It is a remarkable fact, due to Slepian and Pollak [**Slep-Pol**], that the eigenvalues of K_s are all distinct, and that when they are arranged in decreasing order

$$1 > \lambda_0(s) > \lambda_1(s) > \lambda_2(s) > \lambda_3(s) > \lambda_4(s) > \cdots > 0$$

it is the eigenvalues of even index $\lambda_{2j}(s)$ which have an even eigenfunction, and those of odd index $\lambda_{2j+1}(s)$ which have an odd eigenfunction.

AG.0.2. The eigenvalues $\lambda_j(s)$ for $0 \leq j \leq 20$ and $0 \leq s \leq 4$ were computed numerically by Steve Miller and Peter Doyle independently, both using the computer program of Van Buren [**VB**] as modified by Lloyd and Odlyzko. Using their data, Michael Rubinstein computed numerically the CDF's. He then computed, by numerical differentiation (cf. [**Fuchs**, Lemma 1, page 321] for another approach) the densities of $\nu(1), \nu(-, 1)$ and $\nu(+, 1)$, which are displayed in Figures 2 to 4. The density of $\mu_0 := \mu(\text{univ, sep. } 0)$ is (by 7.6.4) minus the derivative of the density of $\nu(1)$, but the graph of the density of μ_0 shown in Figure 1 was not calculated this way, but rather plotted from data kindly provided by Odlyzko (cf. [**Odl-Distr**, Section 6]), who used [**Mehta-Des Cloiz**, 2.34].

AG.0.3. The power series of these densities at $s = 0$ can be computed from the power series of $E_0(s)$ and of $E_{\pm,0}(s)$ at $s = 0$. Thanks to 7.2.2 and 7.3.5.2, these in turn can be computed to any desired order, say $\bmod s^{l+1}$, by computing $E(T, s)$ and $E_{\pm}(T, s) \bmod (T^{l+1}, s^{l+1})$, which amounts to computing the multiple integrals $e_n(s)$ and $e_{\pm,n}(s) \bmod s^{l+1}$ for $1 \leq n \leq l$. Mehta [**Mehta-PS**] uses a more efficient method and gives many more terms. [Our $E_0(s)$ is his $E(0, s)$, but our $E_{\pm,0}(s)$ is his $E_{\pm}(0, 2s)$.] We have

$$E_0(s) = 1 - s + (\pi^2/36)s^4 - (\pi^4/675)s^6 + O(s^8),$$
$$E_{+,0}(s) = 1 - 2s + (2\pi^2/9)s^3 - (2\pi^4/75)s^5 + (64\pi^4/8100)s^6 + O(s^7),$$
$$E_{-,0}(s) = 1 - (2\pi^2/9)s^3 + (2\pi^4/75)s^5 + O(s^7).$$

AG.0.4. The density of μ_0 is $(d/ds)^2 E_0(s) = (\pi^2/3)s^2 - (2\pi^4/45)s^4 + \cdots$, and is displayed in Figure 1. The second order vanishing at $s = 0$ is often referred to as the "quadratic repulsion" between eigenvalues in GUE. The tail of μ_0 as $s \to \infty$

is $O(e^{-s^2/8})$, as was shown in 6.13.4. The "unitary Wigner surmise" is the unique probability measure on $\mathbb{R}_{\geq 0}$ of mean 1 (recall from 7.5.15 that μ_0 has mean 1) of the form $As^2e^{-Bs^2}\,ds$. This forces A to be $32/\pi^2$ and B to be $4/\pi$. Its density is also displayed in Figure 1. The third density displayed in Figure 1 is that of the rescaling $\tilde{\nu}(-,1)$ of $\nu(-,1)$ which has mean 1, i.e., if we denote $\alpha := \int s\,d\nu(-,1) \sim 0.782$, then $\tilde{\nu}(-,1)$ is the measure on $\mathbb{R}_{\geq 0}$ defined by

$$\int f(s)\,d\tilde{\nu}(-,1) := \int f(s/\alpha)\,d\nu(-,1).$$

That these three densities are so near each other is a remarkable coincidence, and a source of possible confusion in interpreting numerical data. [They are in fact all distinct, as one sees from comparing their power series at $s = 0$.]

AG.0.5. The density of $\nu(-,1)$, the large N limit of the measures

$$\nu(1, USp(2N)),$$

is shown in Figure 2. Its series expansion at $s = 0$ is

$$(-d/ds)E_{-,0}(s) = (2\pi^2/3)s^2 - (2\pi^4/15)s^4 + \cdots.$$

The second order zero at $s = 0$ shows that the eigenvalues of elements in $USp(2N)$ for large N tend to repel (quadratically) the point 1. There is no such repulsion for $\nu(1, SO(2N))$ or for $\nu(1, U(N))$, whose large N limits, $\nu(+,1)$ and $\nu(1)$, have densities $2 + O(s^2)$ and $1 + O(s^3)$ respectively.

AG.0.6. The densities of $\nu(+,1)$ and $\nu(1)$ are displayed in Figures 3 and 4. The mean of $\nu(+,1)$ is $0.321\ldots$, while that of $\nu(1)$ is $0.590\ldots$.

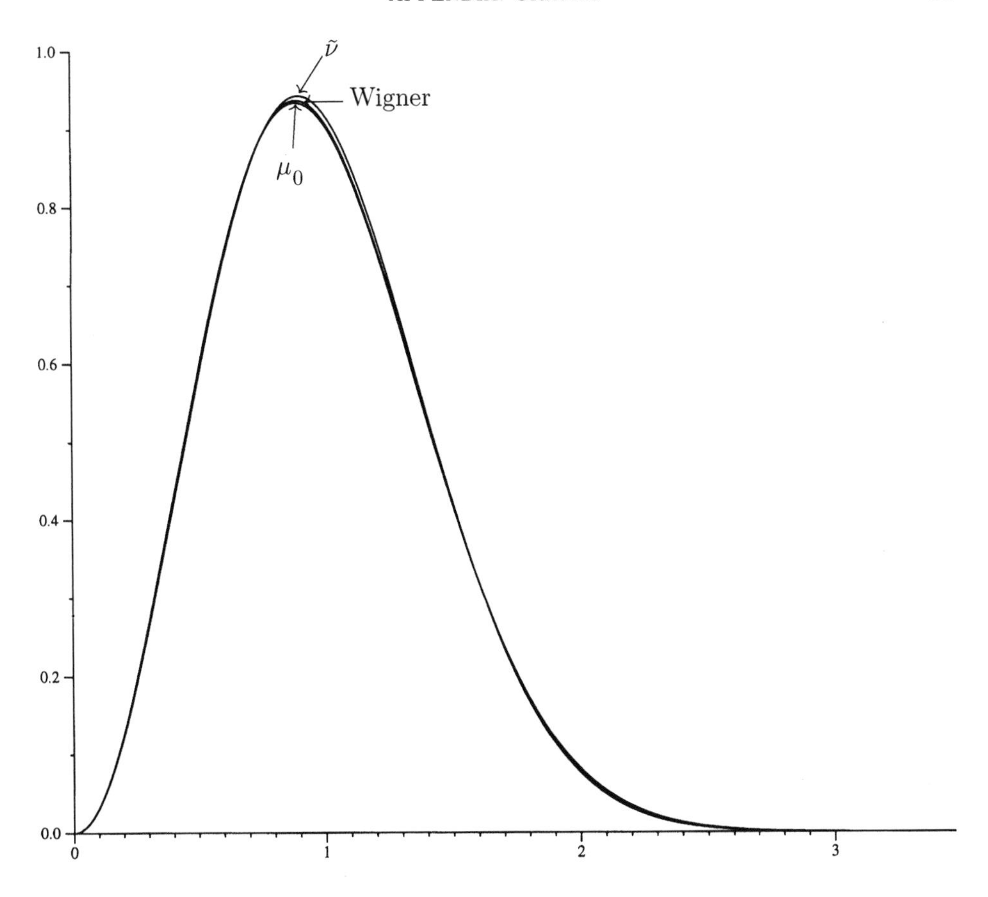

FIGURE 1. Densities of $\mu_0, \tilde{\nu}(-, 1)$, and the unitary Wigner surmise

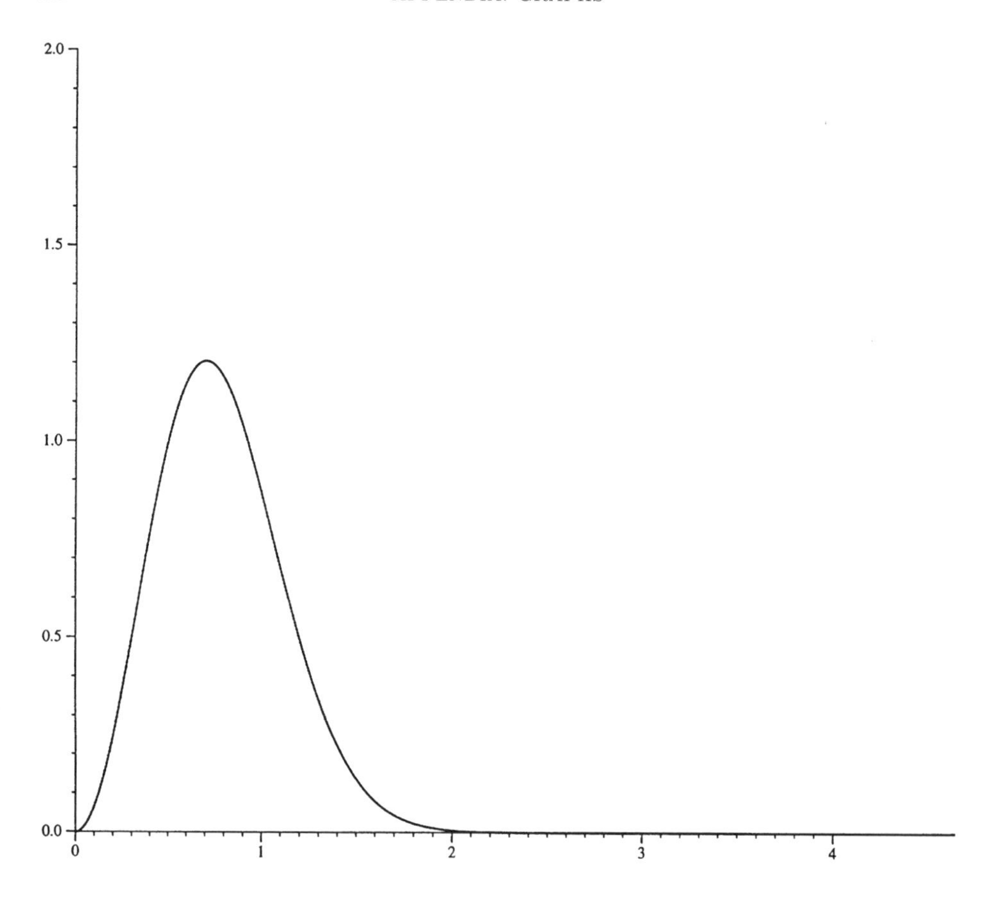

FIGURE 2. Density of $\nu(-,1)$

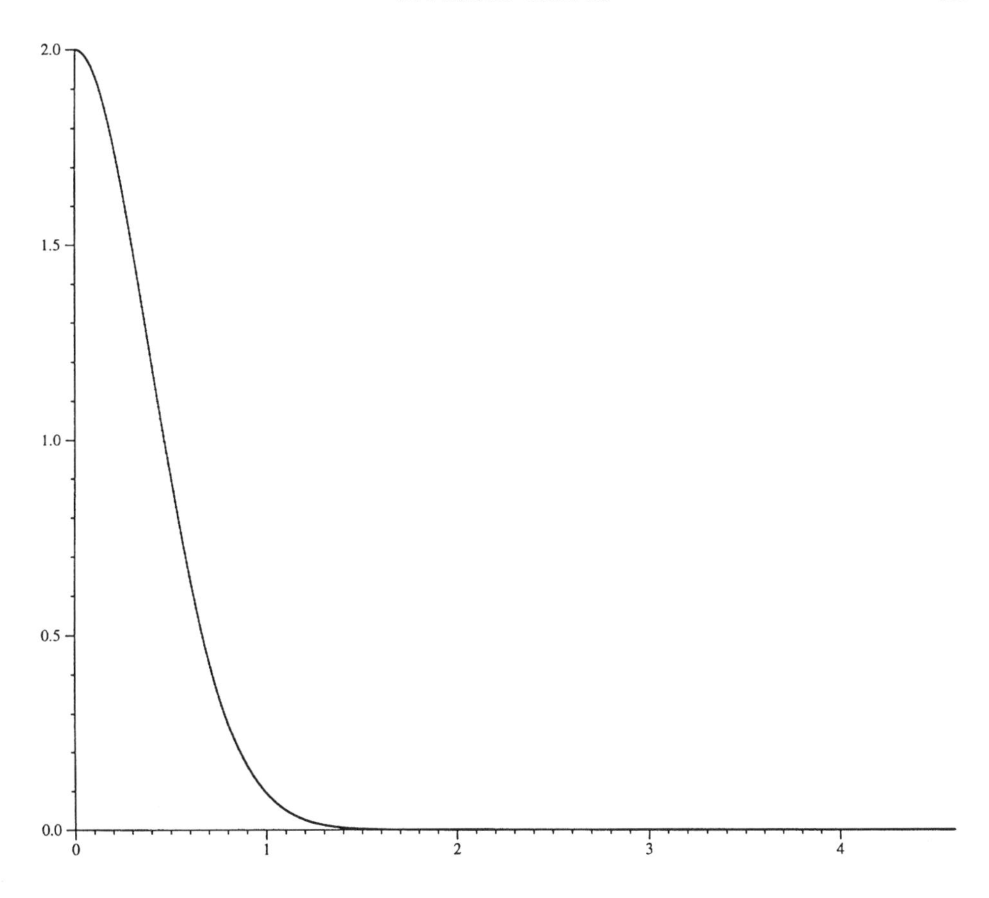

FIGURE 3. Density of $\nu(+,1)$

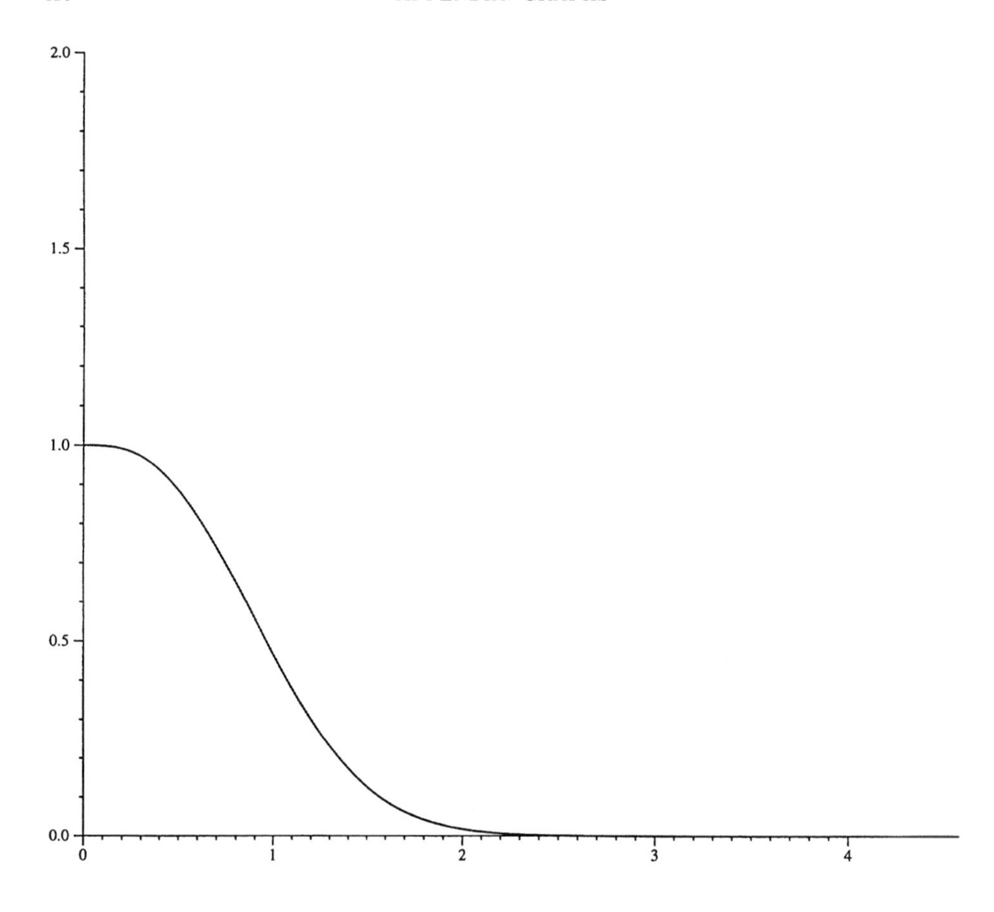

FIGURE 4. Density of $\nu(1)$

References

[A-K] Altman, A. and Kleiman, S., *Introduction to Grothendieck duality theory*, Lecture Notes in Math., vol. 146, Springer-Verlag, 1970.

[Artin] Artin, E., *Quadratische Körper in Gebiete der höheren Kongruenzen*. I, II, Math. Z. **19** (1924), 153–246.

[B-T-W] Basor, E., Tracy, C., and Widom, H., *Asymptotics of level-spacing distributions for random matrices*, Phys. Rev. Lett. **69** (1992), 5–8; Errata, Phys. Rev. Lett. **69** (1992), 2880.

[Bott] Bott, R., *Homogeneous vector bundles*, Ann. of Math. (2) **66** (1957), 203–248.

[Bour-L9] Bourbaki, N., *Groupes et algèbres de Lie*, Chapitre 9, Masson, Paris, 1982.

[C-F] Chai, C. L. and Faltings, G., *Degeneration of abelian varieties*, Springer-Verlag, 1990.

[Chav] Chavdarov, Nick, *The generic irreducibility of the numerator of the zeta function in a family of curves with large monodromy*, Duke Math. J. **87** (1997), 151–180.

[De-AFT] Deligne, P., *Application de la formule des traces aux sommes trigonométriques*, Cohomologie Étale (SGA 4 1/2), Lecture Notes in Math., vol. 569, Springer-Verlag, 1977, pp. 168–232.

[De-CCI] Deligne, P., *Cohomologie des intersections complètes*, Groupes de Monodromie en Géométrie Algébrique (SGA 7, Part II) Lecture Notes in Math., vol. 340, Springer-Verlag, 1973, pp. 39–61.

[De-CEF] Deligne, P., *Courbes elliptiques: formulaire (d'après J. Tate)*, Modular Functions of One Variable. IV, Lecture Notes in Math., vol. 476, Springer-Verlag, 1975, pp. 57–73.

[De-Mum] Deligne, P. and Mumford, D., *Irreducibility of the space of curves of given genus*, Inst. Hautes Études Sci. Publ. Math. **36** (1969), 75–109.

[De-Weil I] Deligne, P., *La conjecture de Weil*. I, Inst. Hautes Études Sci. Publ. Math. **48** (1974), 273–308.

[De-Weil II] Deligne, P., *La conjecture de Weil*. II, Inst. Hautes Études Sci. Publ. Math. **52** (1981), 313–428.

[Dw] Dwork, B., *On the rationality of the zeta function of an algebraic variety*. Amer. J. Math. **82** (1960), 632–648.

[Fel] Feller, W., *An introduction to probability theory and its applications*. Vol. II, Wiley, 1966.

[Fer] Fermigier, S., *Étude expérimentale du rang de famillies de courbes elliptiques sur* $\mathbb{Q}$, Exper. Math. **5** (1966), 119–130.

[FGA] Grothendieck, A., *Fondements de la géométrie algébrique* (a collection of his Bourbaki talks), Secrétariat Math., Paris, 1962.

[Fuchs] Fuchs, W. H. J., *On the eigenvalues of an integral equation arising in the theory of band-limited signals*, J. Math. Anal. Appl. **9** (1964), 317–330.

[Gaudin] Gaudin, M., *Sur la loi limite de l'espacement des valeurs propres d'une matrice aléatoire*, Nuclear Phys. **25** (1961), 447–458.

[Gro-FL] Grothendieck, A., *Formule de Lefschetz et rationalité des fonctions L*, Séminaire Bourbaki 1964–65, Exposé 279, reprinted in *Dix exposés sur la cohomologie des schémas*, North-Holland, 1968.

[Hasse] Hasse, H., *Zur Theorie der abstrakten elliptischen Funktionenkorper*. I, II, III, J. Reine Angew. Math. **175** (1936), 55–62, 69–88, 193–208.

[Ig] Igusa, J., *Fibre systems of Jacobian varieties.* III, Amer. J. Math. **81** (1959), 561–577.

[Ill-DFT] Illusie, L., *Deligne's l-adic Fourier transform*, Algebraic Geometry: Bowdoin 1985 (Bloch, S., ed.), Proc. Sympos. Pure Math., vol. 46, part 2, Amer. Math. Soc., 1987, pp. 151–163.

[Ill-Ord] Illusie, L., *Ordinarité*, The Grothendieck Festschrift. Vol. II (P. Cartier et al., eds.), Birkhäuser, 1990, pp. 375–405.

[Ka-ACT] Katz, N., *Affine cohomological transforms, perversity and monodromy*, J. Amer. Math. Soc. **6** (1993), 149–222.

[Ka-BTBM] Katz. N., *Big twists have big monodromy*, in preparation.

[Ka-ESDE] Katz., N., *Exponential sums and differential equations*, Ann. of Math. Studies, vol. 124, Princeton Univ. Press, 1990.

[Ka-GKM] Katz, N., *Gauss sums, Kloosterman sums, and monodromy groups*, Ann. of Math. Studies, vol. 116, Princeton Univ. Press, 1988.

[Ka-Lang] Katz, N., and Lang, S., *Finiteness theorems in geometric classfield theory*, L'Enseignment Math. (2) **27** (1981), 285–314.

[Ka-Maz] Katz, N., and Mazur, B., *Arithmetic moduli of elliptic curves*, Ann. of Math. Studies, vol. 108, Princeton Univ. Press, 1985.

[Ka-MG] Katz, N., *On the monodromy groups attached to certain families of exponential sums*, Duke Math. J. **54** (1987), 41–56.

[Ka-ODW21] Katz, N., *An overview of Deligne's work on Hilbert's twenty-first problem*, Mathematical Developments Arising from Hilbert Problems, Proc. Sympos. Pure Math., vol. 28, Amer. Math. Soc., 1976, pp. 537–557.

[Ka-RLS] Katz, N., *Rigid local systems*, Ann. of Math. Studies, vol. 138, Princeton Univ. Press, 1995.

[Ka-Sar] Katz, N., and Sarnak, P., *Zeros of zeta functions and symmetry*, Bull. Amer. Math. Soc. (to appear).

[Ka-SE] Katz, N., *Sommes exponentielles*, rédigé par G. Laumon, Astérisque, vol. 79, Soc. Math. France, 1980.

[Ka-TA] Katz, N., *On a theorem of Ax*, Amer. J. Math. **93** (1971), 484–499.

[Ka-TL] Katz, N., *Travaux de Laumon*, Séminaire Bourbaki 1987–88, Astérisque, vol. 161–162, Soc. Math. France, 1988, pp. 105–132.

[K-S] Kodaira, K., and Spencer, D. C., *On deformations of complex structures.* II, Ann. of Math. (2) **67** (1958), 403–466.

[Kra-Zag] Kramarz, G., and Zagier, D., *Numerical investigations related to the L-series of certain elliptic curves*, J. Indian Math. Soc. **97** (1987), 313–324.

[Lang-LSer] Lang, S., *Sur les séries L d'une variété algébrique*, Bull. Soc. Math. France **84** (1956), 385–407.

[Lang-Weil] Lang, S., and Weil, A., *Number of points of varieties in finite fields*, Amer. J. Math. **76** (1954), 819–827.

[Lau-TF] Laumon, G., *Transformation de Fourier, constantes d'équations fonctionnelles et conjecture de Weil*, Inst. Hautes Études Sci. Publ. Math. **65** (1987), 131–210.

[Mat-Mon] Matsumura, H., and Monsky, P., *On the automorphisms of hypersurfaces*, J. Math. Kyoto Univ. **3** (1964), 347–361.

[Mehta] Mehta, M. L., *Random matrices*, Academic Press, 1991.

[Mehta-PS] Mehta, M. L., *Power series for level spacing functions of random matrix ensembles*, Z. Phys. B: Condensed Matter **86** (1992), 285–290.

[Mehta-Des Cloiz] Mehta M. L., and Des Cloizeaux, J., *The probabilities of several consecutive eigenvalues of a random matrix*, Indian J. Pure Appl. Math. **3** (1972), 329-351.

[Messing] Messing, W., *The crystals associated to Barsotti-Tate groups; with applications to abelian schemes*, Lecture Notes in Math., vol. 264, Springer-Verlag, 1972.

[Mon] Montgomery, H., *The pair correlation of zeros of the zeta function*, Analytic Number Theory (H. G. Diamond, ed.), Proc. Sympos. Pure Math., vol. 24, Amer. Math. Soc., 1973, pp. 181–193.

[Mum-AV] Mumford, D., *Abelian varieties*, Oxford Univ. Press, 1970.

[Mum-GIT] Mumford, D., *Geometric invariant theory*, Springer-Verlag, 1965.

[Oda] Oda, T., *The first de Rham cohomology groups and Dieudonné modules*, Ann. Sci. École Norm. Sup. (4) **2** (1969), 63–135.